Study Guide and Student Solutions Manual
to accompany

PRINCIPLES OF
MODERN
CHEMISTRY

FOURTH
EDITION

OXTOBY/GILLIS/NACHTRIEB

WADE A. FREEMAN
University of Illinois
at Chicago

SAUNDERS GOLDEN SUNBURST SERIES

Saunders College Publishing
Harcourt Brace College Publishers

Fort Worth   Philadelphia   San Diego   New York   Orlando   Austin
San Antonio   Toronto   Montreal   London   Sydney   Tokyo

FOREWORD

The fourth edition of *Principles of Modern Chemistry*, by David W. Oxtoby, H. P. Gillis, and Norman H. Nachtrieb, presents a thorough introduction to University chemistry organized in seven major parts, or units:

1. The first three chapters cover macroscopic principles and the classical description of chemical bonding;

2. Chapters 4 through 6 cover kinetic molecular theory as it explains the existence and behavior of the different states of matter;

3. Chapters 7 through 12 treat chemical equilibrium and thermodynamics;

4. Chapters 13 and 14 concern the rates of chemical and physical processes;

5. Chapters 15-19 treat the current theory of atomic and moleular structure and modern spectroscopy;

6. Chapters 20 through 22 take up important and useful chemical processes;

7. The final three chapters explain the chemistry of both traditional and modern materials.

All 25 chapters include extensive problem sets. The first portion of each problem set is organized according to the topical sections of the chapter. The problems are paired—they treat the same or closely related material. The final portion of the problem sets consists of unpaired problems in two categories: "Additional Problems" and "Cumulative Problems".

• **Purpose of this Guide.** This book is designed to help students study more conveniently and effectively. It summarizes definitions, concepts, and equations, gives additional insights into the material presented in the text, and shows the way in problem-solving. Each chapter corresponds to the same-numbered chapter in the text. The first part of each chapter reviews concepts and skills with particular emphasis on the type and point of the problems that arise from them. Problems using specific concepts are referenced in boldface type. The second part consists of detailed solutions of the odd-numbered problems.

• **Problem Solving.** Students cannot expect to succeed in a serious chemistry course without solving problems. The best way to learn to solve problems is to work through detailed examples and then to try similar problems for which the solutions are available. The text furnishes a total of 1647 problems, half of which are solved in detail in this Guide.

• **How to Study Using this Guide.** Students should use an electronic calculator featuring scientific notation, logarithms and a y^x key. They should read over the chapter and then try some odd-numbered problems at the end, devoting five to ten minutes to an earnest attempt on each. They should write things down as they go back over the related passages in the text and in this Guide. If *no* progress is made, they should turn to the "Detailed Solutions to the Odd-Numbered Problems" in this Guide. Students should get a start from the "Detailed Solution" and then complete a written solution on their own. After working several odd-numbered problems, they should rest. Later, they should try the related even-numbered problems to see whether the freshly acquired knowledge has been retained. The solutions to the even-numbered problems are given in the "Instructor's Manual" available to instructors from the publisher.

What is a Detailed Solution? The "Detailed Solutions" give the answers, most of which are enclosed in $\boxed{\text{boxes}}$ to make them easier to spot, but are intended as illustrations of problem-solving techniques as well as a source of answers. They therefore show the set-up of all required calculations. Moreover, they give alternative methods of attack, point out common pitfalls, and offer additional insights. **Tip.** This heading identifies such additional material.

Extra Problems as a Source for Exam Questions. The large number of problems in the text makes it unlikely that any student will write out solutions to them all. Students should not however ignore unassigned or apparently duplicative problems, but instead study the solutions given here to confirm their understanding.

Queries. Send queries about solutions to the problems and report difficulties in using this text directly to the author via Internet. The address is WFreeman@uic.edu.

Wade A. Freeman

June, 1998

Contents

Chapter 1

The Nature and Conceptual Basis of Modern Chemistry

This chapter initiates a three-chapter introduction to modern chemistry. It covers:

1. Evidence for the existence of atoms and molecules;

2. The use of atomic theory in understanding chemical transformations;

3. Simple bonding theory and the determination of molecular shapes.

The Nature of Modern Chemistry

Chemistry is the scientific study of the properties of substances and the reactions that create new substances from old. Progress in chemistry (and other sciences) requires a blend of theory and experiment. A series of related experimental results may reveal patterns that allow assertion of a **scientific law**, which is a statement of expectation based on empirical facts. **Hypotheses** are theoretical models to explain experimental results. Hypotheses (and theories) are always subject to revision in the light of newly discovered facts. Experimental facts, once confirmed by repetition, endure. Scientific experimentation amounts to asking nature a question about the validity of a hypothesis. A carefully crafted question gets an unambiguous reply.

Chemical events occur in the ordinary, visible world of daily life and have enormous practical importance. Chemical events are however best understood by reference to a rarely seen, microscopic world of atoms and molecules. The two aspects (macroscopic and microscopic) interact constantly in modern chemistry.

Two laws, or principles, are essential in modern chemical theory:

- **Matter is neither created nor destroyed in chemical reactions.**

The relationships among the particles that make up matter change profoundly during chemical reactions but the total quantity of matter does not change.

- **Energy is neither created nor destroyed in chemical reactions.**

Energy is converted in form or transported from one region to another in space during chemical change, but the total quantity of energy remains unaltered.

These principles and their consequences are central themes in Chapter 1. Conservation of mass is discussed in text Section 1-3; the conservation of energy is discussed in text Section 1-7.

Macroscopic Methods

Experimental chemistry consists traditionally of **analysis** and **synthesis**. Analysis means taking things apart; synthesis means putting them together. An analytical chemist separates mixtures into their components and then subjects these pure substances to further analysis to learn what elements are present and the details of how they are combined. A synthetic chemist makes new substances that have theoretical or practical interest and subjects them to analysis to verify their composition and learn their properties.

Analysis shows that most materials encountered in the course of a day are **mixtures**. Mixtures are materials that can be separated into two or more pure **chemical substances** by ordinary physical means. Physical means are means that rely on differences in physical properties (such as melting point, solubility, or magnetism).

Mixtures are classified as either **homogeneous** (having properties that are uniform from region to region throughout the sample) or **heterogeneous** (having identifiable regions with different properties). Materials are often obviously heterogeneous and therefore mixtures. If a material is homogeneous, it may be a mixture but may also be a substance. More information is required. If all efforts by physical means fail to separate a homogeneous material into portions having different properties, then the material is a chemical substance or, more simply, a substance.[1] This definition relies on a negative experimental result. This is less desirable than discovering a positive indication and can have awkward consequences (see **1-4**).

Some chemical substances can be decomposed into new substances by chemical means. These are means (such as heating) that involve or induce chemical reactions. Substances that can be broken down further into new substances with different properties are **compounds**. Substances that cannot be broken down by chemical

---

[1]The term "pure substance" is often encountered. Logically, "pure" is redundant because the essence of a substance, by the definition just given, is that it is not mixed with anything. If it is not pure it is not a substance. However, saying "pure substance" rarely does any harm.

means are chemical **elements**. Thus, compounds must contain two or more different elements.

Many practical situations require the classification of a material as a mixture or substance and the further classification of the mixture (as heterogeneous or homogeneous) or of the substance (as compound or element). See **1-1**. Text Figure 1-2 assists in the classification of materials. This classification scheme is not perfect. Physical and chemical means of separation are not always distinct. For example, one substance may be separated from another because its molecules stick slightly better in cavities provided by the molecules of a third substance. The interaction is neither entirely physical nor entirely chemical.

All e 112 known chemical elements have names and symbols. These are given on the inside back cover of the text. A proper chemical symbol consists either of a single capital letter or a capital letter followed by a lower-case letter.[2]

• **Learn the correctly spelled names and symbols of the chemical elements in proper correspondence.**

This task cannot be completed overnight. It is easier when undertaken as part of a study of the structure of the periodic table of the elements (Section 1-5). Learn the names and symbols of the *common* elements first. See text Figure 1-11[3] and the inside front cover of the text for versions of the modern periodic table.

The Laws of Chemical Combination

The Law of Conservation of Mass

Carefully designed experiments on all sorts of chemical systems show that matter is neither created nor destroyed in chemical change. Rather, the mass of the starting materials always equals the mass of the products in chemical change. Many reactions *appear* to cause losses or gains in mass. Burning for example changes a massive pile of wood to a scant heap of ashes. In fact, in such cases the visible reactant picks up mass from the surroundings (for example: oxygen from the air) or evolves mass into the surroundings. When proper experimental design eliminates these confounding factors, the conclusion stands: mass is conserved in chemical reactions. Also, one chemical element is never transmuted into another in chemical reactions.

These empirical facts lead to the principle or law[4] of conservation of mass. Its most useful form for chemistry is:

---

[2]Elements that have not yet received a permanent names and symbols have temporary symbols consisting of a capital letter followed by two lower-case letters.

[3]Text page 21.

[4]The term law is used when the emphasis is on the empirical, experimental basis of the statement.

• **In chemical change, the mass of the elements is conserved, element by element.**

The law of conservation of mass helps in problem-solving. See **1-57**, **1-65**, **1-89**, and **2-95**.

The Law of Definite Proportions

This important result derived from early synthetic and analytical studies that tracked the masses of the different substances participating in chemical reactions:

• **In a pure chemical compound the proportions by mass of the constituent elements are fixed and definite, independent of the history of the sample.**

Pure synthetic vitamin C (a compound of carbon, hydrogen, and oxygen) is indistinguishable from pure vitamin C derived from a natural source. Both have the same proportions of the three constituent elements (see **1-5**). The law of definite proportions has direct use in solving difficult problems. See **2-31, 2-33**.

Modern work shows that the compositions of some solids, **nonstoichiometric** compounds,[5] violate the law of definite proportions. For example, the solid compound of nominal formula $K_2Pt(CN)_4$ has been isolated as $K_{1.3}Pt(CN)_4$. The law of definite proportions, however, applies rigorously to all gaseous and liquid compounds.

The Atomic Theory of Dalton

In 1808 John Dalton advanced the first modern atomic theory. Part of his motivation was to explain the law of definite proportions. Learn Dalton's original postulates:

1. Matter consists of indivisible atoms.

2. All of the atoms of a chemical element are identical in their mass of other properties.

3. The atoms of different elements differ in all respects.

4. Atoms are indestructible and retain their identities in chemical reactions.

5. Compounds form from the combination of atoms of unlike elements in small whole-number ratios.

This essence of the theory that the subdivisibility of matter has a lower limit, called the **atom**. The above 19th-century postulates are discussed in the light of modern understanding in **1-45**.

---

[5]See text Section 19-5.

The Law of Multiple Proportions

A **chemical formula** specifies the definite composition of a compound by telling the relative number and kind of atoms that it contains. In chemical formulas, chemical symbols specify the kind of atoms. Subscripts following the symbols specify the relative numbers of atoms.[6] Thus the formulas NO_2 and N_2O_4 specify the same composition.

Often, a given pair of elements combines to form two or more compounds of differing definite compositions. That is, the same two elements can combine in different proportions to give more than one compound (example: NO_2 and N_2O). Studies of such cases led to the law of multiple proportions:

- **When two elements form a series of compounds, the masses of the one element that combine with a fixed mass of the other stand to each other in the ratio of small whole numbers.**

Problem **1-7** traces the relationship between the multiple proportions in a pair of compounds and their chemical formulas. Also, see **2-41**. The mass ratios of the elements in the four oxides of vanadium (VO, V_2O_3, VO_2, V_2O_5), given in **1-9**, provide yet another example of the law.

The Law of Combining Volume

During the first decade of the 19th century, Joseph Gay-Lussac established experimentally that:

- **In gas-phase reactions, the volumes that combine and the volumes that form stand in the ratio of small integers as long as the volumes are measured at the same temperature and pressure.**

The theme of small whole-number ratios introduced with the law of definite proportions recurs in Gay-Lussac's work.

Avogadro's Hypothesis

Amedeo Avogadro offered an explanation of Gay-Lussac's results. He hypothesized:

- **Equal volumes of different gases at the temperature and pressure contain equal numbers of particles.**

Thus, 1 L of oxygen contains the same number of molecules as 1 L of argon provided the volumes are measured under the same conditions. See **1-11**, **1-13**, and

---

[6]The subscript "1" in chemical formulas is however omitted, by convention.

2-15.

The particles in Avogadro's hypothesis are not necessarily the same as the atoms of Dalton's atomic theory. They may be atoms (as in gaseous Ar and He), or they may be **molecules** (such as CH_4, NO, or H_2O).

• **Molecules are groups of two or more atoms bound together by attractive forces that are strong enough to maintain the grouping for a reasonable period of time.**

Molecules form from identical atoms as well as from atoms of different elements. Thus, the elements hydrogen, oxygen, nitrogen, fluorine, chlorine, bromine and iodine, which are all gases at or near room conditions, exist as diatomic molecules: H_2, O_2, N_2, F_2, Cl_2, Br_2, I_2. Memorize these seven "diatomic elements." Their names all end in "-gen" or "-ine." Numerous elements that are solids at room temperature have their atoms organized as molecules as well. Some examples are S_8, P_4, and C_{60}).[7]

Avogadro's hypothesis affords a method (Cannizzaro's method) for the determination of the relative masses of the molecules of gases. The densities of a reference gas and an unknown gas are measured at identical conditions of temperature and pressure. The ratio of the density of the unknown gas to the density of the reference gas equals the ratio of the molecular masses of the two. See **2-15** and **2-17**.

Elemental analysis of pure compounds cannot reveal the relative atomic masses of the constituent elements unless the relative numbers of atoms of each element in the compounds are known (**1-51**). This fact was a big problem for early chemists, who had no way to know the subscripts in the formulas of compounds. The advent of Cannizarro's method gave reliable estimates of relative molecular masses. This allowed the derivation of approximate formulas. Approximate formulas are as good as exact because atoms combine in small whole-number ratios, and errors vanish in the required process of rounding to whole numbers.

The Physical Structure of Atoms

By the end of the 19th century, evidence had accumulated that atoms are not indivisible, but are composed of yet smaller particles, now called **protons**, **neutrons**, and **electrons**. The following experiments provided knowledge of subatomic particles:

• **The Thomson Experiment.** This experiment determined the charge-to-mass ratio of the electron. It was known that an electrical current flows across the gap between two electrodes inserted in an evacuated tube and held at a large difference in electrical potential. The current flows even if little or no air or other gas remains in

---

[7]Some elements may have their atoms organized into molecules in more than one way. For example: O_2 and O_3; S_8 and S_6.

the tube as a conductor. The current is carried by **cathode rays**, so called because they emanate from the negatively charged cathode. Cathode rays are streams of **electrons**, the fundamental particles of electricity.

An electric field deflects cathode rays that are passing through it in a direction parallel to the direction of the field. A magnetic field deflects cathode rays passing through it in a direction perpendicular to the direction of the field. J. J. Thomson passed a beam of cathode rays in an evacuated tube simultaneously through an electric and magnetic field oriented at right angles to each other. By adjusting the strengths of the two fields (E and H), Thomson attained a *balance* between their deflecting effects and caused the beam to pass undeviated through the tube. Knowing the strengths of the two fields allowed calculation of the velocity of the cathode rays ($v = E/H$). Turning off the magnetic field allowed the electric field alone to deflect the cathode rays. Measuring their displacement s as they left the field and the length ℓ of the region of the field allowed calculation of the ratio of charge to mass:

$$\frac{e}{m_e} = \frac{2sE}{\ell^2 H^2} = 1.7588196 \times 10^{11} \text{ C kg}^{-1}$$

The SI units[8] of E are N C^{-1}, and the SI units of H are N s m^{-1}C^{-1} (where N stands for the newton, the unit of force, and C stands for the coulomb, the unit of charge). Naturally, s and ℓ are in meters. Cancel units in the preceding to verify that e/m_e indeed has units of C kg^{-1}. Practice first by doing the problems in text Appendix B.

• **The Millikan Oil-Drop Experiment.** the large size of the ratio e/m_e indicates that electrons are either very highly charged or very light. Determination of either e or m_e tells which. The Millikan experiment determined the charge of the electron.

The oil-drop experiment starts with a spray of oil droplets. Electrons are lost or gained by some droplets, either by friction or other means, giving them an electrostatic charge. The droplets fall under the influence of gravity, but are caught and stopped by a properly arranged electrical field of variable strength.

When a droplet is held motionless, the electrical and gravitational forces acting on it balance each other. The gravitational force depends on the mass of the droplet, which can be determined from its terminal velocity as it falls through the air in the absence of the electric field, and g, the acceleration of the earth's gravity (a known). The electrical force depends on the known strength of the electric field and the charge on the oil droplet. The only unknown, the charge on the oil droplet, is thus determined.

Millikan determined the electrical charge on many different droplets.[9] The results were always an integral multiple of the same basic value. He suggested that different

---

[8]See text Appendix B for a discussion of units.
[9]Some typical data appear in text Appendix A-2 on text page A-3.

droplets carry different integral numbers of electrons, all with the same fundamental charge, and computed that charge. The modern value equals $1.6021773 \times 10^{-19}$ C. Using e/m_e and this value in a quick computation gives the mass m_e of the electron. It is 9.109390×10^{-31} kg.

● **The Rutherford Experiment.** The experiment led to the nuclear model of the structure of the atom. Rutherford and his co-workers investigated the scattering of alpha particles (helium atoms with both electrons removed[10]) as beams of them impinged on thin foils of gold and other metallic elements. Most of the alpha-particles passed through the foils undeflected or only slightly deflected. Significantly, however, the foils deflected some alpha particles through large angles.

This unexpected result implied that the mass of the foils was concentrated in small, dense, positively-charged **nuclei**. The small nuclei were missed by most of the alpha particles. Occasional close approaches caused a few to be scattered through large angles. A detailed analysis based on this model predicted the number of scattered alpha particles as a function of angle. The prediction matched with the experimental results. Hence:

● **Atoms consist of massive nuclei of charge $+Ze$ and radius about 10^{-15} m surrounded by Z electrons at distances on the order of 10^{-10} m.**

Protons, Neutrons, and Isotopes

An atomic nucleus consists of Z protons which account for the nuclear charge by contributing one unit of positive charge ($+1.602 \times 10^{-19}$ C) each. The lightest atom (hydrogen) has $Z = 1$. The nucleus of hydrogen consists of a single proton. The mass of the proton is 1.672623×10^{-27} kg.

For elements other than hydrogen, the atomic mass always *exceeds* the mass contributed by Z protons. Rutherford accounted for the discrepancy by assuming the existence of additional, electrically neutral particles, **neutrons**, in the nuclei of non-hydrogen atoms. Later experiments[11] detected such particles. The neutron mass is 1.674929×10^{-27} kg, which is quite close to the proton mass. The neutron charge is zero.

Chemistry focuses on an atom's electrons and the way they are affected by the approach of other atoms. Events among the electrons are however profoundly influenced by the nature of the nucleus:

● The positive charge on the atomic nucleus dictates the number of negatively-charged electrons in the atom. In this way, the number of protons Z, the **atomic**

---

[10]See text Section 14-2 on text page 498.
[11]Conducted by James Chadwick.

number, determines the chemical identity of an atom. See **1-15**. Each of the 112 different chemical symbols is synonymous with a different Z. The values of Z run from 1 to 112.

- Neutrons as well as protons occur in the nucleus. The **neutron number N** equals the number of neutrons in the nucleus. Although neutrons are required for the stability of the nucleus, the neutron number has little influence on the chemistry of the atom. See **1-15** and **1-17**.

- The **mass number A** of an atom equals the number of protons plus the number of neutrons in the nucleus—it equals the sum of the atomic number and the neutron number: $A = Z + N$. See **1-17**.

- A **nuclide** is an atom of specific Z and A. Avoid confusing the term nuclide (plural: nuclides) with the term nucleus (plural: nuclei). A nucleus is a part of an atom; a nuclide *is* an atom.

- **Isotopes** are nuclides that have the same Z but different A's. Because isotopic atoms have the same number of protons in their nuclei, they have the same chemical symbol. Isotopes differ only very slightly in chemical behavior.

A full symbol for an atom or ion consists of the chemical symbol augmented by a left superscript, a left subscript and possibly a right superscript. The left superscript is A, the mass number, the left subscript is Z, the atomic number, and the right superscript is the electrical charge on the particle, if any. For example:

$$^{16}_{8}O \quad \text{and} \quad ^{17}_{8}O \quad \text{and} \quad ^{18}_{8}O^{+}$$

represent three isotopic atoms of oxygen, one of which has lost one electron. Chemists often omit the left subscript because the chemical symbol conveys the same information. The above species can be called "oxygen-16," "oxygen-17," and "O-18 +1 ion."

The Periodic Table of the Elements

The periodic table provides an organizing framework for the understanding of chemical facts and relationships. The table was originally derived from the study of patterns in the chemical and physical properties of the elements when they listed in order of relative atomic mass. The patterns are now ascribed to underlying similarities in the arrangement of the electrons within the atoms of the elements[12] The **periodic law** is:

---

[12]See text Section 15-8 on text page 560.

• **The properties of the elements are periodic functions of their atomic number** Z.

Periodic trends in dozens of properties, both physical and chemical, have been established. Numerical values of several physical properties are given in Appendix F.[13] The graphing of periodic trends in physical properties and the prediction of both chemical and physical properties using the periodic table are standard problems (see **1-19** and **1-21**. Periodic trends are found up and down the columns of the table, which define the **groups**, and across the rows of the table, which are **periods**.

The Representative Elements

The following is a periodic table of the representative elements:

I	II	III	IV	V	VI	VII	VIII
H							He
Li	Be	B	C	N	O	F	Ne
Na	Mg	Al	Si	P	S	Cl	Ar
K	Ca	Ga	Ge	As	Se	Br	Kr
Rb	Sr	In	Sn	Sb	Te	I	Xe
Cs	Ba	Tl	Pb	Bi	Po	At	Rn
Fr	Ra						

In this table, the 44 known representative elements are organized into eight groups, each designated by a Roman numeral at the top of its column. Elements in the same group have similarities in their properties. Group I is the **alkali metals**; Group II is the **alkali earths**; Group VI is the **chalcogens**; Group VII is the **halogens**; Group VIII is the **noble gases**. Groups may also be named after the top-most element in the column (**Example:** the Nitrogen Group). The eight groups of representative elements are often called the **main groups**.

The periodic table of the representative (main-group) elements has some symmetry that the full table lacks. Compare it to the full table in text Figure 1-11.[14] The

---

[13]Text page A-41.
[14]Text page 21.

heavy stepped line divides the elements into two near-equal sets. Those to the left of and down from the line are **metals**; the rest (including H) are **nonmetals**. The metal/nonmetal distinction is based on a constellation of physical properties such as metallic luster, electrical and thermal conductivity, and malleability. Some elements fit in one category with respect to one property and in the other with respect to a second. Elements in boxes bordering the heavy step line often have such intermediate character. Such elements are **metalloids**.

Metals tend to combine chemically with non-metals. Thus, Group I elements form 1 : 1 binary compounds with Group VII elements (examples: NaCl, KBr, CsF). Group I elements form 2 : 1 compounds with the Group VI elements (Na_2O, K_2S, Rb_2Te) and form 3 : 1 compounds with the Group V elements (Na_3N, K_3P). Group II elements form 2 : 1 compounds with Group VII elements, 1 : 1 compounds with Group VI elements, and 3 : 2 compounds with Group V elements. The pattern "metal plus non-metal yields compound" is so common that the metal/nonmetal distinction is used in naming compounds.

The elements that are not representative elements are all metals. They fall into these classes:

- **The Transition Elements.** These consist of 10 groups (columns in the table) of three or four metals each. The first transition series consists of Sc (element 21) through Zn (element 30) in the 4th row of the periodic table. The second, third and fourth transition series are located below the first series in the 5th, 6th and 7th row respectively. The transition elements are also called **transition metals**. Their chemistry is discussed in text Section 18-1.

- **The Lanthanides.** These elements include lanthanum, element 57 in the table, and the 14 elements immediately subsequent. The lanthanide elements (also called the **rare earths**) are similar in their chemistries. They are all metals.

- **The Actinides.** The elements include actinium, element 89 in the table, and the 14 elements that follow. They are similar in their chemistries and are also all radioactive. Many are produced artificially and available only in small amounts. All are metals.

The Mole Concept

The masses of atoms are much too small to be measured with a balance. Although absolute masses are now known (**1-27**), traditional laboratory chemistry requires only the relative masses of atoms. It is possible to arrive at a list of the relative atomic masses of the elements by strictly chemical methods. This is accomplished by synthesizing compounds of known chemical formula, purifying them, and analyzing to determine the relative masses of the different elements that they contain.

Relative masses are compiled on a scale defined by assigning an arbitrary relative mass to some reference element:

- **The current accepted scale of relative atomic mass assigns a relative mass of exactly 12 to $^{12}_{6}C$ atoms.**

On this scale, relative atomic masses range from approximately 1 to 250.

Periodic tables of the elements very often cite relative atomic masses. Modern tables all use this scale.[15] They are very important in solving chemistry problems. As ratios, they do not have units.

Mass Spectrometry

Mass spectrometry, a physical method, is the most accurate method available to determine relative atomic masses. It has supplanted chemical methods entirely. In a mass spectrometer, the atoms under study are converted to positively charged ions, accelerated by an electric field, and then passed through a magnetic field. More massive ions are deflected less by the magnetic field, and the ions in a mixture are separated according to mass. The inclusion of a supply of reference atoms allows very accurate measurement of relative masses by comparison of the relative amounts of deflection experienced by the unknown and the reference atom.

Mass spectrometry confirms the presence of isotopes. Because $^{17}O^+$ ions, for example, are heavier than $^{16}O^+$ ions, the mass spectrometer separates them, something a chemical method does not do. In addition, a mass spectrometer measures the **fractional abundance** of each isotope. The fractional abundance of a particular isotope of an element equals the number of atoms of that isotope divided by the total number of atoms of all the isotopes of the element.

The masses of the isotopes of an element, when combined in a **weighted average** (see **1-23**) using fractional abundances as weighting factors, give the **chemical relative atomic mass** of an element. When working problems, always use fractional abundances. Convert percent abundances, if they are given, to fractions by dividing them by 100. Problems like **1-23**, **1-25**, and **1-46** are frequent on examinations.

The **relative molecular mass** of a compound is the sum of the relative atomic masses of the elements that make it up, each one multiplied by the number of atoms of that element in a molecule. When the atoms of a substance are not organized as molecules or when the nature of the organization at the atomic level is unknown, chemists refer to a **relative formula mass** instead of a relative molecular mass. This avoids the implication that molecules are present. Computing a formula mass is identical to computing a molecular mass.

---

[15]The inside front cover of the text presents atomic masses organized according to the periodic table. The inside back cover presents relative masses alphabetically by element name.

Mass spectrometer can be used for the direct measurement of relative molecular masses if the molecules of a compound can be introduced into a mass spectrometer and caused to acquire an electric charge without experiencing decomposition.

Avogadro's Number

The connection between the macroscopic scale of masses used in the laboratory and the microscopic scale of individual atoms and molecules is provided by Avogadro's number (N_0):

- **Avogadro's number equals the number of atoms in exactly 0.012 kg (12 g) of ^{12}C (carbon-12).**

Avogadro's number cannot be deduced from any theory. It is strictly an experimental result. Its currently accepted value is $N_0 = 6.022137 \times 10^{23}$.

The Mole Concept

When factories must inventory large numbers of small parts (like screws and washers), they determine the mass of some convenient number (for example, 1000) of the parts and then weigh the whole batch. The number of thousands of washers in a bin of washers equals the total mass of washers divided by the mass per thousand. The concept of the mole allows an identical procedure in chemistry. The "convenient number" in chemistry is Avogadro's number.

- **Avogadro's number of "elemental entities" equals one mole of those entities.**

Working with moles allows chemists to weigh collections of atoms, molecules, or other entities instead of counting them. The mole is essential because the small size of chemistry's elementary entities makes counting them an impossibility (see **1-31**).

Thus, the mole is a unit of **amount of substance,** or **chemical amount.** Its formal definition is:

One mole of a substance equals the amount of the substance that contains the same number of elementary entities as exactly 0.012 kg of ^{12}C.

Clearly, one mole of a substance contains Avogadro's number of the atoms or molecules or formula units that make it up. The mole is a "collective unit." Familiar units of this type are the dozen (12), the score (20), and the pair (2). A mole contains 6.022×10^{23} items, which is a lot more than 12, 20, or 2, but no different in principle.

Chemical reactions consist of the combination or reorganization of different types of elementary entities in small whole-number ratios. This means that chemical reactions always involve small whole-number ratios of number of *moles* of various substances. This simplicity is obscured by mass, volume, and all other measures of amount.

The mole is recognized as a fundamental unit in the International System (SI) of units.[16]

It is acceptable to modify the term "mole" by adding prefixes to specify various powers of 10; the millimole (1 mmol = 10^{-3} mol) and the kilomole (1 kmol = 10^3 mol) are common. See **1-95** and **1-97**.

The Molar Mass

The **molar mass** of an element or compound equals the mass of exactly one mole of its atoms, molecules, or formula units. The following statements follow from the preceding discussion and help greatly in using molar masses:

- **The molar mass of an element is its relative atomic mass expressed in grams per mole.**

- **The molar mass of a compound is its relative molecular mass (or formula mass) expressed in grams per mole.**

Use these two statements to get from the mass of a sample of a substance to its chemical amount or vice versa. To convert mass in grams to chemical amount in moles, *divide* by the molar mass in grams per mole; to convert chemical amount in moles to mass in grams, *multiply* by the molar mass in grams per mole.

The **unit-factor method**[17] is an additional guide to deciding when to divide and when to multiply by the molar mass. Close study of the detailed solution to the formidable **1-47** on page 23 of this Guide will help in understanding the units of molar mass and the mole concept. The most common units of molar mass are grams (of substance) per mole: g mol^{-1}, but units such as kilograms per mole or milligrams per millimole are also often encountered.

Density and Molecular Size

The **density** of a sample equals its mass divided by its volume. This definition holds for solids, liquids, and gases. Possible units of density include pounds per cubic foot, pounds per gallon, kilograms per liter, and kilograms per cubic meter.

---

[16]See Appendix B, text page A-10.

[17]Appendix B, text page A-12. See **1-27** for the use of unit-factors in a simple case.

Chemists frequently use the gram per cubic centimeter and gram per milliliter to express density.[18] Common symbols for density are rho (ρ) and d. Many problems require manipulations with density (see **1-35** and **1-37**). Densities provide a unit-factor "bridge" that makes it possible to measure either volume or mass, whichever is more convenient, and immediately know the other.

- **The densities of materials change with their temperature and pressure.**

The dependence is very strong for gases. It is much less strong but still significant for solids and liquids (see **4-61** and text Figure 5-13).

When the molar mass of a substance is divided by its density, the result has the units of volume per mole. This is the **molar volume**

$$\text{molar volume} = V_\text{m} = \frac{\text{molar mass}}{\text{density}}$$

The molar volumes of gases are much larger than those of liquids and solids. The large molar volume (low density) of gases implies that their molecules are separated by large distances; the smaller molar volumes in solids and liquids suggest that the molecules nearly touch each other.

At a given temperature and pressure, all gases have approximately the same molar volume. The molar volumes of gases, like their densities, depend strongly on the temperature and pressure:

The Concept of Energy

Energy is neither created nor destroyed in chemical reactions, but it is very readily transformed from one form to another.

Forms of Energy

The following are some important forms of energy:

- The **kinetic energy** of a moving object is defined as

$$E_\text{k} = KE = \tfrac{1}{2}mv^2$$

  where m is the mass of the object and v is its speed.

- The **potential energy**, V or E_p, of an object is the energy stored in an object by virtue of its location relative to another object that repels or attracts it.

---

[18]These two units are identical in size: $1 \text{ g cm}^{-3} = 1 \text{ g mL}^{-3}$.

Each of these forms of energy may be categorized further. For example, "gravitational potential energy" refers to the energy that an object has by being high rather than low in a gravitational field. "Electrical potential energy" would mean the same for a charged particle in an electric field. "Chemical potential energy" refers to the potential energy of a mixture of substances when poised to begin a reaction that gives off energy into its surroundings. The particular type of potential energy lies in the relative positions of the particles that make up the reacting substances. Once the chemical reaction starts, the chemical potential energy is converted to other varieties of potential energy and to kinetic energy of the particles themselves or of nearby bodies. See **1-39**.

Potential Energy Curves

The potential energy of objects that attract or repel each other depends on the relative positions of the objects. The potential energy can be graphed as a function of the distance between two such objects. The result is a **potential energy curve**. Often, potential energy curves can be modeled fairly precisely by relatively simple mathematical functions. The potential energy curve for a two-particle system often dips and then rises again, corresponding to net attraction at some distances and net repulsions on others. The inter-particle distance corresponding to the bottom of the dip is the distance at which attractions and repulsions balance. The particles settle at this distance, if left alone, because it gives the lowest energy arrangement. Potential energy curves are used in the description of chemical bonds and other attractions among atoms, molecules, or ions. See text Sections 3-2, 4-7, and 5-3.[19]

Detailed Solutions to Odd-Numbered Problems

1-1 Table salt consists of sodium chloride plus additives; the additives make it a heterogeneous mixture. Sodium chloride however is a substance (a compound, NaCl). Wood is a heterogeneous mixture; air is a homogeneous mixture of several gases. Mercury is a substance (in fact, it is an elemental substance), and water is a substance (a compound, H_2O), but seawater is a homogeneous mixture of many compounds. Mayonnaise is a heterogeneous mixture (of egg and oil, which are themselves also mixtures).

1-3 The chemist is writing about $\boxed{\text{substances.}}$ Mixtures of substances can be separated (resolved) into the individual compounds by physical means.

----

[19]Figures 3-1 (text page 64), 4-20 (text page 126), and 5-9 (text page 146) are potential energy curves.

1-5 According to the law of definite proportions, a compound such as ascorbic acid has the same chemical composition regardless of source (as long as it is pure). Therefore, the ratio of carbon to oxygen in the natural sample (from lemons) must equal the ratio in the laboratory sample. The laboratory sample contains 40.00 g of O for every 30.00 g of C. The mass of oxygen in the sample isolated from lemons is accordingly:

$$m_{\text{oxygen}} = \left(\frac{40.00 \text{ g O}}{30.00 \text{ g C}}\right) \times 12.7 \text{ g C} = \boxed{16.9 \text{ g O}}$$

1-7 a) 100.00 g of compound 1 contains 66.72 g of Si and 33.28 g of N. It follows that compound 1 contains 66.72/33.28 g of Si per 1.000 g of N. This equals $\boxed{2.005 \text{ g}}$ of Si per 1.000 g of N. This ratio does *not* depend on the amount of compound 1 considered.

Compound 2 contains 60.06/39.94 g of Si per 1.000 g of N. This equals $\boxed{1.504 \text{ g}}$ of Si per 1.000 g of N.

b) To test the law of multiple proportions, compare the masses of Si associated with 1.000 g of N in the two compounds. The best comparison of the two quantities is their ratio:

$$\frac{2.005 \text{ g Si}/1.000 \text{ g N}}{1.504 \text{ g Si}/1.000 \text{ g N}} = 1.333$$

According to the law of multiple proportions, this ratio should equal a ratio of small whole numbers. Recognizing that $1.333 = 4/3$ (to four significant figures) confirms the law applies in this case.

Compound 1 has more Si per gram of N than compound 2; it is richer in Si by the factor 4/3. To obtain the formula of compound 1, take the formula of compound 2 (given as Si_3N_4) and multiply the subscript on the Si by this "richness factor." The result is Si_4N_4. When rewritten using the smallest possible whole-numbers subscripts, Si_4N_4 becomes "Si_1N_1". Subscripts equal to 1 are customarily omitted in chemical formulas, so the answer is $\boxed{SiN}$. Integral multiples (such as Si_2N_2, Si_3N_3) are also correct.

Tip. Learn the decimal equivalents of small whole-number ratios like 2/3, 3/4, 4/5, 5/8, and so forth.

1-9 The problem asks for the *relative* number of atoms of oxygen for a given mass of vanadium in four compounds. "Relative" means "take a ratio," that is, divide. The first compound contains 23.90 g of O for every 76.10 g of V. Take a ratio of these two masses:

$$\frac{23.90 \text{ g O}}{76.10 \text{ g V}} = \frac{0.3140 \text{ g O}}{1.000 \text{ g V}}$$

Compute similar ratios for the second, third and fourth compounds:

$$\frac{0.4710 \text{ g O}}{1.000 \text{ g V}} \quad \text{for cmpd 2} \qquad \frac{0.6281 \text{ g O}}{1.000 \text{ g V}} \quad \text{for cmpd 3} \qquad \frac{0.7851 \text{ g O}}{1.000 \text{ g V}} \quad \text{for cmpd 3}$$

The increasing size of the ratios shows an increasing richness of oxygen in the compounds. Next, compare the ratios. For example, divide the second by the first: $0.4710/0.3140 = 1.500$. This means the second compound has 1.500 times more oxygen per given quantity of vanadium than the first. The 1.500 : 1 ratio holds right down to a single atom of vanadium. Compare the third and fourth compounds to the first in the same way. The ratios are $0.6281/0.3140$ and $0.7851/0.3140$, which equal 2.000 : 1 and 2.500 : 1 respectively. The relative numbers of atoms of oxygen for a given amount of vanadium in these four compounds are therefore 1 to $1\frac{1}{2}$ to 2 to $2\frac{1}{2}$. This is the same as $\boxed{2 \text{ to } 3 \text{ to } 4 \text{ to } 5}$.

1-11 a) The law of combining volumes states that at constant temperature and pressure the volumes of gases combining to form a substance are in the ratio of small whole numbers. The problem describes the reverse of combination, namely the breakdown of a compound into two gases. Still, a law of *un*-combining volumes clearly must apply. Also, it is reasonable to assume that conditions of temperature and pressure are the same at the two electrodes. Therefore, the ratio of the number of particles of gaseous hydrogen to the number of particles of gaseous oxygen equals the ratio of the volume of gaseous hydrogen to the volume of gaseous oxygen. This ratio equals 14.4 mL/14.4 mL or 1.00 : 1.00. The particles of gaseous hydrogen (H_2 molecules) contain the same number of hydrogen atoms as the particles of gaseous oxygen (O_2 molecules) contain of oxygen atoms. Therefore, the simplest chemical formula is H_1O_1, or $\boxed{\text{HO}}$.

b) The formula HO is just one possible answer. That is, all formulas in which H and O have equal subscripts are also correct. Complete decomposition of any compound having a formula of this type gives gaseous H_2 and gaseous O_2 in the same 1-to-1 ratio of volumes.

1-13 From the mention of "pure nitrogen dioxide", it is clear that the reaction between the dinitrogen oxide (N_2O) and oxygen (O_2) generates nitrogen dioxide (NO_2) exclusively and also goes to completion (does not stop as long as both reactants are available). According to the law of combining volumes, the volumes of gases taking part in this reaction are in the ratio of small whole numbers. These small whole numbers are just the ratios obtained by balancing the chemical equation that represents the reaction:

$$2\,N_2O + 3\,O_2 \rightarrow 4\,NO_2$$

Thus $\boxed{2.0 \text{ L of } N_2O}$ and $\boxed{3.0 \text{ L of } O_2}$ react to form 4.0 L of NO_2.

Tip. In this reaction, 2.0 L of one gas combine with 3.0 L of a second gas to give 4.0 L of a third gas. Clearly, no "principle of conservation of volume" exists.

1-15 a) The atomic number Z of Pu equals 94. Hence, an atom of Pu has 94 protons in its nucleus. An atom of ^{239}Pu has a mass number A of 239, that is, a total of 239 protons and neutrons in its nucleus. Since the neutron number N equals $A - Z$, the atom has 145 neutrons. The requested ratio is 145/94, which equals $\boxed{1.54}$.

b) Because the Pu atom is electrically neutral, its extranuclear electrons contribute exactly enough negative charge to balance the positive charge of the 94 protons in the nucleus. The charges on the electron and proton are equal in magnitude, so the atom has $\boxed{94 \text{ electrons}}$.

1-17 The atomic number of americium is 95; americium has $\boxed{95 \text{ protons}}$ in its nucleus. In the neutral atom there are also exactly $\boxed{95 \text{ electrons}}$ because the negative charge of the electrons balances the positive charge of the protons. Of the 241 nucleons, those that are not protons are neutrons. There are accordingly $\boxed{146 \text{ neutrons}}$.

1-19 According to the periodic law, the properties of scandium should be intermediate between those of calcium and titanium. Simply average the numerical data:

Property	Predicted	Observed
Melting point	1250°C	1541°C
Boiling point	2386°C	2831°C
Density	3.02 g cm^{-3}	2.99 g cm^{-3}

The numbers in the "observed" column come from text Appendix F.

1-21 Antimony is in Group V: $\boxed{SbH_3}$; bromine is in Group VII: $\boxed{HBr}$; tin is in Group IV: $\boxed{SnH_4}$; selenium is in Group VI: $\boxed{H_2Se}$.

1-23 The atomic mass of naturally-occurring Si is the *weighted* mean (weighted average) of the atomic masses of the three isotopes listed. What does weighting an average imply? The *un*-weighted mean of the masses of the three isotopes would be:

$$\text{regular mean} = \bar{n} = \frac{1}{3}(27.97693) + \frac{1}{3}(28.97649) + \frac{1}{3}(29.97376)$$

Weighting corresponds to replacing the $\frac{1}{3}$'s in this expression with values telling each isotope's *true* contribution to the total. These values are the abundances. Fractional abundances (which add up to exactly 1.00) rather than percent abundances (which add up to 100.0) must be used:

$$\text{weighted mean} = 0.9221(27.97693) + 0.0470(28.97649) + 0.0309(29.97376) = \boxed{28.086}$$

1-25 The relative atomic mass of natural boron is the weighted mean of the relative masses of the two isotopes:

$$A_{\text{boron}} = A_{^{10}\text{B}}\, p_{^{10}\text{B}} + A_{^{11}\text{B}}\, p_{^{11}\text{B}}$$

where the A's represent relative atomic masses and the p's represent fractional abundances. With one exception, all of the quantities in this equation are known:

$$10.811 = (10.013)(0.1961) + A_{^{11}\text{B}}(0.8039) \qquad \text{hence} \qquad A_{^{11}\text{B}} = \boxed{11.01}$$

1-27 The mass of a single iodine atom is:

$$\left(\frac{126.90447 \text{ g}}{1 \text{ mol I}}\right) \times \left(\frac{1 \text{ mol I}}{6.022137 \times 10^{23} \text{ atom I}}\right) = \boxed{2.107300 \times 10^{-22} \frac{\text{g}}{\text{atom I}}}$$

1-29 Use the relative atomic masses from the inside back cover of the text.
a) P_4O_{10}: $4(30.974) + 10(15.999) = \boxed{283.886}$.
b) $BrCl$: $79.904 + 35.453 = \boxed{115.357}$.
c) $Ca(NO_3)_2$: $40.08 + 2\left(14.01 + 3(16.00)\right) = \boxed{164.09}$.
d) $KMnO_4$: $39.098 + 54.938 + 4(15.999) = \boxed{158.032}$.
e) $(NH_4)_2SO_4$: $2\left(14.007 + 4(1.0079)\right) + 32.06 + 4(15.999) = \boxed{132.13}$.

1-31 Set up the entire computation as a single string of unit-factors:

$$m_{\text{Au}} = 80 \text{ yr} \times \left(\frac{365.25 \text{ day}}{1 \text{ yr}}\right) \times \left(\frac{24 \text{ h}}{1 \text{ d}}\right) \times \left(\frac{3600 \text{ s}}{1 \text{ h}}\right) \times \left(\frac{1 \text{ atom Au}}{1 \text{ s}}\right)$$

$$\times \left(\frac{1 \text{ mol Au}}{6.022 \times 10^{23} \text{ atoms Au}}\right) \times \left(\frac{197 \text{ g Au}}{1.00 \text{ mol Au}}\right) = \boxed{8.3 \times 10^{-13} \text{ g Au}}$$

Advanced microbalances can detect down to about 10^{-10} g. Even after a lifetime of counting, the mass of the counted atoms is much too small to detect.

1-33 According to the formula, 51 atoms of all kinds are contained in a single molecule of vitamin A. Use this with a series of unit-factors to find out how many atoms there are in 1.000 mol of vitamin A:

$$N_{\text{atoms}} = 1.000 \text{ mol vit A} \times \left(\frac{N_0 \text{ molecule}}{1 \text{ mol vit A}}\right) \times \left(\frac{51 \text{ atoms}}{1 \text{ molecule}}\right) = 51.00 N_0 \text{ atoms}$$

Now compute the chemical amount of vitamin A_2 that contains this number of atoms:

$$n_{A_2} = 51.00 N_0 \text{ atoms} \times \left(\frac{1 \text{ molecule } A_2}{49 \text{ atoms}}\right) \times \left(\frac{1 \text{ mol } A_2}{N_0 \text{ molecule}}\right) = \boxed{1.041 \text{ mol } A_2}$$

Tip. The N_0's cancel out. The numerical value of Avogadro's number is not needed to complete the problem, just the concept that such a number exists.

1-35 The volume of a "flask" of mercury equals the volume per unit mass of mercury multiplied by the mass of mercury contained in a flask. The volume per unit mass is the reciprocal of the density (the density divided into one):

$$V_{\text{flask}} = 34.5 \times 10^3 \text{ g} \times \left(\frac{1 \text{ cm}^3 \text{ Hg}}{13.6 \text{ g Hg}}\right) \times \left(\frac{1 \text{ L}}{1000 \text{ cm}^3}\right) = \boxed{2.54 \text{ L}}$$

1-37 Use unit-factors to progress from volume of Al_2O_3 to the number of atoms of Al. The correct answer must be on the order of 10^{23} atoms because the amount of corundum is on the ordinary human scale:

$$N_{\text{Al}} = 15.0 \text{ cm}^3 \text{ Al}_2\text{O}_3 \times \left(\frac{3.97 \text{ g Al}_2\text{O}_3}{1 \text{ cm}^3 \text{ Al}_2\text{O}_3}\right) \times \left(\frac{1 \text{ mol Al}_2\text{O}_3}{101.96 \text{ g Al}_2\text{O}_3}\right)$$

$$\times \left(\frac{6.022 \times 10^{23} \text{ Al}_2\text{O}_3 \text{ units}}{1 \text{ mol Al}_2\text{O}_3}\right) \times \left(\frac{2 \text{ atom Al}}{1 \text{ Al}_2\text{O}_3 \text{ unit}}\right) = \boxed{7.03 \times 10^{23} \text{ atom Al}}$$

Tip. Unit-factors can be "flipped over" (numerator and denominator exchanged) at will. To make progress with a chain of unit-factors, arrange each one so the desired unit is in the top (numerator) and the unit to be canceled away is in the bottom (denominator). Note the rather creative last factor in the preceding.

1-39 The mass of the volleyball is not given. Make a reasonable guess, such as 0.5 pounds.[20] Substitute 0.5 lb and the estimated speed of the spiked ball into the formula for kinetic energy:

$$KE = 1/2\, mv^2 = 1/2\,(0.5 \text{ lb})(100 \text{ mi h}^{-1})^2 = 2500 \text{ lb mi}^2 \text{ h}^{-2}$$

The arithmetic is correct, but the units are non-standard. Convert to joules, the SI units of energy. This requires knowing how many meters in a mile, how many kilograms in a pound, and how many seconds in an hour. Some memorize such numbers. Others use calculators that are programmed with such conversion factors or look them up in tables. The arithmetic proceeds as follows:

$$KE = 2500\frac{\text{lb mi}^2}{\text{h}^2} \times \left(\frac{1 \text{ h}}{3600 \text{ s}}\right)^2 \times \left(\frac{1609 \text{ m}}{1 \text{ mi}}\right)^2 \times \left(\frac{0.454 \text{ kg}}{1 \text{ lb}}\right) = \boxed{2 \times 10^2 \text{ J}}$$

The single significant figure reflects the use of estimates used in the calculation: the kinetic energy of the spiked volleyball is between 100 and 300 J.

---

[20]The rules of volleyball require a ball with mass between 260 and 280 g, which is more like 0.6 pounds.

Tip. Make reasonable estimates when hard data are unavailable. Also, squaring a unit-factor (such as 1 hr/3600 s) must give a new unit-factor, because $1^2 = 1$. Finally, Appendix B confirms that a joule equals a kilogram meter squared per second squared.

1-41 a) Soft-wood chips: wood is a $\boxed{\text{mixture}}$ of many substances. Water: H_2O is a $\boxed{\text{pure compound}}$. Sodium hydroxide: NaOH is a $\boxed{\text{pure compound}}$.

b) Because the iron vessel was sealed, nothing was able to enter or escape, including gases. Therefore, all of the original mass is still contained in the vessel—no more, no less. The total mass is $17.2 + 150.1 + 22.43 = \boxed{189.7 \text{ kg}}$.

1-43 The density of the nucleus of ^{127}I equals its mass divided by its volume. The problem gives the nuclear mass explicitly and provides a route to the nuclear volume. Start with the formula for the volume of a sphere in terms of its radius r and substitute with the formula for the radius of a nucleus in terms of the mass number A:

$$V_{^{127}I} = \frac{4}{3}\pi r^3 = \frac{4}{3}\pi(kA^{\frac{1}{3}})^3 = \frac{4}{3}\pi k^3 A$$

$$= \frac{4}{3}\pi(1.3 \times 10^{-13} \text{ cm})^3(127) = 1.17 \times 10^{-36} \text{ cm}^3$$

The density of the iodine nucleus then equals:

$$\rho_{^{127}I} = \frac{m_{^{127}I}}{V_{^{127}I}} = \frac{2.1 \times 10^{-22} \text{ g}}{1.17 \times 10^{-36} \text{ cm}^3} = \boxed{1.8 \times 10^{14} \text{ g cm}^{-3}}$$

This is billions of times more dense than solid iodine!

1-45 Dalton's postulates were:

1. Matter consists of indivisible atoms. We now know that atoms are not indivisible. Some elements (such as uranium and radium) are radioactive, and the atoms spontaneously decompose to different atoms and subatomic particles (see Chapter 12).

2. All atoms of a given chemical element are identical in mass and in all other properties. Dalton had no way of knowing about isotopes. Atoms of a given chemical element can have different masses. For example, the element hydrogen has three isotopes. Isotopes have virtually identical chemical properties.

3. Different chemical elements have different kinds of atoms, and in particular, such atoms have different masses. This statement (so far) needs no modification or extension.

4. Atoms are indestructible and retain their identity in chemical reactions. Atoms are not indestructible. They can be split to give new kinds of atoms in particle accelerators. No instances of atoms changing their identity in chemical reactions are known.

5. The formation of a compound from its elements occurs through combining atoms of unlike elements in small whole-number ratios. Certain compounds have compositions that vary within a range. They are nonstoichiometric compounds.[21] The law of definite proportions is strictly true for gaseous and liquid compounds but not for all solid compounds.

1-47 a) No matter what units are used, the masses of atoms of ^{32}S and of atoms of P are the same in the distant galaxy as here. With this firmly in mind, the problem of evaluating the exotic Ordagova's number (symbolized N_{or}) can be treated as an exercise in the conversion of units. In earthly usage, there are N_0 (6.022137×10^{23}) atoms per mole of ^{32}S. Therefore:

$$N_{or} = \frac{N_0 \text{ atom } ^{32}S}{31.972 \text{ g } ^{32}S} \times \frac{4.8648 \text{ g } ^{32}S}{1 \text{ marg } ^{32}S} \times \frac{32.000 \text{ marg } ^{32}S}{1 \text{ elom of } ^{32}S} = N_0 \frac{(4.8648)(32.000) \text{ atom}}{31.972 \text{ elom}}$$

$$= (6.022137 \times 10^{23}) \frac{(4.8648)(32.000) \text{ atom}}{31.972 \text{ elom}} = \boxed{2.9322 \times 10^{24} \text{ atom elom}^{-1}}$$

b) On earth, the molar mass of phosphorus is 30.9738 g mol^{-1}. String together unit-factors to convert this mass to marg elom^{-1}:

$$\text{Molar Mass of P} = \frac{30.9738 \text{ g P}}{1 \text{ mol P}} \times \frac{1 \text{ marg P}}{4.8648 \text{ g P}} \times \frac{1 \text{ mol P}}{N_0 \text{ atom P}} \times \frac{N_{or} \text{ atom P}}{1 \text{ elom P}}$$

$$= N_{or} \left(\frac{1}{N_0}\right) \frac{30.9738}{4.8648} \text{ marg elom}^{-1}$$

From the previous part, it is true that:

$$N_{or} = N_0 \frac{(4.8648)(32.000) \text{ atom}}{31.972 \text{ elom}}$$

Substituting this expression gives:

$$\text{Molar Mass of P} = \left(N_0 \frac{(4.8648)(32.000)}{31.972}\right) \left(\frac{1}{N_0}\right) \frac{30.9738}{4.8648} \text{ marg elom}^{-1}$$

$$= (32.000) \frac{30.9738}{31.972} = \boxed{31.001 \text{ marg elom}^{-1}}$$

Tip. N_0 and 4.8648 cancel away. This means that the answer does not depend on the definition of a "marg" in terms of a gram or an "elom" in terms of a mole. Such a result makes sense because aliens in a distant galaxy would hardly need our definitions to do their chemistry.

---

[21]See text Section 17-5.

Chapter 2

Chemical Equations and Reaction Yields

Chemical substances are designated in two ways. They are assigned names, which may be either trivial or systematic,[1] and they are assigned chemical formulas, of which several varieties exist. Chemical change consists of the transformation of a beginning set of chemical substances to a final set. A specific chemical change is represented by means of a chemical equation in which the formulas of the starting ("input") substances appear on one side and the formulas of the ending ("output") substances appear on the other.

Empirical and Molecular Formulas

A molecule is a set of two or more atoms joined in a persistent combination. In a **molecular formula,** the subscripts that follow the symbols of the elements state the exact number of atoms of each element present in one molecule of the substance. **Examples:**

Names	Molecular Formula	Atoms in one Molecule
water	H_2O	two H and one O
hydrogen peroxide	H_2O_2	two H and two O
dioxygen, oxygen	O_2	two O
trioxygen, ozone	O_3	three O
glucose	$C_6H_{12}O_6$	six C, twelve H, six O

In an **empirical formula** the subscripts give the correct *relative* numbers of atoms of every kind in a substance. By convention, the subscripts are chosen as the smallest set of whole numbers that expresses the required ratios. **Examples:**

---

[1]Many substances have several names, some derived historically, others arrived at according to rival naming systems.

Names	Empirical Formula	Relative Numbers of Atoms
water	H_2O	2 H per 1 O
hydrogen peroxide	HO	1 H per 1 O
dioxygen, oxygen	O	O only
trioxygen, ozone	O	O only
glucose	CH_2O	1 C per 2 H per 1 O

If a list of the relative atomic masses of the elements is available, obtaining an empirical formula requires nothing more than the accurate determination of the relative mass of each element present in a sample of substance.

Obtaining a molecular formula requires information about the way the atoms of the substance are organized. Substances with their atoms organized in distinct molecules are called molecular substances. A meaningful molecular formula be written only for a molecular substance. Many substances are *not* molecular substances. Chemists use the term **formula unit** to refer to the grouping of atoms indicated by the empirical formula of a non-molecular substance or a substance that has an unknown microscopic organization.

Here are some points about chemical formulas:

- The subscript 1 is omitted in chemical formulas. See **2-1** and **2-3**.

- The molecular formula of a substance is always some whole-number multiple of its empirical formula. The whole number may equal one.

- The symbol of one or more elements may be repeated.

- Parentheses and nested parentheses work the same way as in mathematical formulas. For example, the formula $C(NO_2)_4$ specifies the same number and kind of atoms as CN_4O_8. See **2-1**.

- The *order* of symbols follows no hard-and-fast rule. However, order is frequently used to tell which atoms are bonded to which; such information is often very helpful. When no special reason exists to do otherwise, list the elements in this order: C, H, followed by other elements in alphabetic order of their symbol. See **2-4** and **2-39**.

Common problems based on chemical formulas include:

- Calculation of the mass percentage of an element in a compound given the chemical formula. Add up the relative mass contributed by the element in question, divide by the relative mass of the formula unit, and convert to a percentage. **Example:** Calculate the mass percent of N in NH_4NO_3 **Solution:** The relative mass of NH_4NO_3 equals $2(14.01) + 4(1.008) + 3(16.00) = 80.05$. The mass *fraction* of nitrogen is $2(14.01)/80.05 = 0.3500$. The mass percentage of nitrogen is 100 times larger: 35.00%. Compare to **1-29**, and **2-1**. It is not necessary to know what the symbols stand for in the formula to do this problem. Just find the symbols on a copy of the periodic table and read off the atomic masses. Sometimes these problems can involve a lot of arithmetic (see **2-39**).

● Calculation of the mass percentage of a *group* of elements in a compound. **Example:** Calculate the mass percent of water in $CuSO_4 \cdot 5H_2O$. **Solution:** The "dot 5" in the formula means that the compound contains 5 molecules of loosely-bound water per $CuSO_4$ formula unit. The percentage of water is then $(5 \times 18.015/249.69) \times 100\% = 36.07\%$, where the 249.69 is the relative mass of the $CuSO_5H_{10}$ formula unit and 18.015 is the relative mass of H_2O.

● Determination of the elemental composition of a compound that has a complex formula. **Example:** Simplify $(NO_2)_2C_6H_3CH_2CH_2COOH$. **Solution:** Add up the subscripts for each element with due respect for the parentheses to obtain $C_9H_8N_2O_6$. The example formula conveys structural information (connections among atoms) in addition to compositional information. See **2-23**.

Chemical Formulas and Percentage Composition

One of the first analytical procedures with any new compound is **elemental analysis**, which consists of breaking down a weighed sample of the compound into its constituent elements and weighing them separately. The aim is to determine the new compound's empirical formula. The calculation proceeds as follows:

1. Divide the mass of each element by its molar mass. This converts from mass of each (in, for example, grams) to chemical amount of each (in moles).

2. Divide each chemical amount by the smallest chemical amount in the group of data. Dividing is the same as taking a ratio. Such a set of divisions gives the ratios of the chemical amounts of the different elements to the chemical amount of the least prevalent element. (Obviously, the ratio for the least prevalent element is 1 : 1.)

3. Multiply the fractional ratios by the lowest factor that converts them all to whole numbers.

4. Apply the whole numbers as subscripts to the symbols of their corresponding elements in the formula of the substance. See **2-7**, **2-9** and **2-11**.

Practical elemental analysis of compounds often gives stable compounds of the constituent elements instead of the pure elements themselves. For example, analysis of C_6H_{12} might give CO_2 and H_2O instead of carbon and hydrogen. Additional procedures could be carried out to break down such compounds and isolate pure elements, but a computation is easier–the chemical amount of the compound is determined, and the chemical amount of the element in question is obtained by examining the formula of the compound that contains it. bf Example: Elemental analysis of a sample of a compound containing iron gives 1.000 g of Fe_3O_4 instead of the elemental Fe itself. How many moles of Fe was present? **Solution:** The chemical amount of Fe is 0.0130 mol, which is three times the chemical amount of Fe_3O_4. This reflects the subscript 3 in the formula Fe_3O_4). See **2-13**.

Writing Balanced Chemical Equations

A **chemical equation** represents in succinct form the start and end points of a chemical change. People write chemical equations; the symbols on the paper obviously cannot influence real events in a beaker or flask:

> • **Equations are only models of the reality of chemical reactions.**

A chemical equation gives the formulas of the **reactants** on the left and the formulas of the **products** on the right, linking them by an arrow to indicate the change. In a **balanced chemical equation**, the number of atoms of every element represented on the left-hand side equals the number of atoms of the same element on the right-hand side. This **material balance** follows from the law of conservation of mass.

Ions are atoms or groups of atoms that have gained a non-zero electric charge. An ions is represented by a chemical formulas to which is added a right superscript that states the net electric charge in units of the charge on the electron. Thus the formula X^{2+} means a deficiency of two electrons from number of electrons that neutral X requires, and the formula X^{3+} means an excess of three electrons. When ions appear as reactants or products in a chemical equation, the sum of all of the electrical charges represented on the left-hand side must equal the sum of the charges on the right-hand side. This is **charge balance.**[2] When no right superscripts appear on any of the formulas in a chemical equation, the equation is automatically balanced as to charge (zero charge on the left; zero charge on the right).

> • **In chemical reactions, mass is conserved, element by element; charge is conserved.**

Balancing equations is essential in chemistry.

---

[2]Material and charge balances emerge as powerful tools in text Section 10-8.

- Balance is achieved using **coefficients** in front of the formulas representing the reactants and products.

- Arbitrarily changing the chemical formulas of reactants and products is not allowed in balancing equations, but is a common error.

- Misreading chemical formulas, altering or omitting subscripts and superscripts, omitting one or more compounds entirely, and incorporation of coefficients into formulas are common errors in balancing equations. For example the formula $[Co(NH_3)_6]^{3+}$ includes $3 \times 6 = 18$ H atoms and specifies a net $+3$ electrical charge on the 19-atom group. Taking the formula to represent fewer H atoms or a different electrical charge leads to disaster in balancing equations.

- Many chemical equations can be balanced **by inspection**:

  1. Assign 1 as the coefficient of one species. Choose the most complicated species that contains the largest number of different elements.

  2. Identify, in sequence, elements that appear in only one chemical species whose coefficient is not yet determined. Assign that coefficient so as to balance the number of moles of atoms of that element. Continue until all coefficients are identified.

  3. If desired, eliminate fractions by multiplying by a suitable integer. See **2-19**, **2-49**, and **6-23** for simple examples of balancing by inspection. Problem **2-45** presents a more challenging case.

- The coefficients in chemical equations refer either to moles or molecules of substances, depending on context. Because specifying a fraction of a molecule or atom is chemically improper, many chemists eliminate fractional coefficients from their balanced equations, writing, for example:

$$H_2O + N_2O_5 \rightarrow 2\,HNO_3 \text{ instead of } \tfrac{1}{2}H_2O + \tfrac{1}{2}N_2O_5 \rightarrow HNO_3$$

However both versions represent the necessary balance and are acceptable.

Mass Relationships in Chemical Reactions

Balanced chemical equations provide relationships among the amounts of reactants and products in a reaction; chemical formulas provide similar relationships among the amounts of the elements comprising a compound. **Stoichiometry** concerns the use of these relationships. Typical practical problems in stoichiometry include:

- **Computation of a Molar Mass From a Formula.** To compute the molar mass of any substance from its formula, add up the relative atomic masses of all of the atoms represented in the formula. See **1-29**. Then affix the unit grams per mole ($g \, mol^{-1}$). The result can also be expressed in $kg \, kmol^{-1}$ (see **2-49a** or $kg \, mol^{-1}$ or $ng \, mmol^{-1}$ or any other unit that represents mass divided by chemical amount by use of appropriate unit-factors. See **4-41**.

- **Yield Problems.** Given the mass of a reactant and a balanced chemical equation, the task is to determine the maximum possible amount of one or more products if all of the reactant is consumed by the reaction. First calculate the chemical amount (moles) of the given reactant; from this calculate the chemical amounts of all the other substances by setting up **chemical unit-factors** and multiplying with them in such a way as to change units from moles of the known reactant to moles of the target reactant or product. Chemical unit-factors come from the coefficients in the balanced chemical equation representing the reaction. For example, the balanced chemical equation $2 \, N_2 + O_2 \rightarrow 2 \, N_2O$ gives six factors:

$$\frac{2 \, mol \, N_2}{1 \, mol \, O_2} \quad \frac{1 \, mol \, O_2}{2 \, mol \, N_2} \quad \frac{2 \, mol \, N_2}{2 \, mol \, N_2O} \quad \frac{2 \, mol \, N_2O}{2 \, mol \, N_2} \quad \frac{1 \, mol \, O_2}{2 \, mol \, N_2O} \quad \frac{2 \, mol \, N_2O}{1 \, mol \, O_2}$$

Observe that the second factor is the reciprocal of the first, the fourth the reciprocal of the third, and the sixth the reciprocal of the fifth. The different balanced equation $2 \, N_2 + 3 \, O_2 \rightarrow 2 \, N_2O_3$ gives six different factors:

$$\frac{2 \, mol \, N_2}{3 \, mol \, O_2} \quad \frac{3 \, mol \, O_2}{2 \, mol \, N_2} \quad \frac{2 \, mol \, N_2}{2 \, mol \, N_2O_3} \quad \frac{2 \, mol \, N_2O_3}{2 \, mol \, N_2} \quad \frac{3 \, mol \, O_2}{2 \, mol \, N_2O_3} \quad \frac{2 \, mol \, N_2O_3}{3 \, mol O_2}$$

Once the chemical amount (moles) of a substance is determined, then multiplication by the molar mass (in $g \, mol^{-1}$) quickly gives its mass (grams). See **2-21** and **2-53**.

Variations of this problem recognize that in real experiments some product may be lost to side-reactions or during purification. The stoichiometric yield of a product is a **theoretical yield.** The **actual yield** refers to the weighed quantity of product experimentally isolated from a reaction. It is possible to perform a chemical reaction with great success but have an actual yield of zero (by accidentally throwing away the product). **Percent yield** is the actual yield of product divided by its theoretical yield then multiplied by 100 percent. See **2-37**.

- **Determination of Empirical Formulas.** The results of elemental analysis of a compound often appear as a list of percentages by mass of the several elements in the compound. A common task is to compute the empirical formula from such a list. To do this:

1. Imagine some convenient amount of the compound (usually 100 g).

2. Use the percentage data to compute the mass of the several elements in this sample.

3. Divide each element's mass by the molar mass of that element.

4. Figure out the ratio of these chemical amounts. The way to do this is shown in **2-5** and **2-43**.

It is worth the effort to confirm that the amount of compound amount of compound imagined has no effect on the answer. To do this, rework **2-11** assuming 200 g of the compounds.

- **Limiting Reactant Problems.** In general a chemical reaction uses up one reactant before the others. The first reactant to run out is the **limiting reactant**. An error-proof method of determining which reactant is limiting follows:

    1. Compute the yield of a product (it does not matter which) assuming that all reactants but one are present in *unlimited supply*.

    2. Repeat the computation making the same assumption for each of the different reactants.

    3. Identify the reactant that gives the smallest yield of product. This reactant is the limiting reactant. All others are **in excess**.

The solution to **2-35** chooses a conceptual method to identify the limiting reactant but also gives a short-cut.

Detailed Solutions to Odd-Numbered Problems

2-1 The repetition of F in the formula implies that this compound contains F in at least two different settings. Such a distinction does not matter in obtaining mass percentages, so rewrite the formula as $PtClO_2F_8$. The mass of exactly one mole of this substance equals 414.52 g, a result obtained by multiplying the molar masses (in g mol^{-1}) of the various elements by their subscripts in the formula and adding the results (as in **1-29**). The mass percentage of each element is the mass that it contributes divided by the mass of the whole and multiplied by 100%:

$$\text{for Cl:} \quad \frac{1 \text{ mol}(35.453 \text{ g mol}^{-1})}{414.52 \text{ g}} \times 100\% = \boxed{8.553\% \text{ Cl}}$$

$$\text{for F:} \quad \frac{8 \text{ mol}(18.998 \text{ g mol}^{-1})}{414.52 \text{ g}} \times 100\% = \boxed{36.67\% \text{ F}}$$

$$\text{for O:} \quad \frac{2 \text{ mol}(15.999 \text{ g mol}^{-1})}{414.52 \text{ g}} \times 100\% = \boxed{7.720\% \text{ O}}$$

$$\text{for Pt:} \quad \frac{1 \text{ mol}(195.08 \text{ g mol}^{-1})}{414.52 \text{ g}} \times 100\% = \boxed{47.06\% \text{ Pt}}$$

Although Pt ties with Cl as the least prevalent element in the compound on the basis of number of atoms, it is by far the most prevalent on the basis of mass.

2-3 The mass percentage of hydrogen in each of the compounds can be calculated,[3] and the resulting numbers used to get the required order. The results are 11.19% for H_2O, 15.35% for $C_{12}H_{26}$, 9.742% for N_4H_6, and 12.68% for LiH, Therefore:

$$\boxed{N_4H_6 \ < \ H_2O \ < \ \text{LiH} \ < \ C_{12}H_{26}}$$

Tip. This method takes a lot of computation. A faster way is to settle for estimates of the hydrogen content of each compound. Get the estimates by adding up the masses of the non-H atoms and dividing by the number of H's. Exact arithmetic is not necessary. Thus, in H_2O, there are $16/2 = 8$ units of non-H mass per hydrogen atom. In $C_{12}H_{26}$ there are $144/26 \approx 6$ such units; in N_4H_6 there are $56/6 \approx 9$ such units; in LiH there are 7.9 such units. The compound that is richest in H has the smallest amount of non-H mass per hydrogen atom.

2-5 Calculate the fraction (not percentage) by mass of hydrogen (H) in the compound C_4H_{10} (butane) by the method of **2-1** and multiply the result by 0.0130, the fraction of butane in "Q-gas". This fraction-of-a-fraction method works because helium, the other component of Q-gas, contains no hydrogen:

$$\left(\frac{10 \times (1.008) \text{ g H}}{(4 \times 12.011) + (10 \times 1.008) \text{ g butane}} \right) \times \left(\frac{0.0130 \text{ g butane}}{1 \text{ g Q gas}} \right) = \frac{0.00225 \text{ g H}}{1 \text{ g Q gas}}$$

Multiply by 100% to obtain the desired percentage: $\boxed{0.225\% \text{ H}}$ by mass.

2-7 The empirical formula of a compound is the smallest whole-number ratio of atoms of different kinds (or moles of atoms of different kinds) in the compound. First,

---

[3]As in **2-1**.

calculate the chemical amount of each element from the given masses:

$$n_O = 16.58 \times 10^{-3} \text{ g O} \times \left(\frac{1 \text{ mol O}}{15.999 \text{ g O}} \right) = 1.036 \times 10^{-3} \text{ mol O}$$

$$n_P = 8.02 \times 10^{-3} \text{ g P} \times \left(\frac{1 \text{ mol P}}{30.97 \text{ g O}} \right) = 2.59 \times 10^{-4} \text{ mol P}$$

$$n_{Zn} = 25.40 \times 10^{-3} \text{ g Zn} \times \left(\frac{1 \text{ mol Zn}}{65.38 \text{ g Zn}} \right) = 3.885 \times 10^{-4} \text{ mol Zn}$$

Next, divide through by the smallest number of moles. This puts the quantities on a basis of 1 for comparison:

$$\frac{\text{mol O}}{1 \text{ mol P}} = \frac{1.036 \times 10^{-3}}{2.59 \times 10^{-4}} = 4.00 \quad \text{and} \quad \frac{\text{mol Zn}}{1 \text{ mol P}} = \frac{3.885 \times 10^{-4}}{2.59 \times 10^{-4}} = 1.50$$

This gives the formula $Zn_{1.5}PO_4$. But the empirical formula is defined as the smallest whole-number ratio of moles of elements in a compound. Simply multiply all subscripts by two to get rid of the fractions. The result is $\boxed{Zn_3P_2O_8}$, which is often written $Zn_3(PO_4)_2$ to show how the atoms are organized in the compound.

2-9 The percentages of Fe and Si in the crystalline grain in the fulgurite apply to any arbitrary amount of the compound. A 100.0-g sample would contain 46.01 g Fe and 53.99 g Si. Compute the chemical amounts of Fe and Si in such a sample:

$$n_{Fe} = 46.01 \text{ g Fe} \times \left(\frac{1 \text{ mol Fe}}{55.847 \text{ g Fe}} \right) = 0.8239 \text{ mol Fe}$$

$$n_{Si} = 53.99 \text{ g Si} \times \left(\frac{1 \text{ mol Si}}{28.086 \text{ g Si}} \right) = 1.922 \text{ mol Si}$$

The two chemical amounts have the ratio 2.333 mol of Si to 1.000 mol of Fe. This is expressed by the formula $FeSi_{2.333}$. Multiplying through by 3 to eliminate fractions gives the empirical formula $\boxed{Fe_3Si_7}$.

2-11 Consider the two cases separately. 100.000 g of the first compound has 90.745 g of Ba and, by subtraction, 9.255 g of N. Compute the chemical amounts of the two elements:

$$n_{Ba} = 90.745 \text{ g Ba} \times \left(\frac{1 \text{ mol Ba}}{137.33 \text{ g Ba}} \right) = 0.66078 \text{ mol Ba}$$

$$n_N = 9.255 \text{ g N} \times \left(\frac{1 \text{ mol N}}{14.007 \text{ g N}} \right) = 0.6607 \text{ mol N}$$

The two elements are present in equal chemical amounts, that is, they are present in a 1-to-1 molar ratio. Thus, the empirical formula is $\boxed{\text{BaN}}$.

100.000 g of the second compound has 93.634 g of Ba and, by subtraction, 6.366 g of N. The chemical amounts are:

$$n_{\text{Ba}} = 93.634 \text{ g Ba} \times \left(\frac{1 \text{ mol Ba}}{137.33 \text{ g Ba}} \right) = 0.68182 \text{ mol Ba}$$

$$n_{\text{N}} = 6.366 \text{ g N} \times \left(\frac{1 \text{ mol N}}{14.007 \text{ g N}} \right) = 0.4545 \text{ mol N}$$

Dividing both these chemical amounts by the smaller establishes that the two elements are in a 1.500-to-1 molar ratio. Thus, the empirical formula is $\boxed{\text{Ba}_3\text{N}_2}$.

2-13 a) Burning the compound in oxygen gives 0.692 g of H_2O and 3.381 g of CO_2. Determine the masses of elemental H and C in these amounts of H_2O and CO_2:

$$m_{\text{H}} = 0.692 \text{ g H}_2\text{O} \times \left(\frac{2.016 \text{ g H}}{18.015 \text{ g H}_2\text{O}} \right) = \boxed{0.0774 \text{ g of H}}$$

$$m_{\text{C}} = 3.381 \text{ g CO}_2 \times \left(\frac{12.01 \text{ g C}}{44.01 \text{ g CO}_2} \right) = \boxed{0.9226 \text{ g of C}}$$

b) The masses of C and H in the CO_2 and H_2O add up to 1.000 g. The compound therefore contains $\boxed{\text{no other elements}}$.

c) The compound is $\boxed{7.74\% \text{ H}}$ and $\boxed{92.26\% \text{ C}}$ by mass.

d) To determine the empirical formula of the compound, convert the masses of C and H in the sample to chemical amounts (in moles) and determine their ratio:

$$n_{\text{H}} = 0.0774 \text{ g H} \times \left(\frac{1 \text{ mol H}}{1.008 \text{ g H}} \right) = 0.0767 \text{ mol H}$$

$$n_{\text{C}} = 0.9226 \text{ g C} \times \left(\frac{1 \text{ mol}}{12.01115 \text{ g C}} \right) = 0.0768 \text{ mol C}$$

The C and H are present in a 0.0768/0.0767 molar ratio, which is a 1.00-to-1.00 molar ratio. The empirical formula is therefore $\boxed{\text{CH}}$.

2-15 a) Imagine that some fluorine and some of the fluorocarbon are confined side-by-side in identical vessels that hold exactly 1 L. The two vessels hold the same number of molecules because equal volumes of gases under the same conditions contain the same number of molecules (Avogadro's hypothesis). The sample of fluorocarbon has a mass of 8.93 g whereas the sample of fluorine has a mass of only 1.70 g. It follows that the molecules of the fluorocarbon are 8.93/1.70 times more massive than those of fluorine (F_2). The relative mass of F_2 equals 38.0. Therefore:

$$\text{Relative mass of fluorocarbon} = 38.0 \times \left(\frac{8.93}{1.70}\right) = 200$$

A relative mass of 200 requires four CF_2 units (each of which contributes a relative mass of 50. Hence the molecular formula of the fluorocarbon is $(CF_2)_4$, or $\boxed{C_4F_8}$.

2-17 a) The unknown binary compound is gaseous. Although it is 1.94 times more dense than gaseous O_2, the unknown has the same number of molecules per liter (Avogadro's hypothesis). Therefore, its molecules have a mass that is 1.94 times larger than the mass of an O_2 molecule. An O_2 molecule has a relative molecular mass of 32.0. This makes the relative molecular mass of the unknown $1.94 \times 32.0 = \boxed{62.1}$.

b) The unknown compound contains hydrogen (H) and one other element. Burning 1.39 g of it in oxygen gives 1.21 g of water. This amount of water contains all of the H present in the unknown. Compute the chemical amount of this H:

$$n_H = 1.21 \text{ g } H_2O \times \left(\frac{1 \text{ mol } H_2O}{18.0153 \text{ g } H_2O}\right) \times \left(\frac{2 \text{ mol}}{1 \text{ mol } H_2O}\right) = 0.1343 \text{ mol H}$$

Compare this to the chemical amount of the unknown compound. First obtain the number of moles of the unknown:

$$n_{\text{unknown}} = 1.39 \text{ g unknown} \times \left(\frac{1 \text{ mol unknown}}{62.08 \text{ g unknown}}\right) = 0.0224 \text{ mol}$$

Then do the comparison by dividing the chemical amount of H by the chemical amount of unknown:

$$\frac{n_H}{n_{\text{unknown}}} = \frac{0.1343 \text{ mol H}}{0.0224 \text{ mol unknown}} = \frac{6.00 \text{ mol H}}{1 \text{ mol unknown}}$$

There are 6 mol of H per mole of the unknown, and therefore $\boxed{6}$ atoms of H per molecule of unknown.

c) The unknown contains H and one other element, call it Z. Its molecular mass is 62.08. The maximum possible atomic mass of Z is $62.08 - 6(1.00794) = \boxed{56.03}$. This is the atomic mass of Z if exactly one atom of Z is present per molecule of unknown, that is, if the molecular formula of the unknown is ZH_6.

d) The molecule may contain more than one atom of Z. Two atoms of Z implies a relative atomic mass for Z of 28.02; Three atoms of Z implies a relative atomic mass for Z of 18.68. As the subscript of Z gets larger, the atomic mass of Z gets smaller:

Formula	Atomic Mass of Z	Formula	Atomic Mass of Z
Z_1H_6	56.0	Z_2H_6	28.02
Z_3H_6	18.7	Z_4H_6	14.01
Z_5H_6	11.21	Z_6H_6	9.34
Z_9H_6	6.23	$Z_{14}H_6$	4.00
$Z_{56}H_6$	1.00		

Compare the atomic masses in this list with the real atomic masses of the elements. If the subscript of Z equals 2 or 4, values quite close to the atomic masses of $\boxed{\text{Si}}$ and $\boxed{\text{N}}$ come out. Most other subscripts give atomic masses not corresponding to atomic masses of authentic elements. A subscript of 56 gives an atomic mass of 1.00, which is close to the atomic mass of H, but the unknown would then be H_{62}, which is not a binary compound. A subscript of 14 gives an atomic mass of 4.00, but helium (atomic mass 4.00) forms no compounds.

e) The compound is either Si_2H_6 (silane, molecular mass 62.2186) or N_4H_6 (tetrazane, molecular mass 62.0746). Both substances exist, but $\boxed{Si_2H_6}$ is more stable.

2-19 Balance the equations by inspection. For example, in part a, assign 1 as the coefficient of NH_3. This obliges a coefficient on $\frac{1}{2}$ for the N_2 because $\frac{1}{2}$ mol of N_2 furnishes the whole 1 mol of N that is signified in "1 NH_3." Similarly, it obliges a coefficient of $\frac{3}{2}$ for the H_2 because $\frac{3}{2}$ mol of H_2 contains exactly the same number of atoms of H as 1 mol of NH_3. The answer to part a in the following table clears the fractions from this set of coefficients by multiplying them all by 2.

a) $3\,H_2 + N_2 \rightarrow 2\,NH_3$
b) $2\,K + O_2 \rightarrow K_2O_2$
c) $PbO_2 + Pb + 2\,H_2SO_4 \rightarrow 2\,PbSO_4 + 2\,H_2O$
d) $2\,BF_3 + 3\,H_2O \rightarrow B_2O_3 + 6\,HF$
e) $2\,KClO_3 \rightarrow 2\,KCl + 3\,O_2$
f) $CH_3COOH + 2\,O_2 \rightarrow 2\,CO_2 + 2\,H_2O$
g) $2\,K_2O_2 + 2\,H_2O \rightarrow 4\,KOH + O_2$
h) $3\,PCl_5 + 5\,AsF_3 \rightarrow 3\,PF_5 + 5\,AsCl_3$

2-21 a) According to the equation $Mg + 2\,HCl \rightarrow H_2 + MgCl_2$, the reaction produces 1 mol of H_2 for every 1 mol of Mg consumed. Diatomic hydrogen has a relative molecular mass of $2 \times 1.00797 = 2.01594$, and Mg has a relative atomic mass of 24.305. Therefore, "1 mol Mg $\rightarrow$ 1 mol H_2" implies "24.305 g Mg $\rightarrow$ 2.01594 g H_2". This fact provides a unit-factor to compute the mass of Mg that yields 1.000 g of H_2:

$$m_{Mg} = 1.000 \text{ g } H_2 \times \left(\frac{24.305 \text{ g Mg}}{2.01594 \text{ g } H_2} \right) = \boxed{12.06 \text{ g Mg}}$$

b) The equation states that 1 mol of I_2 arises from every 2 mol of $CuSO_4$:

$$2\,CuSO_4 + 4\,KI \rightarrow 2\,CuI + I_2 + 2\,K_2SO_4$$

Write down unit-factors to create a train of conversions, starting from the 1.000 g of I_2:

$$m_{CuSO_4} = 1.000 \text{ g } I_2 \times \left(\frac{1 \text{ mol } I_2}{253.809 \text{ g } I_2}\right) \times \left(\frac{2 \text{ mol } CuSO_4}{1 \text{ mol } I_2}\right) \times \left(\frac{159.602 \text{ g } CuSO_4}{1 \text{ mol } CuSO_4}\right)$$

$$= \boxed{1.258 \text{ g } CuSO_4}$$

c) According to the balanced equation, 1 mol of $NaBH_4$ yields 4 mol of H_2. Some might argue that such a reaction is not possible, reasoning that no chemical reaction can transform the 4 H atoms of $NaBH_4$ into the 8 H atoms of 4 H_2. In fact, the extra H comes from the other reactant, water. Write a series of unit-factors:

$$m_{NaBH_4} = 1.000 \text{ g } H_2 \times \left(\frac{1 \text{ mol } H_2}{4 \text{ mol } H_2}\right) \times \left(\frac{37.833 \text{ g } NaBH_4}{1 \text{ mol } NaBH_4}\right) = \boxed{4.692 \text{ g } NaBH_4}$$

2-23 One mole of $K_2Zn_3[Fe(CN)_6]_2$ contains 12 moles of C. Since carbon has no other place to go, 12 mol of K_2CO_3 must form per mole of $K_2Zn_3[Fe(CN)_6]_2$. This fact provides the second unit-factor in the following. The other factors are routine:

$$m = 18.6 \text{ g } K_2CO_3 \times \left(\frac{1 \text{ mol } K_2CO_3}{138.2 \text{ g } K_2CO_3}\right) \times \left(\frac{1 \text{ mol } K_2Zn_3[Fe(CN)_6]_2}{12 \text{ mol } K_2CO_3}\right)$$

$$\times \left(\frac{698.2 \text{ g } K_2Zn_3[Fe(CN)_6]_2}{1 \text{ mol } K_2Zn_3[Fe(CN)_6]_2}\right) = \boxed{7.83 \text{ g } K_2Zn_3[Fe(CN)_6]_2}$$

2-25 Balance a chemical equation to learn the relationship between the chemical amount of Si_2H_6 consumed and the chemical amount of SiO_2 formed. By inspection:

$$6 \, Si_2H_6 + 21 \, O_2 \rightarrow 12 \, SiO_2 + 18 \, H_2O$$

Next, use the density and volume of the gaseous Si_2H_6 to obtain its chemical amount. Then obtain the number of moles of the SiO_2 product and finally the mass of the SiO_2. The following does this all in a single series of unit-factors:

$$m_{SiO_2} = 25.0 \text{ cm}^3 \times \left(\frac{2.78 \times 10^{-3} \text{ g}}{1.00 \text{ cm}^3}\right) \times \left(\frac{1 \text{ mol } Si_2H_6}{62.2186 \text{ g } Si_2H_6}\right) \times \left(\frac{12 \text{ mol } SiO_2}{6 \text{ mol } Si_2H_6}\right)$$

$$\times \left(\frac{60.084 \text{ g } SiO_2}{1 \text{ mol } SiO_2}\right) = \boxed{0.134 \text{ g } SiO_2}$$

2-27) Use a series of unit-factors to pass from grams of Al_2O_3 to grams of cryolite:

$$m_{Na_3AlF_6} = 287 \text{ g } Al_2O_3 \times \left(\frac{1 \text{ mol } Al_2O_3}{101.962 \text{ kg } Al_2O_3} \right) \times \left(\frac{2 \text{ mol } Na_3AlF_6}{1 \text{ mol } Al_2O_3} \right)$$

$$\times \left(\frac{209.94 \text{ g } Na_3AlF_6}{1 \text{ mol } Na_3AlF_6} \right) = \boxed{1.18 \times 10^3 \text{ g } Na_3AlF_6}$$

2-29) It does not matter whether the substance in question is a product or reactant. The form of the unit-factors is similar:

$$567 \text{ g } KNO_3 \times \frac{1 \text{ mol } KNO_3}{101.103 \text{ g } KNO_3} \times \frac{1 \text{ mol } KCl}{1 \text{ mol } KNO_3} \times \frac{74.551 \text{ g } KCl}{1 \text{ mol } KCl} = \boxed{418 \text{ g } KCl}$$

For the mass of the by-product Cl_2, switch direction after the first unit-factor:

$$567 \text{ g } KNO_3 \times \frac{1 \text{ mol } KNO_3}{101.103 \text{ g } KNO_3} \times \frac{2 \text{ mol } Cl_2}{4 \text{ mol } KNO_3} \times \frac{70.906 \text{ g } Cl_2}{1 \text{ mol } Cl_2} = \boxed{199 \text{ g } Cl_2}$$

2-31 a) The small whole-number ratios in chemical formulas and balanced chemical equations refer to chemical amount, never to mass. The balanced equation given in this problem assures that the chemical amounts of XCl_2 and XBr_2 are equal. To use this fact, convert the mass of XBr_2 to a chemical amount of XBr_2. Also, convert the mass of XCl_2 to a chemical amount of XCl_2. The conversions require the molar masses of the two compounds, which require the unknown molar mass of X. Call this quantity x:

$$\text{molar mass } XBr_2 = (2 \times 79.909 + x) \text{ g mol}^{-1}$$
$$\text{molar mass } XCl_2 = (2 \times 35.453 + x) \text{ g mol}^{-1}$$

The chemical amounts of the two compounds are:

$$n_{XBr_2} = 1.500 \text{ g } XBr_2 \times \left(\frac{1 \text{ mol } XBr_2}{(159.818 + x) \text{ g } XBr_2} \right) = \frac{1.500}{159.818 + x} \text{ mol } XBr_2$$

$$n_{XCl_2} = 0.890 \text{ g } XCl_2 \times \left(\frac{1 \text{ mol } XCl_2}{(70.906 + x) \text{ g } XCl_2} \right) = \frac{0.890}{70.906 + x} \text{ mol } XCl_2$$

But the chemical amounts of the XBr_2 and XCl_2 are equal: Hence:

$$\frac{1.500}{159.818 + x} = \frac{0.890}{70.906 + x}$$

This equation is easily solved for x, the molar mass of the unknown element. It equals 58.8 g mol^{-1}; the relative atomic mass of the element is $\boxed{58.8}$.

b) Checking the relative atomic masses in the periodic table shows that the unknown element is most likely $\boxed{\text{nickel}}$.

2-33 There seems to be insufficient information. What is overlooked is that both NaCl and KCl furnish only a set, definite fraction of their mass to the total mass of chlorine. These two fractions are readily available from the atomic masses and formulas of KCl and NaCl. Thus, in addition to the obvious relationship: $x + y = 1.0000$ g where x and y are the masses of the NaCl and KCl respectively, we can write:

$$\text{mass of Cl from NaCl} + \text{mass of Cl from KCl} = \text{mass of Cl in AgCl}$$

Compute the molecular masses of NaCl, KCl and AgCl and use them to get the fraction of each compound that is Cl:

$$\left(\frac{35.4527}{58.4425}\right) x + \left(\frac{35.4527}{74.5510}\right) y = \left(\frac{35.4527}{143.3209}\right) 2.1476$$

Divide both sides of this equation by 35.4527:

$$\frac{x}{58.4425} + \frac{y}{74.5510} = \left(\frac{1}{143.3209}\right) 2.1476$$

Combining this equation with $x + y = 1.0000$ and solving gives:

$$x = 0.4249 \qquad \text{and} \qquad y = 0.5751 \text{ g}$$

The mass percentages of NaCl and KCl in the original mixture of NaCl and KCl are then $\boxed{42.49\%}$ and $\boxed{57.51\%}$ respectively.

2-35 Write the balanced chemical equation for the reaction:

$$\text{HCl}(g) + \text{NH}_3(g) \rightarrow \text{NH}_4\text{Cl}(s)$$

One mole of HCl gas weighs 36.46 g, and one mole of NH_3 weighs only 17.03 g. It takes fewer heavy molecules than light molecules to make up a specific mass. Therefore, equal masses of HCl and NH_3 contain more molecules of NH_3. When the two react in a 1-to-1 molar ratio, the HCl is used up first. This means HCl is the limiting reactant. When the HCl is used up, the reaction stops, leaving excess NH_3. The mass of NH_4Cl is:

$$m_{\text{NH}_4\text{Cl}} = 2.00 \text{ g HCl} \times \left(\frac{1 \text{ mol HCl}}{36.46 \text{ g HCl}}\right) \times \left(\frac{1 \text{ mol NH}_4\text{Cl}}{1 \text{ mol HCl}}\right) \times \left(\frac{53.49 \text{ g NH}_4\text{Cl}}{1 \text{ mol NH}_4\text{Cl}}\right)$$

$$= \boxed{14.7 \text{ g NH}_4\text{Cl}}$$

Since 20.0 g of matter was present originally, the mass of left-over NH_3 is $(20.0 - 14.7) = \boxed{5.3 \text{ g } NH_3}$. The above is a conceptual method to identify the limiting reactant. The recommended method is to compute that 10.0 g of NH_3 and unlimited HCl would give 31.4 g of NH_4Cl, whereas 10.0 g of HCl and unlimited NH_3 would give only 14.7 g of NH_4Cl.

Tip. A quick way to identify the limiting reactant is to divide the chemical amount of each reactant by the coefficient that the reactant has in the balanced chemical equation. The smallest answer identifies the limiting reactant.

2-37 Compute the maximum possible yield (the theoretical yield) of Fe as follows:

$$433.2 \text{ g } Fe_2O_3 \times \left(\frac{1 \text{ mol } Fe_2O_3}{159.69 \text{ g } Fe_2O_3} \right) \times \left(\frac{2 \text{ mol } Fe}{1 \text{ mol } Fe_2O_3} \right) \times \frac{55.87 \text{ g } Fe}{1 \text{ mol } Fe} = \boxed{303.0 \text{ g } Fe}$$

Compare this result to the actual yield. The percent yield is:

$$\frac{\text{Actual Yield}}{\text{Theoretical Yield}} \times 100\% = \frac{254.3 \text{ g Fe actual}}{303.0 \text{ g Fe theoretical}} \times 100\% = \boxed{83.93\%}$$

2-39 a) The problem requires more significant figures than the computations in **1-29** but follows the same principles. The relative molecular mass of the human parathormone is $13,932.24$. The mass percentages are: C, 59.571%; H, 6.4967%; N, 12.5668%; O, 18.833%; S, 2.532%.

2-41 a) The word "binary" means that the only elements in the three compounds are oxygen and the metal (label it M). The first compound contains 13.38 g of O for every 86.62 g of M, which is $\boxed{0.15457 \text{ g}}$ of O per gram of M. The second and third compounds have $\boxed{0.1029 \text{ g}}$ of O and $\boxed{0.07721 \text{ g}}$ of O per gram of M, respectively.

b) If the first compound is MO_2, then the second is $MO_{4/3}$ because the second compound has 2/3 as much oxygen per given quantity of M as the first. This formula is improved by multiplying both subscripts by 3 to clear the fraction. The answer is $\boxed{M_3O_4}$. The third compound is $\boxed{MO}$ if the first is MO_2 because it has almost exactly 1/2 as much oxygen per quantity of M.

c) Continue the assumption that the first compound is MO_2. Compute the amount of metal that combines with 2 mol of O (2×15.9994 g of O) in each of the first compounds. It is 207.2 g. Because the formula tells us that there is 1 mol of M per 2 mol of O, the relative atomic mass of M equals 207.2. Consulting the periodic table establishes that M is $\boxed{\text{lead}}$.

2-43 Imagine 1.000 g of the first oxide. It contains 0.6960 g of Mn and 0.3040 g of O. Dividing these masses by the relative atomic masses of Mn and O gives the relative numbers of moles of the two elements. The smallest whole-number ratio of these chemical amounts gives the subscripts in the compound's empirical formula:

$$\text{Mn}_{\frac{0.6960}{54.94}}\text{O}_{\frac{0.3040}{16.00}} \implies \text{Mn}_{0.01267}\text{O}_{0.01900} \implies \text{Mn}_{1.000}\text{O}_{1.500} \implies \boxed{\text{Mn}_2\text{O}_3}$$

Repeat the procedure for the second oxide:

$$\text{Mn}_{\frac{0.6319}{54.94}}\text{O}_{\frac{0.3681}{16.00}} \implies \text{Mn}_{0.01150}\text{O}_{0.02301} \implies \text{Mn}_{1.000}\text{O}_{2.000} \implies \boxed{\text{MnO}_2}$$

Tip. Getting from the second to the third formula in the preceding requires dividing each subscript by the smallest of the subscripts.

2-45 a) The balanced equations for the conversion of cyanuric acid to isocyanuric acid and the reaction of isocyanuric acid with nitrogen dioxide are:

$$\boxed{\text{C}_3\text{N}_3(\text{OH})_3 \rightarrow 3\,\text{HNCO}} \quad \text{and} \quad \boxed{8\,\text{HNCO} + 6\,\text{NO}_2 \rightarrow 7\,\text{N}_2 + 8\,\text{CO}_2 + 4\,\text{H}_2\text{O}}$$

Balancing by inspection in the second equation works requires some care. Assign 1 as the coefficient of HNCO. Then, focus on C and O because these elements occur in only one compound on each side of the equation. If HNCO on the left has a coefficient of 1, CO_2 on the right must have a coefficient of 1 and H_2O on the right must have a coefficient of 1/2 to achieve balance in C and H. These two coefficients on the right imply a total of $2\frac{1}{2}$ mol of O on the right. The "1 HNCO" on the left supplies only 1 mol of O, and its coefficient must not be changed. The NO_2 must supply the other 3/2 mol of O. To do this, its coefficient must be 3/4. The left side now has 7/4 mol of N. The coefficient of N_2 on the right must therefore equal 7/8 because $7/8 \times 2 = 7/4$ (the 2 comes from the subscript in N_2). Multiplying all five coefficients by 8 eliminates fractional coefficients.

b) Use a series of unit-factors constructed from the molar masses of the compounds and the coefficients of the two balanced equations:

$$m_{\text{C}_3\text{N}_3(\text{OH})_3} = 1.70 \times 10^{10}\text{ kg NO}_2 \times \left(\frac{1\text{ mol NO}_2}{0.046\text{ kg NO}_2}\right) \times \left(\frac{8\text{ mol HNCO}}{6\text{ mol NO}_2}\right)$$

$$\times \left(\frac{1\text{ mol C}_3\text{N}_3(\text{OH})_3}{3\text{ mol HNCO}}\right) \times \left(\frac{0.129\text{ kg C}_3\text{N}_3(\text{OH})_3}{1\text{ mol C}_3\text{N}_3(\text{OH})_3}\right) = \boxed{2.1 \times 10^{10}\text{ kg C}_3\text{N}_3(\text{OH})_3}$$

2-47 The only product of the reaction that contains nitrogen is *m*-toluidine; the only reactant that contains nitrogen is 3′-methylphthalanilic acid. The mass of nitrogen coming from the reactant must equal the mass of nitrogen ending up in the product by the law of conservation of mass. The m-toluidine (empirical formula C_7H_9N) is 13.1% nitrogen by mass (calculated as in **2-1**). The 5.23 g of m-toluidine therefore contains 0.685 g of nitrogen. The $3'$-methylphthalanilic acid consists of 5.49% nitrogen by mass (as given in the problem). The issue thus becomes finding the mass of $3'$-methylphthalanilic acid that contains 0.685 g of nitrogen. Let this mass be x. Then $0.0549x = 0.685$ g. Solving gives x equal to 12.5 g $3'$-methylphthalanilic acid. This wordy analysis is equivalent to the following:

$$5.23 \text{ g toluidine} \times \left(\frac{13.1 \text{ g N}}{100 \text{ g toluidine}} \right) \times \left(\frac{100 \text{ g } 3'\text{-methyl}}{5.49 \text{ g N}} \right) = \boxed{12.5 \text{ g } 3'\text{-methyl}}$$

2-49 a) Write an unbalanced equation to represent what the problem tells about the process:

$$C_{12}H_{22}O_{11} + O_2 \rightarrow C_6H_8O_7 + H_2O$$

Balance this equation as to carbon by inserting the coefficient 2 in front of the citric acid. Then balance the hydrogens by putting a 3 in front of the water (of the 22 H's on the left, 16 appear in the citric acid and the rest appear in the water). Next, consider the oxygen. The right side has $(2 \times 7) + (3 \times 1) = 17$ O's. On the left side, the sucrose furnishes 11 O's so the remaining 6 must come from 3 molecules of oxygen. The balanced equation is

$$\boxed{C_{12}H_{22}O_{11} + 3\,O_2 \rightarrow 2\,C_6H_8O_7 + 3\,H_2O}$$

b) The balanced equation provides the information to write the second unit-factor in the following:

$$m_{C_6H_8O_7} = 15.0 \text{ kg sucrose} \times \left(\frac{1 \text{ kmol sucrose}}{342.3 \text{ kg sucrose}} \right) \times \left(\frac{2 \text{ kmol citric acid}}{1 \text{ kmol sucrose}} \right)$$
$$\times \left(\frac{192.13 \text{ kg citric acid}}{1 \text{ kmol citric acid}} \right) = \boxed{16.8 \text{ kg citric acid}}$$

Tip. Save time by creating factors such as "1 kilomole sucrose / 342.3 kg sucrose." Also, only *part* of the balanced equation is needed, namely the 2 : 1 molar ratio of citric acid from sucrose. The O_2 and H_2O could have been left unbalanced.

2-51 a) Compute the chemical amount of XBr_2 that is present and recognize that two moles of AgBr appear for every one mole of XBr_2 present in the 5.00 g sample. This fact appears in the second unit-factor in the following:

$$1.0198 \text{ g AgBr} \times \left(\frac{1 \text{ mol AgBr}}{187.77 \text{ g AgBr}} \right) \times \left(\frac{1 \text{ mol } XBr_2}{2 \text{ mol AgBr}} \right) = 0.002716 \text{ mol } XBr_2$$

The molar mass $\mathcal{M}$ of any substance equals its mass divided by its chemical amount:

$$\mathcal{M}_{XBr_2} = \frac{m_{XBr_2}}{n_{XBr_2}} = \frac{0.500 \text{ g}}{0.002716 \text{ mol}} = \boxed{184.1 \text{ g mol}^{-1}}$$

b) The atomic mass of X equals the molecular mass of the compound minus the contribution of the bromine:

$$\text{atomic mass of X} = 184.1 - 2(79.9) = \boxed{24.3}$$

Checking the atomic masses in the periodic table shows that X is magnesium, $\boxed{\text{Mg}}$.

2-53 In the Solvay process, one mole of NH_3 takes up one mole of CO_2, holds it for NaCl to attack, and departs as NH_4Cl. Each mole of the $NaHCO_3$ that results then goes on to give 1/2 mol of Na_2CO_3. Thus, for each mole of NH_3 1/2 mol of Na_2CO_3 forms. In the following equation the prefix "M" in a unit stands for mega:[4]

$$m_{Na_2CO_3} = 1 \text{ metric ton NH}_3 \times \left(\frac{1 \text{ Mmol NH}_3}{17.03 \text{ metric ton NH}_3} \right)$$

$$\times \left(\frac{1 \text{ Mmol Na}_2CO_3}{2 \text{ Mmol NH}_3} \right) \times \left(\frac{105.98 \text{ Mg Na}_2CO_3}{1 \text{ Mmol Na}_2CO_3} \right) = 6.22 \text{ Mg Na}_2CO_3$$

A metric ton equals a megagram, so $\boxed{6.22 \text{ metric tons}}$ of sodium carbonate is produced per metric ton of ammonia.

2-55 Assume that the limestone raw material is pure calcium carbonate. Then compute the theoretical yield of ethylene in the three-step process. An essential unit-factor comes from viewing the equations as a group: it takes 1 mol of $CaCO_3$ to produce 1 mol of C_2H_2. It is *not* necessary to compute the theoretical yield of the lime (CaO) and calcium carbide (CaC_2) produced and later consumed on the way to the final product. The following conversions use an additional short-cut—just as a metric ton (1000 kg) is a million times bigger than a gram, a megamole (Mmol) is a million time bigger than a mole:

$$m_{C_2H_2} = 10.0 \text{ metric ton CaCO}_3 \times \left(\frac{1 \text{ Mmol CaCO}_3}{100.1 \text{ metric ton CaCO}_3} \right)$$

$$\times \left(\frac{1 \text{ Mmol C}_2H_2}{1 \text{ Mmol CaCO}_3} \right) \times \left(\frac{26.03 \text{ metric ton C}_2H_2}{1 \text{ Mmol C}_2H_2} \right) = 2.60 \text{ metric ton C}_2H_2$$

The percent yield equals the actual yield divided by the theoretical yield and multiplied by 100%:

$$\text{percent yield C}_2H_2 = \frac{2.32 \text{ metric ton C}_2H_2}{2.60 \text{ metric ton C}_2H_2} \times 100\% = \boxed{89.2\%}$$

---

[4]See text Table B-3, text page A-11.

Chapter 3

Chemical Bonding: The Classical Description

Atoms consist of a small positively charged nucleus surrounded by a swarm of negatively charged electrons. Chemical bonds result from the redistribution of electrons among atoms as they approach each other. Bonds persist because the new distribution of electrons has lower energy than the old, unbonded one. The description of chemical bonding begins by distinguishing two idealized types of bonds:

Covalent. The redistribution of electrons leads to a sharing between the atoms.

Ionic. The redistribution of electrons leads to a transfer of electrons from one atom to another.

Intermediate redistributions of electrons (neither fully ionic nor fully covalent) are exceedingly common. The partial transfer of electrons (but with some sharing as well) gives **polar covalent** bonds.

Electronegativity

The character (ionic versus covalent) of a bond is determined by the relative electronegativity of the atoms that are bonded.

Electronegativity is a measure of the power of an atom to attract electrons to itself in a bonding situation.

If two bonded atoms have a large difference in electronegativity, then the bond between them is ionic; if the difference is small, then the bond is covalent (see **3-39**).

43

Ionization Energies and Electron Affinities

The ionization energy of an atom is the energy required to detach one electron from that atom. The electron affinity is the energy released when an atom attaches an electron:

$$X(g) \rightarrow X^+(g) + e^- \qquad \text{electron detachment}$$
$$X(g) + e^- \rightarrow X^-(g) \qquad \text{electron attachment}$$

The ionization energy and electron affinity condition the tendency of atoms to share or transfer electrons. Recognize the following points:

- Ionization energies and electron affinities are precisely measurable properties of single atoms.

- Ionization energies are positive for the atoms of all the elements in the periodic table. It always requires energy to detach electrons from atoms.

- Electron affinities are sometimes positive (meaning that the attachment of the electron releases energy) and sometimes negative.

- Attachment of an electron by an atom is *not* the reverse of the detachment of an electron from that same atom—the two preceding equations are not each other's reverses.

Ionization energies are measured by photoelectron spectroscopy (see text Section 15-3[1]). In general the ionization energy *increases* from left to right across a period (row) in the periodic table and *decreases* from top to bottom within a group (column). The periodic trends in *EA* parallel the changes in *IE*. Both atomic properties tend to increase from left to right in a period and from bottom to top in a group.

Mulliken Electronegativities

Robert Mulliken observed that atoms with low *IE*'s and low *EA*'s readily lose their own electrons and are poor at accepting new electrons while atoms with high *IE*'s and high *EA*'s strongly resist giving up their own electrons and are good at accepting new electrons. He therefore defined electronegativity as:

$$\text{Electronegativity} \propto \tfrac{1}{2}(IE + EA)$$

Mulliken electronegativities are invariant atomic properties.

The electronegativities of the elements increase to the right across the periodic table and decreases going down the table in the groups to the extreme left and extreme right. In the center of the table, electronegativities increase going down the table. Elements of low electronegativity are said to be **electropositive**.

---

[1]On text page 537.

Ionic Bonding

Not all of an atom's electrons participate significantly in chemical bonding. Electrons are organized in **shells** surrounding the nucleus. Electrons in inner shells (**core electrons**) are not directly involved in interactions with other atoms. The outermost, partially filled shell (**the valence shell**) of an atom holds the electrons that are most active in chemical bonding. These are the **valence electrons**.

Lewis Diagrams

The **Lewis diagram** of an atom represents its valence electrons explicitly with dots. The dots go around the chemical symbol at its four sides. If there are more than four valence electrons, dots for them are paired with those already present. See **3-1**. It does not matter which location gets the first dot or the first pair of dots. The maximum number of dots surrounding an atom is usually eight, but exceptions are common (see below).

  • **The number of valence electrons (dots) to use in a Lewis diagram of a main-group element equals the group number of the element.**

Atoms often gain or lose one or more valence electrons. A positively charged ion results when electrons are lost. A negatively charged ion results when electrons are gained. These ions are called **cations** and **anions** respectively. Lewis dot symbols for ions are written by removing or adding the proper number of dots from the symbol for the atom. Then the net charge is indicated by a right superscript. For a $+1$ cation and a -1 anion, the superscripts are a simple plus and minus sign. For other cations and anions, the superscript is written as in Ca^{2+}. Avoid putting the sign first in the superscript ("Ca^{+2}" is incorrect).

The Formation of Ionic Compounds

The formulas of many compounds can be understood on this basis:

  • **Atoms tend to lose or gain electrons to form ions that have the same number of valence electrons as a noble-gas atom.**

The noble-gas atoms, with the exception of helium, have eight electrons in their valence shells. The above statement is therefore abbreviated: "atoms tend to lose or gain electrons to attain a **completed valence octet** of electrons". This is the **octet rule**.

The octet rule, in combination with the principle of charge neutrality, explains the formulas of binary **ionic compounds** between the electropositive, metallic elements on the left side of the periodic table and the electronegative, nonmetallic elements on the right side. **Charge neutrality** requires that the total positive charge on the cations must equal the total negative charge on the anions. See **3-3**. Thus, the formation of salt, sodium chloride, is represented:

$$\text{Na·} + :\ddot{\text{C}}\text{l·} \rightarrow (\text{Na})^+ \, (:\ddot{\text{C}}\text{l}: \,)^-$$

Sodium chloride is the archetypal ionic compound. Accordingly, simple ionic compounds often called **salts**. Compounds such as NH_4Cl or $CaSO_4$ in which the cation or anion is a molecular ion (see below) are also salts.

Names and Formulas of Ionic Compounds

Ionic compounds are named by giving the name of the cation followed by a space and then the name of the anion. See **3-5**. Monatomic cations, which derive from metallic elements, simply take the name of the element. **Example:** Na is sodium; Na^+ is sodium ion. Some metals form more than one cation. If so, Roman numerals are put in parentheses after the name of the metal to indicate the charge. **Examples:** Cu^+ and Cu^{2+} are copper(I) and copper(II); Fe^{2+} and Fe^{3+} are iron(II) and iron(III).

Monatomic anions are derived from nonmetals. They are named by adding the suffix *-ide* to the stem of the name of the parent element. **Examples:** Cl^-, O^{2-}, and P^{3-} are chloride, oxide and phosphide respectively.

There are many **polyatomic** anions, particularly oxoanions. See text Table 3-1.[2]

• **Memorize the names of the common inorganic polyatomic anions.**

Naming compounds from formula and writing formulas from names require this information (see **3-5**, **3-7**, and **3-9**.

Coulomb Stabilization Energy in Ionic Bonding

Removal of an electron from a gaseous atom requires the input of energy, the ionization energy *IE*, for every element. Many atoms release energy, the electron affinity (*EA*), as they accept an electron. For all possible pairs of atoms:

$$\Delta E_\infty = (IE - EA) > 0$$

It always costs energy to transfer an electron from one neutral atom to another.

---

[2]Text page 62.

Ionic bonds form because of the **Coulomb attraction** (electrostatic attraction) between charges of unlike sign. The change in energy that occurs as two ions approach each other is:

$$\Delta E_{\text{Coulomb}} = \frac{Q_1 Q_2}{4\pi\epsilon_0 R}$$

where R is the distance between the ions, the Q's are the charges of the ions, and ϵ_0 is a constant $(8.854 \times 10^{-12} \text{ C}^2 \text{ J}^{-1} \text{m}^{-1})$. According to this formula, the energy of a plus/minus pair of ions goes to negative infinity as R goes to zero. But ions are not point charges. At very short distances the core electrons of the ions start to repel each other, and the energy rises steeply. See text Figure 3-1.[3] The distance at which the short-range repulsions balance the Coulomb attraction is R_e, the equilibrium bond distance.

The energy of dissociation of the ion-pair is symbolized ΔE_d. It is the energy required to unmake the bond between the ions and restore them as separated neutral atoms. It equals the negative of the sum of $\Delta E_{\text{Coulomb}}$ and ΔE_∞:

$$\Delta E_d = -\left(\frac{Q_1 Q_2}{4\pi\epsilon_0 R} + \Delta E_\infty\right)$$

See **3-11** and **3-71** for the use of this formula.

The preceding applies to bonds between single pairs of gaseous ions. A collection of such bonded pairs of ions in the same region of space would release additional energy as it condensed to an ionic crystal, in which every ion interacts with all others. This "lattice energy" is treated in text Section 19-4.[4]

The Structures of Isolated Molecules

Chemical bonds are characterized by these physical measures:

• **Bond Lengths.** The distance from one nucleus to another is an interatomic distance. Interatomic distances have no upper limit. Interatomic distances are **bond lengths** if the two atoms under consideration are connected by a chemical bond. Bond lengths are on the order of 1×10^{-10} m, which is on the order of 1 Ångstrom (Å):

1 Ångstrom $= 1 \times 10^{-10}$ m $= 0.1$ nanometer (nm) $= 100$ picometer (pm)

Bond lengths are symbolized R or R_e. The subscript (e for equilibrium) emphasizes that molecular vibrations can temporarily lengthen and shorten bonds. Get some typical bond lengths in mind: C-to-C bonds range from 1.20 to 1.54 Å; O-to-H bonds

---

[3]Text page 64.

[4]Text page 713.

in various compounds range from 0.94 to 1.09 Å. Text Table 3-2[5] gives the bond length in many diatomic molecules. Bond lengths range up to about 3 Å. Within a group in the periodic table, bond lengths usually increase with increasing atomic number. See **3-13**. Bond lengths for a given type of bond do not change much from molecule to molecule. See Table 3-3.[6]

• **Bond Energies.** Chemical bonds form because the atoms involved have a lower energy when close together than when far apart. Breaking a bond means moving the bonded atoms apart. It requires energy. This energy is ΔE_d, the **dissociation energy**, of the bond. It measures the strength of the bond.

Bond energies are always positive and are on the order of 100 to 1000 kJ mol^{-1}. They grow less with increasing atomic number of one or both of the bonded atoms (see **3-13**). Bond energies are reproducible within about 10% for a given pair of bonded atoms as long as the bonds being compared have the same order (single, double, or triple).

• **Bond Angles.** Any three atoms in a molecule define an angle, which is measured in degrees.[7] **Bond angles** are the angles defined by bonded atoms. They are always taken in the sense that makes them less than or equal to 180°.

• **Bond Order.** Sometimes the length and energy of a bond between the same two kinds of atoms (for example C and N atoms) are sharply different from one compound to another. This breakdown in reproducibility is ascribed to the bonds have different **bond orders**. The most common bond orders are single (as in C—N), double (as in C=N), and triple (as in C≡N). Fractional bond orders are also possible. See **3-29**.

Covalent Bonding: Electron Pair Model

Many elements have no strong tendency to form either anions or cations. Atoms of these elements attain valence octets by *sharing* electrons with other atoms. Electrons have a negative charge. If an electron shared by two nuclei spends most of its time *between* the two, it tends to lower the repulsion between the positively-charged nuclei because it is attracted to both.

The Lewis model of the covalent bond represents covalent bonds as shared pairs of valence electrons positioned between nuclei.

Compounds in which the bonding is predominately covalent are called **covalent compounds.** Covalent bonds are strong and hold atoms close together. They maintain groups of atoms in distinct molecules or **molecular ions.** The latter are bound

---

[5]Text page 67.
[6]Text page 67.
[7]In fact, three points in space define *two* angles, the second being equal to 360° minus the first.

groups of atoms that have an overall positive on negative charge. Molecular ions are the same as polyatomic ions.

Lewis Diagrams for Covalent Bonding

The **Lewis structure** or Lewis diagram of a molecule or molecular ion displays the sharing of valence electrons by positioning the shared pairs between the symbols of the atoms. Valence electrons that are not shared are **lone pairs.** They make no direct contribution to the bonding, but stay on their original atoms. Lewis structures are written so as to follow the octet rule, which holds that atoms of the main-group elements should be surrounded by (should "see") eight electrons. An important exception is hydrogen, which should "see" two electrons, according to the octet rule.

Whenever bonds connect atoms in molecules, the question of **connectivity** (the order of linkages) arises. A Lewis structure *always* states the connectivity of atoms. The Lewis structure of water shows that the order of the three atoms is H—O—H and not H—H—O. Writing three atoms in a straight line is equivalent in a Lewis structure to writing them with a bend at the middle atom. One does *not* read bond angles from a Lewis structure. Lewis structures are formulated in the following steps:

1. Sum up the total number of valence electrons in the molecule or molecular ion under consideration. **Example:** ClO_2^+. Each O atom contributes 6 valence electrons (oxygen is in Group VI) and the Cl atom contributes 7 valence electrons (chlorine is in Group VII). The +1 charge on the group as a whole removes one electron: $2(6) + 7 - 1 = 18$ valence electrons.

2. Count the number of connections among the atoms. Note that joining p atoms together requires at least $(p - 1)$ connections. Structures may use more than $p - 1$ links but can never use fewer.

3. Determine the "skeleton" of the molecule or molecular ion. This is the pattern by which the atoms are joined together. Neither a molecular nor an empirical formula reveals which atoms are located next to which. Indeed some sets of atoms may arrange themselves in more than one skeleton, resulting in **isomers,** molecules that contain the same number and kind of atoms in different orders (see **3-17**, **3-31**, and **3-49**).[8] However, there are ways to get a skeleton in most cases:

    (a) The chemical formula as written may specify the order of the connections among the atoms. **Example:** The connections between atoms in Cl_2O_7 could be anything. The formula $O_3ClOClO_3$ suggests:

---

[8]For more on isomers, see pages 488 and 507 of this Guide.

(b) Hydrogen atoms are nearly always on the periphery in Lewis structures because hydrogen rarely forms more than one covalent bond. **Example:** H_2O_2. The skeleton is [H O O H], not [H O H O] or anything else with interior hydrogen atoms. Fluorine atoms are also always on the periphery.

(c) Molecules and molecular ions tend to be clusters with the unique atom (if any) at the center of the cluster, rather than chains. **Example:** SF_4 (see **3-33b**). Place the single sulfur at the center of a cluster with all four fluorine atoms bonded to it. Avoid S—F—F—F—F and other chains.

(d) The skeleton may be described explicitly. See **3-65b**.

4. Put the valence electrons around the symbols for the elements in the skeleton in such a way that all the elements see a noble-gas electron configuration, if possible. For most representative elements this means that the atoms should see an octet of electrons—four pairs. For hydrogen, each atom should see one pair of electrons.

Sometimes the last step cannot be completed. To understand why, analyze what it means to satisfy the octet rule on every atom. If there is no sharing of electrons, satisfying the octet rule requires eight times the number of non-hydrogen atoms plus twice the number of hydrogen atoms. Sharing reduces the number of electrons required:

- **Every single bond reduces the required number of electrons by two.**

This is because shared electrons count in the octets of two atoms. Since it takes $p-1$ links to connect p atoms, the number of electrons for "octets all around" equals:

$$N_{e-} = [8 \times N_{non-H}] + [2 \times N_H] - [2 \times (p-1)]$$

Example: HNO_3. Substitution in the formula gives:

$$N_{e-} = [8 \times 4] + [2 \times 1] - [2 \times (4-1)] = 26$$

Only when N_{e-} equals the actual number of electrons available is it possible to put a shared pair of electrons at every connection between atoms, to fill in octets all around (but duets on H), and to arrive at a satisfactory Lewis structure. Two other cases arise:

- N_{e^-} *exceeds* the actual number of valence electrons available.

  When this happens, invoke **multiple bonding**, the sharing of multiple pairs of electrons between atoms. The sharing of two pairs is a **double bond**; the sharing of three pairs is a **triple bond**. A double bond has bond order two; a triple bond has bond order three. More sharing means that fewer valence electrons are required to give every atom an octet. Most often, double bonds involve a connection to a C, N, O or S atom, and triple bonds involve a connection to a C or N atom. Double and triple bonds are shorter (and stronger) than single bonds connecting the same atoms.[9]

  The formation of *rings* of atoms also allows the use of fewer electrons. See **3-25**.

- N_{e^-} may *be less than* the actual number of valence electrons. When this happens, the octet rule fails. There are too many electrons to be accommodated according to the rule, yet the electrons must be accommodated. Invoke **valence-shell expansion** on at least one atom. An atom with an expanded valence shell sees more than eight valence electrons. When resorting to valence-shell expansion, abandon the octet rule for the central atom first. Thus, in SF_6 the central S atom sees 12 electrons, but the F atoms have octets. For other examples, see **3-33** and **3-67**. Valence-shell expansion is common for elements from the third and subsequent period, that is, the octet rule is well obeyed only for second-row elements.

A Lewis structure that satisfies the octet rule strongly suggests stability for the molecule; the unavailability of such a structure suggests instability.

No certainties exist because of valence-shell expansion and these further deviations from the octet rule:

Octet-Deficient Molecules. A few molecules (BF_3 and BeF_2 for examples) do *not* have octets on the central atom, despite the fact that multiple bonding would satisfy the octet rule.

Odd-Electron Molecules. Many "odd molecules" (those with an odd number of valence electrons) exist. Yet, satisfactory Lewis structures cannot be drawn unless the number of valence electrons is even. The Lewis approach is clearly incomplete. See **3-65**.

Formal Charges

The calculation of a quantity called **formal charge (f.c.)** helps in the selection of the best Lewis structure when more than one is possible. A formal charge is a whole

---

[9]See text Table 3-3, text page 68.

number associated with a specific atom in a Lewis structure. Every atom in a Lewis structure has a formal charge; the best Lewis structures put formal charges of zero or near zero on all atoms. See **3-61** and **3-65a**.

The formal charge does *not* equal the actual electrical charge residing on an atom in a molecule or molecular ion, but is determined by the following formal procedure:

1. Draw the Lewis structure.

2. For each atom apply the formula:

$$\text{f.c.} = \left(\text{Group No.}\right) - \left(\text{No. } e^-\text{'s in lone pairs}\right) - \frac{1}{2}\left(\text{No. } e^-\text{'s in bonds}\right)$$

This formula states that the formal charge on an atom equals the number of valence electrons it starts with minus the number it retains for itself (in lone pairs) minus the number it would get if all bonds were broken with a 50:50 distribution of the shared electrons. **Example:** Compute the f.c.'s on the atoms in a Lewis structure for SO_2: **Solution:**

$$\overset{0}{:}\overset{\;}{\ddot{O}}\overset{+1}{=}\overset{\;\;}{\ddot{S}}\overset{-1}{-}\ddot{O}:$$

The formula works this way for the oxygen atom on the left:

$$\text{f.c.} = \left(\text{Group No.}\right) - \left(\text{No. } e^-\text{'s in lone pairs}\right) - \frac{1}{2}\left(\text{No. } e^-\text{'s in bonds}\right)$$
$$= 6 - 4 - 2 = 0$$

In words: the O atom on the left has 4 electrons in lone pairs. An even split of its double bond gives it 2 more for a total of 6. This equals its original 6 valence electrons, so its f.c. equals zero.

The O on the right has 6 electrons in three lone pairs. An even break of its single bond gives it 1 more electron. It originally had 6 electrons, so its f.c. is -1. The S atom has 1 single bond and 1 double bond. If these bonds are broken, the S atom gets 3 electrons to go with the 2 in its lone pair. Because it is 1 electron short of the 6 it started with, the S has f.c. $6 - 2 - 3 = +1$.

The algebraic sum of the formal charges on all of the atoms in a molecule or molecular ion must equal the actual charge on that species. This provides a useful check against errors in writing Lewis structures and in computing formal charge. For this reason:

- **Compute formal charges routinely when writing Lewis structures.**

Resonance

Often, two or more Lewis structures can be drawn on the same skeleton but depicting different distributions of electrons. Lewis structures related in this way are **resonance structures**. The rules for constructing Lewis structures require a specific location for every valence electron. Resonance is invoked when there is no reason to prefer one distribution of locations over another. The structures:

$$\overset{0}{:\ddot{O}}=\overset{+1}{\ddot{O}}-\overset{-1}{\ddot{O}:} \qquad \longleftrightarrow \qquad \overset{-1}{:\ddot{O}}-\overset{+1}{\ddot{O}}=\overset{0}{\ddot{O}:}$$

are resonance structures for the O_3 molecule. Both localizations of the double bond (to the right and to the left) are incomplete representations of the bonding. The actual bonding is an equal mixture of the resonance contributors. There is a 3/2 bond (the average of a single and double) on each side.

Do not misunderstand "resonance" to mean that the electrons flip rapidly back and forth between the two sites. The electron distribution is a stationary intermediate hybrid of the extremes represented by the resonance structures. See **3-27** and **3-63**.

Resonance structures do not in general contribute equally to the final picture of the bonding in a molecule or molecular ion. See **3-31**. Structures that violate the octet rule and build up large formal charges on several atoms are then not wrong but rather "low-percentage resonance contributors". Every Lewis structure that displays the correct number of valence electrons, however distributed, is defensible as a low-percentage contributor. Low-percentage contributors are usually marked wrong on examinations unless accompanied by high-percentage contributors and an explanation.

Bonding in Polar Covalent Molecules

Pauling's Electronegativity Scale

The symbol for electronegativity is the Greek chi (χ)—the electronegativity of atom A is represented χ_A. Numerical electronegativities are derived in at least two different ways: the method of Mulliken (see page 44 in this Guide) and the method of Pauling.

Linus Pauling defined the difference in electronegativity between two atoms as:

$$\chi_A - \chi_B = 0.102\sqrt{\Delta}$$

where Δ equals the bond energy of the A—B bond minus the geometric mean[10] of the bond energies of the pure covalent A—A and B—B bonds:

$$\Delta = \Delta E_{d(A-B)} - \sqrt{\Delta E_{d(A-A)} \times \Delta E_{d(B-B)}}$$

---

[10] Recall that the geometric mean of two numbers equals the square root of their product.

The Pauling scale sets the electronegativity of fluorine equal to 3.98. Other χ's are then arrived at by measuring bond energies and calculating $\chi_F - \chi_A$. Cesium has the smallest χ, so cesium is the most electropositive element. Slightly different χ's are obtained for the different elements using bond energy data for different sets of compounds. Tables[11] of Pauling electronegativities list averages based on measured bond energies from many compounds.

Dipole Moment and Percent Ionic Character

Molecules in which the center of the distribution of positive charge does not coincide with the center of the distribution of negative charge possess an electric **dipole moment**. In effect they have an electrically positive end and an electrically negative end. They are **polar**. When placed in an electric field, polar molecules tend to align themselves with the direction of the field. Their negative ends are attracted by the positive side of the field and repelled by the negative; the reverse holds for their negative ends. Molecules not aligned with the field experience a torque, or twist, which tends to align them. For a given electric field, the size of the torque depends on the magnitude of the molecule's dipole moment. Numerically, the dipole moment is:

$$\mu = \mathcal{Q}R$$

where R, is the distance separating charges of opposite sign and equal magnitude, $\mathcal{Q}$.

A difference in electronegativity between two bonded atoms means indicates a separation of charge across the length of the bond—a dipole moment is associated with the bond.

For diatomic molecules, which have only one connection between atoms, this dipole moment of the bond equals the dipole moment of the molecule. In molecules that contain more than one bond, the overall molecular dipole moment depends both on bond polarity and molecular geometry (see below).

Dipole moments are experimentally measurable. The magnitude of a dipole moment in a polyatomic molecule provide important information about the molecule's geometry. Absence of a dipole moment argues for some kind of symmetrical molecular structure. See **3-47** and **3-51**.

The Coulomb meter (C m) is the SI unit for dipole moment, but dipole moments are nearly always measured in a non-SI unit, the **Debye**.

$$1 \text{ D} = 3.336 \times 10^{-30} \text{ C m}$$

The Debye is preferred by chemists because most molecular dipole moments are between 0 and 10 Debye.

---

[11]Such as in text Figure 3-7, text page 78.

For diatomic molecules (and only for diatomic molecules), experimental dipole moments allow estimates of the ionic character of the bond. The greater the dipole moment for a given bond distance, then the greater is the bond's ionic character. If δ is the fraction of a unit charge on each atom in a diatomic molecule then:

$$\delta = \frac{\mu}{eR}$$

When μ is in Debye and R is in Ångstrom:

$$\delta = 0.2082 \text{ Å D}^{-1} \left(\frac{\mu}{R}\right)$$

The quantity δ equals the **fractional ionic character** of a bond. It ranges from 0.0 to 1.0. The **fractional covalent character** of a bond equals $1.0 - \delta$. The equation is straightforward to use in problems. The only complication is the factor of 0.2082 which arises from the use of the non-SI unit for dipole moment. See **3-37**. Estimates of fractional ionic character from dipole moments agree fairly well with estimates from electronegativities. See **3-39**.

The Shapes of Molecules

Molecules having more than two atoms have bond angles as well as distances and energies. Changes in bond angles affect the energies of molecules strongly, and certain geometrical arrangements of atoms are associated with the greater stability than others.

The **valence shell electron pair repulsion (VSEPR)** theory holds that valence electron pairs distribute themselves about atoms in such a way as to minimize electron pair repulsions. Electron pairs are either lone pairs, not directly involved in bonding, or bonding pairs, which are pairs shared between atoms.

The **steric number SN** of an atom in a molecule equals the number of other atoms bonded to it plus the number of lone pairs on it. To get the steric number of an atom in a molecule, draw a valid Lewis structure for the molecule and then count the atom's neighbors and electron pairs. See **3-41**, **3-43**, and **3-73**. A double or triple bond contributes only one to the steric number of an atom (**3-43d**).

In VSEPR theory, lone pairs and bonding pairs repel each other with different strengths. The relative strengths are:

lone pair-lone pair > lone pair-bonding pair > bonding pair-bonding pair

Different geometries minimize the electron pair repulsions for the different steric numbers. The shapes are shown in text Figure 3-9.[12] Knowing these shapes and the

---

[12]Text page 81.

repulsion order just quoted allows construction of a table of the molecular shapes for steric numbers from 2 to 6. In this table, X is an atom bonded to the central atom A, and E stands for an electron pair.

SN of A	Molecular Type X is a bonded atom; E is a lone pair	Predicted Shape
2	AX_2	linear
3	AX_3	trigonal planar
	AX_2E	bent
4	AX_4	tetrahedral
	AX_3E	trigonal pyramidal
	AX_2E_2	bent
5	AX_5	trigonal bipyramidal
	AX_4E	distorted see-saw
	AX_3E_2	distorted T
	AX_2E_3	linear
6	AX_6	octahedral
	AX_5E	square pyramidal
	AX_4E_2	square planar

Problems **3-41** and **3-43** show the use of VSEPR. The theory can be extended to *SN*'s exceeding 6 (see **3-73**).

Dipole Moments

In polyatomic molecules, the dipole moment of the molecule equals the vector sum of the dipole moments of all of the bonds. Consider CCl_4. Chlorine is more electronegative than carbon. The four C−Cl bonds are therefore polar. According to VSEPR theory, the four equal bonds point from the central C atom to Cl atoms at the corners of a tetrahedron. This symmetrical arrangement means that the vector sum of the four C—Cl bond dipoles equals zero. See text Figure 3-13d.[13] Also, see **3-77** and **3-47**.

Inorganic Nomenclature and Oxidation Numbers

Oxidation Numbers

The **oxidation number** of an atom in a compound equals the charge that the atom would have if all of the shared electrons in its covalent bonds suddenly belong either entirely to it or else entirely to its neighbors. In ionic compounds, nonmetals have

---

[13]Text page 85.

oxidation numbers equal to their negative charge and metals have oxidation numbers equal to their positive charge.

Oxidation numbers are an artificial device devised to help keep track of electrons. The following points apply to oxidation numbers:

1. The oxidation numbers of the atoms in a neutral molecule must add up to zero; those in an molecular ion must add up to the charge on the ion.

2. Alkali metal atoms have oxidation number +1; alkali earth atoms have oxidation number +2.

3. Fluorine always has an oxidation number of −1 in its compounds. Other halogens have an oxidation number of −1 in their compounds except in compounds with oxygen or with other halogens, in which they may have positive oxidation numbers.

4. Hydrogen has an oxidation number of +1 in its compounds except those with metals from Group I and Group II, in which it has an oxidation number of −1.

5. Oxygen has an oxidation number of −2 in compounds except those with fluorine and those containing O—O bonds.

These rules allow the assignment of oxidation numbers in a very large number (but not all) compounds and molecular ions. See **3-53**.

Oxidation numbers are used in chemical nomenclature (**3-55** and **3-57**) and in discussing oxidation-reduction reactions. They are emphatically not the same as formal charges, which find use only in conjunction with Lewis structures.

The term **oxidation state** is synonymous with oxidation number.

Naming Binary Covalent Compounds

The naming of binary covalent compounds resembles the naming of ionic compounds. If a pair of elements forms only one covalent compound, the name of the compound consists of the name of the more electropositive followed by the name of the more electronegative element with *-ide* added to its root. Often, a given pair of elements forms more than one binary covalent compound. In such cases either:

- Specify the number of atoms of each element in the molecular formula with Greek prefixes. *Di-* stands for two. *Tri, tetra, penta, hexa, hepta* stand for three through seven, in order.

- Treat the compound as if it were ionic and place the oxidation number of the first-named element as a Roman numeral in parentheses after the name of that element. See **3-57**.

Detailed Solutions to Odd-Numbered Problems

3-1 a) An atom of radon has 86 electrons. Of these, 78 are core electrons, and 8 are valence electrons. The Lewis diagram is $\ddot{:Rn:}$

b) The monopositive strontium ion has a total of 37 electrons. Of these, 36 are core electrons and 1 is a valence electron. The diagram is $(\cdot Sr)^+$

c) The selenide(2−) ion has a total of 36 electrons. Of these 28 are core electrons and 8 are valence electrons. The Lewis diagram is $(:\ddot{Se}: \,)^{2-}$

d) The antimonide(1−) ion has a total of 52 electrons. Of these 46 are core electrons and 6 are valence electrons. The Lewis diagram is $(\cdot\ddot{Sb}\cdot)^-$

3-3 The theme here is how simple rules on the transfer of valence electrons govern the formation of ionic compounds.

a) Cesium chloride (CsCl) is a compound between chlorine and cesium. The Lewis symbols for the elements before and after reaction are $Cs\cdot \,+\, :\ddot{Cl}\cdot \,\rightarrow\, (Cs)^+ \,(:\ddot{Cl}: \,)^-$.

b) Calcium and astatine form $CaAt_2$, calcium astatide. Each At atom gains an electron and the Ca atom loses two: $\cdot Ca\cdot \,+\, 2\,(:\ddot{At}\cdot) \,\rightarrow\, (:\ddot{At}: \,)^- \,(Ca)^{2+} \,(:\ddot{At}: \,)^-$.

c) Aluminum and sulfur form Al_2S_3, aluminum sulfide. Each S atom gains two electrons and each Al atom loses three: $2\cdot\dot{Al}\cdot \,+\, 3\,(\cdot\ddot{S}\cdot) \,\rightarrow\, (Al^{3+})_2 \,(:\ddot{S}: {}^{2-})_3$.

d) Potassium and tellurium form K_2Te, potassium telluride. Each Te atom gains two electrons and each K atom loses one: $2\,K\cdot \,+\, (\cdot\ddot{Te}\cdot) \,\rightarrow\, (K^+)(:\ddot{Te}: {}^{2-})(K^+)$.

3-5 It is best just to memorize the patterns of the nomenclature of simple inorganic compounds.

a) Al_2O_3	aluminum oxide	**b)** Rb_2Se	rubidium selenide
c) $(NH_4)_2S$	ammonium sulfide	**d)** $Ca(NO_3)_2$	calcium nitrate
e) Cs_2SO_4	cesium sulfate	**f)** $KHCO_3$	potassium hydrogen carbonate

Another name for the last item is potassium bicarbonate.

3-7 The formulas of the anions come from text Table 3-1.

a) Silver cyanide	AgCN	**b)** Calcium hypochlorite	$Ca(OCl)_2$
c) Potassium chromate	K_2CrO_4	**d)** Gallium oxide	Ga_2O_3
e) Potassium superoxide	KO_2	**f)** Barium hydrogen carbonate	$Ba(HCO_3)_2$

3-9 The phosphate ion has the formula PO_4^{3-} (text Table 3-1). The systematic name for this ionic compound is $\boxed{\text{sodium phosphate}}$. Trisodium phosphate is therefore $\boxed{Na_3PO_4}$.

3-11 The problem is similar to Example 3-2.[14] The gaseous KCl molecule is treated as two point charges separated by 2.67 Å. The energy of dissociation equals the negative of the change in energy arising from the Coulomb (electrostatic) attraction between the potassium and chloride ions minus the change in energy consumed in forming the ions by transferring an electron from a chlorine atom to the potassium ion:

$$\Delta E_d = -\Delta E_{\text{Coulomb}} - \Delta E_\infty$$

The Coulomb potential energy of a pair of charges separated by distance R_e is:

$$\Delta E_{\text{Coulomb}} = \frac{Q_1 Q_2}{4\pi \epsilon_0 R_e}$$

In the case of KCl(g), Q_1 and Q_2 equal $+1.602 \times 10^{-19}$ C (K$^+$ ion) and -1.602×10^{-19} C (Cl$^-$ ion), and R_e equals 2.67×10^{-10} m. Substituting these values (and ϵ_0 and π) gives:

$$\Delta E_{\text{coulomb}} = \frac{(+1.602 \times 10^{-19} \text{ C})(-1.602 \times 10^{-19} \text{ C})}{4(3.1416)(8.854 \times 10^{-12} \text{ C}^2 \text{ J}^{-1}\text{m}^{-1})(2.67 \times 10^{-10} \text{ m})}$$
$$= -8.64 \times 10^{-19} \text{ J}$$

This is the Coulomb potential energy change in a *single* K$^+$ to Cl$^-$ attraction. For a mole of such pair-wise attractions, multiply by Avogadro's number:

$$\Delta E_{\text{Coulomb}} = -8.64 \times 10^{-19} \text{ J pair}^{-1} \times 6.022 \times 10^{23} \text{ pair mol}^{-1} = -520 \text{ kJ mol}^{-1}$$

It requires $+520$ kJ to dissociate one mole of the KCl molecules to ions, that is, to move all the Cl$^-$ ions an infinite distance away from the K$^+$ ions to which they are bonded and give a collection of K$^+(g)$ ions all infinitely distant one from another and a second collection of Cl$^-(g)$ ions, also infinitely distant one from another. Removing the electrons from the Cl$^-(g)$ ions would *consume* 349 kJ mol^{-1} (based on the electron affinity of Cl(g) compiled in text Appendix F). Feeding the electrons into the K$^+(g)$ ions would release 419 kJ mol^{-1} (based on the ionization energy of potassium). The energy change for the electron transfer accordingly is:

$$\Delta E_\infty = 419 - 349 = 70 \text{ kJ mol}^{-1}$$

---

[14]Text page 64.

Substituting in the original formula gives:

$$\Delta E_d = -\Delta E_{\text{Coulomb}} - \Delta E_\infty$$
$$= -(-520 \text{ kJ mol}^{-1}) - 70 \text{ kJ mol}^{-1}$$
$$= \boxed{450 \text{ kJ mol}^{-1}}$$

3-13 The As—H bond length should lie between the 1.42 Å of P—H and the 1.71 Å of Sb—H. A length of $\boxed{1.56 \text{ Å}}$, (the average) is a reasonable guess. The experimental bond length is 1.519 Å. The X—H bond will be weakest in $\boxed{SbH_3}$, which has the longest bonds.

3-15 a) SO_4^{2-} In this structure the four O atoms have formal charges (f.c.'s) of -1, and the central S has f.c. $+2$.

Tip. Oxygen atoms with a single bond always have f.c. -1.

b) $S_2O_3^{2-}$ In this structure for the thiosulfate ion, the central sulfur has f.c. $+2$. The three oxygen atoms and the peripheral sulfur have f.c. -1.

c) SbF_3 All atoms have f.c. zero in this structure.

d) SCN^- The sulfur atom in this structure has f.c. -1; the C and N have f.c. zero. **Tip.** Whenever C has four covalent bonds and no lone pairs in a Lewis structure, its formal charge is zero; whenever N has three covalent bonds and one lone pair, its formal charge is zero.

3-17 a) The structure H—$\ddot{\text{N}}$=$\ddot{\text{O}}$: has f.c. $\boxed{\text{zero}}$ on all atoms. The isomeric structure H—$\ddot{\text{O}}$=$\ddot{\text{N}}$: has f.c.'s of $\boxed{\text{zero}}$ on the H, $\boxed{+1}$ on the O and $\boxed{-1}$ on the N. The first of the isomers is favored because no formal charge builds up on any of the atoms.

Tip. The structures for HNO and HON are *not* resonance structures because they use different skeletons.

3-19 a) In this structure of "ZO_2" the oxygen atoms both possess a formal charge of 0. Since there is no net charge on the molecule, Z must also have an f.c. of 0. Therefore, it has 4 valence electrons and belongs to $\boxed{\text{Group IV}}$. CO_2 is an example.

b) In this Lewis structure of "Z_2O_7" each of the six peripheral O atoms has f.c. -1, but the bridging O atom has f.c. zero. Since the molecule has no net charge and is symmetrical, the f.c. on each Z is $+3$. Therefore, Z has 7 valence electrons and is in $\boxed{\text{Group VII}}$. An example is Cl_2O_7 (dichlorine heptaoxide).

c) In "ZO_2^-" one of the oxygen atoms has f.c. zero, but the other has f.c. -1. Since the species has a -1 net charge, Z must have f.c. zero. Therefore, it comes from Group V (5 valence electrons). An example is NO_2^- (nitrite ion).

d) In "$HOZO_3^-$" three of the O atoms have f.c. -1, and the other O atom and the H atom have f.c. zero. Since the ion has a -1 net charge, Z must have f.c. $+2$. Therefore Z has 6 valence electrons and comes from Group VI. An example is $HOSO_3^-$ (hydrogen sulfate ion).

3-21 The octet rule is satisfied for all atoms in these structures. Non-zero formal charges are indicated near the atom, and the non-zero overall charges of molecular ions are shown outside of enclosing brackets:

a) H $\ddot{\cdot}\,\overset{\cdot}{\underset{\cdot\cdot}{As}}\,\ddot{\cdot}$ H

        H

b) H $\colon \overset{\cdot\cdot}{\underset{\cdot\cdot}{O}} \,\colon\, \overset{\cdot\cdot}{\underset{\cdot\cdot}{Cl}}\colon$

c) $\left[\colon\!\overset{\cdot\cdot}{Kr}\,\colon\,\overset{\cdot\cdot}{\underset{\cdot\cdot}{F}}\!\colon \right]^{+}$ (with $+1$ on Kr)

d) $\left[\overset{-1}{\colon\!\overset{\cdot\cdot}{\underset{\cdot\cdot}{O}}} - \overset{\overset{\displaystyle :\overset{\cdot\cdot}{Cl}:}{|}}{\underset{\underset{\displaystyle :\overset{\cdot\cdot}{Cl}:}{|}}{P^{+1}}} - \overset{-1}{\overset{\cdot\cdot}{\underset{\cdot\cdot}{O}}\!\colon} \right]^{-}$

Tip. In Lewis structures, a line equals a pair of dots, but lines are rarely used for lone pairs.

3-23 Referring to text Table 3-5,[15] the bond lengths should be: N—H, 1.01×10^{-10} m; N—C, 1.47×10^{-10} m; C=O, 1.20×10^{-10} m. The following Lewis structure for urea has f.c. zero on all atoms:

3-25 A Lewis structure for S_8:

$$\begin{array}{ccccc}
:\!\ddot{S}\!&-&\!\ddot{S}\!&-&\!\ddot{S}\!: \\
| & & \ddot{\,} & & | \\
:\!\underset{\cdot}{S}\!: & & & & :\!\underset{\cdot}{S}\!: \\
| & & & & | \\
:\!S\!&-&\!\ddot{S}\!&-&\!S\!:
\end{array}$$

Each sulfur obeys the octet rule and has f.c. zero.

---

[15]Text page 68.

3-27 a) The nitrogen and boron atoms would both get 4 valence electrons if all bonds were broken and the two electrons from each pair parcelled out evenly between the two atoms that share them. The nitrogen atom is supposed to have 5 valence electrons, so its formal charge is $\boxed{+1}$. The f.c. is $\boxed{-1}$ on the boron atom, and $\boxed{\text{zero}}$ on the rest of the atoms:

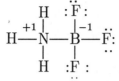

b) The single-bonded O atom has $\boxed{\text{f.c. } -1}$. All other f.c.'s are $\boxed{\text{zero}}$. A double-headed arrow indicates resonance structures:

c) The hydrogen carbonate ion has 24 valence electrons. The C atom contributes 4, the H atom contributes 1, and the 3 O atoms contribute 6 each; a final electron comes from outside to make the overall charge −1. The resonance Lewis structures are:

Formal charges on all atoms are $\boxed{\text{zero}}$ except as indicated.

Tip. Resonance structures differ only in the positions of the electrons. A common error is to include a third structure in which the oxygen atom on the upper left shares two pairs of electrons with the central carbon atom and the H atom is moved to avoid putting a +1 formal charge onto that oxygen atom. Such a structure is *not* a resonance structure, because an atom as well as electrons has moved. Resonance structures always use the same atomic skeleton.

3-29 The main resonance structures (others break the octet rule) for NO_2^- ion are:

$$\left[\ddot{\underset{..}{O}} - \ddot{N} = \ddot{O} \quad \longleftrightarrow \quad \ddot{O} = \ddot{N} - \ddot{\underset{..}{O}} \right]^-$$

The two N-to-O bonds in this ion should be equivalent and intermediate in length between 1.43 Å (single bond) and 1.10 Å (double bond), that is, about $\boxed{1.27 \text{ Å}}$.

3-31 Resonance structures of methyl isocyanate include:

$$\left[\overset{0}{H_3C}-\overset{0}{\overset{..}{N}}=\overset{0}{C}=\overset{..}{O}: \longleftrightarrow \overset{-1}{H_3C}-\overset{0}{\overset{..}{N}}-\overset{+1}{C}\equiv O: \longleftrightarrow \overset{+1}{H_3C}-\overset{0}{N}\equiv\overset{-1}{C}-\overset{..}{\underset{..}{O}}: \right]$$

The left structure has f.c. zero on all atoms. The center structure has f.c. -1 on the N, f.c. $+1$ on the O and f.c. zero on the other atoms. The right structure has f.c. $+1$ on the N, -1 on the O and zero on all of the other atoms. The predominant resonance contributor is the left structure, which has f.c. zero on all atoms.

Tip. The "iso" in methyl isocyanate suggests that the compound is an isomer of something. Here are Lewis structures for methyl cyanate, which contains the same atoms arranged differently. These Lewis structures are not resonance structures of the preceding three:

$$\left[\overset{+1}{H_3C}-\overset{0}{\overset{..}{O}}=\overset{-1}{C}=\overset{..}{N}: \longleftrightarrow \overset{0}{H_3C}-\overset{0}{\overset{..}{O}}-\overset{0}{C}\equiv N: \longleftrightarrow \overset{+2}{H_3C}-\overset{0}{O}\equiv\overset{-2}{C}-\overset{..}{\underset{..}{N}}: \right]$$

3-33 a) In the structure for PF_5, each atom has f.c. zero. Bonding on the F atoms obeys the octet rule but the P atom has an expanded octet (10 electrons).

b) In the structure for SF_4, bonding on all F atoms obeys the octet rule. The S has an expanded octet (10 electrons). All f.c.'s equal 0.

c) In the structure of XeO_2F_2, the F atoms and the O atoms obey the octet rule. The Xe has an expanded octet. All atoms have f.c. zero.

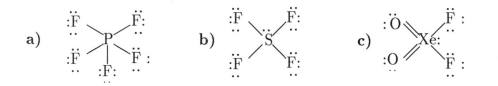

3-35 Binary ionic compounds have large differences in electronegativity between their elements whereas binary covalent compounds have small differences. High vapor pressure is associated with relatively weak intermolecular attractions and therefore with molecular (covalent) compounds. The correct choice in each case is the compound with the smaller difference in electronegativity:

a) $\boxed{CI_4}$ b) $\boxed{OF_2}$ c) $\boxed{SiH_4}$.

3-37 In diatomic molecules, the fractional ionic character δ is:

$$\delta = (0.2082 \text{ Å D}^{-1}) \left(\frac{\mu}{R}\right)$$

when the dipole moment μ is in Debye and the bond distance R is in Ångstroms. The percent ionic character is just 100 times the fractional ionic character. Substitute in the formula to obtain:

Compound	Bond Length	Dipole Moment	(δ)	% Ionic Character
ClO	1.573 Å	1.239 D	0.16	16
KI	3.051	10.82	0.74	74
TlCl	2.488	4.543	0.38	38
InCl	2.404	3.79	0.33	33

3-39 The Δ in the expression $16\Delta + 3.5\Delta^2$ in this problem is *not* the Δ in the Pauling definition of electronegativity on text page 77. Rather it is $\chi_A - \chi_B$, the difference in electronegativity. Substitute in the formula to obtain:

Compound	$|\chi_A - \chi_B|$	Calc. % Ionic	Dipole % Ionic
HF	1.78	40	41
HCl	0.96	19	18
HBr	0.76	14	12
HI	0.46	8	6
CsF	3.19	87	70

Tip. The point of the problem is the generally good agreement between the ionic character calculated from differences in electronegativity and that based on dipole moment.

3-41 a) The molecule of CBr_4 is tetrahedral. The central C atom has SN 4. There are no lone pairs on the central carbon atom so this is an AX_4 case.[16]

b) In SO_3, the central S atom has SN 3 and no lone pairs. The molecule is trigonal planar with an O—S—O angle of 120°. The fact that one or more of the S-to-O bonds can be shown in a Lewis structure as a double bond does *not* affect the steric number.

c) In SeF_6, the central Se atom has SN 6. There are no lone pairs on the central Se atom, so the expected geometry of the F atoms about the Se is octahedral.

[16]See the table on page 56 of this Guide.

d) In $SOCl_2$ the central S has SN $\boxed{4}$ and one lone pair. It is a AX_3E case. The disposition of the electron pairs about the S is approximately tetrahedral. The actual geometry of the molecule is $\boxed{\text{pyramidal}}$.

e) In ICl_3, the central I atom has SN $\boxed{5}$. It is surrounded by 3 Cl atoms and 2 lone pairs. The disposition of the electron pairs is trigonal bipyramidal. The true molecular geometry (which only considers atoms) of this AX_3E_2 case is a $\boxed{\text{distorted T-shape}}$.

3-43 a) The molecular ion ICl_4^- is square planar. The central I atom has SN $\boxed{6}$, which means the geometry of the electron pairs about the central I is octahedral. The 2 lone pairs lie opposite each other on the octahedral pattern, minimizing lone-pair to lone-pair interactions. The 4 Cl atoms surround the central I atom in a $\boxed{\text{planar square}}$.

b) In OF_2, the central O atom has SN $\boxed{4}$. The molecule is of the type AX_2E_2.[17] The molecule is $\boxed{\text{bent}}$ to accommodate the two lone pairs on the O atom. The F—O—F angle is less than $109.5°$.

c) In BrO_3^-, the central Br atom has SN $\boxed{4}$ making the molecular type of the molecular ion AX_3E. The single lone pair on the central Br atom occupies one corner of a tetrahedron about the Br atom. The resulting molecule is $\boxed{\text{pyramidal}}$. The presence of the lone pair forces the O atoms together slightly, so that the O—Br—O angle is less than $109.5°$.

d) In CS_2, the central C atom has SN $\boxed{2}$. Both of the C-to-S bonds are double bonds, but this plays no part in figuring the SN of the C atom. The molecule, which is of the type AX_2, is $\boxed{\text{linear}}$.

3-45 a) Planar AB_3: BF_3, BH_3, SO_3. **b)** Pyramidal AB_3: NH_3, NF_3. **c)** Bent AB_2^-: ClO_2^-, NO_2^-. **d)** Planar AB_3^{2-}: CO_3^{2-}.

3-47 A molecule or molecular ion has a dipole moment when the center of its spatial distribution of positive charge does not coincide with the center of its distribution of negative charge. The *bonds* in the listed compounds are all polar. The symmetry of certain molecular shapes however causes the vector sum of the individual bond dipoles to equal zero. Thus, SeF_6 (octahedral), CBr_4 (tetrahedral), and SO_3 (trigonal planar) have $\boxed{\text{non-polar molecules}}$.

The molecules, ICl_3 (distorted T) and $SOCl_2$ (pyramidal), are less symmetrical, and the vector sums of their bond dipoles are not zero. They are $\boxed{\text{polar}}$.

---

[17]See page 56 of this Guide.

3-49 The fact that the molecule is bent is unhelpful in deciding between the formulations:

$$\overset{-1}{:\ddot{N}} = \overset{+1}{\ddot{O}} -\overset{0}{\ddot{F}}: \qquad \text{and} \qquad \overset{0}{:\ddot{O}} = \overset{0}{\ddot{N}} -\overset{0}{\ddot{F}}:$$

because both structures feature a central atom having *SN* 3 with 2 bonds and 1 lone pair. VSEPR theory predicts a bent molecule in both cases.

Tip. The above structures include the formal charges. Considerations of formal charge (as in **3-65**) favor the isomer on the right. Both isomers exist. The one on the right (nitrosyl fluoride) is a colorless gas of reasonable stability, but the one on the left is highly unstable and has been characterized only spectroscopically (by methods described in Chapter 16).

3-51 a) The resonance structures

$$\left[\overset{-1}{:\ddot{N}} = \overset{+1}{N} = \overset{0}{\ddot{O}}: \quad \longleftrightarrow \quad \overset{0}{:N} \equiv \overset{+1}{N} - \overset{-1}{\ddot{O}}: \right]$$

can be written for the NNO molecule. When either is considered, the *SN* of the central nitrogen atom equals two. The predicted molecular geometry is $\boxed{\text{linear}}$.

b) The linear geometry in NNO would cause the N=O and N=N bond dipoles to add vectorially to zero if they were equal in magnitude. The observed net dipole moment means that the two bond dipoles differ in magnitude. The N=O bond should be more polar than the N=N bond because O is more electronegative than N. The $\boxed{\text{N end}}$ of the molecule is therefore expected to have the positive partial charge.

3-53 The oxidation numbers are determined by the standard rules:

$SrBr_2$	Sr +2	Br −1	$Zn(OH)_4^{2-}$	Zn +2	O −2	H +1	
SiH_4	Si −4	H +1	$CaSiO_3$	Ca +2	Si +4	O −2	
$Cr_2O_7^{2-}$	Cr +6	O −2	KO_2	K +1	O −1/2		
CsH	Cs +1	H −1	$Ca_5(PO_4)_3F$	Ca +2	P +5	O −2	F −1

3-55 a) SiO_2 **b)** $(NH_4)_2CO_3$ **c)** PbO_2 **d)** P_2O_5 **e)** CaI_2 **f)** $Fe(NO_3)_3$.

3-57
a) Copper(I) sulfide and copper(II) sulfide **b)** Sodium sulfate
c) Tetraarsenic hexaoxide[18] or arsenic(III) oxide **d)** Zirconium(IV) chloride
e) Dichlorine heptaoxide or chlorine(VII) oxide **f)** Gallium(I) oxide

3-59 a) Ammonium phosphate is $(NH_4)_3PO_4$; potassium nitrate is KNO_3; ammonium sulfate is $(NH_4)_2SO_4$.

b) $(NH_4)_3PO_4$. First calculate the molecular masses by summing up the individual atomic masses of each element in the compound multiplied by the number of times it appears in the compound. Next, compute ratios of the individual masses of each element in the compound to the total molecular mass and convert to percentages:

Compound	Molecular Mass	Percent N	Percent P	Percent K
$(NH_4)_3PO_4$	149.09	28.18	20.78	0
KNO_3	101.10	13.85	0	38.67
$(NH_4)_2SO_4$	132.14	21.20	0	0

3-61 a) A Lewis structure for OPCl in which the octet rule is obeyed for all atoms and in which all atoms have a formal charge of zero is $: \ddot{O} = \ddot{P} - \ddot{C}l :$

Tip. In this compound, the oxidation number of O is -2, of Cl -1, and of P $+3$. Oxidation numbers are *not* the same as formula charge.

b) Three resonance Lewis structure for O_2PCl are given below. In the structure on the left, all atoms have f.c. zero, but the octet rule is violated on the P atom, which sees 10 electrons. In the two structures to the right, the octet rule is obeyed for every atom, but formal charges build up as shown. All three contribute to the "true" bonding:

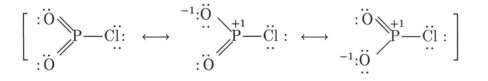

3-63 The following resonance structures of nitryl chloride together imply equivalent N-to-O bonds:

Tip. Nitryl chloride is a reactive gas that decomposes readily to nitrogen dioxide (see **3-65**) and chlorine. Despite this, it is named as a salt: nitryl ion, a $+1$ cation, in combination with chloride ion, a -1 anion. The following Lewis structure takes up the suggestion of this name by showing ionic bonding to the Cl. Note that the two N-to-O bonds are still equivalent, but the pair of electrons previously between N and Cl now belongs entirely to the Cl. The octet rule is satisfied for all atoms:

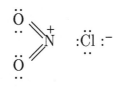

3-65 a) The molecule of nitrogen dioxide has 17 valence electrons. At least one atom cannot achieve an octet. The following four resonance structures are the candidates for best single Lewis structure:

$$\left[\ \overset{-1}{:\ddot{O}}-\overset{+1}{\dot{N}}=\overset{0}{\ddot{O}}\quad\longleftrightarrow\quad \overset{0}{\ddot{O}}=\overset{+1}{\dot{N}}-\overset{-1}{\ddot{O}}:\quad\longleftrightarrow\quad \overset{0}{\cdot\ddot{O}}-\overset{0}{\dot{N}}=\overset{0}{\ddot{O}}\quad\longleftrightarrow\quad \overset{0}{:\ddot{O}}=\overset{0}{\dot{N}}-\overset{0}{\ddot{O}}\cdot\ \right]$$

Other candidate structures break the octet rule on more than one atom or use octet expansion rather than octet deficiency. If the N atom is octet-deficient (left two structures), then formal charges build up as shown. If an O is octet-deficient (right two structures), all atoms have f.c. zero. The best structure puts the odd electron on the ⬚O atoms⬚.

b) Four resonance structure can be drawn varying the relative positions of the double and single bonds:

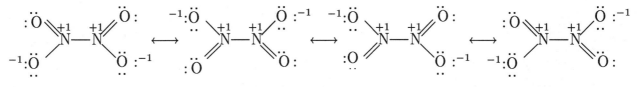

3-67 Xenon compounds always have an expanded octet:

a) $\left(:\overset{+1}{\ddot{X}e}-\ddot{F}:\right)^{+}$ **b)** $:\ddot{F}-\ddot{X}e-\ddot{F}:$

3-69 The difference in electronegativities of the atoms in HF is 1.78; the difference in LiCl is 2.18. The two compounds ⬚differ⬚ greatly in their bonding according to the evidence of their physical properties. Lithium chloride is an ionic compound (high boiling, high melting); hydrogen fluoride is a covalent compound (low melting, low boiling).

3-71 a) The energy ΔE_{∞}[19] equals the difference between the *IE* of M(*g*) and the *EA* of X(*g*). It equals the energy required to extract an electron from an isolated M(*g*)

---

[19]See text page 63.

and place it on an isolated $X(g)$. At the critical distance R_c, the decrease in energy from the Coulomb attraction between the $M^+(g)$ and $X^-(g)$ ions just compensates for ΔE_∞:[20]

$$\Delta E_\infty = IE - EA = -\Delta E_{Coulomb} = -\frac{Q_1 Q_2}{4\pi\epsilon_0 R_c}$$

The constant ϵ_0 is 8.854×10^{-12} $C^2\,J^{-1}m^{-1}$, and, in alkali halides, Q_1 and Q_2 are $+1.60218\times10^{-19}$ C and -1.60218×10^{-19} C. Solve the preceding for R_c and substitute the various values:

$$R_c = -\frac{Q_1 Q_2}{4\pi\epsilon_0 \Delta_\infty} = \frac{(2.3071 \times 10^{-28}\text{ J m})}{(IE - EA)\text{ J}}$$

Text Appendix F and other sources give IE's and EA's on a per mole basis. Revise the preceding equation, which applies to a single pair of particles, to allow the use of IE's and EA's in joule per mole. Do this by multiplying the numerator by N_0:

$$R_c = \frac{(6.0221 \times 10^{23}\text{ mol}^{-1})(2.3071 \times 10^{-28}\text{ J m})}{(IE - EA)\text{ J mol}^{-1}} = \frac{1.3894 \times 10^{-4}\text{ J m mol}^{-1}}{(IE - EA)\text{ J mol}^{-1}}$$

b) From Appendix F for LiF, $(IE - EA)$ is $520.2 - 328.0 = 192.2 \times 10^3$ J mol^{-1}. Substitution in the preceding expression gives R_c equal to $\boxed{7.229 \times 10^{-10}\text{ m}}$.

For KBr, $(IE - EA)$ equals 94.1×10^3 J mol^{-1} giving R_c equal to $\boxed{14.8 \times 10^{-10}\text{ m}}$.

For NaCl, $(IE - EA)$ equals 149.8 kJ mol^{-1}, making R_c equal to $\boxed{9.275 \times 10^{-10}\text{ m}}$.

3-73 a) In $SbCl_5^{2-}$ ion, the central Sb atom has SN 6. It is surrounded by five bonding pairs (single bonds to the five fluorine atoms) and one lone pair. As an AX_5E case,[21] the molecular ion is $\boxed{\text{square pyramidal}}$.

b The central Sb atom in $SbCl_6^{3-}$ ion has SN $\boxed{7}$. This steric number is rare. The required extension of VSEPR theory is inclusion of SN 7 and other higher SN's. In fact, three geometries have been observed for SN 7: pentagonal bipyramidal, capped octahedral (in which the seventh atom occupies one face of an octahedron surrounding the central atom) and capped trigonal prism (in which the seventh atom occupies one rectangular face of a trigonal prism about the central atom).

3-75 The central S atom in F_4SO has SN 5 and falls into the class AX_5. The geometry of the molecule is therefore based on the $\boxed{\text{trigonal bipyramid}}$. The real question is whether the oxygen atom is equatorial or axial. Putting the double-bonded oxygen

---

[20]Obviously at this same distance ΔE_d equals zero.
[21]See the table on page 56 of this Guide.

atom at an equatorial position minimizes 90° interactions with the four fluorine atoms and should be preferred, according to VSEPR theory. Also, the F—S—F angles will be slightly less than 90° and 120°.

3-77 Set up a coordinate system and position the oxygen atom at its origin. Let the y-axis bisect the angle θ defined by H—O—H:

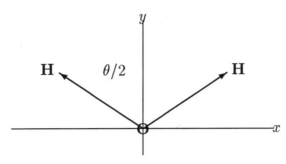

The dipole moments of the two O—H bonds are symbolized μ_{OH}. They parallel the two bonds. The x-components of these two vectors oppose each other and cancel. The y-components point in the same direction and add. The magnitude of the y-components is $\mu_{OH} \cos(\theta/2)$, and the sum of the two y-components equals the dipole moment of the molecule as a whole. Therefore:

$$\mu(H_2O) = 1.86 \text{ D} = 2\mu_{OH} \cos\left(\frac{\theta}{2}\right) = 2\mu_{OH} \cos\left(\frac{104.5°}{2}\right)$$

Solving for μ_{OH} gives $\boxed{1.52 \text{ D}}$.

3-79 From the formula Bi_5^{3+} we must conclude that the oxidation number (or at least the average oxidation number) of the bismuth is $\boxed{+3/5}$. Because the oxidation number of F is -1 by convention, the oxidation number of the As is $\boxed{+5}$. The elevated dot in the formula means that SO_2 is loosely associated with this salt in its solid state. The oxidation number of S in SO_2 is $\boxed{+4}$; the oxidation number of O is $\boxed{-2}$.

3-81 a) The oxidation number of lead in PbO is $+2$; in PbO_2 it is $+4$; in Pb_2O_3 it is $+3$; in Pb_3O_4 it is $+8/3$.

b) Note that the formula "Pb_2O_3" is the sum of PbO and PbO_2. Perhaps in Pb_2O_3 half of the lead is Pb(II) and half is Pb(IV), as in "(PbO)(PbO$_2$)." Similarly, the lead in Pb_3O_4 may be 2/3 Pb(II) and 1/3 Pb(IV) as in "(PbO)$_2$(PbO$_2$)."

3-83 a) The element M loses two electrons in forming compounds, since it forms compounds MCl_2 and MO. It belongs to $\boxed{\text{group II}}$, the alkaline-earth metals.

b) The relative molecular mass of MCl_2 equals the sum of the relative masses of the three constituent atoms: $x + 2(35.453) = x + 70.906$ where x stands for the relative atomic mass of the element M. The fraction of mass that is chlorine is 0.447. So:

$$0.447 = \frac{70.906}{(x + 70.906)}$$

Solving gives $x = 87.7$. Reference to a list of relative atomic masses identifies the element M as $\boxed{\text{strontium}}$.

3-85 a) The elemental analysis reveals that the compound contains only F, Cl, and O. Divide the respective mass percentages by the molar masses of the elements to obtain the relative number of moles of each.[22] The ratios correspond to the empirical formula $\boxed{ClO_3F}$.

b) The central atom in the molecule ClO_3F is certainly the Cl. One good Lewis structure is:

This structure shows an expanded octet on the central Cl, but the formal charges equal zero on all atoms. Resonance structures in which electrons are shifted from the double bonds to reside completely on the O's can also be written.[23]

c) The central Cl has a steric number of four. VSEPR theory predicts a structure based on the tetrahedron. Because F is more electronegative than O, it tends to attract electrons away from the central Cl, reducing the electron-pair repulsion and causing the F—Cl—O angles to become smaller than tetrahedral while the O—Cl—O angles become larger than tetrahedral.

---

[22]See **2-7**, page 31 of this Guide, if necessary.
[23]See **3-61b** on page 67 of this Guide.

Chapter 4

The Gaseous State

The Chemistry of Gases

Unlike solids or liquids, gases expand to fill the container they occupy. The molecules of gases interact with each other only weakly; the average distances between them are large in comparison to their diameters. The atmosphere is a mixture of gases, principally nitrogen and oxygen. Gases form chemically:

- By thermal decomposition of many solids (such as mercury(II) oxide, calcium carbonate, potassium chlorate (see **4-29**) and ammonium hydrogen carbonate). Look for the formula or partial formula of a known stable gas embedded in the formula of a solid. The solid may well decompose when heated to give the gas represented by this formula. See **4-1**.

- By the direct reaction of nonmetallic elements with oxygen to give oxides.

- By the action of acids on many ionic solids. See **4-67**.

- By the action of acids on active metals. Such reactions generate gaseous hydrogen and a salt of the metal.

- By the action of bases on ammonium salts. Such reactions generate ammonia and a new salt.

Pressure and Temperature of Gases

The three physical variables that are most used in describing the physical state of gases are the pressure temperature and volume:

- **Pressure** is force divided by area. Intuitively, a force is a push. Pressure takes into account both the magnitude of the push and the size of the region upon which

72

it pushes. A weak push driving a sharp pin against the skin hurts because the area of the point is small—enormous pressure is applied directly under the point.

The SI unit of force is the **newton**, which equals a kg m s^{-2}. From the definition of pressure, the SI unit of pressure is the newton per square meter or newton·meter^{-2}, also called the **pascal**.[1] The newton is a conveniently sized unit of force for chemical applications, but the square meter is quite a large area. Consequently, the pascal (Pa) is too small for common applications. Many other pressure units are defined and in use. Some important ones are related to the pascal as follows:

$$\text{1 atmosphere (atm)} = 101325 \text{ Pa exactly} \quad \text{1 torr} = \frac{101325}{760} \text{ Pa exactly}$$

$$\text{1 lb in}^{-2} = 6.894757 \times 10^3 \text{ Pa} \quad \text{1 bar} = 10^5 \text{ Pa exactly}$$

Many unit factors can be constructed from these equations. For example:

$$\frac{1 \text{ atm}}{760 \text{ torr}} = 1 \quad \frac{1 \text{ atm}}{14.69595 \text{ lb in}^{-2}} = 1 \quad \frac{1 \text{ atm}}{0.9869233 \text{ bar}} = 1$$

Such unit factors are used in **4-7** and **4-9**.

At one time the torr was defined as the pressure exerted at its base by a column of mercury exactly one millimeter high when the temperature of the mercury was 0°C. The current definition of the torr in terms of the pascal maintains this equality:

$$\text{1 torr} = 1 \text{ mm Hg} \quad \text{at } 0°\text{C}$$

The millimeter of mercury (mm Hg) appears as a unit of pressure because reading a mercury barometer consists of measuring the height of a column of the liquid metal mercury (Hg) in a glass tube. The diameter of the column does not matter. A broad column exerts a large force but spreads it over a large area. A thin column exerts less force, but because the area at its base is proportionately less, the pressure at its base is the same. This fact is the basis for the operation of a barometer and is summarized:

$$P = \rho g h$$

where ρ is the density of the liquid in the column, g is the acceleration of gravity, and h is the height of the column. The acceleration of gravity at the surface of the earth is 9.80665 m s^{-2}. See **4-5**, **4-7**, and **4-61**.

The sea-level pressure of the earth's atmosphere fluctuates in the neighborhood of one standard atmosphere (atm). A psi equals a pound of force per square inch.

The pressure *of* a gas is equal to the pressure *on* that gas. When the internal pressure of a sample of gas differs from the external pressure, then the gas either expands or is compressed.

---

[1] See text Table B-2, text page A-11.

• **Volume** is the three-dimensional region available to the sample of gas under consideration. The obvious SI unit for volume is the cubic meter. Because this unit is inconveniently large, many other units of volume are used. The most important in chemistry are the cubic decimeter, which is named the **liter**, and the cubic centimeter. The following should be memorized:

$$\frac{1}{1000} \ \text{m}^3 = 1 \ \text{dm}^3 = 1 \ \text{liter (L)} \qquad \text{and} \qquad \frac{1}{1000} \ \text{L} = 1 \ \text{mL} = 1 \ \text{cm}^3$$

The volumes of samples of gas are easily measured. Consequently volume is often studied as a function of other variables.

Boyle's law states that the pressure of a quantity of gas is inversely proportional to its volume when the gas is held at a constant temperature:

$$P = C\left(\frac{1}{V}\right)$$

A useful operational version of Boyle's law is:

$$P_1 V_1 = P_2 V_2 \quad \text{for a fixed amount of gas at constant temperature}$$

The subscripts refer to the variables before (state 1) and after (state 2) some change.

• **Temperature** is connected with **Charles's law**. Charles's law states that the volume of a quantity of gas is linearly dependent on its temperature when the gas is held at a constant pressure. All gases (at low enough pressures) expand by $1/273.15$ of V_0, their volume at $0°C$, for every degree Celsius that they are heated:

$$V = V_0\left(1 + \frac{t}{273.15°C}\right)$$

This finding by Charles suggested the definition of the absolute scale of temperature upon which the volume of the gas is directly proportional to the temperature:

$$V \propto T \quad \text{(fixed pressure and fixed amount of gas)}$$

See **4-13** and **4-15**.

The **Kelvin temperature scale** is an absolute scale having divisions (named kelvins) equal in magnitude to Celsius degrees. The difference is that the Kelvin scale has its zero at $-273.15°C$. Hence:

$$T\,(\text{Kelvin}) = t\,(\text{Celsius})\left(\frac{1 \ \text{K}}{1°C}\right) + 273.15 \ \text{K}$$

Another scale of temperature is the **Fahrenheit scale**. The Fahrenheit degree is exactly 5/9 the size of the Celsius degree (or kelvin). More Fahrenheit degrees than Celsius degrees (or kelvins) are needed to cover a given difference in temperature. Also, the zero of the Fahrenheit scale is offset by 32°F. relative to the zero of the Celsius scale. To convert from Fahrenheit to Celsius, first adjust for the offset by subtracting 32°F, and then use a unit factor:

$$t\,(\text{Celsius}) = \left(t\,(\text{Fahrenheit}) - 32°\text{F}\right)\left(\frac{5°\text{C}}{9°\text{F}}\right)$$

Problem **4-15** shows conversion from one temperature scale to another.

The Ideal-Gas Law

At sufficiently low densities, all gases follow the **ideal-gas law**, an equation combining Boyle's law, Charles's law, and the observation that the volume of a gas depends directly on its amount at a given temperature and pressure. The mathematical form of ideal-gas law is:

$$PV = nRT$$

In this equation, n equals the chemical amount of gas, V, P, and T are the volume, pressure and absolute temperature, and R is a constant.

● **Use the absolute temperature whenever using the ideal-gas law.**

See **4-19** and **4-21**. The constant R is the **universal gas constant**. It is a fundamental constant of nature, like the speed of light. Two numerical values of R find frequent use:

$$R = 8.31451 \text{ J K}^{-1}\text{mol}^{-1} = 0.0820578 \text{ L atm mol}^{-1}\text{K}^{-1}$$

The first is mostly used in calculations involving the energy of molecular motion (see **4-41**) and in thermodynamics. The second is common in calculations involving different states of gases. It is legitimate and often advantageous to recast the units of R. For example:

$$R = 8.315 \times 10^{-3} \text{ kJ K}^{-1}\text{mol}^{-1} = 82.057 \text{ cm}^3 \text{ atm K}^{-1}\text{mol}^{-1}$$

See **4-39** and **4-53**.

A comparison of the different units of R shows that the L atm must be a unit of energy. It is (force/area) × volume which equals force × distance (see **7-1**). Think of a piston being forced along a cylinder by the expansion of a gas.

Problems Based on the Ideal-Gas Law

Many problems use the ideal-gas equation. Some have practical importance; others are more fanciful. Most fall into two broad categories:

Problems Involving Changes of State. The T, P and V and n of a gas are stated or implied. One or more of the variables is then changed, and the final value of another variable must be found. To solve such problems, make a list of n_1, P_1, V_1, T_1 and n_2, P_2, V_2, T_2. Then substitute into the following "before/after" relationship:

$$\frac{P_1 V_1}{n_1 T_1} = \frac{P_2 V_2}{n_2 T_2}$$

Cancellation of variables that are not known but that stay constant between state 1 and state 2 is frequent. See **4-11**, **4-19**, **4-21**, and **4-67**. Situations in which n changes along with or instead of P, V and T strike students as more difficult. See **4-73**.

Problems Involving an Unknown in the Gas Law. The problem gives or implies the values of three out of four of the variables in the ideal-gas equation and asks for the fourth. A suitable value for R, the universal gas constant and due care with substitution and cancellation of the units does the rest. See **4-29** and **4-89**. Chemical reactions consume or generate gases, which means they cause changes in n, the number of moles of gas.

Two problem themes have particular practical importance: the calculation of n from measurements of T and P and V of a gas; the calculation of V when a gas is generated by reaction of a known amount of reactants and collected at a known T and P. See **4-31** and **4-67**.

Some problems do not give explicit values of variables but instead give a quantity relating two of them. For example, knowing the ratio n/V, the molar density of a gas, is just as good as knowing both n and V if P is known and T must be computed, or vice versa. See **4-21** and **5-95**.

Mixtures of Gases

According to **Dalton's law of partial pressures**, each gas in a mixture of gases exerts a **partial pressure** equal to the pressure it would exert if it were alone under the same conditions. The sum of the partial pressures of the components of a mixture of gases equals the total pressure:

$$P_{\text{tot}} = P_{\text{A}} + P_{\text{B}} + P_{\text{B}} + \cdots$$

where P_{A} is the partial pressure of the first component, P_{B} is the partial pressure of the second, and so forth. Dalton's law is understood by reasoning that a molecule in a mixture of gases must interact with *unlike* molecules to exactly the same extent it interacts with *like* molecules. In other words, the molecules of the several gases in a mixture that follows Dalton's law effectively interpenetrate each other. See **4-71** for a use of Dalton's law.

The partial pressure of a component gas in a mixture of ideal gases equals the total pressure of the mixture multiplied by the mole fraction of that component. The **mole fraction** (symbolized X with a subscript identifying the gas) equals the number of moles of the gas in the mixture divided by the total number of moles of all gases present:

$$X_A = \frac{n_A}{n_A + n_B + n_C + \cdots} = \frac{n_A}{n_{tot}}$$

The mole fraction of a substance in a mixture always equals the fraction of the *molecules* in the mixture that are molecules of the substance. See **4-33** and **4-37**.

The composition of gaseous mixtures can be given using volume fractions (or volume percentages, or parts per million by volume) of the components. For example, text Table 4-1[2] gives the composition of the atmosphere in percentages by volume. By Boyle's law, the volume fraction of a gas in a mixture equals its mole fraction. See **4-35**.

A common application of Dalton's law occurs when gases are collected over volatile liquids, *e.g.,* the collection of oxygen gas by bubbling it through a trough collecting water and into the mouth of an inverted bottle. In this experiment, the O_2 gas gets unavoidably "wet" (mixed with water vapors). The total pressure of the sample is then:

$$P_{tot} = P_{oxygen} + P_{water\ vapor}$$

To find the pressure that the oxygen would exert if it were pure, the contribution of the water vapor must be subtracted. See **4-3**. The vapor pressure of water depends only on the temperature and is widely tabulated.[3]

The Kinetic Theory of Gases

The postulates of the **kinetic theory of gases** are brief:

1. A pure gas has many identical molecules[4] of negligible size. That is, the molecules are approximately point masses.

2. The molecules of a gas are in chaotic motion with a distribution of speeds.

---

[2]Text page 99.

[3]As in text Table 5-1, text page 151.

[4]A few gases (such as He) consist of atoms not molecules. In kinetic theory, the atoms of monatomic gases are also called molecules.

3. There are no interactions among the molecules except during elastic collisions. (An elastic collision is a collision in which no translational energy is lost.)

4. The collisions of molecules with the walls of the container are elastic.

From these postulates, the ideal-gas law, and Newton's laws of motion, it is possible to derive[5] a relationship that links the speeds of molecules and the temperature. The derivation results in this expression:

$$\tfrac{1}{3} N_0 m \bar{u^2} = RT$$

where $\bar{u^2}$ is the **mean-square speed** of the molecules of the gas and m is the mass of the individual molecule. The mean-square speed equals the average of the squares of the speeds of the molecules. According to the equation, the mean-square speed of the molecules in an ideal gas depends solely on the temperature and mass of the molecules. The mass of a single molecule of a substance multiplied by Avogadro's number equals the molar mass $\mathcal{M}$ of the substance. Therefore:

$$\bar{u^2} = \frac{3RT}{\mathcal{M}}$$

An important aspect of the derivation is the difference between **speed** and **velocity**. The velocity v of an object is a vector. It has both magnitude and direction. Speed has magnitude alone. A car proceeding too fast gets a speeding ticket. A car proceeding (even slowly) the wrong way on a one-way street gets a velocity ticket.

• The addition of velocities must take their directions into account.

The **momentum** of a particle equals its velocity times its mass. Therefore, the addition of momenta also must take directions in account.

Distribution of Molecular Speeds

The speeds of the molecules in a gaseous substance at room temperature are distributed across a wide but predictable range. The prediction is made by the **Maxwell-Boltzmann speed distribution**:

$$f(u) = 4\pi \left(\frac{m}{2\pi k_{\mathrm{B}} T}\right)^{3/2} u^2 \exp(-mu^2/2k_{\mathrm{B}}T)$$

where k_{B} is a new constant (**Boltzmann's constant**) equal to R/N_0, m is the mass of the molecules (which are all identical), and T is the absolute temperature.

To learn what $f(u)$ means, take the function apart. Note first the "exp" means to raise the base e to a power equal to what follows in parentheses. Now, suppose that

---

[5]Text page 114–6.

m and T are fixed. Two factors involve the molecular speed u: $\exp(-mu^2/2k_{\mathrm{B}}T)$ and u^2. The quantity $-mu^2/2k_{\mathrm{B}}T$ gets more and more negative as u increases (k_{B} is constant). Therefore $\exp(-mu^2/2k_{\mathrm{B}}T)$ gets smaller and smaller as u increases If this exponential part alone controlled the distribution of molecular speeds, the molecules in a gas would scarcely be moving. But the exponential part is multiplied by u^2, which increases as u increases. As a result, $f(u)$ grows as u rises from zero, reaches a maximum, and then diminishes as the effect of the exponential term becomes dominant. The following table shows this for nitrogen gas (N_2) at T equal to 273.15 K:

u (m s^{-1})	u^2 (m^{2}s^{-2})	$\exp\left(-mu^2/2k_{\mathrm{B}}T\right)$	$f(u)$ (s m^{-1})
0	0	1.000	0.0
50	2.5×10^3	0.985	8.5×10^{-5}
100	1.0×10^4	0.940	3.3×10^{-4}
200	4.0×10^4	0.781	1.1×10^{-3}
300	9.0×10^4	0.574	1.8×10^{-3}
400	1.6×10^5	0.373	2.1×10^{-3}
500	2.5×10^5	0.214	1.9×10^{-3}
1000	1.0×10^6	2.10×10^{-3}	7.3×10^{-5}
3000	9.0×10^6	7.81×10^{-25}	2.4×10^{-25}

At low speeds u^2 dominates. At high speeds the exponential term crushes $f(u)$ toward zero.

The units of $f(u)$ are s m^{-1}, the reciprocal of the units of speed. When $f(u)$ is multiplied by Δu, a range of speed, all of the units cancel out. This reveals the quantity $f(u)\Delta u$ as a **probability distribution**. It is a pure number telling the fraction of the molecules having a speed between u and $(u + \Delta u)$.

Example: What fraction of the molecules in a sample of gaseous N_2 held at 273.15 K have a speed between 390 and 410 m s^{-1}?

Solution: Δu is 20 m s^{-1}, $f(u)$ at 400 m s^{-1} is 2.1×10^{-3} s m^{-1} (from the table). Therefore, $f(u)\Delta u = 0.042$. This is also the chance that an individual N_2 molecule has a speed in the specified range.

The answer makes the approximation that $f(u)$ is constant across the 390 to 410 m s^{-1} range. It can be improved by computing $f(u)$ at various speeds between 390 and 400 m s^{-1}, using shorter Δu's that bracket these speeds, and adding up the subtotals. See text Appendix C[6] for details on this operation, which amounts to estimating an area under the Maxwell-Boltzmann speed distribution curve. Also, see **4-45** and **4-77**.

The Maxwell-Boltzmann distribution is scaled in such a way that the sum of all possible probabilities equals 1—the molecule must have *some* speed.

---

[6]Text page A-22.

The **root-mean-square speed** u_{rms} equals the square root of the mean of the squares of the speeds of a collection of molecules. It is defined by the equation:

$$u_{rms} = \sqrt{\frac{3k_B T}{m}}$$

The **Boltzmann constant** equals 1.38×10^{-23} J K^{-1}. As always, T and m stand for the absolute temperature of the gas and the mass of the molecules. If m is in kilograms, then u_{rms} comes out in meters per second. Common gases at room temperature have root-mean-square molecular speeds on the order of 10^3 m s^{-1}. At high temperatures, average speeds are higher. See **4-75**.

Since the Boltzmann constant equals the universal gas constant divided by Avogadro's number: ($k_B = R/N_0$), k_B/m can be replaced by $R/\mathcal{M}$ where $\mathcal{M}$ is the mass of a mole (Avogadro's number) of molecules (see **4-43**). Therefore the root-mean-square speed is also:

$$u_{rms} = \sqrt{\frac{3RT}{\mathcal{M}}}$$

If R in this equation is 8.315 J K^{-1}mol^{-1} then $\mathcal{M}$ must have the units kg mol^{-1} (and not the more usual g mol^{-1}) in order to come out in m s^{-1}. To prove this, recall that a joule is a kg m^{2}s^{-2} and see **4-41**.

The Maxwell-Boltzmann distribution curve is not symmetrical about its maximum. Consequently other representative speeds arise:

$$\bar{u} = \sqrt{\frac{8k_B T}{\pi m}} = \sqrt{\frac{8RT}{\pi \mathcal{M}}} \qquad \text{and} \qquad u_{mp} = \sqrt{\frac{2k_B T}{m}} = \sqrt{\frac{2RT}{\mathcal{M}}}$$

The **average speed** $\bar{u}$ is the arithmetic mean of the speeds of the molecules in a sample of gas. It is always less than u_{rms} because squaring the speeds, the procedure used in computing a root-mean-square speed, gives extra emphasis to larger speeds. The **most probable speed** u_{mp} is always less than both u_{rms} and $\bar{u}$. It is the speed at which the Maxwell-Boltzmann distribution curve reaches its maximum.

Applications of the Kinetic Theory

Wall Collisions, Effusion and Diffusion

The rate of collisions of the molecules of a gas with an area of a wall of its container is:

$$Z_w = \frac{1}{4}\frac{N}{V}\bar{u}A = \frac{1}{4}\frac{N}{V}\sqrt{\frac{8RT}{\pi \mathcal{M}}}A$$

where A is the wall area under consideration and N/V is the **number density** of the gas. The number density of a sample of gas equals its molar density (in, for example, mol L^{-1}) multiplied by Avogadro's number (N_0). The units of Z_w are time^{-1}, that is, Z_w gives the number of collisions per unit time. See **4-47**.

Graham's law of effusion follows from this expression. Effusion is the escape of a gas from a container through a small hole into a vacuum. Graham's law states that the rate of effusion of a gas at constant temperature and pressure is inversely proportional to the square root of its molar mass:

$$\text{rate of effusion} \propto \sqrt{\frac{1}{\mathcal{M}}} \quad \text{(constant } T \text{ and } P\text{)}$$

Graham's law is often applied to mixtures of two gases effusing through a small orifice. The ratio of the rate of effusions is then:

$$\frac{\text{rate of effusion of A}}{\text{rate of effusion of B}} = \frac{N_A}{N_B}\sqrt{\frac{\mathcal{M}_B}{\mathcal{M}_A}}$$

The gas that emerges is enriched in the lighter component by the **enrichment factor** $\sqrt{\mathcal{M}_B/\mathcal{M}_A}$ if B is heavier than A. The rate of **diffusion**, which is physically different from effusion, is also inversely proportional to the square root of the molar mass of the diffusing gas (see **4-49** and **4-51**).

The typical problem involving Graham's law requires the computation of the molar mass of an unknown gas by comparison of its rate of effusion to the rate of effusion of a known gas under the same conditions. Trouble arises in problems based on Graham's law because the "rate of effusion" of a gas may be stated in a variety of ways: the number of molecules that escape per unit time, the volume of gas that escapes per unit time, and the mass of gas that escapes per unit time. See **4-79**. Effusion data may also be given in terms of the inverses of the preceding: the time it takes a fixed number of molecules to escape, the time it takes a fixed volume of gas to escape, and the time it takes a fixed mass of gas to escape. Rates of diffusion can be stated in all of these units and also in terms of the distance a gas diffuses down a tube per unit time and its inverse, the time it takes a gas to diffuse a fixed distance.

Frequency of Molecular Collisions

If the molecules of a gas were mathematical points, they would always miss each other and collide only with the walls of the container. In real gases, the molecules have a real effective diameter d and the rate at which a typical molecule experiences collisions becomes of interest. It is:

$$Z_1 = \sqrt{2}\frac{N}{V}\pi d^2 \bar{u} = 4\frac{N}{V}d^2\sqrt{\frac{\pi RT}{\mathcal{M}}}$$

The usual units of Z_1 are s^{-1}. The derivation of this expression assigns a diameter d to the molecules of the gas and imagines an individual molecule sweeping out a cylinder over some fixed time. The frequency of collisions of this molecule with others varies with the square of the diameter of its cylinder: a bigger molecule is harder to avoid. Because a faster moving molecule sweeps out a longer cylinder in the fixed time, the frequency of collisions is also directly proportional to the root-mean-square speed of the molecules. Finally, the greater the density of the gas, then the more encounters the test molecule experiences per unit time. This explains each of the three factors that are multiplied together to give Z_1.

Mean Free Path and Diffusion

The **mean free path λ** of a gas is the average distance that its molecules travel between collisions. It is the product of the average speed of the molecules ($\bar{u}$) and the average time between their collisions. The average time between collisions is the reciprocal of the rate of collision. Therefore:

$$\lambda = \bar{u} \left(\frac{1}{Z_1} \right)$$

Upon substituting for Z_1, $\bar{u}$ cancels out:

$$\lambda = \frac{1}{\sqrt{2}\pi d^2 N/V}$$

Thus, the mean free path of a gas is independent of its temperature and its molar mass, a surprising result. See **4-53** and **4-81** for calculations using this formula. At higher densities, the mean free path become comparable to the diameter of the molecules. At this point, deviations from ideality (next section) become important.

Diffusion in a gas can be described in terms of the mean-square displacement $\overline{\Delta r^2}$ and the root-mean-square displacement $\sqrt{\overline{\Delta r^2}}$ of the molecules over the course of time. A molecule follows a chaotic path as the random buffeting of other molecules constantly redirects it. The mean-square displacement is the average of the squares of such displacements over many molecules. In the absence of gas currents, the mean-square displacement equals a constant D multiplied by the elapsed time:

$$\overline{\Delta r^2} = 6Dt$$

where D is the **diffusion constant**. The diffusion constant depends on the temperature, density, molar mass, and molecular diameter of the gas:

$$D = \frac{3}{8} \sqrt{\frac{RT}{\pi \mathcal{M}}} \frac{1}{d^2 N/V}$$

As in the formulas for the root-mean-square speed, average speed, and most probable speed of the molecules in gases, the molar mass $\mathcal{M}$ must be in kg mol^{-1} if R in J K^{-1}mol^{-1} is used. Also, the number density should have units of m^{-3}, and the molecular diameter should be in meters. See **4-53**. Then, the units of D, which have the form (length)^2·(time)$^{-1}$, come out to m^{2}s^{-1}. When this unit is multiplied by a time, the result is a distance squared.

Real Gases

The ideal-gas equation is an equation of state. It relates the state variables that describe a sample of gas. The behavior of real gases deviates from the predictions of the ideal-gas equation of state. The deviations worsen at low temperature and high pressure. Deviations are revealed in the **compressibility factor z** of a gas:

$$z = \frac{PV}{nRT}$$

For an ideal gas, z equals 1 under all conditions. For real gases, z deviates from 1. Other equations of state make better predictions by explicitly accounting for the effects of intermolecular attractions and the definite, if small, size of the molecules.

The Van der Waals Equation of State

One comparatively simple equation of state is the **van der Waals equation**:

$$\left(P + a\frac{n^2}{V^2}\right)(V - nb) = nRT$$

The van der Waals equation is the ideal-gas equation modified by two correction factors. An *additive* correction (an^2/V^2) is applied to the observed pressure of the gas. Intermolecular attractions tend to reduce pressures from what they would be in the ideal case (no attractions) and the addition of a-factor makes up for the reduction.

A *subtractive* correction (nb) is applied to the actual volume available to the motion of the molecules of the gas. This correction accounts for the volume of the gas that is excluded to the motion of molecules because it is taken up by the molecules themselves.

The strength of the intermolecular attractions and the size of the excluded volume depend on the chemical nature and physical size of the molecules. Therefore the values of the van der Waals constants a and b differ from substance to substance. They are experimentally determined. See text Table 4-3.[7]

In working the van der Waals equation, it is easy to compute P and T given everything else (see **4-55** and **4-57**), but hard to compute V. The iterative approach[8]

---

[7]Text page 124.
[8]Text Appendix C, text page A-17.

is the most practical means to obtain V.

Intermolecular Forces

As two real molecules approach each other, they at first attract each other (the a term in the van der Waals equation). As R, the distance between them, grows even shorter, they begin to repel (the b term in the van der Waals equation). The potential energy of the pair of particles drops as a function of R, reaches a minimum and starts to rise again When the potential energy is plotted against R, the result has aa characteristic dip. See text Figure 4-20.[9] The minimum of the curve is the bottom of a "potential-energy well" of the system.

The **Lennard-Jones potential** is a mathematical function that approximates the shape of the potential energy curve for non-bonding interactions between two identical atoms and molecules. It consists of a repulsion term (the term involving the twelfth power) and an attraction term:

$$V_{\mathrm{LJ}} = 4\epsilon \left[\left(\frac{\sigma}{R} \right)^{12} - \left(\frac{\sigma}{R} \right)^{6} \right]$$

where ϵ and σ are constants that differ for various atoms and molecules (see text Table 4-4). It is instructive to graph the intermolecular potential energy curve for one or two different gases, given ϵ and σ. Thus, the parameters for argon in text Table 4-4 can be used to draw the blue curve in text Figure 4-20. Problem **4-87** lends additional physical meaning to the Lennard-Jones parameters by relating them to the van der Waals constants.

Detailed Solutions to Odd-Numbered Problems

4-1 The decomposition of ammonium hydrosulfide produces ammonia NH_3 and hydrogen sulfide H_2S:

$$\boxed{NH_4HS(s) \rightarrow NH_3(g) + H_2S(g)}$$

Tip. To predict the course of chemical reactions, look for the formulas of stable gaseous compounds embedded in more complicated formulas. Heating often drives out such gases. Look for water (H_2O), carbon dioxide (CO_2), carbon monoxide (CO), ammonia (NH_3), hydrogen chloride and the other hydrogen halides (HCl, HBr, HI, HF), oxygen (O_2), nitrogen (N_2), and hydrogen sulfide (H_2S).

4-3 Generate ammonia from ammonium bromide by dissolving the ammonium bromide in water and adding a strong base, such as sodium hydroxide. NH_4Br dissolves

---

[9]Text page 126.

in water to form ammonium ion $NH_4^+(aq)$ and bromide ion $Br^-(aq)$. The $NH_4^+(aq)$ ion donates a hydrogen ion to $OH^-(aq)$ to form gaseous ammonia $NH_3(g)$:

$$\boxed{NH_4^+(aq) + OH^-(aq) \rightarrow NH_3(g) + H_2O(l)}$$

Tip. Heating helps to liberate the ammonia (gases are usually less soluble in hot water than in cold).

4-5 Because water is less dense than mercury, a longer column of water is required to balance the pressure of the atmosphere. A pressure of 1.00 atm is balanced in a barometer by a column of mercury 76.0 cm high. The density of mercury is 13.6 g cm^{-3}, whereas the density of water is only 1.00 g cm^{-3}. The column of water must be longer in inverse proportion to the ratio of the densities:

$$76.0 \text{ cm} \times \left(\frac{13.6 \text{ g cm}^{-3}}{1.00 \text{ g cm}^{-3}}\right) = 1.03 \times 10^3 \text{ cm} = \boxed{10.3 \text{ m}}$$

This is nearly 34 feet. Still, water barometers have been built.

Tip. The problem can also be solved by substitution in $P = \rho g h$:

$$h = \frac{P}{\rho g} = \frac{101\,325 \text{ Pa}}{(1.00 \times 10^3 \text{ kg m}^3)(9.807 \text{ m s}^{-2})} = \boxed{10.3 \text{ m}}$$

Note the conversion of the pressure to pascals, the SI unit, and the conversion of the density of water to kilograms per cubic meter, a combination of SI base units. These conversions guarantee an answer in meters, the SI base unit of length.

4-7 Convert the 414 atm pressure to pascals (Pa), and then use the formula $P = \rho g h$ to compute the depth (h) of sea-water that exerts the same pressure. In this computation, g is the acceleration of the earth's gravity (9.807 m s^{-2}) and ρ is the density of water. Take the density of sea-water as a constant 1.00×10^3 kg m^{-3} throughout the oceans.[10] The pressure is:

$$P = 414 \text{ atm} \times \left(\frac{1.01325 \times 10^5 \text{ Pa}}{1 \text{ atm}}\right) = 4.19 \times 10^7 \text{ Pa}$$

The depth of water that exerts this pressure is:

$$h = \frac{P}{\rho g} = \frac{4.19 \times 10^7 \text{ kg m}^{-1}\text{s}^{-2}}{(1.0 \times 10^3 \text{ kg m}^{-3})\, 9.807 \text{ m s}^{-2}} = 4.3 \times 10^3 \text{ m}$$

---

[10]The density of the deep sea is however somewhat affected by dissolved salts and the compression of overlying layers.

Text Table B-2[11] confirms that a pascal equals a $kg\,m^{-1}s^{-2}$. One meter equals 3.28 feet, so the depth of 4300 m equals $\boxed{14\,000}$ feet.

4-9 The pascal (Pa) is a newton per square meter ($N\,m^{-2}$). One standard atmosphere is defined as 1.01325×10^5 Pa. Convert using unit factors as follows:

$$172.00 \text{ MPa} \times \left(\frac{10^6 \text{ Pa}}{\text{MPa}}\right) \times \left(\frac{1 \text{ atm}}{1.01325 \times 10^5 \text{ Pa}}\right) = \boxed{1.6975 \times 10^3 \text{ atm}}$$

Convert the pressure from pascals to bars by multiplying by the proper unit factor:

$$1.7200 \times 10^8 \text{ Pa} \times \left(\frac{1 \text{ bar}}{10^5 \text{ Pa}}\right) = \boxed{1.7200 \times 10^3 \text{ bar}}$$

All of the unit factors in the preceding conversions come from definitions and therefore have an unlimited number of significant digits.

4-11 Assume that the N_2 in the tank behaves ideally. Since neither the temperature nor the amount of the N_2 changes during the expansion, use Boyle's law in the form $P_1V_1 = P_2V_2$. The initial pressure P_1 is 3.00 atm, the initial volume V_1 is 2.00 L, P_2 is the final pressure (this is the desired answer), and V_2 is the final volume. If the volumes of the valve and associated plumbing are negligibly small, $V_2 = 2.00 + 5.00 = 7.00$ L. Then:

$$P_2 = \frac{P_1V_1}{V_2} = (3.00 \text{ atm})\left(\frac{2.00 \text{ L}}{7.00 \text{ L}}\right) = \boxed{0.857 \text{ atm}}$$

4-13 Apply Charles's law in the form $V_1T_2 = V_2T_1$. The V_1 is given (4.00 L), and the absolute temperature is doubled, that is, $T_2 = 2T_1$. Accordingly, $V_2 = \boxed{8.00 \text{ L}}$.

4-15 Use Charles's Law. In this problem, $V_1 = 17.4$ gill and V_2 is required. The temperatures are given on the Fahrenheit scale and must be converted to an absolute scale for use with Charles's law. The following shows these as unit conversions:

$$T_1 = (100°F - 32°F)\left(\frac{5°C}{9°F}\right)\left(\frac{1 \text{ K}}{1°C}\right) + 273.15 \text{ K} = 310.9 \text{ K}$$

$$T_2 = (0°F - 32°F)\left(\frac{5°C}{9°F}\right)\left(\frac{1 \text{ K}}{1°C}\right) + 273.15 \text{ K} = 255.4 \text{ K}$$

Insert these T's into the equation for Charles's law:

$$V_2 = \left(\frac{T_2}{T_1}\right)V_1 = \left(\frac{255.4 \text{ K}}{310.9 \text{ K}}\right) 17.4 \text{ gill} = \boxed{14.3 \text{ gill}}$$

Tip. Do not worry about converting gills to more familiar units.

---

[11]Text page A-11.

4-17 The complete reaction of a set mass of CaC_2 produces a set mass of $C_2H_2(g)$ regardless of T and P. Since the pressure is the same (1 atm) in both cases, this is a Charles's law problem with $V_1 = 64.5$ L, $t_1 = 50°C$, $t_2 = 400°C$, and V_2 unknown. Convert the temperatures to kelvins. Then:

$$V_2 = \left(\frac{T_1}{T_2}\right) V_1 = \left(\frac{(400 + 273.15)\ \text{K}}{(50 + 273.15)\ \text{K}}\right) 64.5\ \text{L} = \boxed{134\ \text{L}}$$

4-19 The true pressure inside the bicycle tire is 14.7 psi (1 atm) more than the gauge pressure since the gauge reads zero when the pressure is 1 atm.[12] P is $30.0 + 14.7 = 44.7$ psi. Assuming that the air in the tire behaves ideally:

$$\frac{P_1 V_1}{n_1 T_1} = \frac{P_2 V_2}{n_2 T_2}$$

where the subscripts refer to the variables before and after warming from $0°C$ to $32°C$. Because the tire does not expand, V_1 equals V_2. Also, heating does not change the quantity of air inside the tire, so n_1 equals n_2. Converting the temperatures to the Kelvin scale gives $T_1 = 273$ K and $T_2 = 305$ K, and $P_1 = 44.7$ psi, as just established. Cancel the V's and n's, rearrange and substitute:

$$P_2 = P_1 \frac{T_2}{T_1} = P_1 \left(\frac{305\ \text{K}}{273\ \text{K}}\right) = 44.7\ \text{psi} \left(\frac{305}{273}\right) = 49.9\ \text{psi}$$

Gauge pressure is always 1 atm (14.7 psi) less than the actual pressure. The gauge pressure of the tire at $32°C$ is thus $49.9 - 14.7 = \boxed{35.2\ \text{psi}}$.

Tip. The conversion to absolute temperature was necessary. Using temperatures in degrees Celsius leads to division by zero in this problem.

4-21 a) Let state 1 be the original state of the air, and let state 2 be the state of the air after the compression. Assume that the ideal-gas law applies. Neither the chemical amount of air nor its temperature changes between state 1 and state 2. Hence Boyle's law applies:

$$P_2 = \frac{P_1 V_1}{V_2} = \frac{(1.01\ \text{atm})(20.6\ \text{L})}{1.05\ \text{L}} = \boxed{19.8\ \text{atm}}$$

b) Let state 3 be the state of the bottled air in the European laboratory. The pressure in the bottle is bigger because T_3 (294 K) exceeds T_2 (253 K). Note the conversion

---

[12]A flat tire still has 1 atm of air pressure inside it.

of the temperatures to an absolute scale. Solve the ideal-gas equation for n/V and write it for states 2 and 3:

$$\frac{P_2}{RT_2} = \frac{n_2}{V_2} \quad \text{and} \quad \frac{P_3}{RT_3} = \frac{n_3}{V_3}$$

Neither the volume of the bottle nor the chemical amount of the air it contains changes during the trip to Europe. This means

$$\frac{n_2}{V_2} = \frac{n_3}{V_3} \quad \text{from which} \quad \frac{P_2}{T_2} = \frac{P_3}{T_3}$$

Solve this last equation for P_3, and substitute the known values of the other quantities:

$$P_3 = P_2 \frac{T_3}{T_2} = (19.8 \text{ atm}) \frac{294 \text{ K}}{253 \text{ K}} = \boxed{23.0 \text{ atm}}$$

4-23 The information about the density of the gas is nearly worthless because gas densities depend strongly on P and T, but neither is specified. One might assume room temperature and pressure, but other assumptions are plausible.

Using the fact that $n = m/\mathcal{M}$, the ideal-gas equation can be rewritten as:

$$PV = \frac{m}{\mathcal{M}} RT \quad \text{which gives} \quad \frac{m}{V} = \frac{P\mathcal{M}}{RT} \quad \text{which becomes} \quad \rho = \frac{P\mathcal{M}}{RT}$$

because the density ρ equals mass divided by volume. Solve the last equation for T and insert the stated density and the molar mass of the H_2Te, which are 6.234 g L^{-1} and 129.615 g mol^{-1} respectively:

$$T = \frac{P\mathcal{M}}{R\rho} = \frac{(1.00 \text{ atm})(129.615 \text{ g mol}^{-1})}{(0.08206 \text{ L atm mol}^{-1}\text{K}^{-1})(6.234 \text{ g L}^{-1})} = 253.4 \text{ K} = \boxed{-19.8°\text{C}}$$

4-25 a) The other product of the reaction is sodium chloride:

$$\boxed{2\,Na(s) + 2\,HCl(g) \rightarrow H_2(g) + 2\,NaCl(s)}$$

b) First, calculate the chemical amount of $H_2(g)$ produced by the complete reaction of 6.24 g of $Na(s)$:

$$n_{H_2} = 6.24 \text{ g Na} \times \left(\frac{1 \text{ mol Na}}{22.99 \text{ g Na}}\right) \times \left(\frac{1 \text{ mol H}_2}{2 \text{ mol Na}}\right) = 0.1357 \text{ mol H}_2$$

Next, convert the temperature into kelvins (323 K), rearrange the ideal-gas equation to give V explicitly, and substitute the known values:

$$V_{H_2} = \frac{n_{H_2}RT}{P} = \frac{(0.1357 \text{ mol})(0.08206 \text{ L atm mol}^{-1}\text{K}^{-1})(323 \text{ K})}{0.850 \text{ atm}} = \boxed{4.23 \text{ L}}$$

4-27 Calculate the chemical amount of NaCl being reacted:

$$2.5 \times 10^6 \text{ g NaCl} \times \left(\frac{1 \text{ mol NaCl}}{58.44 \text{ g NaCl}} \right) = 4.28 \times 10^4 \text{ mol NaCl}$$

According to the balanced equation, 1 mol of HCl forms per 1 mol of NaCl consumed. Therefore, 4.28×10^4 mol NaCl in theory produces 4.28×10^4 mol HCl. Use the ideal-gas equation to compute the volume of this amount of gaseous HCl under the specified conditions. Note the conversion of the temperature from Celsius to absolute:

$$V_{\text{HCl}} = \frac{n_{\text{HCl}} RT}{P} = \frac{(4.28 \times 10^4 \text{ mol})(0.08206 \text{ L atm mol}^{-1}\text{K}^{-1})(823 \text{ K})}{0.970 \text{ atm}}$$

$$= \boxed{3.0 \times 10^6 \text{ L}}$$

4-29 According to the balanced chemical equation, there is a 3-to-2 molar ratio between the O_2 formed and $KClO_3$ consumed. This fact furnishes a crucial unit factor in the following series:

$$n_{O_2} = 87.6 \text{ g KClO}_3 \times \left(\frac{1 \text{ mol KClO}_3}{122.54 \text{ g KClO}_3} \right) \times \left(\frac{3 \text{ mol O}_2}{2 \text{ mol KClO}_3} \right) = 1.072 \text{ mol O}_2$$

Now, use the ideal-gas equation to compute the volume, not forgetting to convert the temperature from Celsius to absolute:

$$V_{O_2} = \frac{n_{O_2} RT}{P} = \frac{(1.072 \text{ mol})(0.08206 \text{ L atm mol}^{-1}\text{K}^{-1})(286.4 \text{ K})}{1.04 \text{ atm}} = \boxed{24.2 \text{ L}}$$

4-31 a) The problem is similar to text Example 4-5[13] and to **4-25**. Calculate the theoretical amount of H_2S needed in this reaction to give 2.00 kg of S:

$$n_{H_2S} = (2.00 \times 10^3 \text{ g S}) \times \left(\frac{1 \text{ mol S}}{32.066 \text{ g S}} \right) \times \left(\frac{2 \text{ mol H}_2\text{S}}{3 \text{ mol S}} \right) = 41.58 \text{ mol H}_2\text{S}$$

Use the ideal-gas equation to compute the volume the gaseous H_2S occupies under the stated conditions:

$$V_{H_2S} = \frac{n_{H_2S} RT}{P} = \frac{(41.58 \text{ mol})(0.08206 \text{ L atm mol}^{-1}\text{K}^{-1})(273.15 \text{ K})}{1.00 \text{ atm}} = \boxed{932 \text{ L}}$$

Note the conversion of the temperature into kelvins. The final answer has three significant figures because the mass of sulfur was given to only three significant figures.

---

[13]Text page 111.

b) Copy the approach that worked in the preceding part:

$$n_{SO_2} = (2.00 \times 10^3 \text{ g S}) \times \left(\frac{1 \text{ mol S}}{32.066 \text{ g S}}\right) \times \left(\frac{1 \text{ mol SO}_2}{3 \text{ mol S}}\right) = 20.79 \text{ mol SO}_2$$

The mass of this chemical amount of sulfur dioxide is:

$$m_{SO_2} = 20.79 \text{ mol SO}_2 \times \left(\frac{64.06 \text{ g SO}_2}{1 \text{ mol SO}_2}\right) = \boxed{1.33 \times 10^3 \text{ g SO}_2}$$

Obtain the volume occupied by the SO_2 by substituting n_{SO_2}, P, and T into the ideal-gas equation:

$$V_{SO_2} = \frac{n_{SO_2}RT}{P} = \frac{(20.79 \text{ mol})(0.08206 \text{ L atm mol}^{-1}\text{K}^{-1})(273.15 \text{ K})}{1.00 \text{ atm}} = \boxed{466 \text{ L}}$$

Another way to get the volume of SO_2 is to recall Avogadro's hypothesis,[14] which requires that V_{SO_2} equal half of V_{H_2S}.

4-33 Apply the definition of mole fraction to the SO_3:

$$X_{SO_3} = \frac{\text{mol SO}_3}{\text{total moles}} = \frac{17.0 \text{ mol}}{26.0 \text{ mol} + 83.0 \text{ mol} + 17.0 \text{ mol}} = \frac{17.0 \text{ mol}}{126.0 \text{ mol}} = \boxed{0.135}$$

The partial pressure of the SO_3 equals its mole fraction times the total pressure:

$$P_{SO_3} = X_{SO_3}P_{tot} = (0.135)(0.950 \text{ atm}) = \boxed{0.128 \text{ atm}}$$

4-35 Ideal gases under a given set of conditions take up volume in direct proportion to the number of molecules that comprise them. Hence the mole percentage (or fraction) of N_2 in Martian air equals its volume percentage (or fraction): $X_{N_2} = \boxed{0.027}$. The partial pressure of N_2 equals the total pressure multiplied by the mole fraction of N_2:

$$P_{N_2} = X_{N_2}P_{tot} = (0.027)(5.92 \times 10^{-3} \text{ atm}) = \boxed{1.6 \times 10^{-4} \text{ atm}}$$

4-37 a) The mole fraction of CO in the mixture equals:

$$X_{CO} = \frac{\text{mol CO}}{\text{total moles}} = \frac{10.0 \text{ mol}}{10.0 \text{ mol} + 12.5 \text{ mol}} = \frac{10.0 \text{ mol}}{22.5 \text{ mol}} = \boxed{0.444}$$

b) The balanced chemical equation shows that formation of 3.0 mol of CO_2 requires consumption of 3.0 moles of CO. Therefore, the chemical amount of CO in the mixture at this point in the reaction is 7.0 mol. The chemical amount of O_2 at this point equals

---

[14]Text page 13.

12.5 mol $-$ (3.0 mol/2) $=$ 11.0 mol where the factor 2 arises from the 2-to-1 reaction stoichiometry between the CO_2 and O_2. The mixture of gases consists of 7.0 mol of CO, 11.0 mol of O_2, and 3.0 mol of CO_2. The mole fraction of CO is therefore:

$$X_{CO} = \frac{\text{mol CO}}{\text{total moles}} = \frac{7.0 \text{ mol}}{7.0 \text{ mol} + 11.0 \text{ mol} + 3.0 \text{ mol}} = \frac{7.0 \text{ mol}}{21.0 \text{ mol}} = \boxed{0.33}$$

4-39 a) Treat the saturated air as a mixture of ideal gases that contribute to the total pressure according to Dalton's law. The volume of the mixture is 1.0 cm³, its temperature is $(20 + 273.15)$ K, and the partial pressure of water vapor equals 0.0230 atm. Solve the ideal-gas equation for n and applying it to the water vapor:

$$n_{H_2O} = \frac{P_{H_2O}V}{RT} = \frac{(0.0230 \text{ atm})(1.0 \text{ cm}^3)}{(82.057 \text{ cm}^3 \text{ atm K}^{-1}\text{mol}^{-1})(293.15 \text{ K})} = 9.56 \times 10^{-7} \text{ mol}$$

A mole of H_2O contains N_0 molecules. Hence: so the 1.0 cm³ of saturated air contains:

$$\frac{9.56 \times 10^{-7} \text{ mol H}_2\text{O}}{1.0 \text{ cm}^3 \text{ air}} \times \left(\frac{6.022 \times 10^{23} \text{ molecules}}{1 \text{ mol}}\right) = \boxed{\frac{5.8 \times 10^{17} \text{ molecules H}_2\text{O}}{\text{cm}^3 \text{ air}}}$$

b) 1.0 cm³ of air holds only about 10^{-6} mol of water. Hence it requires more than 1.0 cm³ to hold 0.50 mol of water. Use unit factors as follows:

$$0.50 \text{ mol H}_2\text{O} \times \left(\frac{1.0 \text{ cm}^3 \text{ sat. air}}{9.56 \times 10^{-7} \text{ mol H}_2\text{O}}\right) \times \left(\frac{1 \text{ L}}{1000 \text{ cm}^3}\right) = \boxed{520 \text{ L sat. air}}$$

4-41 a) The root-mean-square speed of the molecules in a gas is given by:

$$u_{rms} = \sqrt{\frac{3k_BT}{m}} = \sqrt{\frac{3RT}{\mathcal{M}}}$$

where k_B is the Boltzmann constant, T is the absolute temperature, m is the mass of the molecules, R is the gas constant, and $\mathcal{M}$ is the molar mass of the gas. For H_2 at 300 K:

$$u_{rms} = \sqrt{\frac{3RT}{\mathcal{M}}} = \sqrt{\frac{3(8.315 \text{ J K}^{-1}\text{mol}^{-1})(300 \text{ K})}{0.002016 \text{ kg mol}^{-1}}} = \boxed{1.93 \times 10^3 \text{ m s}^{-1}}$$

b) For sulfur hexafluoride SF_6 at 300 K:

$$u_{rms} = \sqrt{\frac{3RT}{\mathcal{M}}} = \sqrt{\frac{3(8.315 \text{ J K}^{-1}\text{mol}^{-1})(300 \text{ K})}{0.14607 \text{ kg mol}^{-1}}} = \boxed{226 \text{ m s}^{-1}}$$

The heavier SF_6 molecules have a root-mean-square speed about 8.5 times slower than the lighter H_2 molecule at the same temperature. The ratio of the speeds equals the square root of the reciprocal of the ratio of the molar masses.

Tip. The analysis of the units is worth separate study:

$$\sqrt{\frac{\text{J K}^{-1}\text{mol}^{-1}\ \text{K}}{\text{kg mol}^{-1}}} = \sqrt{\frac{\text{kg m}^2\ \text{s}^{-2}\ \text{K}^{-1}\ \text{mol}^{-1}\ \text{K}}{\text{kg mol}^{-1}}} = \sqrt{\text{m}^2\ \text{s}^{-2}} = \text{m s}^{-1}$$

4-43 The rms speed of helium atoms at the surface of the sun (6000 K) is:

$$u_{\text{rms}} = \sqrt{\frac{3RT}{\mathcal{M}}} = \sqrt{\frac{3(8.315\ \text{J K}^{-1}\text{mol}^{-1})(6000\ \text{K})}{0.004003\ \text{kg mol}^{-1}}} = \boxed{6100\ \text{m s}^{-1}}$$

In an interstellar cloud at 100 K:

$$u_{\text{rms}} = \sqrt{\frac{3RT}{\mathcal{M}}} = \sqrt{\frac{3(8.315\ \text{J K}^{-1}\text{mol}^{-1})(100\ \text{K})}{0.004003\ \text{kg mol}^{-1}}} = \boxed{790\ \text{m s}^{-1}}$$

"Comparison" may mean to take the ratio of the two rms speeds rather than to calculate the actual speeds. Getting the ratio is simpler that the previous calculation because $3R$ and $\mathcal{M}$ cancel out:

$$\frac{u_{\text{rms}}\ (\text{near sun})}{u_{\text{rms}}\ (\text{interstellar})} = \sqrt{\frac{6000\ \text{K}}{100\ \text{K}}} = 7.7$$

The rms speed of the molecules of a gas goes up by a factor of only about eight between 100 K and 6000 K.

4-45 The molecular speeds in ClO_2 follow the Maxwell-Boltzmann distribution because the gas is at thermal equilibrium. If 35.0% of the molecules have speeds exceeding 400 m s^{-1}, then obviously 65.0% have speeds between 0 and 400 m s^{-1}. Since over half of the molecules have speeds less than 400 m s^{-1}, that speed lies well on the high side of the hump in the Maxwell-Boltzmann distribution.[15] An increase in temperature shifts the hump toward higher speeds and also flattens it out. Because the temperature increase is slight, the first effect predominates. The percentage of molecules having speeds in excess of 400 m s^{-1} will $\boxed{\text{increase}}$.

Tip. This answer can be confirmed numerically, but not very easily. The area under this Maxwell-Boltzmann curve from $u = 0$ to $u = 400$ m s^{-1} equals 0.65 of the total area under the curve. Integrate the Maxwell-Boltzmann function from $u = 0$ to

---

[15]See graph on text page 118.

$u = 400$, set the integral equal to 0.65, and determine the T that makes the integral equal to 0.65. The answer is 397 K. If the temperature is raised slightly, the area under the curve between 0 and 400 m s^{-1} changes as the hump in the distribution shifts to the right. Suppose the slightly higher new temperature is 407 K. The integral of the $T = 407$ Boltzmann-Maxwell function from 0 to 400 m s^{-1} is only 0.64. Hence, the percentage of molecules having speeds exceeding 400 m s^{-1} rises from 35% to about 36% when ClO_2 is heated from 397 to 407 K.

4-47 Compute the chemical amount of air that leaked into the 500 cm^3 bulb during the one-hour period immediately after it is sealed by inserting the pressure observed at the one-hour time and the known volume and temperature into the ideal-gas equation:

$$n_{air} = \frac{PV}{RT} = \frac{(1.00 \times 10^{-7} \text{ atm})(0.500 \text{ L})}{(0.082057 \text{ L atm mol}^{-1}\text{K}^{-1})(300 \text{ K})} = 2.03 \times 10^{-9} \text{ mol}$$

The rate of the leak was:

$$\frac{2.03 \times 10^{-9} \text{ mol}}{1 \text{ hr}} \times \left(\frac{6.022 \times 10^{23} \text{ molec.}}{\text{mol}}\right) \times \left(\frac{1 \text{ hr}}{3600 \text{ s}}\right) = \frac{3.4 \times 10^{11} \text{ molec.}}{\text{s}}$$

Outside air entered the vessel when its molecules "collided" with the area of the tiny hole. Therefore, work with the formula for the rate of wall collisions by a gas.[16] The rate at which molecules exit the bulb through the hole is surely negligible, so the observed rate of the leak equals Z_w in this formula. Compute the density of the outside air using the ideal-gas equation and its known temperature and pressure:

$$\left(\frac{n}{V}\right)_{air} = \left(\frac{1.00 \text{ atm}}{(0.082057 \text{ L atm mol}^{-1}\text{K}^{-1})(300 \text{ K})}\right) = 4.062 \times 10^{-2} \text{ mol L}^{-1}$$

Multiplying this result by N_0 converts it to a number density:

$$\frac{4.062 \times 10^{-2} \text{ mol}}{1 \text{ L}} \times \left(\frac{6.022 \times 10^{23} \text{ molec.}}{\text{mol}}\right) \times \left(\frac{1000 \text{ L}}{1 \text{ m}^3}\right) = 2.46 \times 10^{25} \text{ molecule m}^{-3}$$

Solve text equation 4-13 for A, the area of the wall:

$$A = 4\,\frac{1}{N/V}\sqrt{\frac{\pi \mathcal{M}}{8RT}}\, Z_w$$

Insert numbers for the several quantities on the right, taking care with units:

$$A = 4\left(\frac{1}{2.46 \times 10^{25} \text{ m}^{-3}}\right)\sqrt{\frac{3.1416(0.0288 \text{ kg mol}^{-1})}{8(8.315 \text{ J K}^{-1}\text{mol}^{-1})(300 \text{ K})}}(3.4 \times 10^{11} \text{ s}^{-1})$$

$$= 1.18 \times 10^{-16} \text{ m}^2$$

---

[16]Equation 4-13, text page 119.

Because the hole is circular, its radius is:

$$r = \sqrt{\frac{A}{\pi}} = \sqrt{\frac{1.18 \times 10^{-16} \text{ m}^2}{3.1416}} = \boxed{6.1 \times 10^{-9} \text{ m}}$$

4-49 The ratio of the rates of effusion of two gases is given by Graham's Law:[17]

$$\frac{\text{rate of effusion of A}}{\text{rate of effusion of B}} = \frac{N_A/V}{N_B/V}\sqrt{\frac{\mathcal{M}_B}{\mathcal{M}_A}}$$

Let the methane be A and the unknown be B. The ratio of N_A/V to N_B/V equals 1 because the gases in the two effusion experiments are held under identical conditions in the same container, and Avogadro's principle therefore applies. Insert the molar mass of methane and the observed rates of effusion: of methane (gas A) and the unknown (gas B):

$$\frac{1.30 \times 10^{-8} \text{ mol s}^{-1}}{5.41 \times 10^{-9} \text{ mol s}^{-1}} = \sqrt{\frac{\mathcal{M}_B}{16.04 \text{ g mol}^{-1}}}$$

Solving for $\mathcal{M}_B$ gives $\boxed{92.6 \text{ g mol}^{-1}}$.

4-51 One pass of a mixture of $^{235}\text{UF}_6$ and $^{238}\text{UF}_6$ through a diffusion apparatus enriches the product mixture by a factor of $\sqrt{352.038/349.028}$ or 1.0043 in the lighter gas, $^{235}\text{UF}_6$. This is computed in text Examples 4-9 and 4-10.[18] This problem calls for enrichment from 0.72 percent $^{235}\text{UF}_6$ to 95 percent $^{235}\text{UF}_6$. It is understood that these percentages are number percentages, not mass percentages. Let (n_{235}/n_{238}) equal the ratio of the number of molecules of the lighter gas $^{235}\text{UF}_6$ to the number of molecules of the heavier gas $^{238}\text{UF}_6$ after any number of diffusion passes. Then:

$$\left(\frac{n_{235}}{n_{238}}\right)_{\text{before}} = \frac{0.72}{99.27} = 0.0007253 \qquad \left(\frac{n_{235}}{n_{238}}\right)_{\text{after}} = \frac{95}{5} = 19$$

Each pass multiplies the light-heavy ratio from the previous pass by 1.0043. Let x equal the number of passes. Then:

$$(0.007253)(1.0043)^x = 19$$

Dividing through by 0.007253 and taking the logarithm of both sides[19] gives:

$$x \log(1.0043) = \log\left(\frac{19}{0.007253}\right) \quad \text{from which} \quad x = \boxed{1830}$$

---

[17]Equation 4-14, text page 120.
[18]Text page 120-21.
[19]See text Appendix C.

4-53 The key is to recognize that the pressure of the krypton is directly proportional to its number density. This follows from the ideal gas law:

$$P = \left(\frac{n}{V}\right) RT \qquad \text{from which} \qquad P = \frac{1}{N_0}\left(\frac{N}{V}\right) RT$$

because N, the number of molecules, divided by N_0, Avogadro's number, equals the number of moles of any substance. The mean free path (λ) of the molecules is:

$$\lambda = \frac{1}{\sqrt{2}\pi d^2 (N/V)} \qquad \text{from which} \qquad \frac{N}{V} = \frac{1}{\sqrt{2}\pi d^2 \lambda}$$

where d is the molecular diameter. Substituting this equation into the preceding gives:

$$P = \frac{1}{N_0}\left(\frac{1}{\sqrt{2}\pi d^2 \lambda}\right) RT$$

Next, obtain the diameter of the spherical vessel. The volume equals 1.00 L (1.00×10^{-3} m^3). For a sphere, $V = (4/3)\pi r^3$. Solving for r gives the radius as 0.0620 m. Its diameter is therefore 0.124 m.

Set λ equal to 0.124 m, because the mean free path must be comparable to the diameter of the vessel. From the statement of the problem, T is 300 K and d is 3.16×10^{-10} m. The gas constant equals 8.206×10^{-5} m^3 atm mol^{-1}K^{-1} (note the carefully chosen units of R). Substitution gives:

$$P = \frac{1}{6.022 \times 10^{23}\ \text{mol}^{-1}} \left(\frac{1}{\sqrt{2}\pi(3.16 \times 10^{-10}\ \text{m})^2(0.124\ \text{m})}\right)$$
$$\times\ (8.206 \times 10^{-5}\ \text{m}^3\,\text{atm mol}^{-1}\ \text{K}^{-1})(300\ \text{K}) = \boxed{7.4 \times 10^{-7}\ \text{atm}}$$

The number density (N/V) of the krypton is needed to calculate the diffusion constant. At this P and T, the ordinary density of krypton is:

$$\rho_{Kr} = \frac{n}{V} = \frac{P}{RT} = \frac{7.4 \times 10^{-7}\ \text{atm}}{(0.08206\ \text{L atm mol}^{-1}\text{K}^{-1})(300\ \text{K})} = 3.00 \times 10^{-8}\ \text{mol L}^{-1}$$

Convert this to a number density:

$$\frac{3.00 \times 10^{-8}\ \text{mol}}{\text{L}} \times \left(\frac{6.022 \times 10^{23}\ \text{molec.}}{\text{mol}}\right) \times \left(\frac{1000\ \text{L}}{1\ \text{m}^3}\right) = \frac{1.81 \times 10^{19}\ \text{molec.}}{\text{m}^3}$$

Substitute into the formula for the diffusion constant of a gas:

$$D = \frac{3}{8}\sqrt{\frac{RT}{\pi\mathcal{M}}}\left(\frac{1}{d^2 N/V}\right)$$

$$= \frac{3}{8}\sqrt{\frac{8.315 \text{ J K}^{-1}\text{mol}^{-1}(300 \text{ K})}{\pi(0.08380 \text{ kg mol}^{-1})}}\left(\frac{1}{(3.16\times10^{-10} \text{ m})^2(1.81\times10^{19} \text{ m}^{-3})}\right)$$

$$= \boxed{20 \text{ m}^2\,\text{s}^{-1}}$$

Tip. Check the units separately:

$$\sqrt{\frac{\text{J K}^{-1}\text{mol}^{-1}\ \text{K}}{\text{kg mol}^{-1}}}\left(\frac{1}{\text{m}^2\ \text{m}^{-3}}\right) = \sqrt{\frac{\text{kg m}^2\ \text{s}^{-2}\ \text{K}^{-1}\ \text{mol}^{-1}\ \text{K}}{\text{kg mol}^{-1}}}\left(\frac{1}{\text{m}^{-1}}\right) = \text{m}^2\,\text{s}^{-1}$$

4-55 The problem is to use the van der Waals equation to estimate the pressure of 6.80 kg of gaseous oxygen in a 28.0-L container at 20°C. First, solve the van der Waals equation for P. Then substitute for n, V, P a and b and complete the arithmetic. The values $a = 1.360$ atm L^{2}mol^{-2} and $b = 0.031834$ L mol^{-1} come from text Table 4-3.[20] The chemical amount of O_2 equals 212.5 mol, which is its mass divided by its molar mass. The details of the substitution follow:

$$P = \frac{nRT}{V-nb} - a\frac{n^2}{V^2}$$

$$= \frac{(212.5 \text{ mol})(0.08206 \text{ L atm mol}^{-1}\text{K}^{-1})(293 \text{ K})}{28.0 \text{ L} - (212.5 \text{ mol})(0.031834 \text{ L mol}^{-1})} - a\frac{n^2}{V^2}$$

$$= 240.6 \text{ atm} - (1.360 \text{ atm L}^2\text{mol}^{-2})\frac{(212.5)^2 \text{ mol}^2}{(28.0)^2 \text{ L}^2}$$

$$= 240.6 \text{ atm} - 78.33 \text{ atm} = \boxed{162 \text{ atm}}$$

This pressure is equivalent to $\boxed{2380 \text{ psi}}$, since 14.696 psi equals 1 atm.

4-57 The problem provides a comparison between the ideal-gas pressure and the van der Waals pressure of a typical gas under ordinary conditions. The data for this sample of CO_2 are:

$$n = 50.0 \text{ g}/44.0 \text{ g mol}^{-1} = 1.136 \text{ mol} \quad T = 298.15 \text{ K}$$
$$R = 0.08206 \text{ L atm mol}^{-1}\text{K}^{-1} \quad\quad V = 1.00 \text{ L}$$

---

[20]Text page 124.

a) Solve the ideal-gas equation for P and substitute:

$$P = \frac{(1.136 \text{ mol})(0.08206 \text{ L atm mol}^{-1}\text{K}^{-1})(298.15 \text{ K})}{1.00 \text{ L}} = \boxed{27.8 \text{ atm}}$$

b) The van der Waals equation includes terms (a and b) that depend on the identity of the gas. For CO_2, a is 3.592 atm $L^2\text{mol}^{-2}$ and b is 0.04267 L mol^{-1}. Solving the van der Waals equation for P and substitution give:

$$\begin{aligned}
P &= \frac{nRT}{V - nb} - a\frac{n^2}{V^2} \\
&= \frac{(1.136 \text{ mol})(0.08206 \text{ L atm mol}^{-1}\text{K}^{-1})(298.15 \text{ K})}{1.00 \text{ L} - (1.136 \text{ mol})(0.04267 \text{ L mol}^{-1})} - a\frac{n^2}{V^2} \\
&= 29.2 \text{ atm} - (3.592 \text{ atm L}^2\text{mol}^{-2})\frac{(1.136)^2 \text{ mol}^2}{(1.00)^2 \text{ L}^2} \\
&= 29.2 \text{ atm} - 4.64 \text{ atm} = \boxed{24.6 \text{ atm}}
\end{aligned}$$

Thus the van der Waals P is less than the ideal-gas P. The effect of the b term in the van der Waals equation was to increase P from 27.8 to 29.2 atm; the effect of a was to decrease P by 4.64 atm. $\boxed{\text{The attractive forces dominate}}$ in this case. They are manifested in the a term.

Tip. The corrections oppose. Consequently the ideal-gas equation gives fair approximations for P over a larger range of conditions that it would otherwise.

4-59 The solid earth is covered by an ocean of air that exerts an average pressure of 730 mm Hg all over its surface. Imagine the air replaced by an ocean of liquid mercury. To exert an equal pressure the mercury would need to be only 730 mm deep. The volume of the mercury would equal, to a close approximation, the surface area of the earth times its depth d. Insert $r = 6.370 \times 10^6$ m (converted from km) and $d = 730 \times 10^{-3}$ m into the formula for this area:

$$V = 4\pi r^2 d = 4(3.14159)(6.370 \times 10^6 \text{ m})^2(730 \times 10^{-3} \text{ m}) = 3.72 \times 10^{14} \text{ m}^3$$

Compute the mass of the ocean of mercury by multiplying its volume by its density, which is the density of mercury at ordinary temperatures:

$$3.72 \times 10^{14} \text{ m}^3 \times \left(\frac{10^6 \text{ cm}^3}{1 \text{ m}^3}\right) \times \left(\frac{13.6 \text{ g}}{\text{cm}^3}\right) \times \left(\frac{1 \text{ kg}}{1000 \text{ g}}\right) = \boxed{5.06 \times 10^{18} \text{ kg}}$$

This mass equals the mass of the atmosphere because the atmosphere exerts the same pressure at the surface of the earth as the hypothetical ocean of mercury.

4-61 From the equation $P = \rho gh$ the height h of the column of liquid in a barometer is inversely proportional at a given pressure P to the density of the liquid that is in it. Hence, the height that the column of Hg would have at 0.0°C is:

$$h_{\text{at 0°C}} = h_{\text{at 35°C}} \times \left(\frac{13.5094 \text{ g cm}^{-3}}{13.5955 \text{ g cm}^{-3}} \right) = 760.0 \text{ mm} \times 0.993667 = 755.19 \text{ mm}$$

This height in a mercury barometer at 0.0°C means a pressure of 755.19 torr, which equals $\boxed{0.9937 \text{ atm}}$.

Tip. Substitution into the formula $P = \rho gh$ gives the same answer.

4-63 Reasoning by analogy to the textbook statements of Charles's law and Boyle's law, we conclude that Amontons's law must state that at constant volume, the pressure of a sample of a gas is directly proportional to its absolute temperature.

4-65 a) 1005 mol of helium displaces 1005 mol of air since the pressure and temperature of the gases inside and outside of the balloon are the same.[21] The masses of these gases are:

$$1005 \text{ mol He} \times \left(\frac{4.003 \text{ g He}}{1 \text{ mol He}} \right) = 4.023 \times 10^3 \text{ g He}$$

$$1005 \text{ mol air} \times \left(\frac{29.0 \text{ g air}}{1 \text{ mol air}} \right) = 2.91 \times 10^4 \text{ g air}$$

The answer is the difference between these two masses, which equals $\boxed{2.51 \times 10^4 \text{ g}}$.

b) The balloon still contains 1005 mol of He after the ascent to 10 miles, and the temperature and pressure inside the balloon still equal the temperature and pressure outside of it. Despite the change in T and P from their ground-level values, the balloon still displaces 1005 mol of air. Therefore, the answer is again $\boxed{2.51 \times 10^4 \text{ g}}$.

4-67 One way to solve this problem is to think in terms of proportions. The ideal-gas law states that a given volume contains moles of gas in *inverse* proportion to their absolute temperature, as long as the pressure is constant. This means the higher the temperature, the lower the amount of gas. Also, the amount of products of a chemical reaction is in direct proportion to the amount of reactants. Raising the temperature at which HCl is collected from 323.15 to 773.15 K (from 50 to 500°C) multiplies T by 2.392. The number of moles of HCl produced in the high-temperature experiment therefore equals the number produced in the low-temperature experiment divided by this factor. Since the number of moles of $HCl(g)$ equals the number of moles of

---

[21]This follows from Avogadro's hypothesis, text page 14.

NaCl reacted, the amount of NaCl used in the high-temperature experiment is less in the same proportion. The answer is simply 10.0 kg divided by 2.392. It equals $\boxed{4.18 \text{ kg NaCl}}$.

Another approach is to compute the "certain volume" and then find the chemical amount (number of moles) of gas it contains at both 323.15 K and 773.15 K. First, find the theoretical yield of HCl from 10.0 kg of NaCl:

$$10.0 \times 10^3 \text{ g NaCl} \times \left(\frac{1 \text{ mol NaCl}}{58.44 \text{ g NaCl}} \right) \times \left(\frac{1 \text{ mol HCl}}{1 \text{ mol NaCl}} \right) = 171.1 \text{ mol HCl}$$

The "certain volume" occupied by this HCl at 50°C (323.15 K) is:

$$V = \frac{nRT}{P} = \frac{171.1 \text{ mol}(0.08206 \text{ L atm mol}^{-1}\text{K}^{-1})(323.15 \text{ K})}{1 \text{ atm}} = 4537 \text{ L}$$

At 500°C (773.15 K) this volume contains fewer moles:

$$n_{500°C} = \frac{PV}{RT} = \frac{(1 \text{ atm})(4537 \text{ L})}{(0.08206 \text{ L atm mol}^{-1}\text{K}^{-1})(773.15 \text{ K})} = 71.51 \text{ mol HCl}$$

Finally, compute the mass of NaCl required to produce this amount of HCl:

$$n_{\text{NaCl}} = 71.51 \text{ mol HCl} \times \left(\frac{1 \text{ mol NaCl}}{1 \text{ mol HCl}} \right) \times \left(\frac{58.44 \text{ g NaCl}}{1 \text{ mol NaCl}} \right)$$
$$\times \left(\frac{1 \text{ kg NaCl}}{1000 \text{ g NaCl}} \right) = \boxed{4.18 \text{ kg NaCl}}$$

Tip. The second calculation is slower, but attractive in its concreteness. But suppose that the problem had simply stated that the pressure was the same in the two experiments (not telling what it was). None of the intermediate numbers in the second method could be computed, but the first method would work the same.

4-69 a) Balance the equations for the two reactions:

$$CaCO_3(s) \rightarrow CaO(s) + CO_2(g)$$
$$CaO(s) + H_2O(l) \rightarrow Ca(OH)_2(s)$$

Determine the chemical amount of CO_2 to produce 8.47 kg of $Ca(OH)_2$:

$$8.47 \text{ kg Ca(OH)}_2 \times \left(\frac{1 \text{ mol Ca(OH)}_2}{0.074093 \text{ kg Ca(OH)}_2} \right) \times \left(\frac{1 \text{ mol CO}_2}{1 \text{ mol Ca(OH)}_2} \right) = 114.32 \text{ mol CO}_2$$

Then, use the ideal-gas law to find the volume of the gaseous CO_2:

$$V_{CO_2} = \frac{nRT}{P} = \frac{(114.3 \text{ mol})(0.08206 \text{ L atm mol}^{-1}\text{K}^{-1})(1223 \text{ K})}{0.976 \text{ atm}} = \boxed{1.18 \times 10^4 \text{ L}}$$

The final answer is rounded to three significant figures because the mass of $Ca(OH)_2$ was given to only three significant figures.

4-71 The total volume of the system is 12.00 L, the sum of the volumes of the three containers (assuming that the volume of the connecting tube is negligible). If the three gases behave ideally in their containers, their chemical amounts are:

$$n_{O_2} = \frac{2.51 \times 5.00 \text{ L atm}}{RT} \qquad n_{N_2} = \frac{0.792 \times 4.00 \text{ L atm}}{RT} \qquad n_{Ar} = \frac{1.23 \times 3.00 \text{ L atm}}{RT}$$

The total pressure after the gases mix is also given by the ideal-gas equation, assuming Dalton's law holds. In the expression for P_{tot}, the total chemical amount of the mixed gas equals the sum of the chemical amounts of the three components:

$$P_{tot} = n_{tot}\frac{RT}{V} = n_{O_2} + n_{N_2} + n_{Ar}\frac{RT}{V}$$
$$= \left(\frac{12.7 \text{ L atm}}{RT} + \frac{3.168 \text{ L atm}}{RT} + \frac{3.69 \text{ L atm}}{RT}\right)\left(\frac{RT}{12.00 \text{ L}}\right) = \boxed{1.63 \text{ atm}}$$

4-73 Assume the gases behave ideally before the catalyst is introduced and after the reaction is finished. The reaction itself is profoundly non-ideal behavior. Imagine that the temperature and volume of the system are such that the total chemical amount of gases equals 1.00 mol at the start. The reaction:

$$C_2H_2(g) + 2\,H_2(g) \rightarrow C_2H_6(g)$$

then must decrease the total chemical amount of gas to 0.42 mol. This is true because chemical amount is directly proportional to pressure if T and V do not change. Let x represent the original chemical amount of $C_2H_2(g)$ and y the original chemical amount of $H_2(g)$. Before the reaction starts, there is no $C_2H_6(g)$ so:

$$x + y = 1.00 \text{ mol}$$

The reaction produces x mol of $C_2H_6(g)$ as it consumes $2x$ mol of $H_2(g)$ and x mol of $C_2H_2(g)$. Since $C_2H_2(g)$ is the limiting reagent, the reaction stops when the x mol of C_2H_2 gas is gone. At the end of the reaction, the vessel contains x mol of $C_2H_6(g)$, the product, and $(y - 2x)$ mol of left-over $H_2(g)$. Hence:

$$x + (y - 2x) = 0.42 \text{ mol}$$

Solving the two simultaneous equations gives $x = 0.029$. The original amount of $C_2H_2(g)$ is 0.29 mol; the original mole fraction of $C_2H_2(g)$ is $0.29/1.00 = \boxed{0.29}$.[22]

4-75 a) The average kinetic energy of an atom of deuterium, which is given in the problem, depends only on the absolute temperature (assuming ideal-gas behavior): $KE_{avg} = 3/2k_BT$. Solve for T and substitute:

$$T = \frac{2KE_{avg}}{3k_B} = \frac{2(8 \times 10^{-16} \text{ J})}{3(1.38 \times 10^{-23} \text{ J K}^{-1})} = \boxed{3.86 \times 10^7 \text{ K}}$$

The atomic mass of ^2H is not needed in this part of the problem.

b) The average kinetic energy of the particles in a gas equals $1/2m\bar{u}^2$. Write this relationship for ^1H and divide it by a similar relationship for ^2D:

$$\frac{1/2m_H\bar{u}_H^2}{1/2m_D\bar{u}_D^2} = \frac{32 \times 10^{-16} \text{ J}}{8 \times 10^{-16} \text{ J}}$$

Solve for the ratio of the rms speeds:

$$\frac{u_{rms,H}}{u_{rms,D}} = \sqrt{\frac{\bar{u}_H^2}{\bar{u}_D^2}} = \sqrt{\frac{32}{8}}\sqrt{\frac{2.015}{1.0078}} = \boxed{2.83}$$

4-77 For a Maxwell-Boltzmann speed distribution, the quantity $f(u)\Delta u$ equals the fraction of the molecules in a gas having speeds between u and $u + \Delta u$.[23] This fraction equals the desired probability and is:

$$\frac{\Delta N}{N} = f(u)\Delta u = 4\pi\left(\frac{\mathcal{M}}{2\pi RT}\right)^{3/2}u^2\exp(-\mathcal{M}u^2/2RT)\Delta u$$

The gas in this case is O_2, for which $\mathcal{M} = 0.0320$ kg mol^{-1}. The temperature equals 300 K, and Δu equals 10 m s^{-1} (from 500 to 510 m s^{-1}). Evaluate $f(u)$ at $u = 500$ m s^{-1}:

$$f(u) = 4\pi\left(\frac{0.0320 \text{ kg mol}^{-1}}{2\pi(8.315 \text{ J K}^{-1}\text{mol}^{-1})(300 \text{ K})}\right)^{3/2} \times (500 \text{ m s}^{-1})^2$$

$$\times \exp\left(\frac{-0.032 \text{ kg mol}^{-1}(500 \text{ m s}^{-1})^2}{2(8.315 \text{ J K}^{-1}\text{mol}^{-1})(300 \text{ K})}\right) = 1.843 \times 10^{-3} \text{ s m}^{-1}$$

---

[22]Had we assumed the system to be big enough to hold, say, 100 mol of gases, all of the numbers in this computation, except the answer, would have been 100 times bigger.

[23]Text page 108.

The value of $f(u)$ changes over the range of u. The hint proposes a way to deal with this change, which amounts to a 2.5 percent decrease as u rises from 500 to 510 meters per second. This decrease is shown in the following table:

u (m s^{-1})	$f(u)$ (s m^{-1})	u (m s^{-1})	$f(u)$ (s m^{-1})
500	1.843×10^{-3}	506	1.816×10^{-3}
501	1.839×10^{-3}	507	1.812×10^{-3}
502	1.834×10^{-3}	508	1.807×10^{-3}
503	1.830×10^{-3}	509	1.802×10^{-3}
504	1.825×10^{-3}	510	1.797×10^{-3}
505	1.821×10^{-3}	–	–

The desired probability $f(u)\Delta u$ equals the area under the distribution curve between 500 and 510 m s^{-1}.[24] This area has a width of 10 m s^{-1} and a smoothly changing height. Approximate it by 10 narrow columns of width 1 m s^{-1} and heights in s m^{-1} given by the first ten values of $f(u)$ in the table. The sum of these ten areas is 1.823×10^{-2} (no units). The desired probability is thus $\boxed{1.82 \text{ percent}}$.

4-79 Call the unknown gas Z. Convert the rates of effusion of the oxygen and Z from g min^{-1} to mol min^{-1} so that Graham's law can be applied. Use 32.0 g mol^{-1} as the molar mass of $O_2(g)$ and $\mathcal{M}_Z$ as the molar mass of Z. The two rates equal:

$$\text{rate}_{O_2} = \frac{3.25 \text{ g min}^{-1}}{32.0 \text{ g mol}^{-1}} = 0.1016 \text{ mol min}^{-1} \qquad \text{rate}_Z = \frac{1.96 \text{ g min}^{-1}}{\mathcal{M}_Z \text{ g mol}^{-1}} = \frac{1.96}{\mathcal{M}_Z} \text{ mol min}^{-1}$$

Write Graham's law as a comparison of the two gases, as in text equation 4-14:[25]

$$\frac{\text{rate of effusion of } O_2}{\text{rate of effusion of Z}} = \frac{N_{O_2}}{N_Z}\sqrt{\frac{\mathcal{M}_Z}{\mathcal{M}_{O_2}}}$$

The numbers of molecules in the vessel are the same in the two experiments because the experiments are carried out as the same temperature and pressure. Hence N_{O_2} and N_Z cancel out of the preceding equation. Inserting the two rates gives:

$$\frac{0.1016 \text{ mol min}^{-1}}{(1.96/\mathcal{M}_Z) \text{ mol min}^{-1}} = \frac{0.1016 \mathcal{M}_Z}{1.96} = \sqrt{\frac{\mathcal{M}_Z}{32.0 \text{ g mol}^{-1}}}$$

Solution of the last equation gives $\mathcal{M}_Z$ equal to $\boxed{11.6 \text{ g mol}^{-1}}$.

Tip. It is not necessary to know T and P as long as they were the same in the two experiments. A smaller mass of the unknown gas than of oxygen effuses in the given

---

[24]See text Appendix C, text page A-26.
[25]Text page 120.

time. Despite this, the molar mass of the unknown is less than the molar mass of oxygen. The result is easier to accept when it is noted that 0.169 mol of the unknown effuses, but only 0.102 mol of oxygen effuses.

4-81 The mean free path is given by:

$$\lambda = \frac{1}{\sqrt{2}\pi d^2 (N/V)}$$

Substituting the ideal-gas law in the form $N/V = N_0 P / RT$[26] gives:

$$\lambda = \frac{1}{\sqrt{2}\pi d^2} \frac{(R/N_0)T}{P} = \frac{1}{\sqrt{2}\pi d^2} \frac{k_B T}{P}$$

And solving for P gives:

$$P = \frac{k_B T}{\sqrt{2}\pi d^2 \lambda}$$

All the quantities on the right side of this equation are known. Substitute them:

$$P = \frac{(1.38 \times 10^{-23} \text{ J K}^{-1})(300 \text{ K})}{\sqrt{2}\pi(3.1 \times 10^{-10} \text{ m})^2(0.1 \text{ m})} = 0.097 \text{ J m}^{-3} = 0.097 \text{ Pa} = \boxed{9.7 \times 10^{-7} \text{ atm}}$$

4-83 Molecules of UF_6 are much heavier than those of H_2, but the rms speed of UF_6 molecules is much slower than those of H_2 molecules. The pressure of a gas comes from the force exerted by its molecules hitting the walls of the container. This force depends not only on the mass of the molecules, but also on their speed, that is, upon their momenta.

4-85 Assume that the 2.00-mol sample of argon behaves ideally. This is reasonable because the pressure is low and the temperature range is well above the boiling point of argon. Equations in text Sections 4-5 and 4-6 give the dependence of the pressure P, the average energy per atom KE_{avg}, the root-mean-square speed u_{rms}, the rate of collisions per area of wall Z_{wall}, the frequency of Ar-Ar collisions Z_1, and the mean free path λ) upon T, V and n. The following table states the effects of the proposed changes as multiplying factors. Thus, the entry 1 means no change. It is assumed that T, V and n change singly.

Change	P	KE_{avg}	u_{rms}	Z_{wall}	Z_1	λ
a) T from 50 to $-50°$C	$223/323$	$223/323$	$\sqrt{223/323}$	$\sqrt{223/323}$	$\sqrt{223/323}$	1
b) V doubled	$1/2$	1	1	$1/2$	$1/2$	2
c) n_{Ar} from 2 to 3 mol	$3/2$	1	1	$3/2$	$3/2$	$2/3$

[26]See the solution to **4-53**.

4-87 Data for both plots are in the following table:

Gas	b (L mol^{-1})	$N_0\sigma^3$ (L mol^{-1})	a (J m^3 mol^{-2})	$\epsilon\sigma^3 N_0^2$ (J m^3 mol^{-2})
Ar	0.03219	0.0237	0.13628	0.0236
H$_2$	0.02661	0.0151	0.02476	0.00466
CH$_4$	0.04278	0.0336	0.22829	0.0414
N$_2$	0.03913	0.0305	0.14084	0.0241
O$_2$	0.03183	0.0273	0.13780	0.0270

a) This part is concerned with b versus $N_0\sigma^3$, which are in the second and third columns of the table. The b's are copied directly from text Table 4-3.[27] The values of $N_0\sigma^3$ are computed from the Lennard-Jones σ's in text Table 4-4.[28] For example, for argon:

$$\left(N_0\sigma^3\right)_{\text{Ar}} = (6.022 \times 10^{23} \text{ mol}^{-1})(3.40 \times 10^{-10} \text{ m})^3 \times \left(\frac{10^3 \text{ L}}{\text{m}^3}\right) = 0.0237 \text{ L mol}^{-1}$$

Having $N_0\sigma^3$ in the same units as b allows easy comparison of the two. The table shows a strong correlation; the ratio of b to $N_0\sigma^3$ lies mostly in a range centered at 1.25.

b) The units of a in text Table 4-3 are atm L^2 mol^{-2}. Convert to the SI units J m^3 mol^{-2}. For example, for argon:

$$a_{\text{Ar}} = \frac{1.345 \text{ atm L}^2}{\text{mol}^2} \times \left(\frac{101.325 \text{ J}}{\text{L atm}}\right) \times \left(\frac{10^{-3} \text{ m}^3}{\text{L}}\right) = 0.1363 \text{ J m}^3 \text{ mol}^{-2}$$

The rest of the results are in the fourth column of the table. Next, combine the Lennard-Jones constants of each gas with Avogadro's number to obtain $\epsilon\sigma^3 N_0^2$. The motivation is that the units of this particular combination are J m^3 mol^{-2}, which are the same units a has. For example, for argon:

$$(\epsilon\sigma^3 N_0^2)_{\text{Ar}} = (1.654 \times 10^{-21} \text{ J})(3.40 \times 10^{-10} \text{ m})^3 (6.022 \times 10^{23} \text{ mol}^{-1})^2$$

$$= 0.0236 \text{ J m}^3 \text{ mol}^{-2}$$

The rest of the results are in the fifth column in the table. The a's and $\epsilon\sigma^3 N_0^2$'s correlate strong. The ratio of the two stays within the range from 5.1 to 5.8 for the five gases.

[27]text page 124.
[28]Text page 127.

4-89 A sample of a gaseous hydrocarbon burns in excess oxygen to give 47.4 g of H_2O and 231.6 g of CO_2. Assume that these are the *only* compounds resulting from the combustion. Compute the chemical amount of the hydrocarbon from the *P-V-T* data:

$$n_{\text{hydrocarbon}} = \frac{PV}{RT} = \frac{(3.40 \text{ atm})(25.4 \text{ L})}{(0.08206 \text{ L atm mol}^{-1}\text{K}^{-1})(400 \text{ K})} = 2.63 \text{ mol}$$

Compute the chemical amounts of the carbon dioxide and water:

$$n_{CO_2} = 231.6 \text{ g } CO_2 \times \left(\frac{1 \text{ mol } CO_2}{44.01 \text{ g } CO_2}\right) = 5.262 \text{ mol } CO_2$$

$$n_{H_2O} = 47.4 \text{ g } H_2O \times \left(\frac{1 \text{ mol } H_2O}{18.015 \text{ g } H_2O}\right) = 2.63 \text{ mol } H_2O$$

2.63 mol of the hydrocarbon reacts to give 5.26 mol of C (tied up in the form of carbon dioxide) and 5.26 mol of H (tied up in the form of water). This can happen only if there are 2 mol of C per mole of the hydrocarbon and 2 mol of H per mole of the hydrocarbon. The molecular formula of the hydrocarbon is therefore $\boxed{C_2H_2}$.

4-91 Let x equal the mass of barium carbonate and y equal the mass of calcium carbonate in the mixture. The chemical amounts of $BaCO_3$ and $CaCO_3$ in the mixture equal:

$$n_{BaCO_3} = \frac{x \text{ g}}{181.34 \text{ g mol}^{-1}} = \frac{x}{181.34} \text{ mol} \qquad n_{CaCO_3} = \frac{y \text{ g}}{100.08 \text{ g mol}^{-1}} = \frac{y}{100.08} \text{ mol}$$

One mole of $BaCO_3$ generates one mole of CO_2 in reaction with the hydrochloric acid; one mole of $CaCO_3$ generates one mole of CO_2. Accordingly:

$$n_{CO_2} = \left(\frac{x}{181.34} + \frac{y}{100.08}\right) \text{ mol}$$

Compute n_{CO_2} by substitution of the given *V-P-T* data into the ideal-gas equation:

$$n_{CO_2} = \frac{PV}{RT} = \frac{(0.904 \text{ atm})(1.39 \text{ L})}{(0.08206 \text{ L atm mol}^{-1}\text{K}^{-1})(323.15 \text{ K})} = 0.04738 \text{ mol}$$

Combining the two preceding equations gives:

$$0.04738 = \frac{x}{181.34} + \frac{y}{100.08}$$

Also, $x + y = 5.40$. Solving the two simultaneous equations gives:

$$x = 1.47 \text{ g} \qquad \text{and} \qquad y = 3.93 \text{ g}$$

This means that the $CaCO_3$ comprises $\boxed{72.8\%}$ of the mixture and the $BaCO_3$ comprises $\boxed{27.2\%}$.

Chapter 5

Solids, Liquids, and Phase Transitions

The bulk properties of substances are observed for large assemblies of identical atoms or molecules and not for the individual particles. Collections of atoms or molecules behave as solid, liquids or gases, but single atoms or molecules do not. Similarly, the reddish-brown color of solid copper is a bulk property, and color has no meaning for single atoms.

A **phase** is a region of a sample of material having uniform bulk properties. It is separated from other regions by boundaries across which bulk properties change sharply. A solid , liquid and gas in the same sample typically establish themselves as three separate phases. Changing the temperature or pressure (or other conditions) acting on a sample of matter often causes a given phase in the sample to change into a new phase with different properties. This is a **phase transition**. Phase transitions are associated with changes in the *organization* of the particles comprising a substance; they do *not* involve changes in the nature or identity of those particles.

Bulk Properties of Gases, Liquids, and Solids

A pure substance consists of a large assembly of identical particles of fixed properties. The bulk properties of the substance depend on the average distances among the particles and on the nature of the forces acting among them. These two factors determine the physical state (solid, liquid or gaseous) of the substance at a given temperature and pressure, which is one of the most important of all bulk properties. Certain other easily observed bulk properties have widely differing values for solids, liquids, and gases. Substances can be classified as solid, liquid, or gas according to these properties. See **5-1** and **5-5**.

Classification by checking bulk properties cannot succeed if the bulk properties vary greatly from region to region within a sample, that is, when the sample is heterogeneous. Many common materials (such as smokes, pastes, and foams) are heterogeneous mixtures of solids, liquids and gases dispersed through each other. Such materials cannot be meaningfully classified as solid, liquid or gaseous.

The following bulk properties have both practical and theoretical importance. Learn to explain, on a molecular basis, why these differ from solid to liquid to gas:

• **Molar volume.** This bulk property is defined early in the text[1] It is:

$$\text{molar volume} = V_\mathrm{m} = \frac{\text{molar mass}}{\text{density}} = \frac{\mathcal{M}}{\rho}$$

Liquids and solids, the two **condensed phases** have much smaller molar volumes than gases. At ordinary temperatures:

$$V_\mathrm{m} = 10 \text{ to } 100 \text{ cm}^3 \text{ mol}^{-1} \text{ for liquids and solids}$$
$$= 24,000 \text{ cm}^3 \text{ mol}^{-1} \text{ for gases}$$

Liquids and solids are called **condensed** states of matter because they occupy so much less volume per mole.

• **Compressibility.** Compression forces the molecules of a substance into a smaller volume. The density of the substance increases, and its molar volume decreases if it is compressed. The compressibility (symbolized β) gives the *rate* at which these changes occur. It equals the fractional change in the volume of a substance observed for a given change in pressure. The fractional change in volume equals $\Delta V / V$, which is the change in volume divided by the original volume; the change in pressure equals ΔP. Therefore

$$\text{compressibility} = \beta = -\frac{1}{V}\frac{\Delta V}{\Delta P}$$

The minus sign is needed to make the compressibility a positive quantity. The units of the compressibility are atm^{-1}.

The compressibilities of gases are large because the molecules of a gas are widely separated. The compressibilities of liquids and solids are much smaller than those of gases because their molecules are in contact or nearly in contact.

• **Thermal Expansion.** The coefficient of thermal expansion (symbolized α) gives the rate at which the volume of a substance changes when its temperature is raised or lowered. It is

$$\alpha = \frac{1}{V}\frac{\Delta V}{\Delta T}$$

Both liquids and solids typically expand with increasing temperature. this means that α is typically positive. The fractional change in volume is roughly ten times greater for liquids than for solids. Thus, the coefficient of thermal expansion of copper is

---

[1]See text page 29 and page pagerefmolarV of this manual.

1.65×10^{-5} K^{-1} at room conditions, but that of water is 21×10^{-5} K^{-1}. The α's of gases at room conditions are bigger yet. They are three to ten times greater than those of liquids. Many condensed materials exhibit preferential expansion in a certain direction or directions.

• **Fluidity and Rigidity.** Substances possess an inherent resistance to change of form. This resistance, which is called the **shear viscosity** of the substance, arises from internal "friction" among the molecules that comprise the substance. It is a consequence of attractive forces operating among the molecules. Shear viscosity opposes the sliding (shearing) motion of one layer of molecules across another that must take place if a substance is to flow. The shear viscosity of solids is very great; they are rigid and flow very poorly. It is small in fluids (liquids and gases), which flow readily. High shear viscosity is associated with **hardness** (resistance to indentation) and **elasticity** (capacity to recover shape when a deforming stress is removed).

• **Diffusion.** Substances in contact tend to mix by diffusion, which is the translation of one substance through the medium of a dissimilar substance based on the random chaotic motion of the molecules of both. The **diffusion constant** of a substance is a measure of its tendency to diffuse. Diffusion constants of gases at a given temperature and pressure exceed those of liquids, which greatly exceed those of solids.

• **Surface Tension.** If two fluids (such as air and water) come in contact and do not mix, then they are separated by a boundary surface. Surface tension is a characteristic of this surface. It arises from the difference in the cohesive forces among the molecules on the one side of the surface and those on the other side. Water in contact with air at room temperature has a surface tension of 0.073 N m^{-1}. Liquid metallic mercury under the same conditions has a surface tension of 0.49 N m^{-1}. The difference reflects the fact that the intermolecular forces in metals are about an order of magnitude larger than those in water. A tension is a pull, or force, but surface tension is actually a force per unit length (as shown by its unit, the newton per meter). Surface tension causes suspended drops of a liquid to acquire a spherical shape. This shape minimizes the area of surface exposed to a surrounding fluid.

Types of Intermolecular Forces

Intermolecular forces are generally weaker than covalent chemical bonds, which are *intra*molecular forces. They are less directional in character than covalent chemical bonds and operate at longer range. The following list includes ion-ion forces (Coulomb forces) for completeness. Ion-ion attractions are responsible for ionic bonds. They are as strong or stronger than covalent chemical bonds.

• **Ion-ion** forces occur between electrically charged particles. Unlike charges attract; like charges repel. Ion-ion forces are responsible for ionic bonding. Text Figure 5-9 shows a potential energy diagram for the approach of a positive ion to a negative ion. The attraction is strong and long-range. The potential energy of two interacting ions goes as to $1/R$ where R is the distance between them. See **5-19** and **5-20**.

• **Dipole-dipole** forces occur between polar molecules. In such molecules, the center of positive charge and center of negative charge do not coincide. The molecules consequently have a positive end and a negative end. The positive end of one attracts the negative end of a second. The strength of interaction depends on orientation of the molecules, and the magnitude of the dipoles. The potential energy of a pair of dipoles depends on $1/R^3$. This means that dipole-dipole forces are inherently shorter range than ion-ion forces.

• **Ion-dipole** forces operate when an ion interacts with a polar species. See **4-55**. Ion-dipole interactions are similar to ion-ion interactions except that they are more sensitive to distance. Their potential energy goes as to $1/R^2$ rather than $1/R$.

• **Induced dipole** forces operate when an ion distorts the electron distribution in an otherwise nonpolar species, making it polar and attracting it. Induced dipole forces are weak and have shorter range than dipole-dipole forces: the potential energy of an ion/molecule pair depends on $1/R^4$.

• **London dispersion** forces occur even between nonpolar species. They arise as fluctuations in the distribution of electrons on one species induce a dipole in a second. The newly-induced neighbor dipole then induces a dipole on the first species. The strength of dispersion forces depends on the polarizability of the species, that is, on how readily electrons are influenced from their symmetrical distribution. Dispersion forces are the weakest and shortest range of all intermolecular attractions . The potential energy of two molecules attracted only by dispersion forces goes as to $1/R^6$. Dispersion forces are always attractive. They increase rapidly with the number of electrons in the molecule and the volume that they occupy and are in principle always present (see bf 5-15).

• **Repulsive** forces between species are short range and grow sharply with decreasing distance. They arise from the strong reluctance of the core electrons of neighboring atoms to share the same volume of space. The potential energy due to repulsive forces depends on $1/R^n$ where the exponent n is as large as 12. These forces are exceedingly short-range. They cause atoms to behave as if they were hard, incompressible spheres. The characteristic size of such a sphere is its **van der Waals radius**. Non-bonded attractions cannot bring two atoms closer together than the sum of their van der Waals radii.

Intermolecular Forces in Liquids

Intermolecular forces cause gases to behave nonideally and, when strong enough, cause the condensation of gases to liquids and solids. Attractive forces vary greatly in strength, and liquids with stronger intermolecular forces have higher boiling points

(**5-23**). Liquids can be categorized as to the nature of the forces that hold their particles together:

• **Ionic Liquids.** The attractions arise from plus-to-minus (Coulomb) interactions between ions. The forces are quite strong (they are the same strength as ionic bonds).

• **Polar Liquids.** Polar molecules have electron distributions such that one end of the molecule attains a small positive charge and the other end a small negative charge. Polar molecules have dipole moments. They align themselves with an outside electric field. One polar molecule attracts another thanks to plus-to-minus electrical interactions between the positive end of the first molecule and the negative end of the second. These are dipole-dipole forces.

• **Non-Polar Liquids.** The molecules in these liquids are held close together by dispersion forces only. Within a series of non-polar liquids of similar structure, boiling point increases with increasing molar mass because more massive molecules have more electrons and generally larger volumes.

Hydrogen Bonds

Hydrogen bonds are a special case of the dipole-dipole force; they are stronger than other dipole-dipole attractions. They are found only in compounds containing hydrogen chemically bonded to nitrogen, oxygen or fluorine. See **5-27.** The hydrogen bond links the hydrogen atom of one molecule to an N, O, or F atom of a neighboring molecule. The strength of hydrogen bonds depends on the identity of the H-donor, the type of atom chemically bonded to the hydrogen, and the H-acceptor, the atom with lone pairs of electrons which interact with the hydrogen. Hydrogen bonds vary, but have roughly 10 percent the strength of regular chemical bonds.

Special Properties of Water

Hydrogen bonds cause peculiarities in the properties of substances in which they are prevalent, such as HF, NH_3 and, most importantly, H_2O. Hydrogen bonds are responsible for the open, low-density structure of ice, for the remarkably high melting point of ice and boiling point of water, and for the very high specific heat capacity[2] of water. These properties are profoundly influential in the biosphere. In addition, water is an excellent solvent for substances having polar molecules, particularly substances that can hydrogen-bond to its molecules.

Phase Equilibrium and Phase Transitions

A **phase** is a sample of matter that is uniform throughout in both chemical composition and physical state. **Phase transitions** include the six possible changes linking

---

[2]See Section 7-3, text page 213.

the three states of matter:

$$\text{solid} \rightleftharpoons \text{liquid} \qquad \text{solid} \rightleftharpoons \text{gas} \qquad \text{liquid} \rightleftharpoons \text{gas}$$

Some of the names for these six types of phase transition are familiar, but others are not. Melting and freezing refer to the two directions of the solid-liquid transition; **sublimation** and condensation refer to solid-gas transitions; vaporization and condensation refer to gas-liquid transitions. Many substances have more than one solid phase, but none has more than one liquid or gaseous phase.[3] Depending on conditions two or even three phases of a pure substance may coexist indefinitely.

Consider a quantity of a volatile solid or liquid (such as water) confined in a vessel but not completely filling the available volume. The solid or liquid evaporates until the pressure of the vapors in the space above the liquid reaches a characteristic value, the equilibrium **vapor pressure** of the substance. At this point the molecules of the vapor return to the condensed phase just as fast as molecules in the condensed phase escape into the vapor, and no further chance is perceptible. The vapor pressure of the solid or liquid depends only on the temperature, and not on the size of the container See text Figure 5-15.[4] Problem **5-59** uses the fact that at 100% relative humidity air is saturated with water vapor. At saturation the partial pressure of the water vapor is close to the vapor pressure of water. A **phase equilibrium** exists when the system persists indefinitely with no further net transport of matter from one phase to another.

Phase equilibria have these characteristics:

- They are dynamic on the molecular level; individual molecules continue to move from one phase to the other despite the absence of any visible sign of change.

- Their characteristics do not depend on their history (the way in which they were attained).

Phase Diagrams

A **phase diagram** of a pure substance is a plot with T on the horizontal axis and P on the vertical axis. Lines (generally curved) on this plot define combinations of temperature and pressure at which two phases coexist at equilibrium. If two phases are at equilibrium, then no further macroscopic changes are detectable, although the exchange of molecules between the phases continues. A generic phase diagram:

---

[3]See however the discussion of liquid crystals in text Section 19-6, text page 719.
[4]Text page 151.

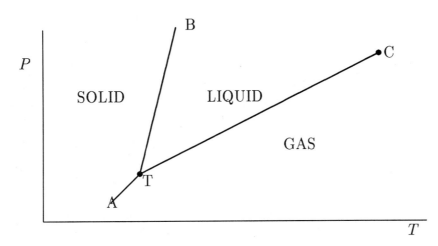

This diagram displays the combinations of P and T that favor the solid, liquid or gaseous phase of a pure substance at equilibrium. Phase diagrams have these features:

- Three different **coexistence lines** radiate from a **triple point** (labeled T). At a triple point, three phases coexist in equilibrium.

- All along the coexistence lines (TA, TB, and TC), *two* phases coexist at equilibrium.

- At combinations of P and T not on a line in the diagram, only **one** phase can exist at equilibrium. Labels identify the favored phase.

- The liquid-gas coexistence line terminates at the **critical point** (labeled C).

- The solid-liquid and solid-gas coexistence lines have no distinct terminus.

- The slope of a coexistence line tells the relative densities of the two phases it separates. Line TB above has a positive slope, which means the solid is more dense than the liquid.[5]

Use this diagram as a model to solve **5-45** and **5-66**. Phase diagrams using variables other than P and T can also be plotted (see **5-69**).

Freezing, Boiling and Sublimation Points

The **normal freezing point** of a pure substance is the temperature at which solid and liquid are in equilibrium when the pressure equals 1 atm. Similarly, the **normal boiling point** is the temperature at which liquid and gas are in equilibrium when

---

[5]Also, the solid is more dense than the gas and the liquid is more dense than the gas because lines AT and TC have positive slopes.

the pressure is 1 atm. Finally, **the normal sublimation point** of a pure substance is the temperature at which solid and gas are in equilibrium at 1 atm. All three, in diagrammatic terms, are intersections of the $P = 1$ atm line and the appropriate coexistence line on a phase diagram (see **5-45**). No pure substance can have all three of these points. After all, a solid heated slowly at a constant pressure of 1 atm either melts or sublimes, but not both (see **5-49**). Some substances have *none* of these points. They simply decompose chemically before the temperature is high enough for them to melt, boil, or sublime.

Phase Equilibria and Triple Points

Consider just the three phases: solid, liquid, and gas. There are exactly three possible equilibrium lines (solid-liquid, solid-gas and liquid-gas) along which two of these three phases coexist. Regions between these lines are sets of P-T values at which *one* phase exists at equilibrium. If two phase coexistence lines intersect, then, at the point of intersection, *three* phases coexist. This means that the third coexistence line must pass through the intersection of the first two. In general the coexistence lines on the P versus T diagram do intersect. The resulting specific combination of pressure and temperature at which three phases coexist is a **triple point.**

The Critical Point

The essence of a phase transition is an abrupt, discontinuous change in physical properties (*e.g.* density or viscosity) as the transition occurs. The distinction between liquid and gas *no longer exists* at temperatures and pressures that exceed the T and P of the **critical point,** a certain P-T combination on the phase diagram. In experimental terms, the **meniscus**, which is the boundary separating the more-dense liquid from the less-dense gas, disappears. A substance held at conditions beyond its critical point is a **supercritical fluid.** Distinguishable, abrupt phase transitions do not take place in the supercritical region. The critical point is defined on a graph of P versus T. The temperature at the critical point is the **critical temperature**; the pressure at the critical point is the **critical pressure**:

At temperatures above its critical temperature, a gas cannot be liquefied no matter how great the pressure.

The critical pressure is the pressure just necessary to liquefy a gas at its critical temperature.

The critical pressure exceeds normal atmospheric pressure for all known substances. The observed behavior of substances at and beyond their critical points seems paradoxical.

Many thought-provoking problems are constructed around the facts presented by a phase diagram.

- Description of the state of a substance at specified conditions or of the events within a container of a pure substance when the temperature is changed at constant pressure or when the pressure is changed at constant temperature. See **5-48**, **5-49b**, **5-61**, and **5-65**. A pressure change at constant temperature corresponds to motion along a vertical line on a phase diagram. A temperature change at constant pressure corresponds to a motion along a horizontal line. When such lines cut through phase coexistence lines, then phase transitions occur.

- Construction of a phase diagram given the temperature and pressure at the triple point and critical point and perhaps the densities of the solid and liquid phases. See **5-45** and **5-66**.

- Determination of the equation defining one or more of the phase coexistence lines. Although these lines are in general *not* straight lines, segments of them (over limited T and P ranges) are often approximated as straight. The determination of the slopes of the coexistence lines on phase diagrams requires use of the Clapeyron equation[6].

The solid-liquid coexistence line on a *P-T* phase diagram usually goes almost straight up. This means that freezing points are seldom strongly dependent on pressure and that the densities of liquid and solid are nearly equal. The slight slope that solid-liquid lines have is generally positive because, for most substances, the density of the solid phase exceeds the density of the liquid phase. Water is an exception. Ice floats in liquid water because it is less dense than the liquid. The solid-liquid coexistence line for H_2O has a slight negative slope. For another exception see **5-43**.

The scales on the T and P axes in phase diagrams are often deliberately distorted.[7] Diagrams constructed using regularly spaced increments of T and P on the axes usually omit some area of interest (triple point, critical point) or lack essential detail.

Detailed Solutions to Odd-Numbered Problems

5-1 The substance is most likely to be a ⌐gas⌐. Its large compressibility and large coefficient of thermal expansion are typical of gaseous materials.

---

[6]This equation is discussed on text page 819 in text Section 23-2.

[7]An example is text Figure 5-19, text page 154.

5-3 a) Compute the density of the material

$$d = \left(\frac{2.71 \text{ kg}}{258 \text{ cm}^3}\right) \times \left(\frac{1000 \text{ g}}{1 \text{ kg}}\right) = 10.5 \text{ g cm}^{-3}$$

The material is $\boxed{\text{condensed}}$, and not a gas because of its high density, which exceeds the density of water considerably.

b)

$$\text{molar volume} = V_\text{m} = \left(\frac{1 \text{ cm}^3}{10.5 \text{ g}}\right) \times \left(\frac{108 \text{ g}}{1 \text{ mol}}\right) = \boxed{10.3 \text{ cm}^3 \text{ mol}^{-1}}$$

5-5 The volume increase amounts to only 0.3%. Warming a gas from 293 K to 313 K would, according to Charles's law, cause an expansion of:

$$\frac{(313 - 293) \text{ K}}{293 \text{ K}} \times 100\% = 6.8\%$$

A non-ideal gas would also expand by several percent during this change in temperature. The substance is accordingly $\boxed{\text{condensed}}$.

5-7 The huge increase in volume is due to the transition of water from a condensed phase (liquid water) to a gaseous phase (steam) at 100°C.

5-9 Solid sodium chloride has an extended structure maintained by ion-ion interactions that are non-directional. Solid carbon tetrachloride, by contrast, is maintained as a solid by weaker dispersion forces. Indentation requires breaking bonds in the solid. Solid NaCl is $\boxed{\text{harder}}$ than solid CCl_4 because it is harder to overcome the ion-ion interactions than the dispersion interactions.

5-11 In all three phases, the diffusion constant should $\boxed{\text{decrease}}$ as the density of the phase is increased. At higher densities molecules are closer to each other. In gases they will collide more often and travel shorter distances between collisions. In liquids and solids, there will be less space for molecules to move around each other.

5-13 Ion-dipole forces arise from molecular or ionic properties that are always present, as in the interaction of the permanent charge of the potassium ion and the permanent dipole of a water molecule. Ion-induced dipole forces arise when the charge on an ion induces a temporary change in another species. An induced dipole does not exist in the absence of the ion that induces it. An example is the attraction between an (Fe^{3+}) ion and the temporary dipole that it induces in a molecule of oxygen.

5-15 a) Potassium and fluorine differ considerably in electronegativity. As explained in text Section 3-1, the bonding in the compound potassium fluoride is therefore expected to be ionic. These ⸢ion-ion⸣ attractions predominate in this compound; dispersion forces are also present.

b) ⸢Dipole-dipole⸣ attractions predominate in the interactions between molecules in hydrogen iodide. The positive (H) end of one molecule is attracted by the negative (I) end of another, but repelled by its positive (H) end. Dispersion forces are also present.

c) ⸢Dispersion forces⸣ are the only forces operating among the atoms in a sample of radon. Single Rn atoms have completely symmetrical (spherically symmetrical) distributions of charge. Two neighboring Rn atoms induce temporary dipoles in each other that cause them to attract each other.

d) ⸢Dispersion forces⸣ are the only intermolecular forces possible between molecules of N_2.

5-17 A sodium ion should be most strongly attracted to a ⸢bromide ion⸣. The attraction between ions of unlike charge such as these is stronger than the ion-dipole attraction between Na^+ and HBr and the ion-induced dipole attraction between Na^+ and Kr.

5-19 a) Read along the horizontal axis in Figure 5-9[8] to find the locations of the minima in the potential energy curves. These locations give the bond distances. in Cl_2 the bond is about ⸢2.0×10^{-10} m⸣ long; in molecular KCl, the bond is about ⸢2.5×10^{-10} m⸣ long.

b) The bond in Cl_2 is shorter than the bond in KCl, but is distinctly weaker, as shown by the depth of the minimum in the potential energy curve for Cl—Cl (about -225 kJ mol^{-1}) compared to the depth for K—Cl (about -490 kJ mol^{-1}).

Tip. Potential energy curves such as the ones in Figure 5-9 are much more informative than the type of over-general statement quoted in the problem.

5-21 The boiling point and melting point of a substance depend on the strength of the attractive forces operating among the particles that comprise the liquid or solid substance. These forces tend to increase with increasing molar mass in a group of related substances. The halogens are certainly closely related chemically. Therefore their boiling and melting points tend to rise with increasing molar mass.

Tip. The observed melting points and boiling points of the halogens are:

---

[8]Text page 146.

Substance	$\mathcal{M}$ (g mol^{-1})	m.p. (°C)	b.p. (°C)
F$_2$	38	−219.6	−187.9
Cl$_2$	71	−101	−34.05
Br$_2$	160	−7.2	58.2
I$_2$	254	113.6	184.5

5-23 Substances with the strongest intermolecular forces require the highest temperature to make them boil. Liquid RbCl has strong Coulomb forces holding its ions together. It has the highest boiling point. Liquid NH$_3$ has dipole-dipole attractions, as does liquid NO. In liquid NH$_3$, these are particularly strong. They are hydrogen bonds. In liquid NO, the dipole-dipole attractions are weaker. Liquid NH$_3$ boils at a higher temperature than liquid NO. Induced dipole-induced dipole forces are the only intermolecular attractions in liquid neon. Consequently, it has the lowest boiling point of all: $\boxed{\text{Ne} < \text{NO} < \text{NH}_3 < \text{RbCl}}$.

5-25 The facts suggest the (CH$_3$OH)$_4$ molecule has a $\boxed{\text{cyclic}}$ structure. Why would a straight chain of molecules stop at exactly four links? The ring is probably maintained by hydrogen bonding between the —OH hydrogen atom and the O of a neighboring molecule. It would consist of a total of eight atoms—four H's and 4 O's in alternation:

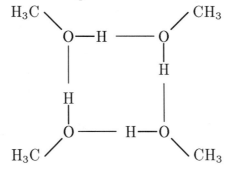

5-27 Although the two substances have comparable molar masses, the boiling point of hydrazine N$_2$H$_4$ should $\boxed{\text{exceed}}$ the boiling point of ethylene C$_2$H$_4$ because N—H$\cdots$N hydrogen bonds are present in hydrazine but no hydrogen bonds occur in ethylene.

5-29 Compute the number of molecules of water present in the sample:

$$N_{\text{H}_2\text{O}} = 1.0 \text{ kg H}_2\text{O} \times \left(\frac{1 \text{ mol H}_2\text{O}}{0.018 \text{ kg H}_2\text{O}} \right) \times \left(\frac{6.022 \times 10^{23} \text{ molecules}}{1 \text{ mol H}_2\text{O}} \right)$$

$$= 3.35 \times 10^{25} \text{ molecules}$$

Each water molecule is involved in a maximum of 4 hydrogen bonds. Also, each bond has 1 water at each end. Therefore the maximum number of H-bonds in the sample is twice the number of molecules: $\boxed{6.7 \times 10^{25}}$.

5-31 Rearrange the ideal-gas equation and use it as follows

$$V_{H_2} = \frac{n_{H_2}RT}{P} = \frac{(1.00 \text{ mol})(0.08206 \text{ L atm mol}^{-1}\text{K}^{-1})(16.0 \text{ K})}{0.213 \text{ atm}} = \boxed{6.16 \text{ L}}$$

This volume is 27% of the volume of one mole of gaseous hydrogen at STP.

5-33 First compute the number of moles of Hg per unit volume in the space above the surface of the mercury.

$$\frac{n}{V} = \frac{P}{RT} = \frac{2.87 \times 10^{-6} \text{ atm}}{(0.08206 \text{ L atm mol}^{-1}\text{K}^{-1})(300.15 \text{ K})} = 1.165 \times 10^{-7} \text{ mol L}^{-1}$$

Multiply by Avogadro's number to find the number of Hg atoms per unit volume:

$$N_{Hg} = 1.165 \times 10^{-7} \text{ mol L}^{-1} \times \left(\frac{6.022 \times 10^{23} \text{ molecules}}{1 \text{ mol}}\right) \times \left(\frac{1 \text{ L}}{1000 \text{ cm}^3}\right)$$

$$= \boxed{7.02 \times 10^{13} \text{ atom cm}^{-3}}$$

5-35 The pressure on the interior walls of the vessel containing the collected acetylene comes from collisions by molecules of H_2O molecules as well as C_2H_2. Therefore, subtract the partial pressure of the water vapor from the total pressure inside the container. This gives the pressure that the acetylene would exert if it were present by itself. The partial pressure of water vapor depends on solely on the temperature. The required value is given, so:

$$P_{C_2H_2} = P_{\text{total}} - P_{\text{water}} = 0.9950 - 0.0728 = 0.9222 \text{ atm}$$

Use the ideal gas law to compute the chemical amount of acetylene per unit volume of collected gases that exerts a pressure of 0.9222 atm at a T of 40°C (313.15 K):

$$\frac{n_{C_2H_2}}{V} = \frac{P_{C_2H_2}}{RT} = \frac{0.9222 \text{ atm}}{(0.08206 \text{ L atm mol}^{-1}\text{K}^{-1})(313.15 \text{ K}} = 0.03589 \text{ mol L}^{-1}$$

The molar mass of acetylene equals 26.038 g mol^{-1}. Multiplying the chemical amount by this molar mass gives $\boxed{0.9345 \text{ g L}^{-1}}$ as the mass of acetylene present per unit volume of gas.

5-37 Determine the partial pressure of the CO_2, and then use the ideal gas law to obtain the chemical amount of CO_2. From the balanced equation, the number of moles of $CaCO_3$ that reacts equals the number of moles of CO_2 that forms. Convert this amount of $CaCO_3$ to a mass.

$$P_{CO_2} = P_{total} - P_{water}$$
$$= 0.9963 - 0.0231 = 0.9732 \text{ atm}$$
$$n_{CO_2} = \frac{P_{CO_2} V_{gases}}{RT} = \frac{0.9732 \text{ atm}(0.722 \text{ L})}{(0.08206 \text{ L atm mol}^{-1}\text{K}^{-1})(293.15 \text{ K})} = 0.0292 \text{ mol } CO_2$$

$$0.0292 \text{ mol } CO_2 \times \left(\frac{1 \text{ mol } CaCO_3}{1 \text{ mol } CO_2}\right) \times \left(\frac{100.09 \text{ g } CaCO_3}{1 \text{ mol } CaCO_3}\right) = \boxed{2.92 \text{ g } CaCO_3}$$

5-39 Reading from text Figure 5-15,[9] the vapor pressure of water at 90°C is approximately $\boxed{0.70 \text{ atm}}$. Because the whole of the earth's atmosphere exerts a pressure of 1 atm, this means that 70% of the earth's is above the level of the explorer's camp; $\boxed{30\%}$ of the atmosphere is below the level of the camp.

5-41 The trends in boiling and melting points indicate that the interatomic attractions are much stronger in iridium than in sodium. The surface tension of molten $\boxed{\text{iridium}}$ should therefore be higher than that of molten sodium.

5-43 Compression favors the denser phase, but the denser phase (the liquid) is already present. The sample of Pu $\boxed{\text{stays liquid}}$.

5-45

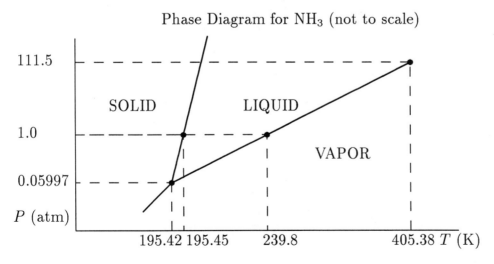

Phase Diagram for NH_3 (not to scale)

---

[9]Text page 151.

5-47 Check whether the specified point is within the boundaries of the solid, liquid, or gas region on the phase diagram of argon that appears in text Figure 5-21.[10]

a) liquid   **b)** gas   **c)** solid   **d)** gas.

5-49 a) The temperature of the triple point of acetylene $\boxed{\text{lies above}}$ $-84.0°C$. For a pure substance, T's at which liquid and gas are in equilibrium equal or exceed T's at which liquid and solid are in equilibrium.

b) Note that 0.8 atm is *less* than 760 torr, which is the vapor pressure of solid acetylene at $-84°C$. If solid acetylene is heated at $P = 0.8$ atm, it therefore passes directly into the vaporous (gaseous) state without ever existing as a liquid. It $\boxed{\text{sublimes}}$ at some temperature below $-84.0°C$.

5-51 The nitrogen confined in the glass tube becomes supercritical as it is heated past the critical temperature. This is assured because the size of the container and the amount of nitrogen that it contains assure that the critical density of nitrogen is exceeded. Originally, a meniscus separates the liquid from gaseous nitrogen within the tube. The $\boxed{\text{meniscus disappears}}$ at 126.19 K as the two phases merge into a single fluid phase of uniform density; the distinction between liquid and gas ceases to exist.

5-53 Candle wax at room temperature is a $\boxed{\text{solid}}$, although a low-melting one. Natural rubber, a polymer of an organic compound called isoprene (C_8H_8), is a $\boxed{\text{solid}}$ at room conditions. Both are incompressible compared to a gas. Neither flows readily, compared to ordinary liquids and gases.

5-55 The HCl molecule is polar. When liquid hydrogen chloride dissolves an ionic compound, its molecules tend to line up with their positive ends (the H ends) closest to negatively charged ions and with their negative ends (the Cl atoms) closest to positively charged ions.

$$\overset{\delta-}{\text{Cl}}\!-\!\overset{\delta+}{\text{H}} \searrow \quad \nearrow \overset{\delta+}{\text{H}}\!-\!\overset{\delta-}{\text{Cl}} \qquad \overset{\delta+}{\text{H}}\!-\!\overset{\delta-}{\text{Cl}} \searrow \quad \nearrow \overset{\delta-}{\text{Cl}}\!-\!\overset{\delta+}{\text{H}}$$

$$\overset{\delta-}{\text{Cl}}\!-\!\overset{\delta+}{\text{H}} \nearrow \quad \searrow \overset{\delta+}{\text{H}}\!-\!\overset{\delta-}{\text{Cl}} \qquad \overset{\delta+}{\text{H}}\!-\!\overset{\delta-}{\text{Cl}} \nearrow + \searrow \overset{\delta-}{\text{Cl}}\!-\!\overset{\delta+}{\text{H}}$$

Tip. The diagram shows four H—Cl molecules surrounding each ion, but the actual number is often more than four, depending on the size and charge of the ion.

5-57 Water at $T = 10$ K ($-263°C$) and $P = 1$ atm exists as a solid (ice). As the temperature of this very cold ice is raised, the average internal kinetic energy, which is the energy associated with the movement of its molecules, rises in (approximately)

---

[10]Text page 155.

direct proportion. Its average internal potential energy, which is the energy that its molecules have by virtue of their relative positions with respect to the attractions that link them, increases slightly (as the ice expands against the constant outside pressure of 1 atm).

At 273.15 K (and $P = 1$ atm), the ice starts to melt During this phase change the temperature stays constant and so does the average internal of the molecules despite the continuing addition of energy continues in the form of heat. The energy goes to (partially) overcome attractions among the molecules. It increases the average internal potential energy. Once the sample is entirely liquefied, more added energy goes mostly to raising the internal kinetic energy. The internal potential energy increases only slightly as the liquid is heated. At 373.15 K the liquid water starts to boil. During boiling, the average internal kinetic energy of of the molecules stays constant. The average internal potential energy of the molecules increases considerably as the molecules, which attract each other, are repositioned from being fairly close neighbors to locations remote from each other.

5-59 "Saturation" means that the vapor pressure of the water in the room has reached its maximum—the humidity is 100% and drops of water are condensing on the walls. The partial pressure of water vapor in the room is very close to the vapor pressure of pure water at the given temperature.[11] This vapor pressure equals 0.03126 atm. The rest of the pressure in the room comes from oxygen, nitrogen, and the other components of the air. Compute the volume of the room in liters:

$$V = 110 \text{ m}^3 \times \left(\frac{10^3 \text{ L}}{\text{m}^3}\right) = 1.10 \times 10^3 \text{ L}$$

Combine this with the known partial pressure of water to obtain the chemical amount and the mass of the water vapor in the air in the room:

$$n_{H_2O} = \frac{PV}{RT} = \frac{0.03126 \text{ atm}(1.10 \times 10^3 \text{ L})}{(0.08206 \text{ L atm mol}^{-1}\text{K}^{-1})(298.15 \text{ K})} = 140.5 \text{ mol}$$

$$m_{H_2O} = 140.5 \text{ mol H}_2\text{O} \times \left(\frac{18.02 \text{ g H}_2\text{O}}{1 \text{ mol H}_2\text{O}}\right) = \boxed{2530 \text{ g H}_2\text{O}}$$

This is roughly 2.5 L of liquid water, or almost 3 quarts.

Tip. Converting R to m^3 atm mol^{-1} K^{-1} (to fit the units of volume as given) leads to the same answer. See **4-53**. A standard reference[12] gives the measured mass of water in one cubic meter of saturated air at 1 atm and 298.15 K as 23.14 g m^{-3}. This corresponds to 2545 g of water in 110 m^3 of air, which is quite close to the above answer.

---

[11]The difference derives from the interaction of the water with the air.
[12]Kaye and Laby, *Tables of Physical and Chemical Constants*, 16th edition, page 38.

5-61 The only substance within the sealed can is water. If liquid water and water vapor coexist at 60°C, the pressure of the water vapor[13] must equal $\boxed{0.20 \text{ atm}}$.

5-63 Why don't spacecraft just boil away in the vacuum of space? The term "boiling" implies an active or even violent event. Spacecraft do tend to lose individual atoms from their hulls into the vacuum of space. This process is however so very slow that no change is apparent.

5-65 When chunks of solid CO_2 are added to room-temperature ethanol in an open beaker, portions of the solid sublime off as gaseous CO_2. Much bubbling and roiling accompany the escape of this gas. The process will in theory chill the ethanol to $\boxed{-78.5°C}$, the sublimation temperature of the solid CO_2, but no lower. Since ethanol at $P = 1$ atm requires a temperature below $-114.5°C$ to freeze, it stays liquid in this experiment. Once the ethanol is good and cold, small amounts of further sublimation take place to counteract heat flowing in from the surroundings; the mixture will always fizz a little.

5-67 Substitution of the appropriate "VDW" (van der Waals) a's and b's in the three equations quoted in the problem gives the results that appear in columns four through six of the following table. Because the different VDW constants are in atm L^2 mol^{-2} and L mol^{-1} respectively, a value of 0.08257 L atm $mol^{-1}K^{-1}$ should be used for R. The answer columns include the observed values (in parentheses) for comparison.

Gas	a (atm L^2mol^{-2})	b (L mol^{-1})	T_c (K)	P_c (atm)	$(V/n)_c$ (L mol^{-1})
O_2	1.360	0.03183	154.3(154.6)	49.72 (49.8)	0.09549(0.0734)
CO_2	3.592	0.04267	303.9(304.2)	73.07 (72.9)	0.1280 (0.0940)
H_2O	5.464	0.03049	647.1(647.1)	217.7(217.6)	0.09147(0.0567)

Tip. The observed values T_c and P_c for the gases are taken from experimental reports and are more precise than can be obtained by reading the scale in text Figure 5-21. Molar volumes cannot be read from P-T phase diagrams at all.

---

[13]See text Figure 5-15, text page 151.

5-69 a)

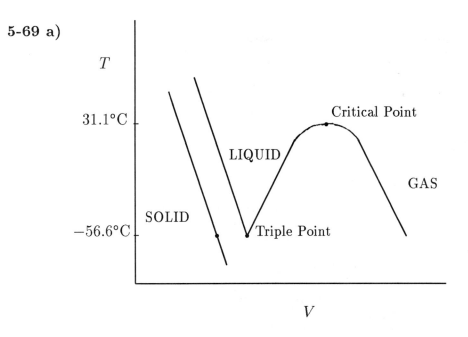

b) Such a diagram cannot be drawn because two different phases (liquid and solid) can have the same temperature and molar volume.

5-71 The problem gives the boiling points of fluorides of elements in the second row of the periodic table. The high boiling points of LiF and BeF_2 result from the strong ion-ion attractions in the liquids. The large decrease in boiling point going from BeF_2 to BF_3 suggests a changeover from ion-ion forces to much weaker dipole-dipole or induced-dipole forces. The continued decrease in boiling point of the chlorides across the rest of the second period corresponds to decreasing ionic character in the bonds and a parallel decrease in the strength of the intermolecular attractions.

5-73 The bonds in SbF_5 have less ionic character than in SbF_3. Ionic character tends to decrease with increasing oxidation number of the metal. The bonds in AsF_5 should have less ionic character than those in SbF_5 because Sb lies below As in the periodic table. The bond in F_2 has the least ionic character of all, since it is a covalent bond. Therefore the trend in ionic character is:

$$\text{least} \quad F_2 < AsF_5 < SbF_5 < SbF_3 \quad \text{most}$$

If the boiling points increase with increasing ionic character, then:

$$\boxed{F_2 < AsF_5 < SbF_5 < SbF_3}$$

Tip. The experimental boiling points of the first three substances equal: -188.14, -53, and $149.5°C$. SbF_3 sublimes at a temperature exceeding $319°C$.

Chapter 6

Solutions

A **solution** is a homogeneous mixture of two or more substances. When substances dissolve, they disperse (spread out) into each other at the level of individual atoms, molecules, or ions. We see formerly distinct phases merge into a new single phase. The major component in a solution is called the **solvent;** the minor component or components are the **solutes.** Solutions vary in composition. For each solute added to solvent, another variable, a composition variable, is needed fully to describe the system.

The Composition of Solutions

The following measures are used to describe the composition of solutions:

- **Mass Fraction/Mass Percentage.** The mass fraction (also called weight fraction) of a component equals the mass of the component divided by the total mass of the solution. The mass percentage equals the mass fraction multiplied by 100%. See **6-3** and **6-59**. The sum of the mass fractions of all the components in a solution equals 1; the sum of the mass percentages equals 100%.

- **Mole fraction.** The mole fraction X_A of component A in a solution equals the number of moles of A divided by the total number of moles of all the different components: $X_A = n_A/n_{\text{tot}}$. See **6-7**. The only facts needed to convert freely between mass percentages and mole fractions are the molar masses of the components.

  The sum of the mole fractions of all of the components of a solution must equal 1. Two-component systems are common in problems, and the relationship $X_A + X_B = 1$ is often crucial in solving these problems. See **6-3** and **6-7**.

- **Molality.** The molality m_A of solute A in a solution equals the number of moles of A divided by the number of kilograms of solvent. When more than

124

one solute is present, then the solution has m_A, its molality in solute A, m_B, its molality in solute B, and so forth. The solvent is the component that is present in the greatest chemical amount. The units of molality are mol kg^{-1}.

The only fact needed to convert from mole fractions to molalities and vice versa is the molar mass of the solvent. See **6-3**.

Knowledge of both the molality of a solute and the composition by mass of the solution allows computation of the molar mass of the solute. See **6-39**. This situation is common in practical problems.

- **Molarity.** The molarity of solute A in a solution equals the number of moles of A per liter of the solution. Measures of composition defined per unit volume of solution are **concentrations**. Molarity is thus a unit of concentration. The symbol for molarity is M, which stands for mol L^{-1}. See **6-3** and **6-11**. The molarity differs from the preceding units by having a volume in its denominator. Conversions from molarity to molality or mole fraction (and the reverse) require the density of the *solution* (not of the pure solvent or solutes). See **6-5**. It follows that if the molarity and molality of a two-component solution are both known, then the density of the solution can be computed. Also, if the mass percentage and molarity are known, the density can be calculated. See **6-9**.

  Densities vary with temperature and consequently so do molarities. See **6-71**. In *dilute* aqueous solutions the molarity is approximately the same as the molality because the density of the solution is approximately 1 g cm^{-3}. See **6-1b**.

Many problems require conversions among the different units of composition. It is not wise to memorize formulas. Instead learn the definitions of the different units of composition and apply them (see **6-3**).

Units of composition fall into two categories, those having a *mass* in the denominator (mole fraction, molality, mass percent) and those having a *volume* in the denominator (molarity). Avoid confusion between mola*li*ty (with *mass* of sol*vent* in the denominator) and mola*ri*ty (with *volume* of sol*ution* in the denominator). The similarity in the names is unfortunate.

The Preparation of Solutions

The preparation of solutions of known composition is an important laboratory skill. Suppose that 1 liter of 1.20 M aqueous NaCl solution is needed. Weigh out 1.20 mol of NaCl (this equals 70.13 g). Add enough water to make the volume of the mixture equal 1.00 L. This is *not* the same as adding 1.00 L of water: the solute occupies volume too. If only 456 mL of the same solution is needed, then weigh out (0.456 ×

1.20) mol of NaCl, and add enough water to bring the total volume exactly to 456 mL. See **6-3**. Special **volumetric flasks** are calibrated to contain precisely known volumes at specified temperatures (**6-9**). Volumetric flasks are used to mix solutions when good accuracy is required.

Doubling the total volume of a solution by adding more solvent obviously cuts the concentrations of all solutes in half. It does not change the chemical amounts of the solutes that are present. For a dilution:

$$c_f = \frac{\text{moles solvent}}{\text{final solution volume}} = \frac{c_i V_i}{V_f} = c_i \frac{V_i}{V_f}$$

where c and V stand for concentration and volume respectively and the subscripts refer to the final and initial values.

The preparation and dilution of solutions are the subjects of many exercises. Mixing of solutions of differing concentrations of a single solute presents an important variation (**6-11**). Another variation of "dilution problems" involves *concentrating* a solution from a lower to a higher concentration. The above formula applies except that V_f is smaller than V_i instead of larger.

The Nature of Dissolved Species

In dissolution, new chemical species form at the expense of old ones. In this respect, dissolution is a chemical change. However, the components of a solution are present in indefinite (variable) proportions. In this respect, dissolution resembles the mixing of (for example) flour and sugar, which can be combined in any proportion. Dissolution is thus intermediate between a physical and chemical process.

Chemical equations to represent dissolution do not show the solvent as a reactant. Instead, abbreviations are added in parentheses to the right of the chemical formulas of the solute and its products. The following table shows the pattern of these abbreviations:

State	Abbreviation
Solid	(s)
Liquid	(l)
Gaseous	(g)
Dissolved in water	(aq)
Dissolved in solvent X	(X)

Thus, the dissolution of the liquid compound ethanol in water is represented

$$C_2H_5OH(l) \rightarrow C_2H_5OH(aq)$$

A dissolved species is surrounded by molecules of its solvent, or **solvated** by the solvent. Solvation by water is called **aquation.**

Mechanisms of dissolution vary depending on the nature of the solute and the solvent. Water dissolves polar molecular species (such as ethanol) by offering strong dipole-dipole interactions between H_2O molecules and the solute molecules to replace the solute-solute attractions (and some of the H_2O-H_2O attractions) previously existing.

Dissolution of most ionic compounds by water occurs with **dissociation** of the solute into two or more ions. For example:

$$NaCl(s) \rightarrow Na^+(aq) + Cl^-(aq)$$

These ions interact with H_2O molecules by means of ion-dipole attractions. Shells of solvating (aquating) water molecules surround the ions, which are free to move independently in solution.

Solutions containing large concentrations of mobile ions (arising from complete or near-complete dissociation) conduct electricity well and are called **strong electrolytes.** Some substances dissociate partially when dissolved in water. Their solutions are poor conductors of electricity and are called **weak electrolytes.** Aqueous solutions of solutes that do not dissociate do not conduct electricity and are called **non-electrolytes.**

Solubility and Precipitation

The **solubility** of a solute in a solvent is the maximum amount of the solute that a fixed quantity (such as 100 g) of the solvent can dissolve. When the maximum is reached, the solution is **saturated.** No further dissolution occurs. If the solubility of a solute is exceeded, then the solution is supersaturated. The solute then tends to separate from solution until the point of saturation is again reached. When solids separate from a liquid they generally settle or precipitate to the bottom of the container. For this reason, the reverse of dissolution is known as **precipitation.**

Many poorly soluble ionic substances can be precipitated by mixing solutions that each contain just one of their component ions. Precipitation takes place as the ions in the insoluble combination encounter each other. A **net ionic equation** shows such encounters. It omits the **spectator ions** that maintain electrical neutrality in the solutions that are mixed but play no role in the reaction. See **6-13.** Net ionic equations are used in all types of equations to focus attention on the particles that are actually reacting.

The Stoichiometry of Reactions in Solutions

The molar concentration of a solution furnishes a new type of unit factor to use in stoichiometric calculations. Concentrations relate the chemical amount of a substance with the volume of the solution. For example, if an aqueous solution of alcohol ($C_2H_5OH(aq)$) has a concentration of 0.575 mol L^{-1}, the factors:

$$\left(\frac{0.575 \text{ mol } C_2H_5OH}{1 \text{ L solution}}\right) \quad \text{and} \quad \left(\frac{1 \text{ L solution}}{0.575 \text{ mol } C_2H_5OH}\right)$$

can immediately be written. Problems **6-15**, **6-17**, and **6-31** show the technique, which should be studied as an extension of the methods of Chapter 2. Problem **6-59** is a good example of the combined use of several types of unit factors.

Titration

A **titration** is the addition of a solution of known concentration c_1 in solute 1 to a measured volume V_2 of a second solution of unknown concentration c_2 in solute 2. The solutes react as the addition proceeds. In a titration:

- The aim is to determine the unknown concentration c_2.

- The equation for the reaction taking place between the two substances must be known.

- The reaction must go to completion.

- A way must exist to determine the **equivalence point** of the titration. At the equivalence point, the last bit of solute 2 has just been reacted away by the addition of solution 1. Disappearance of the color of solute 2 (if it has a color and no interfering colors are present) indicates this point.

- The presence of an **indicator**, an additional substance that changes color at the equivalence point, is often necessary in a titration.

- If solutes 1 and 2 react in a 1-to-1 molar ratio, then at the equivalence point:

$$\boldsymbol{V_1 c_1 = n_1 = n_2 = V_2 c_2} \qquad \textbf{at equivalence}$$

This relationship applies to **6-25**.

- If solutes 1 and 2 do not react in a 1-to-1 molar ratio, then the stoichiometry of their reaction must be taken into account (**6-31**).

- The volume V_2 of the unknown solution is not always measured. The titration then gives the quantity $V_2 c_2$, which equals n_2, the chemical amount of solute 2.

Titrations are classified according to the type of reaction they employ.

Acid-base Titrations. These employ a **neutralization** reaction. The Arrhenius definition of acid and base states that acids increase the amount of $H^+(aq)$ ion in water and bases increase the amount of $OH^-(aq)$ ion in water.[1] The reaction underlying acid-base titrations in water is then the combination of the extra hydrogen ion with the extra hydroxide ion to form water ("hydrogen hydoxide"):

$$H^+(aq) + OH^-(aq) \rightarrow H_2O(l)$$

This is a net ionic equation. In an actual neutralization, negative ions are present to balance the charge of the H^+ ions in the acid, and positive ions are present to balance the charge of the OH^- ions in the base. These ions combine to form the second product of acid-base neutralization in water, a **salt.** If the salt is poorly soluble in water it precipitates. If it is soluble in water it remains in solution.

Redox Titrations. These employ a **redox** reaction. In redox, electrons are transferred. The species that experiences a loss of electrons is **oxidized**; the species that experiences a gain of electrons is **reduced.** In oxidation, the oxidation number of one or more atoms is raised. In reduction, the oxidation number of one or more atoms is lowered. Reactions can be classified as redox or non-redox by checking oxidation numbers because any change in oxidation number from one side of an equation to the other signifies redox. See **6-27.**

An atom, molecule, or ion that experiences oxidation is a **reducing agent.** Its loss of electrons enables another species to gain electrons. A species that experiences reduction is, by similar reasoning, an **oxidizing agent.** Oxidation cannot occur without reduction (and vice versa) In redox, the number of electrons lost by the reducing agent must equal the number gained by the oxidizing agent.

Colligative Properties of Solutions

Colligative properties of solutions depend on the number of solute particles but not on the identity of the solute. To understand these properties, first understand **Raoult's law** which states that the vapor pressure of a component in a solution depends on its mole fraction times the vapor pressure exerted by the pure component:

$$P_1 = X_1 P_1^\circ$$

If P_1° equals zero, then component 1 has zero vapor pressure and is **non-volatile.**

---

[1]The Arrhenius definitions of acid and base are less general that subsequent definitions, but still quite useful for aqueous titrations.

Solutions that follow Raoult's law are **ideal solutions.** Compare them with ideal gases:

In an ideal gas, forces of attraction among the molecules of the gas are negligibly weak;

In an ideal solution, intermolecular attractions are strong, but the solvent-to-solute attractions are the *same* as the solvent-solvent interactions.

Vapor Pressure Lowering. Mixtures of a volatile solvent (component 1) and a non-volatile solute (component 2) are common. In such cases, the presence of the solute reduces the total vapor pressure of the solution. The reduction is easily measured and turns out to be proportional to the mole fraction of the solute.

$$\Delta P = \text{vapor pressure change} = -P^{\circ}_{(solvent)} X_2$$

Thus, measurements of vapor pressure can give the composition of solutions **6-33**. An observed vapor-pressure lowering, taken in combination with the molar mass of the solvent and the masses of the solute and solvent, gives the molar mass of the solute **6-61**.

Boiling Point Elevation. The boiling point of a volatile solvent is raised by the presence of a non-volatile solute. The (approximate) equation for this change is:

$$\Delta T_b = K_b m$$

where K_b is a constant that depends only on the solvent, m is the molality of the solute, and ΔT is the final boiling temperature minus the original boiling temperature. The quantity ΔT_b is always positive, by the nature of the phenomenon. Tables of K_b values are available.[2] The values in such tables come from experimental determinations of boiling points of solutions of known concentration (**6-35**). The units of K_b are K kg mol^{-1} where the kg refers to the mass of the solvent, and the mol refers to the chemical amount of the solute.

The simple form of the boiling point elevation formula lends itself to many practical problems involving the determination of the molar mass of unknown non-volatile solutes (**6-37**). All kinds of solutes, not just water, have their boiling points raised by the presence of non-volatile solutes (**6-35**).

Freezing Point Depression. Freezing point depression is analogous to boiling point elevation. The change in the freezing point of a volatile solvent caused by a non-volatile solute is:

$$\Delta T = -K_f m$$

---

[2]Text Table 6-2, text page 179.

where K_f is the solvent's freezing point depression constant. The constants K_f and K_b differ from solvent to solvent. The negative sign appears because the final T is always less than the initial T in a temperature lowering and all K_f's are positive.[3] Typical applications again involve the determination of the molar masses of unknown solutes (**6-39**).

Osmotic Pressure. The colligative property that is most used in biochemical applications is the osmotic pressure. All solutions have an osmotic pressure. In the case of dilute solutions:

$$\pi = cRT$$

where π is the osmotic pressure, c is the concentration of the solution, T is its absolute temperature, and R is the gas constant. Osmotic pressures are measured using semi-permeable membranes, materials that allow passage of molecules of solvent but not of solute. As is the case with the other colligative properties, observations of osmotic pressure are useful in determinations of molar mass (**6-45** and **6-47**).

The colligative properties are closely related. This allows the computation of one colligative property based on the observation of another (**6-65** and **6-69**).

Solutes That Dissociate

A complication arises when solutes break down into smaller pieces in solution.. Most salts dissolve in water by dissociating into ions, so this complication is frequent. Dissociating solutes give rise to two or more particles in solution for every particle added. This affects the magnitude of the several colligative properties. If a colligative property is enhanced in this way, the apparent or **effective molality**, not the actual molality, is increased (**6-43**). The number of particles formed upon dissociation is often evident from the formula or name of the solute: Na^+Cl^- has two; $Ca^{2+}(Cl^-)_2$ has three; $(K^+)_2(SO_4^{2-})$ has three; and so forth. But this is not always so (see **problem 6-67**). Dissociation of the solute affects all four colligative properties (**6-69**).

Mixtures and Distillation

Solutions of two (or more) components have phase diagrams, just as do pure substances. A complete phase diagram of a two-component solution requires three axes, one for pressure, one for temperature, and one for composition. A three-component solution requires *four* axes, including two composition axes. Evidently, complete phase diagrams for many-component solutions are difficult to draw. Instead cross-sections of phase diagrams are sketched. Text Figure 6-14[4] is a constant-temperature cross-section of the phase diagram of a two-component ideal solution. Pressure and

---

[3]See text Table 6-2, text page 179.
[4]Text page 189.

composition vary on the vertical and horizontal axes respectively. Text Figure 6-15 shows composition on the horizontal axis but temperature on the vertical axis. The pressure is constant.

Phase diagrams of real solutions are complex and differ sharply from the simple predictions of Raoult's law. Nevertheless, at a sufficiently low concentration of solute (even in nonideal solutions) there is still some simplicity:

$$P_{\text{solute}} = kX_{\text{solute}}$$

This is **Henry's law**. The constant k differs depending on the solute-solvent combination. Henry's law often occurs in problems involving solutions of gases in liquids. See **6-49**, **6-51**, and **6-71**.

Distillation

Heating a solution with volatile components increases the vapor pressure above the solution. The vapors in equilibrium with the liquid are in general richer in the more volatile components than the liquid. This enrichment is the basis for the separation of mixtures by **distillation**.

If the solution is ideal, the degree of enrichment can be computed using Raoult's law for the vapor pressures of the components and Dalton's law to figure the mole fractions of the components in the vapor. See **6-53**, **6-55**, and **6-72**.

Fractional distillation provides means for the rapid separation of the components of a volatile solution. In this technique, an equilibrium between a liquid mixture and its vapors, which are richer in the more volatile component, is established. The vapors are liquefied (condensed), and the new liquid is brought to a new liquid-vapor equilibrium. Numerous repetitions of this cycle can achieve nearly perfect separation of the components.

A nonideal solution deviates from Raoult's law. Nonideal solutions have special compositions called azeotropic compositions. An **azeotrope** is a solution that boils to give vapors having the same composition as the liquid. Azeotropic mixtures cannot be separated by distillation.

Colloidal Suspensions

Colloids are mixtures of two or more substances in which one phase is suspended as small particles in a second. The particles of the **dispersed phase** are larger than single molecules (see **6-73**). They are from 10^{-9} to 10^{-6} m in diameter and too small to be distinguished by eye. The particles are in constant motion (**Brownian motion**). In principle, the dispersed phase in a colloid will settle out—eventually. In practice, sedimentation is slow. When colloids are **flocculated** by adding soluble salts to the dispersing phase, **aggregation** of the particles occurs, accelerating their sedimentation.

Detailed Solutions to Odd-Numbered Problems

6-1 a) The molar mass of cholesterol is 386.64 g mol^{-1}; one liter (L) equals 10 deciliters (dL); one gram (g) equals 1000 milligrams (mg). Use unit factors written from these equivalencies as follows:

$$c_{\text{cholesterol}} = \frac{214 \text{ mg}}{1 \text{ dL}} \times \left(\frac{1 \text{ g}}{1000 \text{ mg}}\right) \times \left(\frac{10 \text{ dL}}{1 \text{ L}}\right) \times \left(\frac{1 \text{ mol}}{386.64 \text{ g}}\right)$$

$$= \boxed{0.00553 \text{ mol L}^{-1}}$$

b) Assume that blood, like water, has a density of 1.0 g mL^{-1}. One liter (1000 mL) of blood then has a mass of 1.0 kg, of which 2.14 g (0.00214 kg) is cholesterol. Assume that the remaining 0.998 kg is solvent water. Insert these numbers into the definition of molality:

$$m = \frac{\text{moles of cholesterol}}{\text{kilograms of solvent}} = 2.14 \text{ g} \times \left(\frac{1 \text{ mol}}{386.64 \text{ g}}\right) \times \frac{1}{0.998 \text{ kg}} = \boxed{0.0055 \text{ mol kg}^{-1}}$$

c) If there is 2.14 g of cholesterol per liter of blood, then one liter of blood contains 2.14 g of cholesterol. This simple turn-about gives the unit factor in the following:

$$V_{\text{blood}} = 8.10 \text{ g cholesterol} \times \left(\frac{1 \text{ L blood}}{2.14 \text{ g cholesterol}}\right) = \boxed{3.79 \text{ L blood}}$$

Tip. The second answer has only two significant figures because the assumption about the density of blood is weak. Whole blood in fact has a density of 1.06 g mL^{-1}. This means that 1000 mL of blood weighs 1.06 kg. Taking this into account alters the calculation of the molality of the cholesterol:

$$m = \frac{0.00553 \text{ mol cholesterol}}{1.0578 \text{ kg solvent}} = 0.0052 \text{ mol kg}^{-1}$$

Clearly, the subtraction of the 2.14 g is superfluous in both calculations.

6-3 To compute the various quantities, obtain the masses and chemical amounts of HCl and H_2O in some set quantity of solution. Then use the definitions. Exactly 100.0 g of solution contains 38.00 g of HCl and 62.00 g of H_2O. Its volume is:

$$V = 100 \text{ g solution} \times \left(\frac{1 \text{ mL solution}}{1.1886 \text{ g solution}}\right) = 84.133 \text{ mL solution}$$

Use the molar masses of HCl and H_2O to convert their masses to chemical amounts:

$$\frac{38.00 \text{ g HCl}}{36.4606 \text{ g mol}^{-1}} = 1.0422 \text{ mol HCl} \qquad \frac{62.00 \text{ g H}_2\text{O}}{18.0153 \text{ g mol}^{-1}} = 3.4415 \text{ mol H}_2\text{O}$$

The molarity of the HCl equals the number of moles of HCl divided by the number of liters of solution. The 84.133 mL (0.084133 L) sample of solution contains 1.0422 mol of HCl, for a molarity of $\boxed{12.39 \text{ mol L}^{-1}}$.

The molality of the HCl equals the number of moles of HCl divided by the mass of solvent in kilograms. 1.0422 mol of HCl is dissolved in 62.00 g (0.06200 kg) of water. The molality of the HCl is therefore $\boxed{16.81 \text{ mol kg}^{-1}}$.

The mole fraction of water is the number of moles of water divided by the total number of moles of all components of the solution:

$$X_{\text{H}_2\text{O}} = \frac{3.4415 \text{ mol}}{1.0422 + 3.4415 \text{ mol}} = 0.7676$$

Because there are two components, X_{HCl} is simply $1 - 0.7676 = \boxed{0.2324}$.

6-5 Compute the mass of acetic acid in 1 kg of the 6.0835 M solution. To do this, use the molarity of the acetic acid and the density of the solution in unit factors:

$$1 \text{ kg sol'n} \times \left(\frac{1 \text{ L sol'n}}{1.0438 \text{ kg sol'n}}\right) \times \left(\frac{6.0835 \text{ mol C}_2\text{H}_4\text{O}_2}{1 \text{ L sol'n}}\right) \times \left(\frac{60.052 \text{g C}_2\text{H}_4\text{O}_2}{\text{mol C}_2\text{H}_4\text{O}_2}\right)$$
$$= 350.00 \text{ g C}_2\text{H}_4\text{O}_2$$

By subtraction, 1 kg of solution contains 650.00 g (0.65000 kg) of water. The 350.00 g of acetic acid equals 5.8282 mol of acetic acid. The molality of the acetic acid equals the number of moles of acetic acid divided by the mass of the solvent in kilograms:

$$m_{\text{C}_2\text{H}_4\text{O}_2} = \frac{5.8282 \text{ mol C}_2\text{H}_3\text{O}_2}{0.65000 \text{ kg}} = \boxed{8.9665 \text{ mol kg}^{-1}}$$

6-7 Water is the solute, and liquid nitrogen is the solvent, but the definition of mole fraction works the same. Use it to obtain the number of moles of water:

$$X_{\text{H}_2\text{O}} = 1.00 \times 10^{-5} = \frac{n_{\text{H}_2\text{O}}}{n_{\text{N}_2} + n_{\text{H}_2\text{O}}} = \frac{n_{\text{H}_2\text{O}}}{35.6972 + n_{\text{H}_2\text{O}}}$$

where 35.6972 is the number of moles of N_2 in 1.00 kg of N_2 (non-significant figures are carried along deliberately). Solving for $n_{\text{H}_2\text{O}}$ becomes simpler when it is realized that $n_{\text{H}_2\text{O}}$ can be neglected in the denominator in the preceding. Thus, the 1.00 kg of N_2 contains 3.5697×10^{-4} mol of H_2O. This amounts to $\boxed{0.00643 \text{ g}}$ of H_2O.

6-9 a) According to the problem, 100.00 g of commercial $H_3PO_4(aq)$ contains 90.00 g of pure H_3PO_4 and 10.00 g of H_2O. This much H_3PO_4 is 0.9184 mol of H_3PO_4 ($\mathcal{M} = $ 97.995 g mol^{-1}). Then:

$$\frac{100.00 \text{ g solution}}{0.9184 \text{ mol } H_3PO_4} \times \left(\frac{12.2 \text{ mol } H_3PO_4}{1 \text{ L solution}}\right) = \frac{1.33 \times 10^3 \text{ g solution}}{1 \text{ L solution}}$$

This is a correct answer, but densities are usually given in gram per milliliter. In those units, the answer is $\boxed{1.33 \text{ g mL}^{-1}}$.

b) The 2.00 L of 1.00 M $H_3PO_4(aq)$ must contain 2.00 mol of H_3PO_4. The volume of the 12.2 M $H_3PO_4(aq)$ solution that provides 2.00 mol of H_3PO_4 is:

$$2.00 \text{ mol } H_3PO_4 \times \left(\frac{1 \text{ L solution}}{12.2 \text{ mol } H_3PO_4}\right) = \boxed{0.164 \text{ L solution}}$$

To make 2.00 L of a 1.00 M $H_3PO_4(aq)$ solution, put 0.164 L (164 mL) of 12.2 M H_3PO_4 in a 2-liter volumetric flask and then add water to bring the total volume up the 2.00 L mark.

6-11 First calculate how many moles of NaOH were in the solution before any solid NaOH was added:

$$n_{NaOH} = 1.50 \text{ L solution} \times \left(\frac{2.40 \text{ mol NaOH}}{1 \text{ L solution}}\right) = 3.60 \text{ mol NaOH}$$

The molar mass of NaOH is 40.00 g mol^{-1}, so the added 25.0 g of NaOH equals 0.625 mol. After the addition, the total amount of NaOH in the container equals $3.60 + 0.625 = 4.23$ mol. Meanwhile, the final volume of the solution is 4.00 L. The final concentration of NaOH is therefore:

$$c_{NaOH} = \frac{4.23 \text{ mol NaOH}}{4.00 \text{ L}} = \boxed{1.06 \text{ mol L}^{-1}}$$

6-13 Break up each soluble reactant and product into its component ions; cancel out the spectator ions appearing on both sides of the equation:
a) $Ag^+(aq) + Cl^-(aq) \rightarrow AgCl(s)$
b) $K_2CO_3(s) + 2H^+(aq) \rightarrow 2K^+(aq) + CO_2(g) + H_2O(l)$
c) $2Cs(s) + 2H_2O(l) \rightarrow 2Cs^+(aq) + 2OH^-(aq) + H_2(g)$ (unchanged)
d) $2MnO_4^-(aq) + 16H^+(aq) + 10Cl^-(aq) \rightarrow 5Cl_2(g) + 2Mn^{2+}(aq) + 8H_2O(l)$

Tip. The product in part a is labeled (s) because CaF_2 precipitates from the solution as the reactants are mixed. How do you know this? You learn the table of solubilities of ionic compounds given in the text.[5]

---

[5]Table 11-2, text page 372.

6-15 The balanced chemical equation given in the problem indicates a 4-to-2 molar relationship between HNO_3 and PbO_2 in the reaction. Use this fact to construct a unit factor. The given molarity also gives unit factor, and the molar mass of PbO_2 (239.2 g mol^{-1}) furnishes another. The computation using these three factors goes as follows:

$$15.9 \text{ g } PbO_2 \times \left(\frac{1 \text{ mol } PbO_2}{239.2 \text{ g } PbO_2}\right) \times \left(\frac{4 \text{ mol } HNO_3}{2 \text{ mol } PbO_2}\right) \times \left(\frac{1 \text{ L } 7.91 \text{ M } HNO_3 \text{ sol'n}}{7.91 \text{ mol } HNO_3}\right)$$

$$= \boxed{0.0168 \text{ L } 7.91 \text{ M } HNO_3 \text{ sol'n}}$$

6-17 The carbon dioxide in this problem is a gas (with volume measured in liters), and the potassium carbonate is in aqueous solution (with volume measured in liters).[6] According to the balanced equation, the $CO_2(g)$ and $K_2CO_3(aq)$ react in a 1-to-1 molar ratio. The following uses this fact together with the concentration of the $K_2CO_3(aq)$ to determine the chemical amount of $CO_2(g)$ that reacts:

$$n_{CO_2} = 187 \text{ L solution} \times \left(\frac{1.36 \text{ mol } K_2CO_3}{1 \text{ L solution}}\right) \times \left(\frac{1 \text{ mol } CO_2}{1 \text{ mol } K_2CO_3}\right) = 254.3 \text{ mol } CO_2$$

The volume occupied by this much $CO_2(g)$ depends on T and P. Insert the given T and P in the ideal-gas equation:

$$V_{CO_2} = \frac{nRT}{P} = \frac{(254.3 \text{ mol})(0.08206 \text{ L atm mol}^{-1}\text{K}^{-1})(323.15 \text{ K})}{1.00 \text{ atm}} = \boxed{6.74 \times 10^3 \text{ L}}$$

6-19 When a salt forms in an acid-base reaction, the cation derives from the base and the anion from the acid:
a) $Ca(OH)_2(aq) + 2\,HF(aq) \rightarrow CaF_2(s) + 2\,H_2O(l)$
calcium hydroxide + hydrofluoric acid $\rightarrow$ calcium fluoride + water
b) $2\,RbOH(aq) + H_2SO_4(aq) \rightarrow Rb_2SO_4(aq) + 2\,H_2O(l)$
rubidium hydroxide + sulfuric acid $\rightarrow$ rubidium sulfate + water
c) $Zn(OH)_2(aq) + 2\,HNO_3(aq) \rightarrow Zn(NO_3)_2(aq) + 2\,H_2O(l)$
zinc hydroxide + nitric acid $\rightarrow$ zinc nitrate + water
d) $KOH(aq) + HCH_3COO(aq) \rightarrow KCH_3COO(aq) + H_2O(l)$
potassium hydroxide + acetic acid $\rightarrow$ potassium acetate + water

---

6-21 The reaction is $H_2S + 2\,NaOH \rightarrow Na_2S + 2\,H_2O$. $\boxed{\text{Sodium sulfide}}$ is the salt produced by this neutralization reaction.

Tip. Without the hint, NaHS (sodium hydrogen sulfide) is a possible answer.

---

[6]The use of the same unit can cause confusion between the two.

6-23 a) Phosphorus trifluoride is PF_3; phosphorous acid is H_3PO_3, and hydrofluoric acid is HF. The equation is easily balanced by inspection. Assign 1 as the coefficient for PF_3. All of the fluorine ends up in HF. This means the coefficient for HF is 3. All of the oxygen ends up in H_3PO_3, making 3 the coefficient for the H_2O:

$$\boxed{PF_3 + 3\,H_2O \rightarrow H_3PO_3 + 3\,HF}$$

b) First determine the chemical amount of $PF_3(g)$ in 1.94 L of gaseous PF_3 at 25° C (298 K) and 0.970 atm:

$$n_{PF_3} = \frac{PV}{RT} = \frac{(0.970 \text{ atm})(1.94 \text{ L})}{(0.08206 \text{ L atm mol}^{-1}\text{K}^{-1})(298 \text{ K})} = 0.07695 \text{ mol}$$

According to the equation, 1 mol of PF_3 reacts to give 1 mol of H_3PO_3 and 3 mol of HF. This means 0.07695 mol of H_3PO_3 and 0.2309 mol of HF are produced from 0.07695 mol of PF_3. Both acids dissolve as they are formed. enough water is present to give a final volume of 872 mL (0.872 L). The acids are mixed with each other, but their respective concentrations are computed by *separately* dividing chemical amount by the final volume:

$$[H_3PO_3] = \frac{0.07695 \text{ mol}}{0.872 \text{ L}} = \boxed{0.0882 \text{ M}} \qquad [HF] = \frac{0.2309 \text{ mol}}{0.872 \text{ L}} = \boxed{0.265 \text{ M}}$$

6-25 The problem is very similar to text Example 6-6.[7] Each mole of KOH dissolves in water to give one mole of $K^+(aq)$ ion and one mole of $OH^-(aq)$ ion. Therefore, the chemical amount of $OH^-(aq)$ in 37.85 mL of 0.1279 M aqueous KOH is:

$$n_{OH^-} = 37.85 \text{ mL} \times \left(\frac{0.1279 \text{ mmol}}{1 \text{ mL}}\right) = 4.841 \text{ mmol}$$

Nitric acid furnishes one mole of $H^+(aq)$ ion per mole dissolved. Also, the stoichiometric ratio in the acid-base reaction between HNO_3 and KOH is 1-to-1. Thus, the chemical amount of HNO_3 in the 100.0 mL sample before the reaction was also 4.841 mmol. The concentration of HNO_3 in the original solution was:

$$[HNO_3] == \frac{4.841 \text{ mmol}}{100.0 \text{ mL}} = \frac{0.04841 \text{ mmol}}{1 \text{ mL}} = \boxed{0.04841 \text{ mol L}^{-1}}$$

Tip. Note the use of the convenient unit, the millimole (mmol). The concentration of a solution in mmol mL^{-1} equals its concentration in mol L^{-1}.

---

[7]Text page 174.

6-27 a) $2 \overset{+3}{P} F_2 I(l) + 2 \overset{0}{Hg}(l) \rightarrow \overset{+2}{P_2} F_4(g) + \overset{+1}{Hg_2} I_2(s)$

b) $2 K \overset{+5}{Cl} \overset{-2}{O_3}(s) \rightarrow 2 K \overset{-1}{Cl}(s) + 3 \overset{0}{O_2}(g)$

c) $4 \overset{-3}{N} H_3(g) + 5 \overset{0}{O_2}(g) \rightarrow 4 \overset{+2}{N} \overset{-2}{O}(g) + 6 H_2 \overset{-2}{O}(g)$

d) $2 \overset{0}{As}(s) + 6 NaO \overset{+1}{H}(l) \rightarrow 2 Na_3 \overset{+3}{As} O_3(s) + 3 \overset{0}{H_2}(g)$

6-29 Refer to the rules on oxidation numbers. Neither hydrogen (oxidation state $+1$) nor oxygen (oxidation state -2) changes oxidation state in this reaction. The gold loses $3\,e^-$ per atom, passing from the zero to the $+3$ oxidation state; $\boxed{Au}$ is oxidized. The Se atom in H_2SeO_4 passes from the $+6$ to the $+4$ oxidation state; it gains $2\,e^-$ and so $\boxed{H_2SeO_4}$ is reduced. Note that only half of the H_2SeO_4 reacting is actually reduced.

6-31 The potassium dichromate solution contains 5.134 g of the solute per 1000 mL of solution. 34.26 mL of it brings the titration to the endpoint. The chemical amount of $K_2Cr_2O_7$ that reacts is:

$$n_{K_2Cr_2O_7} = 34.26 \text{ mL sol'n} \times \left(\frac{5.134 \text{ g } K_2Cr_2O_7}{1000 \text{ mL sol'n}} \right) \times \left(\frac{1 \text{ mol } K_2Cr_2O_7}{294.18 \text{ g } K_2Cr_2O_7} \right)$$

$$= 5.979 \times 10^{-4} \text{ mol}$$

In aqueous solution, 1 mol of $Cr_2O_7^{2-}(aq)$ forms for every 1 mol of $K_2Cr_2O_7$ that dissolves. Also, according to the balanced equation (which is a net ionic equation), 1 mol of $Cr_2O_7^{2-}(aq)$ reacts with 6 mol of $Fe^{2+}(aq)$. Cast these facts as unit factors to compute the chemical amount of Fe^{2+}:

$$n_{Fe^{2+}} = 5.979 \times 10^{-4} \text{ mol } K_2Cr_2O_7 \times \left(\frac{1 \text{ mol } Cr_2O_7^{2-}}{1 \text{ mol } K_2Cr_2O_7} \right) \times \left(\frac{6 \text{ mol } Fe^{2+}}{1 \text{ mol } Cr_2O_7^{2-}} \right)$$

$$= 0.003587 \text{ mol}$$

This is the amount of Fe^{2+} in 500.0 mL of solution. The amount per liter (1000.0 mL) is twice as much. The concentration of Fe^{2+} in the sample is $\boxed{0.007175 \text{ mol L}^{-1}}$.

6-33 Add up the molar masses of acetone and benzophenone, and use them to compute the chemical amounts of the two in the solution:

$$n_{C_3H_6O} = 15.0 \text{ g } C_3H_6O \times \frac{1 \text{ mol } C_3H_6O}{58.08 \text{ g } C_3H_6O} = 0.86087 \text{ mol } C_3H_6O$$

$$n_{C_{13}H_{10}O} = 15.0 \text{ g } C_{13}H_{10}O \times \frac{1 \text{ mol } C_{13}H_{10}O}{182.22 \text{ g } C_{13}H_{10}O} = 0.0823 \text{ mol } C_{13}H_{10}O$$

The mole fraction of benzophenone in the solution equals:

$$X_{C_{13}H_{10}O} = \frac{0.0823}{0.0823 + 0.86087} = 0.08727$$

The change in vapor pressure of the acetone due to the presence of the benzophenone is:

$$\Delta P_{C_3H_6O} = -X_{C_{13}H_{10}O}P^\circ_{C_3H_6O} = -0.08727(0.3270 \text{ atm}) = -0.02853 \text{ atm}$$

The final vapor pressure of the acetone equals its P° plus the change. This is 0.3270 atm minus 0.02853 atm or $\boxed{0.2985 \text{ atm}}$.

6-35 The boiling-point elevation of a solvent caused by a single nonvolatile solute is proportional to the molality of the solute:

$$\Delta T_b = K_b m \qquad \text{hence} \qquad K_b = \frac{\Delta T_b}{m}$$

To obtain m, use the definition of molality. The chemical amount of anthracene equals its mass divided by its molar mass, which is 178.2 g mol^{-1}:

$$n_{\text{anthracene}} = \frac{7.80 \text{ g}}{178.2 \text{ g mol}^{-1}} = 0.04376 \text{ mol}$$

The molality of the anthracene in the toluene is this number of moles divided by the mass of the toluene in kilograms:

$$m_{\text{anthracene}} = \frac{0.4376 \text{ mol}}{0.1000 \text{ kg}} = 0.4376 \text{ mol kg}^{-1}$$

The change in the boiling temperature is clearly $112.06 - 110.60 = 1.46°C$. Then:

$$K_b = \frac{\Delta T_b}{m} = \frac{1.46°C}{0.4376 \text{ mol kg}^{-1}} = 3.34°C \text{ mol kg}^{-1}$$

This can also be expressed as $\boxed{3.34 \text{ K mol kg}^{-1}}$.

Tip. Convert both temperatures in the problem from Celsius to absolute, and subtract T_1 from T_2 to check this last statement.

6-37 Compute the molality of the aqueous solution of the nonvolatile, nondissociating solute sugar using the formula for boiling-point elevation and taking K_b for water from text Table 6-2:[8]

$$m_{\text{sugar}} = \frac{\Delta T_b}{K_b} = \frac{0.30 \text{ K}}{0.512 \text{ K kg mol}^{-1}} = \frac{0.5859 \text{ mol}}{\text{kg}}$$

This means that 200.0 g of water contains 0.1172 mol of sugar. The mass of the sugar is given as 39.8 g. Its molar mass equals 39.8 mol/0.1172 mol = $\boxed{340 \text{ g mol}^{-1}}$.

[8]Text page 179.

6-39 Assume that the unknown is nonvolatile. The molality of the unknown then is related to the change in the freezing point of the camphor as follows:

$$m = -\frac{\Delta T_f}{K_f} = -\frac{(170.8 - 178.4)°C}{37.7°C \text{ kg mol}^{-1}} = 0.20 \text{ mol kg}^{-1}$$

Note the switch in the temperature units in the freezing-point depression constant from K to °C.[9] There is 0.20 mol of unknown per kilogram of camphor, but the problem deals with 25.0 g (0.0250 kg) of camphor. Compute the amount of unknown in the 25.0 g of camphor:

$$0.20 \text{ mol kg}^{-1} \times 0.0250 \text{ kg} = 0.0050 \text{ mol}$$

The 0.840 g of unknown dissolved in the 25.0 g of camphor also equals 0.0050 mol of unknown. The molar mass of a substance equals its mass divided by its chemical amount: $0.840 \text{ g}/0.0050 \text{ mol} = \boxed{1.7 \times 10^2 \text{ g mol}^{-1}}$.

6-41 The ice-cream mixture contains 34 g of sucrose for every 66 g of water. Use this fact in a unit factor to figure out how much sucrose is present in 1000 g (1 kilogram) of water:

$$1000 \text{ g water} \times \left(\frac{340 \text{ g sucrose}}{660 \text{ g water}}\right) = 515 \text{ g sucrose}$$

Divide this mass by 342.3 g mol^{-1}, the molar mass of sucrose, to convert to moles. The result is 1.50 mol. Since there is 1.50 mol of sucrose per 1000 g of water, the molality of the aqueous sucrose is 1.50 mol kg^{-1}. The change in the freezing point is:

$$\Delta T = -K_f m = (-1.86 \text{ K mol kg}^{-1})(1.50 \text{ kg mol}^{-1}) = -2.8 \text{ K}$$

The freezing point of the mixture equals this change added to the freezing point of the pure solvent (0°C). It is $\boxed{-2.8°C}$.

Tip. As pure ice freezes out, the remaining solution becomes more concentrated in sucrose and the freezing point is depressed further.

6-43 Calculate the effective molality from the change in the freezing point:

$$m = -\frac{\Delta T}{K_f} = -\frac{(-4.218 \text{ K})}{1.86 \text{ K kg mol}^{-1}} = 2.268 \text{ mol kg}^{-1}$$

The ratio of the effective molality to the actual molality is $2.268/0.8402 = 2.70$. Thus each Na_2SO_4 unit dissociates effectively into $\boxed{2.70 \text{ particles}}$.

---

[9]The switch is legitimate since the problem concerns a *change* in temperature. See **6-35**.

Tip. This is somewhat less than the theoretical value of 3.00 (corresponding to two Na^+ and one SO_4^{2-} per formula unit) because the positive and negative ions in this rather concentrated solution tend to associate, reducing the effective number of free ions.

6-45 The osmotic pressure π of this solution is related to the concentration of the unknown solute by the equation:

$$\pi = cRT$$

Measurements of osmotic pressure therefore give the concentration:

$$c_{unknown} = \frac{\pi}{RT} = \frac{0.0105 \text{ atm}}{(0.08206 \text{ L atm mol}^{-1}\text{K}^{-1})(300 \text{ K})} = 4.265 \times 10^{-4} \text{ mol L}^{-1}$$

This solution was obtained by dissolving 200 mg (0.200 g) of the solute in 25.0 mL of solution, a procedure that gives the same concentration as dissolving 8.00 g solute in 1.00 L of water. Thus 8.00 /Xof solute equals 4.265×10^{-4} mol of solute. Accordingly. the molar mass is:

$$\mathcal{M}_{unknown} = \frac{8.00 \text{ g}}{4.265 \times 10^{-4} \text{ mol}} = \boxed{1.88 \times 10^4 \text{ g mol}^{-1}}$$

6-47 Text Figure 6-12[10] shows the experimental set-up. The difference h between the level of the solution in the tube and the level outside the tube is proportional to the osmotic pressure of the solution. The problem gives h as 15.2 cm (0.152 m) of solution. To get the osmotic pressure, substitute the density ρ of the solution and the acceleration g of gravity in the formula $\pi = \rho g h$.[11] The density of the solution is 1.00 g cm^{-3}, which is equivalent to 1.00×10^3 kg m^{-3}. Then:

$$\pi = \rho g h = (1.00 \times 10^3 \text{ kg m}^{-3})(9.807 \text{ m s}^{-2})(0.152 \text{ m}) = 1.49 \times 10^3 \text{ kg m}^{-1}\text{s}^{-2}$$

This equals 1.49×10^3 Pa.[12] Converting to atm:

$$\pi = 1.49 \times 10^3 \text{ Pa} \times \left(\frac{1 \text{atm}}{101\,325 \text{ Pa}}\right) = 0.0147 \text{ atm}$$

Now, calculate the concentration of the polymer in the solution:

$$c = \frac{\pi}{RT} = \frac{0.0147 \text{ atm}}{(0.08206 \text{ L atm mol}^{-1}\text{K}^{-1})(288.15 \text{ K})} = 6.22 \times 10^{-4} \text{ mol L}^{-1}$$

---

[10]Text page 185.

[11]Text page 184. Compare this formula to the formula for the "regular" pressure exerted by a gas that appears on text page 102.

[12]Text Table B-2 in text Appendix B.

The solution holds 6.22×10^{-4} mol of polymer per liter. There are also 4.64 g of polymer per liter. Therefore:

$$\mathcal{M} = \frac{4.64 \text{ g}}{6.22 \times 10^{-4} \text{ mol}} = \boxed{7.46 \times 10^3 \text{ g mol}^{-1}}$$

Tip. It is possible to compute the concentration of the polymer without changing the pressure to atmospheres. Use SI units as follows:

$$c = \frac{\pi}{RT} = \frac{1.49 \times 10^3 \text{ Pa}}{(8.315 \text{ J K}^{-1}\text{mol}^{-1})(288.15 \text{ K})} = 0.622 \text{ mol m}^{-3}$$

This answer is the same because there are 1000 L in a cubic meter. Confirm the cancellation of units as in **6-33**, using the equivalencies in text Appendix B.

6-49 a) The partial pressure of gaseous CO_2 above the aqueous solution of CO_2 is 5.0 atm. Henry's law relates the mole fraction the dissolved CO_2 to this partial pressure:

$$P_{CO_2} = k_{CO_2} X_{CO_2} \qquad \text{from which} \qquad X_{CO_2} = \frac{P_{CO_2}}{k_{CO_2}} = \frac{5.00 \text{ atm}}{(1.65 \times 10^3 \text{ atm})} = 0.0030$$

This fraction means that there is 0.0030 mol of CO_2 in solution for every 0.9970 mol of water. But there is 55.5 mol of water per liter of solution if the solution (which is dilute) has the same density as water. Hence:

$$\frac{0.0030 \text{ mol CO}_2}{0.9970 \text{ mol H}_2\text{O}} \times \left(\frac{55.5 \text{ mol H}_2\text{O}}{1.00 \text{ L solution}} \right) = 0.17 \text{ mol CO}_2 \text{ L}^{-1}$$

The amount of CO_2 dissolved in a liter of water held under a pressure of 5.00 atm of gaseous CO_2 is therefore $\boxed{0.17 \text{ mol}}$.

b) Before the cap is removed, gaseous CO_2 in the small space above the liquid is in equilibrium with the dissolved CO_2. This means that CO_2 molecules are constantly moving from the gas phase to the dissolved phase and back. The rate of movement of CO_2 out of solution equals the rate of movement into solution. When the cap is removed, gaseous CO_2 escapes from the bottle because the partial pressure of CO_2 in the atmosphere is far less than 1 atm. Equilibrium is re-established with a far smaller concentration of CO_2 in the solution.

6-51 Determine the chemical amount of the methane that was dissolved in the 1.00 kg of solution before the boiling. Assume that all the methane was expelled and that the expelled gas behaves ideally:

$$n_{CH_4} = \frac{PV}{RT} = \frac{(1.00 \text{ atm})(3.01 \text{ L})}{(0.08206 \text{ L atm mol}^{-1}\text{K}^{-1})(273.15 \text{ K})} = 0.1343 \text{ mol}$$

The molar mass of CH_4 is 16.04 g mol^{-1}, so the expelled methane has a mass of 2.154 g. The 1.00 kg of solution contained only water and methane. The mass of the water left after removal of methane is therefore 0.9978 kg. This mass of H_2O equals 55.39 mol of water.[13] Calculate the mole fraction of CH_4:

$$X_{CH_4} = \frac{0.1343 \text{ mol}}{0.1343 + 55.39 \text{ mol}} = 0.002419$$

This fraction of CH_4 was present with 1.00 atm of CH_4 above the solution. Henry's law applies, and the Henry's law k equals:

$$k_{CH_4} = \frac{P_{CH_4}}{X_{CH_4}} = \frac{1.00 \text{ atm}}{0.002419} = \boxed{413 \text{ atm}}$$

6-53 Write Raoult's law for both the benzene and toluene:

$$P_{benz} = X_{benz} P^\circ_{benz} \quad \text{and} \quad P_{tol} = X_{tol} P^\circ_{tol}$$

Since equal numbers of moles of benzene and toluene were mixed, $X_{benz} = X_{tol} = 0.500$. Use these mole fractions and the vapor pressures of the pure substances to compute the partial pressure of each substance above the mixture:

$$P_{benz} = 0.500(0.0987 \text{ atm}) = 0.04935 \text{ atm}$$
$$P_{tol} = 0.500(0.0289 \text{ atm}) = 0.01445 \text{ atm}$$

The number of moles of a particular gas in a gaseous mixture at constant temperature is directly proportional to its partial pressure:

$$n_{benz} = P_{benz} \frac{V}{RT} \quad \text{and} \quad n_{tol} = P_{tol} \frac{V}{RT}$$

The mole fraction of benzene in the vapor is

$$X_{benz,vap} = \frac{n_{benz}}{n_{benz} + n_{tol}} = \frac{P_{benz}(V/RT)}{P_{benz}(V/RT) + P_{tol}(V/RT)} = \frac{P_{benz}}{P_{benz} + P_{tol}}$$
$$= \frac{0.04935 \text{ atm}}{(0.04935 + 0.01445) \text{ atm}} = \boxed{0.774}$$

Tip. Half of the molecules in the liquid are benzene, but over three quarters of the molecules in the vapors above the liquid are benzene. Such enrichment in the more volatile component is the basis for separation by distillation.

---

[13]Using $\mathcal{M} = 18.0153$ g mol^{-1} for water.

6-55 a) Convert the masses of CCl_4 and $C_2H_4Cl_2$ to chemical amount by dividing by their respective molar masses:

$$n_{CCl_4} = 30.0 \text{ g} \times \left(\frac{1 \text{ mol}}{153.82 \text{ g}}\right) = 0.1950 \text{ mol} \qquad n_{C_2H_4Cl_2} = 20.0 \text{ g} \times \left(\frac{1 \text{ mol}}{98.96 \text{ g}}\right) = 0.2021 \text{ mol}$$

Compute the mole fraction of CCl_4 in the solution from these values:

$$X_{CCl_4} = \frac{0.1950 \text{ mol}}{0.1950 + 0.2021 \text{ mol}} = \boxed{0.491}$$

b) The total vapor pressure above the solution equals the sum of the partial pressures of the two components in the vapors above the solution. Raoult's law gives these partial pressures. Therefore:

$$P_{tot} = P_{CCl_4} + P_{C_2H_4Cl_2} = X_{CCl_4} P^\circ_{CCl_4} + X_{C_2H_4Cl_2} P^\circ_{C_2H_4Cl_2}$$
$$P_{tot} = (0.491)(0.293 \text{ atm}) + (1 - 0.491)(0.209 \text{ atm}) = \boxed{0.250 \text{ atm}}$$

where the mole fraction of CCl_4 comes from the preceding part, and the vapor pressures of the pure components are given in the problem. The two mole fractions add up to one because there are only two components in the solution.

c) The mole fraction of a component in a gaseous mixture equals its partial pressure divided by the total pressure:

$$X_{CCl_4,vap} = \frac{P_{CCl_4}}{P_{tot}}$$

According to Raoult's law, the partial pressure of $CCl_4(g)$ in the vapor above the solution equals its mole fraction in the solution times its vapor pressure when pure:

$$X_{CCl_4,vap} = \frac{P_{CCl_4}}{P_{tot}} = \frac{X_{CCl_4} P^\circ_{CCl_4}}{P_{tot}}$$

But $P^\circ_{CCl_4}$ is given in the problem and X_{CCl_4} was found in part a). Substitution gives:

$$X_{CCl_4,vap} = \frac{0.491(0.293 \text{ atm})}{0.250 \text{ atm}} = \boxed{0.575}$$

Tip. The mole fraction of CCl_4 is 0.491 in the solution but 0.575 in the vapors above the solution. As in **6-53**, the vapors are enriched in the more volatile component.

6-57 a) The element iodine exists in the Donovan's solution in different chemical forms. However, the total amount of this element in the solution, which is prepared by mixing pure compounds that dissolve completely, depends only on the amounts of the compounds and the fraction of each comprised by iodine. The total mass of iodine in 100 mL of Donovan's solution is thus the fraction by mass of elemental iodine in AsI_3 ($\mathcal{M} = 455.6$ g mol^{-1}) multiplied by the 1.00 g of AsI_3 plus the fraction by mass of iodine in HgI_2 ($\mathcal{M} = 454.4$ g mol^{-1}) multiplied by the 1.00 g of HgI_2:

$$m_I = \left(\frac{(3)(126.9) \text{ g I}}{455.6 \text{ g } AsI_3}\right) 1.00 \text{ g } AsI_3 + \left(\frac{(2)(126.9 \text{ g I})}{454.4 \text{ g } HgI_2}\right) 1.00 \text{ g } HgI_2 = 1.39 \text{ g}$$

where 126.9 is the atomic mass of iodine. The mass of iodine per liter (which is 10 times 100 mL) of solution is 10 times this answer or $\boxed{13.9 \text{ g L}^{-1}}$.

b) The 0.100 M AsI_3 solution contains 45.56 g of AsI_3 per liter and therefore furnishes 4.556 g of AsI_3 per 100 mL. To make 3.50 L of Donovan's solution, 35.0 g of AsI_3 is needed. Measure out $(35.0/4.556) \times 100$ mL $= 768$ mL of the AsI_3 solution. Add to it 35.0 g of $HgI_2(s)$ and 31.5 g of $NaHCO_3(s)$. Then add enough water to bring the total volume to 3.50 L.

6-59 The careful wording of the problem assures the reader that no Cl is lost at any point during the transformation $NaCl \rightarrow Cl_2 \rightarrow HCl$. For every mole of Cl present in the original 150 mL of 10.00% aqueous NaCl, one mole of Cl is formed in the 250 mL of $HCl(aq)$. Compute this number of moles:

$$150 \text{ mL sol'n} \times \left(\frac{1.0726 \text{ g sol'n}}{1 \text{ mL sol'n}}\right) \times \left(\frac{10.0 \text{ g NaCl}}{100 \text{ g sol'n}}\right) \times \left(\frac{1 \text{ mol NaCl}}{58.44 \text{ g NaCl}}\right)$$

$$= 0.2753 \text{ mol NaCl}$$

Note the third term, in which the mass percentage of NaCl in the solution is used as a unit factor. The 0.2753 mol of NaCl implies 0.2753 mol of Cl, because each mole of NaCl contains one mole of Cl. All of the Cl ends up in the form of HCl, so 0.2753 mol of HCl is present in the 250 mL of solution that is formed. The concentration of the HCl is:

$$c_{HCl} = \frac{0.2753 \text{ mol}}{0.250 \text{ L}} = \boxed{1.10 \text{ mol L}^{-1}}$$

6-61 The change in the vapor pressure ΔP is $0.3868 - 0.3914 = -0.0046$ atm. The mole fraction of the sulfur present in the sulfur-CS_2 system is therefore:

$$X_{sulfur} = -\frac{\Delta P}{P^\circ_{CS_2}} = -\frac{-0.0046 \text{ atm}}{0.3914 \text{ atm}} = 0.0117$$

The solution contains 1.00 kg of CS_2 which is 13.13 mol of CS_2.[14] Hence:

$$X_{sulfur} = 0.0117 = \frac{n_{sulfur}}{n_{sulfur} + n_{CS_2}} = \frac{n_{sulfur}}{n_{sulfur} + 13.13}$$

---

[14] $\mathcal{M} = 76.14$ g mol^{-1}

Solving for the chemical amount of sulfur gives 0.155 mol. Because this amount of sulfur is simultaneously 40.0 g of sulfur, the molar mass of sulfur as it exists in this solution is 40.0 g/0.155 mol = $\boxed{257 \text{ g mol}^{-1}}$. This is almost exactly eight times larger than 32 g mol^{-1}, the molar mass of S The molecular formula of the sulfur in the solution must be $\boxed{S_8}$.

6-63 The soft drink is a solution of CO_2 (and other solutes) in water. When the cap is on, gaseous CO_2 in the space above the fluid is held at a pressure exceeding 1 atm. Henry's law requires a higher concentration of dissolved CO_2 than if the pressure of CO_2 were only 1 atm. The dissolved CO_2 depresses the freezing point of the solution. When the cap is popped, the partial pressure of CO_2 over the solution suddenly drops to far less than 1 atm. Gaseous CO_2 bubbles out of solution. The freezing point of the soft drink rises. If the temperature of the soft drink is colder than the elevated freezing point caused by loss of the gaseous solute, the solution will freeze.

6-65 Raoult's law allows calculation of the effective mole fraction of $CaCl_2$ in the solution at 25°C using the vapor-pressure lowering:

$$X_2 = -\frac{P_1 - P_1^0}{P_1^0} = -\frac{0.02970 - 0.03126 \text{ atm}}{0.03126 \text{ atm}} = 0.0499$$

$CaCl_2$ dissociates in water to form one mole of $Ca^{2+}(aq)$ cations and two moles of $Cl^-(aq)$ anions per mole dissolved. The X_2 just calculated is therefore not the true mole fraction of the solute, but is an *effective* mole fraction that is larger that the true mole fraction because of the dissociation of $CaCl_2$ into ions. Now, calculate the effective molality of the $CaCl_2$:

$$\frac{0.0499 \text{ mol solute}}{(1.000 - 0.0499) \text{ mol H}_2\text{O}} \times \left(\frac{1 \text{ mol H}_2\text{O}}{0.018015 \text{ kg H}_2\text{O}}\right) = 2.92 \text{ mol kg}^{-1}$$

Put this effective molality into the standard formula for freezing-point depression:

$$\Delta T = -K_f m = -(1.86 \text{ K kg mol}^{-1})(2.92 \text{ mol kg}^{-1}) = -5.43 \text{ K} = -5.43°C$$

Recall that the kelvin and the degree Celsius are equal in size. The freezing point of the solution is the original freezing point plus the change:

$$0.00°C + (-5.43°C) = \boxed{-5.43°C}$$

Tip. It is assumed that the effective number of particles of solute is unchanged by cooling from 25°C, where the vapor pressure was recorded, to −5.43°C.

6-67 The salt $GaCl_2$ would be expected to dissociate in water according to:

$$GaCl_2(aq) \rightarrow Ga^{2+}(aq) + 2\,Cl^-(aq)$$

If the "$GaCl_2$" were actually $Ga[GaCl_4]$,[15] then the dissociation would be:

$$Ga[GaCl_4](aq) \rightarrow Ga^+(aq) + [GaCl_4]^-(aq)$$

In the first case, dissociation gives three ions; in the second case, it gives only two ions. We can therefore measure a colligative property to distinguish between the two cases. For example, imagine that enough compound is dissolved in water to make a solution that is 0.0100 mol kg^{-1} in $GaCl_2$. This solution would have a freezing point of $-0.056°C$ if the formula $GaCl_2$ were correct. This freezing point is predicted using an effective molality of 0.0300 mol kg^{-1} in the formula for freezing-point depression. The effective molality is triple m_{GaCl_2} because three moles of ions are formed by dissociation of one mole of $GaCl_2$. Now try the formula $Ga[GaCl_4]$. The identical aqueous solution has a $m_{Ga[GaCl_4]}$ of 0.00500 mol kg^{-1}, because the new formula for the solute corresponds to a molar mass that is twice as large, and an effective molality of 0.0100 mol kg^{-1}, because two ions are formed upon dissociation. The predicted freezing point of $-0.0186°C$ is measurably different.

4-69 Determine the effective molality of the NaCl solution that freezes at $-0.406°C$:

$$m_{NaCl} = -\frac{\Delta T_f}{K_f} = -\frac{-0.406\ K}{1.86\ K\ kg\ mol^{-1}} = 0.218\ mol\ kg^{-1}$$

This is also the effective molality of the contents of the red blood cell. It is assumed in the problem that the molality inside the cell equals the molarity. Then, 0.218 M can be used to calculate the osmotic pressure:

$$\pi = cRT = (0.218\ mol\ L^{-1})(0.08206\ L\ atm\ mol^{-1}K^{-1})(298.15\ K) = \boxed{5.33\ atm}$$

6-71 Write Henry's law for a solution of benzene in water:

$$P_{benz} = k_{benz}X_{benz} = (301\ atm)X_{benz}$$

The chemical amount of benzene ($\mathcal{M} = 78.11$ g mol^{-1}) in the solution described in the problem is 2.0 g/78.11 g $mol^{-1} = 0.0256$ mol. The mole fraction of benzene is:

$$X_{benz} = \frac{0.0256\ mol}{0.0256 + (55.5 \times 10^3)\ mol} = 4.6 \times 10^{-7}$$

---

[15]The bracketed group is a complex ion. See text page 665.

The large amount of water completely drowns out the contribution of the benzene to the denominator of this fraction. Insert the mole fraction of benzene and the given Henry's law constant into the equation for Henry's law:

$$P_{\text{benz}} = k_{\text{benz}} X_{\text{benz}} = (301 \text{ atm})(4.6 \times 10^{-7}) = \boxed{1.4 \times 10^{-4} \text{ atm}}$$

Then substitute this pressure and the temperature in kelvins into the rearranged ideal-gas equation:

$$\frac{n_{\text{benz}}}{V} = \frac{P_{\text{benz}}}{RT} = \frac{1.4 \times 10^{-4} \text{ atm}}{(0.082057 \text{ L atm K}^{-1}\text{mol}^{-1})(298.15 \text{ K})} = 5.7 \times 10^{-6} \text{ mol L}^{-1}$$

This result is the concentration of the benzene in the vapor above the solution. Convert to molecules per cubic centimeter as follows:

$$\frac{5.7 \times 10^{-6} \text{ mol}}{1 \text{ L}} \times \left(\frac{1 \text{ L}}{1000 \text{ cm}^3}\right) \times \left(\frac{6.022 \times 10^{23} \text{ molecule}}{1 \text{ mol}}\right) = \boxed{3.4 \times 10^{15} \frac{\text{molecule}}{\text{cm}^3}}$$

6-73 The difference between a solution and a colloidal suspension lies with the size of the dispersed particles. In a solution, the solute is dispersed at the molecular (or ionic) level. Each particle is surrounded by a cage of several solvent molecules. Examples are solutions of NaCl or alcohol in water. In a colloidal suspension, the dispersed particles are aggregates of hundreds to thousands of solute molecules. The aggregates are frequently surrounded by interacting solvent molecules that prevent them from sticking together to form a visible precipitate. The particles do not settle on the bottom of the container because the agitation caused by collisions of neighboring molecules is strong enough to keep them up. An example of a colloid is homogenized milk. The white opacity of milk is caused by tiny particles of fat that are too small to be filtered. In some cases, it is difficult to classify a mixture definitively as solution or suspension. If the particles are aggregates of only small numbers of molecules, the properties of the mixture will be similar to those of a solution, but deviate somewhat toward those of a colloid.

6-75 First obtain the empirical formula of the compound. Compute the chemical amounts of C and H in the sample:

$$n_{\text{C}} = 5.46 \text{ g CO}_2 \times \left(\frac{1 \text{ mol CO}_2}{44.0 \text{ g CO}_2}\right) \times \left(\frac{1 \text{ mol C}}{1 \text{ mol CO}_2}\right) = 0.1241 \text{ mol}$$

$$n_{\text{H}} = 2.23 \text{ g H}_2\text{O} \times \left(\frac{1 \text{ mol H}_2\text{O}}{18.015 \text{ g H}_2\text{O}}\right) \times \left(\frac{2 \text{ mol H}}{1 \text{ mol H}_2\text{O}}\right) = 0.2476 \text{ mol}$$

These amounts correspond to 0.2495 g of H and 1.490 g of C, so the mass of oxygen in the combustion sample equals:

$$m_{\text{O}} = m_{\text{tot}} - m_{\text{H}} - m_{\text{C}} = 2.40 - 0.2495 - 1.490 = 0.6605 \text{ g}$$

The chemical amount of O is:

$$n_O = 0.6611 \text{ g O} \times \left(\frac{1 \text{ mol O}}{15.9994 \text{ g O}} \right) = 0.0413 \text{ mol}$$

The three elements are present in the molar ratio $C_{0.124}H_{0.248}O_{0.0413}$, which gives the empirical formula C_3H_6O.

The observed depression of the freezing point gives the molality of the solution:

$$m = -\frac{\Delta T_f}{K_f} = -\frac{-0.97 \text{ K}}{1.86 \text{ K kg mol}^{-1}} = 0.522 \text{ mol kg}^{-1}$$

The 0.281 kg of solvent therefore contains 0.146 mol of the compound. The mass of this 0.146 mol of compound equals 8.69 g. Hence the molar mass of the compound is approximately 59 g mol^{-1}. The molecular formula is clearly $\boxed{C_3H_6O}$, which has a molar mass of 58.08 g mol^{-1}.

Chapter 7

Thermodynamic Processes and Thermochemistry

Thermodynamics is the study of the transfers of energy that accompany physical and chemical changes. The goals of chemical thermodynamics are to make predictions. Is given process possible? Under what conditions can a desired process occur?

Systems, States, and Processes

The basic terms in thermodynamics require careful definition:

• **System.** A real or imaginary portion of the universe that is confined by physical boundaries or mathematical constraints. In essence, a thermodynamic system is that part of the universe which we decide to study. In problem-solving, a shrewd choice of system clarifies difficult situations. A useful tactic is to define a system as the sum of a set of sub-systems (see **7-11**). One then deals with each sub-system in turn. In complex problems the system should be fully defined *in writing* before trying any computations (see **7-69**).

• **Closed System.** A system that has boundaries that do not allow the transfer of matter. Closed systems are common in problems.

• **Open System.** A system that *does* allow the transfer of matter across its boundaries. The typical chemical reaction, performed in an open flask in the laboratory, takes place in an open system.

• **Surroundings.** The portion of the universe that lies outside the system. The system plus its surroundings equal the **thermodynamic universe** for a given situation or process.

• **Extensive Property.** Extensive properties depend on the *extent*, or size of a system. The value of an extensive property for the whole system is the sum of the values for the individual sub-systems. The volume is an extensive property. Two

150

sub-systems of volume 3.0 and 7.0 L make a system of volume 10.0 L. The energy is another important extensive property.

- **Intensive Property.** A property of a system that does not depend on how big the system is. Examples are the pressure and temperature. If a system at 298 K is divided into two sub-systems, both are at 298 K (and not at 149 K).
- **Thermodynamic State.** A system is in a unique thermodynamic state when each of its properties (pressure, temperature, volume, energy, and so forth) has a definite, time-independent value. Left to itself such a system remains unchanged, in a state called *equilibrium,* indefinitely. A laboratory apparatus is often arranged to exert **constraints** on a system. Constraints maintain constant values for the thermodynamic properties of a system. **Example:** A glass vessel constrains the volume of a gaseous system to a certain fixed value. If the vessel is removed, the volume of the system can change.
- **Thermodynamic Process.** Such processes lead to changes in the thermodynamic states of systems. The **path** of a process is the sequence of intermediate conditions occurring in the system during a change.

An **irreversible** process passes through intermediate conditions that are not thermodynamic states. If the walls of a vessel constraining the volume of a sample of gas are suddenly removed, then the gas expands irreversibly. A **reversible** process proceeds through a continuous series of equilibrium states. At any moment during a reversible change, an infinitesimal alteration in external conditions can reverse the direction of the process. Reversible expansion of a sample of gas would require an infinite series of infinitely small increases in the volume of the container. A reversible decrease in temperature would require a similar infinite series of temperature reductions.

- **State Function.** A state function is a property of a system that depends only on the thermodynamic state of the system. The value of a state function does not depend on the way in which the state was achieved. State functions are symbolized in thermodynamic equations by capital letters. A change in a state function is indicated by a Δ. Thus:

$$\Delta P = P_2 - P_1$$

means the difference in pressure between the final state (state 2) and the initial state (state 1). In many applications the *change* in a state function is much more important than the actual final and initial values. Hence thermodynamic equations feature many Δ's.

In any process, the change in a state function depends only on the initial and final states and not on the path by which the change occurred.

In problems, adhere strictly to the convention that Δ applied to a function means the *final* value minus the *initial* value. Carelessness about this convention leads to tiresome sign errors.

The First Law of Thermodynamics

Work (w) and heat (q) are the two different means by which energy is transferred into and out of a system. The signs of w and q tells the direction of the transfer:

$+q$	the system gains heat
$-q$	the system loses heat
$+w$	work is done on the system
$-w$	the system performs work

Strict adherence to this sign convention reduces errors in problem solving. See the solutions to **7-17** and **7-57**.

Work

Work is the product of an external force acting on a body times the distance through which the force acts.

Work is the transfer of energy by oriented, non-random macroscopic motions.

A transfer of work into a system goes to increase the internal kinetic energy of the system or the internal potential energy or both. Work is measured in joules, the unit used to measure energy, since work is a manifestation of energy. Force is measured in newtons (N), and distance in meters (m). Work is therefore also measured in newton meters (N m). The units check because a newton is a kg m s^{-2}, which makes a newton-meter a kg m^2 s^{-2}, which is a joule.[1]

Of major concern in chemistry are electrical work (discussed in Chapter 12) and mechanical work. The transfer of mechanical work requires mechanical contact between a system and its surroundings. In the absence of levers, wheels, pulleys, and other mechanisms, the only type of mechanical work that is possible in chemical processes is **pressure-volume work**. Pressure-volume work results from the change in volume (compression or expansion) of a system against a resisting pressure. If a system changes volume against an constant external pressure, then work flows:

$$w = -P_{\text{ext}}\Delta V$$

The negative sign must not be omitted. It maintains the convention that positive work is work done on the system. For an expansion, ΔV is positive because V_2 is larger

---

[1]See text Appendix B.

than V_1. The system pushes back its surroundings; it does work on the surroundings. Under the sign convention this fact is stated as follows: negative work is done on the system. If the pressure is in atmospheres and the volume is in liters, then the unit of pressure-volume work is the L atm: **1 L atm is equal to 101.325 J**.

Heat

Heat, like work, is a way in which energy is exchanged between a system and its surroundings. As a manifestation of energy, heat has the same units (joules) as energy. The transfer of heat requires thermal contact (rather than mechanical contact) across a system boundary. Transfer of heat occurs because of a temperature difference.

Heat is the transfer of energy by random, incoherent, non-directed microscopic motions.

Heat lacks the large-scale organization of motion that is characteristic of work.

Internal Energy

Two types of energy are distinguished. The **kinetic energy** of a system is its energy of motion:

$$KE = E_{K} = \tfrac{1}{2}Mv^2$$

where M is its mass and v its velocity. An object has kinetic energy based on its own motion as a whole. A system has *internal* kinetic energy based on the relative motions of the atoms that compose it.

The **potential energy** of an object is its energy of position. An object has gravitational potential energy relative to the center of the earth. If its height h near the surface of the earth is changed, then the change in its potential energy is:

$$\Delta(PE) = V = Mg\Delta h$$

where g is the acceleration of gravity (9.81 m s^{-2}) and M is the mass of the object (see **7-3**). *Internal* potential energy is potential energy that a system has due to the relative positions of the atoms that compose it. A pound of TNT for instance has much internal potential energy relative to the internal potential energy of the mixtures of gases that it forms when it explodes.

The **internal energy E** of a system equals the sum of its internal kinetic energy and internal potential energies. Changes in the internal energy of systems (symbolized ΔE) are very important in chemical thermodynamics.

Heat Capacity, Enthalpy, and Calorimetry

Calorimetry is the measurement of quantities of heat. The name of an important non-SI unit of heat, the **calorie**, reflects this. **One calorie is 4.184 J exactly**. The fundamental method in calorimetry is to let some heat flow into a system, checking the temperature before and after. As the system gains heat its temperature increases. If the transfer is at constant pressure in a system that stays in the same state throughout the process, then the specific heat capacity of the system can be measured:

• The **specific heat capacity** c_s of a system is the amount of heat needed to raise the temperature of a specified mass of material through a given temperature interval.

This quantity has many different units. The text uses J $(°C)^{-1}$ g^{-1}. (or J $K^{-1}g^{-1}$, see **7-11**). Once a specific heat capacity is known, it can be used to measure flows of heat in practical experiments. In such cases:

$$q = Mc_s\Delta T$$

where M is the mass of the system. See **7-65**.

• The **heat capacity** of a system is the quantity of heat necessary to raise its temperature of the entire system through a given temperature interval.

The text uses the unit J K^{-1} for heat capacity. In this definition, the focus is not on some one-pound or a one-gram portion of the system, but rather on the whole. The heat capacity of a system is an extensive property.

The **molar heat capacity** is a heat capacity put on a per-mole basis. Like the specific heat capacity, it is an intensive property. It equals the amount of heat necessary to raise the temperature of one mole of a substance through a given temperature interval. Molar heat capacities have the unit J $K^{-1}mol^{-1}$. Molar heat capacities cannot be used unless the system is a substance or known mixture of substances (see **7-69**).

Heat capacities depend whether the heating is carried out under conditions of constant volume or constant pressure. Therefore, a **constant-pressure molar heat capacity** c_p and **constant-volume molar heat capacity** c_v are defined. The difference exists because a system at constant pressure may change volume as it absorbs heat. If it does, the system exchanges some work with its surroundings. This affects its internal energy and final temperature. At constant volume, there can be no such pressure-volume work.

At constant volume: $q_v = nc_v\Delta T$ at constant pressure: $q_p = nc_p\Delta T$

The difference between c_v and c_p is small for liquids and solids and is often ignored. Thus **7-5** and **7-7** ask for molar heat capacities of several solids without stating whether volume or pressure was constant. The distinction between c_v and c_p becomes important with gases and figures strongly in many problems (see below).

• **Always check units particularly carefully in calorimetry problems.**

Mixing up the heat capacity, an extensive variable, with the molar heat capacity or specific heat capacity, which are intensive variables, is a common error. Although the kelvin temperature scale has its zero point 273.15° lower than the zero on the Celsius scale, 1 K is still exactly the same size as 1°C. Heat capacities in J $(°C)^{-1}$ are for this reason numerically equal to heat capacities in J K^{-1}. Similarly, the J $(°C)^{-1}g^{-1}$ is the same as the J $K^{-1}g^{-1}$. See **7-11**.

The First Law of Thermodynamics

The first law of thermodynamics is a process-oriented re-statement of the principle of conservation of energy. Any change in the internal energy of a system during a process must equal the work *done on* the system plus the heat *absorbed* by the system:

$$\Delta E = w + q$$

In a general process, both heat and work cross the boundaries of the system. The two are positive when they flow into the system. This explains the emphasis on *absorption* in the statement of the first law.

What if a system is carefully insulated so that heat cannot flow across its boundaries? Heat is neither absorbed nor lost so q equals 0. If the system now does some work, it does so entirely at the expense of its internal energy. When the internal energy diminishes in a process, E_2 is smaller than E_1. This means that ΔE, which equals $E_2 - E_1$ is negative. According to the first law, if q is zero and ΔE is negative, then w has to be negative. See **7-53**.

Although both q and w depend on the path along which a change occurs, their sum ΔE is a state function and does *not* depend on the path. See **7-9** for an interesting use of the first law.

Heat Capacities of Ideal Gases

At constant volume the molar heat capacity of a monatomic ideal gas equals $3/2R$. The gas constant R is equal to 8.315 J $K^{-1}mol^{-1}$ so this c_v is 12.472 J $K^{-1}mol^{-1}$. At constant pressure, the same monatomic ideal gas has a molar heat capacity of $5/2R$, which equals 20.788 J $K^{-1}mol^{-1}$. It is 1.666 times larger. Problem **7-19b** requires these facts as do **7-17**, **7-21**, **7-51**, and **7-53**.

In the first two of these problems, the solver must recall that argon and neon are monatomic. The qualification "monatomic" is important because diatomic and polyatomic gases have larger c_v's and c_p's than monatomic gases. The quantities c_p and c_v increase with the number of atoms per molecule. Gaseous C_8H_{18} has a c_p of 327 J $K^{-1}mol^{-1}$ compared to only 29.8 J $K^{-1}mol^{-1}$ for $N_2(g)$. See **7-69**. For ideal gases, whether monatomic or polyatomic: $c_p = c_v + R$.

Thermochemistry

Thermochemistry is the study of heat effects in chemical reactions. When a reaction evolves or absorbs heat at constant pressure, then the heat lost or gained is gained the **enthalpy change** or **enthalpy** of that reaction. Conditions of constant pressure are very common in chemistry so changes in enthalpy are very important.

Enthalpy

The enthalpy is an additional, very useful state function. It is defined in terms of other state functions:

$$H = E + PV$$

In principle, the enthalpy of a system is computed by determining its internal energy and adding the product of its pressure and volume. Clearly H has the same units as E (joules). Although PV ordinarily comes out in L atm, an easy conversion (1 L atm = 101.325 J) would allow adding it to E. In practice, one never actually performs such an addition. Absolute values of E are unattainable because of the impossibility of defining a zero for energy. The emphasis is on *changes* in E and H instead. For a process (a change) in a system:

$$\Delta H = \Delta E + \Delta(PV)$$

The definition of enthalpy was deliberately chosen so that $q_p = \Delta H$. This means that the change in the enthalpy of a system during a process equals the amount of heat that the system would absorb if the process were carried out at constant pressure.

A change does *not* actually have to occur at constant pressure for the accompanying change in enthalpy to be meaningful. Enthalpy is a state function. However, the enthalpy is such a very useful function just because many processes *do* occur under constant (atmospheric) pressure. This circumstance makes q_p relatively easy to measure. The similarity of the relationships between q_p and ΔH to that between q_v and ΔE is not accidental:

$$q_v = \Delta E \quad \text{and} \quad q_p = \Delta H$$

These are key relationships in thermochemistry.

Enthalpies of Reaction

Chemical reactions cause large changes in the internal energy E and enthalpy H of systems because reactions involve the making and breaking of chemical bonds, events that greatly alter the internal potential energy. A negative enthalpy change in a reaction means that H_2, the enthalpy of the products, is *less* than H_1, the enthalpy of the reactants:

$$\Delta H = H_2 - H_1 = q_p < 0$$

Because q_p is negative the reaction gives off heat. The reaction is **exothermic**. Reactions in which heat is absorbed are **endothermic** and have a positive ΔH.

The enthalpy change of a reaction is often written immediately after the reaction on the same line. In this position, ΔH refers to the enthalpy change of the balanced chemical reaction with which it is associated. For example, in **7-65**, the enthalpy change for the combustion of isooctane is computed:

$$C_8H_{18}(l) + \tfrac{25}{2}\,O_2(g) \to 8\,CO_2(g) + 9\,H_2O(l) \quad \Delta H = -5530 \text{ kJ}$$

If all of the coefficients in the reaction are doubled, then ΔH is also doubled:

$$2\,C_8H_{18}(l) + 25\,O_2(g) \to 16\,CO_2(g) + 18\,H_2O(l) \quad \Delta H = -11060 \text{ kJ}$$

A ΔH is a change in an extensive property; the coefficients in the balanced equation give the extent (size) of the system. Problems **7-23** and **7-65** show how the quantities of reactants are considered in figuring out how much heat a reaction evolves.

If a reaction is reversed, the roles of products and reactants are exchanged. The exchange makes H_2 into H_1 and H_1 into H_2. The upshot is to change the sign of ΔH, but not its magnitude. Thus, for the reverse of the combustion of isooctane:

$$8\,CO_2(g) + 9\,H_2O(l) \to C_8H_{18}(l) + 25/2\,O_2(g) \quad \Delta H = +5530 \text{ kJ}$$

ΔH is independent of the path used to convert reactants to products. This fact leads to **Hess's law**:

• If two or more chemical reactions are added together to give an overall reaction, their ΔH's add up to the ΔH of the overall equation.

In other words, the ΔH for the conversion of a given set of reactants into a given set of products is the same regardless of whether the conversion occurs in one direct step or circuitously by a series of steps. Experiments are expensive, and calculations are cheap. Hess's law means that if a chemical reaction can be constructed (on paper) as a combination of other reactions with known ΔH's, then the ΔH of the new reaction can be calculated instead of measured. Problem **7-31** is an example showing the use of known ΔH's to get a ΔH for a new reaction.

To compute ΔE for a reaction when ΔH is known, use the relationship:

$$\Delta E = \Delta H - \Delta(PV)$$

For solids and liquids the term $\Delta(PV)$ is small. Consequently, for reactions involving exclusively solids and liquids, the difference between ΔE and ΔH can be safely neglected. For reactions involving gases, neglecting the difference is *not* safe. It is however nearly always acceptable to assume that the ideal-gas equation is followed by the gases. If so, then, at constant temperature:

$$\Delta E = \Delta H - \Delta n_{\mathrm{g}} RT \quad \text{(at constant temperature)}$$

where Δn_{g} is the number of moles of gas among the products minus the number of moles of gas among the reactants. As **7-43** and **7-65** show, the difference between ΔE and ΔH is usually only a small fraction of ΔH. **Warning.** Avoid computing $\Delta n_{\mathrm{g}} RT$ in joules and then subtracting it from a ΔH value that is in kilojoules. Check units before adding or subtracting.

Standard-State Enthalpies

The enthalpy, internal energy, and several other thermodynamic properties of pure substances, depend, often strongly, on the temperature, pressure and physical state of the substance. **Standard states** or reference states must be defined to allow the meaningful compilation and comparison of thermodynamic data. Three differently defined standard states are recognized:

- For solid and liquid substances, the standard state is the thermodynamically stable state at a pressure of 1 atm and at a specified temperature.

- For gases, the standard state is the gaseous phase at a pressure of 1 atm, at a specified temperature, and exhibiting ideal-gas behavior

- For solutes, the standard state is a 1 M solution at a pressure of 1 atm, at a specified temperature, and exhibiting ideal-solution behavior.

A superscript zero ("naught") following the symbol for a thermodynamic value identifies it as a standard-state value. The specified temperature appears as a subscript following the symbol. Thus, the standard-state enthalpy of substance X in the gaseous state at $T = 400$ K is symbolized $H^{\circ}_{400}(\mathrm{X}(g))$. The most frequent choice of temperature is 298.15 K (25°C). When no temperature is given, a temperature of 298.15 K should be assumed.

Gases cannot be prepared in their standards states because real gases do not behave ideally. Similarly, 1 M ideal solutions do not exist. Fortunately, it is always possible computationally to relate thermodynamic values of real substances to those in their idealized gaseous or dissolved states.

The Standard Enthalpy of Formation

Absolute values of H and E (and some other thermodynamic variables as well) cannot be measured because no "natural zero" exists for these functions. An arbitrary zero must be defined. For the enthalpy:

- **$H° = 0$ for each element in its most stable form at 298.15 K.**

In principle, every chemical compound can by synthesized by direct combination of its constituent chemical elements in their most stable forms. Reactions of this type are **formation reactions.** Equations representing formation reactions always have some assortment of pure elements on the left-hand side and a single compound on the right-hand side.

- **The standard molar enthalpy of formation $\Delta H_f°$ of a substance is the enthalpy change of the reaction that produces one mole of the substance from its constituent elements in their standard states.**

The units of standard enthalpies of formation are kJ mol^{-1}. For example, the formation of 1 mol of ammonium iodide is represented:

$$\tfrac{1}{2}\,N_2(g) + 2\,H_2(g) + \tfrac{1}{2}\,I_2(s) \rightarrow NH_4I(s) \quad \Delta H_f° = 201.4\ \text{kJ}$$

To emphasize that this is a formation reaction its $\Delta H°$ is subscripted with an "f" (for formation). As written, the reaction produces 1 mol of $NH_4I(s)$. Hence the $\Delta H_f°$ of ammonium iodide is 201.4 kJ mol^{-1}. This particular ΔH was obtained for reactants and products in standard states at 298.15 K. It is accordingly a standard enthalpy change, $\Delta H_f°$.

- **The standard enthalpy of formation of pure elements equals zero.**

The equation

$$O_2(g) \rightarrow O_2(g)$$

represents the formation of elemental oxygen "from its constituent elements in their standard states." The two sides of the equation are identical. The "change" is no change at all. The $\Delta H_f°$ of $O_2(g)$ is accordingly zero.

Standard enthalpies of formation of compounds are tabulated in long lists. See text Appendix D.[2] Such tables are useful because every reaction, no matter how complex, can be imagined to go by the decomposition of all of the reactants into elements

---

[2]Such data are also available on the Internet. See for example http://www.crct.polymtl.ca/FACT/websites.htm#Online Thermodynamic Data

followed by the formation of all of the products directly from those elements. As a consequence of Hess's law then:

$$\Delta H^\circ = \sum \Delta H_f^\circ \text{ (products)} - \sum \Delta H_f^\circ \text{ (reactants)}$$

To use the above equation:

1. Look up the ΔH_f°'s of all the products and reactants in text Appendix D. Remember that elements in their standard states have ΔH_f°'s of zero.

2. Multiply each substance's ΔH_f°, which is given in kJ mol^{-1}, by the number of moles of that substance shown in the balanced chemical equation.

3. Add up these numbers for the products and separately add them up for the reactants.

4. Subtract the answer for the products from the answer for the reactants.

The result is ΔH° of the reaction. It has units of kJ. Sometimes ΔH's of reactions are quoted in kJ mol^{-1} (or J mol^{-1}). This refers to the enthalpy change *per mole of reaction as written*. See **7-37**. Enthalpy changes can be computed on a per gram basis as well (**7-23**). Problems **7-35**, **7-37**, **7-65**, and **7-73** further illustrate the use of this important equation. Problem **7-41** shows how to use it to obtain a standard enthalpy of formation when an enthalpy of combustion has been measured. This is a very frequent calculation.

Bond Enthalpies

A **bond enthalpy** is the ΔH° of the reaction to break a specific type of bond in a gaseous substance. For example, the ΔH° for the reaction $Cl_2(g) \rightarrow 2\,Cl(g)$ equals the bond enthalpy for one mole of Cl—Cl bonds. This particular reaction is the breakdown of a molecule into atoms. Bond enthalpies of this type are called **enthalpies of atomization.** Bond enthalpies are obtained experimentally. They are always positive.

Bond enthalpies are fairly reproducible from one compound to another. That is, the bond enthalpy for a C—H bond is about the same (within about 10 percent) for the four C—H bonds in CH_4 and the 50 C—H bonds in $C_{30}H_{50}$.

Big differences in the enthalpies of C-to-C, C-to-O, N-to-C, and other bonds do occur, but are attributed to multiple bonding. Thus text Table 7-3[3] gives different enthalpies for C—N, C=N, and C≡N bonds.

---

[3]Text page 228.

Tables of bond enthalpies allow estimates of the ΔH_f°'s of gaseous compounds (**7-45**) and estimates of the ΔH°'s of gas-phase reactions (**7-47**). Imagine that a reaction proceeds by the complete disruption of all chemical bonds in the reactants to give isolated gaseous atoms and that this atomization is followed by recombination of the atoms into products. The total bond enthalpy of the reactants is the enthalpy change in the atomization. The total bond enthalpy of the products is, similarly, the enthalpy to break all the bonds among the products. The *change* in bond enthalpy is the bond enthalpy of the products minus the bond enthalpy of the reactants. It is the negative of ΔH° of the reaction. See **7-77**.

Reversible Processes in Ideal Gases

Changes in state functions depend only on the initial and final state and not at all on the path by which a process or reaction takes place. Sometimes, however, knowledge of the path of a process can help in computing ΔH, ΔE, and the changes in other state variables. To compute values for q and w, which *do* depend on the path, a good specification of the path is needed. Several technical terms describe paths:

• **Isothermal.** An isothermal process goes on at constant temperature; ΔT equals zero in an isothermal change. The internal energy of an ideal gas depends only on its temperature. If T is constant, $\Delta E = 0$ and:

$$q = -w \qquad \text{(isothermal process, ideal gas)}$$

In addition, ΔH is zero in an isothermal change of an ideal gas. For non-ideal gases, liquids, and solids, ΔE and ΔH may be non-zero even if the change is isothermal.

• **Isothermal and Reversible.** An isothermal and reversible process goes on at constant temperature by a series of infinitesimally small steps. The system is always at equilibrium, and at any time the direction of the process can be changed. For an ideal gas in such a process:

$$q = -w = nRT \ln\left(\frac{V_2}{V_1}\right) \qquad \text{(isothermal reversible process, ideal gas)}$$

Boyle's law states that $P_1 V_1 = P_2 V_2$ for an ideal gas at constant temperature. Solve for V_2 and substitute into the previous equation:

$$q = -w = nRT \ln\left(\frac{P_1}{P_2}\right) \qquad \text{(isothermal reversible process, ideal gas)}$$

In an isothermal expansion, V_2 is bigger than V_1. The gas expands and absorbs heat from the surroundings as it does so. It simultaneously performs an exactly equivalent amount of work on the surroundings. In that way ΔE remains zero.

• **Adiabatic.** In an adiabatic process, heat cannot transfer into or out of the system: q equals zero in all adiabatic changes. Any work that the system performs comes at the expense of its internal energy; any work done on the system adds to its internal energy: $\Delta E = w$.

• **Adiabatic and Reversible.** For an ideal gas it can be shown that:

$$P_1 V_1^{\gamma} = P_2 V_2^{\gamma} \quad \text{(adiabatic reversible process, ideal gas)}$$

where γ is the ratio of c_p to c_v. In addition:

$$T_1 V_1^{\gamma-1} = T_2 V_2^{\gamma-1} \quad \text{(adiabatic reversible process, ideal gas)}$$

The solutions to **7-19** and **7-21** show how to avoid a common pitfall with these equations. They apply only to an *ideal* gas undergoing an *adiabatic* and *reversible* change. All three qualifications must be met. Problems **7-53**, **7-69**, and **7-71** show applications of these equations to more complex cases.

Detailed Solutions to Odd-Numbered Problems

7-1 The work done *on* a gas in a change of volume at constant pressure is given by $w = -P_{ext}\Delta V$. The problem states values for the external pressure and for the final and initial volumes. Substituting gives:

$$w = -P_{ext}\Delta V = -(50.0 \text{ atm})(974 \text{ L} - 542 \text{ L}) = \boxed{-2.16 \times 10^4 \text{ L atm}}$$

where, as ever, the change in a quantity (in this case the volume) is the final value minus the initial. To convert to joules, multiply by the proper unit factor:[4]

$$w = -2.16 \times 10^4 \text{ L atm} \times \left(\frac{101.325 \text{ J}}{1 \text{ L atm}}\right) = \boxed{-2.19 \times 10^6 \text{ J}}$$

Tip. Negative work is performed on the nitrogen; the nitrogen does positive work on its surroundings. The difference is a matter of point of view. Do we put ourselves with the gas looking at the surroundings, or in the surroundings looking at the gas?

7-3 A ball of mass M falls a distance Δh under the influence of gravity. It experiences a change in potential energy equal to $Mg\Delta h$, where Δh is the change in height and g is the acceleration of gravity. The (non-bouncing) ball stops dead when it hits the ground. According to the problem, the total energy of the ball does not change at impact. All of the potential energy instead is converted in internal energy that goes to heat up the ball. This is expressed mathematically:

$$Mc_s\Delta T + Mg\Delta h = \Delta(\text{total energy}) = 0$$

---

[4]From text page 208.

where c_s is the specific heat capacity of the ball. Cancel out the M's and solve for Δh:

$$\Delta h = -\frac{c_s \Delta T}{g}$$

In this problem, ΔT equals 1.00°C (which equals 1.00 K) and c_s equals 0.850×10^3 J K^{-1}kg^{-1}. Note that c_s is put on a per-kilogram basis to aid the cancellation of units. Also, g is 9.81 m s^{-2}. Substituting gives:

$$\Delta h = -\frac{(0.850 \times 10^3 \text{ J K}^{-1}\text{kg}^{-1})(1.00 \text{ K})}{9.81 \text{ m s}^{-2}} = -86.6 \text{ J kg}^{-1}\text{m}^{-1}\text{s}^2$$

By its definition a joule equals a kg m^{2}s^{-2}. Therefore, in the above cluster of units all but the meter cancel out: Δh is -86.6 m. The negative sign simply means that the final height of the ball is less than the initial height. The ball falls a distance of $\boxed{86.6 \text{ m}}$ down.

7-5 The molar heat capacity of a substance equals its specific heat capacity multiplied by its molar mass. Here is a sample calculation for lithium:

$$c_p = c_s \mathcal{M} = 3.57 \text{ J K}^{-1}\text{g}^{-1}(6.94 \text{ g mol}^{-1}) = 24.8 \text{ J K}^{-1}\text{mol}^{-1}$$

The full set of values in the group:

Li(s)	Na(s)	K(s)	Rb(s)	Cs(s)	
24.8	28.3	29.6	31.0	32.2	J K^{-1}mol^{-1}

Beyond sodium there is a steady increase of about 1.3 J K^{-1}mol^{-1} for every element. Extrapolation of the trend assigns francium a molar heat capacity of about $\boxed{33.5 \text{ J K}^{-1}\text{mol}^{-1}}$. Although the trend is small, it is distinct. Indeed, the molar heat capacities of the metallic elements are remarkably constant. This constancy is the law of Dulong and Petit (see **7-7**).

7-7 Again, the molar heat capacity of a substance equals its specific heat capacity multiplied by its molar mass. The calculations proceed as in **7-5** with the results:

Ni(s)	Zn(s)	Rh(s)	W(s)	Au(s)	U(s)	
26.1	25.4	25.0	24.3	25.4	27.6	J K^{-1}mol^{-1}

7-9 a) During the heating process, heat flows from the surroundings to the system. Therefore, $\boxed{q \text{ is positive}}$. Since the container is rigid, it neither expands nor contracts. Hence ΔV is zero. Consequently, no pressure-volume work is performed on the system. No other type of work is possible, so $\boxed{w = 0}$. Then, by the first law, $\boxed{\Delta E \text{ is positive}}$.

b) During the cooling process, the heat absorbed by the system is negative $\boxed{q < 0}$. No work can be done on the system $\boxed{w = 0}$. The energy of the system is therefore lowered $\boxed{\Delta E < 0}$.

c) Since no work was done in either step 1 or step 2, $\boxed{w_1 + w_2 = 0}$. Nothing in the problem suggests that the system is in the same thermodynamic state after it is cooled back to its original temperature. All that is stated is that the temperature is the same. Other variables, such as the internal energy, might be greatly affected by the heating-cooling cycle. Hence $\Delta E_1 + \Delta E_2$ is *not* necessarily zero.

Tip. A trap in this problem is to assume, without justification, that the system in the container is ideal gas (for which the internal energy depends only on the temperature). A related trap is to assume that any changes brought on by the heating are exactly reversed by the cooling. This is not the case when an egg is boiled so why should it be true here? All that can be said is $\boxed{\Delta E_1 + \Delta E_2 = q_1 + q_2}$. The two sides of this equation could be positive, negative or zero.

7-11 Let the system under consideration consist of two sub-systems: the metal and the water. If the mixing of hot metal and cool water takes place in a well-insulated container (which prevents leaks of heat), then the heat absorbed by the system equals zero. The system is the sum of the two sub-systems. Therefore:

$$q_{\text{sys}} = 0 = q_{\text{m}} + q_{\text{w}}$$

For both sub-systems, the amount of heat gained equals the specific heat capacity times the mass times the temperature change:

$$q_{\text{m}} + q_{\text{w}} = M_{\text{w}} c_{s,\text{w}} \Delta T_{\text{w}} + M_{\text{m}} c_{s,\text{m}} \Delta T_{\text{m}} = 0$$

Solving for the specific heat capacity of the metal:

$$c_{s,\text{m}} = \frac{-M_{\text{w}} c_{s,\text{w}} \Delta T_{\text{w}}}{M_{\text{m}} \Delta T_{\text{m}}} = -\frac{(100.0 \text{ g}) \, 4.18 \text{ J K}^{-1}\text{g}^{-1}(6.39°\text{C})}{(61.0 \text{ g})(-93.61°\text{C})} = \boxed{0.468 \text{ J K}^{-1}\text{g}^{-1}}$$

Tip. We do not bother to convert °C to K. A change of one degree Celsius is identical to a change of one kelvin. The Kelvin and Celsius scales have the same size increments and differ only in the location of their zeros.

7-13 Body 1 and body 2 are originally at different temperatures. They are brought into thermal contact with each other and held in thermal isolation from other objects. Then:

$$q_1 + q_2 = M_1 c_{s1} \Delta T_1 + M_2 c_{s2} \Delta T_2 = 0$$

If the masses of the two bodies are equal, then $M_1 = M_2$, and:

$$c_{s1} \Delta T_1 = -c_{s2} \Delta T_2$$

$$\boxed{c_{s1}/c_{s2} = -\Delta T_2/\Delta T_1}$$

This equation shows that the specific heat capacities of the two bodies are inversely proportional to the temperature changes they undergo in this experiment.

7-15 The difference in temperature ΔT between water at its boiling point and melting point is 100°C. The heat needed to bring 1.00 g of water at 0°C to 100°C equals:

$$q = Mc_s\Delta T = (1.00 \text{ g})\left(4.18 \text{ J} (°C)^{-1} \text{g}^{-1}\right)(100°C) = 418 \text{ J}$$

The amount of heat needed to melt 1.00 g of ice is, according to the statement of Lavoisier and Laplace, 3/4 of this amount or $\boxed{314 \text{ J}}$. More recent experiments set the amount of heat to melt 1.00 g of ice at 333 J.

7-17 The 0.500 mol of neon expands against a constant pressure of 0.100 atm. Define the system as the neon. Before the expansion, the volume of the system is 11.20 L (calculated using the ideal-gas equation with n equal 0.500 mol at 1.00 atm and 273 K). The expanded volume is 43.08 L (calculated from the ideal-gas equation with $P = 0.200$ atm, $n = 0.500$ mol, and $T = 210$ K). The gas expands against a constant pressure (of 0.100 atm). The work done on the system is:

$$w = -P_{ext}\Delta V = -0.100 \text{ atm}(43.08 - 11.20) \text{ L} = \boxed{-3.19 \text{ L atm}}$$

The gas cools from 273 to 210 K. Since it is the ideal monatomic gas, the change in its internal energy is directly proportional to the change in its temperature; the constant of proportionality is $n(\frac{3}{2})R$, the heat capacity at constant volume:

$$\Delta E = nc_v\Delta T = n\left(\frac{3}{2}R\right)\Delta T$$

Substituting gives:

$$\Delta E = 0.500 \text{ mol} \left(\frac{3}{2} 0.08206 \text{ L atm mol}^{-1}\text{K}^{-1}\right)(-63 \text{ K}) = \boxed{-3.88 \text{ L atm}}$$

By the first law:

$$q = \Delta E - w = -3.88 \text{ L atm} - (-3.19 \text{ L atm}) = \boxed{-0.69 \text{ L atm}}$$

The three answers can also be given in joules (1 L atm = 101.325 J):

$$w = -323 \text{ J} \qquad \Delta E = -393 \text{ J} \qquad q = -70 \text{ J}$$

7-19 a) The statement of the problem gives the initial quantity (2.00 mol), pressure (3.00 atm), and temperature (350 K) of the ideal monatomic gas. The initial volume of the gas is $V = nRT/P = 19.15$ L. The final volume is *twice* this original volume or $\boxed{38.3 \text{ L}}$. The change in volume ΔV equals $38.30 - 19.15 = 19.15$ L.

b) The adiabatic expansion occurs against a *constant* pressure of 1.00 atm. Under that circumstance, the work done on the gas is:

$$w = -P\Delta V = -1.00(19.15) \text{ L atm} \times \left(\frac{101.325 \text{ J}}{1 \text{ L atm}}\right) = \boxed{-1.94 \times 10^3 \text{ J}}$$

The expansion is adiabatic so $\boxed{q = 0}$ by definition, and:

$$\Delta E = q + w = 0 - 1.94 \times 10^3 \text{ J} = \boxed{-1.94 \times 10^3 \text{ J}}$$

c) Any change in the internal energy of an ideal gas causes a change in temperature in direct proportion:

$$\Delta E = nc_v\Delta T$$

Solving for ΔT and substituting the various values:

$$\Delta T = \frac{\Delta E}{nc_v} = \frac{-1.94 \times 10^3 \text{ J}}{2.00 \text{ mol}(3/2)8.315 \text{ J K}^{-1}\text{mol}^{-1}} = -77.8 \text{ K}$$

Thus, T_2, the final temperature, is $T_1 + \Delta T = 350 + (-77.8) = \boxed{272 \text{ K}}$.

7-21 The system consists of the 6.00 mol of argon. The change in internal energy of this monatomic gas (assuming ideality) is:

$$\Delta E = nc_v\Delta T = (6.00 \text{ mol})\left(\frac{3}{2}8.315 \text{ J K}^{-1}\text{mol}^{-1}\right)(150 \text{ K}) = 11.2 \times 10^3 \text{ J}$$

The change is adiabatic which means that $\boxed{q = 0}$. From the first law:

$$w = \Delta E - q = 11.2 \times 10^3 \text{ J} - 0 = \boxed{+11.2 \times 10^3 \text{ J}}$$

The work done on the argon is $\boxed{11.2 \times 10^3 \text{ J}}$, *all* of which goes to increase its internal energy.

7-23 The balanced equation tells the enthalpy change taking place during the production or consumption of a specific number of moles of product or reactant. All that is necessary is to put these enthalpy changes on a basis of mass.

a) $\quad \Delta H = \dfrac{-828 \text{ kJ}}{2 \text{ mol Na}_2\text{O}} \times \left(\dfrac{1.00 \text{ mol Na}_2\text{O}}{62.0 \text{ g Na}_2\text{O}}\right) = \boxed{-6.68 \text{ kJ g}^{-1}}$

b) $\quad \Delta H = \dfrac{302 \text{ kJ}}{1 \text{ mol MgO}} \times \left(\dfrac{1.00 \text{ mol MgO}}{40.31 \text{ g MgO}}\right) = \boxed{7.49 \text{ kJ g}^{-1}}$

c) $\quad \Delta H = \dfrac{33.3 \text{ kJ}}{2 \text{ mol CO}} \times \left(\dfrac{1.00 \text{ mol CO}}{28.01 \text{ g CO}}\right) = \boxed{0.594 \text{ kJ g}^{-1}}$

7-25 Only 119.0 J of the measured 121.3 J of heat comes from the reaction of the 0.00288 mol of $Br_2(l)$. The rest of the heat (2.34 J) is added mechanically[5] by breaking the capsule and stirring the liquid. The amount of heat evolved from 1.00 mol of $Br_2(l)$ is:

$$1.00 \text{ mol} \times \left(\dfrac{119.0 \text{ J}}{2.88 \times 10^{-3} \text{ mol}}\right) = \boxed{41.3 \times 10^3 \text{ J}}$$

7-27 The vaporization is $CO(l) \rightarrow CO(g)$. For this change, ΔH_{vap} is 6.04 kJ mol^{-1}. The following series of conversions provides the answer:

$$2.38 \text{ g CO} \times \left(\dfrac{1 \text{ mol CO}}{28.01 \text{ g CO}}\right) \times \left(\dfrac{6.04 \text{ kJ}}{1 \text{ mol CO}}\right) = \boxed{0.513 \text{ kJ}}$$

7-29 The 36.0 g ice cube contains 2.00 mol of H_2O because 1.00 mol of water equals 18.0 g of water. The cube is put in contact with 20.0 mol of 20°C water. At -10°C, the ice is well below its melting point. It must heat up before it can start to melt. Warming the ice from -10°C to 0°C absorbs:

$$q = nc_p\Delta T = (2.00 \text{ mol})(38 \text{ J K}^{-1}\text{mol}^{-1})(10 \text{ K}) = 760 \text{ J}$$

Melting the ice at 0°C gives water at 0°C and absorbs:

$$q = n\Delta H_{fus} = (2.00 \text{ mol})(6007 \text{ J mol}^{-1}) = 12\,014 \text{ J}$$

On the other hand, cooling 360 g (20.0 mol) of water from 20°C to 0°C would absorb:

$$q = nc_p(T_f - T_i) = (20.0 \text{ mol})(75 \text{ J K}^{-1}\text{mol}^{-1})(-20 \text{ K}) = -3.0 \times 10^4 \text{ J}$$

This result can be rephrased: cooling 20.0 mol of water from 20°C to 0°C requires *removal* of $+3.0 \times 10^4$ J. Since the ice cube absorbs only 12774 J by warming up and

---

[5]See text Figure 7-7, text page 211.

then melting to liquid water at 0°C, T_f, the final temperature of the mixture, must be above 0°C.

No heat is lost to the surroundings. The heat absorbed in warming and melting the ice, and then warming the melt-water to the actual T_f can therefore be added with the heat absorbed in cooling the 20.0° water to T_f to equal zero:

$$\underbrace{12\,774 \text{ J}}_{q \text{ for ice}} + \underbrace{(2.00 \text{ mol})(75 \text{ J K}^{-1}\text{mol}^{-1})(T_f - 0)}_{q \text{ for meltwater}}$$

$$+ \underbrace{(20.0 \text{ mol})(75 \text{ J K}^{-1}\text{mol}^{-1})(T_f - 20.0)}_{q \text{ for warm water}} = 0$$

Solving gives $T_f = \boxed{10.4°C}$.

Tip. A source of difficulty in this problem is the wrong concept that ice is always be at 0°C. Like any other material, ice comes to the temperature of its surroundings.

7-31 Multiply the equation and ΔH for the combustion of methane by 2. Reverse the equation for the combustion of ketene and multiply its ΔH by -1:

$$2\,CH_4(g) + 4\,O_2(g) \rightarrow 2\,CO_2(g) + 4\,H_2O(g) \qquad \Delta H = -1604.6 \text{ kJ}$$
$$2\,CO_2(g) + H_2O(g) \rightarrow CH_2CO(g) + 2\,O_2(g) \qquad \Delta H = 981.1 \text{ kJ}$$

Adding these two equations gives the desired equation:

$$2\,CH_4(g) + 2\,O_2(g) \rightarrow CH_2CO(g) + 3\,H_2O(g) \qquad \Delta H = \boxed{-623.5 \text{ kJ}}$$

By Hess's law, the enthalpy change of the total reaction equals the sum of the enthalpy changes of the two reactions that are added.

Tip. How does one know which equations to reverse or double in problems like this? Manipulate to put the correct number of moles of each substance on the correct side of the final equation. Thus, the ketene equation had to be reversed because ketene is among the products in the target equation.

7-33 The conversion $C(gr) \rightarrow C(dia)$ is endothermic (positive ΔH). Therefore, one pound of diamonds contains more enthalpy than one pound of graphite. Both diamond and graphite give the same product (carbon dioxide) when burned. When burned, the pound of $\boxed{\text{diamonds}}$ will give off more heat.

7-35 A reaction enthalpy is calculated by summing the enthalpies of formation of the products and subtracting the enthalpies of formation of the reactants:

$$N_2H_4(l) + 3\,O_2(g) \rightarrow 2\,NO_2(g) + 2\,H_2O(l)$$

$$\Delta H° = 2 \underbrace{(33.18)}_{NO_2(g)} + 2 \underbrace{(-285.83)}_{H_2O(l)} - 1 \underbrace{(50.63)}_{N_2H_4(l)} - 3 \underbrace{(0)}_{O_2(g)} = \boxed{-555.93 \text{ kJ}}$$

In the preceding equation, all of the $\Delta H_f°$'s are in kJ mol^{-1}. All are multiplied by the number of moles of the substance represented in the balanced equation.

7-37 a) As in the preceding:

$$2\,ZnS(s) + 3\,O_2(g) \to 2\,ZnO(s) + 2\,SO_2(g)$$

$$\Delta H° = 2 \underbrace{(-348.28)}_{ZnO(s)} + 2 \underbrace{(-296.83)}_{SO_2(g)} - 2 \underbrace{(-205.98)}_{ZnS(s)} - 3 \underbrace{(0)}_{O_2(g)} = \boxed{-878.26 \text{ kJ}}$$

b) Compute the chemical amount of ZnS (in moles) and multiply it by the molar $\Delta H°$ to get the amount of heat absorbed in the roasting of the 3.00 metric tons of ZnS. It is known that 2 mol of ZnS(s) has a dHo of -878.26 kJ. Hence:

$$q_p = \Delta H° = 3.0 \text{ metric ton ZnS} \times \left(\frac{10^6 \text{ g}}{1 \text{ metric ton}}\right) \times \left(\frac{1 \text{ mol}}{97.456 \text{ g}}\right)$$

$$\times \left(\frac{-878.26 \text{ kJ}}{2 \text{ mol ZnS}}\right) = \boxed{-1.35 \times 10^7 \text{ kJ}}$$

7-39 a) The balanced equation is $CaCl_2(s) \to Ca^{2+}(aq) + 2\,Cl^-(aq)$. Combine the enthalpies of formation as follows:

$$\Delta H° = 2 \underbrace{(-167.16)}_{Cl^-(aq)} - 1 \underbrace{(542.83)}_{Ca^{2+}(aq)} - 1 \underbrace{(-795.8)}_{CaCl_2(s)} = \boxed{-81.4 \text{ kJ}}$$

b) Compute $\Delta H°$ for the dissolution of 20.0 g of $CaCl_2(s)$:

$$\Delta H° = 20.0 \text{ g CaCl}_2 \times \left(\frac{1 \text{ mol CaCl}_2}{110.98 \text{ g CaCl}_2}\right) \times \left(\frac{-81.35 \text{ kJ}}{1 \text{ mol CaCl}_2}\right) = -14.66 \text{ kJ}$$

The process of dissolution absorbs -14.66 kJ. The immediate surroundings of the dissolution (the water) therefore must absorb $+14.66$ kJ. The temperature change of the water equals the heat it absorbs divided by its heat capacity:

$$\Delta T = \frac{q}{c_p M} = \frac{14.66 \times 10^3 \text{ J}}{418 \text{ J K}^{-1}} = 35.1 \text{ K} = 35.1°C$$

The final temperature is $T_f = 20.0°C + 35.1°C = \boxed{55.1°C}$.

7-41 The balanced equation is $C_6H_{12}(l) + 9\,O_2(g) \rightarrow 6\,CO_2(g) + 6\,H_2O(l)$. Set up a calculation of a standard enthalpy of this combustion reaction in terms of standard enthalpies of formation of the products and reactants. The standard enthalpy of combustion is known, but one of the ΔH_f°'s is not known:

$$\Delta H^\circ = -3923.7 \text{ kJ} = 6 \underbrace{(-393.51)}_{CO_2(g)} + 6 \underbrace{(-285.83)}_{H_2O(g)} - 1\,\Delta H_f^\circ\,(C_6H_{12}(l)) - 9 \underbrace{(0)}_{O_2(g)}$$

The standard enthalpies of formation are all in kJ mol^{-1}. All are therefore multiplied by the number of moles of each substance appearing in the balanced equation. Solving gives the ΔH_f° of liquid cyclohexane as $\boxed{-152.3 \text{ kJ mol}^{-1}}$.

7-43 a) The equation is: $\boxed{C_{10}H_8(s) + 12\,O_2(g) \rightarrow 10\,CO_2(g) + 4\,H_2O(l)}$.

b) The amount of heat evolved $(-q)$ in the combustion of 0.6410 g of naphthalene was observed to equal 25.79 kJ. Since the combustion was performed at constant volume, no work was done on the system $(w = 0)$. Therefore, $\Delta E = q + w = -25.79 \text{ kJ} + 0 = -25.79$ kJ. Put ΔE on a molar basis:

$$\Delta E = \left(\frac{-25.79 \text{ kJ}}{0.6410 \text{ g } C_{10}H_8}\right) \times \left(\frac{128.17 \text{ g } C_{10}H_8}{1 \text{ mol } C_{10}H_8}\right) = -5157 \text{ kJ mol}^{-1}$$

The temperature is essentially 25°C both before and after the reaction. Hence, the ΔE° in the combustion of 1.000 mol of naphthalene is $\boxed{-5157 \text{ kJ}}$.

c) To calculate ΔH° use the definition: $\Delta H^\circ = \Delta E^\circ + \Delta(PV)$. If the gases behave ideally, then $\Delta(PV) = \Delta(nRT)$ so that:

$$\Delta H^\circ = \Delta E^\circ + \Delta(nRT) = \Delta E^\circ + (\Delta n_g)RT$$

for this change at constant temperature. The value of Δn is calculated by making the valid assumption that the change in the amount of gases is the only factor that contributes significantly to $\Delta(PV)$. The combustion of 1.000 mol of naphthalene produces 10.00 mol of gas, but requires 12.00 mol of gas. Hence:

$$(\Delta n_g)RT = (-2.00 \text{ mol})(8.315 \text{ J K}^{-1}\text{mol}^{-1})(298.15 \text{ K}) = -4.96 \text{ kJ}$$

$$\Delta H^\circ = \Delta E^\circ + (\Delta n_g)RT = -5157 - 4.96 = \boxed{-5162 \text{ kJ}}$$

d) Use $\Delta H^\circ = \sum \Delta H_f^\circ \text{ (products)} - \sum \Delta H_f^\circ \text{ (reactants)}$ to write:

$$\Delta H^\circ = -5162 \text{ kJ} = 10 \underbrace{(-393.51)}_{CO_2(g)} + 4 \underbrace{(-285.83)}_{H_2O(l)} - 12 \underbrace{(0)}_{O_2(g)} - 1\,\Delta H_f^\circ\,(\text{naphthalene}(s))$$

where each term on the right consists of a ΔH_f° in kJ mol^{-1} multiplied by the number of moles in the balanced equation. Solving gives the ΔH_f° of naphthalene as $\boxed{+84 \text{ kJ mol}^{-1}}$.

7-45 First, write an equation for the formation of CCl$_3$F(g) from the "naked atoms":

$$C(g) + 3\,Cl(g) + F(g) \rightarrow CCl_3F(g)$$

From the average bond enthalpies[6] estimate the ΔH° for this reaction as:

$$\Delta H^\circ = 1\underbrace{(-441)}_{\text{C-F}} + 3\underbrace{(-328)}_{\text{C-Cl}} = -1425 \text{ kJ}$$

Next, write equations that show the preparation of the naked atoms from the elements in their standard states. Each of these has a corresponding atomization enthalpy derived from the data in the text:[7]

$$
\begin{aligned}
C(s) &\rightarrow C(g) & \Delta H^\circ &= 716.7 \text{ kJ} \\
3/2\,Cl_2(g) &\rightarrow 3\,Cl(g) & \Delta H^\circ &= 365.1 \text{ kJ} \\
1/2\,F_2(g) &\rightarrow F(g) & \Delta H^\circ &= 79.0 \text{ kJ}
\end{aligned}
$$

Combine the four reactions to arrive at the formation reaction of CCl$_3$F(g) from the elements in their standard states:

$$
\begin{aligned}
C(g) + 3\,Cl(g) + F(g) &\rightarrow CCl_3F(g) & \Delta H^\circ &= -1425 \text{ kJ} \\
C(s) &\rightarrow C(g) & \Delta H^\circ &= 716.7 \text{ kJ} \\
3/2\,Cl_2(g) &\rightarrow 3\,Cl(g) & \Delta H^\circ &= 365.1 \text{ kJ} \\
1/2\,F_2(g) &\rightarrow F(g) & \Delta H^\circ &= 79.0 \text{ kJ}
\end{aligned}
$$

$$C(s) + 3/2\,Cl_2(g) + 1/2\,F_2(g) \rightarrow CCl_3F(g) \qquad \Delta H^\circ = \boxed{-264 \text{ kJ}}$$

7-47 The reaction is the combustion of propane in oxygen:

$$C_3H_8(g) + 5\,O_2(g) \rightarrow 3\,CO_2(g) + 4\,H_2O(g)$$

As this reaction proceeds, bonds are both broken and formed. Broken are 2 mol of C—C bonds, 8 mol of C—H bonds, and 5 mol of O=O double bonds. Formed are

---

[6]Text Table 7-3, text page 228.

[7]The atomization enthalpies in Table 7-3 are per mole of atom formed. Accordingly, each is multiplied by the number of moles of the atom involved.

6 mol of C=O double bonds and 8 mol of O—H bonds. The net enthalpy change equals:

$$\Delta H \approx 6 \underbrace{(728)}_{C=O} + 8 \underbrace{(463)}_{O-H} - 5 \underbrace{(498)}_{O=O} - 8 \underbrace{(413)}_{C-H} - 2 \underbrace{(348)}_{C-C} = \boxed{-1.58 \times 10^3 \text{ kJ}}$$

Tip. The answer -1.582×10^3 kJ is correct according to the rules for significant digits.[8] In view of the fact that bond enthalpies are only approximately constant[9] -1.58×10^3 kJ is a more sensible answer.

7-49 The Lewis structures are:

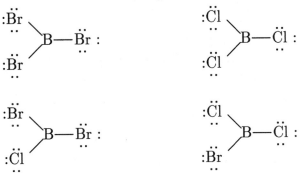

The reaction breaks but then reforms a mole of boron-bromine bonds and a mole of boron-chlorine bonds. Therefore, the sum of the average bond enthalpies in the products equals the sum of the average bond enthalpies in the reactants.

Tip. Boron tribromide and boron trichloride are octet-deficient molecules.

7-51 The system is the 2.00 mol of ideal gas. In an isothermal change, no change of temperature occurs. ($\Delta T = 0$). The internal energy of an ideal gas depends only on its temperature which means that $\boxed{\Delta E = 0}$. As for the enthalpy:

$$\Delta H = \Delta E + \Delta(PV) = 0 + \Delta(nRT) = 0 + nR\Delta T = \boxed{0}$$

The expansion is reversible. Hence:

$$w = -nRT \ln \left(\frac{V_2}{V_1}\right) = -(2.00 \text{ mol}) \left(\frac{8.315 \text{ J}}{\text{mol K}}\right) (298 \text{ K}) \ln \left(\frac{36.00}{9.00}\right) = \boxed{-6.87 \text{ kJ}}$$

The first law requires that $\Delta E = q + w$. Hence q equals $\boxed{+6.87 \text{ kJ}}$.

[8]Text page A-6.
[9]Text page 303.

7-53 During any adiabatic process $q = 0$. During this *reversible* adiabatic expansion of an ideal gas:

$$T_1 V_1^{\gamma-1} = T_2 V_2^{\gamma-1}$$

where γ is c_p/c_v and the subscripts refer the initial and final states of the gas. In this problem, V_1 is 20.0 L, V_2 is 60.0 L, γ is 5/3, and T_1 is 300 K. Solving for T_2 and substituting gives:

$$T_2 = T_1 \left(\frac{V_1}{V_2}\right)^{\gamma-1} = (300 \text{ K}) \left(\frac{20.0 \text{ L}}{60.0 \text{ L}}\right)^{2/3} = 144.22 \text{ K} = \boxed{144 \text{ K}}$$

Meanwhile, the ΔE of the ideal gas depends solely on its change in temperature:

$$\Delta E = n c_v \Delta T = (2.00 \text{ mol}) \left(\frac{3}{2} 8.315 \text{ J K}^{-1}\text{mol}^{-1}\right)(-155.78 \text{ K}) = \boxed{-3.89 \text{ kJ}}$$

This number also equals w, the work done on the gas, because $\Delta E = q + w$ and q is zero in this process. Finally, ΔH of an ideal gas also depends entirely on ΔT:

$$\Delta H = n c_p \Delta T = (2.00 \text{ mol}) \left(\frac{5}{2} 8.315 \text{ J K}^{-1}\text{mol}^{-1}\right)(-155.78 \text{ K}) = \boxed{-6.48 \text{ kJ}}$$

Tip. Notice that $\Delta H = \gamma \Delta E$ for this reversible adiabatic process.

7-55 The law of Dulong and Petit states that all metals have a molar heat capacity of approximately 25 J K^{-1}mol^{-1}. The molar heat capacity equals the specific heat capacity of a substance multiplied by its molar mass. Hence:

$$c = c_s \mathcal{M} \approx 25 \text{ J K}^{-1}\text{mol}^{-1}$$

The experimental specific heat capacity of indium is 0.233 J K^{-1}g^{-1}. A molar mass of 76 g mol^{-1} combines with this number to give a molar heat capacity for indium of only 17.7 J K^{-1}mol^{-1}. This violates the law of Dulong and Petit badly. The modern value of $\mathcal{M}$ for indium (114.8 g mol^{-1}) fits much better in the law of Dulong and Petit.[10]

7-57 a) The system is the 2.00 mol of argon gas. The work done *on* the system is $-P\Delta V$. Since the gas is ideal and P is constant, $P\Delta V = nR\Delta T$ for the system. In this case ΔT is given as -100 K. Then:

$$w = -nR\Delta T = -(2.00 \text{ mol})(8.315 \text{ J K}^{-1}\text{mol}^{-1})(-100 \text{ K}) = \boxed{+1.66 \times 10^3 \text{ J}}$$

---

[10] For solids and liquids the distinction between c_p and c_v is unimportant, especially in an approximate relationship. For this reason there is no p or v subscript on c in this problem.

b) The process goes on at constant pressure so the heat absorbed is q_p:

$$q_p = nc_p\Delta T = (2.00 \text{ mol})\left(\frac{5}{2} \, 8.315 \text{ J K}^{-1}\text{mol}^{-1}\right)(-100 \text{ K}) = \boxed{-4.16 \times 10^3 \text{ J}}$$

c) Use the first law of thermodynamics: $\Delta E = q + w = -4157 + 1663 = -2494$ J. This rounds off to $\boxed{-2.49 \times 10^3 \text{ kJ}}$. Note the use of un-rounded answers from parts a and b in the addition.

d) The ΔH of a system always equals q_p. Hence, ΔH is $\boxed{-4.16 \text{ kJ}}$.

7-59 Because frictional losses and leaks do not occur, the amount of work done by the gas on the paddle mechanism equals the negative of the work absorbed by the gas:

$$w = -(-P\Delta V) = (1.00 \text{ atm})(13.00 - 5.00 \text{ L}) = 8.00 \text{ L atm}$$

This work amounts to 811 J because 1 L atm is 101.325 J. All of this work is converted to heat in the 1.00 L of water. Hence the heat absorbed by the water is +811 J. At the given density, the 1.00 L of water weighs 1.00×10^3 g. Therefore:

$$\Delta T = \frac{q}{c_s M} = \frac{811 \text{ J}}{(4.18 \text{ J K}^{-1}\text{g}^{-1})(1.00 \times 10^3 \text{ g})} = \boxed{0.194 \text{ K}}$$

7-61 Use the molar mass of glucose ($C_6H_{12}O_6$) as a unit factor to obtain the chemical amount of glucose in the candy bar. Then use the molar enthalpy of combustion of glucose as a unit factor to obtain the heat absorbed:

$$q = 14.3 \text{ g } C_6H_{12}O_6 \times \left(\frac{1 \text{ mol } C_6H_{12}O_6}{180.16 \text{ g } C_6H_{12}O_6}\right) \times \left(\frac{-2820 \text{ kJ}}{1 \text{ mol } C_6H_{12}O_6}\right) = -223.8 \text{ kJ}$$

The heat absorbed by the surroundings of the reaction (which are the person's body) therefore equals +223.8 kJ. When this amount of heat is absorbed by 50 kg of water:

$$\Delta T = \frac{q}{c_s M} = \frac{223.8 \times 10^3 \text{ J}}{(4.18 \text{ J K}^{-1}\text{g}^{-1})(50 \times 10^3 \text{ g})} = \boxed{1.1 \text{ K}}$$

7-63 Determine which liquid is a better coolant near the boiling point by comparing their specific heat capacities. $He(l)$ absorbs 4.25 J of heat per gram as it heats up by 1 K. $N_2(l)$ absorbs only 1.95 J of heat per gram as it warms by the same amount. Therefore, $\boxed{He(l)}$ is a better coolant near the boiling point.

At their boiling point, the two liquids cool by vaporization; $\boxed{N_2(l)}$ is better because it absorbs much more heat per gram in vaporization than $He(l)$.

7-65 a) The combustion of isooctane is represented:

$$C_8H_{18}(l) + \tfrac{25}{2}O_2(g) \rightarrow 8\,CO_2(g) + 9\,H_2O(l)$$

b) The combustion of 0.542 g of isooctane is exothermic (isooctane is a fuel) and takes place at constant volume in a bomb calorimeter. This closed system consists of three sub-systems: the combustion reaction, the calorimeter body, and the water inside the calorimeter. As a whole, the system neither gains nor loses heat because the bomb calorimeter is well-insulated. The amounts of heat gained by the three sub-systems add up to zero:

$$q_{sys} = q_{H_2O} + q_{calorimeter} + q_{combustion} = 0 \qquad \text{constant } V$$

The heat absorbed by a system (or sub-system) in a change at constant volume equals:

$$q_v = c_v \Delta T \qquad \text{or} \qquad q_v = M c_s \Delta T$$

depending on whether a heat capacity or specific heat capacity is available. The problem gives the heat capacity c_v of the calorimeter (48 J K^{-1}) and the specific heat capacity of water (4.184 J K^{-1}g^{-1}). The ΔT of the calorimeter equals $28.670 - 20.450 = 8.220°C$, which also equals 8.220 K. The ΔT of the 750 g of water also equals 8.220 K, because the water and calorimeter are in thermal contact. Insert these numbers for the q's of the water and calorimeter:

$$\underbrace{(48 \text{ J K}^{-1})(8.22 \text{ K})}_{\text{calorimeter}} + \underbrace{(750 \text{ g})(4.184 \text{ J K}^{-1}\text{g}^{-1})(8.22 \text{ K})}_{\text{water}} + q_{combustion} = 0$$

Solving for the last q gives -2.62×10^4 J. This equals the heat absorbed by this combustion reaction at constant volume. At constant volume, zero work is done by or upon the combustion reaction. Hence:

$$\Delta E_{combustion} = q + w = q_v + 0 = \boxed{-2.62 \times 10^4 \text{ J}}$$

c) The molar mass of C_8H_{18} is 114.23 g mol^{-1}. The combustion of an entire mole of isooctane absorbs more heat than the combustion of 0.542 g:

$$\Delta E = \frac{-2.62 \times 10^4 \text{ J}}{0.542 \text{ g}} \times 114.23 \text{ g mol}^{-1} = -5.52 \times 10^6 \text{ J mol}^{-1} = \boxed{-5520 \text{ kJ mol}^{-1}}$$

d) By definition, $\Delta H = \Delta E + \Delta(PV)$. If the gases in the combustion reaction are ideal and the liquids have negligible volume, then $\Delta(PV) = (\Delta n_g)RT$, where Δn_g is the change in the number of moles of gas. The balanced equation for the combustion of 1 mol of isooctane shows that $\Delta n_g = 8 - 12.5 = -4.5$ mol. Therefore:

$$(\Delta n_g)RT = (-4.5 \text{ mol})(0.008315 \text{ kJ mol}^{-1}\text{K}^{-1})(298 \text{ K}) = -11.15 \text{ kJ}$$

Although the temperature rises from 20.450° to 28.670°C, taking it as a constant 25°C (298 K) causes little error. Complete the calculation as follows:

$$\Delta H = \Delta E + \Delta(PV) = -5520 + (-11.15) = \boxed{-5530 \text{ kJ}}$$

e) The standard enthalpy change of the combustion reaction written above is:

$$\Delta H^\circ = 8\,\Delta H^\circ_f(CO_2(g)) + 9\,\Delta H^\circ_f(H_2O(l)) - \Delta H^\circ_f(\text{isooctane})$$

Take ΔH° of the reaction as -5530 kJ and insert ΔH°_f's from text Appendix D:

$$-5530 \text{ kJ} = 8\underbrace{(-393.51)}_{CO_2(g)} + 9\underbrace{(-285.83)}_{H_2O(l)} - \Delta H^\circ_f(\text{isooctane})$$

$$\Delta H^\circ_f(\text{isooctane}) = \boxed{-190 \text{ kJ mol}^{-1}}$$

7-67 a) To get ΔH° in the combustion of 1 mol of acetylene, combine ΔH°_f's as follows:

$$\Delta H^\circ = 2\underbrace{(-393.51)}_{CO_2(g)} + 1\underbrace{(-241.82)}_{H_2O(g)} - 1\underbrace{(226.73)}_{C_2H_2(g)} - \frac{5}{2}\underbrace{(0.00)}_{O_2(g)} = \boxed{-1255.57 \text{ kJ}}$$

b) The total heat capacity of the mixture of the two gases equals the molar heat capacity of the first multiplied by the number of moles of the first plus the molar heat capacity of the second multiplied by the number of moles of the second:

$$nc_p = (2.00 \text{ mol})\underbrace{(37 \text{ J K}^{-1}\text{mol}^{-1})}_{CO_2} + (1.00 \text{ mol})\underbrace{(36 \text{ J K}^{-1}\text{mol}^{-1})}_{H_2O} = \boxed{110 \text{ J K}^{-1}}$$

c) Assume for convenience that 1.00 mol of $C_2H_2(g)$ is burned. Then the product gases, which are 2.00 mol of $CO_2(g)$ and 1.00 mol of $H_2O(g)$, absorb 1255.57 kJ of heat. For these gases, which comprise the flame:

$$\Delta T = \frac{q}{nc_p} = \frac{1.25557 \times 10^6 \text{ J}}{110 \text{ J K}^{-1}} = 1.14 \times 10^4 \text{ K} = 11400°C$$

If the temperature before combustion is room temperature (25°C), the maximum flame temperature is $\boxed{11400°C}$.

7-69 Define the system as the contents of the engine cylinder. Before the explosive combustion of the octane, the temperature is 600 K, the volume is 0.150 L, and the pressure is 12.0 atm. Apply the ideal-gas equation to the mixed contents of the cylinder before the combustion:

$$n_{\text{octane}} + n_{\text{air}} = \frac{PV}{RT} = \frac{(12.0 \text{ atm})(0.150 \text{ L})}{(0.08206 \text{ L atm mol}^{-1}\text{K}^{-1})(600 \text{ K})} = 0.03656 \text{ mol}$$

Also, the cylinder holds octane and air in a 1-to-80 molar ratio:

$$80 n_{\text{octane}} = n_{\text{air}}$$

Solving the two simultaneous equations gives:

$$n_{\text{octane}} = 4.514 \times 10^{-4} \text{ mol} \qquad \text{and} \qquad n_{\text{air}} = 0.03611 \text{ mol}$$

According to the problem, the system does not change its volume during the actual combustion of the fuel, so w is zero. Furthermore, q is zero (the combustion happens so fast that there is no time for heat to be lost or gained). Since w and q both equal zero, ΔE of the system equals zero. Imagine the combustion to occur in two stages: a: the reaction goes at a constant temperature of 600 K; b: the product gases heat up at constant volume. The sum of these two changes is the overall change within the cylinder. Therefore:

$$\Delta E_{\text{sys}} = 0 = \Delta E_a + \Delta E_b \quad \text{which means} \quad \Delta E_a = -\Delta E_b$$

The problem offers data pertaining to enthalpy changes, not energy changes, in the two steps. Deal with this by substituting for the ΔE_a and ΔE_b in terms of ΔH's:

$$\Delta H_a - \Delta(PV)_a = -\left(\Delta H_b - \Delta(PV)_b\right)$$

Step a involves ideal gases, takes place at a constant temperature, and involves change in the chemical amount of gas. Therefore $\Delta(PV)_a$ equals $\Delta n_{\text{g}} RT$, where Δn_{g} is the change in the chemical amount of gases during the reaction. Step b is the heating of the ideal gases inside the cylinder. The term $\Delta(PV)_b$ therefore equals $nR\Delta T$ where n is the chemical amount of gases present *after* the reaction. Also, for the change in temperature that comprises step b, ΔH_b is equal to $nc_{\text{p}}\Delta T$, as long as the molar heat capacity c_{p} is independent of temperature. Substitution of these relations gives:

$$\Delta H_a - \Delta n_{\text{g}} RT = -(nc_{\text{p}}\Delta T - nR\Delta T)$$

In this equation T is 600 K, and ΔT is the temperature change during the heating. Compute all the other quantities and then substitute in this equation to get ΔT and, from it, the final temperature.

Air is 80 percent N_2 and 20 percent O_2 on a molar basis. From this fact and the fact that octane burns according to the balanced equation:

$$C_8H_{18}(g) + 12\tfrac{1}{2} O_2(g) \rightarrow 8 CO_2(g) + 9 H_2O(g)$$

the chemical amounts of the different gases within the cylinder can be calculated before the reaction:

$$n_{N_2} = 0.80 \times (0.036109) = 0.02889 \text{ mol}$$
$$n_{O_2} = 0.20 \times (0.036109) = 0.007222 \text{ mol}$$
$$n_{\text{octane}} = 0.0004514 \text{ mol}$$

and after the reaction:

$$n_{N_2} = 0.02889 \text{ mol}$$
$$n_{CO_2} = 8 \times (0.0004514) = 0.003611 \text{ mol}$$
$$n_{H_2O} = 9 \times (0.0004514) = 0.004063 \text{ mol}$$
$$n_{O_2} = (0.007222) - (12.5 \times 0.0004514) = 0.001579 \text{ mol}$$

Note that all the octane burns and that the N_2 does not react. The change in the chemical amount of gases during the reaction is:

$$\Delta n_{\text{g}} = 0.03814 - 0.03656 = +0.001583 \text{ mol}$$

The enthalpy change of combustion of one mole of gaseous octane at 600 K can be computed using the $\Delta H_{\text{f}}^{\circ}$'s of the products and reactants:

$$\Delta H = 9 \underbrace{(-241.8)}_{H_2O(g)} + 8 \underbrace{(-393.5)}_{CO_2(g)} - 1 \underbrace{(-57.4)}_{\text{octane}(g)} - 12.5 \underbrace{(0.00)}_{O_2(g)} = -5266.8 \text{ kJ}$$

This is *not* ΔH_a, the enthalpy change of the combustion inside the cylinder. Because only 0.0004514 mol of octane is in the cylinder:

$$\Delta H_a = -5266.8 \text{ kJ mol}^{-1} \times (4.514 \times 10^{-4} \text{ mol}) = -2.377 \text{ kJ}$$

The composite heat capacity of the contents of the cylinder after the reaction is the sum of the nc_{p} values for the four product gases, as in **7-67b**:

$$nc_{\text{p}} = (0.00158 \text{ mol}) \underbrace{(35.2 \text{ J K}^{-1}\text{mol}^{-1})}_{O_2} + (0.0289 \text{ mol}) \underbrace{(29.8 \text{ J K}^{-1}\text{mol}^{-1})}_{N_2} +$$
$$(0.00406 \text{ mol}) \underbrace{(38.9 \text{ J K}^{-1}\text{mol}^{-1})}_{H_2O} + (0.00361 \text{ mol}) \underbrace{(45.5 \text{ J K}^{-1}\text{mol}^{-1})}_{CO_2} = 1.24 \text{ J K}^{-1}$$

Now, solve the equation derived previously for ΔT and make the various substitutions:

$$\Delta T = \frac{\Delta H_a - \Delta n_g RT}{nR - nc_p}$$

$$\Delta T = \frac{-2377 \text{ J} - (0.001583 \text{ mol})(8.315 \text{ J K}^{-1}\text{mol}^{-1})(600 \text{ K})}{(0.03814 \text{ mol})(8.315 \text{ J K}^{-1}\text{mol}^{-1}) - 1.24 \text{ J K}^{-1}} = 2580 \text{ K}$$

The maximum temperature inside the cylinder is $600 + 2580 = 3180$ K. This equals $\boxed{2910°C}$.

7-71 a) The gases trapped inside the cylinder of the "one-lung" engine have volume V_1 when the piston is fully withdrawn but a smaller volume V_2 when the piston is thrust home. The compression ratio is $8 : 1$ so $V_1 = 8V_2$. The area of the base of the engine's cylinder is πr^2, where r is the radius of the base. The volume of a cylinder is the area of its base times its height h:

$$V_1 = Ah \quad \text{and} \quad V_2 = A(h - 12.00 \text{ cm})$$

which employs the (given) fact that full compression shortens h by 12.00 cm. Because r is 5.00 cm, the area A is 78.54 cm^2. Substituting for V_1 and V_2 in terms of A and h gives:

$$Ah = 8A(h - 12.00 \text{ cm})$$

The A's cancel, allowing solution for h. The result is 13.714 cm. With h known it is easy to compute V_1 and V_2, which equal 1.077 L and 0.1347 L respectively.

The temperature and pressure of the fuel mixture are 353 K (80°C) and 1.00 atm when the mixture enters the cylinder with fully withdrawn piston (V_1). Assuming the fuel mixture is an ideal gas:

$$n_{\text{mixture}} = \frac{(1.00 \text{ atm})(1.077 \text{ L})}{(0.08206 \text{ L atm mol}^{-1}\text{K}^{-1})(353 \text{ K})} = 0.0372 \text{ mol}$$

The molar ratio of air to fuel (C_8H_{18}) is 62.5 to 1. Then:

$$n_{\text{fuel}} + n_{\text{air}} = 0.0372 \text{ mol} \quad \text{and} \quad n_{\text{air}} = 62.5 n_{\text{fuel}}$$

Solving these simultaneous equations establishes that at the start the cylinder contains 0.0366 mol of air and 5.86×10^{-4} mol of octane fuel.

During the compression stroke, the system undergoes an irreversible adiabatic compression to one-eighth of its initial volume. None of the relationships that govern *reversible* adiabatic processes applies here. Assume however, as advised in the problem, that the compression is near to reversible. If it is, then:

$$T_1 V_1^{\gamma-1} \approx T_2 V_2^{\gamma-1} \quad \text{where } \gamma = \frac{35 \text{ J K}^{-1}\text{mol}^{-1}}{26.7 \text{ J K}^{-1}\text{mol}^{-1}} = 1.31$$

The temperature after the compression stroke is:

$$T_2 = T_1 \left(\frac{V_1}{V_2}\right)^{\gamma-1} = (353 \text{ K}) \left(\frac{1.077 \text{ L}}{0.1347 \text{ L}}\right)^{0.31} = (353 \text{ K})(8)^{0.31} = \boxed{673 \text{ K}}$$

b) The compressed gases occupy a volume of $\boxed{0.135 \text{ L}}$ just before they are ignited, as calculated above.

c) The pressure of the compressed fuel mixture just before ignition is P_2. Compute it by applying the ideal-gas equation to the system with $T_2 = 673$ K, $V_2 = 0.1347$ L, and $n = 0.0372$ mol. It equals 15.3 atm. Alternatively, compute P_2 using the formula for a reversible adiabatic change:

$$P_2 = P_1 \left(\frac{V_1}{V_2}\right)^{\gamma} = 1.00 \text{ atm} \left(\frac{1.077}{0.1347}\right)^{1.31} = \boxed{15.3 \text{ atm}}$$

d) ΔH for the combustion of gaseous octane at 600 K equals -5266.8 kJ mol^{-1}.[11] The combustion mixture inside the cylinder contains 5.86×10^{-4} mol of octane. Consequently, the ΔH of combustion in this system equals:

$$\Delta H = \left(\frac{-5266.8 \text{ kJ}}{1 \text{ mol}}\right) \times (5.86 \times 10^{-4} \text{ mol}) = -3.09 \text{ kJ}$$

After the combustion, the cylinder contains CO_2, H_2O and unreacted O_2 and N_2, all gases. The balanced chemical equation shows that the combustion consumes 5.86×10^{-4} mol of octane and $12.5 \times (5.86 \times 10^{-4})$ mol of O_2 to produce $8 \times (5.86 \times 10^{-4})$ mol of CO_2 and $9 \times (5.86 \times 10^{-4})$ mol of H_2O. The effect of the reaction is to increase the chemical amount of gases in the cylinder by $3.5 \times (5.86 \times 10^{-4})$ mol. This is Δn_g for the reaction. The original quantity of gases is 0.0372 mol. After the combustion there is 0.0393 mol of gases. The *energy* (not enthalpy) released from the reaction all goes to heat up the gaseous contents of the cylinder as long as no heat escapes to the cylinder walls and no work is done until the power stroke starts. Therefore:

$$\Delta T = \frac{\Delta H_{\text{react}} - \Delta n_{\text{g}} RT}{nR - nc_{\text{p}}}$$

---

[11] As computed in **7-69**. This number is preferable to $\Delta H° = -5530$ kJ mol^{-1} for the combustion of liquid isooctane that was obtained in **8-69**. The latter is for combustion of a different compound (isooctane) in a different form (liquid, not gaseous) at a different temperature (298 K not 600 K) to give a different product (liquid water, not water vapor).

In this equation, which is derived in **7-69**, every quantity but ΔT is known:

$$\Delta T = \frac{-3090 \text{ J} - (0.002051 \text{ mol})(8.315 \text{ J K}^{-1}\text{mol}^{-1})(673 \text{ K})}{(0.0393 \text{ mol})(8.315 \text{ J K}^{-1}\text{mol}^{-1}) - (0.0393 \text{ mol})(35 \text{ J K}^{-1}\text{mol}^{-1})} = 2960 \text{ K}$$

The temperature inside the cylinder rises by 2960 K to a maximum of $\boxed{3630 \text{ K}}$.

e) Assume that the expansion stroke is not only adiabatic but reversible. Then the formula:

$$T_2 = T_1 \left(\frac{V_1}{V_2}\right)^{\gamma-1}$$

applies. In this case, T_1 is 3630 K. The ratio $V_1 \, / \, V_2$ is 1 to 8 because now the initial state is the *small* volume state just before the expansion stroke of the piston. The exponent $\gamma - 1$ is still 0.31, as previously established. Substituting gives:

$$T_2 = (3630 \text{ K}) \left(\frac{1}{8}\right)^{0.31} = \boxed{1900 \text{ K}}$$

This is the temperature of the exhaust gases.

7-73 The chemical amount of the silane at the T and P stated in the problem equals:

$$n_{\text{SiH}_4} = \frac{PV}{RT} = \frac{(0.658 \text{ atm})(0.250 \text{ L})}{(0.08206 \text{ L atm mol}^{-1}\text{K}^{-1})(298 \text{ K})} = 6.727 \times 10^{-3} \text{ mol}$$

The combustion of this much silane at constant volume (in a bomb calorimeter) absorbs -9.757 kJ of heat at 25°C (which is the same as evolving $+9.575$ kJ). Hence:

$$\Delta E = q_v = -9.757 \text{kJ}$$

For the combustion of 1 mol of silane:

$$\Delta E^\circ = \frac{-9.757 \text{ kJ}}{6.727 \times 10^{-3} \text{ mol}} = -1450 \text{ kJ mol}^{-1}$$

Now, compute the ΔH° of the combustion of 1 mol of silane. The balanced equation given in the problem shows that 3 mol of gaseous reactants gives 0 mol of gaseous products:

$$\begin{aligned} \Delta H^\circ &= \Delta E^\circ + RT\Delta n_g \\ &= -1450.4 \text{ kJ} + (0.008315 \text{ kJ K}^{-1}\text{mol}^{-1})(298.15 \text{ K})(-3 \text{ mol}) = -1458 \text{ kJ} \end{aligned}$$

ΔH° for the combustion of silane equals the sum of the standard enthalpies of formation of the products minus the sum of the standard enthalpies of formation of the reactants. Taking values from Appendix D:

$$\Delta H° = -1458 \text{ kJ} = 1 \underbrace{(-910.94)}_{SiO_2 \text{ quartz}} + 2 \underbrace{(-285.83)}_{H_2O(l)} - 1 \underbrace{(\Delta H_f^\circ)}_{SiH_4(g)}$$

$$\underbrace{(\Delta H_f^\circ)}_{SiH_4(g)} = \boxed{-25 \text{ kJ mol}^{-1}}$$

Compute ΔE_f° of silane from ΔH_f° and the known Δn_g in the formation of 1 mol of silane from its elements:

$$\Delta E_f^\circ = \Delta H_f^\circ - RT\Delta n_g$$
$$= -25 \text{ kJ} - (0.008315 \text{ kJ K}^{-1}\text{mol}^{-1})(298.15 \text{ K})(-1 \text{ mol})$$
$$= \boxed{-23 \text{ kJ}}$$

7-75 Substances with the strongest intermolecular forces have the highest enthalpies of vaporization. Liquid KBr has strong ion-ion forces holding its ions together. It has the highest ΔH_{vap}. NH_3 has dipole-dipole attractions, which are stronger than the weak dispersion (van der Waals) forces that maintain the liquid in Ar and He. The dispersion forces should be stronger in Ar than in He because Ar has a larger molar mass. Therefore: $\boxed{\text{He < Ar < NH}_3 \text{ < KCl}}$.

7-77 a) Lewis structures for carbonic acid show two O—H single bonds, two C—O single bonds and one C=O double bond:

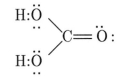

b) Imagine the reaction to proceed by the breaking of the five bonds in H_2CO_3 followed by the making of the four bonds in H—O—H plus O=C=O. The enthalpy change of bond breaking is positive; the enthalpy change of bond making is negative. Take bond enthalpies from text Table 7-3,[12] and combine them accordingly:

$$\Delta H = -2 \underbrace{(463)}_{O—H} -2 \underbrace{(728)}_{C=O} +2 \underbrace{(463)}_{O—H} +2 \underbrace{(351)}_{O—C} +1 \underbrace{(728)}_{C=O} = \boxed{-26 \text{ kJ}}$$

---

[12]Text page 228.

Chapter 8

Second Law of Thermodynamics: Spontaneous Processes and Thermodynamic Equilibrium

A **spontaneous change** is one that occurs by itself, given enough time, without outside intervention. Spontaneous change is directional. If a process is spontaneous then its reverse must be non-spontaneous. When all driving force or tendency for spontaneous change in either direction is expended, a system rests at equilibrium. Neither the forward process nor its reverse occurs at equilibrium. At equilibrium, what can happen has happened.

The **entropy** S a thermodynamic state function, provides a basis for predictions of the direction of spontaneous change. The entropy of the universe always increases in a spontaneous change.

Entropy and Irreversibility: A Statistical Interpretation

The entropy is a measure of the randomness of a system. Spontaneous change occurs from states of low probability (ordered states) to states of high probability (disordered states). **Statistical thermodynamics** identifies and compares such states on a molecular basis.

Consider the expansion of an ideal gas into a vacuum. Originally, the gas is confined behind a closed stopcock in the left-hand bulb (of volume V) of a two-bulb container. The right-hand bulb (of equal volume V) is empty. The stopcock is opened. Very soon, the gas has the volume $2V$. Everybody knows that the gas expands spontaneously. But why? The energy of the ideal gas does not change during the expansion. The gas absorbs neither work nor heat. As far as the *first* law of thermodynamics is concerned, the ideal gas could just as well stay in the left-hand bulk.

To appreciate why the gas expands, suppose for simplicity that there are only six molecules of the gas. How can the six molecules be distributed between the two sides?

Number of Molecules on Left	Number of Molecules on Right	Number of Ways Attainable
6	0	1
5	1	6
4	2	15
3	3	20
2	4	15
1	5	6
0	6	1

There are 64 ways to distribute the six molecules between the two sides. This is the sum of the numbers in the last column of the table. These give only 7 distinguishable states of the gas, corresponding to the 7 lines in the table.

The state that is most likely to occur is the one that can be achieved in the greatest number of ways.

The chance of finding all 6 molecules on the left is only 1/64. The chance of finding 3 on the left and 3 on the right is 20/64. The latter state is 20 times more probable than the former. Left to itself, the gas is 20 times more likely to fill the entire volume evenly than to stay on the left. For a large number of molecules there is a huge number of ways to achieve an even distribution and far fewer ways to achieve a state with all the molecules on the left or right. See **8-7** and **8-49** for related calculations.

Entropy and the Number of Microstates

Nature proceeds spontaneously toward states that have the highest probability of occurring. These states have higher entropies. The entropy of a system in fact depends on the number of microscopically different ways in which the particles of the system can be arranged as it exists in a given state. Each such arrangement is a **microstate** of the system. A microstate is a particular distribution of molecules among the positions and momenta accessible to them. Every microstate is equally likely to be occupied by the system. See **8-3**. The number of microstates of a system is symbolized Ω.

The exact relationship between the entropy of a system and the number of its microstates is:

$$S = k_\mathrm{B} \ln \Omega$$

where k_B is the **Boltzmann constant** (or Boltzmann's constant);

$$k_B = 1.380658 \times 10^{-23} \text{ J K}^{-1}$$

This constant equals the gas constant R divided by Avogadro's number N_0:

$$k_B = \frac{R}{N_0} = \frac{8.31451 \text{ J K}^{-1}\text{mol}^{-1}}{6.0221367 \times 10^{23} \text{ mol}^{-1}}$$

In problems, actual numerical values for Ω are rarely available. Fortunately, ΔS values can be calculated even if values for S are not available. The *change* in the entropy of a system between two states depends on the *ratio* of the number of microstates:

$$\Delta S = k_B \ln \left(\frac{\Omega_2}{\Omega_1}\right)$$

If state 2 of the six-molecule gas is the state with 3 molecules on the left and three on the right and state 1 is the state with all 6 molecules on the left, then Ω_2/Ω_1 is $20/1$ and the difference in entropy between the states is:

$$\Delta S = k_B \ln \left(\frac{20}{1}\right) = (1.381 \times 10^{-23} \text{ J K}^{-1})(2.996) = 4.137 \times 10^{-23} \text{ J K}^{-1}$$

This approach is used in solving **8-7** and **8-49a**.

Entropy and Disorder

The greater the disorder of a system, then the greater its entropy. Disorder should not be conceived of in terms of a motionless array of atoms or molecules, but in terms of all the possible motions of the particles and the ways in which their arrangements change with time. These are the microstates of the system. Their number depends on factors such as the number of atoms per molecule and the strength of the bonds between atoms. As a result two different substances in general have different entropies even if both are solids (or liquids or gases) at the same temperature. Some helpful generalizations:

- The entropy of substances increases when they melt. In solids, every atom or molecule is at or near a prescribed position. In liquids, the particles may move around more. See **8-9** .

- Substances in their gaseous state have more entropy than in their liquid state at the same temperature. There are more microstates for a gas than a liquid. Many more arrangements of atoms or molecules correspond to the same observed state. See **8-29**.

- Gases at low pressure have greater entropy than at high pressure.

- A dilute solution of a given quantity of a substance has greater entropy than a concentrated solution.

- When one substance dissolves in another, ΔS of the system usually is positive.

- The entropy of a system always increases with increasing temperature.

Entropy, Spontaneous Change, and the Second Law

Like other state functions (such as the internal energy, pressure, volume, and temperature), the entropy of a system depends only on the current nature (state) of a system and not on its history. The significance of the entropy is this:

- **All spontaneous change occurs with an increase in the entropy of the universe.**

Clearly, the entropy change of the universe equals the sum of the entropy changes of the system and its surroundings:

$$\Delta S_{\text{univ}} = \Delta S_{\text{sys}} + \Delta S_{\text{surr}}$$

Either of the terms on the right may be negative in a spontaneous process. The sum of the terms, the total entropy change of the universe must be positive in a spontaneous change. Problem **8-19** illustrates this point, as do problems **8-21** and **8-47**, which concern the changes in entropy when a piece of hot iron is plunged into cold water. The iron (the system) of course spontaneously cools. Its entropy *decreases.* The increase in the entropy of the surroundings of the iron (the water) more than compensates.

Calculation of Entropy Changes

To compute the difference in entropy ΔS_{sys} between two states of a system, imagine some **reversible** path. Recall that a reversible process proceeds through a continuous series of thermodynamic states. Equilibrium is attained at each stage.[1] As the system goes along the imagined reversible path, it absorbs heat from its surroundings. ΔS equals the sum of each little quantity of heat dq absorbed divided by the temperature at which it is absorbed:

$$\Delta S_{\text{sys}} = \int \frac{dq_{\text{rev}}}{T}$$

---

[1]See text Section 7-6 on text page 229 or page 151 in the Guide.

This is the formal definition of the entropy. For a change at constant temperature:

$$\Delta S_{\text{sys}} = \frac{q_{\text{rev}}}{T} \qquad \text{at constant temperature}$$

See **8-13**. The q may be positive or negative (recall that a negative q corresponds to the evolution of heat by the system.) The absolute temperature T is always positive. Hence, for a system gaining heat on a reversible path, ΔS_{sys} is positive. For a system losing heat on a reversible path, the ΔS_{sys} is negative. Finally, if a system undergoes a reversible change in which it neither gains nor loses heat, q and ΔS both equal zero.

The same net change between two states can be achieved along an infinite number of different paths. Every different path may have a different total q. The above method of calculation gives ΔS only if reversible paths are chosen. Mapping out a reversible path may seem difficult, especially if the final temperature differs from the original. The best tactic is to break down the total change into a combination of steps that occur either **isobarically** (at constant pressure), **isochorically** (at constant volume), or **isothermally** (at constant temperature). Such an approach is used in **8-45b**. This tactic works because formulas are available to allow calculation of entropy changes along isochoric, isobaric and isothermal path segments. Here is a summary of these formulas:

- **Reversible gain (loss) of heat at constant T.** Direct application of the definition of ΔS is easy because T is a constant:

$$\Delta S = \frac{1}{T} q_{\text{rev}}$$

This situation arises when a system is exchanging work with its surroundings as heat flows in or out. Also, when substances freeze, melt, boil, or condense they stay at a constant temperature while absorbing or emitting heat. For example, boiling water absorbs heat at a constant temperature equal to its boiling point. This heat equals its ΔH of vaporization. The entropy change of vaporization therefore is:

$$\Delta S_{\text{vap}} = \frac{\Delta H_{\text{vap}}}{T_{\text{b}}}$$

This fact is used in **8-31** and, in a way, in **8-13**. Most liquids have nearly the same molar entropy of vaporization. **Trouton's rule** states: $\Delta S_{\text{vap}} = 88 \pm 5 \text{ J mol}^{-1} \text{ K}^{-1}$. This rule in useful in estimates of boiling point and enthalpy of vaporization (see **8-15** and **8-41**).

For fusion:

$$\Delta S_{\text{fus}} = \frac{\Delta H_{\text{fus}}}{T_{\text{f}}}$$

which has the same form as for vaporization. There is no rule like Trouton's for fusion.

• **Reversible heating (cooling) at constant P.** The change in entropy is:

$$\Delta S = n c_p \ln \left(\frac{T_2}{T_1} \right)$$

Whenever a system cools, its entropy diminishes along with its temperature. The entropy change of the system is negative if its temperature drops and positive if it rises. This formula is used in **8-45**. To use it, one must either assume that c_p is not a function of temperature or get information on how c_p varies with temperature.

• **Reversible heating (cooling) at constant V.** The formula is similar to the previous one. Note that at constant volume c_v appears instead of c_p:

$$\Delta S = n c_v \ln \left(\frac{T_2}{T_1} \right)$$

• **Reversible expansion (compression) at constant T.** The entropy of a system always increases with its size as long as the temperature does not change. The entropy change upon expansion *of an ideal gas* is:

$$\Delta S = n R \ln \left(\frac{V_2}{V_1} \right) \qquad \text{ideal gas}$$

The formula:

$$\Delta S = n R \ln \left(\frac{P_1}{P_2} \right) \qquad \text{ideal gas}$$

is an immediate consequence (by combination with the ideal-gas law). If a system that is part solid or liquid and part gaseous expands or contracts, the volume change can, with reasonable safety be assigned entirely to the gas. These formulas can then be used, as long as the gas is near-ideal.

• **Reversible adiabatic changes.** In an adiabatic change, the system is thermally insulated from its surroundings. Therefore, the system absorbs zero heat. Since q_{rev} equals zero, ΔS also equals zero. Reversible adiabatic changes are **isentropic** (keep the entropy constant). Note that a process must reversible as well as adiabatic to be isentropic.

Even the most complex process involving simultaneous changes in T, V, and P can be imagined to proceed along a path involving a change first with T constant, then with P constant, and then with V constant. Coming up with such paths is a major theme in problems involving the entropy function. See **8-43**. When a path is finally sketched, the above formulas allow the computation of the change in entropy.

Example: 10.00 L of an ideal monatomic gas at 10.00 atm and 273.15 K is expanded and cooled to a final volume of 63.93 L and a final temperature of 174.8 K. Compute the entropy change of the gas.

Solution: During the process, T, P, and V all change. Using the ideal-gas equation on the original state shows that the system contains 4.457 mol of gas. None of the gas escapes during the change, so there is 4.457 mol in the final state. Using the ideal-gas equation on the final state gives a final pressure of 1.000 atm. In summary: $P_1 = 10.0$ atm, $P_2 = 1.00$ atm; $V_1 = 10.00$ L, $V_2 = 69.93$ L; $T_1 = 273.15$ K, $T_2 = 174.8$ K; $n = 4.457$ mol. A reversible path connecting these two states is: a) cool the gas reversibly at constant P from T_1 to T_2; b) expand the gas reversibly at constant T from its intermediate volume (the volume attained after the cooling step) to V_2. During the isobaric cooling (step a), the volume of the gas falls to 6.399 L. The entropy change is:

$$\Delta S_a = n c_p \ln \frac{T_2}{T_1} = (4.457 \text{ mol}) \left(\frac{5}{2} 8.315 \text{ J K}^{-1}\text{mol}^{-1}\right) \ln \left(\frac{174.8}{273.15}\right) = -41.35 \text{ J K}^{-1}$$

In step b, the gas expands from 6.3993 L to 63.99 L, at a constant T of 174.8 K. The entropy change is:

$$\Delta S_b = n R \ln \left(\frac{V_2}{V_1}\right) (4.457 \text{ mol})(8.315 \text{ J K}^{-1}\text{mol}^{-1}) \ln \left(\frac{63.99}{6.399}\right) = 85.33 \text{ J K}^{-1}$$

The overall entropy change for the gas in the process is $85.33 + (-41.35) = 43.98$ J K^{-1}.

Other reversible paths connect the final and initial states. The problem could have been solved by imagining an isothermal expansion from 10.0 to 63.99 L followed by an isochoric cooling from 273.15 to 174.8 K.

Irreversible Processes

Reversibility is an idealization. All real processes are more or less **irreversible**. If a comparison is made between an idealized, reversible path connecting two states and a real, irreversible path then:

$$q_{\text{irrev}} < q_{\text{rev}} \quad \text{and} \quad -w_{\text{irrev}} < -w_{\text{rev}}$$

Suppose that as a practical matter one wishes to get work out of a system by some kind of process. The maximum work is extracted (largest $-w$) only if the process proceeds reversibly. The *actual* work performed by any real system during a real process is always less than this maximum: all real changes are irreversible. Similarly, the system will absorb the maximum heat only if the process goes on reversibly. The heat absorbed in a real (irreversible) process will be less than this maximum. Problem **8-47** shows how an irreversible process becomes more nearly reversible as the method by which it is performed is changed. The process in **8-47** approaches reversibility as a limit.

The Second Law

Return to the question of whether a proposed change can occur by itself. As previously mentioned, the entropy function is pivotal in answering this question for all processes. The **second law of thermodynamics** states:

- **In a reversible process the total entropy of a system plus its surroundings is unchanged.**

- **In an irreversible process the total entropy of a system plus its surroundings increases.**

- **In all human experience, processes for which ΔS_{univ} is negative are impossible.**

This all means that in real processes:

$$\Delta S_{\text{univ}} = \Delta S_{\text{sys}} + \Delta S_{\text{surr}} > 0$$

To see if a proposed process can occur, calculate the entropy change of the system. Then calculate the entropy change of the surroundings (regarding them momentarily as a system). If the sum of these two answers exceeds zero, then the proposed change can occur. This is actually done in **8-21**. See also **8-27**.

Carnot Cycles, Efficiency, and Entropy

The Carnot Cycle

A **cyclic process** takes a system through a series of thermodynamic states and returns it finally to the exact state from which it started. In a **Carnot cycle**, an ideal gas undergoes a series of four reversible changes during such an out-and-back excursion:

1. An isothermal reversible expansion. The gas, at its original temperature, T_{h}, expands from V_A to V_B. It performs work on the surroundings. It simultaneously absorbs heat from the surroundings to keep the same internal energy.

2. An adiabatic reversible expansion. The gas continues to expand, going from V_B to V_C. Now, however, q is zero. The continued expansion of the gas performs more work on its surroundings. The energy to do the work comes from the internal energy of the gas. The temperature of the gas drops to T_ℓ.

3. An isothermal reversible compression. At T_ℓ the surroundings compress the gas, from V_C to V_D. Work is performed upon the gas. The temperature is constant and just enough heat flows from the gas to the surroundings that ΔE for this step equals zero.

4. An adiabatic reversible compression. The surroundings continue to do work on the gas. Now however the gas is thermally insulated so that $q = 0$. All the work adds to the internal energy of the gas. The temperature rises toward T_h. The volume diminishes toward V_A. The compression continues until the gas reaches its original state.

In a Carnot cycle the system extracts heat from the surroundings at T_h, converts some of it into work, and dumps the rest back to the surroundings at T_ℓ. The net work in one passage around the cycle is:

$$w_{\text{net}} = -nR(T_h - T_\ell)\ln\left(\frac{V_B}{V_A}\right)$$

The net work is negative, consistent with the convention that work performed *by* a system is negative. The net work depends on how much gas (n is the chemical amount) is taken around the cycle, on the two extremes of temperature and by how much the gas expands in the first step.

Heat Engines, Refrigerators, and Heat Pumps

A **heat engine** is a device that converts the natural heat flow from higher to lower temperature into useful work. A **refrigerator** does the reverse. It destroys work to produce a difference in temperature.

A heat engine operates between a high and low temperature, T_h and T_ℓ. The **efficiency** ϵ of a heat engine is the net work that it performs divided by the heat that it absorbs. Maximum efficiency is obtained only if the engine operates reversibly. It is unattainable in practice. Maximum efficiency depends only on the temperatures:

$$\epsilon = 1 - \frac{T_\ell}{T_h}$$

This equation is derived from consideration of the Carnot cycle. Why does this equation give the theoretical maximum efficiency? The proof consists of assuming the opposite and showing that an engine with greater efficiency would make it possible to transfer heat in a continuous cycle out of a low-temperature reservoir into a high-temperature reservoir without expending any work. In all human experience it is impossible to do this. Heat never spontaneously flows "uphill" (against a temperature gradient).

It is impossible to construct a device that will transport heat from a cold reservoir to a hot reservoir in a continuous cycle without any net expenditure of work.

The theoretical operation of a refrigerator or heat pump is modeled on a Carnot cycle running in reverse. The maximum heat q absorbed from the interior of a refrigerator is:

$$q = w_{\text{net}} \left(\frac{T_\ell}{T_{\text{h}} - T_\ell} \right)$$

where T_ℓ and T_{h} are the low and temperatures and w_{net} is the work during the refrigeration cycle. These equations are used in **8-11**.

Absolute Entropies and the Third Law

Unlike energy and enthalpy, for which there is no natural zero point, the entropy function *does* have an obvious zero. The **third law of thermodynamics** concerns the setting of this zero for the measurement of entropies. The third law states:

In any thermodynamic process involving only pure phases in their equilibrium states, the entropy change ΔS approaches zero as T approaches 0 K.

This law means that the entropy of any pure substance in its equilibrium state approaches zero as T approaches 0 K. It provides a natural reference for the tabulation of entropies. See text Appendix D.

Standard-State Entropies

Appendix D gives the absolute entropies at 1 atm pressure and 298.15 K for a variety of substances. These $S°$ values (note the absence of a Δ) of various pure solids, liquids and gases are all positive because they are all referred to a zero entropy at 0 K. The values are experimental. They were obtained by measuring c_{p} as a function of temperature for each substance. Then the increase from 0 K (T_1) to 298.15 K (T_2) of the entropy of each substance was computed using the formula:

$$\Delta S = n \int_{T_1}^{T_2} \frac{c_{\text{P}}}{T} dT$$

Problem **8-51** displays a plot of c_{P}/T versus T for two substances. The area under such a curve is proportional to ΔS.

In a chemical reaction the entropy change is the sum of the standard entropies of the products minus the sum of the standard entropies of the reactants:

$$\boldsymbol{\Delta S° = \sum S°(\text{products}) - \sum S°(\text{reactants})}$$

The use of this equation to get $\Delta S°$ values is like the use of $\Delta H_{\text{f}}°$'s (tabulated in Appendix D) to get $\Delta H°$'s of reaction. Remember these points:

- The tabulated $S°$ values are *molar* entropies. If a reaction forms, say, 2 mol of $NH_3(g)$ (for which $S°$ is 192.3 J $K^{-1}mol^{-1}$), the contribution of the ammonia to the entropy of the products is 384.6 J K^{-1}. If it forms 3 mol of $NH_3(g)$ then the contribution is 576.9 J K^{-1}. See **8-23** and **8-25**.

- The $S°$ of an element in its standard state is *not* equal to zero. See **8-25** and **8-37**. The mistake of taking elemental $S°$'s to equal zero occurs because the $\Delta H_f°$'s of the elements in their standard states *are* zero.

- Standard entropies are tabulated in *joules* per Kelvin per mole (J $K^{-1}mol^{-1}$). In contrast, $\Delta H_f°$'s are tabulated in kilojoules per mole (kJ mol^{-1}). These are entirely different units. The accidental similarity in the magnitudes of the values in the $S°$ and $\Delta H_f°$ columns in Appendix D sometimes makes users of Appendix D forget this.

- The $\Delta S°$ of a reaction may be positive, negative or even zero. A negative $\Delta S°$ does *not* mean that a reaction is forbidden to occur by the second law of thermodynamics. For example, in **8-37a,** the $\Delta S°$ for the rusting of iron at room conditions comes out to be negative. Iron still rusts spontaneously.

- The solution-phase entropies in Appendix D are not absolute entropies, but are measured relative to an arbitrary standard. Some of them are negative. Use these values in calculations without concern. See **8-57**.

The Gibbs Free Energy

So far, predicting whether a given process can occur requires the calculation of ΔS_{univ}, the sum of ΔS of the surroundings and ΔS of the system. The universe is big and complex, and getting its ΔS for every proposed process is hard. It is possible, at a price, to avoid figuring ΔS_{univ} when evaluating a proposed process. The method requires the definition of a new function, the **Gibbs free energy, G:**

$$G = H - TS$$

The Gibbs free energy has units of energy (such as joules, kilojoules, or calories).

The Gibbs free energy is a state function. Hence, for a change in a system:

$$\Delta G_{sys} = \Delta H_{sys} - \Delta (TS)_{sys}$$

If T does not change it may be placed in front of the Δ:

$$\Delta G_{sys} = \Delta H_{sys} - T\Delta S_{sys} \quad (T \text{ constant})$$

Under an even more restricted set of conditions, *constant temperature and pressure,* a simple criterion for spontaneity emerges:

If ΔG_{sys} for a proposed process at constant temperature and pressure is negative, then the process is spontaneous.

What if ΔG_{sys} is positive or zero?

- If ΔG_{sys} is positive, then the proposed process at constant temperature and pressure is *not* spontaneous, but the reverse process is.

- If ΔG is 0, then the system is at equilibrium at constant temperature and pressure, and no change occurs.

The price paid for the convenience and simplicity of this criterion that it works only for processes at constant temperature and pressure. But, processes at constant T and P are common in chemistry.

The value of ΔG for a reaction at constant T and P represents a compromise between the tendency toward minimum enthalpy in the system and the tendency toward maximum entropy in the system. The following table shows how this compromise works for the four possible combinations of sign of ΔH and ΔS. Also, see **8-37**.

ΔH	ΔS	Outcome
Positive	Negative	Always nonspontaneous
Negative	Positive	Always spontaneous
Positive	Positive	Spontaneous at high T
Negative	Negative	Spontaneous at low T

Free energy is not "no-pay" energy, but energy that is available. The term $T\Delta S$ in the preceding equation is the *unavailable energy.* The unavailable energy is energy that must be discarded in the form of heat, no matter what, for a process to take place. The free energy change ΔG at constant T and P equals the total enthalpy change ΔH minus the unavailable energy. Although the whole ΔH of a reaction at constant T and P can appear in the surroundings as heat, only ΔG is available (free) to do useful work.

Free Energy and Phase Transitions

The phase transitions melting, boiling and sublimation all require the input of heat. Their ΔH's are all positive. For all three transitions, the product phase has higher entropy than the reactant phase. All three have positive ΔS's. At a transition temperature (melting point, boiling point or sublimation point) the G's of two phases are equal. This means that ΔG equals zero at a transition temperature. The change in free energy for a phase transition is given by

$$\Delta G = \Delta H - T\Delta S$$

It follows the transition temperature is:

$$T_{\text{melt}} = \frac{\Delta H_{\text{melt}}}{\Delta S_{\text{melt}}} \qquad T_{\text{boil}} = \frac{\Delta H_{\text{boil}}}{\Delta S_{\text{boil}}} \qquad T_{\text{subl}} = \frac{\Delta H_{\text{subl}}}{\Delta S_{\text{subl}}}$$

Below a transition temperature, the enthalpy term, which favors the lower-enthalpy phase, predominates. Above the transition temperature, the entropy term, which favors the higher entropy phase, predominates. At a transition temperature, both phases are equally favored. Two phases coexist. See **8-31**, **8-33**, and **8-53**.

Standard-State Free Energies

A *standard* free-energy change is the free-energy change measured between reactants and products each in standard states. The superscript "naught" to the right in the $\Delta G°$ designates a standard change. The standard free energy change of a reaction is computed in two ways. The first way is from $\Delta H°$ and $\Delta S°$ values determined at the same temperature and using the equation:

$$\Delta G_T° = \Delta H_T° - T\Delta S_T°$$

See **8-37**. The other way is from tabulated values of standard free energies of formation:

$$\Delta G° = \sum \Delta G_f° \,(\text{products}) - \sum \Delta G_f° \,(\text{reactants})$$

The procedure in this second type of calculation is identical in every respect to the procedure for getting $\Delta H°$'s of reaction from $\Delta H_f°$'s. See **8-55** and page 160 of this Guide. Text Appendix D contains the necessary $\Delta G°$ data.

Warning. The most common error in calculations of $\Delta G°$ is to subtract a $T\Delta S$ term in joules from a ΔH term that is in kilojoules.

Problem **8-57** demonstrates an important point of working with $\Delta H°$, $\Delta S°$ and ΔG. The free-energy change of a process depends strongly on the temperature but the enthalpy change and entropy change depend only weakly. The $\Delta H°$'s associated with changes at temperatures other than 298.15 K can often be approximated by $\Delta H_{298.15}°$; the $\Delta S°$'s associated with changes at temperatures other than 298.15 K can often be approximated by $\Delta S_{298.15}°$. However:

Do not approximate $\Delta G°$ at temperatures other than 25°C by $\Delta G_{298.15}°$.

Assuming temperature-independence for $\Delta S°$ and $\Delta H°$ is common, particularly when the temperature does not change very much. See **8-37**, **8-39**, and **8-57**.

Detailed Solutions to Odd-Numbered Problems

8-1 The definition of a thermodynamical system is ultimately arbitrary. Once a system is defined, the surroundings are then automatically "the rest of the universe." Nevertheless, a wise choice of the system can greatly simplify the analysis of a thermodynamic problem. It also often pays explicitly to recognize the nature of a chosen system's *immediate* surroundings. The following are typical useful choices of system and surroundings.

a) The system is the reaction $NH_4NO_3(s) \rightarrow NH_4^+(aq) + NO_3^-(aq)$. This means that the system includes solid ammonium nitrate, the water in which it dissolves and the aquated ions that are the products of the dissolution process. The inclusion of water in the system is indicated only rather subtly (by the (aq)'s on the formulas of the product ions).

The surroundings include the flask or beaker in which the system is held, the air above the system, and other neighboring materials. The dissolution of ammonium nitrate is $\boxed{\text{spontaneous}}$. Before the process can proceed, any physical separation (such as a glass wall or a space of air) between the water and the ammonium nitrate must be removed. The parts (sub-systems) of a system need not be physically contiguous.

b) The system is the reaction $H_2(g) + O_2(g) \rightarrow$ products. Its surroundings are the walls of the bomb and other portions of its environment that might deliver heat or work to or else absorb heat or work. The reaction of hydrogen with oxygen is $\boxed{\text{spontaneous}}$. Once hydrogen and oxygen are mixed in a closed bomb, $\boxed{\text{no constraint}}$ exists to prevent their reaction. That is, the system just defined is thermodynamically unstable with respect to the explosion. It is found experimentally that this system gives products quite slowly at room temperature (no immediate explosion). It explodes instantly at higher temperatures.

c) The system is the rubber band. The surroundings consist of the weight (visualized as attached to the lower end of the rubber band), a hangar at the top of the rubber band, and the air in contact with the rubber band. The change is $\boxed{\text{spontaneous}}$ once a constraint such as a stand or support underneath the weight is removed.

d) The system is the gas contained in the chamber. The surroundings are the walls of the chamber and the moveable piston head. The process is spontaneous if the force exerted by the weight on the piston exceeds the force exerted by the collisions of the molecules of the gas on the bottom of the piston.[2] Because slow compression of the gas is observed, the change is $\boxed{\text{spontaneous}}$.

---

[2]The forces due to the mass of the piston itself and friction between the piston and the walls within which it slides are neglected.

e) The system is the drinking glass in the "reaction"

$$\text{glass} \rightarrow \text{fragments}$$

The surroundings are the floor, the air, and the other materials in the room. The change is $\boxed{\text{spontaneous}}$. It occurs when the constraint, which is whatever portion of the surroundings holds the glass above the floor, is removed.

8-3 a) The number of available microstates equals the number of possible ways for a number to come up on one die times the number of possible ways for a number to come up on the other die. Each die has six faces and therefore 6 available microstates. The total number of available microstates is $\boxed{36}$.

b) The probability that one die will show a six is 1/6. The same is true for the other die. Thus, the probability that two sixes show up at the same time is $(1/6)(1/6) = \boxed{1/36}$.

8-5 The driving force for

$$\text{H}_2\text{O}(l) + \text{D}_2\text{O}(l) \rightarrow 2\,\text{HOD}(l)$$

is the $\boxed{\text{tendency for the entropy to increase}}$. Two moles of HOD have a larger entropy than a mixture of one mol of H_2O and one mol of D_2O because there is a much larger number of ways for the available H's, D's and O's to be assembled into a collection of HOD molecules than into a collection of H_2O's and D_2O's. The change occurs spontaneously, once the reactants are mixed, even if the system is completely separated from its surroundings.

Tip. A mixture of HOD, H_2O, and D_2O will have an even larger entropy than either pure products or pure reactants. The reaction comes to equilibrium in an intermediate state consisting of just such a mixture.

8-7 Before the stopcock is opened, the number of microstates available to a single H_2 (or He) is proportional to the volume of the glass bulb: $\Omega = cV$ where c is a constant. There are N_0 molecules of H_2 and N_0 atoms of He. The number of possible microstates for each gas is:

$$\Omega_{\text{H}_2} = (cV)^{N_0} \quad \text{and} \quad \Omega_{\text{He}} = (cV)^{N_0}$$

The number of microstates of the entire system, still before the valve is opened, is the product of the Ω's:

$$\Omega_{\text{sys}} = \Omega_{\text{H}_2}\Omega_{\text{He}} = (cV)^{2N_0}$$

This is the number of microstates that have all of the H_2 in the first bulb and all of the He in the second. By symmetry it is also the number of microstates that have all of the H_2 in the *second* bulb and all the He in the first. *After* the stopcock is opened, $2N_0$ molecules occupy a volume of $2V$ and:

$$\Omega_{\text{sys}} = (c\,2V)^{2N_0}$$

The probability p of the "cross-diffused" result, the state in which the H_2 and He trade places, is the number of ways in which it can be constituted divided by the number of ways in which the mixed system can be constituted:

$$p = \frac{(cV)^{2N_0}}{(c2V)^{2N_0}} = 2^{-2N_0}$$

Take the logarithm of both sides of this equation:

$$\log p = -2N_0 \log 2 = -2N_0(0.301) = -3.62 \times 10^{23}; \quad \text{hence}: \quad p = \boxed{10^{-3.62 \times 10^{23}}}$$

8-9 If the amount of disorder in a system increases when a process occurs, the change in entropy ΔS of that system is positive.

a) When NaCl melts it goes from a highly ordered solid to a relatively disordered liquid state. This means an increase in disorder: $\boxed{\Delta S > 0}$.

b) When a building is demolished it goes from a ordered state to a highly disordered state: $\boxed{\Delta S > 0}$.

c) In this case a highly disordered system (air) has order imposed on it by the separation into three distinct subsystems. An increase in order means $\boxed{\Delta S < 0}$.

8-11 a) The maximum theoretical efficiency ϵ of an engine operating between two temperatures is attained when the engine operates reversibly. This maximum efficiency is, according to text equation 8-4:[3]

$$\epsilon = \frac{T_{\text{h}} - T_\ell}{T_{\text{h}}} = 1 - \frac{T_\ell}{T_{\text{h}}}$$

In this problem, T_ℓ is 300 K and T_{h} is 450 K so ϵ is $\boxed{0.333}$.

b) The efficiency of the engine is the ratio of the net work it *performs* to the heat that it *absorbs*:

$$\epsilon = \frac{-w_{\text{net}}}{q}$$

---

[3]Text page 254.

The minus sign is necessary to adhere to the convention that $+w$ is work absorbed. If 1500 J of heat is absorbed per cycle from the 400 K reservoir and ϵ is 0.333, then w_{net} is -500 J in each turn of the cycle. It follows from the first law that the engine discards 1000 J of heat $\boxed{q = -1000 \text{ J}}$ into the low-temperature reservoir during each cycle.

c) The engine has absorbed 1500 J of heat during one portion of the cycle of operation. It must lose this amount of energy by the time it completes the cycle (for which ΔE is zero). Of the 1500 J, 1000 J goes to the 300 K reservoir as heat. Accordingly, 500 J appears as work done by the engine. That is, $\boxed{w = -500 \text{ J}}$.

8-13 Consider a system consisting of solid tungsten at its melting point of 3410°C (3683 K). Imagine supplying 35.4 kJ of heat infinitely slowly and in such a way that the temperature stays constant but 1 mol of tungsten melts. 35.4 kJ then equals q_{rev} for the melting. Substitute this value and T in the definition of entropy. Since the change is at a constant temperature, T can be taken outside the integral sign:

$$\Delta S = \int \frac{dq_{\text{rev}}}{T} = \frac{1}{T} q_{\text{rev}} = \frac{35.4 \times 10^3 \text{ J mol}^{-1}}{3683 \text{ K}} = \boxed{9.61 \text{ J K}^{-1}\text{mol}^{-1}}$$

Tip. The temperature must be an absolute temperature (in kelvins, for example).

8-15 Trouton's rule states that the molar entropy of vaporization ΔS_{vap} of most liquids equals 88 ± 5 J mol^{-1}K^{-1}. Use the rule to estimate the molar enthalpy of vaporization of acetone as follows:

$$\Delta H_{\text{vap}} \approx (329.35 \text{ K})\left(88 \text{ J K}^{-1}\text{mol}^{-1}\right) = \boxed{29 \times 10^3 \text{ J mol}^{-1}}$$

The experimental ΔH_{vap} of acetone equals 30.2×10^3 J mol^{-1}.

8-17 Assume that the 4.00 mol of hydrogen behaves ideally. The internal energy of an ideal gas depends solely on its absolute temperature T. In an isothermal process, T does not change. Hence, ΔE equals $\boxed{\text{zero}}$.

To evaluate ΔH, use its definition:

$$\Delta H = \Delta E + \Delta(PV) \qquad \text{which implies:} \qquad \Delta H = \Delta E + nR\Delta T$$

since $PV = nRT$. But ΔT and ΔE equal zero. Hence, ΔH equals $\boxed{\text{zero}}$.

The work done *on* the gas during the reversible isothermal expansion from 12.0 L to 30.0 L is:

$$w = -nRT \ln\left(\frac{V_2}{V_1}\right) = -4.00 \text{ mol}\left(\frac{8.315 \text{ J}}{\text{mol K}}\right)(400 \text{ K}) \ln\left(\frac{30.0}{12.0}\right) = \boxed{-12.2 \text{ kJ}}$$

The first law requires that if $\Delta E = 0$, then $q = -w$. This means the gas absorbs 12.2 kJ of heat during its expansion, just enough to balance off the 12.2 kJ of work that it performs: $q = \boxed{+12.2 \text{ kJ}}$.

Finally, $\Delta S = q_{rev}/T$ for an isothermal process, and q_{rev} is the q just computed:

$$\Delta S = \frac{q_{rev}}{T} = \frac{+12.2 \times 10^3 \text{ J}}{400 \text{ K}} = \boxed{+30.5 \text{ J K}^{-1}}$$

8-19 Break down the overall process to the three steps described in the problem and calculate ΔS_{sys} for each. Then add up the three contributions. The steps are: I, warming of ice; II, melting of ice; III, warming of melted ice. As the text shows,[4] ΔS for any temperature change at constant pressure is given by the equation:

$$\Delta S = nc_p \ln\left(\frac{T_2}{T_1}\right)$$

Use this formula to obtain ΔS of the system for the first and third steps:

$$\Delta S_I = (1.00 \text{ mol})(38 \text{ J K}^{-1}\text{mol}^{-1})\ln(273.15/253.15) = 2.9 \text{ J K}^{-1}$$
$$\Delta S_{III} = (1.00 \text{ mol})(75 \text{ J K}^{-1}\text{mol}^{-1})\ln(293.15/273.15) = 5.3 \text{ J K}^{-1}$$

In the second step, T stays at 273.15 K, and ΔS equals the quantity of heat absorbed reversibly by the system (q_{rev}) divided by this temperature:

$$\Delta S_{II} = \frac{6007 \text{ J}}{273.15 \text{ K}} = 21.99 \text{ J K}^{-1}$$

The *total* ΔS of the system equals:

$$\Delta S_{sys} = \Delta S_I + \Delta S_{II} + \Delta S_{III} = \boxed{+30.2 \text{ J K}^{-1}}$$

The entire process is reversible so the entropy of the universe remains constant: $\Delta S_{univ} = \Delta S_{sys} + \Delta S_{surr} = \boxed{0}$, which means $\Delta S_{surr} = \boxed{-30.2 \text{ J K}^{-1}}$.

8-21 Hot iron is plunged into cool water. The final temperature is 16.5°C (289.65 K). This process is far from reversible. Nevertheless, the ΔS of the iron and the ΔS of the water may be computed using the equation:

$$\Delta S = nc_p \ln\left(\frac{T_2}{T_1}\right)$$

---

[4]Text page 258.

This works because entropy is a state function. Its change depends only on the original and final states of the system, not on the path by which the change occurs.

As computed in text Example 7-3[5], the iron cools from 373.15 K to 289.65 K. A 72.4 g mass of iron amounts to 1.296 mol, which equals 72.4 g divided by 55.847 g mol^{-1}, the molar mass of iron. Hence:

$$\Delta S_{\text{Fe}} = nc_{\text{p}} \ln\left(\frac{T_2}{T_1}\right)$$

$$= (1.296 \text{ mol})(25.1 \text{ J K}^{-1}\text{mol}^{-1}) \ln\left(\frac{289.65 \text{ K}}{373.15 \text{ K}}\right) = \boxed{-8.24 \text{ J K}^{-1}}$$

The 100.0 g of water equals 5.55 mol. The c_{p} of water is 75.3 J K^{-1}mol^{-1}, and the water is warmed from 283.15 K to 289.65 K. Substituting as before:

$$\Delta S_{\text{H}_2\text{O}} = nc_{\text{p}} \ln\left(\frac{T_2}{T_1}\right) = 5.55 \text{ mol}(75.3 \text{ J K}^{-1}\text{mol}^{-1}) \ln\left(\frac{289.65}{283.15}\right) = \boxed{+9.49 \text{ J K}^{-1}}$$

The overall ΔS of the system is $\boxed{+1.25 \text{ J K}^{-1}}$, which is the sum of the ΔS's of the water and the iron.

8-23 a) The $\Delta S°$ of the reaction as written equals the standard molar entropies (the $S°$'s) of the products, each multiplied by its chemical amount in the balanced equation, minus the $S°$'s of the reactants, each multiplied by its chemical amount in the balanced equation:

$$\Delta S° = 2\underbrace{(239.95)}_{\text{NO}_2(g)} + 2\underbrace{(69.91)}_{\text{H}_2\text{O}(l)} - 1\underbrace{(121.21)}_{\text{N}_2\text{H}_4(l)} - 3\underbrace{(205.03)}_{\text{O}_2(g)} = \boxed{-116.58 \text{ J K}^{-1}}$$

where the numbers in parenthesis are standard molar entropies (in J K^{-1}mol^{-1}) from text Appendix D and the coefficients are the numbers of moles from the balanced chemical equation.

b) In the process $\text{N}_2\text{H}_4(l) \rightarrow \text{N}_2\text{H}_4(g)$, the disorder in the N_2H_4 increases, and therefore its ΔS exceeds 0. In other words, $S°$ for $\text{N}_2\text{H}_4(g)$ is more positive than $S°$ for $\text{N}_2\text{H}_4(l)$. This will cause $\Delta S°$ of the reaction of gaseous N_2H_4 with oxygen to be $\boxed{\text{algebraically smaller}}$ than when liquid N_2H_4 reacts.

8-25 The computations use the method of **8-23**.

$$\text{For LiCl:} \quad \Delta S° = 2\underbrace{(59.33)}_{\text{LiCl}(s)} - 2\underbrace{(29.12)}_{\text{Li}(s)} - 1\underbrace{(222.96)}_{\text{Cl}_2(g)} = \boxed{-162.54 \text{ J K}^{-1}}$$

---

[5]Text page 214.

For NaCl: $\Delta S° = 2\underbrace{(72.13)}_{\text{NaCl}(s)} -2\underbrace{(51.21)}_{\text{Na}(s)} -1\underbrace{(222.96)}_{\text{Cl}_2(g)} = \boxed{-181.12 \text{ J K}^{-1}}$

For KCl: $\Delta S° = 2\underbrace{(82.59)}_{\text{KCl}(s)} -2\underbrace{(64.18)}_{\text{K}(s)} -1\underbrace{(222.96)}_{\text{Cl}_2(g)} = \boxed{-186.14 \text{ J K}^{-1}}$

For RbCl: $\Delta S° = 2\underbrace{(95.90)}_{\text{RbCl}(s)} -2\underbrace{(76.78)}_{\text{Rb}(s)} -1\underbrace{(222.96)}_{\text{Cl}_2(g)} = \boxed{-184.72 \text{ J K}^{-1}}$

For CsCl: $\Delta S° = 2\underbrace{(101.17)}_{\text{CsCl}(s)} -2\underbrace{(85.23)}_{\text{CsCl}(s)} -1\underbrace{(222.96)}_{\text{Cl}_2(g)} = \boxed{-191.08 \text{ J K}^{-1}}$

The $\Delta S°$'s grow increasingly negative moving down the group, but RbCl is an exception.

8-27 By the second law, $\Delta S_{\text{univ}} = \Delta S_{\text{sys}} + \Delta S_{\text{surr}} > 0$. In this example, $\Delta S_{\text{sys}} = -44.7$ J K^{-1}. Thus, $\boxed{\Delta S_{\text{surr}} > +44.7 \text{ J K}^{-1}}$.

8-29 The change is the breakdown of $SiO_2(s)$ to solid silicon and gaseous oxygen. The products consist of a mole of solid and a mole of gas, but the reactant is simply a mole of solid. The products have many more possible microstates both because there are more particles and because the particles are less well-organized.

8-31 a) Solid ammonia is held at a constant temperature of 170 K. It is implied that the pressure is a constant 1 atm.

$$\Delta G = \Delta H - T\Delta S$$
$$= 5.65 \text{ kJ mol}^{-1} - (170 \text{ K})(0.0289 \text{ kJ K}^{-1}\text{mol}^{-1}) = \boxed{0.74 \text{ kJ mol}^{-1}}$$

b) This case differs from previous only in the amount of ammonia; the T and P are the same. The change in free energy is larger because more ammonia is involved. For 1.00 mol of NH_3, ΔG equalled 0.74 kJ. The ΔG for 3.60 mol of NH_3 is simply 3.60 times this value, which is $\boxed{2.65 \text{ kJ}}$.

c) At 170 K, $\Delta G > 0$. Hence the melting of ammonia is $\boxed{\text{not spontaneous}}$ at 170 K (and 1 atm pressure).

d) If solid and liquid NH_3 are in equilibrium, then ΔG equals zero for the process solid $\rightleftharpoons$ liquid. Calculate the T that makes this true using the molar enthalpy and the molar entropy changes quoted in the problem:

$$\Delta G = \Delta H - T\Delta S = 0 \quad \text{hence:} \quad T = \frac{\Delta H}{\Delta S} = \frac{5.65 \times 10^3 \text{ J mol}^{-1}}{28.9 \text{ J K}^{-1}\text{mol}^{-1}} = \boxed{196 \text{ K}}$$

8-33 When 1.00 mol of ethanol is vaporized at its normal boiling point, ΔH equals 38.7 kJ. The vaporization goes on at constant pressure, so $q_p = \Delta H$ and q is $\boxed{38.7 \text{ kJ}}$. The vaporization is isothermal and reversible, so q is also q_{rev}. Hence:

$$\Delta S = \frac{q_{rev}}{T} = \frac{38.7 \text{ kJ}}{351.1 \text{ K}} = 0.110 \text{ kJ K}^{-1} = \boxed{110 \text{ J K}^{-1}}$$

Now for the calculation of ΔE. From the definition of enthalpy:

$$\Delta E = \Delta H - \Delta(PV)$$

At constant pressure $\Delta(PV) = P\Delta V = P(V_2 - V_1)$. In this case, V_2 is the volume of one mole of vaporous ethanol at 351.1 K and V_1 is the volume of one mole of liquid ethanol, also at 351.1 K. The vapor behaves ideally:

$$V_2 = \frac{nRT}{P} = \frac{1.00 \text{ mol}(0.08206 \text{ L atm mol}^{-1}\text{K}^{-1})(351.15 \text{ K})}{1.00 \text{ atm}} = 28.8 \text{ L}$$

The volume of one mole of liquid ethanol (V_1) is less than 0.1 L, which makes it negligibly small compared to 28.8 L. Therefore:

$$P\Delta V = P(V_2 - V_1) = (1.00 \text{ atm})(28.8 \text{ L}) = 28.8 \text{ L atm} \times \left(\frac{0.101325 \text{ kJ}}{1 \text{ L atm}}\right) = 2.92 \text{ kJ}$$

Substitute these values into the expression for ΔE:

$$\Delta E = \Delta H - \Delta(PV) = 38.7 \text{ kJ} - 2.92 \text{ kJ} = \boxed{35.8 \text{ kJ}}$$

By expanding against a constant pressure, the system performs +2.92 kJ of pressure-volume work on its surroundings. This is the only kind of work possible. The total work done on the system is $\boxed{-2.92 \text{ kJ}}$.

For any reversible processes at constant T and P, $\Delta G = 0$. This can be verified in this case:

$$\Delta G = \Delta H - T\Delta S = 38.7 \text{ kJ} - (351.15 \text{ K})(0.110 \text{ kJ K}^{-1}) = \boxed{0.0 \text{ kJ}}$$

8-35 Add the two reactions given in the problem and their ΔG's:

$$2\,Fe_2O_3(s) \rightarrow 4\,Fe(s) + 3\,O_2(g) \qquad \Delta G = +840 \text{ kJ}$$
$$3\,C(s) + 3\,O_2(g) \rightarrow 3\,CO_2(g) \qquad \Delta G = -1200 \text{ kJ}$$

$$2\,Fe_2O_3(s) + 3\,C(s) \rightarrow 4\,Fe(s) + 3\,O_2(g) \qquad \Delta G = -360 \text{ kJ}$$

The last reaction is spontaneous because it has a negative ΔG. The removal of O_2 by reaction with C drives the decomposition of the Fe_2O_3.

8-37 a) Calculate $\Delta H^\circ_{298.15}$ and $\Delta S^\circ_{298.15}$ from the data in Appendix D:

$$\Delta H^\circ_{298.15} = 2\underbrace{(-824.2)}_{Fe_2O_3(s)} - 4\underbrace{(0.00)}_{Fe(s)} - 3\underbrace{(0.00)}_{O_2(g)} = -1648.4 \text{ kJ}$$

$$\Delta S^\circ_{298.15} = 2\underbrace{(87.40)}_{Fe_2O_3(s)} - 4\underbrace{(27.28)}_{Fe(s)} - 3\underbrace{(205.03)}_{O_2(g)} = -549.41 \text{ J K}^{-1}$$

The problem asks for the temperature range in which the reaction is spontaneous. The changeover from spontaneity to non-spontaneity occurs at $\Delta G^\circ = 0$. Use the relationship $\Delta G^\circ = \Delta H^\circ - T\Delta S^\circ$ to obtain the temperature T° that makes ΔG° equal zero. Take $\Delta H^\circ_{298.15}$ and $\Delta S^\circ_{298.15}$ as close approximations to the actual values of ΔH° and ΔS° at whatever T° turns out to be. Remember to convert $\Delta S^\circ_{298.15}$ to kJ K^{-1} or $\Delta H^\circ_{298.15}$ to J. The units must cancel:

$$T^\circ \approx \frac{\Delta H^\circ_{298.15}}{\Delta S^\circ_{298.15}} = \frac{-1648.1 \text{ kJ}}{-0.54941 \text{ kJ K}^{-1}} = 3000 \text{ K}$$

Because ΔH° and ΔS° are both negative, the reaction is $\boxed{\text{spontaneous below 3000 K}}$. Above 3000 K the ever-growing $-T\Delta S^\circ$ term finally makes ΔG° positive.

b) Perform similar calculations:

$$\Delta H^\circ_{298.15} = \underbrace{(-395.72)}_{SO_3(g)} - \underbrace{(-296.83)}_{SO_2(g)} - 0.5\underbrace{(0.00)}_{O_2(g)} = -98.89 \text{ kJ}$$

$$\Delta S^\circ_{298.15} = \underbrace{(256.65)}_{SO_3(g)} - \underbrace{(248.11)}_{SO_2(g)} - 0.5\underbrace{(205.03)}_{O_2(g)} = -93.98 \text{ J K}^{-1}$$

$$T^\circ \approx \frac{\Delta H^\circ_{298.15}}{\Delta S^\circ_{298.15}} = \frac{-98.89 \text{ kJ}}{-0.09398 \text{ kJ K}^{-1}} = 1052 \text{ K}$$

Since ΔH° and ΔS° are both negative, the reaction is $\boxed{\text{spontaneous below 1050 K}}$.

c)

$$\Delta H^\circ_{298.15} = \underbrace{(82.05)}_{N_2O)(g)} + 2\underbrace{(-241.82)}_{H_2O(g)} - \underbrace{(-365.56)}_{NH_4NO_3(s)} = -36.03 \text{ kJ}$$

$$\Delta S^\circ_{298.15} = \underbrace{(219.74)}_{N_2O)(g)} + 2\underbrace{(188.72)}_{H_2O(g)} - \underbrace{(151.08)}_{NH_4NO_3(s)} = 446.10 \text{ J K}^{-1}$$

$$T^\circ \approx \frac{\Delta H^\circ_{298.15}}{\Delta S^\circ_{298.15}} = \frac{-36.03 \text{ kJ}}{0.44610 \text{ kJ K}^{-1}} = -80.7 \text{ K}$$

A negative absolute temperature is physically meaningless. In this calculation it signals that the reaction is either never spontaneous ($\Delta S°$ positive and $\Delta H°$ negative) or always spontaneous ($\Delta S°$ negative and $\Delta H°$ positive).[6] Since $\Delta S°$ is positive, this reaction is $\boxed{\text{spontaneous at all temperatures.}}$

8-39 The reduction reaction is:

$$\boxed{WO_3(s) + 3\,H_2(g) \to W(s) + 3\,H_2O(g)}$$

Calculate $\Delta H°$ and $\Delta S°$ for this process using the data in Appendix D. The set-up is the same as in **8-37**:

$$\Delta H°_{298.15} = 1\underbrace{(0.00)}_{W(s)} + 3\underbrace{(-241.82)}_{H_2O(g)} - 1\underbrace{(-842.87)}_{WO_3(s)} - 3\underbrace{(0.00)}_{H_2(g)} = +117.41 \text{ kJ}$$

$$\Delta S°_{298.15} = 1\underbrace{(32.64)}_{W(s)} + 3\underbrace{(188.72)}_{H_2O(g)} - 1\underbrace{(75.90)}_{WO_3(s)} - 3\underbrace{(130.57)}_{H_2(g)} = +131.19 \text{ J K}^{-1}$$

$\boxed{\text{Because } \Delta H° \text{ and } \Delta S° \text{ are both positive}}$, the reaction becomes spontaneous at high enough temperature. The changeover temperature is:

$$T° = \frac{\Delta H°}{\Delta S°} = \frac{117.41 \times 10^3 \text{ J}}{131.19 \text{ J K}^{-1}} = \boxed{895 \text{ K}}$$

Tip. The reaction *does* proceed to some extent at temperatures below 895 K, but reactants predominate.

8-41 The liquid and gaseous forms of a substance are in equilibrium at its normal boiling point. "Normal" means that the pressure equals 1.000 atm. It follows that ΔG for boiling equals zero as long as both liquid and vapor are present at 1 atm.

$$\text{If} \qquad \Delta G = \Delta H_{vap} - T\Delta S_{vap} = 0 \qquad \text{then} \qquad \Delta S_{vap} = \frac{\Delta H_{vap}}{T}$$

Substitution of the values from the problem gives:

$$\Delta S_{vap} = \frac{38.74 \times 10^3 \text{ J}}{351.6 \text{ K}} = \boxed{110.2 \text{ J K}^{-1}\text{mol}^{-1}}$$

The computation is identical to **8-33** except that an additional significant figure is available.

Trouton's rule states that ΔS_{vap} is close to 88 J K^{-1}mol^{-1} for all liquids. The ΔS_{vap} for ethanol is 25% higher than predicted by Trouton's rule.

---

[6]See the discussion on text page 271.

Tip. The ΔS_{vap} and ΔH_{vap} in the preceding computation are *not* the same as $\Delta H^{\circ}_{298.15}$ and $\Delta S^{\circ}_{298.15}$ for the vaporization of ethanol. Computing these two values from the 298.15 K data on $C_2H_5OH(l)$ and $C_2H_5OH(g)$ in Appendix D confirms this:

$$\text{For } C_2H_5OH(l) \rightarrow C_2H_5OH(g) \qquad \Delta H^{\circ}_{298.15} = 42.59 \text{ kJ} \quad \Delta S^{\circ}_{298.15} = 121.89 \text{ J K}^{-1}$$

The reason for the difference is that boiling takes place at the boiling point and not at 25°C. This distinction is the point of problem **8-58**. Using the 298.15 K data gives 76.3°C as the boiling point of ethanol, which is more than 2°C low.

8-43 a) The compression of the oxygen is reversible and adiabatic. This means $q_{rev} = 0$. Therefore, ΔS_{sys} equals $\boxed{\text{zero}}$.

b) When an ideal gas is compressed reversibly and adiabatically from an initial (P_1, V_1) to a final state (P_2, V_2) then:

$$\frac{P_1}{P_2} = \left(\frac{V_2}{V_1}\right)^{\gamma}$$

where γ is the ratio of c_p to c_v of the gas. For oxygen γ equals 29.4 J $K^{-1}mol^{-1}$ divided by 21.09 J $K^{-1}mol^{-1}$ or 1.394. The original volume (V_1) of the 2.60 mol of oxygen in this problem is 64.0 L, as computed using the ideal-gas equation with $T_1 = 300$ K. Substitute this V_1, P_2 (8.00 atm), P_1 (1.00 atm) and γ into the preceding:

$$\frac{1.00 \text{ atm}}{8.00 \text{ atm}} = \left(\frac{V_2}{64.0 \text{ L}}\right)^{1.394} \qquad \text{so that} \qquad V_2 = 14.4 \text{ L}$$

Inserting P_2 and V_2 in the ideal-gas equation gives T_2. In summary, states 1 and 2 of the 2.60 mol of O_2 are:

$$P_1 = 1.00 \text{ atm} \qquad V_1 = 64.0 \text{ L} \qquad T_1 = 300 \text{ K}$$
$$P_2 = 8.00 \text{ atm} \qquad V_2 = 14.4 \text{ L} \qquad T_2 = 540 \text{ K}$$

The problem traces an alternative path from state 1 to state 2. The oxygen is first heated to T_2 at constant pressure and then compressed reversibly and isothermally to P_2. Compute all of the state variables in the *intermediate* state (subscripted i), after the isochoric heating but before the isothermal compression:

$$P_i = 1.00 \text{ atm} \qquad T_i = 540 \text{ K} \qquad V_i = 115.2 \text{ L}$$

The volume comes from the ideal-gas law, with $n = 2.60$ mol. The entropy change during the constant-pressure heating is:

$$\Delta S_{1 \rightarrow i} = nc_p \ln\left(\frac{T_i}{T_1}\right) = (2.60 \text{ mol})(29.4 \text{ J K}^{-1}mol^{-1}) \ln\left(\frac{540}{300}\right) = 44.9 \text{ J K}^{-1}$$

The entropy change during the constant-temperature compression is:

$$\Delta S_{i \to 2} = nR \ln \left(\frac{V_2}{V_i} \right) = (2.60 \text{ mol})(8.315 \text{ J K}^{-1}\text{mol}^{-1}) \ln \left(\frac{14.40}{115.2} \right) = -45.0 \text{ J K}^{-1}$$

The ΔS for the overall process is the sum of these two values. It equals $\boxed{\text{zero}}$, allowing for round-off errors.

8-45 a) If the motion of air masses through the atmosphere is adiabatic and reversible, then q_{rev} equals zero, and ΔS equals $\boxed{\text{zero}}$.

b) Upward displacement of an air mass causes its temperature and pressure to drop concurrently. Break down this overall process into two parts: a temperature change at constant pressure (step I) and a pressure change at constant temperature (step II). The initial values of temperature and pressure are T_0 and P_0 and the final values are T and P. For the two steps:

$$\Delta S_{\text{I}} = nc_{\text{p}} \left(\frac{T}{T_0} \right) \qquad \Delta S_{\text{II}} = nR \ln \left(\frac{P_0}{P} \right)$$

In the first step, ΔS is *less* than zero. because cooling a system reduces its entropy. In the second step, ΔS is *greater* than zero. This step is the expansion of the air mass. The sum of the ΔS's must be zero because the overall process, the sum of the two steps, is isentropic:

$$\boxed{c_{\text{p}} \ln \left(\frac{T}{T_0} \right) + R \ln \left(\frac{P_0}{P} \right) = 0}$$

c) If $\ln(P/P_0)$ is approximately equal to $-\mathcal{M}gh/RT$, then:

$$-\ln \left(\frac{P}{P_0} \right) = \ln \left(\frac{P_0}{P} \right) \approx \frac{+\mathcal{M}gh}{RT}$$

Substitute this result into the final expression in part b and rearrange:

$$T \ln \left(\frac{T}{T_0} \right) \approx \frac{-\mathcal{M}gh}{c_{\text{p}}}$$

All of the quantities on the right side are given in the problem:

$$T \ln \left(\frac{T}{T_0} \right) \approx \frac{-(0.029 \text{ kg mol}^{-1})(9.8 \text{ m s}^{-2})(5.9 \times 10^3 \text{ m})}{29 \text{ J K}^{-1}\text{mol}^{-1}} = -57.8 \text{ K}$$

This means that T, the temperature on top of the mountain, fulfills the equation:

$$T \ln \left(\frac{T}{311 \text{ K}} \right) \approx -57.8 \text{ K}$$

where the sea-level temperature (311 K) has replaced T_0. Obtain T by guessing a few trial values and using a calculator: $T = 246$ K or $\boxed{-27°C}$.

8-47 In **8-22**, a 1.000 mol piece of iron at 100°C is plunged into a large reservoir of water at 0°C. It loses 2510 J to the water as its temperature falls from 373 K to 273 K. Its entropy decreases. The change is:

$$\Delta S_{Fe} = nc_p \ln \frac{T_2}{T_1} = (1.00 \text{ mol})(25.1 \text{ J K}^{-1}\text{mol}^{-1}) \ln \left(\frac{373.15}{273.15}\right) = \boxed{-7.83 \text{ J K}^{-1}}$$

a) The piece of iron is first cooled from 100 to 50°C and then from 50 to 0°C using two water reservoirs. It loses 1255 J of heat to the first reservoir and 1255 J of heat to the second. The entropy change of the first reservoir, which *absorbs* 1255 J of heat and which is so big it stays at 323.15 K, is:

$$\Delta S_I = \frac{q}{T} = \frac{1255 \text{ J}}{323.15 \text{ K}} = 3.88 \text{ J K}^{-1}$$

The entropy change of the second reservoir, which also absorbs 1255 J of heat but at 273.15 K, is larger:

$$\Delta S_{II} = \frac{q}{T} = \frac{1255 \text{ J}}{273.15 \text{ K}} = 4.60 \text{ J K}^{-1}$$

These two reservoirs comprise the surroundings of the iron:

$$\Delta S_{surr} = \Delta S_I + \Delta S_{II} = 3.88 + 4.60 = 8.48 \text{ J K}^{-1}$$

The ΔS of the iron is still -7.83 J K^{-1} because only the path by which it cooled has changed. It still ends up in the same final state. Therefore:

$$\Delta S_{univ} = \Delta S_{surr} + \Delta S_{Fe} = 8.48 \text{ J K}^{-1} - 7.83 \text{ J K}^{-1} = 0.65 \text{ J K}^{-1}$$

b) Each of the four reservoirs absorbs 627.5 J, one-fourth of the total given up by the iron. The entropy changes of the four reservoirs are:

$$\Delta S_I = \frac{627.5 \text{ J}}{348.15 \text{ K}} = 1.80 \text{ J K}^{-1} \qquad \Delta S_{II} = \frac{627.5 \text{ J}}{323.15 \text{ K}} = 1.94 \text{ J K}^{-1}$$

$$\Delta S_{III} = \frac{627.5 \text{ J}}{298.15 \text{ K}} = 2.11 \text{ J K}^{-1} \qquad \Delta S_{IV} = \frac{627.5 \text{ J}}{273.15 \text{ K}} = 2.30 \text{ J K}^{-1}$$

ΔS_{surr} is the sum of the ΔS's of the four reservoirs. It is 8.15 J K^{-1}. The ΔS of the iron is still -7.83 J K^{-1}. Therefore:

$$\Delta S_{univ} = \Delta S_{surr} + \Delta S_{Fe} = 8.15 - 7.83 \text{ J K}^{-1} = \boxed{0.32 \text{ J K}^{-1}}$$

c) Using four reservoirs makes the process more nearly reversible as evidenced by the smaller ΔS_{univ}. Making the process exactly reversible would require an infinite series of reservoirs each one absorbing an infinitesimal quantity of heat from the iron at a temperature infinitesimally less than the previous reservoir. The ΔS's of all the reservoirs would add up to $+7.83$ J K^{-1}, and ΔS_{univ} would be zero.

8-49 a) Several different ideal gases each occupy their own original volumes and all at the same temperature and pressure. Constraints are removed (for example, valves between the containers are opened) and the gases mix. Clearly ΔS is positive for the mixing. To get an expression for ΔS compute ΔS for each of the gases *separately*, and then add up the several contributions. Work with the i-th gas. This gas starts at V_1 and expands to V_2. In both state 1 and state 2, the entropy of the gases depends on its number of microstates Ω:

$$S_1 = k_B \ln \Omega_1 \qquad S_2 = k_B \ln \Omega_2$$

The *change* in entropy of the i-th gas is:

$$\Delta S_i = S_2 - S_1 = k_B \ln \Omega_2 - k_B \ln \Omega_1 = k_B \ln \left(\frac{\Omega_2}{\Omega_1}\right)$$

In both state 1 and state 2, the number of microstates available to one molecule of the gas is proportional to the volume ($\Omega = cV$). The number of microstates available to *all* the molecules of this gas is proportional to the volume raised to the power $n_i N_0$, the total number of molecules of the i-th gas (N_0 is Avogadro's number and n_i is the number of moles). The change in entropy for the i-th gas now equals:

$$\Delta S_i = k_B \ln \left(\frac{\Omega_2}{\Omega_1}\right) = k_B \ln \left(\frac{(cV_2)^{n_i N_0}}{(cV_1)^{n_i N_0}}\right) = n_i N_0 k_B \ln \left(\frac{V_2}{V_1}\right)$$

Now, focus on the term (V_2/V_1). By Boyle's law, it equals (P_1/P_2), the ratio of the original pressure of the i-th gas to the final partial pressure of the i-th gas in the mixture. This latter pressure is, by Dalton's law:

$$P_2 = X_i P_{tot}$$

where X_i is the mole fraction of the i-th gas. But P_1, the original pressure of the i-th gas, *equals* P_{tot}, because all of the gases started at the same pressure.[7] Therefore:

$$\frac{V_2}{V_1} = \frac{P_1}{P_2} = \frac{P_{tot}}{X_i P_{tot}} = \frac{1}{X_i}$$

---

[7]Opening a valve between a container of gas 1 at 2.0 atm and a container of gas 2 also at 2.0 atm gives a mixture at 2.0 atm. Pressure is an intensive property.

Substituting this result into the expression for ΔS_i gives

$$\Delta S_i = n_i N_0 k_B \ln \left(\frac{1}{X_i} \right) = -n_i N_0 k_B \ln X_i$$

Next, substitute R for $N_0 k_B$ and insert $X_i n$, where n is the total number of moles of gas, for n_i:

$$\Delta S_i = -(X_i n)(N_0 k_B) \ln X_i = -nRX_i \ln X_i$$

Finally, add up the contributions of all of the gases to get the overall ΔS:

$$\boxed{\Delta S = \sum_i \Delta S_i = -nR \sum_i X_i \ln X_i}$$

b) Calculate the mole fractions of O_2, N_2, and Ar in the mixture and substitute into the preceding formula. Divide 50 g by the respective molar masses of the three gases to obtain the chemical amount of each. The mole fraction of each is its chemical amount divided by the total chemical amount in the mixture. The results of these calculations are:

Gas	Chemical Amount	X (Mole Fraction)	$X \ln X$
O_2	1.563 mol	0.3399	−0.3668
N_2	1.784	0.3879	−0.3673
Ar	1.252	0.2722	−0.3542
	4.596	1.0000	−1.0883

The last row in the table contains the sums of the columns. Then:

$$\Delta S = -nR \sum_i X_i \ln X_i = -4.596 \text{ mol}(8.315 \text{ J K}^{-1}\text{mol}^{-1})(-1.0883) = \boxed{42 \text{ J K}^{-1}}$$

This is the entropy change of mixing at any temperature and pressure as long as the assumption of ideal-gas behavior holds.

c) Separating the components of air is the reverse of mixing them. The entropy change of separation is therefore the negative of the entropy change of mixing, assuming ideal-gas behavior. To solve the problem, compute ΔS_{sys} for the process of mixing and then change its sign. Table 4-1[8] gives the volume percentages of the various gases in the air. The mole fractions (X's) of the gases equal these numbers divided by 100. The following table give these mole fractions, and the quantity $X \ln X$ for each gas:

---

[8]Text page 99.

Gas	X (Mole Fraction)	$X \ln X$
N_2	0.78110	-0.19297
O_2	0.20953	-0.32747
Ar	0.00934	-0.04365
Ne	0.00001818	-0.0001976

Continuing the table to include more trace gases does not provide $X \ln X$ values significantly different from zero. The sum of the numbers in the last column equals -0.56429. Hence:

$$\Delta S = -nR \sum_i X_i \ln X_i$$

$$= -(4.09 \text{ mol})(8.315 \text{ J K}^{-1}\text{mol}^{-1})(-0.56429) = 19.2 \text{ J mol}^{-1}$$

where n was obtained using the ideal-gas law with V equal 100 L, T equal 298.15 K, and P equal 1 atm. The entropy change of separation of the components *of the system* is $\boxed{-19.2 \text{ J K}^{-1}}$.

Tip. The problem asks simply for the "entropy change." The entropy change of the universe cannot be calculated because there is no information about the surroundings of the system. It is of course certain that ΔS_{univ} exceeds zero.

8-51 The absolute entropy is proportional to the area under such a curve. Gold has the $\boxed{\text{higher}}$ absolute entropy at 200 K.

8-53 a) Higher temperature makes the conversion of rhombic to monoclinic sulfur a spontaneous process. From the discussion on text page 271, both $\Delta H°$ and $\Delta S°$ must then both be positive: if the two had unlike signs, then $T°$, the changeover temperature, would be negative (meaning no changeover). If both were negative, then the conversion would be favored by lower temperature.

b) Equilibrium between rhombic and monoclinic sulfur at constant temperature and pressure means $\Delta G° = \Delta H° - T\Delta S° = 0$. Rearranging and substituting the values from the problem gives:

$$\Delta S° = \frac{\Delta H°}{T} = \frac{400 \text{ J}}{368.5 \text{ K}} = \boxed{1.09 \text{ J K}^{-1}}$$

8-55 The reaction is $3\,CO_2(g) + Si_3N_4(s) \rightarrow 3\,SiO_2(s) + 2\,N_2(g) + 3\,C(s)$. The problem and text Appendix D supply the necessary $\Delta G_f°$ data:

$$\Delta G° = 3\,\underbrace{(-856.67)}_{SiO_2(s)} + 2\,\underbrace{(0)}_{N_2(g)} + 3\,\underbrace{(0)}_{C(s)} - 3\,\underbrace{(-394.36)}_{CO_2(g)} - 1\,\underbrace{(-642.6)}_{Si_3N_4(s)} = \boxed{-744.3 \text{ kJ}}$$

8-57 a) The reaction of interest is $2\,CuCl_2(s) \rightarrow 2\,CuCl(s) + Cl_2(g)$. Text Appendix D supplies ΔH_f° and S° values for the computation of ΔH° and ΔS° of this reaction:

$$\Delta H_{298.15}^\circ = 2\underbrace{(-137.2)}_{CuCl(s)} + 1\underbrace{(0)}_{Cl_2(g)} - 2\underbrace{(-220.1)}_{CuCl_2(s)} = 165.8 \text{ kJ}$$

$$\Delta S_{298.15}^\circ = 2\underbrace{(86.2)}_{CuCl(s)} + 1\underbrace{(222.96)}_{Cl_2(g)} - 2\underbrace{(108.07)}_{CuCl_2(s)} = 179.2 \text{ J K}^{-1}$$

b)

$$\Delta G_{590}^\circ \approx \Delta H_{298.15}^\circ - T\Delta S_{298.15}^\circ$$

$$\approx 165.8 \text{ kJ} - (590 \text{ K})(0.1792 \text{ kJ K}^{-1}) = \boxed{60.1 \text{ kJ}}$$

c) Use the experimental values at 590 K instead of the values at 298.15 K:

$$\Delta G_{590}^\circ = \Delta H_{590}^\circ - T\Delta S_{590}^\circ = 158.36 \text{ kJ} - (590 \text{ K})(0.17774 \text{ kJ K}^{-1}) = \boxed{53.5 \text{ kJ}}$$

The answer using $\Delta H_{298.15}^\circ$ and $\Delta S_{298.15}^\circ$ is about 12% larger than the actual ΔG_{590}°.

8-59 a) Potassium ions tend to transfer so as to equalize the K^+ concentrations on the two sides of the cell wall. Thus K^+ ions will go $\boxed{\text{from inside the muscle cells out}}$ into the surrounding fluids.

b) The problem is to compute ΔG for transporting 1.00 mol of K^+ ions from a concentration c_1 of 0.0050 M to a concentration c_2 of 0.15 M. The answer must be positive because the separation of a uniform concentration of K^+ ions into two regions of differing concentration is clearly non-spontaneous. Assume that the two solutions are ideal solutions and that T is the normal human body temperature of 37°C (310 K). Consult Section 9-2[9] where the text states the following for the change in the Gibbs free energy of the system as an ideal solution goes from c_1 to c_2:

$$\Delta G = nRT \ln\left(\frac{c_2}{c_1}\right)$$

Substitute the appropriate values:

$$\Delta G = (1.00 \text{ mol})(8.315 \text{ J K}^{-1}\text{mol}^{-1})(310 \text{ K}) \ln\left(\frac{0.15}{0.0050}\right) = \boxed{8770 \text{ J}}$$

---

[9]Specifically, see Text page 289.

Chapter 9

Chemical Equilibrium: Principles and Applications to Gas-Phase Reactions

The Nature of Chemical Equilibrium

A chemical reaction left to itself ultimately comes to a state of **equilibrium.** At equilibrium, all tendency for observable change has been exhausted, and the amounts of the reactants and products no longer change. Mechanical analogies are easy to find: the unwinding of a spring, the slump of a pile of gravel to a final angle of repose. Such analogies have some merit but most are imperfect. True chemical equilibria display *all* of the following characteristics:

- They show no macroscopic evidence of change.

- They are reached through spontaneous processes.

- A dynamic balance of forward and reverse processes exists within them.

- They are the same regardless of the direction from which they were approached.

Although visible signs of change cease once a reaction is at equilibrium, continuous exchange between the reactants and products nonetheless continues on the molecular level. Most analogies miss this.

An absence of observable change in a reaction system does *not* prove that the system is at equilibrium because:

- A system may be far from equilibrium, but drifting toward equilibrium so slowly that no change is observable.

- A **steady state** may exist. In a steady state, a dynamic balance exists between a process that supplies reactants to a system and a second process that removes products. A steady state is not an equilibrium state.

The Law of Mass Action

For a chemical reaction involving only gases such as:

$$a\,A(g) + b\,B(g) \rightleftharpoons c\,C(g) + d\,D(g)$$

the expression:

$$\frac{(P_C)^c_{eq}(P_D)^d_{eq}}{(P_A)^a_{eq}(P_B)^b_{eq}}$$

is very close to constant as long as the temperature is held constant. The subscript "eq" means that equilibrium partial pressures of the reactants and products must be used in the expression. Expression of this form are called **mass-action expressions** or **equilibrium expressions.** The numerical value of the mass-action expression is the **empirical equilibrium constant K_P:**

$$K_P = \frac{(P_C)^c_{eq}(P_D)^d_{eq}}{(P_A)^a_{eq}(P_B)^b_{eq}}$$

If arbitrary partial pressures of the gases A, B, C, and D are mixed, substitution in the mass-action expression gives a number that is not equal to K_P. The arbitrary mixture would tend to react in one direction or the other (as suggested by the double arrow in the balanced equation) until the value of the mass-action expression becomes equal to K_P.

For a reaction in solution, the mass-action expression has the same form but uses concentrations instead of partial pressures. The empirical equilibrium constant for a reaction among solutes is subscripted C (for concentration) instead of P:

$$K_C = \frac{[C]^c_{eq}[D]^d_{eq}}{[A]^a_{eq}[B]^b_{eq}}$$

In this equation, the concentration of a dissolved reactant or product (c_A for example) is indicated by a set of brackets around its formula. Thus, $[CO_2]_{eq}$ stands for a number, the equilibrium concentration of carbon dioxide, but CO_2 stands for a compound.

The observed constancy of K_P and K_C is the **law of mass action.** Note these points:

- Empirical equilibrium constants depend on the temperature.

- At a given temperature the value of K_P or K_C is characteristic of the reaction itself. It does not depend on the history or surroundings of the reaction. In particular, K_P or K_C is not affected by the presence of concurrent or competing reactions.

- A very large K_P or K_C mean that the products are strongly favored at equilibrium; a very small K_P or K_C mean that the reactants are strongly favored.

- Numerical values of K_P or K_C depend on the units chosen to express the equilibrium partial pressures or concentrations. The units of K_P are (press.)$^{c+d-a-b}$ where "press." stands for the selected unit of pressure. The units of K_C are (conc.)$^{c+d-a-b}$ where "conc." stands for the chosen unit of concentration.

- K_P and K_C have units except in the special case that $c + d = a + b$.

- The law of mass action applies to equilibria involving any combination of solids, liquids and gases. Equilibria across phase boundaries (such as a liquid-gas boundary) are *heterogeneous equilibria*. They are the topic of text Chapter 11.[1]

Writing mass-action expressions is an essential skill. See **9-1**, **9-3**, **9-5**, and **9-59b**.

Thermodynamic Description of Equilibrium

A reaction system that is undergoing spontaneous chemical change at constant T and P has a negative ΔG. Otherwise the change would not be taking place. The progress of the reaction raises ΔG upward toward zero. When ΔG finally equals zero, the reaction stops. Equilibrium has been reached. Part of the cause for changes in ΔG are changes in the partial pressures or concentrations of the reactants and products:

The free energies of gases and solutes depend on their respective partial pressures and concentrations.

The following equations give the change (at constant T and P) in the free energy of a substance if its partial pressure P or its concentration c is changed from value 1 to value 2:

$$\Delta G = nRT\left(\frac{P_2}{P_1}\right) \qquad \Delta G = nRT\left(\frac{c_2}{c_1}\right)$$

From these equations, the text shows that for gas-phase reactions (and for reactions in solution):

$$\boldsymbol{\Delta G^\circ = -RT \ln K} \qquad \text{or} \qquad \boldsymbol{\frac{-\Delta G^\circ}{RT} = \ln K}$$

---

[1]See text page 364 for a statement of the law of mass action for heterogeneous equilibria.

where K is not K_P or K_C, which have units, but rather a **thermodynamic equilibrium constant** that has no units.

Suppose that you have the partial pressures of gases A, B, C, and D at equilibrium in the reaction:

$$a\,A(g) + b\,B(g) \rightleftharpoons c\,C(g) + d\,D(g)$$

To obtain K for this system, measure the equilibrium partial pressures of the four gases, set up the mass-action expression, divide each equilibrium partial pressure by a reference partial pressure expressed in the same units, and insert all the now-unitless numbers into the expression. In brief:

$$K = \left[\frac{(P_C/P_{ref})_{eq}^c (P_D/P_{ref})_{eq}^d}{(P_A/P_{ref})_{eq}^a (P_B/P_{ref})_{eq}^b}\right]$$

Completing the arithmetic gives a pure number K.

The natural logarithm of this number equals $-\Delta G^\circ / RT$ *if* the reference pressure P_{ref} equals the standard-state pressure that was used in the compilation of the ΔG° values. The text uses a standard-state pressure of 1 atm exclusively.

Similar points apply for the K of a reaction taking place in solution. The only difference is that the division is by a reference concentration that equals the standard-state concentration used in the complication of the ΔG° values. The text uses a standard-state concentration of 1 M exclusively.

Calculation of K's from Calorimetric Data

The equation

$$\ln K = \frac{-\Delta G^\circ}{RT}$$

allows calculation of the thermodynamic equilibrium constant of a chemical reaction from tabulated calorimetric data. Text Appendix D is a source of such data.

Problems fall into these types:

• **Calculate K for a reaction at 25°C.** Balance the chemical equation, and locate the ΔG_f°'s of the reactants and products in Appendix D. These are 298.15 K (25°C) values. Multiply each substance's ΔG_f° by the coefficient that it has in the balanced reaction. Subtract the total for the reactants from the total for the products. The answer is $\Delta G_{298.15}^\circ$ for the reaction, in kilojoules. Regard $\Delta G_{298.15}^\circ$ as having the units "kilojoules per mole of the chemical reaction as written." Multiplication by 1000 J kJ^{-1} converts the units to J mol^{-1}. Subsequent division by 8.315 J mol^{-1}K^{-1} and 298.15 K gives $-\ln K$. See **9-7** and **9-9**.

The same procedure works to obtain a K for a reaction taking place in solution. See **9-10b**.

● **Calculate $\Delta G°$, given K.** This requires use of the same equation, but in the reverse direction. See **9-49a** and **22-7**. The K may be given for any temperature (not necessarily for 298.15 K). The resulting $\Delta G°$ is correct for temperature at which K is correct, and not at other temperatures.

Sometimes either $\Delta H°$ or $\Delta S°$ is also given, and the problem is extended to the calculation of whichever one is missing. See **9-49**. Simply use $\Delta G° = \Delta H° - T\Delta S°$.

● **Calculate K at T not equal to 25°.** Obtain $\Delta H°$ and $\Delta S°$ for the reaction by looking up $\Delta H_f°$ and $S°$ data in Appendix D and combining them according to Hess's law. Do not calculate $\Delta G°$ from the $\Delta G_f°$ data in Appendix D. $\Delta G°$'s depend strongly on temperature. The $\Delta G_{298.15}°$, which is what the $\Delta G_f°$ data provide, has nothing to do with K's at temperatures other than 298.15 K. By contrast, $\Delta H°$'s and $\Delta S°$'s depend only weakly on temperature. The thermodynamic equilibrium constant is calculated using:

$$\ln K_T = \frac{-\Delta G_T°}{RT} \approx \frac{-(\Delta H_{298.15}° - T\Delta S_{298.15}°)}{RT}$$

Equilibrium Calculations for Gas-Phase Reactions

Evaluating K's from Reaction Data

The thermodynamic equilibrium constant K is easily obtained if all the equilibrium partial pressures or all the equilibrium concentrations are available:

1. Write a balanced equation and set up the mass-action expression.

2. Express the pressures in atm and the concentrations in mol L^{-1}.

3. Divide equilibrium pressures by the reference pressure (1 atm) or the equilibrium concentrations by the reference concentration (1 mol L^{-1}). This amounts to discarding their units.

4. Substitute the resulting numbers into the mass-action expression.

5. Complete the arithmetic.

See **9-5b, 9-11,** and **9-65.**

Often, equilibrium partial pressures are not given explicitly. Instead, one or more *initial* partial pressures is either supplied explicitly or else can be obtained from other information (see **9-15**). Other times, quantities that are proportional to the equilibrium partial pressures are stated (**9-13**). Yet other times, a total pressure is given (**9-67**). The key to such problems is to use the known stoichiometry of the reaction to generate relationships that yield the required partial pressures. For example, if a balanced equation shows 2 mol of A giving 3 mol of B, and if neither A nor B is taking part in other reactions, then a decrease in P_A by $2x$ atm causes an increase in P_B by $3x$ atm. Tracing such relationships is assisted by drawing up a three-line "table of changes" of the form:

	reactants	$\rightleftharpoons$	products
Initial pressure (atm)	—		—
Change in pressure (atm)	—		—
Equilibrium pressure (atm)	—		—

Relationships Among Equilibrium Expressions

When the coefficients of a balanced chemical equation are all multiplied by a constant, the equation is still balanced. The corresponding mass-action expression is raised to a power equal to the multiplying constant. This holds if the multiplying constant is negative or fractional. Multiplying the coefficients by a negative number corresponds to reversing the direction of reaction. Multiplying by a fraction corresponds to dividing all the coefficients by a constant. Consider these two chemical equations and their associated mass-action expressions:

$$A(g) \rightleftharpoons B(g) + 2\,C(g) \qquad \left(\frac{P_B/P_{ref})_{eq}(P_C/P_{ref})^2_{eq}}{(P_A/P_{ref})_{eq}}\right) = K_1$$

$$\tfrac{1}{2}\,A(g) \rightleftharpoons \tfrac{1}{2}\,B(g) + C(g) \qquad \left(\frac{P_B/P_{ref})^{1/2}_{eq}(P_C/P_{ref})_{eq}}{(P_A/P_{ref})^{1/2}_{eq}}\right) = K_2$$

The second equation equals the first multiplied by one-half. Hence, $K_2 = \sqrt{K_1}$, because raising to the $1/2$ power means taking the square root.

- If two balanced chemical equations are added, then the corresponding mass-action expressions are multiplied. See **9-17**, **9-19**, and **9-55**.

- If one chemical equation is subtracted from a second, then the mass-action expression of the resulting equation equals the mass-action expression of the second *divided* by the mass-action expression of the first.

Whenever a number is quoted as an equilibrium constant, it should be accompanied by a chemical equation. For example, it is ambiguous to say "K for the synthesis of ammonia at $25°$ equals 6.78×10^5," Is this K for $N_2(g) + 3\,H_2(g) \rightleftharpoons 2\,NH_3(g)$ or perhaps $\tfrac{1}{2}\,N_2(g) + \tfrac{3}{2}\,H_2(g) \rightleftharpoons NH_3(g)$? See also the solution to **9-3**.

Calculating Equilibrium Compositions

The law of mass action relates the equilibrium partial pressures of several substances taking part in a reaction. It provides only one relationship, and several gases might join the reaction. Obtaining the partial pressures of all the gases in an equilibrium mixture almost always requires finding additional relationships among the partial pressures. Note these points:

- The partial pressures substituted into a mass-action expression must be *equilibrium* partial pressures. Problems often quote initial partial pressures; final (equilibrium) partial pressures usually differ sharply. See **9-33a.**

- The units of all partial pressures must be the same. Pressure is measured in many different units If a problem uses mixed units, convert to a single unit, preferably atmospheres.

- The initial partial pressures in a reaction mixture are determined by the person setting up the experiment. They may be given specifically in a problem as in **9-25** and **9-37.** When not specified they often equal zero (**9-26** and **9-37**) or a value implied by the wording of the problem (**9-21**, **9-23**, and **9-29**).

- The stoichiometry of the equation determines the *changes* in the partial pressures of reactants and products as a single chemical reaction moves toward equilibrium.

- The sum of the partial pressures of all of the gases in a system equals the total pressure. It often appears in problems because it is usually easy to measure in experiments. Knowing it provides an additional relationship among the several partial pressures. See **9-15**, **9-23**, and **9-67.**

To solve gas-phase equilibrium problems:

1. Write down the balanced equation and the corresponding mass-action expression.

2. Write down the initial partial pressures of all reactants and products. Include zero partial pressures for gases not initially present.

3. Determine the changes required in the partial pressures to reach equilibrium. If a change is not known, symbolize it with x or y. Changes occurring as the reaction goes to equilibrium are related by the coefficients in the balanced

equation. Keep track in a three-line table of the form:

	reactants $\rightleftharpoons$ products	
Initial pressure (atm)	—	—
Change in pressure (atm)	—	—
Equilibrium pressure (atm)	—	—

For each product and reactant, the entries on the third line equal the sum of the entries on the first and second lines. If changes are positive for the products, they are negative for the reactants. This format is used repeatedly in the text and in this Guide, starting with **9-15.** For contrast, the solution to **9-23** does not use this format, but does use the ideas behind it.

4. Substitute the equilibrium partial pressures into the mass-action expression and solve for the unknowns. Use the quadratic formula[2] if quadratic equations arise (they often do). Higher-order equations also develop. Numerical considerations can often lead to big simplifications (**9-27**), although analytical solution is always an option. See **9-29** and **9-61.**

5. Remember the x's or y's are related to the changes; they are seldom the desired answer. Calculate the equilibrium partial pressures by adding the changes to the initial partial pressures.

6. Examine the answer critically. Negative partial pressures are impossible. Negative *changes* in partial pressure correspond to a substance being used up in the reaction. They are possible, but getting them unexpectedly can be upsetting. In **9-27**, a negative change in partial pressure is avoided by a suitable definition of the direction of change.

Concentrations of Gases in Equilibrium Calculations

The concentration in mol L^{-1} of a substance is indicated by a set of brackets around its formula. This applies to gases as well as to solutes.

The concentration and partial pressure of a gas are related. For an ideal gas A:

$$[A] = \frac{n_A}{V} = \frac{P_A}{RT}$$

See **9-55, 9-57,** and **9-59.** Given this equation, it is easy to recast a mass-action expression from partial pressures to concentrations. This can often simplify calculations (see **9-31**), but obviously never influences the course of the reaction.

---

[2]Text page A-21.

For the general reaction:

$$a\,A(g) + b\,B(g) \rightleftharpoons c\,C(g) + d\,D(g)$$

the law of mass action in terms of concentrations of the gases is:

$$\frac{[C]_{eq}^c [D]_{eq}^d}{[A]_{eq}^a [B]_{eq}^b} = K \left(\frac{RT}{P_{ref}}\right)^{a+b-c-d} = K \left(\frac{RT}{P_{ref}}\right)^{-\Delta n_g}$$

The expression on the left has the form of the mass-action expression but uses equilibrium concentrations instead of equilibrium partial pressures. The quantity $(c+d-a-b)$ is symbolized Δn_g because it equals the change in the number of moles of gas between the two sides of the reaction.

In terms of empirical equilibrium constants, the above relationship is

$$\boldsymbol{K_C = K_P(RT)^{-\Delta n_g}}$$

The Reaction Quotient

The **reaction quotient** of a gas-phase reaction such as:

$$a\,A(g) + b\,B(g) \rightleftharpoons c\,C(g) + d\,D(g)$$

is defined as

$$Q = \frac{(P_C/P_{ref})^c (P_D/P_{ref})^d}{(P_A/P_{ref})^a (P_B/P_{ref})^b}$$

The form of the expression is the same as in the law of mass action, but the several partial pressures can have any arbitrary values and do not have to be equilibrium partial pressures. (Note the absence of subscript eq's.) For any reaction at constant T and P:

$$\Delta G = \Delta G° + RT \ln Q$$

Analysis of this expression confirms that at equilibrium, where ΔG equals zero, Q equals K. Away from equilibrium, ΔG is not equal to zero, and Q differs from K. Under all circumstances of constant T and P:

The reaction quotient adjusts to become equal to the equilibrium constant. That is, Q chases K.

The reaction quotient helps in determining the direction a reaction will take to come to equilibrium. If $Q < K$, then the reaction proceeds to the right. The change increases the numerator of Q and decreases its denominator, both of which make the quotient bigger. If $Q > K$, then the proceeds to the left. The change decreases the numerator of Q and increases the denominator. See **9-33**, **9-35**, and **9-63**.

An examination problem might give a balanced equation and ask for a calculation of remaining free energy that the reaction has to offer at some T and P. This would be ΔG in the equation:

$$\Delta G_T = \Delta G_T^\circ + RT \ln Q$$

Substituting for ΔG° gives

$$\Delta G_T = -RT \ln K_T + RT \ln Q = RT \ln \left(\frac{Q}{K_T} \right)$$

If $Q < K$, ΔG is negative, and the reaction proceeds from left to right. If $Q > K$ ΔG is positive and the reaction proceeds in reverse (from right to left). The problem would give T explicitly and then Q and K either explicitly or in some form of words. It might add nuance by furnishing information to compute Q and K_T (for example, by listing partial pressures at equilibrium and away from equilibrium).

Substitution of Q and K_T into the equation gives ΔG in units of kilojoules per mole of the reaction as written.

As Q chases K, their ratio in the preceding equation gets closer and closer to 1. Because $\ln 1$ equals zero, ΔG becomes equal to zero when Q catches K, that is, at equilibrium.

External Effects and LeChatelier's Principle

LeChatelier's principle states that if a stress is applied to a system at equilibrium, then the position of the equilibrium will shift in the direction that counteracts the stress. Many kinds of stresses occur:

- **Product removed.** The equilibrium shifts to the right partially to remedy the loss.

- **Reactant removed.** The equilibrium shifts to the left partially to compensate.

- **Product added.** The equilibrium shifts to the left, consuming some of the newly added product.

- **Reactant added.** The equilibrium shifts to the right, consuming some of the new reactant.

- **Temperature raised.** An exothermic equilibrium shifts to the left; an endothermic equilibrium shifts to the right. These shifts occur because the K of an exothermic reaction decreases with increasing T, but the K of an endothermic reaction increases with increasing T. **Exothermic** means generating heat, and **endothermic** means absorbing heat. Think of heat as a product in an exothermic reaction and a reactant in an endothermic reaction. Raising T adds "product heat" to an exothermic equilibrium, causing it to shift left, but adds "reactant heat" to an endothermic equilibrium, causing it to shift right. See **9-41**.

- **Temperature lowered.** An exothermic equilibrium shifts to the right; an endothermic equilibrium shifts to the left.

- **Volume decreased.** Shrinking the volume causes an increase in pressure; equilibria respond by shifting to the side having less volume. If gases are part of the reaction, this is the side have fewer moles of gas.

- **Volume increased.** Expanding the volume shifts an equilibrium toward the side having a large volume (more moles of gas). See **9-63**.

Remember: the response to a stress can never completely nullify the stress.[3]

LeChatelier's principle helps to figure out how to change conditions to drive a desired reaction to give more products (**9-39** and **9-43**) and in setting up calculations (**9-27**).

The Temperature Dependence of Equilibrium Constants

The equation

$$-\Delta G° = RT \ln K$$

relates the equilibrium constant to the temperature at which that reaction is carried out. Write the equation twice, once for the reaction at constant temperature T_1 and once for the reaction at constant temperature T_2:

$$-RT_1 \ln K(T_1) = \Delta G°_{T_1} = \Delta H° - T_1 \Delta S°$$

$$-RT_2 \ln K(T_2) = \Delta G°_{T_2} = \Delta H° - T_2 \Delta S°$$

Combine the two equations by eliminating $\Delta S°$ between them:

$$\ln \frac{K(T_2)}{K(T_1)} = \frac{-\Delta H°}{R} \left(\frac{1}{T_2} - \frac{1}{T_1} \right)$$

This is the **van't Hoff equation.** It states the dependence of K on the temperature as long as $\Delta H°$ and $\Delta S°$ remain constant between T_1 and T_2. In fact, $\Delta H°$ and $\Delta S°$

---

[3]See text Figure 9-6, text page 301.

do change somewhat with temperature (see **9-71** and **See 9-47b**). To the extent that they change, the van't Hoff equation is an approximation.

The following helpful statements follow from the van't Hoff equation:

When $\Delta H°$ is positive, a reaction is endothermic, and its equilibrium constant increases with increasing temperature.

When $\Delta H°$ is negative, a reaction is exothermic, and its equilibrium constant decreases with increasing temperature. Compare to **9-41**.

Problems often combine the preceding thermodynamic relationship with concentration and partial pressure data. Thus, an elaborate problem might give enough partial pressure (or concentration) data to calculate K at two different temperatures. Then it would ask for $\Delta H°$ and $\Delta S°$ for the reaction. Remember that the standard state (reference state) for all of the thermodynamic values in Appendix D is either a pressure of 1 atm or a concentration of 1 M.

Review of the Meaning of Different Symbols

ΔG The change in the free energy of a process. At any stage in a chemical reaction, ΔG equals the free energy of the products minus the free energy of the reactants. It changes as a reaction progresses.

$\Delta G°$ The standard free-energy change. It equals the change in free energy when the reaction transforms pure reactants in standard states into pure products in standard states and in the amounts (number of moles) indicated by the coefficients in the balanced equation. It does *not* change as a reaction progresses. It is *strongly* dependent on the temperature, which is consequently often given as a subscript. If a subscript is not given, a temperature of 298.15 K should be assumed.

$\Delta G°_{298.15}$ The standard free-energy change at 298.15 K (25°C exactly).

$\Delta G°_T$ The standard free-energy change at some general temperature T.

K The thermodynamic equilibrium constant of a reaction. K has no units. Values are meaningful only with reference to a specific balanced equation. K is strongly dependent on T.

$K(T)$ The thermodynamic equilibrium constant of a reaction at temperature T. This symbol is used when it is desired to emphasize the temperature dependence of K.

K_T The same as $K(T)$.

$K_{298.15}$ The thermodynamic equilibrium constant of a reaction at the temperature 298.15 K (25°C exactly).

K_P The empirical (as opposed to thermodynamic) equilibrium constant. It is obtained by substituting observed equilibrium partial pressures into a mass-action expression. It has units and is strongly dependent on the temperature.

K_C The empirical equilibrium constant in terms of concentrations of substances. It has units and is strongly dependent on the temperature.

Q The reaction quotient. It changes as a reaction progresses from pure reactants ($Q = 0$) to pure products ($Q = \infty$). It is unitless and is independent of temperature.

$S°$ The standard entropy. It equals the amount of entropy that a substance has under standard conditions. It is strongly dependent on the temperature.

$S°_{298.15}$ The standard entropy at a temperature of 298.15 K.

ΔS The difference in the entropy between the products and reactants in a reaction. ΔS changes as a reaction progresses. It is only weakly dependent on the temperature.

$\Delta S°$ The standard entropy change. It equals the change in entropy when a reaction transforms pure reactants in standard states into pure products in standard states in the amounts indicated by the coefficients in the balanced equation. It does *not* change as a reaction progresses. It is only weakly dependent on the temperature at which the change occurs.

$\Delta S°_{298.15}$ The standard entropy change in a reaction proceeding at 298.15 K. $\Delta S°_{298.15}$) is often used as an approximation for $\Delta S°$'s at other temperatures ($\Delta S°_T$'s).

ΔH The change in the enthalpy of a reaction. At any stage in a chemical reaction, ΔH equals the enthalpy of the products minus the enthalpy of the reactants. It changes as a reaction progresses. It is weakly dependent on the temperature at which the reaction takes place.

$\Delta H°$ The standard enthalpy change. It is the change in enthalpy if a reaction transforms pure reactants in standard states into pure products in standard states in the amounts indicated by the coefficients in the balanced equation. It does *not* change as a reaction progresses. It is weakly dependent on the temperature at which the reaction takes place.

$\Delta H°_{298.15}$ The standard enthalpy change of a reaction at 298.15 K. It is often used as an approximation for $\Delta H°$ at other temperatures ($\Delta H°_T$).

Detailed Solutions to Odd-Numbered Problems

9-1 An equilibrium expression for a reaction requires a balanced equation. The partial pressures of the products are raised to powers equal to their coefficients in the balanced equation and multiplied together to form numerator of a fraction. The denominator of this fraction is formed by the partial pressures of the reactants raised to powers equal to *their* coefficients and similarly multiplied together.[4] Apply this to the three

---

[4]See page 215 of this Guide.

specific examples in the problem:

a) $\dfrac{(P_{H_2O})_{eq}^2}{(P_{H_2})_{eq}^2(P_{O_2})_{eq}}$ b) $\dfrac{(P_{XeF_6})_{eq}}{(P_{Xe})_{eq}(P_{F_2})_{eq}^3}$ c) $\dfrac{(P_{CO_2})_{eq}^{12}(P_{H_2O})_{eq}^6}{(P_{C_6H_6})_{eq}^2(P_{O_2})_{eq}^{15}}$

Tip. If equilibrium partial pressures are inserted directly into these expressions, the result in each case is a K_P. If all P_{eq}'s are divided first by a reference pressure (1 atm, for example), the result is a thermodynamic K.

9-3 One balanced equation and its associated mass-action expression are:

$$P_4(g) + 2\,O_2(g) + 6\,Cl_2(g) \rightleftharpoons 4\,POCl_3(g) \qquad \dfrac{P_{POCl_3}^4}{P_{P_4} P_{O_2}^2 P_{Cl_2}^6}$$

Tip. The answer omits the subscript eq's, a common practice.
Other answers are possible. For example, if the coefficients in the balanced equation are doubled, then the exponents in the mass-action expression are doubled, and the overall expression is squared:

$$2\,P_4(g) + 4\,O_2(g) + 12\,Cl_2(g) \rightleftharpoons 8\,POCl_3(g) \qquad \dfrac{P_{POCl_3}^8}{P_{P_4}^2 P_{O_2}^4 P_{Cl_2}^{12}}$$

9-5 a) The law of mass action applied to the reaction

$$CO_2(g) + H_2(g) \rightleftharpoons CO(g) + H_2O(g)$$

states that the following equation, in which K is a constant, holds at equilibrium:

$$\boxed{\dfrac{(P_{CO_2}/P_{ref})(P_{H_2}/P_{ref})}{(P_{CO}/P_{ref})(P_{H_2O}/P_{ref})} = K}$$

b) The problem gives K and three of the four equilibrium partial pressures. Substitute in the mass-action expression and solve:

$$\dfrac{(0.70)(P_{H_2}/P_{ref})}{(0.10)(0.10)} = 3.9$$

In this equation, each numerical partial pressure has already been divided by the reference pressure of 1 atm. Solving for P_{H_2} gives $\boxed{0.056 \text{ atm}}$.

9-7 Insert data from text Appendix D into the usual form:

$$\Delta G^{\circ}_{298.15} = 2 \underbrace{(51.29)}_{NO_2(g)} + 3 \underbrace{(-228.59)}_{H_2O(g)} - 2 \underbrace{(-16.48)}_{NH_3(g)} - 7/2 \underbrace{(0)}_{O_2(g)} = \boxed{-550.23 \text{ kJ}}$$

Substitute this answer in the equation:

$$\ln K_{298.15} = \frac{-\Delta G^{\circ}_{298.15}}{RT} = \frac{-(-550.23 \times 10^3 \text{ J mol}^{-1})}{8.315 \text{ J K}^{-1}\text{mol}^{-1}(298.15 \text{ K})} = 221.95$$

Hence, $K = e^{221.95} = \boxed{2.5 \times 10^{96}}$.

Tip. The $\Delta G^{\circ}_{298.15}$ is properly reported in kJ, but the value used in the calculation of K is in J mol^{-1}. The extra per mole refers to "per mole of the reaction as it is written." If the equation were rewritten with all the coefficients doubled, then ΔG° would double, and the equilibrium constant K would be squared.

9-9 Calculate the standard free energy change at 25° ($\Delta G^{\circ}_{298.15}$) for the reaction of 1 mol of SO_2 with $\frac{1}{2}$ mol of O_2 to give 1 mol of SO_3. This requires a table of molar free energies of formation (Appendix D). Then compute the equilibrium constant K using the relationship $\Delta G^{\circ} = -RT \ln K$:

$$\Delta G^{\circ}_{298.15} = 1 \underbrace{(-371.08)}_{SO_3(g)} - 1 \underbrace{(-300.19)}_{SO_2(g)} - 1 \underbrace{(0.00)}_{O_2(g)} = -70.89 \text{ kJ}$$

$$\ln K_{298.15} = \frac{-\Delta G^{\circ}_{298.15}}{RT} = \frac{-(-70.89 \times 10^3 \text{ J mol}^{-1})}{8.315 \text{ J K}^{-1}\text{mol}^{-1}(298.15 \text{ K})} = 28.6$$

$$K = e^{28.6} = \boxed{2.6 \times 10^{12} = \frac{P_{SO_3}}{P_{SO_2} P_{O_2}^{1/2}}}$$

It is understood that the partial pressures in the boxed equation must be divided by a reference pressure of 1 atm because the standard-state pressure for the ΔG° was 1 atm.

Tip. The form of the equilibrium expression derives from the set of coefficients that is used in the computation of the ΔG°. In the above, the set was "$1 + 1/2 \rightarrow 1$" and not "$2 + 1 \rightarrow 2$" or any of the other sets that balances the equation.

9-11 Write the mass-action expression that corresponds to the equation given in the problem. Then substitute the equilibrium partial pressures and calculate K:

$$3 \, Al_2Cl_6(g) \rightleftharpoons 2 \, Al_3Cl_9(g) \qquad \frac{P^2_{Al_3Cl_9}}{P^3_{Al_2Cl_6}} = K = \frac{(1.02 \times 10^{-2})^2}{(1.00)^3} = \boxed{1.04 \times 10^{-4}}$$

9-13 The 1,3-di-*t*-butylcyclohexane is a gas at 580 K. In a collection of, say, 10,000 molecules, 642 would be in the chair form and 9358 would be in the boat form at equilibrium. The partial pressures of gases are proportional to the number of molecules present, assuming ideality. That is, $P_{gas} = kN$ where k is a constant of proportionality that depends on the temperature and volume only. Therefore:

$$K = \frac{P_{boat}}{P_{chair}} = \frac{kN_{boat}}{kN_{chair}} = \frac{9358}{642} = \boxed{14.6}$$

9-15 a) Calculate the chemical amount (in moles) of SO_2Cl_2 from the mass of the compound that was loaded into the flask:

$$n_{SO_2Cl_2} = 3.174 \text{ g } SO_2Cl_2 \times \left(\frac{1 \text{ mol } SO_2Cl_2}{135.0 \text{ g } SO_2Cl_2} \right) = 0.02351 \text{ mol } SO_2Cl_2$$

Imagine that the SO_2Cl_2 vaporizes in one step and then reacts in a second distinct step. Use the ideal-gas law to compute the partial pressure of the $SO_2Cl_2(g)$ after it fills the flask at 100°C but before it has a chance to react:

$$P_{SO_2Cl_2} = n_{SO_2Cl_2} \left(\frac{RT}{V} \right)$$

$$= 0.02351 \left(\frac{(0.08206 \text{ L atm mol}^{-1}K^{-1})(373.15 \text{ K})}{1.000 \text{ L}} \right) = 0.7199 \text{ atm}$$

The partial pressures of both products equal zero at this point. As the reaction advances toward equilibrium, the three partial pressures change. The SO_2Cl_2 decomposes to generate SO_2 and Cl_2 in equal chemical amounts:

	$SO_2Cl_2(g)$	$\rightleftharpoons$	$SO_2(g)$	+	$Cl_2(g)$
Init. pressure (atm)	0.7199		0		0
Change in pressure (atm)	$-x$		$+x$		$+x$
Equil. pressure (atm)	$0.7199 - x$		x		x

The total pressure in the flask at equilibrium is the sum of the three equilibrium partial pressures:

$$P_{tot} = 1.30 \text{ atm} = P_{SO_2Cl_2} + P_{Cl_2} + P_{SO_2} = (0.7199 - x) + x + x$$

Solving gives x equal to 0.5801 atm. The equilibrium partial pressures of the two products are accordingly both $\boxed{0.58 \text{ atm}}$, and the equilibrium partial pressure of the reactant is $\boxed{0.14 \text{ atm}}$.

b) K is computed by substituting equilibrium partial pressures into the appropriate mass-action expression:

$$K = \frac{(P_{SO_2}/1 \text{ atm})(P_{Cl_2}/1 \text{ atm})}{(P_{SO_2Cl_2}/1 \text{ atm})} = \frac{(0.58)(0.58)}{(0.14)} = \boxed{2.4}$$

9-17 The reaction is the combustion of carbon disulfide to give carbon dioxide and sulfur dioxide. Equation 2 represents this as meaningfully as equation 1 but has all of the coefficients divided by three. The mass-action expression for equation 2 (K_2) therefore equals that for equation 1 (K_1) except with all of the exponents divided by 3. Dividing exponents by 3 corresponds to taking the cube root: $\boxed{K_2 = \sqrt[3]{K_1}}$.

9-19 When chemical equations 1 and 2 are added to obtain equation 3, the equilibrium constant associated with equation 3 equals the product of the equilibrium constants associated with equations 1 and 2. Also if a equation is reversed, the new equilibrium constant is the reciprocal of the original. Writing the first equation in this problem in reverse and adding the second equation gives the equation of interest. Thus, K for the reaction of interest equals $K_2 \times (1/K_1)$ or $\boxed{K_2/K_1}$.

9-21 a) Calculate the chemical amount n of the gaseous $C_6H_5CH_2OH$. Then use the ideal-gas law to calculate its initial partial pressure: Refer to this compound as "BzOH."

$$n_{BzOH} = 1.20 \text{ g BzOH} \times \left(\frac{1 \text{ mol BzOH}}{108 \text{ g BzOH}}\right) = 0.0111 \text{ mol BzOH}$$

$$P_{BzOH} = \frac{n_{BzOH}RT}{V} = \frac{(0.0111 \text{ mol})(0.08206 \text{ L atm mol}^{-1}\text{K}^{-1})(523 \text{ K})}{2.00 \text{ L}} = 0.238 \text{ atm}$$

The following three-line table shows how the partial pressures of benzyl alcohol and its products change as equilibrium is approached:

	$C_6H_5CH_2OH(g) \rightleftharpoons$	$C_6H_5CHO(g) +$	$H_2(g)$
Init. pressure (atm)	0.238	0	0
Change in pressure (atm)	$-x$	$+x$	$+x$
Equil. pressure (atm)	$0.238 - x$	x	x

Substitute the final pressures in the mass-action expression:

$$\frac{P_{C_6H_5CHO}P_{H_2}}{P_{C_6H_5CH_2OH}} = 0.558 = \frac{(x)(x)}{(0.238 - x)} \qquad \text{from which} \qquad x^2 + 0.558x - 0.133 = 0$$

Use the quadratic formula[5] to solve for x:

$$x = \frac{-(0.558) \pm \sqrt{(0.558)^2 - 4(1)(0.133)}}{2(1)} = \frac{-0.558 \pm 0.918}{2}$$

$$x = 0.180 \quad \text{and} \quad x = -0.738$$

Disregard the solution $x = -0.738$ because it leads to negative partial pressures for all three gases. The answer is $P_{C_6H_5CHO} = \boxed{0.180 \text{ atm}}$.

b) The fraction of the benzyl alcohol dissociated at equilibrium equals the amount dissociated divided by the initial amount. These amounts are respectively proportional to the decrease in partial pressure of the benzyl alcohol and the initial partial pressure of the benzyl alcohol. Hence:

$$f = \frac{\text{amount of BzOH dissociated}}{\text{original amount of BzOH}} = \frac{0.180 \text{ atm}}{0.238 \text{ atm}} = \boxed{0.756}$$

9-23 At equilibrium, the bulb contains a mixture of $PCl_3(g)$ and $Cl_2(g)$, the two products, and whatever $PCl_5(g)$, the sole reactant, remains. The total pressure of this mixture is given as 0.895 atm. If this gaseous mixture follows Dalton's law, then its final total pressure is equal to the sum of the partial pressures of the components:

$$P_{tot} = P_{PCl_3} + P_{Cl_2} + P_{PCl_5} = 0.895 \text{ atm}$$

Assume that the reaction

$$PCl_5(g) \rightleftharpoons Cl_2(g) + PCl_5(g)$$

is the only reaction taking place. It then follows that the partial pressure of PCl_3 and the partial pressure of Cl_2 always remain equal. These two pressures equal x atm. Then the equilibrium partial pressure of PCl_5 equals $(0.895 - 2x)$ atm. The contents of the bulb are at equilibrium, so the partial pressures satisfy the equation:

$$K = 2.15 = \frac{P_{Cl_2} P_{PCl_3}}{P_{PCl_5}} = \frac{(x)(x)}{0.895 - 2x}$$

Solving (using the quadratic formula) gives $x = 0.40866$ and a physically meaningless root ($x = -4.7087$). The equilibrium partial pressures of the Cl_2 and the PCl_3 are both $\boxed{0.409 \text{ atm}}$, and the equilibrium partial pressure of the PCl_5 is $\boxed{0.078 \text{ atm}}$.

---

[5]Text page A-21.

9-25 Write the equation for the reaction, and insert the given data in the usual table:

	$Br_2(g)$	$+\ I_2(g)$	$\rightleftharpoons\ 2\,IBr(g)$
Init. pressure (atm)	0.0500	0.0400	0
Change in pressure (atm)	$-x$	$-x$	$+2x$
Equil. pressure (atm)	$0.0500 - x$	$0.0400 - x$	$2x$

$$\frac{P_{IBr}^2}{P_{Br_2}P_{I_2}} = 322 = \frac{(2x)^2}{(0.0500-x)(0.0400-x)}$$

Rearrangement leads to the quadratic equation: $x^2 - 0.09113x + 2.025 \times 10^{-3} = 0$. Solving (using the quadratic formula) gives $x = 0.0384$ and $x = 0.0527$. The second root is "unphysical" because there was only 0.0400 atm of I_2 at the start. Going down by 0.0527 atm is impossible. The correct partial pressures come from the first root:

$$P_{IBr} = \boxed{0.0768\ \text{atm}} \qquad P_{I_2} = \boxed{0.0016\ \text{atm}} \qquad P_{Br_2} = \boxed{0.0116\ \text{atm}}$$

9-27 Write the equation, the initial partial pressures, the change in the pressures required to reach equilibrium, and the final partial pressures:

	$N_2(g) +$	$O_2(g)$	$\rightleftharpoons\ 2\,NO(g)$
Init. pressure (atm)	0.41	0.59	0.21
Change in pressure (atm)	$+x$	$+x$	$-2x$
Equil. pressure (atm)	$0.41 + x$	$0.59 + x$	$0.21 - 2x$

Note the sign of x. A positive x in this set-up corresponds to the loss of product and the formation of reactants as the reaction comes to equilibrium. The mass-action expression relates the equilibrium partial pressures to K:

$$K = 4.2 \times 10^{-31} = \frac{P_{NO}^2}{P_{N_2}P_{O_2}} = \frac{(0.21 - 2x)^2}{(0.41 + x)(0.59 + x)}$$

Before attempting to solve for x, consider the relative sizes of the numbers. This saves getting bogged down with algebra. The equilibrium constant is very small. Hence, the numerator of the equilibrium expression must be very small—almost all of the $NO(g)$ is consumed at equilibrium. Suppose therefore that *all* of the $NO(g)$ reacts. Then, $2x = 0.21$ and $x = 0.105$. Using this value of x gives an equilibrium pressure of N_2 of $0.41 + 0.105$ or $\boxed{0.52\ \text{atm}}$. The equilibrium pressure of O_2 is, by a similar computation, $\boxed{0.70\ \text{atm}}$. To get the true (non-zero) equilibrium partial pressure of NO, substitute these two pressures back into the original expression:

$$4.2 \times 10^{-31} = \frac{P_{NO}^2}{P_{N_2}P_{O_2}} = \frac{(P_{NO})^2}{(0.52)(0.70)}$$

Solving for P_{NO} gives $\boxed{3.9 \times 10^{-16} \text{ atm}}$. The reaction lies *quite* far toward the reactants at equilibrium.

9-29 The equation for the synthesis of ammonia that is given in the problem and the corresponding equilibrium expression are:

$$N_2(g) + 3\,H_2(g) \rightleftharpoons 2\,NH_3(g) \qquad \frac{P_{NH_3}^2}{P_{H_2}^3 \, P_{N_2}} = K = 6.78 \times 10^5$$

This can be written almost effortlessly. The difficulty starts with translating the other statements in the problem into mathematical terms. If the H to N atom ratio is 3 to 1 then:

$$P_{H_2} = 3P_{N_2}$$

because the third component, NH_3, maintains, within itself, the required 3 to 1 ratio of atoms. The fact that the total pressure is 1.00 atm means:

$$P_{N_2} + P_{H_2} + P_{NH_3} = 1.00 \text{ atm}$$

Let x equal the equilibrium partial pressure of N_2, and substitute in the mass-action expression:

$$6.78 \times 10^5 = \frac{(1.00 - 4x)^2}{(3x)^3 x} = \frac{(1.00 - 4x)^2}{27x^4}$$

This equation is not too hard to solve analytically (see below), but is it necessary to even try? The x is expected to be small because at equilibrium the mixture is mostly NH_3 (K is big). Using this idea simplifies the algebra. Suppose $4x \ll 1.00$. Then:

$$6.78 \times 10^5 \approx \frac{1.00}{27x^4} \quad \text{from which} \quad x \approx 0.0153$$

Improve this answer by doing successive approximations.[6] The idea is to guess values of x and compute the right-hand side of the equation for each. Improve each new guess based on the results of the previous computation. Soon, the computed value becomes sufficiently close to 6.78×10^5. The following table maps such a process. It starts with the x from the rough solution:

x	$(1.00 - 4x)^2$	$27x^4$	$(1.00 - 4x)^2/27x^4$
0.0153	0.8813	1.480×10^{-6}	5.96×10^5
0.0145	0.8874	1.194×10^{-6}	7.43×10^5
0.0149	0.8843	1.331×10^{-6}	6.64×10^5
0.0147	0.8859	1.261×10^{-6}	7.03×10^5
0.0148	0.8851	1.295×10^{-6}	6.83×10^5
0.01483	0.8849	1.306×10^{-6}	6.776×10^5

---

[6]See text Appendix C, text page A-17.

Therefore x equals 0.01483 to four significant digits. Then:

$$P_{N_2} = \boxed{0.0148 \text{ atm}} \quad P_{H_2} = \boxed{0.0445 \text{ atm}} \quad P_{NH_3} = \boxed{0.941 \text{ atm}}$$

Tip. Analytical solution of the equation:

$$6.78 \times 10^5 = \frac{(1.00 - 4x)^2}{27x^4}$$

proceeds by multiplying both sides of the equation by 27 and then taking the square root of both sides. This gives:

$$4.278 \times 10^3 = \frac{1.00 - 4x}{x^2} \quad \text{which gives} \quad (4.278 \times 10^3)x^2 + 4x - 1.00 = 0$$

Substitution in the quadratic formula affords two roots:

$$x = \frac{-b \pm \sqrt{b^2 - 4ac}}{2a} = \frac{-4 \pm \sqrt{16 + 17122}}{8556} = +0.0148 \text{ and } -0.0158$$

The positive root is the same answer obtained by approximation. The negative root is physically meaningless.

9-31 The reaction and corresponding mass-action expression in terms of pressures are:

$$2CO(g) + Cl_2(g) \rightleftharpoons COCl_2(g) \qquad \frac{P_{COCl_2}}{P_{CO} P_{Cl_2}} = K = 0.20$$

where the 0.20 comes from **9-6**. Write the mass-action expression in terms of concentrations. This expression is related to the above K as follows:[7]

$$\frac{[COCl_2]}{[CO][Cl_2]} = 0.20 \left(\frac{RT}{P_{ref}}\right)^{-\Delta n_g}$$

where Δn_g equals the change in the number of moles of gas from left to right in the reaction (-1 in this case because there is 1 mol of gas on the right and 2 mol on the left of the equation), and P_{ref} is the reference pressure used in obtaining the given K (1.00 atm in this case). Insert these values along with the gas constant R and the absolute temperature T:

$$\frac{[COCl_2]}{[CO][Cl_2]} = 0.20 \left(\frac{(0.08206 \text{ L atm mol}^{-1}\text{K}^{-1})(873.15 \text{ K})}{1 \text{ atm}}\right)^{-(-1)} = 14.3$$

Finally, substitute the equilibrium concentrations of the CO and Cl_2:

$$\frac{[COCl_2]}{[2.3 \times 10^{-4}][1.7 \times 10^{-2}]} = 14.3$$

and solve for $[COCl_2]$. The answer is $\boxed{5.6 \times 10^{-5} \text{ mol L}^{-1}}$.

---

[7]Text page 297.

9-33 a) The reaction quotient Q has the form of a mass-action expression. A K is computed only by substitution of *equilibrium* partial pressures into this mathematical form. Computations of Q on the other hand may employ whatever partial pressures might temporarily prevail. In the following, the subscript zero means initial P's were used:

$$Q_0 = \frac{(P_{Al_3Cl_9})_0^2}{(P_{Al_2Cl_6})_0^3} = \frac{(1.02 \times 10^{-2})^2}{(0.473)^3} = \boxed{9.83 \times 10^{-4}}$$

b) From **9-11**, $K_{454} = 1.04 \times 10^{-4}$. The initial reaction quotient Q_0 exceeds this K so the reaction approaches equilibrium by "shifting" from the right to the left (generating reactants at the expense of problems). The process $\boxed{\text{consumes } Al_3Cl_9}$ and produces Al_2Cl_6.

9-35 The fading of color means that the reaction consumes $Br_2(g)$ as it goes toward equilibrium: change goes from left to right in the reaction:

$$H_2(g) + Br_2(g) \rightleftharpoons 2\,HBr(g)$$

The initial reaction quotient Q_0 is accordingly less than K. The data given in the problem allow computation of Q_0:

$$Q_0 = \frac{(P_{HBr})_0^2}{(P_{H_2})_0(P_{Br_2})_0} = \frac{(0.90 \text{ atm})^2}{(0.40 \text{ atm})(0.40 \text{ atm})} = 5.1$$

Thus $\boxed{K \text{ must exceed } 5.1}$.

9-37 a) Let P_{di} stand for the partial pressure of gaseous diphosphorus $P_2(g)$ and P_{tet} for the partial pressure of gaseous tetraphosphorus $P_4(g)$. For the process:

$$\frac{P_{di}^2}{P_{tet}} = Q$$

Initially, $P_{di} = 2.00$ atm and $P_{tet} = 5.00$ atm making $Q_0 = \boxed{0.800}$. Because $Q_0 > K$ the equilibrium shifts to the $\boxed{\text{left}}$ (reducing the numerator and increasing the denominator in the above expression) until Q equals K.

b) Let x equal the increase in the pressure of $P_4(g)$ during the change. Then:

$$K = \frac{(2.00 - 2x)^2}{(5.00 + x)} = 0.612$$

The equation can be solved with the quadratic formula. It is instructive however to get x numerically.[8] Construct a table:

x	$(2.00 - 2x)^2$	$5.00 + x$	Q
0.00	4.00	5.00	0.800
0.100	3.24	5.10	0.635
0.120	3.10	5.12	0.605
0.115	3.13	5.115	0.612

As x increases Q decreases from 0.800. As the table shows, progress to a good answer is quick. Electronic calculators make it painless. At $x = 0.120$, Q is only slightly less than K. If $x = 0.115$:

$$P_{di} = 2.00 - 2(0.115) = \boxed{1.77 \text{ atm}} \quad \text{and} \quad P_{tet} = 5.00 + 0.115 = \boxed{5.12 \text{ atm}}$$

c) If the volume of the system is increased, then, by LeChatelier's principle, there will be net $\boxed{\text{dissociation}}$ of P_4. The system responds to its forced rarefaction by producing more molecules to fill the larger volume.

9-39 Chemical systems always tend toward equilibrium. If a stress is applied to a system at equilibrium, the system reacts to minimize the stress.

a) The stress is the addition of $N_2O(g)$. The system reacts to decrease the concentration of N_2O. It does this by proceeding from $\boxed{\text{right to left}}$ until a new equilibrium is reached.

b) The stress is the reduction in volume. The partial pressures of all the compounds will momentarily rise. The equilibrium will then shift in such a way as to reduce the number of molecules of gas (chemical amount of gas) in the container and reduce the total pressure. There are three moles of gas on the reactant side of the equation and two moles of gas on the product side. The equilibrium will thus shift from $\boxed{\text{left to right}}$.

c) The reaction is exothermic. Cooling the equilibrium mixture therefore shifts the reaction from $\boxed{\text{left to right}}$ (to favor the products).

d) In order to maintain a constant pressure, the volume of the system must have increased. Thus, the reaction will shift from $\boxed{\text{right to left}}$.

e) The partial pressures of the gases are unchanged by the addition of an inert gas, and the equilibrium law is independent of total pressure. Consequently, there is $\boxed{\text{no effect}}$ on the position of the equilibrium.

---

[8]See text Appendix C.

9-41 a) The equilibrium constant increases with decreasing temperature. This means that decreasing the temperature favors the products. Because removing heat shifts the reaction to the right, heat must be generated on the right—the reaction is exothermic .

b) Reducing the volume shifts this particular equilibrium to the right. Shrinking favors the side of the reaction with fewer moles of gas. Hence, there is a net decrease in the number of gas molecules from left to right in the reaction.

9-43 Good design would provide for transferring the heat generated in the chlorination of the ethylene (the first reaction) into the dehydrochlorination of the dichloroethane (the second reaction). Removal of the product heat from the first reaction would tend to drive the first reaction toward the right; input of this heat to the second reaction would drive it toward its products. Good design would also arrange for the continuous removal of the gaseous products of both reactions. This would shift them in the desired direction.

9-45 The ΔH of the reaction is clearly less than zero. The hydration of ethylene to give ethanol is thus exothermic. Think of the heat as a reaction product:

$$C_2H_4(g) + H_2O(g) \rightarrow C_2H_5OH(g) + \text{ heat}$$

By LeChatelier's principle, the equilibrium production of ethanol is maximized by running the reaction at low temperature (which allows the product heat to escape better). There are two moles of gas on the reactant side and only one mol of gas of the product side; high pressure will force the equilibrium from left to right, increasing the yield of ethanol.

9-47 Use the van't Hoff equation to calculate $\Delta H°$ for the reaction from the two K's and the temperatures at which the K's were measured:

$$\ln \frac{K(T_2)}{K(T_1)} = \frac{-\Delta H°}{R} \left(\frac{1}{T_2} - \frac{1}{T_1} \right)$$

$$\ln \left(\frac{0.00121}{6.8} \right) = -8.634 = \frac{-\Delta H°}{8.315 \text{ J K}^{-1}\text{mol}^{-1}} \left(\frac{1}{473.15 \text{ K}} - \frac{1}{298.15 \text{ K}} \right)$$

$$\Delta H° = \boxed{-5.8 \times 10^4 \text{ J mol}^{-1}}$$

9-49 a) Use the equation $\Delta G° = -RT \ln K$:

$$\Delta G°_{298.15} = -RT \ln K = (-8.315 \text{ J K}^{-1}\text{mol}^{-1})(298 \text{ K}) \ln(9.3 \times 10^9)$$

$$= \boxed{-56.9 \times 10^3 \text{ J mol}^{-1}}$$

b) Use the van't Hoff equation and the two values of K to obtain $\Delta H°$. Then $\Delta S°$ can be calculated from $\Delta G°_{298.15}$ (above) and the equation $\Delta G° = \Delta H° - T\Delta S°$. The van't Hoff equation gives:

$$\ln\left(\frac{3.3 \times 10^7}{9.3 \times 10^9}\right) = \frac{-\Delta H°}{8.315 \text{ J K}^{-1}\text{mol}^{-1}}\left(\frac{1}{398 \text{ K}} - \frac{1}{298 \text{ K}}\right)$$

Completion of the arithmetic reveals that $\Delta H°$ equals -55.6 kJ mol^{-1}. This means that $\Delta H°$ in the reaction of 1/2 mol of Cl_2 and 1/2 mol of F_2 to give 1 mol of ClF is $\boxed{-55.6 \text{ kJ}}$. Finally: .

$$\Delta S° = \frac{\Delta H° - \Delta G°}{T} = \frac{-55.63 \times 10^3 \text{ J} - (-56.88 \times 10^3 \text{ J})}{298 \text{ K}} = \boxed{4.2 \text{ J K}^{-1}}$$

9-51 Use the van't Hoff equation to calculate the equilibrium constant at 600 K (K_{600}) from the constant K_{298} and the standard enthalpy change $\Delta H°$, both of which are given:

$$\ln\left(\frac{K_{600}}{K_{298}}\right) = \frac{-\Delta H°}{R}\left(\frac{1}{T_2} - \frac{1}{T_1}\right)$$

$$\ln\left(\frac{K_{600}}{5.9 \times 10^5}\right) = \frac{-(-92.2 \times 10^3 \text{ J mol}^{-1})}{8.315 \text{ J K}^{-1}\text{mol}^{-1}}\left(\frac{1}{600 \text{ K}} - \frac{1}{298 \text{ K}}\right)$$

$$K_{600} = \boxed{4.3 \times 10^{-3}}$$

9-53 a) Use the van't Hoff equation to calculate ΔH_{vap}. The equilibrium constants at T_1 and T_2 equal the vapor pressure of the liquid at T_1 and T_2:

$$\ln\left(\frac{P_2}{P_1}\right) = -\frac{\Delta H_{vap}}{R}\left(\frac{1}{T_2} - \frac{1}{T_1}\right)$$

$$\ln\left(\frac{4.2380 \text{ atm}}{0.4034 \text{ atm}}\right) = \frac{-\Delta H_{vap}}{8.315 \text{ J K}^{-1}\text{mol}^{-1}}\left(\frac{1}{273.15 \text{ K}} - \frac{1}{223.15 \text{ K}}\right)$$

$$\Delta H_{vap} = \boxed{23.8 \text{ kJ mol}^{-1}}$$

b) The normal boiling point T_b of a liquid is defined as the temperature at which the vapor pressure of the liquid equals 1 atm exactly. Therefore, set P_1 equal to 1.000 atm and T_1 equal to T_b in the previous equation. Set $T_2 = 273.15$ K and $P_2 = 4.2380$ atm because 4.2380 atm is the vapor pressure at 273.15 K:

$$\ln\left(\frac{4.2380 \text{ atm}}{1.000 \text{ atm}}\right) = \frac{-2.38 \times 10^3 \text{ J}}{8.315 \text{ J K}^{-1}\text{mol}^{-1}}\left(\frac{1}{273.15} - \frac{1}{T_b}\right)$$

$$T_b = \boxed{240 \text{ K}}$$

Using 0.4034 atm for P_2 along with 223.15 K for T_2 gives the same answer.

9-55 a) The partial pressures of the gases are in direct proportion to their chemical amounts (assuming that Dalton's law and the ideal-gas law apply):

$$P_{\text{gas}} = n_{\text{gas}} \left(\frac{RT}{V} \right)$$

Suppose that the volume of the container is such that the actual chemical amounts of the four gases are 90, 470, 200, and 45 mol. Then, the partial pressure of the $BCl_3(g)$ is $P_{BCl_3} = 90(RT/V)$. Expressions for the partial pressures of the other three gases are similar. The equilibrium expression for reaction 1 is:

$$\frac{P_{BFCl_2}^3}{P_{BCl_3}^2 P_{BF_3}} = K_1$$

Substituting the three partial pressures gives:

$$\frac{(45RT/V)^3}{(90RT/V)^2 (470RT/V)} = K_1 = \frac{(45)^3}{(90)^2(470)} = \boxed{0.024}$$

The RT/V's canceled out! The cancellation means the volume of the container has no effect on the position of this particular equilibrium. We were therefore justified in making any assumption we wanted about the volume. A similar procedure with the equilibrium expression for reaction 2 gives $K_2 = \boxed{0.40}$.

b) The given equation (equation 3) equals the sum of equations 1 and 2 divided by three. Obtain the required equilibrium constant by taking the cube root of the product of K_1 and K_2. This gives $\boxed{0.21}$.

Tip. The same answer is obtained by substituting the original partial pressures into the mass-action expression for reaction 3, but the preceding emphasizes how K_3 derives entirely from K_1 and K_2; K_3 is a new number but adds no new information.

9-57 a) If we assume that 1.000 mol of the *cis* form is initially present, then the equilibrium chemical amounts of each species are:

	cis	$\rightleftharpoons$	*trans*
Initial amount	1.000		0
Change	$-x$		$+x$
Equilibrium amount	$1.000 - x$ mol		x mol

At equilibrium, 73.6% of the *cis* has been converted to the *trans* form. Therefore x is 0.736 mol. The equilibrium chemical amounts are:

$$n_{\text{cis}} = 1.000 - x = 1.000 - 0.736 = 0.264 \text{ mol} \quad \text{and} \quad n_{\text{trans}} = x = 0.736 \text{ mol}$$

Assuming ideal-gas behavior, the equilibrium partial pressures of the two forms are:

$$P_{cis} = \frac{n_{cis}RT}{V} = \frac{(0.264)RT}{V} \quad \text{and} \quad P_{trans} = \frac{n_{trans}RT}{V} = \frac{(0.736)RT}{V}$$

Substitution in the mass-action expression gives K:

$$K = \frac{P_{trans}}{P_{cis}} = \frac{(0.736)RT/V}{(0.264)RT/V} = \boxed{2.79}$$

b) From the preceding it is clear that at equilibrium:

$$P_{trans} = 2.788(P_{cis}) \quad \text{implying:} \quad n_{trans} = 2.788(n_{cis})$$

Also, the combined chemical amount of the two forms of the compound is 0.525 mol because one mole of the *cis* is consumed for every mole of the *trans* created:

$$n_{trans} + n_{cis} = 0.525$$

Solving the two equations in two unknowns gives n_{trans} equal to 0.386 mol. The partial pressure of the *trans* form is:

$$P_{trans} = n_{trans}\left(\frac{RT}{V}\right) = 0.386 \text{ mol}\left(\frac{(0.08206 \text{ L atm mol}^{-1}\text{K}^{-1})(698.75 \text{ K})}{15.00 \text{ L}}\right)$$
$$= \boxed{1.48 \text{ atm}}$$

9-59 Equilibrium constants depend strongly on temperature. The equation in the problem gives the experimentally determined T-dependence of the K of the hydrogenation reaction:

$$C_5H_5N(g) + 3\,H_2(g) \rightleftharpoons C_5H_{11}N(g)$$

a) Substitute $T = 500$ K into the expression:

$$\log_{10} K = -20.281 + \frac{10560}{T} = -20.281 + \frac{10560}{500} = 0.839$$

Taking the antilog of both sides gives $K = 10^{0.839} = \boxed{6.90}$.

b) Assume that the hydrogenation reaction is at equilibrium. The fraction f of nitrogen in the form of C_5H_5N (pyridine or "py") equals the chemical amount of pyridine divided by the sum of the chemical amounts of pyridine and $C_5H_{11}N$ (piperidine or "pip"). If Dalton's law holds, these chemical amounts are directly proportional to the partial pressures of the two compounds. Hence:

$$f = \frac{n_{py}}{n_{py} + n_{pip}} = \frac{P_{py}}{P_{py} + P_{pip}}$$

Also, the partial pressures must satisfy the law of mass action:

$$\frac{P_{\text{pip}}}{P_{\text{py}} P_{\text{H}_2}^3} = K \qquad \text{which gives} \qquad \frac{P_{\text{pip}}}{P_{\text{py}}} = 6.90(1.00)^3$$

when $K = 6.90$ and $P_{\text{H}_2} = 1.00$ atm are put in. Let the partial pressure of the pyridine be y atm. The partial pressure of the piperidine is then, by the preceding equation, $6.90y$ atm. Insert the P's into the equation for f:

$$f = \frac{P_{\text{py}}}{P_{\text{py}} + P_{\text{pip}}} = \frac{y}{y + 6.90y} = \frac{1}{1 + 6.90} = \boxed{0.127}$$

9-61 The set-up of the problem is exactly as given for **9-29** but now K is different, and the total pressure is 100 atm. The set-up yields the equation:

$$3.19 \times 10^{-4} = \frac{(100 - 4x)^2}{(3x)^3 x}$$

where x is the equilibrium partial pressure of nitrogen. Factor the numerator and multiply out the denominator to obtain:

$$3.19 \times 10^{-4} = \frac{16(25.0 - x)^2}{27x^4} \quad \text{which gives} \quad 5.383 \times 10^{-4} = \frac{(25.0 - x)^2}{x^4}$$

Taking the square root of both sides of this equation leads to a quadratic equation in x that can be solved routinely. It is also quite quick to solve by successive approximations. First, observe that x must be less than 25. If P_{N_2} exceeded 25 atm, the combined pressure of just two (the nitrogen and hydrogen) out of the three gases would exceed the total pressure in the container. Now, guess an x between 0 and 25 and compute the right-hand side of the equation. Revise the guess based on experience until the computed value becomes sufficiently close to 5.383×10^{-4}. The following table shows the process. The first guess is near the middle of the range:

x	$(25.0 - 4x)^2/x^4$	Comment
15	1.97×10^{-3}	x is too small
20	1.56×10^{-4}	x is too large
18	4.67×10^{-4}	x is too large
17.5	6.00×10^{-4}	x is too small
17.7	5.43×10^{-4}	x is a bit too small
17.72	5.375×10^{-4}	x acceptable

The answer is $x = 17.72$. If $x = 17.72$ then:

$$P_{\text{N}_2} = \boxed{17.7 \text{ atm}} \qquad P_{\text{H}_2} = \boxed{53.2 \text{ atm}} \qquad P_{\text{NH}_3} = \boxed{29.1 \text{ atm}}$$

Tip. Check the answers by confirming that the sum of the three partial pressures is 100 atm and that:

$$\frac{P_{NH_3}^2}{P_{H_2}^3 P_{N_2}} = \frac{(29.1)^2}{(53.2)^3(17.7)} = 3.18 \times 10^{-4} = K$$

9-63 a) The reaction and reaction-quotient expression are:

$$PCl_5(g) \rightleftharpoons PCl_3(g) + Cl_2(g) \qquad \frac{P_{Cl_2} P_{PCl_3}}{P_{PCl_5}} = Q$$

The problem gives three initial partial pressures. Substitution gives an initial reaction quotient Q_0 of $\boxed{120}$. Since Q_0 exceeds K, the reaction proceeds to the $\boxed{\text{left}}$.

b) As originally mixed, the system has $Q = 120$ and $K = 11.5$. Let y equal the partial pressure of $Cl_2(g)$ consumed by the right-to-left reaction. An equal pressure of $PCl_3(g)$ is also consumed, and an equal pressure of $PCl_5(g)$ is created. This follows from the stoichiometry of the equation. Therefore:

$$K = 11.5 = \frac{(6.0 - y)(2.0 - y)}{(0.10 + y)} \qquad \text{from which} \quad y^2 - 19.5y + 10.85 = 0$$

The roots of the quadratic equation are 0.573 and 18.93. Only the first makes physical sense. The second corresponds to negative partial pressures. Complete the computation by finding the various partial pressures:

$$P_{PCl_3} = 2.0 - 0.573 = \boxed{1.4 \text{ atm}} \qquad P_{PCl_2} = 6.0 - 0.573 = \boxed{5.4 \text{ atm}}$$
$$P_{PCl_5} = 0.10 + 0.573 = \boxed{0.67 \text{ atm}}$$

c) By LeChatelier's principle, an increase in volume causes the reaction to shift to the side having *more* moles of gas. The amount of $PCl_5(g)$ will $\boxed{\text{decrease}}$.

9-65 The reaction is the splitting of tetraphosphorus (P_4) into diphosphorus (P_2).[9]

a) The pressure of a sample of $P_4(g)$ will always exceed the pressure predicted by the ideal-gas law, because the reaction to give $P_2(g)$ furnishes extra molecules to strike the sides of the container. Consider a *P-V* experiment performed on a sample of $P_4(g)$ at constant temperature. In the experiment, the pressure is tracked as the volume of the sample is changed, and the results are plotted with P on the vertical axis and $1/V$ on the horizontal axis. Boyle's law predicts a straight-line plot. According to

---

[9]The effect of a change in volume on this reaction is shown in text Figure 9-7, text page 302.

Boyle's law, P rises in direct proportion as $1/V$ rises. The actual experimental plot will be a line that curves toward $\boxed{\text{higher}}$ pressure at low values of $1/V$.

b) Consider a V-T experiment in which the volume of a sample is $P_4(g)$ is tracked as the temperature is changed and the pressure is kept constant. Charles's law predicts that the volume should rise linearly with the temperature. But increasing the temperature shifts the endothermic equilibrium $P_4 \rightleftharpoons 2P_2$ toward the right, generating more molecules in the sample. The "extra" molecules cause the sample to expand an extra amount with temperature. Thus, a plot of the volume of the P_4 sample as a function of T curves toward $\boxed{\text{higher}}$ volumes than predicted by Charles's law as T increases.

9-67 Let the partial pressure of $N_2(g)$ before any of its molecules break down equal P_0. Let the fraction that has broken down at equilibrium equal α. The value of α is 0.0065 at 5000 K but rises to 0.116 at 6000 K. Represent the approach to equilibrium at either temperature in the usual way:

	$N_2(g)$	$\rightleftharpoons$	$2N(g))$
Init. Pressure (atm)	P_0		0
Change in Pressure (atm)	$-\alpha P_0$		$+2\alpha P_0$
Equil. Pressure (atm)	$P_0 - \alpha P_0$		$2\alpha P_0$

The total pressure of the equilibrium mixture equals 1.000 atm at both T's:

$$P_{N_2} + P_N = (P_0 - \alpha P_0) + 2\alpha P_0 = P_0(1 + \alpha) = 1.000 \text{ atm}$$

Since α is given at both temperatures, the original pressure of N_2 at both temperatures is readily computed from the preceding equation: P_0 was 0.9935 atm at 5000 K and 0.8961 atm at 6000 K. Combine these original pressures with α as indicated above to get the equilibrium partial pressures of N and N_2 at the two temperatures:

Temperature	5000 K	6000 K
Equil. P_{N_2} (atm)	$P_0(1 - \alpha) = 0.9871$	$P_0(1 - \alpha) = 0.7921$
Equil. P_N (atm)	$2\alpha P_0 = 0.0129$	$2\alpha P_0 = 0.2079$

It is easy to verify that the partial pressures add up to 1.000 atm at both temperatures. Substitute these equilibrium partial pressures in the mass-action expression to obtain K's at the two temperatures:

$$K_{5000} = \frac{P_N^2}{P_{N_2}} = \frac{(0.0129)^2}{0.9871} = 1.69 \times 10^{-4} \qquad K_{6000} = \frac{P_N^2}{P_{N_2}} = \frac{(0.2079)^2}{0.7921} = 5.46 \times 10^{-2}$$

Next, use the van't Hoff equation to estimate $\Delta H°$ for the reaction from the two K's and their temperatures:

$$\ln \frac{K_{6000}}{K_{5000}} = \frac{-\Delta H°}{R} \left(\frac{1}{6000} - \frac{1}{5000} \right)$$

Substitute:

$$\ln\left(\frac{5.46 \times 10^{-2}}{1.69 \times 10^{-4}}\right) = 5.78 = \frac{-\Delta H^\circ}{8.315 \text{ J K}^{-1}\text{mol}^{-1}}\left(\frac{1}{6000 \text{ K}} - \frac{1}{5000 \text{ K}}\right)$$

Solving for ΔH° gives $\boxed{1440 \text{ kJ mol}^{-1}}$.

Tip. The answer is only an estimate because ΔH° and ΔS° are not constant over the 1000 K range. The answer exceeds the $N\equiv N$ bond enthalpy[10] of 945 kJ mol^{-1}, which is a $\Delta H^\circ_{298.15}$, by over 50%.

---

[10]See text Table 7-3, text page 228.

Chapter 10

Acid-Base Equilibria

The concepts of chemical equilibrium (Chapter 9) apply to all types of chemical reactions. This chapter focuses on equilibria in acid-base reactions.

Classification of Acids and Bases

Acids react with bases; bases react with acids. Three definitions of acid and base find common use. Each is more general than the previous.

- **Arrhenius Acids and Bases.** Arrhenius acids increase the concentration of the aquated hydrogen ion ($H^+(aq)$) when put in water. Arrhenius bases increase the concentration of the aquated hydroxide ion ($OH^-(aq)$) when put in water.

  When an Arrhenius acid and an Arrhenius base are mixed, **neutralization** occurs. For example:

  $$HCl(aq) + NaOH(aq) \rightarrow H_2O(l) + NaCl(aq)$$

  In general, an Arrhenius acid reacts with an Arrhenius base to give water and a salt. The essence of Arrhenius neutralization is given by a single net ionic equation:

  $$\mathbf{H^+(aq) + OH^-(aq) \rightarrow H_2O(l)}$$

- **Brønsted-Lowry Acids and Bases.** A Brønsted-Lowry acid can donate hydrogen ions; a Brønsted-Lowry base can accept such an ion. The equation:

  $$acid \rightleftharpoons base^- + H^+$$

  summarizes the definition. The acid and base in this equation make a **conjugate acid-base pair.** In such a pair the acid becomes the base by the act of donating a hydrogen ion; the base becomes the acid by the act of accepting a hydrogen ion.

A standard task is to write the formulas of the conjugate acids or conjugate bases of a list of chemical species. See **10-1** and **10-71**. The conjugate acid is obtained by adding H^+ to the formula of the species; the conjugate base is obtained by subtracting H^+ from the formula. If there is no H in a compound's formula, then that compound has no Brønsted-Lowry conjugate base.

Transfer of a hydrogen ion from one species to another can take place in non-aqueous solvents, such as liquid ammonia, or even in the gas phase. Such reactions qualify as Brønsted-Lowry acid-base reactions.

- **Lewis Acids and Bases.** A Lewis base is any species that donates electrons through coordination to its lone pairs of electrons; A Lewis acid is any species that accepts such electron pairs. See **10-7**. "Coordination" means that the donor shares one (or more) of its lone pair and does not give them up. Thus neutralization between a Lewis base and Lewis acid proceeds with the formation of a covalent bond (called a coordinate covalent bond). See **10-5**

The Lewis definition is more general than the Brønsted-Lowry because it does not require the presence of hydrogen or any other element—all substances contain electrons. It applies to non-aqueous systems and allows convenient use of Lewis structures. The formation of coordination complexes can be regarded as the interaction between Lewis bases and acids.[1] Applications of the Lewis acid-base concept appear in **22-25** and **22-27** in this Guide.

The Properties of Acids and Bases in Aqueous Solution

The Brønsted-Lowry definition provides a good basis for a general understanding of acid-base reactions in water solution. The study of these reactions is simplified by the fact that they are fast. They generally come to equilibrium as rapidly as the reacting solutions can be mixed.

Autoionization of Water

The conjugate acid and base of H_2O are the **hydronium ion** (H_3O^+) and the **hydroxide ion** (OH^-) respectively. Remarkably, water establishes an acid-base equilibrium with itself:

$$2\,\mathbf{H_2O}(l) \rightleftharpoons \mathbf{H_3O^+}(aq) + \mathbf{OH^-}(aq)$$

---

[1]See text Section 18-2 on text page 665.

This reaction is called the **autoionization** of water.

In pure water, the concentrations of hydronium and hydroxide ion are equal:

$$[H_3O^+] = [OH^-] \qquad \text{in pure water}$$

When a solute that donates or accepts hydrogen ions (that is, an acid or base) dissolves in water, this equality is destroyed. However, the following equation, which derives from the mass-action expression for autoionization, holds at equilibrium in all aqueous solutions:

$$[H_3O^+][OH^-] = K_w$$

The *product* of the hydronium ion and the hydroxide ion concentrations in water is a constant. If one goes up, the other comes down. The thermodynamic equilibrium constant for the autoionization of water K_w equals 1.0×10^{14} at room temperature. Like all equilibrium constants, it is strongly dependent on the temperature. Raising the temperature to 60°C increases K_w nearly ten-fold. Many calculations emphasize other aspects of acid-base equilibria, assuming tacitly that the temperature is 25°C. Problem **10-81** considers the temperature dependence of K_w. Also see **10-17**.

Strong Acids and Bases

A **strong acid** in aqueous solution is one that donates hydrogen ion essentially completely to the solvent; a **strong base** accepts hydrogen ion essentially completely from the solvent.

The seven common strong acids in water are:

$$HClO_4, \quad HClO_3, \quad HNO_3, \quad HCl, \quad HBr, \quad HI, \quad H_2SO_4$$

Each contains hydrogen ion and donates essentially all of it. The second hydrogen in sulfuric acid is not completely donated. See text Table 10-2, text page 323. Any acid not in this group may be assumed to be a weak acid (see below).

Some strong bases in water are the amide ion NH_2^- and the hydride ion H^-, both of which strongly accept hydrogen ions essentially completely from water (to form NH_3 and H_2, their respective conjugate acids, and OH^- ion. The most common strong bases in water are:

$$NaOH, \quad KOH, \quad RbOH, \quad CsOH, \quad Ca(OH)_2, \quad Sr(OH)_2, \quad Ba(OH)_2$$

These bases dissolve completely in water to give $OH^{(}aq)$.

The strongest possible acid in water is the H_3O^+ ion; the strongest possible base is the OH^- ion. All acids stronger than hydronium ion donates effectively all their available hydrogen ions to H_2O molecules when mixed with water. Hydronium ion (H_3O^+) results. Two strong acids (such as HNO_3 and HCl) therefore have the same apparent strength when dissolved in water. This is called the **leveling effect.** All bases stronger than hydroxide ion take hydrogen ions from H_2O when placed in water. Hydroxide ion (OH^-) results. Thus bases as well as acids are leveled.

Acidic hydrogens are susceptible to donation by a Brønsted-Lowry acid. They are often segregated to the extreme left or extreme right when chemical formulas are written. Thus, C_6H_5COOH (in **10-29**) donates the H^+ at the extreme right to give the conjugate base $C_6H_5COO^-$.

Many Brønsted-Lowry acids and bases have non-zero charges. Hydrogen ions can be donated by a positively charged species (*e.g.* NH_4^+), by a neutral species (*e.g.* HCN) or by a negatively charged species (*e.g.* HCO_3^-). Each of these serves as an Brønsted-Lowry acid when it donates a hydrogen ion. Similarly, bases can be negatively charged (like CN^-), neutral (like NH_3) and positively charged (like $NH_2NH_3^+$). Probably the most common positively charged acid in problems is the NH_4^+ ion (the ammonium ion). Aqueous ammonium ions are obtained by dissolving a salt like NH_4Cl or NH_4NO_3. Many other cationic acids can be viewed as derived from the NH_4^+ ion. For example, in **10-69,** thiamine hydrochloride, $C_{12}H_{17}ON_4SCl \cdot HCl$, dissolves to give $thiH^+$ (where "thi" equals $C_{12}H_{17}ON_4SCl$) and Cl^- ions. The $thiH^+$ ion is analogous to the NH_3H^+ ion. Seventeen of the H's in thiamine hydrochloride are non-acidic. Because they have no important in the compounds acid-base chemistry they are "buried" in the symbol thi.

The pH Function

The pH of an aqueous solution is defined as:

$$pH = -\log_{10}[H_3O^+]$$

Logarithms of numbers less than 1 are negative. The minus sign in the definition makes the pH of commonly-encountered solutions, in which $[H_3O^+]$ is less than one, positive. Further important points about pH:

- The pH goes *down* as the concentration of hydronium ion goes *up*. This is a major source of confusion.

- Negative pH's are uncommon but possible. **Example:** Find the pH of a solution with a hydronium ion concentration of 1.5 M. **Solution:** -0.18.

- pH's greater than 14 are possible.

- If a solution is 1 M in hydronium ion, it has pH 0. Standard acid conditions (used extensively in Chapter 10) are pH 0; standard basic conditions are pH 14.

The "p" operator says "take the negative of the logarithm of what follows". It is frequently applied to quantities other than $[H_3O^+]$. Thus, pOH is the negative logarithm of the concentration of hydroxide ion, and pK_w is the negative logarithm of the autoionization constant of water. Applying the "p" operator to both sides of $[H_3O^+] \times [OH^-] = K_w$ gives:

$$pH + pOH = pK_w$$

At 25°C, pK_w is 14.0. If the pH of an (exceedingly) acidic solution at 25°C is for instance -1, then its pOH is 15. See **10-15** and **10-75**. If the pH of a solution at 25°C is 7, then the pOH is also 7. Such a solution is **neutral.**

The computation of pH is as easy as pushing the "log" button on an electronic calculator. Getting from pH to $[H_3O^+]$ requires taking an antilogarithm, also called an inverse logarithm. There are two kinds of logarithm in ordinary use. The pH function uses "$\log_{10}$", logarithm to the base 10, not "ln", logarithm to the base e.

Acid and Base Strength

Free hydrogen ions do not exist in solution. To be donated, a hydrogen ion must simultaneously be accepted. This "push-pull" phenomenon prevents the determine the intrinsic tendencies of acids to donate H^+ (or bases to accept H^+) in aqueous solution. Instead, all evaluations of acid strength or base strength are relative to the strength of the H_3O^+ or OH^- ions. These evaluations consist of the measurement of equilibrium constants (K_a's and K_b's). These constants are called the **acid ionization constant** and **base ionization constant** respectively.

- A K_a tells the outcome of a *competition* between an acid and the reference acid H_3O^+ in donating hydrogen ions.

- A K_b tells the outcome of a competition between a base and the reference base OH^- in accepting hydrogen ions.

Consider the behavior of the acid HA in water:

$$HA(aq) + H_2O(l) \rightleftharpoons A^-(aq) + H_3O^+(aq) \qquad K_a = \frac{[A^-][H_3O^+]}{[HA]}$$

The competition is between HA and H_3O^+. A large K_a means that HA wins. The equilibrium lies far to the right (and the HA is nearly all gone). The HA is a strong acid. A small K_a means that H_3O^+ wins. The equilibrium then lies far to the left. Most of the HA remains, and little H_3O^+ ion is present. The HA is a weak acid. There are of course gradations of strength among weak acids. The larger the K_a, then the stronger the acid (**10-23**).

What about the base strength of A^-, the conjugate base of HA? The strength of A^- relative to OH^- is judged in the equilibrium:

$$A^-(aq) + H_2O(l) \rightleftharpoons HA(aq) + OH^-(aq) \qquad K_b = \frac{[HA][OH^-]}{[A^-]}$$

Do *not* identify this reaction as the reverse of the K_a reaction of HA. It shows A^- acting as a base to seize hydrogen ions from H_2O and generate OH^-. Its equilibrium constant is symbolized K_b:

- **K_b's refer to equilibria that form OH^- from a base plus water; the size of K_b tells the relative strength of the base.**

- **K_a's refer to equilibria that form H_3O^+ from an acid plus water; the size of K_b tells the relative strength of the acid.**

Write the K_b equilibrium of $A^-(aq)$ and the K_a equilibrium of $HA(aq)$:

$$A^-(aq) + H_2O(l) \rightleftharpoons HA(aq) + OH^-(aq) \qquad \frac{[HA][OH^-]}{[A^-]} = K_a$$

$$HA(aq) + H_2O(l) \rightleftharpoons A^-(aq) + H_3O^+(aq) \qquad \frac{[A^-][H_3O^+]}{[HA]} = K_b$$

Adding the two equilibria (and multiplying their K's gives:

$$2\,H_2O(l) \rightleftharpoons OH^-(aq) + H_3O^+(aq) \qquad [H_3O^+][OH^-] = K_a K_b$$

which is the autoionization of water. Therefore, for any acid-base conjugate pair:

$$\boldsymbol{K_a K_b = K_w}$$

Weaker acids (smaller K_a's) have stronger conjugate bases (larger K_b's) and vice versa. See **10-79**.

This relationship is very useful (as in **10-21**, **10-43**, and **10-95**). Because of it, text Table 10-2[2] provides the K_b's of the conjugate bases of the listed acids as well as the K_a's of the acids. For example, the nitrite ion (NO_2^-), which is the conjugate base of HNO_2, has K_b equal to K_w divided by 4.6×10^{-4}, which is the listed K_a of HNO_2. See **10-13** and **10-65**.

A strong acid or base is not necessarily either concentrated or hazardous. The term "strong" refers only to the ability of the acid or base to donate or accept hydrogen ions. Organic acids are prominent weak acids. Examples are acetic acid, CH_3COOH (**10-45**), and formic acid, $HCOOH$ (**10-49**). Acetic acid is a typical weak acid. It is so common that the special formula "HOAc" is often used to represent it. Important weak bases are NH_3 (ammonia) and $C_2H_3O_2^-$ (the acetate ion).

---

[2]Text page 323.

Electronegativity and Oxoacid Strength

Oxoacids contain groups of the type —X—O—H where the central atom X may be bonded to additional —OH groups, to O atoms, or to H atoms. If the electronegativity of X is low, then the X to O bond is almost completely ionic, which puts a -1 charge on the O—H group and makes the species basic (an OH^- donor). This is the case for X equal to Ca, as in $Ca(OH)_2$. A more electronegative X withdraws more electron density from the O atom bonded to it. This causes the H atom to be more easily released, since there is less electron density to hold it: the X—O—H group gains acidity when X is more electronegative. This correlation is consistent with the finding that metal oxides are base anhydrides and nonmetal oxides are acid anhydrides.[3] See text Section 21-5,[4] which also gives K_a data on a range of oxoacids, for further discussion.

Amphoterism

A species capable of acting as both acid and base is **amphoteric.** Under the Brønsted-Lowry definition, *every* species containing hydrogen is in principle amphoteric because formulas of both a conjugate acid and base can be written. The idea is that any substance containing H can be forced to act as an acid by a super-strong base that rips away a hydrogen ion, and that any substance can be forced by a super-strong acid to accept hydrogen ions.

Common amphoteric species are H_2O, HCO_3^- (see **10-3**), $H_2PO_4^-$, and HSO_4^-. The predominant reaction of an amphoteric species depends on its surroundings. In the presence of stronger acids, an amphoteric molecule or ion *accepts* hydrogen ions. In the presence of stronger bases, the same species *donates* hydrogen ions. Both behaviors however can take place to a measurable extent at once. For example, sodium hydrogen carbonate ($NaHCO_3$) dissolves in water to give $Na^+(aq)$ and $HCO_3^-(aq)$ ions. The $HCO_3^-(aq)$ ions accept H^+ ions from water:

$$HCO_3^-(aq) + H_2O(l) \rightleftharpoons H_2CO_3(aq) + OH^-(aq)$$

Simultaneously they donate H^+ ions to water:

$$HCO_3^-(aq) + H_2O(l) \rightleftharpoons CO_3^{2-}(aq) + H_3O^+(aq)$$

Thus, carbonic acid, hydrogen carbonate ion and carbonate ion are all present as equilibrium (although not in equal amount). See **10-67**.

---

[3]See text Section 10-1, text page 318.
[4]Text page 776.

Indicators

An **indicator** is a weak acid (or base) whose conjugate base (or acid) has a different color. In a good indicator, the colors are intense enough to be visible even when the indicator concentration is too low to affect the pH. The acid-base equilibrium of an indicator HIn and its (slightly rearranged) K_a expression are:

$$HIn(aq) + H_2O(l) \rightleftharpoons H_3O^+(aq) + In^-(aq) \qquad [H_3O^+] = K_a\frac{[HIn]}{[In^-]}$$

At high concentrations of hydronium ion (low pH's), most of the indicator is in the HIn form. This corresponds to a large numerator and small denominator in the preceding K_a equation. The solution takes on the color of the acid form of the indicator. If the hydronium ion concentration is decreased, the ratio $[HIn]/[In^-]$ must get smaller, in order to maintain the equality. Eventually, the base form of the indicator predominates, and the solution takes on the color of that form.

If $[H_3O^+]$ equals the indicator's K_a, then the acid and base forms of the indicator have equal concentrations. The color of the solution is a mixture of the colors of the two forms. Experience shows that this intermediate range of color exists when the $[HIn]$ to $[In^-]$ ratio is between about 1/10 and 10/1. In other words, a complete color change requires about a hundred-fold change in $[H_3O^+]$ (two pH units). See **10-25**.

There are dozens of organic acids and bases having different colors in acid and base forms and covering a wide range of K_a's and K_b's. All are potential indicators. See text Figure 10-8.[5]

One can estimate the pH of an unknown solution by adding a different indicator to each of several small portions and noting the colors. Finding that a solution is acidic to one indicator and basic to another brackets the pH of the solution. See **10-25**.

Equilibria Involving Weak Acids and Bases

Suppose c_a mol L^{-1} of the acid HA dissolves in water. It interacts with H_2O, which accepts some hydrogen ions. Equilibrium is quickly established:

$$HA(aq) + H_2O(l) \rightleftharpoons A^-(aq) + H_3O^+(aq)$$

If the reaction goes to completion (all the way to the right), then the final concentration of HA is zero, and the final concentrations of A^- and H_3O^+ both equal c_0. A large value of K_a corresponds to an equilibrium shifted *almost* all the way to the right. A large K_a makes HA a strong acid. Therefore:

---

[5]Text page 328.

• **If the acid HA is strong, the concentration of H_3O^+ in its solution equals c_a.**

If K_a is less than 1, then HA scarcely reacts. It is a weak acid. If y mol L^{-1} of it reacts, then the equilibrium concentrations of A^- and H_3O^+ are both y, and the concentration of the $HA(aq)$ left unreacted is $c_a - y$. The mass-action expression becomes:

$$K_a = \frac{y^2}{(c_a - y)}$$

Many pH calculations (such as **10-29a**, **10-33**, and **10-37**) require nothing more than setting up and solving an equation just like this.

The analysis for solutions of weak bases is exactly similar, except that K_b replaces K_a and y represents the concentration of OH^-. See **10-35**.

The dissolved acid or base is usually the main source of H_3O^+ (or OH^-), but aqueous solutions always have a second source—the autoionization of water. Most of the time, the concentration of H_3O^+ (or OH^-) coming from autoionization is negligible compared to the concentration coming from the weak acid (or base). But sometimes it is not. See **10-69**.

In Section 10-8, the text derives an *exact* equation for the H_3O^+ concentration of an aqueous solution of a monoprotic weak acid having an original concentration of weak acid equal to c_a. The weak acid donates hydrogen ions, changing the hydrogen-ion concentration such that:

$$[H_3O^+]^3 + K_a[H_3O^+]^2 - (K_w + c_aK_a)[H_3O^+] - K_aK_w = 0$$

The derivation includes no approximations.[6] In principle, the pH of any aqueous solutions prepared by dissolving a single weak acid can be computed using this equation.

Memorizing this equation is *not* recommended. For one thing, relying on a memorized equation leaves one powerless to deal with obvious variations, such as a mixture of two different weak acids. What counts is chemical insight. Therefore, study the equation for insight into acid-base equilibria:

• **What if K_a is very small?** The acid is vanishingly weak. All the terms in which K_a multiplies another quantity are nearly zero. The cubic equation devolves to:

$$[H_3O^+]^2 = K_w \qquad K_a \text{ very small}$$

The hydronium ion concentration becomes equal to $\sqrt{K_w}$, as in pure water. An infinitely weak acid (K_a zero) has no effect on pH.

---

[6]Actually, the derivation is for a more general case, the mixture of a weak acid with its conjugate base. The cubic equation given here is a rearrangement of equation **c′** on text page 349 with c_b set to zero.

• **What if K_a is very large?** The acid is very strong. The terms in the cubic equation containing K_a then overshadow the terms not containing K_a. Neglect the small terms. The equation becomes:

$$K_a[H_3O^+]^2 - c_aK_a[H_3O^+] - K_aK_w = 0 \qquad K_a \text{ large}$$

But K_a appears in each term and can be divided out:

$$[H_3O^+]^2 - c_a[H_3O^+] - K_w = 0 \qquad K_a \text{ large}$$

Dividing out K_a means this: once K_a is big it does not matter how big. If an acid is already essentially 100 percent effective in donating hydrogen ions, an increase in K_a achieve no more. Two strong acids cannot be distinguished in strength. They are leveled to the same strength.[7]

Recall that K_w is on the order of 10^{-14}. Unless c_a is quite small, the third term in the preceding is much smaller than the second and can be neglected. Doing so gives:

$$[H_3O^+]^2 - c_a[H_3O^+] = 0 \qquad K_a \text{ large, } c_a \text{ not tiny}$$

This means in practice that $[H_3O^+]$ equals c_a when K_a is large, and c_a is anything but tiny. "Tiny" is less than about 10^{-5} M.

• **What if c_a is very small?** Chemically, this corresponds to the removal of the weak acid from the solution. If c_a equals zero, the cubic equation becomes:

$$[H_3O^+]^3 + K_a[H_3O^+]^2 - K_w[H_3O^+] - K_aK_w = 0$$

This equation has only one physically meaningful root *regardless* of the value of K_a. That root is $+\sqrt{K_w} = 10^{-7}$. The "solution" (it is really just pure water) becomes neutral.

• **What if K_a and c_a have practical values?** The relative importance of the terms in the cubic equation depends on c_a. Unless the solution is quite dilute (c_a tiny), the term K_aK_w is much smaller than $(K_w + c_aK_a)[H_3O^+]$. Furthermore, K_w is much smaller than c_aK_a. Neglecting the small quantities gives the tractable quadratic:

$$[H_3O^+]^2 + K_a[H_3O^+] - c_aK_a = 0 \qquad c_a \text{ and } K_a \text{ practical}$$

The absence of K_w in this equation means that the autoionization of water has been neglected. The equation is equivalent to the quadratic (in y) worked out previously.

Worry about K_w when the acid is very weak (K_a less than about 10^{-10}), or the acid is very dilute (c_a less than about 10^{-5}).

Example: Calculate the pH of a 1×10^{-7} M solution of HCl at 25° C. **Solution:** Not 7. Making pure water (pH 7) even a little bit acidic lowers the pH. The autoionization of water makes a significant contribution to the H_3O^+ concentration. The correct answer is 6.79, obtained by substituting in: $[H_3O^+]^2 - c_a[H_3O^+] - K_w = 0$.

---

[7]See page 248 of this Guide.

Hydrolysis

Hydrolysis refers to the splitting apart ("lysis") of a substance by reaction with water. Every K_a and K_b reaction has H_2O among the reactants and is, in a way, a hydrolysis. In a more restricted sense, hydrolysis refers to those acid-base reactions that occur when a salt dissolves in water to give ions with large enough K_a or K_b to change the pH of the solution (see **10-83**). For example, sodium acetate dissolves in water to give Na^+ ions and OAc^- ions. Acetate ion has a K_b of about 10^{-10}. Sodium acetate raises the pH of water by the reaction:

$$CH_3COO^-(aq) + H_2O(l) \rightleftharpoons CH_3COOH(aq) + OH^-(aq)$$

If NH_4Cl is dissolved in water, the NH_4^+ ion hydrolyzes to give H_3O^+ and NH_3:

$$NH_4^+(aq) + H_2O(l) \rightleftharpoons NH_3(aq) + H_3O^+(aq)$$

This is the K_a reaction of the NH_4^+ ion. It lowers the pH. In these two examples, the "other" ions (Na^+ and Cl^-) do nothing.[8] However, both cation and anion in a salt can react significantly with water. For example, in ammonium acetate, (NH_4OAc) both ions hydrolyze, and the final pH is a compromise between the acidity of the first and the basicity of the second.

Buffer Solutions

Buffer solutions resist changes in pH. A typical buffer solution is made by mixing a weak acid with its conjugate base (see **10-45**). A solution of a weak base and its conjugate acid is likewise a buffer solution (see **10-43**). The weak base tends to neutralize any acid added to the buffer, and the weak acid tends to neutralize any added base. Buffers are not magic. Acids or bases do lower or raise their pH, even if only slightly. Also, enough acid or base always overwhelms their **buffer capacity** and lowers or raises the pH substantially.

Calculations of the pH of Buffer Solutions

The weak acid HA mixed in water with some of the salt NaA comprises a buffer because NaA dissociates to furnish A^-, the conjugate base of HA. The general problem is to compute the concentration of H_3O^+ (and then the pH). Let the initial concentration of HA equal c_a and the initial concentration of A^- equal c_b. It is known that:

$$[H_3O^+] = K_a \frac{c_a - [H_3O^+] + [OH^-]}{c_b + [H_3O^+] - [OH^-]}$$

---

[8] In more elegant terms: the K_a and K_b of the Na^+ and Cl^- ions are effectively zero.

This equation is derived in the text.[9] Study it for chemical insight:

- As long as the concentrations of the weak acid and its conjugate base are big they drown out the terms added to and subtracted from them. The equation becomes:

$$[H_3O^+] \approx K_a \frac{c_a}{c_b}$$

  The concentration *ratio,* not the concentrations, governs the hydrogen ion concentration! A little acid from outside converts some base to acid—it slightly reduces c_b and increases c_a. The ratio c_a/c_b hardly changes. A little outside base reduces c_a and increases c_b but again changes their ratio only a little. This is the crux of the solution's buffer action. This preceding approximate equation works to solve a majority of practical buffer problems (such as **10-45**), but thoughtless use leads to trouble (see **problem 10-85**).

- If either c_a or c_b goes to zero, their ratio goes either to zero or infinity. Either way, the ratio loses meaning in determining $[H_3O^+]$. Buffer action no longer occurs. Both acid and conjugate base must be present for buffer action.

- A buffer is most effective (resists changes in pH the best) when c_a equals c_b. This is sometimes called the **buffer point** (see **10-47**) of the buffer.

- Rewriting the equation with "a" for "b" (and "b" for "a"), OH$^-$ for H_3O^+ (and the reverse) and K_b for K_a, describes a buffer solution of a weak base and its conjugate acid. The symmetry of this transformation echoes the chemical symmetry between acid and base.

- Taking K_a to zero transforms the exact equation to $[OH^-] = c_b$. This is an expression for the OH$^-$ concentration in a solution of a strong base of concentration c_b. The insight is that as K_a gets very small the acid HA becomes no real acid at all, and the conjugate base A$^-$ automatically becomes a strong base.

- Making K_a very large transforms the exact equation to $[H_3O^+] = c_a$. This is the hydronium-ion concentration in a solution of a strong acid of concentration c_a. As the acid becomes very strong its conjugate base becomes so weak it no longer exerts any effect on the pH.

Acid-Base Titration Curves

A acid-base titration is an analytical procedure for determining the amount of an acid or base in a solution. During a titration, acid and base are mixed in a controlled

---

[9]Text page 241.

fashion. The progress of the neutralization is followed by measuring the pH of the solution as a function of the volume of titrant added. A **titration curve** is a plot of pH versus volume of added titrant. Text Figures 10-12 and 10-13[10] show typical titration curves.

Calculation of Points on a Titration Curve

During the course of a titration, two factors affect the pH of the solution:

- The acid and base neutralize each other. The neutralization reaction is:

$$H_3O^+(aq) + OH^-(aq) \rightleftharpoons 2\,H_2O(l)$$

  Naturally, the pH goes up when an acid is titrated with a base and down when a base is titrated with acid.

- The two solutions dilute each other. Simple dilution affects pH. For example, 10 mL of 0.1 M HCl has a pH of 1. Diluting it to 100 mL lowers the concentration to 0.01 M HCl and thereby raises the pH to 2.

In calculations concerning titrations consider the two factors separately:

1. Imagine that no reaction occurs until mixing is complete. Compute the effect of dilution on the concentrations of all species of importance.

2. Allow the neutralization reaction to proceed 100 percent to the right until the limiting reactant, whether acid or base, is entirely consumed.

3. Identify the important equilibria, and use them with appropriate K's to compute $[H_3O^+]$.

Suppose a strong base is being titrated with a strong acid. The course of the titration has four regions.

The Starting Point. No titrant has been added. The H_3O^+ concentration equals the concentration of the strong acid.

Approach to Equivalence. Some base has been added. The reaction:

$$H_3O^+(aq) + OH^-(aq) \rightarrow 2\,H_2O(l)$$

is assumed to go to completion. The amount (not concentration) of H_3O^+ that remains in solution equals the amount originally present minus the amount of OH^- added. The concentration of H_3O^+ equals this answer divided by the volume of the solution. The volume of the solution is the sum of the original volume of acid solution and the volume of titrant added.

---

[10]Text pages 340 and 342.

Equivalence Point. The amount of base that has been added exactly equals the amount of H_3O^+ originally present. The $[H_3O^+]$ in the solution comes entirely from the autoionization of water. The pH is 7.

Beyond the Equivalence Point. The excess strong base builds up in solution. The concentration of OH^- is the molar amount of excess OH^- divided by the total volume of the solution.

The titration of a weak acid with a strong base has the same four regions. Problems **10-53** and **10-71** show the calculation of points on the titration curve when a weak acid is titrated with a strong base. Problem **10-55** covers the conjugate case—the titration of a weak base with a strong acid. Also, **10-53** identifies the **half-equivalence point.**

Polyprotic Acids

Acids that contain two or more acidic H atoms are **polyprotic** acids. Sulfuric acid (H_2SO_4) is a diprotic acid; phosphoric acid (H_3PO_4) is a triprotic acid. Although acetic acid (CH_3COOH) contains four hydrogen atoms per molecule, it is only a monoprotic acid. The three H atoms bonded to the C atom are not acidic (are not donated to H_2O).

Polyprotic acids donate H^+ ions in two or more stages. Typically the equilibrium constant for the first stage (K_{a1}) is about 10^5 times larger than for the second stage (K_{a2}). A crucial point comes up in working with equilibria of polyprotic acids (or their conjugate bases):

The different stages in the reaction of a polyprotic acid with water interact with each other.

This means that when two or more acid-ionization reactions go on simultaneously, each contributes hydronium ion, but the hydronium ion concentration used in all the mass-action expressions is one and the same.

Exact calculations of the pH of solutions of polyprotic acid therefore require solving simultaneous equations. These systems of equations are often hard to solve. In practical problems, especially those dealing with inorganic polyprotic acids, K_{a1}, K_{a2} and other K_a's almost always differ so much in magnitude that the equilibria can be treated as if they were separate. Thus, in **10-63**, the second donation of H^+ ion by arsenic acid adds only negligibly to the hydronium ion concentration coming from the first.

At high pH, most of the concentration of a polyprotic acid will be in the form of the most negative conjugate base; at low pH most of the concentration will be in the form of the most acidic form. Problems **10-67** and **10-91** show how to compute the fraction of each of the forms of a polyprotic acid that is present at any pH.

Exact Treatment of Acid-Base Equilibria

Principle of Electrical Neutrality

Every solution must be electrically neutral.

- When a solution contains ions, the total amount of positive charge equals the total amount of negative charge. This fact furnishes an important mathematical relationship among the concentrations of the ions in a solution.

For example, in an aqueous solution of NaCl:

$$[Na^+] + [H_3O^+] = [Cl^-] + [OH^-]$$

The principle of electrical neutrality applies to the number of electrical charges and not the number of ions. In a solution of sulfuric acid in water:

$$[H_3O^+] = [OH^-] + [HSO_4^-] + 2\,[SO_4^{2-}]$$

where the coefficient of 2 for the sulfate concentration reflects the contribution by each sulfate of two negative charges to the total negative charge in the solution. See **10-69** for the use of this principle in analyzing a problem.

Material Balance

Substances often, when dissolved in water, end up partially converted to new forms. A substance may distribute itself among several different forms. See **10-49** and **10-69**.

- The sum of the concentrations of the derivative forms and the unconverted original form always equals the number of moles per liter originally added.

For example, if 0.1 mol of H_2SO_4 is placed in a liter of solution, some of it reacts with the solvent to produce HSO_4^- and SO_4^{2-} ions. The material balance condition is:

$$0.1 = [H_2SO_4] + [HSO_4^-] + [SO_4^{2-}]$$

How to Do an Exact Treatment

The exact treatment of acid-base equilibria consists of identifying and counting all the species in the equilibrium solution, coming up with as many equations relating their concentrations as there are species, and then solving the system of equations. Specifically:

1. List all of the species, both neutral and charged, that can be present in solution at equilibrium. In aqueous solutions this includes H_3O^+, OH^- and H_2O.

2. Employ the principle of electrical neutrality to write a single equation relating the concentrations of the ions in the solution.

3. Apply the principle of material balance to each substance that was placed in solution. If, for example, five weak acids are mixed in the solution, then five material balance equations, one for each acid, can be written.

4. Write a mass-action expression for every equilibrium taking place in the solution, including the autoionization of water.

5. Look up the equilibrium constants (at the proper temperature) for all of the equilibria that have been identified.

6. Verify that the number of equations equals the number of different chemical species in solution.

7. Solve the system of simultaneous equations.

Exact solutions to sets of simultaneous equations often require daunting algebraic effort. Use chemical insight to simplify the equations. Neglect terms in the equations which make only tiny contributions when added to or subtracted from other terms. (Never neglect terms that are multipliers or divisors even if they are small!) play the kind of "what-if" game illustrated earlier.

Amphoteric Equilibria

One class of problem requires frequent use of simultaneous equations involving K_{a1} and K_{a2}. The product after the first step of reaction of a diprotic acid with water is amphoteric. This species is ready to donate H^+ (according to K_{a2}), but also ready to accept H^+ to give back the original diprotic acid (according to $K_{b2} = K_w/K_{a1}$). When such an amphoteric species is put in water (in the form for example of its sodium salt), the hydrogen-ion concentration is related to the K's as follows:

$$[H_3O^+] \approx \sqrt{\frac{K_{a1}K_{a2}c_0 + K_{a1}K_w}{K_{a1} + c_0}}$$

where c_0 is the original concentration of the amphoteric species. If K_{a1} is negligible compared to c_0, and if K_w is negligible compared to $K_{a2}c_0$ then this equation becomes:

$$[H_3O^+] \approx \sqrt{K_{a1}K_{a2}}$$

These equations are used in **10-71** and **10-95**, and are discussed further there. An interesting variant of amphoteric equilibria (perfect for an examination question) appears in **25-27**.

Detailed Solutions to Odd-Numbered Problems

10-1 a) The chloride ion Cl^- cannot act as a Brønsted-Lowry acid because it has no hydrogen.

b) The hydrogen sulfate ion HSO_4^- can act as a Brønsted-Lowry acid; its conjugate base is $\boxed{SO_4^{2-}}$ (the sulfate ion).

c) The ammonium ion NH_4^+ can act as a Brønsted-Lowry acid; its conjugate base is $\boxed{NH_3}$ (ammonia).

d) Ammonia NH_3 can act as a Brønsted-Lowry acid; its conjugate base is $\boxed{NH_2^-}$ (the amide ion).

e) Water H_2O can act as a Brønsted-Lowry acid; its conjugate base is $\boxed{OH^-}$ (the hydroxide ion).

10-3 Citric acid donates hydrogen ion to the hydrogen carbonate ion, $\boxed{HCO_3^-(aq)}$, which serves as a base:

$$H_3C_5H_5O_7(aq) + HCO_3^-(aq) \rightleftharpoons H_2C_5O_7^-(aq) + H_2CO_3(aq)$$

The H_2CO_3 that is produced quickly decomposes to water and gaseous carbon dioxide, which causes the cookies to rise. A preliminary step is the dissolution of the solid sodium hydrogen carbonate in the lemon juice.

10-5 a) The slaking of lime: $\boxed{CaO(s) + H_2O(l) \rightarrow Ca(OH)_2(s)}$.

b) The reaction can be seen as a Lewis acid-base reaction. The $\boxed{CaO}$ is the Lewis base. It donates a pair of electrons (located on the oxide ion) to a hydrogen atom in the H_2O molecule. The result is a new O—H bond.

10-7 a) The fluoride ion (F^-) has a negative charge. In the Brønsted-Lowry system, an acid is a donor of a positive entity (the H^+ ion). By plus-minus symmetry then, an acid in this scheme is a $\boxed{\text{fluoride acceptor}}$.

b) In $ClF_3O_2 + BF_3 \rightarrow ClF_2O_2 \cdot BF_4$, BF_3 accepts a F^- ion from ClF_3O_2, so BF_3 is the acid, and ClF_3O_2 is the base.
In $TiF_4 + 2\,KF \rightarrow K_2[TiF_6]$, TiF_4 accepts an F^- ion from KF, so TiF_4 is the acid, and KF is the base.

10-9 Oxides of metals are base anhydrides; oxides of nonmetals are acid anhydrides.

a) MgO is the base anhydride of magnesium hydroxide $Mg(OH)_2$.
b) Cl_2O is the acid anhydride of hypochlorous acid $HOCl$.
c) SO_3 is the acid anhydride of sulfuric acid H_2SO_4.
d) Cs_2O is the base anhydride of cesium hydroxide $CsOH$.

10-11 The equations illustrating the amphoteric behavior of $SnO(s)$ are:

$$\overset{\text{base}}{SnO}(s) + 2\,HCl(aq) \rightarrow Sn^{2+}(aq) + 2\,Cl^-(aq) + H_2O(l)$$

$$\overset{\text{acid}}{SnO}(s) + NaOH(aq) + H_2O(l) \rightarrow Sn(OH)_3^-(aq) + Na^+(aq)$$

10-13 The pH of an aqueous solution equals the negative logarithm of the hydronium-ion concentration: $pH = -\log[H_3O^+] = -\log(2.0 \times 10^{-4}) = \boxed{3.70}$.

10-15 The extremes of the pH range for urine each give a H_3O^+ concentration:

$$[H_3O^+]_{\text{high}} = 10^{-5.5} = \boxed{3 \times 10^{-6} \text{ M}} \qquad [H_3O^+]_{\text{low}} = 10^{-6.5} = \boxed{3 \times 10^{-7} \text{ M}}$$

The pOH comes from the pH using the relation $pOH = pK_w - pH$.

$$pOH = 14.0 - 5.5 = 8.5 \qquad pOH = 14.0 - 6.5 = 7.5$$

where the use of $pK_w = 14.0$ assumes a temperature of 25°C. Then:

$$[OH^-]_{\text{low}} = 10^{-8.5} = \boxed{3 \times 10^{-9} \text{ M}} \qquad [OH^-]_{\text{high}} = 10^{-7.5} = \boxed{3 \times 10^{-8} \text{ M}}$$

10-17 The pH of the seawater equals 8.00. Using the definition of pH:

$$[H_3O^+] = 10^{-8.00} = \boxed{1.0 \times 10^{-8} \text{ M}}$$

Use 13.776 instead of 14.00 as pK_w when calculating pOH:

$$pOH = pK_w - pH = 13.776 - 8.00 = 5.78 \qquad [OH^-] = 10^{-5.78} = \boxed{1.7 \times 10^{-6} \text{ M}}$$

10-19 The equation $2\,K(s) + 2\,H_2O(l) \rightarrow 2\,KOH(aq) + H_2(g)$ is the better representation of what really happens. The reaction starts fast and continues with great vigor. If the equation involving H_3O^+ applied, one would expect a slow process since H_3O^+ is only 10^{-7} M in pure water. Even if the reaction were vigorous at low concentrations of H_3O^+, the reaction generates $OH^-(aq)$, which lowers the concentration of H_3O^+. Loss of H_3O^+ would quickly cause progress to flag. The other equation represents a direct interaction between $K(s)$ and $H_2O(l)$. The concentration of H_2O remains relatively constant and high (about 56 M).

10-21 a) A base (Brønsted-Lowry) is a hydrogen-ion acceptor. In ephedrine and many other organic bases the site of attachment of the hydrogen ion is a nitrogen atom:

$$\boxed{C_{10}H_{15}ON(aq) + H_2O(l) \rightleftharpoons C_{10}H_{15}ONH^+(aq) + OH^-(aq)}$$

b) Use the equation $K_a K_b = K_w$, which relates the strength of an aqueous acid and its conjugate base. Substitution gives:

$$K_a = \frac{K_w}{K_b} = \frac{1.0 \times 10^{-14}}{1.4 \times 10^{-4}} = \boxed{7.1 \times 10^{-11}}$$

c) According to text Table 10-2,[11] the K_a for ammonium ion NH_4^+ equals 5.6×10^{-10}. The K_b for ammonia NH_3 is calculated as follows:

$$K_b = \frac{K_w}{K_a} = \frac{1.0 \times 10^{-14}}{5.6 \times 10^{-10}} = 1.8 \times 10^{-5}$$

The larger the K_b the stronger the base. Because K_b for ephedrine (1.4×10^{-4}) exceeds K_b for ammonia (1.8×10^{-5}), ephedrine is $\boxed{\text{stronger}}$ as a base than ammonia.

10-23 The equation given in the problem can be obtained by combining two equations referenced in text Table 10-2:

$$H_2O(l) + HClO_2(aq) \rightleftharpoons H_3O^+(aq) + ClO_2^-(aq) \qquad K_a = 1.1 \times 10^{-2}$$
$$H_2O(l) + HNO_2(aq) \rightleftharpoons H_3O^+(aq) + NO_2^-(aq) \qquad K_a = 4.6 \times 10^{-4}$$

If the second is subtracted from the first, the equilibrium constant of the first is divided by the equilibrium constant of the second:

$$HClO_2(aq) + NO_2^-(aq) \rightleftharpoons HNO_2(aq) + ClO_2^-(aq) \qquad K = \frac{1.1 \times 10-2}{4.6 \times 10^{-4}} = \boxed{24}$$

In this case the constant exceeds 1, which means that equilibrium favors the products. The relatively small concentration of $HClO_2$ at equilibrium means that $\boxed{HClO_2}$ is a $\boxed{\text{stronger acid}}$ than HNO_2; the comparatively large concentration of HNO_2 at equilibrium means $\boxed{NO_2^-}$ is a $\boxed{\text{stronger base}}$ than ClO_2^-.

10-25 a) The color changes of indicators occur over ranges of 1 to 1.9 pH units.[12] Assume that the center of the range is roughly equal to the pK_a of the indicator. Then:

bromocresol green	$pK_a = 4.6$	$K_a = 2.51 \times 10^{-5}$
methyl orange	$pK_a = 3.8$	$K_a = 1.58 \times 10^{-4}$

---

[11]Text page 323.
[12]See text Figure 10-8, text page 328.

The acid form of methyl orange is the stronger acid because it has the larger K_a (smaller pK_a).

b) Text Figure 10-8 shows that both of these indicators are in their transition ranges. This means that the pH of the solution must simultaneously lie between pH 3.8 and pH 5.4 (bromocresol green) and 3.2 and 4.4 (methyl orange). Therefore, the pH of the solution lies in the range 3.8–4.4.

10-27 Abbreviate $HC_9H_7O_4$ (aspirin) and its conjugate base as aspH and asp⁻ respectively. Compute the concentration of aspH when 0.65 g of it dissolves in 50.0 mL of water:

$$n_{aspH} = 0.65 \text{ g aspH} \times \left(\frac{1 \text{ mol aspH}}{180.16 \text{ g aspH}} \right) = 0.00361 \text{ mol}$$

$$[aspH] = \frac{0.00361 \text{ mol}}{0.0500 \text{ L}} = 0.0722 \text{ mol L}^{-1} = 0.0722 \text{ M}$$

This result equals the "initial" concentration of aspH (*after* dissolution but *before* reaction with water). Next, consider the reaction with water, which is:

$$aspH(aq) + H_2O(l) \rightleftharpoons asp^-(aq) + H_3O^+(aq) \qquad \frac{[H_3O^+][asp^-]}{[aspH]} = K_a = 3.0 \times 10^{-4}$$

Call the concentration of H_3O^+ at equilibrium x:

	aspH(aq)	+H_2O(aq) $\rightleftharpoons$	asp⁻(aq) +	H_3O^+(aq)
Init. Conc. (M)	0.0722	—	0	small
Change in Conc. (M)	$-x$	—	$+x$	$+x$
Equil. Conc. (M)	$0.0722 - x$	—	x	x

Substitute the equilibrium concentrations into the mass-action expression:

$$\frac{[asp^-][H_3O^+]}{[aspH]} = K_a = \frac{x^2}{0.0722 - x} = 3.0 \times 10^{-4}$$

Solve for x using the quadratic formula or by successive approximation. The answer is $x = 4.5 \times 10^{-3}$. A hydrogen-ion concentration of 4.5×10^{-3} M translates to a pH of 2.35.

Tip. Neglecting x compared to 0.0722 gives a hydrogen-ion concentration that is over 3% too high and an answer of 2.33, which is close, but incorrect.

10-29 a) Benzoic acid is C_6H_5COOH.[13] Represent it as HOBz. As the 0.20 M HOBz reacts with water, it generates H_3O^+:

	HOBz(aq)	$+H_2O(l) \rightleftharpoons$	$OBz^-(aq) +$	$H_3O^+(aq)$
Init. Conc. (M)	0.20	−	0	small
Change in Conc. (M)	$-x$	−	$+x$	$+x$
Equil. Conc. (M)	$0.20 - x$	−	x	x

The concentration of H_3O^+ arising from the autoionization of water is very small compared to the concentration from the reaction of benzoic acid, so x equals the concentration of H_3O^+ at equilibrium. Insert x in the mass-action expression:

$$K_a = 6.46 \times 10^{-5} = \frac{[OBz^-][H_3O^+]}{[HOBz]} = \frac{x^2}{0.20 - x}$$

Rearrange to obtain: $x^2 + 6.46 \times 10^{-5}x - 1.29 \times 10^{-5} = 0$. Apply the quadratic formula:

$$x = \frac{-6.46 \times 10^{-5} \pm \sqrt{4.17 \times 10^{-9} + 5.16 \times 10^{-5}}}{2}$$

$$x = \frac{-6.46 \times 10^{-5} \pm 7.18 \times 10^{-3}}{2} = -0.00352 \text{ and } 0.00356$$

The negative root has no physical meaning. Using the positive root gives $[H_3O^+] = 0.00356$ M. The pH is $-\log[H_3O^+] = \boxed{2.45}$.

Tip. If x is neglected in comparison to 0.020 in the mass-action equation, then the very simple equation $x = \sqrt{(0.20)(6.46 \times 10^{-5})}$ results. The positive root of this equation is $x = 0.0036$. This equals, to two significant figures, the answer obtained using the quadratic equation. To get a feel for how approximation works, carry out a few computations both exactly and approximately, and compare the results.

b) The equilibrium concentration of $[H_3O^+]$ must equal 3.56×10^{-3} M. Start as in the preceding part, but now x is known and $[HOAc]_0$, the initial concentration of acetic acid is the goal:

$$K_a = \frac{x^2}{[HOAc]_0 - x} \qquad \text{hence} \qquad 1.76 \times 10^{-5} = \frac{(3.54 \times 10^{-3})^2}{[HOAc]_0 - 3.56 \times 10^{-3}}$$

Solving gives $[HOAc]_0 = 0.72$ M. This means $\boxed{0.72 \text{ mol}}$ of acetic acid must be dissolved per liter of solution.

---

[13]Again, the single H segregated in the formula is an acidic hydrogen.

10-31 Use the approach of **10-29a**:

$$HIO_3(aq) \ + H_2O(aq) \rightleftharpoons \ IO_3^-(aq) + \ H_3O^+(aq)$$

Init. Conc. (M)	0.100	—	0	small
Change in Conc. (M)	$-x$	—	$+x$	$+x$
Equil. Conc. (M)	$0.100 - x$	—	x	x

$$\frac{[IO_3^-][H_3O^+]}{[HIO_3]} = K_a \quad \text{hence} \quad \frac{x^2}{0.100 - x} = 0.16$$

This rearranges to: $x^2 + 0.16x - 0.016 = 0$. The quadratic formula gives:

$$x = \frac{-0.16 \pm \sqrt{0.0256 - 4(-0.016)}}{2} = 0.0697 \text{ and} - 0.230$$

If $[H_3O^+] = 0.0697$ M, then pH = $\boxed{1.16}$. The pH is well on the acid side because HIO_3 is a rather strong weak acid.

Tip. The short-cut of neglecting x in comparison to 0.100 gives an H_3O^+ concentration of 0.126 M and a pH of 0.90, which is seriously wrong.

10-33 The $papH^+Cl^-$, a salt, dissolves completely in water to give $papH^+$ ion and Cl^- ion. The Cl^- ion does not react significantly with the water, but the $papH^+$ ion reacts as a weak acid:

$$papH^+(aq) + H_2O(l) \rightleftharpoons pap(aq) + H_3O^+(aq)$$

Assume that this reaction is the sole source of H_3O^+. The concentration of H_3O^+ is 4.90×10^{-4} M (calculated from the pH of 3.31), and the concentration of the conjugate base pap is also 4.90×10^{-4} M. The concentration of $papH^+$ equals its original concentration minus the portion converted into pap. This is $(0.205 - 4.90 \times 10^{-4})$ M. Substitute these equilibrium values into the K_a expression to compute K_a:

$$K_a = \frac{(4.90 \times 10^{-4})(4.90 \times 10^{-4})}{(0.205 - 4.90 \times 10^{-4})} = \boxed{1.2 \times 10^{-6}}$$

Tip. "Forgetting" to subtract in the denominator has a negligible effect.

10-35 The K_b applies to the reaction:

$$morph(aq) + H_2O(l) \rightleftharpoons morphH^+(aq) + OH^-(aq)$$

where "morph" stands for morphine. The initial concentration of morphine (*after* dissolution but *before* reaction with water) is:

$$[morph]_0 = \frac{0.0400 \text{ mol}}{0.600 \text{ L}} = 0.0667 \text{ mol L}^{-1} = 0.0667 \text{ M}$$

Let x equal the loss in the concentration of morph in coming to equilibrium:

	morph(aq)	+H$_2$O(aq) $\rightleftharpoons$	morphH$^+$(aq) +	OH$^-$(aq)
Init. Conc. (M)	0.0667	—	0	small
Change in Conc. (M)	$-x$	—	$+x$	$+x$
Equil. Conc. (M)	$0.0667 - x$	—	x	x

Substitute the equilibrium concentrations into the mass-action expression:

$$\frac{[\text{morphH}^+][\text{OH}^-]}{[\text{morph}]} = K_b = \frac{x^2}{0.0667 - x} = 8 \times 10^{-7}$$

This equation can be solved using the quadratic formula. But, observe that x must be quite small compared to 0.0667. Neglecting the x in the denominator allows the quick conclusion that $x = 2.301 \times 10^{-4}$. This makes the equilibrium concentration of OH$^-$ equal 2.3×10^{-4} M. The pOH is $-\log[\text{OH}^-] = 3.64$, and the pH is this number subtracted from 14.0, which is $\boxed{10.4}$. Use of one significant figure (the 4) in the final answer reflects the fact that K_b has only one significant figure.

10-37 Hydrofluoric acid is a weak acid in water:

$$\text{HF}(aq) + \text{H}_2\text{O}(l) \rightleftharpoons \text{F}^-(aq) + \text{H}_3\text{O}^+(aq) \quad K_a = 6.6 \times 10^{-4}$$

Because the pH of the HF solution at 25°C is 2.13:

$$[\text{H}_3\text{O}^+] = \text{antilog}(-2.13) = 10^{-2.13} = 7.41 \times 10^{-3} \text{ M}$$

If the hydrofluoric acid is the only important source of H$_3$O$^+$, then [F$^-$] is also 7.41×10^{-3} M. The existence of the equilibrium guarantees that:

$$\frac{[\text{H}_3\text{O}^+][\text{F}^-]}{[\text{HF}]} = K_a = 6.6 \times 10^{-4}$$

All of the concentrations expect [HF] are known. Solve for [HF]:

$$[\text{HF}] = \frac{[\text{H}_3\text{O}^+][\text{F}^-]}{K_a} = \frac{(7.41 \times 10^{-3})^2}{6.6 \times 10^{-4}} = \boxed{0.083 \text{ M}}$$

10-39 Aqueous NaOH contains Na$^+$ ions and OH$^-$ ions. The latter react with acetic acid to produce water and acetate ions. Acetate ion is a weak base in its own right. Its reaction (as a base) with water causes the solution of sodium acetate that results from treating acetic acid with an equal chemical amount of NaOH (a procedure called neutralization) to be basic, and have $\boxed{\text{pH greater than 7}}$.

Tip. One might think that "neutralization" would give solutions that are neutral. Wrong. The exact chemical neutralization of an acid by a base does *not* in most cases give a neutral (pH 7) solution. When a weak acid is reacted with a strong base, the pH at the equivalence point exceeds 7. When a weak base is reacted with a strong acid, the pH at the equivalence point is less than 7. Exact neutralization of a strong acid by a strong base does give a solution of pH 7.

Similarly, a soluble salt of a strong acid and a weak base gives an acidic solution, and a soluble salt of a weak acid and a strong base gives a basic solution.

10-41 The ammonium bromide dissolves to give $NH_4^+(aq)$, a weak acid, and $Br^-(aq)$, which is not active as a base. The $NH_4^+(aq)$ ion reacts weakly with water to generate $H_3O^+(aq)$. The NH_4Br solution is therefore somewhat acidic. The hydrogen chloride reacts completely with water to give $Cl^-(aq)$ and $H_3O^+(aq)$, thereby creating a strongly acidic solution. The sodium hydroxide, a strong base, dissolves to $Na^+(aq)$ and $OH^-(aq)$; the latter makes the solution highly basic. The $NaCH_3COO$ gives $Na^+(aq)$ and $CH_3COO^-(aq)$, which reacts weakly as a base. The KI dissolves to $K^+(aq)$ and $I^-(aq)$, neither of which reacts as either acid or base. The order of increasing pH is therefore:

$$\boxed{HCl < NH_4Br < KI < NaCH_3COO < NaOH}$$

10-43 Compute the pK_a of the conjugate acid of "tris" from the pK_b of tris itself:

$$pK_a = 14.00 - pK_b = 14.00 - 5.92 = 8.08$$

The addition of HCl converts some tris to its conjugate acid:

$$tris(aq) + HCl(aq) \rightleftharpoons trisH^+(aq) + Cl^-(aq)$$

The resulting solution is a mixture of a weak acid ($trisH^+$) and its conjugate base (tris). It is a buffer by virtue of the reaction:

$$trisH^+(aq) + H_2O(l) \rightleftharpoons tris(aq) + H_3O^+(aq) \qquad K_a = \frac{[tris][H_3O^+]}{[trisH^+]}$$

Compute the concentrations of the tris and $trisH^+$ after complete reaction with the HCl but before the preceding equilibrium is established:

$$[tris]_0 = \frac{(0.050 - 0.025 \text{ mol})}{2.00 \text{ L}} = 0.0125 \text{ M}$$

$$[trisH^+]_0 = 0.025 \text{ mol}/2.00 \text{ L} = 0.0125 \text{ M}$$

The equilibrium now reduces the concentration of the $trisH^+$ as it forms H_3O^+ and tris in equal amounts. If x is the equilibrium concentration of $H_3O^+(aq)$ ion, then:

	$trisH^+(aq)$	$+H_2O(aq) \rightleftharpoons$	$tris(aq) +$	$H_3O^+(aq)$
Init. Conc. (M)	0.0125	—	0.0125	small
Change in Conc. (M)	$-x$	—	$+x$	$+x$
Equil. Conc. (M)	$0.0125 - x$	—	$0.0125 + x$	x

$$K_a = \frac{[tris][H_3O^+]}{[trisH^+]} = \frac{(0.0125 + x)x}{(0.0125 - x)}$$

Assume that x is small compared to 0.0125. Then the 0.0125's cancel out and:

$$[H_3O^+] = K_a \quad \text{so that} \quad pH = pK_a = \boxed{8.08}$$

Clearly x is less than 10^{-7}, so the assumption was justified.

Tip. The pH equals the pK_a of the weak acid. This is a general result in buffer solutions in which the acid and conjugate base concentrations are equal (and not extremely low).

10-45 a) The problem is exactly like Example 10-7.[14] After mixing and dissolution but before any other chemical change, the solution is 0.10 M in acetic acid and 0.040 M in acetate ion (from the sodium acetate). Then the weak-acid equilibrium comes into play. Let x equal the equilibrium concentration of H_3O^+:

	$HOAc(aq)$	$+H_2O(aq) \rightleftharpoons$	$OAc^-(aq) +$	$H_3O^+(aq)$
Init. Conc. (M)	0.10	—	0.040	small
Change in Conc. (M)	$-x$	—	$+x$	$+x$
Equil. Conc. (M)	$0.10 - x$	—	$0.040 + x$	x

$$K_a = 1.76 \times 10^{-5} = \frac{[OAc^-][H_3O^+]}{[HOAc]} = \frac{(0.040 + x)x}{(0.10 - x)}$$

Assume that x is small compared to 0.10. If so, the x's that are added and subtracted can be neglected in the above expression to give:

$$x = 1.76 \times 10^{-5} \left(\frac{0.10}{0.040} \right) = 4.4 \times 10^{-5}$$

The result clearly justifies the assumption. The pH is $-\log(4.4 \times 10^{-5})$ or $\boxed{4.36}$.

b) The addition of 0.010 mol of $OH^-(aq)$ ion (in the form of NaOH) very quickly converts 0.010 mol of acetic acid (HOAc) to 0.010 mol of acetate ion (OAc^-). The

---

[14]Text page 335.

concentrations of HOAc and OAc⁻ right after the conversion but before the reaction of HOAc as a weak acid are:

$$[\text{HOAc}] = \frac{0.050 - 0.010 \text{ mol}}{0.500 \text{ L}} = 0.080 \text{ M}$$

$$[\text{OAc}^-] = \frac{0.020 - 0.010 \text{ mol}}{0.500 \text{ L}} = 0.060 \text{ M}$$

Now consider the K_a equilibrium. Let y equal the equilibrium concentration of H_3O^+:

	$\text{HOAc}(aq)$	$+H_2O(aq) \rightleftharpoons$	$\text{OAc}^-(aq) +$	$H_3O^+(aq)$
Init. Conc. (M)	0.080	—	0.060	small
Change in Conc. (M)	$-y$	—	$+y$	$+y$
Equil. Conc. (M)	$0.080 - y$	—	$0.060 + y$	y

$$K_a = \frac{[\text{OAc}^-][\text{H}_3\text{O}^+]}{[\text{HOAc}]} = 1.76 \times 10^{-5} = \frac{(0.060 + y)y}{(0.080 - y)}$$

Assume that y is small compared to 0.060 and 0.080. Then:

$$y = 1.76 \times 10^{-5} \left(\frac{0.080}{0.060}\right) = 2.34 \times 10^{-5}$$

The assumption is clearly justified. The pH is the negative logarithm of 2.34×10^{-5} or $\boxed{4.63}$.

Tip. The pH rises from 4.36 to only 4.63 despite addition of a very substantial (20% of the amount of weak acid present) portion of strong base. Resistance to changes in pH characterizes buffered solutions.

10-47 Buffer solutions are most efficient at resisting changes in pH at their **buffer points**. At the buffer point, the concentrations of the conjugate pair are equal, and the pH of the buffer equals the pK_a of the weak acid. The physician should therefore select a weak acid having a pK_a as close as possible to the desired pH. The best choice on the list is $\boxed{m\text{-chlorobenzoic acid}}$, $pK_a = 3.98$.

10-49 500 mL of 0.100 M formic acid is titrated with 0.0500 M NaOH. Before any NaOH is added, the solution contains $0.500 \text{ L} \times 0.100 \text{ mol L}^{-1}$ or 0.0500 mol of formate-containing species. Most is in the form of un-ionized formic acid $\text{HCOOH}(aq)$. There is also a small quantity of $\text{HCOO}^-(aq)$, the formate ion. Adding strong base to the solution does not alter the *total* amount of formate-containing species. It does convert $\text{HCOOH}(aq)$ to $\text{HCOO}^-(aq)$:

$$\text{HCOOH}(aq) + \text{NaOH}(aq) \rightarrow \text{Na}^+(aq) + \text{HCOO}^-(aq) + \text{H}_2\text{O}(l)$$

and thereby raises the pH. The equivalence point for this reaction is the point at which exactly one mole of NaOH has been added for every mole of formic acid originally present. Reaching the equivalence point requires 0.0500 mol of NaOH, an amount provided by 1000 mL of 0.0500 M NaOH. At equivalence, the solution consists of dilute sodium formate. Sodium formate is the salt of a weak acid and strong base; the pH at the equivalence point would exceed 7 (see **10-39**).

To raise the pH to a mere 4.00 clearly requires *less* than 1000 mL of 0.0500 M OH^- ion. At pH = 4.00, $[H_3O^+] = 1.0 \times 10^{-4}$ M. Write the mass-action expression for the acid ionization of formic acid and insert this value into it:

$$1.77 \times 10^{-4} = \frac{[H_3O^+][HCOO^-]}{[HCOOH]} = \frac{(1.0 \times 10^{-4})[HCOO^-]}{[HCOOH]}$$

$$1.77 = \frac{[HCOO^-]}{[HCOOH]}$$

Let the addition of V L of 0.0500 M OH^- bring the 0.500 L of 0.100 M formic acid solution to pH 4.00. The total volume of the mixture at pH 4.00 equals $(0.500 + V)$ L. Each mole of added OH^- converted one mole of $HCOOH(aq)$ to $HCOO^-(aq)$. Assume for the moment that this acid-base reaction is the only source of $HCOO^-(aq)$, although a bit more $HCOO^-(aq)$ does appear from the ionization of HCOOH to $H_3O^+(aq)$ and $HCOO^-(aq)$.[15] Then, because $0.0500V$ mol of OH^- has been added:

$$[HCOO^-] = \frac{0.0500V}{0.500 + V}$$

where the numerator equals the chemical amount of $HCOO^-(aq)$ produced by the acid-base reaction and the denominator equals the total volume of the solution. Dilution reduces the total concentration of formate-containing species from 0.100 M.[16] There are only two formate-containing species, $HCOOH(aq)$ and $HCOO^-(aq)$. After the addition of V liters of NaOH solution:

$$[HCOO^-] + [HCOOH] = \left(\frac{0.500}{0.500 + V}\right) \times 0.100 = \frac{0.0500}{0.500 + V}$$

Solve for [HCOOH] and insert the expression for $[HCOO^-]$

$$[HCOOH] = \frac{0.0500}{0.500 + V} - [HCOO^-]$$
$$= \frac{0.0500}{0.500 + V} - \frac{0.0500V}{0.500 + V} = \frac{0.0500 - 0.0500V}{0.500 + V}$$

---

[15]If only a small amount of $HCOOH(aq)$ reacts this way, then the assumption is *nearly* correct; the question of how near must be answered later.

[16]To review why, see page 126 in this Guide.

Put the equations for [HCOOH] and [HCOO⁻] in terms of V into the equation for their ratio:

$$1.77 = \frac{[HCOO^-]}{[HCOOH]} = \frac{0.0500V}{0.0500 - 0.0500V}$$

It is easy to solve for V, which equals 0.639 L or $\boxed{639 \text{ mL}}$.

The total volume of the solution when the pH reaches 4.00 is 1.139 L. The concentrations of HCOOH(aq) and HCOO⁻(aq) equal 0.0158 and 0.0280 M respectively, at this point. Both values are large in comparison to the [H₃O⁺] concentration, 1.0×10^{-4} M, which justifies having neglected [H₃O⁺] in comparison to the two.

10-51 The substance Ba(OH)₂ is a strong base. It ionizes completely in solution to give one mole of Ba²⁺ ion and two moles of OH⁻ ion per mole dissolved. Before any acid is added, [OH⁻] = 2 × 0.3750 = 0.7500 M. The pOH, which equals the negative logarithm of this number, is 0.1249; pH= 14.000− pOH = 14.00 − 0.1249 = $\boxed{13.88}$.

The chemical amount of OH⁻ ion in the original 100.0 mL of Ba(OH)₂ equals its molarity multiplied by its volume (in liters). It is 0.07500 mol. This means that to reach the equivalence point requires 0.07500 mol of HClO₄. The volume of 0.4540 M HClO₄ that provides this much HClO₄ is:

$$\frac{1 \text{ L}}{0.4540 \text{ mol HClO}_4} \times 0.0750 \text{ mol HClO}_4 = 0.1652 \text{ L} = 165.2 \text{ mL}$$

When the titration is 1.00 mL short of the equivalence point, only 164.2 mL of 0.4540 M HClO₄ has been added for a total of 0.07455 mol of HClO₄. Some OH⁻ ion remains unreacted. Its amount equals the difference between the amount of OH⁻ originally present and the amount reacted. The concentration of OH⁻ equals this same amount divided by the volume of the solution:

$$[OH^-] = \frac{(0.07500 - 0.07455) \text{ mol}}{(0.1000 + 0.1642) \text{ L}} = 0.0017 \text{ M}$$

Note the (correct) use of two significant figures in the answer. The pOH is 2.77, and the pH is therefore $\boxed{11.24}$.

The pH reaches $\boxed{7.00}$ at the equivalence point; this is a titration of a strong base with a strong acid.

When the titration is 1.00 mL past the equivalence point, all of the OH⁻ from Ba(OH)₂ has been reacted away, and excess HClO₄ is present. The amount of excess is:

$$\left(\frac{0.4540 \text{ mol}}{1 \text{ L}} \times 0.1662 \text{ L} \right) - 0.07500 \text{ mol} = 4.5 \times 10^{-4} \text{ mol HClO}_4$$

The excess $HClO_4$, a strong acid, is completely ionized to produce 4.5×10^{-4} mol of H_3O^+ (and, of course, an equal amount of ClO_4^- ion). The concentration of this H_3O^+ is:

$$[H_3O^+] = \frac{4.5 \times 10^{-4} \text{ mol}}{(0.1000 + 0.1662) \text{ L}} = 0.0017 \text{ M}$$

Hence, pH $= -\log(0.0017) = \boxed{2.77}$. Throughout this analysis, the autoionization of water is ignored. Even very small amounts of strong acid or base completely overshadow water as a source of H_3O^+ or OH^- ion.

Tip. The pH plummets dramatically (from 11.24 to 2.77) upon addition of only 2 mL of acid in the range of the equivalence point.

10-53 Hydrazoic acid, a weak acid, reacts with water:

$$HN_3(aq) + H_2O(l) \rightleftharpoons N_3^-(aq) + H_3O^+(aq) \qquad \frac{[N_3^-][H_3O^+]}{[HN_3]} = K_a$$

• *Before Addition of Base.* The initial concentration of the HN_3 is 0.1000 M. Let x equal the concentration of H_3O^+ present at equilibrium in the solution. Then:

$$\frac{[N_3^-][H_3O^+]}{[HN_3]} = K_a = 1.9 \times 10^{-5} = \frac{x^2}{0.1000 - x}$$

where it has been assumed that hydrazoic acid is the only significant source of H_3O^+. Rearranging gives the quadratic equation:

$$x^2 + (1.9 \times 10^{-5})x - 1.9 \times 10^{-6} = 0$$

The positive root of this equation is 1.37×10^{-3}. Therefore:

$$[H_3O^+] = 1.37 \times 10^{-3} \text{ M} \quad \text{and} \quad \text{pH} = \boxed{2.86}$$

• *After Addition of 25.00 mL of Base.* Sodium hydroxide is a strong base. Each added mole of NaOH converts one mole of $HN_3(aq)$ to one mole of $N_3^-(aq)$. The 25.00 mL of 0.1000 M NaOH furnishes:

$$\frac{0.100 \text{ mol } OH^-}{1 \text{ L}} \times 0.250 \text{ L} = 2.500 \times 10^{-3} \text{ mol } OH^-$$

Assume that the added OH^- reacts completely with the HN_3. The reaction produces 2.500×10^{-3} mol of N_3^- and leaves 2.500×10^{-3} mol of HN_3. Exactly half of the hydrazoic acid is reacted—this is the **half-equivalence point** of the titration. The total volume of the solution is 0.0750 L, so the "original" concentrations of the weak acid and its conjugate base are both 0.0333 M. "Original" is in quotation marks because these concentrations are for the state after the mixing of the solutions but before the acid-base equilibrium gets established. As it becomes established, the equilibrium generates H_3O^+ and changes both concentrations slightly. The changes are so slight that the approximate equation:

$$pH \approx pK_a - \log \frac{[HN_3]_0}{[N_3^-]_0}$$

holds at this point in the titration.[17] Also, the calculation of the pH at this stage of a titration is just like the calculation shown in Example 10-10.[18] Substitution gives:

$$pH = 4.72 - \log \frac{0.0333}{0.0333} = \boxed{4.72}$$

The pH equals the pK_a of the weak acid being titrated. This is generally true at half-equivalence in the titration of a weak acid. A titration at half-equivalence is also a buffer at its buffer point. See **10-45**.

• *At the Equivalence Point.* The addition of 50.00 mL of the NaOH solution brings the titration to its equivalence point—the number of moles of OH^- added equals the number of moles of HN_3 originally present. If no hydrazoic acid at all is left, then the concentration of N_3^- ion equals

$$[N_3^-] = \frac{0.00500 \text{ mol}}{0.100 \text{ L}} = 0.0500 \text{ M}$$

In fact, N_3^- ion reacts with water because N_3^- is a weak base:

	$N_3^-(aq)$	$+ H_2O(aq) \rightleftharpoons$	$HN_3(aq) +$	$OH^-(aq)$
Init. Conc. (M)	0.0500	–	0	small
Change in Conc. (M)	$-x$	–	$+x$	$+x$
Equil. Conc. (M)	$0.0500 - x$	–	x	x

The K_b for this reaction is 5.26×10^{-10}, obtained by dividing K_w by the K_a of hydrazoic acid. Use the mass-action expression for the preceding to obtain:

$$\frac{[OH^-][HN_3]}{[N_3^-]} = K_b = 5.26 \times 10^{-10} = \frac{x^2}{0.0500 - x}$$

Solving for x gives $[OH^-] = 5.13 \times 10^{-6}$ M. This corresponds to a pOH of 5.29 and therefore a pH of $\boxed{8.71}$.

---

[17]The equation is derived on text page 338.
[18]Text page 338.

- *Beyond the Equivalence Point.* A total of 51.00 mL of NaOH(aq) has been added. All of the HN_3 has been reacted, and some OH^- remains in excess. The concentration of leftover OH^- equals the amount of OH^- added minus the amount reacted divided by the total volume of the solution:

$$[OH^-] = \frac{(0.05100 \text{ L} \times 0.100 \text{ M}) - 5.000 \times 10^{-3} \text{ mol}}{(0.05000 + 0.05100) \text{ L}} = 9.910 \times 10^{-4} \text{ M}$$

With this much "other" OH^- in solution, the reaction of N_3^- to HN_3 plus OH^- adds just a pittance to the concentration of OH^-. Hence:

$$pOH = -\log(9.901 \times 10^{-4}) = 3.00 \quad \text{hence} \quad pH = \boxed{11.00}$$

10-55 The titration of the weak base ethylamine with the strong acid HCl falls into four ranges: *before* the addition of acid; *between* the first addition of acid and the equivalence point; *at* the equivalence point; *beyond* the equivalence point. The pH of the original 40.00 mL of 0.1000 M ethylamine exceeds 7 because ethylamine is a base. As 0.1000 M HCl is added, the pH falls. Abbreviate ethylamine and its conjugate acid as $EtNH_2$ and $EtNH_3^+$ respectively.

- *Before Addition of Acid.* Ethylamine raises the pH of pure water by the reaction:

$$EtNH_2(aq) + H_2O(l) \rightleftharpoons EtNH_3^+(aq) + OH^-(aq) \qquad \frac{[EtNH_3^+][OH^-]}{[EtNH_2]} = 6.41 \times 10^{-4}$$

Let $[OH^-] = y$, and assume that the concentration of hydroxide ion from the autoionization of water is small. Because no HCl has been added:

$$[OH^-] = [EtNH_3^+] = y \quad \text{and} \quad [EtNH_2] = 0.100 - y$$

$$6.41 \times 10^{-4} = \frac{y^2}{0.1000 - y}$$

This last has the same form as the equation developed in the text for the titration of acetic acid with NaOH. The difference is that y is now $[OH^-]$. Solving gives $y = [OH^-] = 0.00769$ M. The pOH is 2.11, and the pH is $14.00 - 2.114 = \boxed{11.89}$.

Tip. Formulas very similar to the ones developed in the text for the titration of a weak acid with a strong base work to compute the pH along this titration curve. The only difference is that the natural choice for an unknown is $[OH^-]$ rather than $[H_3O^+]$.

• *After First Addition of Acid, Before Equivalence Point.* In this range of the titration:

$$[EtNH_3{}^+] = \frac{c_t V}{V_0 + V} + y$$

where c_t is the concentration of the titrant, V_0 is the original volume of ethylamine solution, V is the volume of titrant added and y is the concentration of OH^-. The numerator $c_t V$ equals the chemical amount of $EtNH_3^+$ generated by the 1-to-1 reaction between the titrant and the $EtNH_2(aq)$, the denominator $(V_0 + V)$ equals the total volume of the solution, and their quotient $c_t V/(V_0 + V)$ equals the concentration of $EtNH_3^+(aq)$ from the neutralization reaction alone. The reaction of $EtNH_2$ to produce $OH^-(aq)$ gives additional $EtNH_3^+(aq)$ and is responsible for the y on the right-hand side of the above equation. Similarly:

$$[EtNH_2] = \frac{c_0 V_0 - c_t V}{V_0 + V} - y$$

where c_0 stands for the original ethylamine concentration. In this titration, c_0 equals 0.1000 M, c_t equals 0.1000 M, and V_0 equals 0.0400 L. If 5.00 mL of HCl has been added, $V = 0.00500$ L. Substitution in the two preceding equations gives:

$$[EtNH_3^+] = 0.01111 + y \qquad \text{and} \qquad [EtNH_2] = 0.07777 - y$$

Put the concentrations into the K_b expression:

$$6.41 \times 10^{-4} = \frac{y(0.01111 + y)}{(0.07777 - y)}$$

In this equation, y is not negligible compared to 0.01111 or 0.07777. Omitting it from the two terms on the right-hand side gives a trial y of 0.00487, which is 43% of 0.01111! Solve the equation by rearranging it and using the quadratic formula. The answer is $[OH^-] = 0.00331$ M; $\boxed{pH = 11.52}$.

At 20.00 mL, the same formulas give $[OH^-] = 6.18 \times 10^{-4}$ M, and a pH of $\boxed{10.79}$; at 39.90 mL, the same formulas give pH $\boxed{8.20}$.

• *At the Equivalence Point.* At equivalence, the reaction mixture consists of 80.00 mL of 0.05000 M ethylammonium chloride, $EtNH_3^+Cl^-$. The cation of this salt reacts as a weak acid:

$$EtNH_3^+(aq) + H_2O(l) \rightleftharpoons EtNH_2(aq) + H_3O^+(aq)$$

The equilibrium constant for this reaction is $K_a = K_w/K_b$. Let $x = [H_3O^+]$. Then:

$$K_a = 1.56 \times 10^{-11} = \frac{x^2}{(0.05000 - x)}$$

Solving gives $x = 8.83 \times 10^{-7}$ so the pH $= -\log(8.83 \times 10^{-7}) = \boxed{6.05}$

Tip. Using the formulas that work in the range before the equivalence point gives a deceptive result at this point:

$$[EtNH_2] = 0 - [OH^-] \quad (?!)$$

This cannot be right since $[OH^-]$ and $[EtNH_2]$ must both be positive.

● *Beyond the Equivalence Point.* In this range, the solution behaves like a simple solution of HCl. Compared to the strong acid HCl, the weakly acidic ethylammonium ion contributes zero to the $H_3O^+(aq)$ concentration. When 40.10 mL of HCl has been added, the first 40.00 mL has gone to produce $EtNH_3^+(aq)$ ion by reacting with all the $EtNH_2(aq)$. The remaining 0.10 mL is free to act as a strong acid. The 0.10 mL of 0.1000 M HCl is of course diluted to 80.10 mL. Every HCl generates one H_3O^+ in aqueous solution: Thus:

$$[H_3O^+] = \frac{0.10}{80.10} \times 0.1000 \text{ M} = 1.25 \times 10^{-4} \text{ M} \qquad \text{pH} = \boxed{3.90}$$

After 50.00 mL of 0.1000 M HCl has been added:
$[H_3O^+] = (10.00/90.00) \times 0.1000 = 1.111 \times 10^{-2}$ M; $\boxed{\text{pH} = 1.95}$.

10-57 Addition of 46.50 mL of the 0.393 M $NaOH(aq)$ solution must bring solution very near to an equivalence point because one more drop of base boosts the pH by more than one entire unit. (At and near the equivalence point in titrations, small additions of titrant cause large changes in pH.) Assume that 46.51 mL of the base brings the titration to the equivalence point. The chemical amount of NaOH added at this point is:

$$n_{NaOH} = 0.04651 \text{ L} \times \left(\frac{0.393 \text{ mol OH}^-}{1 \text{ L}} \right) = 0.01828 \text{ mol}$$

This amount of strong base completes the neutralization of the HCl, but some HCl had previously been neutralized by benzoate ion ($C_6H_5COO^-$, or OBz^-). The benzoate ion, a base, was present in solution from the original sample. Thus:

$$n_{HCl} = n_{NaOH} + n_{OBz^-}$$

The number of moles of HCl is:

$$n_{HCl} = 0.0500 \text{ L} \times \left(\frac{0.500 \text{ mol HCl}}{1 \text{ L}} \right) = 0.0250 \text{ mol}$$

Substitution gives:

$$0.0250 = 0.01828 + n_{OBz^-} \quad \text{hence} \quad n_{OBz^-} = 6.7 \times 10^{-3} \text{ mol}$$

The mass of C_6H_5COONa in the sample is:

$$m_{NaOBz} = 6.72 \times 10^{-3} \text{ mol OBz}^- \times \left(\frac{1 \text{ mol NaOBz}}{1 \text{ mol OBz}^-}\right) \times \left(\frac{144.11 \text{ g NaOBz}}{1 \text{ mol NaOBz}}\right)$$

$$= \boxed{0.97 \text{ g NaOBz}}$$

10-59 Because diethylamine and hydrochloric acid react in a 1-to-1 molar ratio, the chemical amount of HCl to reach the equivalence point equals the chemical amount of diethylamine originally present and is computed as the volume of the added HCl solution multiplied by the molarity of that solution:

$$n_{HCl} = (15.90 \text{ mL}) \times \left(\frac{0.0750 \text{ mmol HCl}}{1 \text{ mL}}\right) = 1.1925 \text{ mmol}$$

Thus there was originally 1.1925 mmol of diethylamine. Convert to grams:

$$1.1925 \times 10^{-3} \text{ mol} \times \left(\frac{73.14 \text{ g } (C_2H_5)_2NH}{1 \text{ mol } (C_2H_5)_2NH}\right) = \boxed{0.0872 \text{ g } (C_2H_5)_2NH}$$

Imagine that at the equivalence point, all of the diethylamine is converted to its conjugate acid, the diethylammonium ion $(C_2H_5)_2NH_2^+$. If so, the concentration of diethylammonium ion equals its chemical amount, 1.1925 mmol, divided by the volume of the solution (115.90 mL):

$$[(C_2H_5)_2NH_2^+] = \frac{1.1925 \text{ mmol}}{115.90 \text{ mL}} = 0.0103 \text{ M}$$

In actuality, some of the diethylammonium ion is reacted away as it donates H^+ ions to increase the H_3O^+ concentration in the solution:

$$(C_2H_5)_2NH_2^+(aq) + H_2O(l) \rightleftharpoons (C_2H_5)_2NH + H_3O^+(aq)$$

The equilibrium expression for this reaction is:

$$K_a = \frac{[H_3O^+][(C_2H_5)_2NH]}{[(C_2H_5)_2NH^+]}$$

where K_a is K_w divided by the K_b of diethylamine. Let x stand for the concentration of H_3O^+, and assume that all H_3O^+ in the solution comes from the reaction of diethylammonium ion as an acid. Then:

$$K_a = \frac{K_w}{K_b} = \frac{1.0 \times 10^{-14}}{3.09 \times 10^{-4}} = 3.236 \times 10^{-11} = \frac{x^2}{0.0103 - x}$$

Solving give $x = 5.77 \times 10^{-7}$. Then $[H_3O^+] = 5.77 \times 10^{-7}$ M for a pH of 6.24.

This concentration of H_3O^+ is only about six times larger than the concentration furnished by autoionization in pure water. How valid then is the assumption that diethylammonium ion furnishes *all* of the H_3O^+? One way to check is to employ the following expression,[19] which takes into account the autoionization of water as a source of H_3O^+ in solutions of weak acids:

$$[H_3O^+]^3 + K_a[H_3O^+]^2 - (K_w + c_a)[H_3O^+] - K_aK_w = 0$$

Inserting $K_w = 1.0 \times 10^{-14}$, $c_a = 0.0103$, and $K_a = 3.24 \times 10^{-11}$ gives the cubic equation:

$$[H_3O^+]^3 + (3.24 \times 10^{-11})[H_3O^+]^2 - (3.433 \times 10^{-13})[H_3O^+] - 3.24 \times 10^{-25} = 0$$

Solving this equation analytically is hard. A better way is to reason that the last term is certainly much smaller than any of the other three because $[H_3O^+]$ can only be larger than 5.77×10^{-7} M. After all, a second source of H_3O^+ ion can only raise the concentration of that ion, never lower it. Omitting the last term on this basis and dividing through by $[H_3O^+]$ gives:

$$[H_3O^+]^2 + (3.24 \times 10^{-11})[H_3O^+] - 3.433 \times 10^{-13} = 0$$

Solution of this quadratic equation is routine (by means of the quadratic formula). The applicable root gives $[H_3O^+] = 5.86 \times 10^{-7}$ M, for a pH of 6.23 . Including the contribution of autoionization lowers the pH only minimally despite the fact that diethylammonium ion is a very weak acid.

Tip. It is *wrong* to account for the autoionization by adding 1.0×10^{-7} to the answer 5.77×10^{-7} M! Autoionization contributes less H_3O^+ in this solution than in pure water. The reason is that the autoionization equilibrium is shifted to the left by H_3O^+ ion from the diethylammonium ion (LeChatelier's principle). Of course, this K_a equilibrium is also shifted (slightly) to the left by hydronium ion from the autoionization, which in turn causes an even slighter secondary effect back on the autoionization, which in turn.... The best understanding starts with two points: 1) only one kind of hydronium ion exists in the solution; 2) the equilibrium concentration of H_3O^+ ion comes as a compromise among all the simultaneous competing tendencies to donate or accept it.

A suitable indicator for the titration is bromothymol blue , which changes color in the pH range centered at 6.24.[20]

---

[19] From text page 350.
[20] See Figure 10-8, text page 328.

10-61 This buffer solution contains N-ethylmorpholine $C_6H_{13}NO$ (call it "M") and its conjugate acid $C_6H_{13}NOH^+$ (MH^+), which forms from the reaction of the N-ethylmorpholine with HCl. The two species are in chemical equilibrium:

$$M(aq) + H_2O(l) \rightleftharpoons MH^+(aq) + OH^-(aq) \quad K_b = \frac{[OH^-][MH^+]}{[M]}$$

The mass-action expression can be used to compute K_b if equilibrium values of all of the concentrations can be determined. The concentration of OH^- is easy because the pH of the solution is given: $[OH^-] = 1.0 \times 10^{-7}$. There was 10.00 mmol of M in the solution before the addition of the HCl, and the added HCl amounts to 8.00 mmol. If the neutralization reaction goes to completion, it forms 8.00 mmol of MH^+ and leaves 2.00 mmol of M unreacted (in excess). The volume of the solution is deliberately brought to 100.0 mL by the addition of water. The equilibrium concentrations of M and MH^+ therefore are:

$$[M] = \frac{2.00 \text{ mmol}}{100.0 \text{ mL}} - 1.0 \times 10^{-7} \text{ M} \qquad [MH^+] = \frac{8.00 \text{ mmol}}{100.0 \text{ mL}} + 1.0 \times 10^{-7} \text{ M}$$

The subtraction and addition of the 1.0×10^{-7} accounts for the slight amount of conversion of M to MH^+ by the action of the equilibrium. Substitution into the K_b expression gives K_b:

$$K_b = \frac{[OH^-][MH^+]}{[M]} = \frac{(1.0 \times 10^{-7})(0.0800 + 1.0 \times 10^{-7})}{(0.0200 - 1.0 \times 10^{-7})} = \boxed{4 \times 10^{-7}}$$

Tip. The 1.0×10^{-7} is so small compared to 0.0200 or 0.0800 that actually subtracting or adding it is not worth the trouble.

10-63 Aqueous arsenic acid donates H^+ ions in three steps. Each has a different K_a:

$$H_3AsO_4(aq) + H_2O(l) \rightleftharpoons H_2AsO_4^-(aq) + H_3O^+(aq) \quad K_{a1} = 5.0 \times 10^{-3}$$
$$H_2AsO_4^-(aq) + H_2O(l) \rightleftharpoons HAsO_4^{2-}(aq) + H_3O^+(aq) \quad K_{a2} = 9.3 \times 10^{-8}$$
$$HAsO_4^{2-}(aq) + H_2O(l) \rightleftharpoons AsO_4^{3-}(aq) + H_3O^+(aq) \quad K_{a3} = 3.0 \times 10^{-12}$$

K_{a2} is thousands of times smaller than K_{a1}, and K_{a3} is thousands of times smaller yet. This means that the first reaction will predominate in producing H_3O^+ and that the subsequent reactions will be negligible sources of H_3O^+. Compute the hydronium-ion concentration as if the first step occurred separately and use the answer in the equilibrium expressions for the following steps. Ignoring the interaction of the equilibria avoids complicated systems of simultaneous equations.

When the first step is considered separately, the problem is just like **10-29a**. Let x be the equilibrium concentration of H_3O^+, which equals the equilibrium concentration of $H_2AsO_4^-$. The mass-action expression becomes:

$$\frac{[H_2AsO_4^-][H_3O^+]}{[H_3AsO_4]} = K_{a1} = 5.0 \times 10^{-3} = \frac{x^2}{(0.1000 - x)}$$

Rearrange and substitute into the quadratic formula to obtain:

$$x = \frac{-5.0 \times 10^{-3} \pm \sqrt{2.50 \times 10^{-5} + 2.0 \times 10^{-3}}}{2}$$

The positive root of the equation is 0.0200. Thus:

$$[H_3AsO_4] = \boxed{0.080 \text{ M}} \qquad [H_2AsO_4^-] = \boxed{0.020 \text{ M}} \qquad H_3O^+] = \boxed{0.020 \text{ M}}$$

Now consider the donation of the second hydrogen ion. Let y equal the concentration of $HAsO_4^{2-}$ produced at equilibrium:

	$H_2AsO_4^-(aq)$	$+H_2O(l) \rightleftharpoons$	$HAsO_4^{2-}(aq) +$	$H_3O^+(aq)$
Init. Conc. (M)	0.0200	—	0	0.020
Change in Conc. (M)	$-y$	—	$+y$	$+y$
Equil. Conc. (M)	$0.020 - y$	—	y	$0.020 + y$

Use of the mass-action expression for K_{a2} gives the equation:

$$\frac{[HAsO_4^{2-}][H_3O^+]}{[H_2AsO_4^-]} = K_{a2} = 9.3 \times 10^{-8} = \frac{y(0.020 + y)}{(0.020 - y)}$$

This equation is easily solved when it is realized that y must be small compared to 0.0200. Then $y = \boxed{9.3 \times 10^{-8} \text{ M}} = [HAsO_4^{2-}]$. Note that $[HAsO_4^{2-}]$ is equal to K_{a2}.

Finally, consider the third of the three reactions. Let z equal the concentration of AsO_4^{3-} produced at equilibrium:

	$HAsO_4^{2-}(aq)$	$+H_2O(l) \rightleftharpoons$	$AsO_4^{3-}(aq) +$	$H_3O^+(aq)$
Init. Conc. (M)	9.3×10^{-8}	—	0	0.0200
Change in Conc. (M)	$-z$	—	$+z$	$+z$
Equil. Conc. (M)	$9.8 \times 10^{-8} - z$	—	z	$0.0200 + z$

Use of the equilibrium expression for K_{a3} gives the equation:

$$\frac{[AsO_4^{3-}][H_3O^+]}{[HAsO_4^{2-}]} = K_{a2} = 3.0 \times 10^{-12} = \frac{z(0.020 + z)}{(9.8 \times 10^{-8} - z)}$$

Solving gives $[AsO_4^{3-}] = \boxed{1.4 \times 10^{-17} \text{ M}}$. This is a very small concentration. One liter of the arsenic acid solution contains fewer than ten million AsO_4^{3-} ions!

10-65 The phosphate ion accepts hydrogen ions from water in three stages:

$$PO_4^{3-}(aq) + H_2O(l) \rightleftharpoons HPO_4^{2-}(aq) + OH^-(aq) \quad K_{b1} = K_w/K_{a3} = 4.55 \times 10^{-2}$$
$$HPO_4^{2-}(aq) + H_2O(l) \rightleftharpoons H_2PO_4^-(aq) + OH^-(aq) \quad K_{b2} = K_w/K_{a2} = 1.61 \times 10^{-7}$$
$$H_2PO_4^-(aq) + H_2O(l) \rightleftharpoons H_3PO_4(aq) + OH^-(aq) \quad K_{b3} = K_w/K_{a1} = 1.33 \times 10^{-12}$$

Treat the successive equilibria independently. Set up a three-line table for the reaction between PO_4^{3-} ion and water in the usual way:

	$PO_4^{3-}(aq)$	$+H_2O(l) \rightleftharpoons$	$HPO_4^{2-}(aq) +$	$OH^-(aq)$
Init. Conc. (M)	0.050	—	0	small
Change in Conc. (M)	$-x$	—	$+x$	$+x$
Equil. Conc. (M)	$0.050 - x$	—	x	x

Writing the mass-action expression then gives:

$$\frac{[HPO_4^{2-}][OH^-]}{[PO_4^{3-}]} = K_{b1} = 4.55 \times 10^{-2} = \frac{x^2}{(0.050 - x)}$$

Rearrange and substitute into the quadratic formula to obtain:

$$x^2 + (4.55 \times 10^{-2})x - 2.27 \times 10^{-3} = 0 \quad \text{which gives} \quad x = 0.0302$$

$$[OH^-] = \boxed{0.030 \text{ M}} \quad [HPO_4^{2-}] = \boxed{0.030 \text{ M}} \quad [PO_4^{3-}] = 0.050 - 0.0302 = \boxed{0.020 \text{ M}}$$

Turn to the second stage. Let y equal the concentration of $H_2PO_4^-$ formed at equilibrium, but use the concentration of OH^- established by the first stage of the reaction in the K_{b2} mass-action expression:

$$\frac{[H_2PO_4^-][OH^-]}{[HPO_4^{2-}]} = K_{b2} = 1.61 \times 10^{-7} = \frac{y(0.0302 + y)}{0.0302 - y}$$

This equation is easily solved because y must be small compared to 0.0302. The result is $y = \boxed{1.61 \times 10^{-7} \text{ M}} = [H_2PO_4^-]$.

Finally, consider the third reaction. The mass-action expression gives:

$$\frac{[H_3PO_4][OH^-]}{[H_2PO_4^-]} = K_{b3} = 1.33 \times 10^{-12} = \frac{z(0.0302 + z)}{1.61 \times 10^{-7} - z}$$

where z equals the equilibrium concentration of phosphoric acid. Solving gives $z = \boxed{7.1 \times 10^{-18} \text{ M}} = [H_3PO_4]$.

10-67 The major natural contributor to the acidity of rainwater is dissolved CO_2, which reacts to form carbonic acid: $CO_2(g) + H_2O(l) \rightleftharpoons H_2CO_3(aq)$. In recent times, $SO_3(g)$ and $NO_2(g)$, which are air pollutants, have joined $CO_2(g)$ as contributors to the acidity of rain. Carbonic acid donates two hydrogen ions:

$$H_2CO_3(aq) + H_2O(l) \rightleftharpoons HCO_3^-(aq) + H_3O^+(aq) \qquad K_{a1} = 4.3 \times 10^{-7}$$
$$HCO_3^-(aq) + H_2O(l) \rightleftharpoons CO_3^{2-}(aq) + H_3O^+(aq) \qquad K_{a2} = 4.8 \times 10^{-11}$$

At equilibrium, the following equations relate the concentrations:

$$K_{a1} = \frac{[H_3O^+][HCO_3^-]}{[H_2CO_3]} \qquad K_{a2} = \frac{[H_3O^+][CO_3^{2-}]}{[HCO_3^-]}$$

The pH of the raindrop is 5.60. It follows that $[H_3O^+] = 2.51 \times 10^{-6}$ M. Substituting this value of $[H_3O^+]$ and the two K_a's into the preceding gives:

$$0.171 = \frac{[HCO_3^-]}{[H_2CO_3]} \qquad 1.91 \times 10^{-5} = \frac{[CO_3^{2-}]}{[HCO_3^-]}$$

It is convenient to recast these equations so that the concentration of the same species, say HCO_3^-, is in the denominator in both. Take the reciprocal of the first and copy the second:

$$5.84 = \frac{[H_2CO_3]}{[HCO_3^-]} \qquad 1.91 \times 10^{-5} = \frac{[CO_3^{2-}]}{[HCO_3^-]}$$

The fraction f of any of the three species present equals its concentration divided by the sum of the concentrations of all three. If the species is H_2CO_3:

$$f_{H_2CO_3} = \frac{[H_2CO_3]}{[H_2CO_3] + [HCO_3^-] + [CO_3^{2-}]}$$

This expression can be simplified by dividing numerator and denominator by $[HCO_3^-]$ and inserting the ratios just calculated:

$$f_{H_2CO_3} = \frac{[H_2CO_3]/[HCO_3^-]}{[H_2CO_3]/[HCO_3^-] + 1 + [CO_3^{2-}]/[HCO_3^-]}$$

$$f_{H_2CO_3} = \frac{5.84}{5.84 + 1 + (1.91 \times 10^{-5})} = \frac{5.84}{6.84} = 0.854$$

Expressions are obtained to compute the fractions of the other three species by changing the numerator as required. The resulting fractions are 0.146 for HCO_3^- and 2.79×10^{-6} for CO_3^{2-}. The total concentration of all three forms of the carbon-containing species is 1.0×10^{-5} M. The answers equal the respective fractions times this total:

$$[H_2CO_3] = \boxed{8.5 \times 10^{-6} \text{ M}} \quad [HCO_3^-] = \boxed{1.5 \times 10^{-6} \text{ M}} \quad [CO_3^{2-}] = \boxed{2.8 \times 10^{-11} \text{ M}}$$

10-69 The molar mass of thiamine hydrochloride is 337.27 g mol^{-1}, according to the molecular formula given in the problem. 3.0×10^{-5} g of this substance in 1.00 L of water makes a solution that is 8.89×10^{-8} M in thiH$^+$ ion[21] and of course 8.89×10^{-8} M in Cl$^-$ ion. The thiH$^+(aq)$ cation is a weak acid:

$$\text{thiH}^+(aq) + \text{H}_2\text{O}(l) \rightleftharpoons \text{thi}(aq) + \text{H}_3\text{O}^+(aq) \qquad K_a = 3.4 \times 10^{-7}$$

This equilibrium produces only small amounts of H$_3$O$^+(aq)$ because K_a is small and the original concentration of thiH$^+(aq)$ is quite small. The simultaneous autoionization of water:

$$2\,\text{H}_2\text{O}(l) \rightleftharpoons \text{H}_3\text{O}^+(aq) + \text{OH}^-(aq)$$

must be reckoned with as a source of H$_3$O$^+(aq)$.
The following four mathematical relationships always hold in this solution:

$$3.4 \times 10^{-7} = K_a = \frac{[\text{thi}][\text{H}_3\text{O}^+]}{[\text{thiH}^+]} \qquad 1.0 \times 10^{-14} = K_w = [\text{H}_3\text{O}^+][\text{OH}^-]$$

$$8.89 \times 10^{-8} = c_a = [\text{thi}] + [\text{thiH}^+] \qquad [\text{H}_3\text{O}^+] + [\text{thiH}^+] = [\text{OH}^-] + [\text{Cl}^-]$$

The last equation follows from the principle of electrical neutrality: for every positive charge in the solution there must be a negative charge. The second-to-last equation represents a material balance. Whatever the distribution between its two forms, the *total* concentration of thiamine-material is known. The first two equations are the usual mass-action expressions. The [Cl$^-$] equals 8.89×10^{-8} M, as stated above, because Cl$^-(aq)$ does not react to any extent with other species. The four simultaneous equations therefore involve four unknowns. It is "merely" a question of algebra to solve for [H$_3$O$^+$]. The details of the algebra are given in the text for a similar case.[22] The result is a cubic equation in [H$_3$O$^+$]:

$$[\text{H}_3\text{O}^+]^3 + K_a[\text{H}_3\text{O}^+]^2 - (K_w + c_a K_a)[\text{H}_3\text{O}^+] - K_a K_w = 0$$

Substitute the numbers specific to this case:

$$[\text{H}_3\text{O}^+]^3 + (3.4 \times 10^{-7})[\text{H}_3\text{O}^+]^2 - (4.03 \times 10^{-14})[\text{H}_3\text{O}^+] - 3.4 \times 10^{-21} = 0$$

The best route to a real root is to guess a value of [H$_3$O$^+$] near 10^{-7} and make successive approximations. The answer is [H$_3$O$^+$] $= 1.367 \times 10^{-7}$ M for a pH of $\boxed{6.86}$.

Tip. If the autoionization of water is (mistakenly) neglected, the result is the quadratic equation:

$$[\text{H}_3\text{O}^+]^2 + (3.4 \times 10^{-7})[\text{H}_3\text{O}^+] - 3.02 \times 10^{-14} = 0$$

---

[21]We let "thi" stand for thiamine C$_{12}$H$_{17}$ON$_4$SCl. Then thiH$^+$ stands for (HC$_{12}$H$_{17}$ON$_4$SCl)$^+$. This is the "thiammonium" cation, the conjugate acid of thiamine.

[22]Text page 350.

Solving gives $[H_3O^+] = 7.31 \times 10^{-8}$ M for a pH of 7.14. But this pH is on the basic side of 7, which is impossible when an acid is dissolved in water.

10-71 Maleic acid is a diprotic acid for which K_{a1} and K_{a2} differ by about four orders of magnitude. Consider the course of the titration region by region.

• *Before Addition of Base.* Before any 0.1000 M NaOH is added, the predominant source of $H_3O^+(aq)$ in the solution is the equilibrium:

$$H_2mal(aq) + H_2O(l) \rightleftharpoons Hmal^-(aq) + H_3O^+(aq)$$

Let y equal $[H_3O^+]$. Then:

$$K_{a1} = 1.42 \times 10^{-2} = \frac{[Hmal^-][H_3O^+]}{[H_2mal]} = \frac{y^2}{0.1000 - y}$$

Solving for y using the quadratic formula gives 0.0313. If $[H_3O^+]$ is 0.0313 mol L^{-1}, the pH is $\boxed{1.51}$.

• *After Addition of Base, But Before the Equivalence Point.* The first 5.00 mL of 0.1000 M NaOH reacts with some of the maleic acid (H_2mal). 5.00 mL of this solution contains only 5.00×10^{-4} mol of NaOH, which is less than the 50.00×10^{-4} mol of H_2mal that is present. The acid is in excess, and the reaction ends when the base runs out. Assuming complete reaction and that no other reactions take place, the yield of $Hmal^-(aq)$ is 5.00×10^{-4} mol, and 45.00×10^{-4} mol of $H_2mal(aq)$ remains. Adding 5.00 mL of dilute aqueous solution raises the volume of the solution to 55.00 mL so:

$$[Hmal^-] = \frac{5.00 \times 10^{-4} \text{ mol}}{0.05500 \text{ L}} = 0.009091 \text{ M} \quad [H_2mal] = \frac{45.00 \times 10^{-4} \text{ mol}}{0.05500 \text{ L}} = 0.08182 \text{ M}$$

The K_{a1} equilibrium now acts to alter these concentrations slightly. It generates hydronium ions. For every $H_3O^+(aq)$ produced, one $H_2mal(aq)$ is consumed and one additional $Hmal^-(aq)$ is generated. Let y equal the concentration of H_3O^+ that is generated. Then:

$$K_{a1} = [H_3O^+]\frac{[Hmal^-]}{[H_2mal]} \qquad 1.42 \times 10^{-2} = y\frac{0.009091 + y}{0.08182 - y}$$

Rearranging and solving this quadratic equation gives $[H_3O^+] = 0.0244$ M. The Ka_2 equilibrium is neglected as a source of H_3O^+. It proceeds to a far lesser extent, and the "starting" concentration of $Hmal^-$ is about nine times smaller than that of H_2mal. The pH equals $\boxed{1.61}$.

• *Halfway to the First Equivalence Point.* At this point, 25.00 mL of titrant has been added raising the total to volume 75.00 mL. Repeating the reasoning just used gives:

$$1.42 \times 10^{-2} = y\frac{0.0333 + y}{0.0333 - y}$$

Again, rearrange and solve by use of the quadratic formula. The answer is $[H_3O^+] = 8.45 \times 10^{-3}$ M and pH = $\boxed{2.07}$.

• *At the First Equivalence Point.* 50.00 mL of 0.1000 M NaOH brings the titration to the *first* equivalence point. The solution consists of 100.00 mL of 0.0500 M NaHmal (sodium hydrogen maleate). The $Hmal^-(aq)$ ion is amphoteric. It behaves as an acid:

$$Hmal^-(aq) + H_2O(l) \rightleftharpoons H_3O^+(aq) + mal^{2-}(aq) \qquad K_{a2} = 8.57 \times 10^{-7}$$

And it behaves as a base:

$$Hmal^-(aq) + H_2O(l) \rightleftharpoons OH^-(aq) + H_2mal(aq) \qquad K_{b2} = \frac{K_w}{K_{a1}} = 7.04 \times 10^{-13}$$

$Hmal^-(aq)$ is just like $HCO_3^-(aq)$ in this respect. Copy the analysis presented for $HCO_3^-(aq)$ in text Section 10-8[23] to obtain:

$$[H_3O^+] \approx \sqrt{\frac{K_{a1}K_{a2}[Hmal^-]_0 + K_{a1}K_w}{K_{a1} + [Hmal^-]_0}}$$

in which $[Hmal^-]_0$ equals 0.0500 M the "original" concentration of $Hmal^-$. Inserting the numbers gives $[H_3O^+] = 9.73 \times 10^{-5}$ M for a pH of $\boxed{4.01}$. Notice that the approximate formula $[H_3O^+] \approx \sqrt{K_{a1}K_{a2}}$ gives a wrong pH of 3.96. The approximate formula fails because K_{a1} of maleic acid is rather large and cannot be neglected compared to $[Hmal^-]_0$.

• *Halfway to the Second Equivalence Point.* After 75.00 mL of 0.1000 M NaOH has been added, the titration is half-way to the *second* equivalence point. In this range, the main source of $H_3O^+(aq)$ is the reaction:

$$Hmal^-(aq) + H_2O(l) \rightleftharpoons mal^{2-}(aq) + H_3O^+(aq) \qquad K_{a2} = 8.57 \times 10^{-7}$$

The concentration of $H_2mal(aq)$ is very small because so much base has been added, and consequently the K_{a1} equilibrium has only a negligible effect on the pH. The titration can be viewed as the addition of 25.00 mL of 0.1000 M NaOH to 100.00 mL of 0.0500 M NaHmal, making a total volume of 125.00 mL. Assuming complete reaction,

---

[23]Text page 350.

this much NaOH converts exactly half of the $Hmal^-(aq)$ to $mal^{2-}(aq)$. Dilution meanwhile reduces the concentrations of both species by the factor $100/125$. Thus *after* completion of the acid-base reaction but *before* any equilibrium starts:

$$[Hmal^-] = 0.025 \times \frac{100}{125} = 0.020 \text{ M} \qquad [mal^{2-}] = 0.025 \times \frac{100}{125} = 0.020 \text{ M}$$

The K_{a2} equilibrium now changes these concentrations slightly. It adds z to the concentration of $mal^{2-}(aq)$ and removes z from the concentration of $Hmal^-(aq)$, where z is the concentration of hydronium ion that it produces. The mass-action expression is:

$$K_{a2} = 8.57 \times 10^{-7} = z \frac{0.020 - z}{0.020 + z}$$

Solving this equation is very easy because z can be neglected compared to 0.020. The answer is $z = 8.57 \times 10^{-7}$ so the pH is $\boxed{6.07}$.

- *Just Short of the Second Equivalence Point.* By similar computations, the pH after the addition of 99.90 mL of 0.1000 M NaOH is $\boxed{8.77}$.

- *At the Second Equivalence Point.* The solution consists of 150.00 mL of 0.0333 M Na_2mal. The *hydrolysis* of $mal^{2-}(aq)$ ion is a major source of OH^- ions:

$$mal^{2-}(aq) + H_2O(l) \rightleftharpoons Hmal^-(aq) + OH^-(aq) \qquad K_{b1} = \frac{K_w}{K_{a2}} = \frac{1.0 \times 10^{-14}}{8.57 \times 10^{-7}}$$

Note how the *first* K_b for the mal^{2-} ion is related to the *second* K_a for H_2mal. Let x be the equilibrium concentration of $OH^-(aq)$. Then:

$$K_{b1} = \frac{1.0 \times 10^{-14}}{8.57 \times 10^{-7}} = \frac{[Hmal^-][OH^-]}{[mal^{2-}]} = \frac{x^2}{0.0333 - x}$$

Solving for x gives 1.97×10^{-5} so the pOH is 4.71 and the pH is $\boxed{9.29}$.

- *Past the Second Equivalence Point.* When excess NaOH has been added, the hydrolysis of mal^{2-} is completely overshadowed by NaOH as a source of $OH^-(aq)$. The first 100.00 mL of NaOH was used up neutralizing the H_2mal. The next 5.00 mL of 0.100 M NaOH makes a solution that is $(5.00/155.00) \times 0.100$ or 3.23×10^{-3} M in $OH^-(aq)$. This corresponds to pOH 2.49 or pH $\boxed{11.51}$.

10-73 The equilibrium constants are for the reaction

$$ClCH_2COOH(aq) + H_2O(l) \rightleftharpoons ClCH_2COO^-(aq) + H_3O^+(aq)$$

Call the first constant (observed at $T_1 = 273.15$ K) K_1 and the second (observed at $T_2 = 313.15$ K) K_2. The hint in the problem strongly suggests the use of these constants in van't Hoff equation (Section 9-5). Proceed as follows:

$$\ln \frac{K(T_2)}{K(T_1)} = \ln \left(\frac{1.230 \times 10^{-3}}{1.528 \times 10^{-3}} \right) = \frac{-\Delta H^\circ}{R} \left(\frac{1}{313.15 \text{ K}} - \frac{1}{273.15 \text{ K}} \right)$$

Substituting $R = 8.315$ J $K^{-1}mol^{-1}$ then gives $\Delta H^\circ = -3.86 \times 10^3$ J mol^{-1}. The ΔH° of the reaction as it is represented above therefore equals $\boxed{-3.86 \text{ kJ}}$.

10-75 Use the relations: $pOH = 14.00 - pH$; $[H_3O^+] = 10^{-pH}$; $[OH^-] = 10^{-pOH}$.

a)	$[H_3O^+] = 1.6 \times 10^{-3}$ M	$pOH = 11.2$	$[OH^-] = 6.3 \times 10^{-12}$ M
b)	$[H_3O^+] = 1.3 \times 10^{-4}$ M	$pOH = 10.1$	$[OH^-] = 7.9 \times 10^{-11}$ M
c)	$[H_3O^+] = 7.9 \times 10^{-5}$ M	$pOH = 9.9$	$[OH^-] = 1.3 \times 10^{-10}$ M
d)	$[H_3O^+] = 3.2 \times 10^{-9}$ M	$pOH = 5.5$	$[OH^-] = 3.2 \times 10^{-6}$ M
e)	$[H_3O^+] = 1.3 \times 10^{-12}$ M	$pOH = 2.1$	$[OH^-] = 7.9 \times 10^{-3}$ M

10-77 The pH should be $\boxed{\text{low}}$. LeChatelier's principle indicates that increasing $[H_3O^+]$ shifts the equilibrium in the problem to the left, favoring $Cl_2(aq)$ at the expense of $Cl^-(aq)$.

10-79 Urea is such a weak base that its conjugate acid is almost a strong acid.

a) The formula of the conjugate acid of urea is obtained by adding H^+ to the formula of urea. The answer is $\boxed{NH_2CONH_3^+}$. This ion is called the urea acidium ion.

b) Let "ureaH$^+$" stand for the urea acidium ion. It reacts with water:

	ureaH$^+(aq)$	$+ H_2O(aq) \rightleftharpoons$	urea(aq) +	$H_3O^+(aq)$
Init. Conc. (M)	0.15	−	0	small
Change in Conc. (M)	$-x$	−	$+x$	$+x$
Equil. Conc. (M)	$0.15 - x$	−	x	x

The K_b for urea is $10^{-13.8} = 1.58 \times 10^{-14}$. Use this value to calculate K_a for ureaH$^+$. Also set up the K_a mass-action expression:

$$K_a = \frac{K_w}{K_b} = \frac{1.0 \times 10^{-14}}{1.58 \times 10^{-14}} = \frac{[\text{urea}][H_3O^+]}{[\text{ureaH}^+]} = \frac{x^2}{(0.15 - x)}$$

$$0.630 = \frac{x^2}{(0.15 - x)}$$

Solving the last equation by means of the quadratic formula gives $x = 0.125$. The equilibrium concentration of the urea is therefore $\boxed{0.125 \text{ M}}$.

10-81 Use the mass-action equation for the K_a equilibrium of acetic acid. Let $[H_3O^+]$ at 25°C be y. Then:

$$K_{a,25} = 1.76 \times 10^{-5} = \frac{y^2}{0.10 - y} \quad \text{which gives} \quad y^2 + (1.76 \times 10^{-5})y - 1.76 \times 10^{-6} = 0$$

The applicable root of the quadratic equation is 1.318×10^{-3}, so at 25°C the pH equals 2.88. Repeat the calculation for the 50°C case. The initial molarity of the acetic acid $[HOAc]_0$ is less at 50°C because the solution expands when heated. The increase in volume is computed from the relative densities at the two temperatures:

$$\rho_{50} = 0.9881\,\rho_{25} \qquad \left(\frac{\text{mass}}{V_{50}}\right) = 0.9881\left(\frac{\text{mass}}{V_{25}}\right) \qquad V_{50} = 1.012\,V_{25}$$

Because V is in the denominator in the definition of molarity;

$$[HOAc]_0 = 0.010 \text{ M} \times \left(\frac{1}{1.012}\right) = 0.09881 \text{ M} \quad \text{at 50°C}$$

Let x represent the equilibrium concentration of H_3O^+ at 50°C. Then:

$$K_{a,50} = 1.63 \times 10^{-5} = \frac{x^2}{0.0988 - x} \quad \text{which gives} \quad x^2 + (1.63 \times 10^{-5})x - 1.61 \times 10^{-6} = 0$$

The applicable root is 0.00126, making the pH at 50°C equal 2.90. The $\boxed{\text{pH increases}}$ slightly when the solution is heated from 25 to 50°C because the acid weakens and the solution becomes more dilute.

Tip. The autoionization of water also changes its extent with temperature. It is not considered because it is a negligible source of hydronium ion at both temperatures.

10-83 The compounds are a strong acid (HCl), a salt of a strong acid and a weak base (NH_4Cl), a salt of a weak acid and a strong base (Na_3PO_4), a salt of a weak acid and a strong base ($NaC_2H_3O_2$), and a salt of a strong acid and a strong base (KNO_3). The 0.100 M solutions of HCl and NH_4Cl are acidic. The 0.100 M solutions of Na_3PO_4 and $NaCH_3COO$ are basic. The 0.100 M solution of KNO_3 is neutral.

10-85 The initial concentrations of C_6F_5COOH and $C_6F_5COO^-$ are:

$$[C_6F_5COOH]_0 = \frac{0.050 \text{ mol}}{2.00 \text{ L}} = 0.025 \text{ M} \qquad [C_6F_5COO^-]_0 = \frac{0.060 \text{ mol}}{2.00 \text{ L}} = 0.030 \text{ M}$$

These are the concentrations *after* the complete mixing of the two solutions, but *before* any reactions involving the pentafluorobenzoic acid and its conjugate base the pentafluorobenzoate ion have a chance to occur. Now set up the problem as in Example 10-7[24] or **10-45**. Let y equal the equilibrium concentration of hydronium ion. Then:

$$K_a = \frac{[H_3O^+][C_6F_5COO^-]}{[C_6F_5COOH]} = \frac{y(0.030 + y)}{(0.025 - y)} = 0.033$$

If y is neglected compared to 0.025 and 0.030, then the equation is easy to solve, and y equals 0.0275. This is obviously *wrong* because 0.0275 is *larger* than 0.025 rather than being a great deal smaller. When such things happen, start over and neglect nothing. The original equation rearranges to:

$$y^2 + 0.063y - 0.000825 = 0$$

Use of the quadratic formula gives the root $y = 0.01113$. It follows that the pH of the buffer is $\boxed{1.95}$.

10-87 $\boxed{\text{Procedure d)}}$ would not make an effective buffer. A good buffer results when substantial amounts of a weak acid and its conjugate base are mixed in solution. Procedure d) yields a solution that is 1.77×10^{-5} M in HCl mixed with some NaCl. The solution would change pH greatly upon addition of either acid or base. All the other solutions are acetate buffers.

10-89 a) The 0.1000 M solution of weak acid HA has a volume of 50.00 mL because 50.00 mL of the 0.1000 M base titrates it to equivalence. The addition of 40.00 mL of the base converts 40.00/50.00 of the acid to its conjugate base and creates a solution with a volume of 90.00 mL. According to the problem, the pH of this 90.00 mL of solution is 4.50. This means:

$$[H_3O^+] = 10^{-4.50} = 3.16 \times 10^{-5} \text{ M}$$

The solution is at equilibrium. The concentration of HA is the amount of unreacted HA divided by the volume of the solution minus the small concentration lost by the donation of H^+ to water:

$$[HA] = \frac{10.00}{90.00}(0.100) - [H_3O^+] = (0.01111 - 3.16 \times 10^{-5}) \text{ M}$$

Similarly, the equilibrium concentration of A^- is the amount of A^- formed divided by the volume of the solution plus the small concentration added by the HA/A^- equilibrium:

$$[A^-] = \frac{40.00}{90.00}(0.100) + [H_3O^+] = (0.04444 + 3.16 \times 10^{-5}) \text{ M}$$

---

[24]Text page 335.

Substitute these values into the K_a expression:

$$K_a = \frac{[H_3O^+][A^-]}{[HA]} = \frac{(3.16 \times 10^{-5})(0.044476)}{0.011079} = \boxed{1.27 \times 10^{-4}}$$

b) The solution at the equivalence point could have been prepared by dissolving 5.000 mmol of NaA in 100.00 mL of water. In such a solution, the initial concentration (before acid-base equilibria) of A^- ion is 0.05000 M. To determine the pH, consider the reaction of A^- ion with water:

$$A^-(aq) + H_2O(l) \rightleftharpoons HA(aq) + OH^-(aq) \qquad \frac{[HA][OH^-]}{[A^-]} = K_b$$

Let x equal the equilibrium concentration of OH^- ion. Then:

$$\frac{x^2}{0.05000 - x} = \frac{K_w}{K_a} = 7.91 \times 10^{-11}$$

Solving for x gives a $[OH^-]$ of 1.99×10^{-6} M. The pOH is accordingly 5.70, and the pH is $\boxed{8.30}$.

10-91 Phosphonocarboxylic acid ("H_3Pho") donates H^+ in three steps:

$$H_3Pho(aq) + H_2O(l) \rightleftharpoons H_2Pho^-(aq) + H_3O^+(aq) \qquad K_{a1} = 1.0 \times 10^{-2}$$
$$H_2Pho^-(aq) + H_2O(l) \rightleftharpoons HPho^{2-}(aq) + H_3O^+(aq) \qquad K_{a2} = 7.8 \times 10^{-6}$$
$$HPho^{2-}(aq) + H_2O(l) \rightleftharpoons Pho^{3-}(aq) + H_3O^+(aq) \qquad K_{a3} = 2.0 \times 10^{-9}$$

The law of mass action gives the following three equations:

$$K_{a1} = \frac{[H_3O^+][H_2Pho^-]}{[H_3Pho]} \qquad K_{a2} = \frac{[H_3O^+][HPho^{2-}]}{[H_2Pho^-]} \qquad K_{a3} = \frac{[H_3O^+][Pho^{3-}]}{[HPho^{2-}]}$$

The pH of the blood is 7.40, and the buffering action of the blood maintains this pH despite the addition of the drug. It follows that $[H_3O^+] = 3.98 \times 10^{-8}$ M. Substituting the known value of $[H_3O^+]$ and the three K's gives:

$$2.510 \times 10^5 = \frac{[H_2Pho^-]}{[H_3Pho]} \qquad 195.927 = \frac{[HPho^{2-}]}{[H_2Pho^-]} \qquad 0.05024 = \frac{[Pho^{3-}]}{[HPho^{2-}]}$$

Recast these equations so that the concentration of the same species, say H_2Pho^-, is in the denominator:

$$3.984 \times 10^{-6} = \frac{[H_3Pho]}{[H_2Pho^-]} \qquad 195.927 = \frac{[HPho^{2-}]}{[H_2Pho^-]} \qquad 9.8430 = \frac{[Pho^{3-}]}{[H_2Pho^-]}$$

The first new equation is the reciprocal of the first of the preceding group. The third comes by multiplying the second and third equations in the preceding group. The fraction f of any of the four Pho-containing species equals its concentration divided by the sum of the concentrations of all four. For example:

$$f_{\text{H}_3\text{Pho}} = \frac{[\text{H}_3\text{Pho}]}{[\text{H}_3\text{Pho}] + [\text{H}_2\text{Pho}^-] + [\text{HPho}^{2-}] + [\text{Pho}^{3-}]}$$

This expression can be evaluated by dividing numerator and denominator by $[\text{H}_2\text{Pho}^-]$ and inserting the ratios just calculated:

$$f_{\text{H}_3\text{Pho}} = \frac{[\text{H}_3\text{Pho}]/[\text{H}_2\text{Pho}^-]}{[\text{H}_3\text{Pho}]/[\text{H}_2\text{Pho}^-] + 1 + [\text{HPho}^{2-}]/[\text{H}_2\text{Pho}^-] + [\text{Pho}^{3-}]/[\text{H}_2\text{Pho}^-]}$$

$$f_{\text{H}_3\text{Pho}} = \frac{3.98 \times 10^{-6}}{3.98 \times 10^{-6} + 1 + 195.927 + 9.8430} = \frac{3.98 \times 10^{-6}}{206.77} = 1.92 \times 10^{-8}$$

The same approach gives the fractions of the other three species. Simply change the numerator as required. The resulting fractions are 0.004836 for H_2Pho^-, 0.94756 for HPho^{2-}, and 0.0476 for Pho^{3-}. The total concentration of all four forms of the drug is 1.0×10^{-5} M. The desired concentrations equal this total multiplied by the respective fractions:

$$[\text{H}_3\text{Pho}] = \boxed{1.9 \times 10^{-13} \text{ M}} \qquad [\text{H}_2\text{Pho}^-] = \boxed{4.8 \times 10^{-8} \text{ M}}$$

$$[\text{HPho}^{2-}] = \boxed{9.5 \times 10^{-6} \text{ M}} \qquad [\text{Pho}^{3-}] = \boxed{4.8 \times 10^{-7} \text{ M}}$$

10-93 As explained in the problem, hyperventilating causes the loss of CO_2 dissolved in the blood. Carbon dioxide is a weak acid in water. Loss of the weak acid causes the blood to $\boxed{\text{rise in pH}}$.

10-95 Represent the amino acid glycine as glyH. Its conjugate acid is then glyH_2^+, and its conjugate base is gly^-. Copy the two equilibria given in the problem:

$$\text{glyH}(aq) + \text{H}_2\text{O}(l) \rightleftharpoons \text{gly}^-(aq) + \text{H}_3\text{O}^+(aq) \qquad K_a = 1.7 \times 10^{-10}$$

$$\text{glyH}(aq) + \text{H}_2\text{O}(l) \rightleftharpoons \text{glyH}_2^+(aq) + \text{OH}^-(aq) \qquad K_b = 2.2 \times 10^{-12}$$

The following equations convey the *same* chemical relationships:

$$\text{glyH}_2^+(aq) + \text{H}_2\text{O}(l) \rightleftharpoons \text{glyH}(aq) + \text{H}_3\text{O}^+(aq) \qquad K_1 = K_w/K_b$$

$$\text{glyH}(aq) + \text{H}_2\text{O}(l) \rightleftharpoons \text{gly}^-(aq) + \text{H}_3\text{O}^+(aq) \qquad K_2 = K_a$$

In this pair of equilibria, the second equation quoted in the problem has been *reversed* and *added* to the water autoionization equation. The change puts the emphasis on $glyH_2^+$ ($^+H_3N\!-\!CH_2\!-\!COOH$) as a diprotic acid rather than on self-neutralization by the glycine. The new equations reveal that solutions of glycine are just like solutions of other amphoteric species—0.10 M aqueous glycine is like 0.10 M $HCO_3^-(aq)$, for example. The similarity is worth spotting because the case of the pH of an aqueous solution of $HCO_3^-(aq)$ is extensively treated in the text.[25] The text develops the approximate formula $[H_3O^+] \approx \sqrt{K_1 K_2}$. According to this formula, the $[H_3O^+]$ in the solution is independent of the concentration of the glycine! Substitution in the formula gives:

$$[H_3O^+] = \sqrt{K_1 K_2} = \sqrt{\left(\frac{K_w}{K_b}\right) K_a} = \sqrt{\left(\frac{1.0 \times 10^{-14}}{2.2 \times 10^{-12}}\right)(1.7 \times 10^{-10})} = 8.8 \times 10^{-7} \text{ M}$$

The pH equals 6.06, as it seems. A more exact analysis[26] gives:

$$[H_3O^+] \approx \sqrt{\frac{K_1 K_2 c_0 + K_1 K_w}{K_1 + c_0}}$$

The treatment leading to this formula includes only one approximation: that the equilibrium concentration of glycine is close to c_0, its original concentration. Substitution of $c_0 = 0.10$ M and the three constants gives $[H_3O^+] = 8.6 \times 10^{-7}$ M, and the pH equals $\boxed{6.07}$.

Tip. If c_0 were 0.01 M, one-tenth of the c_0 given in the problem, the pH of the glycine solution would be 6.14. This deviates substantially from 6.06, the first answer. Thus, as c_0 diminishes it becomes *more* important in determining the pH, at least at first. As c_0 becomes exceedingly small the solution approximates pure water (pH 7.0).

CUMULATIVE PROBLEMS

10-97 a) $HC_3H_5O_3(aq) + NaHCO_3(aq) \rightarrow NaC_3H_5O_3(aq) + H_2O(l) + CO_2(g)$
b) Compute the chemical amount of sodium bicarbonate that is neutralized:

$$\frac{1}{2} \text{ teaspoon } Na_2CO_3 \times \left(\frac{236.6 \text{ cm}^3}{48 \text{ teaspoon}}\right) \times \left(\frac{1 \text{ mL}}{1 \text{ cm}^3}\right) \times \left(\frac{2.46 \text{ g NaHCO}_3}{1 \text{ mL}}\right)$$
$$\times \left(\frac{1 \text{ mol NaHCO}_3}{84.01 \text{ g NaHCO}_3}\right) = 0.0633 \text{ mol NaHCO}_3$$

---

[25]Text Section 10-8, text page 350–51.
[26]Text page 351.

Lactic acid and sodium bicarbonate react in a 1-to-1 molar ratio. Hence the one cup of sour milk contains 0.0633 mol of lactic acid. The concentration of lactic acid is 0.0633 mol/0.2366 L = $\boxed{0.268 \text{ M}}$.

c) The same chemical amount (0.0633 mol) of $CO_2(g)$ will be produced. Its volume is computed using the ideal-gas law:

$$V = \frac{nRT}{P} = \frac{(0.0633 \text{ mol})(0.08206 \text{ L atm mol}^{-1}\text{K}^{-1})(177 + 273.15 \text{ K})}{1 \text{ atm}} = \boxed{2.34 \text{ L}}$$

10-99 Label the gaseous monoprotic acid HY. The data on the temperature, pressure, and density of HY provide its molar mass. As shown in the answer to **4-23**, for an ideal gas:

$$\rho = \frac{m}{V} = \frac{P}{RT}\mathcal{M}$$

Solve for $\mathcal{M}$ and insert the numbers for HY in this case:

$$\mathcal{M}_{HY} = \rho\left(\frac{RT}{P}\right) = \frac{1.05 \text{ g}}{\text{L}}\left(\frac{0.08206 \text{ L atm mol}^{-1}\text{K}^{-1}(313.15 \text{ K})}{1.00 \text{ atm}}\right) = \frac{26.98 \text{ g}}{\text{mol}}$$

The concentration of HY when 1.85 g of it is dissolved in 450 mL of water is:

$$c_{HY} = \frac{n_{HY}}{V} = \frac{m_{HY}}{\mathcal{M}_{HY}V} = \frac{1.85 \text{ g}}{(26.98 \text{ g mol}^{-1})0.450 \text{ L}} = 0.1524 \text{ mol L}^{-1}$$

As the HY reacts with water, it generates H_3O^+. The final concentration of H_3O^+ equals $10^{-5.01} = 9.77 \times 10^{-6}$ M. For HY as an acid:

	HY(aq)	+ H$_2$O(l) ⇌	Y$^-(aq)$ +	H$_3$O$^+(aq)$
Init. Conc. (M)	0.1524	–	0	small
Change in Conc. (M)	-9.77×10^{-6}	–	$+9.77 \times 10^{-6}$	$+9.77 \times 10^{-6}$
Equil. Conc. (M)	0.1524	–	9.77×10^{-6}	9.77×10^{-6}

Now that the equilibrium concentrations of all of the products and reactants in the acid ionization of HY are known, it is easy to compute K_a:

$$K_a = \frac{[H_3O^+][Y^-]}{[HY]} = \frac{(9.77 \times 10^{-6})^2}{0.1524} = 6.27 \times 10^{-10}$$

Inspection of text Table 10-2 shows that HY must be $\boxed{\text{HCN}}$, which has a tabulated K_a of 6.17×10^{-10}. This conclusion is supported by comparing the molar mass of HCN (27.03 g mol^{-1}) to the molar mass of HY from the gas-density data (27.0 g mol^{-1}).

Chapter 11

Heterogeneous Equilibria

The discussion in text Chapter 10 is restricted to **homogeneous equilibria**. These are equilibria among mixed gases and among solutes dissolved in the same solution. General equilibria however involve solids, liquids, solutes, and solvents in all combinations. Equilibria that act across phase boundaries are **heterogeneous equilibria**.

The Concept of Activity

Mass-action expressions written for homogeneous reactions use the partial pressures of gases and the concentrations of solutes. Each is divided by a reference partial pressure or a reference concentration (equal to 1 atm and 1 M respectively). As a consequence, both K, the thermodynamic equilibrium constant, and Q, the reaction quotient, are unitless numbers. This chapter adds two refinements to this story:

1. The use of the thermodynamic **activity a** of a substance to replace the partial pressure divided by reference pressure (or concentration divided by reference concentration).

2. The definition of the thermodynamic activity of pure solids and pure liquids so that such substances can be dealt with in mass-action expressions.

The thermodynamic activity a of a substance is an accurate measure of its influence (how "active" it is) in determining "where" (how far on the road between reactants and products) a reaction comes to equilibrium.

The activity of any substance in its defined standard state equals 1 exactly.

The recognized standard states (reference states) for solids, liquids and gases are given on text page 224 and also on page 158 of this Guide.

295

What about when substances are not in their standard or reference states (which is most of the time)? The activity of the gas A having partial pressure P_A in a system given by the equation:

$$a_A = \frac{\gamma_A P_A}{P_{ref}}$$

The γ_A is an **activity coefficient.** It is a "fix-up factor" that accounts for the fact that the recognized reference state of gases is *as an ideal gas* at a pressure of 1 atm and some selected temperature (usually 25°C). Real gases behave non-ideally. Thus the activity of gas A at a partial pressure of 2.0 atm is not exactly 2.00 because the activity coefficient differs from 1. The activity coefficient of a substance only equals 1 when the substance is exactly in its reference or standard state. This state is unattainable for gases because real gases deviate more or less from ideal-gas behavior. Real gases held at 1 atm exactly and 25°C, have activity coefficients (γ's) and activities (a's) that differ from 1.

Similarly, the activity of solute B in a solution is

$$a_B = \frac{\gamma_B c_B}{c_{ref}} = \frac{\gamma_B [B]}{c_{ref}}$$

The activity coefficient γ_B corrects for the fact that the agreed-upon reference state for solutes is a concentration of 1 M *in an ideal solution*. The concentrations of solutes only approximate their activities. because no real solution is an ideal solution.

The accepted reference state for pure liquids and solids is the most stable form of the solid or liquid at a pressure of 1 atm and a selected temperature (again, the usual selection is 25°C). The reference state of solids and liquids *is* physically attainable.[1] For solids and liquids in their reference states, $a = 1$.

For a general equilibrium involving substances in any combination of phases:

$$a A_{\text{any phase}} + b B_{\text{any phase}} \rightleftharpoons c C_{\text{any phase}} + d D_{\text{any phase}} \qquad K = \frac{(a_C)^a_{eq}(a_D)^d_{eq}}{(a_A)^a_{eq} a_B)^b_{eq}}$$

See **11-1** and **11-3**.

Heterogeneous Equilibria

The mass-action expression for a general reaction starts with the form introduced for gas-only reactions. Then the different possibilities are handled as follows:

- Gases enter the mass-action expression as partial pressures, in atmospheres. Simply discard the atm part of such a pressure and use the numerical part in

---

[1]Depending however on how stringently "pure" is defined.

mass-action expressions. Assume in the absence of contrary information that the activity coefficient γ of the gas equals 1. Activity coefficients of dilute gases in fact are close to 1.

- Dissolved species enter as concentrations, in moles per liter. Merely discard the unit and insert the numerical part of such concentrations into mass-action expressions. The activity coefficient γ for solutes in dilute solutions are also close to 1.

- Pure solids and pure liquids are omitted from mass-action expressions. The activity a of a pure solid or liquid equals 1. Even if raised to a power in a mass-action expression, an activity of 1 makes no numerical difference. See **11-1** and **11-3**. Although the activities of pure solids and liquids does not appear explicitly in mass-action expressions, the presence of the solids or liquids is still essential to the existence of the equilibrium. See **11-9**.

- Solvents are omitted from mass-action expressions. In dilute solutions the activity of the solvent is quite close to 1 because it is nearly pure. See **11-3**.

Extraction and Separation Processes

Partition equilibria are much used in separations. Such equilibria involve the distribution of solutes between two immiscible (mutually insoluble) solvents. Carbon tetrachloride and water are immiscible solvents. They do not stay mixed after being shaken together, but settle into layers. A solute, such as I_2, partitions itself between the layers. The following heterogeneous equilibrium is established:

$$I_2(aq) \rightleftharpoons I_2(CCl_4) \qquad K = \frac{[I_2]_{(CCl_4)}}{[I_2]_{(aq)}}$$

This K is a **partition coefficient**. An impurity in the iodine would partition itself between the two solvents, too, but with a different K. Suppose the impurity has a smaller K than the I_2. Then the impurity concentrates in the water as the I_2 concentrates in the CCl_4. Even small differences in K can lead to effective separations if the equilibria are repeatedly established in a cycled operation. See **11-15**, **11-17**, and **11-83**. **Extraction** takes advantage of the partitioning of a solute between two immiscible solvents to transfer the solute from one solvent into the other. In the above example, I_2 is extracted from the water into the CCl_4 because K for the partitioning process is large.

Various methods of **chromatography** rely on the continuous extraction of a solute from one phase to another. In all forms of chromatography, a **mobile phase** is passed over a **stationary phase**. Solutes are partitioned between the phases with different partition coefficients, and separations are thereby achieved.

The Nature of Solubility Equilibria

The dissolution and subsequent re-precipitation of a soluble substance from a solvent is **recrystallization**. Recrystallization is a powerful method of purification. Successful recrystallizations require adroit manipulation of dissolution-precipitation equilibria. In **dissolution,** a solvent attacks a solute and brings it into solution by **solvating** it at the level of individual molecules or ions. The solute is seen to melt away. In **precipitation**, the reverse occurs. Solid substance is seen to deposit from a solution.

Solvent of crystallization may be incorporated into a precipitated solid. This leads to formulas like $Li_2SO_4 \cdot H_2O(s)$, $CuSO_4 \cdot 5H_2O(s)$, and $Cu(NO_3)_2 \cdot 6H_2O(s)$.[2] See **11-19**.

- Both solid and solution must be present to establish equilibrium in a dissolution-precipitation reaction. When both are present at equilibrium, the solution is **saturated**. The complete dissolution of a solid gives an **unsaturated** solution. Dissolution has occurred, but the reverse reaction (precipitation) cannot be occurring or solid would be visible. These equilibria resemble those between pure liquid and its vapor.[3]

- Dissolution-precipitation reactions generally reach equilibrium slowly. **Supersaturation**, in which the concentration of a solute exceeds its equilibrium value but no precipitate forms, is common. Dissolution to the point of saturation can take days, even with vigorous shaking.

Ionic Equilibria Between Solids and Solution

Ionic solids (salts) typically dissolve in water to give independent separately aquated (hydrated) anions and cations. A saturated solution of a salt is a solution in which equilibrium exists between the solid salt and its dissolved ions. The **solubility** of a salt (or of any substance) equals the number of grams or moles of it present at equilibrium in a given quantity of solvent or a given quantity of solution. The amount of a salt that is dissolved at saturation varies hugely depending on the identity of the cation and anion.

Learn which combinations of ions are insoluble, which are slightly soluble, and which are soluble.

The solubility rules of text Table 11-2[4] present essential information. They provide a basis to predict what happens when solutions of common salts are mixed.

---

[2]See text pages 370, 28, and 6.
[3]Text page 150.
[4]Text page 372.

Dissociation into Ions

Most salts dissociate into ions when they dissolve.[5] Equilibrium then exists between the solid salt and its aquated ions, and not between the solid salt and dissolved molecules of salt. For example:

$$PbSO_4(s) \rightleftharpoons Pb^{2+}(aq) + SO_4^{2-}(aq)$$

This heterogeneous equilibria is described by the mass-action expression:

$$K_{sp} = a_{Pb^{2+}}\, a_{SO_4^{2-}} = [Pb^{2+}][SO_4^{2-}]$$

The K_{sp} is a **solubility-product constant.** It does not differ fundamentally from other equilibrium constants. The special name merely emphasizes that the mass-action expression has the form of a product and not the more usual form of a quotient. See **11-23** and **11-87** for examples of writing a K_{sp} expression. Note that:

- K_{sp} expressions are not written for highly soluble salts because such solutions are too nonideal. The activities of the ions differ too much from their concentrations. Use K_{sp}'s only for slightly soluble salts unless detailed data are available giving the activity coefficients of the dissolved ions.

- K_{sp} values come from experiment (solubility measurements, see below). Compilations of the K_{sp}'s for most slightly soluble salts are widely available, but can disagree, owing to experimental difficulties.

- Numerical values of K_{sp} are never good to more than two significant figures and often good to only one significant figure.

- K_{sp}'s depend strongly on temperature (**11-31** and **11-35**). They usually increase with increasing T, but sometimes diminish.

The solubility and the K_{sp} of a salt are *not* the same constant, but a relationship exists between the two. Simple relationships between the K_{sp} of a salt and S (its solubility in mol L^{-1}) exist if:

- the solution is ideal or near-ideal;

- no side-reactions reduce concentrations of ions once the salt is in solution. For examples of dealing with of side-reactions, see **11-67** and **11-101**.

Some examples of simple K_{sp} to S relationships are:

---

[5]Mercury(II) chloride ($HgCl_2$), lead(II) acetate ($Pb(C_2H_3O_2)_2$), cadmium(II) iodide (CdI_2), and mercury(II) cyanide ($Hg(CN)_2$) are among the few salts that remain undissociated in solution.

For AgCl or BaSO$_4$, $K_{sp} = S^2$ for AgCrO$_4$, $K_{sp} = S(2S)^2 = 4S^3$

Convert solubilities expressed in units other than molarity (such as g L^{-1}, or g/100 mL) to mol L^{-1} before use in K_{sp} expressions. Assume room temperature (25°C) if a solubility is quoted but the temperature is not specified.

Precipitation and the Solubility Product

A salt tends to precipitate from aqueous solution when the activities (concentrations) of its component ions cause Q to exceed K_{sp}. Otherwise, it stays in solution. The concentrations of the component ions need not echo the molar ratio found in the salt. Random ratios are the rule when solutions are mixed. Thus, in **11-37** it is strictly an accident that Ce^{3+} and IO$_3^-$ are mixed in a 1-to-3 ratio to form Ce(IO$_3$)$_3$. A nonstoichiometric ratio (such as the Pb^{2+} to IO$_3^-$ ratio in **11-39**) is more typical. To compute equilibrium concentrations of ions after mixing two solutions:

- Realize that mixing two solutions dilutes both. Use the equation $c_f = c_i(V_i/V_f)$ on page 126 of this Guide to obtain concentrations after dilution.

- Assume that precipitation goes to completion (uses up the ion that is the limiting reactant). The "initial" concentrations in the standard three-line format[6] equal the concentrations *after* this reaction.

- Imagine that equilibrium is then finally attained by back-reaction (dissolution of the precipitate). Changes during back-reaction appear on the second line in the three-line table. See **11-39**. Problem **11-63** takes the same approach with a non-dissolution equilibrium.

The Common-Ion Effect

If a salt is placed in water that contains one of its component ions, then the solubility of the salt is usually *reduced*. This follow from LeChatelier's principle: the common ion increases the concentration of a product of the dissolution equilibrium, which is shifted to the left. See **11-43**. Complexation or other side-reactions can reverse the common-ion effect. See **11-65**. Although the common-ion effect drastically changes the solubilities of salts, it does not change their K_{sp}'s, which are fundamental quantities. See **11-45**.

---

[6]See page 221 of this Guide.

The Effects of pH on Solubility

The pH strongly influences the solubility of many ionic substances. See **11-51**. For hydroxides, which are numerous and important, pH affects solubility directly (see **11-49**). By LeChatelier's principle, metal hydroxides have enhanced solubility at low pH, but depressed solubility at high pH. See **11-47** and **11-97**.

When the cation or anion is a weak acid or base, pH affects solubility indirectly: H_3O^+ or OH^- reacts with the anion or cation to shift the dissolution equilibrium. For example, $AgCH_3COO$ gains solubility at low pH because the acetate ion, a base, accepts H^+ and is removed as $CH_3COOH(aq)$, shifting the dissolution equilibrium to the right.

Typical Problems Involving K_{sp}

- The calculation of K_{sp} from the solubility of a solid and the reverse. See **11-27**, **11-29**, **11-31**, and **11-97**.

- The determination of threshold concentrations for the precipitation of a given salt when its component ions are mixed. Precipitation occurs if the reaction quotient for the dissolution reaction exceeds K_{sp}. See **11-35**, **11-37**, and **11-95**.

- Evaluation of selective precipitations. In these experiments, a mixture of cations (or anions) is separated on the basis of the differing solubilities of their salts with an added anion (or cation). See **11-39**, **11-55**, and **11-57**.

Selective Precipitation of Ions

The metal sulfides are an important group of salts in which the anion is a base. Indeed, the sulfide ion (S^{2-}) does not exist in aqueous solution because it is too strong a base: essentially all S^{2-} ion in water accepts H^+ to give HS^- ion. Thus, the dissolution of the metal sulfide $CdS(s)$ is best represented:

$$CdS(s) + H_2O(l) \rightleftharpoons Cd^{2+}(aq) + HS^-(aq) + OH^-(aq) \qquad [Cd^{2+}][HS^-][OH^-] = K$$

Such equilibria are easily manipulated by changing the pH (using buffers). In computations, remember:

- The concentration of HS^- ion in a solution is linked by a K_a equilibrium to the concentration of the acid H_2S. See **11-57** and **11-59**.

- Hydrogen sulfide H_2S is a somewhat soluble gas. Saturated aqueous H_2S has a maximum concentration of 0.10 M at room temperature. See **11-101**.

Complex Ions and Solubility

In a **coordination complex**, a central metal ion binds to one or more **ligands**. Ligands are molecules or ions, such as Cl^-, NH_3, OH^-, and H_2O, that can share a lone pair of electrons with the metal ion. The interaction between metal ion and ligand is called **coordination**. The **formation constant K_f** of a coordination complex is the equilibrium constant for the direct reaction between the ligands and the central metal ion to form the complex:

$$M^{2+}(aq) + 4\,L(aq) \rightleftharpoons ML_4^{2+}(aq) \qquad K_f = \frac{[ML_4^{2+}]}{[M^{2+}][L]^4}$$

where L stands for ligand and M for metal. The formation of complex ions can affect the solubility of salts profoundly. An example is the dissolution (**11-99**) in aqueous ammonia of insoluble $AgCl(s)$:

$$AgCl(s) + 2\,NH_3(aq) \rightleftharpoons Ag(NH_3)_2^+(aq) + Cl^-(aq)$$

Complex-ion equilibria run simultaneously with the K_{sp} equilibria that govern the solubility of the salt. Think in terms of a competition among different species in binding to the metal ion. In **11-65**, Ag^+ ion is distributed among three competitive forms, $AgCl(s)$, $AgCl_2^-(aq)$ and $Ag^+(aq)$. The factors determining the outcome of the competition are the intrinsic stability of the different forms (as conveyed by the different K's), and the concentration of the $Cl^-(aq)$ ion.

Complex-Ion Equilibria

A metal ion gains ligands one by one just as a base (such as PO_4^{3-} ion in **6-57**) gains H^+ ions one by one. If the full complex has four ligands, then four chemical equilibria giving complexes with one, two, three, and four ligands lead to its formation in solution. Each of the four has a different equilibrium constant, K_1, K_2, and so forth. The product $K_1 \times K_2 \ldots$ equals the formation constant of the complex ion, as defined above. Multiplying the K's corresponds to adding up the step-wise equilibria. For example, the formation reaction:

$$Zn^{2+}(aq) + 4\,NH_3(aq) \rightleftharpoons Zn(NH_3)_4^{2+}(aq) \qquad\qquad 5 \times 10^8 = K_f$$

takes place in a series of four steps:

$$Zn^{2+}(aq) + NH_3(aq) \rightleftharpoons Zn(NH_3)^{2+}(aq) \qquad \frac{[Zn(NH_3)^{2+}]}{[NH_3][Zn^{2+}]} = K_1$$

$$Zn(NH_3)^{2+}(aq) + NH_3(aq) \rightleftharpoons Zn(NH_3)_2^{2+}(aq) \qquad \frac{[Zn(NH_3)_2^{2+}]}{[NH_3][Zn(NH_3)^{2+}]} = K_2$$

$$Zn(NH_3)_2^{2+}(aq) + NH_3(aq) \rightleftharpoons Zn(NH_3)_3^{2+}(aq) \qquad \frac{[Zn(NH_3)_3^{2+}]}{[NH_3][Zn(NH_3)_2^{2+}]} = K_3$$

$$Zn(NH_3)_3^{2+}(aq) + NH_3(aq) \rightleftharpoons Zn(NH_3)_4^{2+}(aq) \qquad \frac{[Zn(NH_3)_4^{2+}]}{[NH_3][Zn(NH_3)_3^{2+}]} = K_4$$

At equilibrium:

$$K_1 K_2 K_3 K_4 = K_f = \frac{[Zn(NH_3)_4^{2+}]}{[Zn^{2+}][NH_3]^4}$$

Difficulties arise often in the calculation of complex-ion equilibria because the K_n values of many complex ions do not differ by much. For example, text Table 11-5[7] gives the four constants for the formation of $Zn(NH_3)_4^{2+}$ as 150, 180, 200, and 90—the four are practically equal. Such is rarely the case with acid-base equilibria. This circumstance obliges us to treat the stages of complexation simultaneously, despite the algebraic difficulties this entails.

Fortunately, there is a common special case in practical problems. **If the ligand is present in excess and the K's are fairly large, then:**

- Most of the metal ion will be tied up in the complex with the most ligands.

- The concentration of the free ligand will approximately equal the concentration of *excess* ligand that is calculated assuming complete complexation.

These are exactly the conditions that prevail in text Example 11-11 and in **11-61** and **11-103**. When these conditions do *not* prevail, set up the problem using the principles of electrical neutrality and material balance.[8] Then apply chemical and mathematical insight to find out what terms (if any) can be neglected. See **10-69** for an example.

Acidity and Amphoterism of Complex Ions

Solutions of many metal ions are acidic because a metal ion coordinated by water can be an effective H^+-donor. **Example:** $Fe(H_2O)_6^{3+}$ has a K_a of 7.7×10^{-5}. A metal ion coordinated by H_2O and by OH^- is amphoteric; it can donate H^+ or accept H^+ depending on other species present. See **11-67**, **11-69**, and **11-105**.

---

[7]Text page 385.
[8]See text Section 10-8.

Detailed Solutions to Odd-Numbered Problems

11-1 The activities of pure solids and liquids equal 1 and may be omitted from the mass-action expressions:

$$\textbf{a)} \ \frac{a_{H_2S}^8}{a_{H_2}^8} = K \qquad \textbf{b)} \ \frac{a_{COCl_2} a_{H_2}}{a_{Cl_2}} = K \qquad \textbf{c)} \ a_{CO_2} = K \qquad \textbf{d)} \ \frac{1}{a_{C_2H_2}^3} = K$$

In all three parts, the partial pressures of the gases divided by the standard-state pressure (1 atm) approximates the activity of the gas. This allows rewriting the answers as follows:

$$\textbf{a)} \ \frac{P_{H_2S}^8}{P_{H_2}^8} = K \qquad \textbf{b)} \ \frac{P_{COCl_2} P_{H_2}}{P_{Cl_2}} = K \qquad \textbf{c)} \ P_{CO_2} = K \qquad \textbf{d)} \ \frac{1}{P_{C_2H_2}^3} = K$$

11-3 Pure solids and liquids are omitted from the expressions because their activities equal 1.

$$\textbf{a)} \ \frac{a_{Zn^{2+}}}{a_{Ag^+}^2} = K \qquad \textbf{b)} \ \frac{a_{VO_3(OH)^{2-}} a_{OH^-}}{a_{VO_4^{3-}}} = K \qquad \textbf{c)} \ \frac{a_{HCO_3^-}^6}{a_{As(OH)_6^{3-}}^2 a_{CO_2}^6} = K$$

In all three parts, the concentrations of the solutes divided by the standard state concentration (1M) approximates the activity of the solute. Also, the activities of gases can be treated just as in **11-1**. This allows rewriting the answers as:

$$\textbf{a)} \ \frac{[Zn^{2+}]}{[Ag^+]^2} = K \qquad \textbf{b)} \ \frac{[VO_3(OH)^{2-}][OH^-]}{[VO_4^{3-}]} = K \qquad \textbf{c)} \ \frac{[HCO_3^-]^6}{[As(OH)_6^{3-}]^2 P_{CO_2}^6} = K$$

11-5a) The graph consists of the values in the second column of the following table plotted (on the y-axis) versus the values in the first column (on the x-axis). The values in the third column are those of the second divided by those of the first:

$[N_2O_4]$	$[NO_2]^2$	$[NO_2]^2/[N_2O_4]$
0.190×10^{-3} M	0.784×10^{-5} M^2	4.12×10^{-2} M
0.686	2.70	3.94
1.54	5.27	3.42
2.55	10.8	4.23
3.75	13.7	3.65
7.86	29.9	3.80
11.9	44.1	3.71

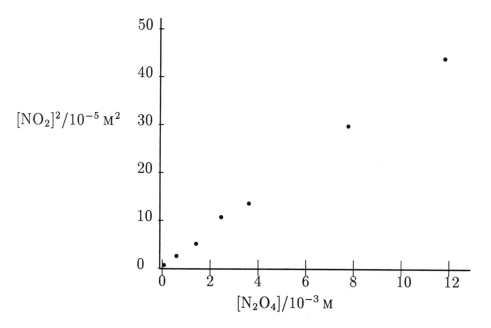

From the mass-action expression for this reaction it follows that:

$$[NO_2]^2 = K[N_2O_4]$$

The equation has the form of the equation for a straight line $y = mx + b$, with $y = [NO_2]^2$, $m = K$, $x = [N_2O_4]$ and $b = 0$. Thus, K equals the slope of the line just plotted.[9]

b) The mean of the values in the last column of the table in the preceding part is the mean experimental K. It is $\boxed{3.83 \times 10^{-2}}$.

11-7 Write equilibrium expressions for the reduction of iron(III) oxide with hydrogen and the reduction of carbon dioxide with hydrogen. Write expressions that correspond to the form of the equation given in the problem:

$$\frac{1}{P_{H_2}^3} = K_1 = 4.0 \times 10^{-6} \qquad \text{and} \qquad \frac{P_{CO}}{P_{CO_2} P_{H_2}} = K_2 = 3.2 \times 10^{-4}$$

If both reactions are simultaneously at equilibrium in the same container, then both equations must be satisfied simultaneously. Compute the partial pressure of the hydrogen from the first equation:

$$P_{H_2} = \sqrt[3]{\frac{1}{K_1}} = \sqrt[3]{\frac{1}{4.0 \times 10^{-6}}} = 63 \text{ atm}$$

---

[9]See text page A-19 for a discussion of the determination of the slope of a straight-line graph.

Insert this partial pressure into the (slightly rearranged) second equation:

$$\frac{P_{CO}}{P_{CO_2}} = K_2 P_{H_2} = (3.2 \times 10^{-4})63 = \boxed{0.020}$$

11-9 a) The reaction and mass-action expression are:

$$NH_3(g) + HCl(g) \rightleftharpoons NH_4Cl(s) \qquad \frac{1}{P_{NH_3} P_{HCl}} = K = 4.0 \quad \text{at } 340°C$$

If P_{NH_3} is 0.80 atm at equilibrium, then, by substitution, the equilibrium partial pressure of HCl(g) is $\boxed{0.31 \text{ atm}}$.

b) Obviously, the equilibrium cannot occur before addition of $NH_4Cl(s)$ to the container filled with ammonia—the ammonium chloride is the only source of HCl(g). For every mole of HCl(g) that is produced, one mole of $NH_3(g)$ joins the quantity of $NH_3(g)$ that was responsible for the original 1.50 atm. The partial pressures of the $NH_3(g)$ and of HCl(g) are directly proportional to their respective chemical amounts (assuming ideal-gas behavior). This means that for every atmosphere of HCl(g) produced one additional atmosphere of $NH_3(g)$ is also produced. Let x equal the equilibrium partial pressure of HCl. Then the equilibrium partial pressure of NH_3 is $1.50 + x$ and:

$$P_{NH_3} P_{HCl} = (1.50 + x)x = 0.25 \qquad \text{which gives} \qquad x^2 + 1.50x - 0.25 = 0$$

Solving the quadratic equation gives x equal to 0.151 (the negative root is rejected). Therefore $P_{HCl} = \boxed{0.15 \text{ atm}}$, and $P_{NH_3} = 1.50 + 0.15 = \boxed{1.65 \text{ atm}}$.

11-11 The "water gas" reaction[10] and its associated mass-action expression:

$$C(s) + H_2O(g) \rightleftharpoons CO(g) + H_2(g) \qquad \frac{P_{CO} P_{H_2}}{P_{H_2O}}$$

If the reaction quotient Q is less than K, the reaction tends to proceed ("shifts") from left to right; if Q is greater than K, the reaction tends to proceed from right to left.

a) A Q is computed by substitution of in the mass-action expression:

$$Q = \frac{P_{CO} P_{H_2}}{P_{H_2O}} = \frac{(1.525)(0.805)}{0.600} = \boxed{2.05}$$

---

[10]This reaction is a possible source of industrial hydrogen. See text page 736.

This Q is less than K (which is 2.6) so the reaction shifts from $\boxed{\text{left to right}}$.

b) All three partial pressures are higher than in the previous part. Use them in the mass-action expression:

$$Q = \frac{P_{CO}P_{H_2}}{P_{H_2O}} = \frac{(1.714)(1.383)}{0.724} = \boxed{3.27}$$

Since Q now exceeds K, the reaction shifts from $\boxed{\text{right to left}}$ to reach equilibrium.

11-13 a) The container holds no gas until some of the $NH_4HSe(s)$ decomposes. Breakdown of the solid is the only source of the two gases than eventually fill the container at equilibrium. The stoichiometry of the decomposition reaction requires that the partial pressure of the H_2Se equal the partial pressure of the NH_3 (assuming ideality). The total pressure equals the sum of the partial pressures of the two gases. In equation form:

$$P_{H_2Se} = P_{NH_3} \quad \text{and} \quad P_{H_2Se} + P_{NH_3} = 0.0184 \text{ atm}$$

Clearly both partial pressures equal 0.00920 atm at equilibrium. The equilibrium constant is:

$$K = (P_{H_2Se})_{eq}(P_{NH_3})_{eq} = (0.00920)^2 = \boxed{8.46 \times 10^{-5}}$$

b) The size of the container has no effect on the K. The two partial pressures can differ, but their product must equal K at equilibrium:

$$8.46 \times 10^{-5} = P_{H_2Se}P_{NH_3} = P_{H_2Se}(0.0252) \quad \text{hence:} \quad P_{H_2Se} = \boxed{0.00336 \text{ atm}}$$

11-15 Assume exactly 1 L each of H_2O and CCl_4. Then at equilibrium the water holds 1.30×10^{-4} mol of I_2, and the remaining I_2 must be in the CCl_4 layer:

$$n_{I_2 (CCl_4)} = 1.00 \times 10^{-2} - 1.30 \times 10^{-4} = 9.9 \times 10^{-3} \text{ mol}$$

The concentration of I_2 in the CCl_4 layer is this amount divided by the volume:

$$[I_2]_{(CCl_4)} = \frac{9.9 \times 10^{-3} \text{ mol}}{1 \text{ L}} = 9.9 \times 10^{-3} \text{ mol L}^{-1}$$

The equilibrium constant equals the ratio of the two concentrations:

$$K = \frac{[I_2]_{(CCl_4)}}{[I_2]_{(aq)}} = \frac{9.9 \times 10^{-3}}{1.30 \times 10^{-4}} = \boxed{76}$$

11-17 a) The mass-action expressions for the dissolution of benzoic acid in water (K_1) and the dissolution of benzoic acid in either(K_2) are quite simple:

$$K_1 = [C_6H_5COOH]_{(aq)} \quad \text{and} \quad K_2 = [C_6H_5COOH]_{(ether)}$$

The concentrations of C_6H_5COOH at saturation (equilibrium) in water and in ether are:

$$c_{water} = \frac{2.00 \text{ g } C_6H_5COOH}{1 \text{ L water}} \times \left(\frac{1 \text{ mol } C_6H_5COOH}{122 \text{ g } C_6H_5COOH} \right) = 0.0164 \text{ M}$$

$$c_{ether} = \frac{660 \text{ g } C_6H_5COOH}{1 \text{ L ether}} \times \left(\frac{1 \text{ mol } C_6H_5COOH}{122 \text{ g } C_6H_5COOH} \right) = 5.4 \text{ M}$$

Hence K_1 is $\boxed{0.0164}$, and K_2 is $\boxed{5.4}$.

b) The partition reaction equals reaction 1 in the previous part subtracted from reaction 2. Subtracting a reaction is the same as reversing it and adding it. The equilibrium constant for the partition reaction (reaction 3) is accordingly:

$$K_3 = K_2 \left(\frac{1}{K_1} \right) = \frac{5.4}{0.0164} = \boxed{330}$$

11-19 Comparison of the two formulas shows that the reaction converting plaster of paris (the half-hydrate of $CaSO_4$) to gypsum (the dihydrate) uses 3/2 mol of water per mole of plaster of paris. Thus:

$$25.0 \text{ kg } CaSO_4 \cdot \tfrac{1}{2}H_2O \times \left(\frac{1 \text{ mol } CaSO_4 \cdot \tfrac{1}{2}H_2O}{0.14515 \text{ kg } CaSO_4 \cdot \tfrac{1}{2}H_2O} \right) \times \left(\frac{3 \text{ mol } H_2O}{2 \text{ mol } CaSO_4 \cdot \tfrac{1}{2}H_2O} \right)$$

$$\times \left(\frac{0.01802 \text{ kg } H_2O}{1 \text{ mol } H_2O} \right) \times \left(\frac{1 \text{ L } H_2O}{1 \text{ kg } H_2O} \right) = \boxed{4.65 \text{ L } H_2O}$$

11-21 On the graph, read across from "80" on the vertical axis until the solubility curve is reached. Then drop down to the horizontal axis and read the temperature. A solubility of 80 g HBr per 100 g H_2O is reached at a temperature of approximately 48°C. The last of the KBr will dissolve at $\boxed{\text{about 48°C}}$.

11-23 Solubility-product constant expressions[11] follow the general rules for heterogeneous equilibria. Pure solids and the solvent do not appear in these expressions. For the dissolution of iron(III) sulfate in water:

$$\boxed{Fe_2(SO_4)_3(s) \rightleftharpoons 2 Fe^{3+}(aq) + 3 SO_4^{2-}(aq)} \qquad \boxed{K_{sp} = [Fe^{3+}]^2[SO_4^{2-}]^3}$$

---

[11]Also called solubility product expressions and K_{sp} expressions.

11-25 The dissolution of thallium(I) iodate is represented:

$$TlIO_3(s) \rightleftharpoons Tl^+(aq) + IO_3^-(aq) \qquad [Tl^+][IO_3^-] = K_{sp} = 3.07 \times 10^{-6}$$

If S mol per liter of $TlIO_3$ dissolves, then S mol per liter of Tl^+ and also S mol per liter of IO_3^- are present at equilibrium, as long as neither ion reacts further in solution. Thus:

$$K_{sp} = 3.07 \times 10^{-6} = [Tl^+][IO_3^-] = S^2 \quad \text{which gives} \quad S = 1.752 \times 10^{-3} \text{ mol L}^{-1}$$

This means that 1.752×10^{-3} mol of thallium(I) iodate saturates a liter of solution at 25°C. This equals the solubility of the thallium(I) iodate in 1000 mL of water because the small amount of solute does not alter the volume of the solution measurably from the volume of the pure solvent. This chemical amount of thallium(I) iodate has a mass of 0.665 g (obtained by multiplying by $\mathcal{M}_{TlIO_3} = 379.3$ g mol^{-1}). The mass of $TlIO_3$ that saturates 100 mL of water is 1/10 as great: $\boxed{0.0665 \text{ g } TlIO_3/100 \text{ mL}}$.

11-27 The equilibrium equation and K_{sp} for the dissolution of potassium perchlorate are:

$$KClO_4(s) \rightleftharpoons K^+(aq) + ClO_4^-(aq) \qquad [K^+][ClO_4^-] = K_{sp} = 1.07 \times 10^{-2}$$

If S mol L^{-1} of $KClO_4$ dissolves, then there must be S mol L^{-1} of K^+ and S mol L^{-1} of ClO_4^- present at equilibrium, as long as neither ion reacts further. Substitute S into the K_{sp} expression:

$$K_{sp} = 1.07 \times 10^{-2} = [K^+][ClO_4^-] = S^2$$

Solving gives $S = 0.103$ mol L^{-1}. Then:

$$\frac{0.103 \text{ mol } KClO_4}{L} \times \left(\frac{138.55 \text{ g } KClO_4}{1 \text{ mol } KClO_4} \right) = \boxed{\frac{14.3 \text{ g } KClO_4}{L}} \text{ at } 25°C$$

11-29 The equilibrium and K_{sp} expressions are:

$$Hg_2I_2(s) \rightleftharpoons Hg_2^{2+}(aq) + 2\,I^-(aq) \qquad [Hg_2^{2+}][I^-]^2 = K_{sp}$$

If S mol L^{-1} of Hg_2I_2 dissolves, then S mol L^{-1} of Hg_2^{2+} and $2S$ mol L^{-1} of I^- are produced. Substituting into the K_{sp} mass-action expression gives:

$$[Hg_2^{2+}][I^-]^2 = S(2S)^2 = 1.2 \times 10^{-28}$$

from which:

$$4S^3 = 1.2 \times 10^{-28} \quad \text{which gives} \quad S = 3.1 \times 10^{-10} \text{ mol L}^{-1}$$

$$[\text{Hg}_2^{2+}] = S = \boxed{3.1 \times 10^{-10} \text{ mol L}^{-1}} \qquad [\text{I}^-] = 2S = \boxed{6.2 \times 10^{-10} \text{ mol L}^{-1}}$$

11-31 As seen in **7-9**, a solubility differs from a solubility-product constant, although there is often a simple relationship between the two. The equilibrium of interest is:

$$\text{Ag}_2\text{CrO}_4(s) \rightleftharpoons 2\,\text{Ag}^+(aq) + \text{CrO}_4^{2-}(aq) \qquad K_{\text{sp}} = [\text{Ag}^+]^2[\text{CrO}_4^{2-}]$$

The problem states that 0.0129 g of silver chromate dissolves in 500 mL (0.500 L) of water. Assume that neither Ag^+ nor CrO_4^{2-} reacts with other species once in the water. Then the chemical amounts of the two ions in solution at equilibrium are:

$$n_{\text{Ag}^+} = 0.0129 \text{ g} \times \left(\frac{1 \text{ mol Ag}_2\text{CrO}_4}{331.7 \text{ g Ag}_2\text{CrO}_4} \right) \times \left(\frac{2 \text{ mol Ag}^+}{1 \text{ mol Ag}_2\text{CrO}_4} \right) = 7.78 \times 10^{-5} \text{ mol}$$

$$n_{\text{CrO}_4^{2-}} = \frac{1}{2} n_{\text{Ag}^+} = 3.89 \times 10^{-5} \text{ mol}$$

The equilibrium concentrations of the two ions equal these chemical amounts divided by 0.500 L, which is the volume of the solution:

$$[\text{Ag}^+] = 15.6 \times 10^{-5} \text{ mol L}^{-1} \qquad [\text{CrO}_4^{2-}] = 7.78 \times 10^{-5} \text{ mol L}^{-1}$$

Substitute into the K_{sp} expression to obtain a numerical K_{sp}:

$$K_{\text{sp}} = [\text{Ag}^+]^2[\text{CrO}_4^{2-}] = (15.6 \times 10^{-5})^2(7.78 \times 10^{-5}) = \boxed{1.9 \times 10^{-12}}$$

Tip. This equals the value in text Table 11-3.

11-33 The dissolution reaction and its K_{sp}-expression are:

$$\text{AgCl}(s) \rightleftharpoons \text{Ag}^+(aq) + \text{Cl}^-(aq) \qquad [\text{Ag}^+][\text{Cl}^-] = K_{\text{sp}}$$

At equilibrium in 1.00 L of the solution at 100°C:

$$n_{\text{Ag}^+} = 0.018 \text{ g} \times \left(\frac{1 \text{ mol AgCl}}{143.3 \text{ g AgCl}} \right) \times \left(\frac{1 \text{ mol Ag}^+}{1 \text{ mol AgCl}} \right) = 1.26 \times 10^{-4} \text{ mol}$$

Since the volume of the solution is 1.00 L, the concentration of Ag^+ ion is 1.26×10^{-4} M. The concentration of Cl^- ion is the same—one mole per liter of chloride ion is produced in solution for every mole per liter of silver ion. Substitute the concentrations into the mass-action expression:

$$K_{sp} = [Ag^+][Cl^-] = (1.26 \times 10^{-4})(1.26 \times 10^{-4}) = \boxed{1.6 \times 10^{-8}}$$

Tip. This K_{sp} is larger (by a factor of 100) than the K_{sp} in text Table 11-3[12] because the temperature is 100°, not 25°C.

11-35 Get the initial concentrations of Ba^{2+} ion and CrO_4^{2-} ion and use them to calculate the initial reaction quotient Q_0 for the reaction:

$$BaCrO_4(s) \rightleftharpoons Ba^{2+}(aq) + CrO_4^{2-}(aq)$$

"Initial" in this case means after the solution cools to 25°C, but before any reaction occurs. Compare Q_0 with K_{sp}. If Q_0 exceeds K_{sp}, then a precipitate will form as the reaction proceeds from right to left, if Q_0 is less than K_{sp}, then there can be no precipitate. The initial amount of dissolved Ba^{2+} is:

$$n_{Ba^{2+}} = 0.0063 \text{ g} \times \left(\frac{1 \text{ mol BaCrO}_4}{253 \text{ g BaCrO}_4}\right) \times \left(\frac{1 \text{ mol Ba}^{2+}}{1 \text{ mol BaCrO}_4}\right) = 2.49 \times 10^{-5} \text{ mol}$$

The initial concentration of Ba^{2+} ion equals 2.49×10^{-5} M, if the volume of the cooled solution is taken as 1.00 L.[13] The initial concentration of the CrO_4^{2-} ion is, by a similar calculation, also equal to 2.49×10^{-5} M. Then:

$$Q_0 = [Ba^{2+}]_0[CrO_4^{2-}]_0 = (2.49 \times 10^{-5})(2.49 \times 10^{-5}) = 6.2 \times 10^{-10}$$

From text Table 11-3, K_{sp} for $BaCrO_4$ equals 2.1×10^{-10}. $\boxed{BaCrO_4 \text{ precipitates}}$ until Q is lowered to equal K_{sp}.

11-37 Calculate the reaction quotient for the reaction:

$$Ce(IO_3)_3(s) \rightleftharpoons Ce^{3+}(aq) + 3\,IO_3^-(aq) \qquad K_{sp} = 1.9 \times 10^{-10}$$

just after the solutions are mixed and compare it to K_{sp}. The desired reaction quotient is $Q = [Ce^{3+}]_0[IO_3^-]_0^3$. where the subscript zero refers to the concentrations prevailing just after mixing. After mixing, the volume of the solution equals 400.0 mL, the sum of the two starting volumes. Mixing dilutes both solutions according to the formula

---

[12]text page 374.

[13]Doing this ignores the contraction of the water as it cools from 100 to 25°C. It also overlooks the very slight change in volume caused by the dissolution of the barium chromate in hot water in the first place.

for dilution given on page 126 of this Guide. Before mixing, the concentration of the Ce^{3+} ion was 0.0020 M. After mixing:

$$[Ce^{3+}]_0 = 0.0020 \text{ M} \times \left(\frac{250.0 \text{ mL}}{400.0 \text{ mL}}\right) = 1.25 \times 10^{-3} \text{ M}$$

Before mixing, the concentration of the IO_3^- ion was 0.010 M. After mixing:

$$[IO_3^-]_0 = 0.010 \text{ M} \times \left(\frac{150.0 \text{ mL}}{400.0 \text{ mL}}\right) = 3.75 \times 10^{-3} \text{ M}$$

$$Q = [Ce^{3+}]_0[IO_3^-]_0^3 = (1.25 \times 10^{-3})(3.75 \times 10^{-3})^3 = 0.66 \times 10^{-10}$$

Q is less than K_{sp}; $\boxed{\text{no precipitate forms}}$.

11-39 The 50.0 mL of 0.0500 M $Pb(NO_3)_2$ dissolves to give 2.50 mmol of $Pb^{2+}(aq)$ ion; the 40.0 mL of 0.200 $NaIO_3$ gives 8.00 mmol of $IO_3^-(aq)$ ion. The two react:

$$Pb^{2+}(aq) + 2\,IO_3^-(aq) \rightleftharpoons Pb(IO_3)_2(s)$$

Assume that this reaction goes to completion. Then zero Pb^{2+} ion remains in solution (it is the limiting reactant), but $8.00 - 2(2.50) = 3.00$ mmol of IO_3^- remains in solution. The concentration of the excess IO_3^- is 3.00 mmol/90.00 mL = 0.0333 M. For a fact, the reaction does not go to completion, but stops short in the equilibrium state. And if it did go to completion, a small amount of the precipitate would soon redissolve anyway:

$$Pb(IO_3^-)_2(s) \rightleftharpoons Pb^{2+}(aq) + 2\,IO_3^-(aq) \qquad K_{sp} = [Pb^{2+}][IO_3^-]^2 = 2.6 \times 10^{-13}$$

The equilibrium state is of course the same regardless of how it is attained. Let S equal the concentration of Pb^{2+} furnished by back-reaction:

	$Pb(IO_3)_2(s) \rightleftharpoons$	$Pb^{2+}(aq) +$	$2\,IO_3^-(aq)$
Init. Conc. (mol L^{-1})	$-$	0.0	0.0333
Change in Conc. (mol L^{-1})	$-$	$+S$	$+2S$
Equil. Conc. (mol L^{-1})	$-$	S	$0.0333 + 2S$

Substitution from the third line of the table into the K_{sp} expression gives:

$$(S)(0.0333 + 2S)^2 = 2.6 \times 10^{-13}$$

This cubic equation is simplified by assuming that $2S << 0.0333$. Then:

$$(0.0333)^2(S) = 2.6 \times 10^{-13} \qquad \text{which gives} \qquad S = 2.3 \times 10^{-10}$$

Hence: $[Pb^{2+}] = \boxed{2.3 \times 10^{-10} \text{ M}}$ and $[IO_3^-] = \boxed{0.033 \text{ M}}$. The assumption that $2S$ was much smaller than 0.0333 was obviously justified.

11-41 Follow the procedure used in **11-39**. The solution contains 5.00 mmol of $AgNO_3$ and 1.8 mmol of Na_2CrO_4 after mixing but before reaction. Assume that the precipitation reaction:

$$2\,Ag^+(aq) + CrO_4^{2-}(aq) \rightleftharpoons Ag_2CrO_4(s)$$

goes to completion. As it does, it consumes 2 mmol of Ag^+ for every 1 mmol of CrO_4^{2-} ion. Because the chemical amount of Ag^+ ion is more than twice that of CrO_4^{2-} ion, CrO_4^{2-} ion is the limiting reactant. The chemical amount of Ag^+ ion left in excess is $5.00 - 2(1.80) = 1.40$ mmol. The concentration of the Ag^+ ion equals this amount divided by the volume of the solution:

$$[Ag^+] = \frac{1.40 \text{ mmol}}{(50.0 + 30.0) \text{ mL}} = 0.01750 \text{ M}$$

Now let some of the precipitate redissolve:

$$Ag_2CrO_4(s) \rightleftharpoons 2\,Ag^+(aq) + CrO_4^{2-}(aq) \qquad K_{sp} = 1.9 \times 10^{-12} = [Ag^+]^2[CrO_4^{2-}]$$

Let S equal the concentration of CrO_4^{2-} in solution after this equilibrium is attained. The concentration of Ag^+ ion is $0.1750 + 2S$ and:

$$K_{sp} = (0.01750 + 2S)^2 S$$

Assume that $2S$ is much smaller than 0.0175. Then:

$$(0.01750)^2 S = 1.9 \times 10^{-12} \quad \text{which gives} \quad S = 6.2 \times 10^{-9}$$

The final concentrations of the two ions are:

$$[CrO_4^{2-}] = \boxed{6.2 \times 10^{-9} \text{ M}} \qquad [Ag^+] = 0.01750 + 2(6.2 \times 10^{-9}) = \boxed{0.018 \text{ M}}$$

11-43 The $F^-(aq)$ ion from the dissolved NaF depresses the solubility of CaF_2. This is the common-ion effect. CaF_2 dissolves according to the equation:

	$CaF_2(s) \rightleftharpoons$	$Ca^{2+}(aq) +$	$2\,F^-(aq)$
Init. Conc. (mol L^{-1})	—	0.0	0.040
Change in Conc. (mol L^{-1})	—	$+S$	$+2S$
Equil. Conc. (mol L^{-1})	—	S	$0.040 + 2S$

where S is the solubility of the CaF_2. For this dissolution reaction:

$$K_{sp} = [Ca^{2+}][F^-]^2 = 3.9 \times 10^{-11}$$

Substitute the equilibrium concentrations from the table, assuming that $2S << 0.040$:

$$S(0.040)^2 = 3.9 \times 10^{-11} \quad \text{which gives} \quad S = 2.4 \times 10^{-8}$$

The computed value is S is indeed much smaller than 0.040, so the assumption is justified. The solubility of the CaF_2 is $\boxed{2.4 \times 10^{-8} \text{ mol L}^{-1}}$.

11-45 a) For every y mol L^{-1} of $Ni(OH)_2$ that is dissolved at equilibrium, y mol L^{-1} of Ni^{2+} and $2y$ mol L^{-1} of OH^- have formed according to the equation:

$$Ni(OH)_2(s) \rightleftharpoons Ni^{2+}(aq) + 2\,OH^-(aq) \qquad [Ni^{2+}][OH^-]^2 = K_{sp} = 1.6 \times 10^{-16}$$

Substituting into the K_{sp} expression gives the equation: $(y)(2y)^2 = 1.6 \times 10^{-16}$. Solving gives $y = 3.4 \times 10^{-6}$. The solubility of $Ni(OH)_2$ is $\boxed{3.4 \times 10^{-6} \text{ mol L}^{-1}}$.

b) The presence of a common ion (the OH^- ion) reduces the solubility of the nickel(II) hydroxide. Set up the usual three-line table:

	$Ni(OH)_2(s) \rightleftharpoons$	$Ni^{2+}(aq)+$	$2\,OH^-(aq)$
Init. Conc. (mol L^{-1})	—	0	0.100
Change in Conc. (mol L^{-1})	—	$+z$	$+2z$
Equil. Conc. (mol L^{-1})	—	z	$0.100 + 2z$

Substitute the equilibrium concentrations into the K_{sp} expression:

$$K_{sp} = [Ni^{2+}][OH^-]^2 = z((0.100 + 2z)^2 = 1.6 \times 10^{-16}$$

If $2z << 0.100$, then:

$$(0.100)^2(z) = 1.6 \times 10^{-16} \quad \text{so that} \quad z = 1.6 \times 10^{-14}$$

Clearly, $2z$ is very small compared to 0.100; the solubility is $\boxed{1.6 \times 10^{-14} \text{ mol L}^{-1}}$.

11-47 As long as the solution is in equilibrium with $Mg(OH)_2(s)$, the following holds:

$$K_{sp} = 1.2 \times 10^{-11} = [Mg^{2+}][OH^-]^2$$

Let S represent the solubility of the $Mg(OH)_2$ *before* any NaOH is added. At this stage:

$$[Mg^{2+}] = S \quad \text{and} \quad [OH^-] = 2S \quad \text{so that} \quad 1.2 \times 10^{-11} = [Mg^{2+}][OH^-]^2 = 4\,S^3$$

Solving gives $S = 1.44 \times 10^{-4}$ mol L^{-1}. Sodium hydroxide dissociates completely to $Na^+(aq)$ and $OH^-(aq)$ ions. Thus, the concentration of OH^- goes up with the addition of NaOH, and the concentration of Mg^{2+} must diminish to maintain the mass-action expression at a constant value. Additional magnesium hydroxide precipitates as NaOH is added. This is the common-ion effect in action. The problem states that the solubility of $Mg(OH)_2$ is reduced to 0.0010 of its original value. This means that after the addition:

$$[Mg^{2+}] = 0.0010(1.44 \times 10^{-4}) = 1.44 \times 10^{-7} \text{ M}$$

Then:

$$[OH^-] = \sqrt{\frac{K_{sp}}{[Mg^{2+}]}} = \sqrt{\frac{1.2 \times 10^{-11}}{1.44 \times 10^{-7}}} = \boxed{9.1 \times 10^{-3} \text{ M}}$$

11-49 If y mol L^{-1} of AgOH(s) dissolves, then y mol L^{-1} of Ag^+ and y mol L^{-1} of OH^- ions are produced. Assume that the concentration of OH^- ion from the dissolution greatly exceeds the concentration of OH^- ion from the autoionization of water:

	AgOH(s) $\rightleftharpoons$	$Ag^+(aq)$ +	$OH^-(aq)$
Init. Conc. (mol L^{-1})	—	0.0	small
Change in Conc. (mol L^{-1})	—	$+y$	$+y$
Equil. Conc. (mol L^{-1})	—	y	y

At equilibrium:

$$K_{sp} = [Ag^+][OH^-] = y^2 = 1.5 \times 10^{-8} \quad \text{so that} \quad y = 1.2 \times 10^{-4} \text{ M}$$

The molar solubility of AgOH in water equals $\boxed{0.00012 \text{ mol } L^{-1}}$.

If the solution is buffered at pH 7.00, then the pH stays at 7.00 even though the dissociation of AgOH produces OH^-. A pH of 7.00 means $[OH^-] = 1.0 \times 10^{-7}$ M. Put this concentration into the K_{sp}-expression to obtain the equilibrium concentration of $Ag^+(aq)$:

$$[Ag^+] = \frac{K_{sp}}{[OH^-]} = \frac{1.5 \times 10^{-8}}{1.0 \times 10^{-7}} = 0.15 \text{ M}$$

The solubility of the AgOH(s) equals the chemical amount of Ag^+ ion in solution per liter; it is therefore $\boxed{0.15 \text{ mol } L^{-1}}$ in water buffered at pH 7. This solubility is 1250 times larger than the solubility of AgOH in pure water.

11-51 a) The solubility of PbI_2 will remain $\boxed{\text{unchanged}}$ as the pH of its solution is lowered. The anion is an exceedingly weak base and has little interaction with H_3O^+ even at a high concentration of H_3O^+.

b) The solubility of AgOH will $\boxed{\text{increase}}$ as the pH of its solution is lowered. The additional H_3O^+ drives additional dissolution by removing OH^- ion.

c) The solubility of $Ca_3(PO_4)_2$ will $\boxed{\text{increase}}$ as the pH of its solution is lowered from 7. A higher concentration of H_3O^+ drives the dissolution by removing product PO_4^{3-} ion as HPO_4^{2-} ion.

11-53 a) Imagine slowly adding oxalate ion to the mixture containing Mg^{2+} and Pb^{2+} ions. Both oxalate salts will stay in solution until their respective reaction quotients exceed their K_{sp}'s. The equilibria and K_{sp} expressions are:

$$PbC_2O_4(s) \rightleftharpoons Pb^{2+}(aq) + C_2O_4^{2-}(aq) \qquad [Pb^{2+}][C_2O_4^{2-}] = K_{sp} = 2.7 \times 10^{-11}$$
$$MgC_2O_4(s) \rightleftharpoons Mg^{2+}(aq) + C_2O_4^{2-}(aq) \qquad [Mg^{2+}][C_2O_4^{2-}] = K_{sp} = 8.6 \times 10^{-5}$$

The concentration of oxalate ion needed to precipitate magnesium oxalate from the 1.0 M Mg^{2+} solution is:

$$[C_2O_4^{2-}] = \frac{K_{sp}}{[Mg^{2+}]} = \frac{8.6 \times 10^{-5}}{0.10} = 8.6 \times 10^{-4} \text{ M}$$

If the $[C_2O_4^{2-}]$ is kept at or below $\boxed{8.6 \times 10^{-4} \text{ M}}$, then magnesium oxalate cannot precipitate. Only the lead oxalate can precipitate. The lead salt precipitates first because its solubility is smaller than that of the magnesium salt.

b) If the $[C_2O_4^{2-}]$ is held at 8.6×10^{-4} M, then:

$$[Pb^{2+}] = \frac{K_{sp}}{[C_2O_4^{2-}]} = \frac{2.7 \times 10^{-11}}{8.6 \times 10^{-4}} = 3.1 \times 10^{-8} \text{ M}$$

$$\text{fraction } Pb^{2+} \text{remaining} = \frac{3.1 \times 10^{-8} \text{ M}}{0.10 \text{ M}} = \boxed{3.1 \times 10^{-7}}$$

11-55 Calculate the $[I^-]$ that just suffices to bring about precipitation of each metal ion. The dissolution reactions and their K_{sp}'s are:

$$Hg_2I_2(s) \rightleftharpoons Hg_2^{2+}(aq) + 2\,I^-(aq) \qquad K_{sp} = 1.2 \times 10^{-28} = [Hg_2^{2+}][I^-]^2$$
$$PbI_2(s) \rightleftharpoons Pb^{2+}(aq) + 2\,I^-(aq) \qquad K_{sp} = 1.4 \times 10^{-8} = [Pb^{2+}][I^-]^2$$

The required concentration is quite small in the case of $Hg_2I_2(s)$ because K_{sp} is very small. In the case of $PbI_2(s)$ the required concentration is:

$$[I^-] = \sqrt{\frac{1.4 \times 10^{-8}}{0.0500}} = 5.3 \times 10^{-4} \text{ M}$$

The optimum $[I^-]$ would be just below $\boxed{5.3 \times 10^{-4} \text{ M}}$. If the concentration of I^- is set to this value, then $PbI_2(s)$ cannot precipitate, but almost all of the $Hg_2^{2+}(aq)$ will precipitate as $Hg_2I_2(s)$.

11-57 Metal ions form sulfides of greatly varying but generally low solubility. Careful control of the pH, which strongly affects these solubilities, allows the separation of the metal ions by differential precipitation of the sulfides. The dissolution of zinc sulfide is represented:

$$ZnS(s) + H_2O(l) \rightleftharpoons Zn^{2+}(aq) + HS^-(aq) + OH^-(aq) \qquad [Zn^{2+}][HS^-][OH^-] = K$$

The equilibrium expression is a triple product, but is otherwise not exceptional. Both the OH^- and HS^- concentrations depend strongly on the concentration of H_3O^+ according to the equations:

$$2\,H_2O(l) \rightleftharpoons H_3O^+(aq) + OH^-(aq) \qquad K_w = [H_3O^+][OH^-]$$

$$H_2S(aq) + H_2O(aq) \rightleftharpoons H_3O^+(aq) + HS^-(aq) \qquad K_{a1} = \frac{[HS^-][H_3O^+]}{[H_2S]}$$

Solve the equations for $[OH^-]$ and $[HS^-]$, and substitute into the triple-product K expression:

$$[Zn^{2+}] \left(\frac{K_{a1}[H_2S]}{[H_3O^+]} \right) \left(\frac{K_w}{[H_3O^+]} \right) = K$$

The K_{a1} of H_2S equals 9.1×10^{-8}; K_w equals 1.0×10^{-14}; the concentration of H_2S equals 0.10 M; the concentration of H_3O^+ equals 1.0×10^{-5} M; and K is 2×10^{-25}.[14] Solve for the concentration of Zn^{2+}, and insert the numbers:

$$[Zn^{2+}] = K \left(\frac{[H_3O^+]}{K_{a1}[H_2S]} \right) \left(\frac{[H_3O^+]}{K_w} \right) = 2 \times 10^{-25} \left(\frac{1.0 \times 10^{-5}}{(9.1 \times 10^{-8})(0.10)} \right) \left(\frac{1.0 \times 10^{-5}}{1.0 \times 10^{-14}} \right)$$

$$= \boxed{2 \times 10^{-13} \text{ M}}$$

11-59 Precipitation of $FeS(s)$ can begin only if the pH is high enough to make the reaction quotient Q for the reaction:

$$FeS(s) + H_2O(l) \rightleftharpoons Fe^{2+}(aq) + HS^-(aq) + OH^-(aq) \qquad [Fe^{2+}][HS^-][OH^-] = 5 \times 10^{-19}$$

exceed the K.[15] Compute the $[H_3O^+]$ that barely causes $FeS(s)$ to precipitate. This concentration is reached when the following equation (obtained as in **11-57**) is satisfied:

$$[Fe^{2+}] \left(\frac{K_{a1}[H_2S]}{[H_3O^+]} \right) \left(\frac{K_w}{[H_3O^+]} \right) = 5 \times 10^{-19}$$

---

[14]See text Table 11-4, text page 383.

[15]The K's in this problem come from text Table 11-4.

Solve for $[H_3O^+]$ and substitute the various numbers. The K_{a1} of H_2S equals 9.1×10^{-8}; K_w equals 1.0×10^{-14}; the concentration of H_2S is 0.10 M; the concentration of Fe^{2+} is 0.10 M:

$$[H_3O^+] = \sqrt{\frac{[Fe^{2+}]K_{a1}[H_2S]K_w}{5 \times 10^{-19}}}$$

$$= \sqrt{\frac{(0.10)(9.1 \times 10^{-8})(0.10)(1.0 \times 10^{-14})}{5 \times 10^{-19}}} = 4.3 \times 10^{-3} \text{ M}$$

This is the minimum concentration of H_3O^+ that keeps FeS in solution. The maximum pH is therefore $\boxed{2.4}$. Higher pH, (implying lower $[H_3O^+]$ and higher $[OH^-]$) shifts the dissolution reaction to the left, causing a precipitate.

The dissolution of $PbS(s)$ is similar to that of $FeS(s)$:

$$PbS(s) + H_2O(l) \rightleftharpoons Pb^{2+}(aq) + HS^-(aq) + OH^-(aq) \quad [Pb^{2+}][HS^-][OH^-] = 3 \times 10^{-28}$$

and the expression:

$$[Pb^{2+}]\left(\frac{K_{a1}[H_2S]}{[H_3O^+]}\right)\left(\frac{K_w}{[H_3O^+]}\right) = 3 \times 10^{-28}$$

is therefore easily written. Substitution of 0.10 for $[H_2S]$, 4.3×10^{-3} for $[H_3O^+]$, and 9.1×10^{-8} for K_{a1} gives $[Pb^{2+}]$ equal to $\boxed{6 \times 10^{-11} \text{ M}}$, which is very small.

Tip. The point is that at pH 2.4, all the Fe^{2+} but (essentially) none of the Pb^{2+} stays in solution.

11-61 The copper(II) nitrate dissolves readily to give $Cu^{2+}(aq)$ and NO_3^- ions. Imagine that the $Cu^{2+}(aq)$ reacts to completion with the $NH_3(aq)$ (which is in excess) to form $Cu(NH_3)_4^{2+}(aq)$:

$$Cu^{2+}(aq) + 4\,NH_3(aq) \rightarrow Cu(NH_3)_4^{2+}(aq)$$

Then:

$$[Cu(NH_3)_4^{2+}] = 0.10 \text{ M} \quad \text{and} \quad [NH_3] = 1.50 - 4(0.10) = 1.10 \text{ M}$$

Now, imagine free Cu^{2+} ion to come from back-reaction (dissociation of $Cu(NH_3)_4^{2+}$). The back-reaction, or dissociation, would proceed in four steps:

$$Cu(NH_3)_4^{2+}(aq) \rightleftharpoons Cu(NH_3)_3^{2+}(aq) + NH_3(aq) \qquad \frac{[Cu(NH_3)_3^{2+}][NH_3]}{[Cu(NH_3)_4^{2+}]} = 1.1 \times 10^{-2}$$

$$Cu(NH_3)_3^{2+}(aq) \rightleftharpoons Cu(NH_3)_2^{2+}(aq) + NH_3(aq) \qquad \frac{[Cu(NH_3)_2^{2+}][NH_3]}{[Cu(NH_3)_3^{2+}]} = 2 \times 10^{-3}$$

$$Cu(NH_3)_2^{2+}(aq) \rightleftharpoons Cu(NH_3)^{2+}(aq) + NH_3(aq) \qquad \frac{[Cu(NH_3)^{2+}][NH_3]}{[Cu(NH_3)_2^{2+}]} = 5 \times 10^{-4}$$

$$Cu(NH_3)^{2+}(aq) \rightleftharpoons Cu^{2+}(aq) + NH_3(aq) \qquad \frac{[Cu^{2+}][NH_3]}{[Cu(NH_3)^{2+}]} = 1.0 \times 10^{-4}$$

where the four constants equal the reciprocals of K_4 through K_1 in text Table 11-5.[16] Label the concentrations of the four Cu-containing products x, y, z, and w respectively, and calculate them in turn. Treat the steps as if they occurred independently. That is, neglect the amount of Cu-containing product reacted away by later steps and assume that NH_3 from later steps adds only negligibly to the 1.10 M NH_3 present when dissociation starts. For the first step:

	$Cu(NH_3)_4^{2+}(aq) \rightleftharpoons$	$Cu(NH_3)_3^{2+}(aq)$	$+ NH_3(aq)$
Init. Conc. (mol L^{-1})	0.10	0	1.1
Change in Conc. (mol L^{-1})	$-x$	$+x$	$+x$
Equil. Conc. (mol L^{-1})	$0.10 - x$	x	$1.1 + x$

$$\frac{[Cu(NH_3)_3^{2+}][NH_3]}{[Cu(NH_3)_4^{2+}]} = 1.1 \times 10^{-2} = \frac{x(1.1 + x)}{0.10 - x} \quad \text{which gives} \quad x = 9.9 \times 10^{-4}$$

The set-up for the second step is similar. It gives

$$\frac{[Cu(NH_3)_2^{2+}][NH_3]}{[Cu(NH_3)_3^{2+}]} = 2 \times 10^{-3} = \frac{y(1.1 + y)}{9.9 \times 10^{-4} - y} \quad \text{from which:} \quad y = 1.8 \times 10^{-6}$$

For the third step:

$$\frac{[Cu(NH_3)^{2+}][NH_3]}{[Cu(NH_3)_2^{2+}]} = 5 \times 10^{-4} = \frac{z(1.1 + z)}{1.8 \times 10^{-6} - z} \quad \text{which gives} \quad z = 8.2 \times 10^{-10}$$

For the fourth step:

---

[16]See text page 385. The product of K_1 through K_4 equals 0.9×10^{12}, but K_f equals 1.1×10^{12} in the table. This inconsistency arises from rounding off the step-wise K's and is unimportant.

$$\frac{[\text{Cu}^{2+}][\text{NH}_3]}{[\text{Cu(NH}_3)^{2+}]} = 1.0 \times 10^{-4} = \frac{w(1.1 + w)}{8.2 \times 10^{-10} - w} \quad \text{which gives} \quad w = 7.4 \times 10^{-14}$$

This w equals the concentration of free Cu^{2+}. The concentrations of the partially dissociated complexes are low. Indeed, all dissociation reduces the concentration of $\text{Cu(NH}_3)_4^{2+}$ by less than 1%. Therefore:

$$[\text{Cu}^{2+}] = \boxed{7 \times 10^{-14} \text{ mol L}^{-1}} \qquad\qquad [\text{Cu(NH}_3)_4^{2+}] = \boxed{0.10 \text{ mol L}^{-1}}$$

11-63 Assume that the reaction between $\text{K}^+(aq)$ and "crown" goes 100% to completion and that free K^+ then comes from back-dissociation of $(\text{Kcrown})^+(aq)$. The equation for this dissociation is the reverse of the equation in the problem:

	$(\text{Kcrown})^+(aq) \rightleftharpoons$	$\text{K}^+(aq) +$	$\text{crown}(aq)$
Init. Conc. (mol L^{-1})	0.0080	0	0
Change in Conc. (mol L^{-1})	$-x$	$+x$	$+x$
Equil. Conc. (mol L^{-1})	$0.0080 - x$	x	x

The reaction has the mass-action expression and K:

$$\frac{[\text{K}^+][\text{crown}]}{[(\text{Kcrown})^+]} = \frac{1}{111.6} \qquad \text{hence:} \qquad \frac{1}{111.6} = \frac{x^2}{0.080 - x}$$

Rearranging gives the quadratic equation:

$$x^2 + (8.961 \times 10^{-3})x - 7.169 \times 10^{-5} = 0$$

Solving for x gives 0.0051; the equilibrium concentration of $\text{K}^+(aq)$ is $\boxed{0.0051 \text{ M}}$. Calculation of the concentration of free Na^+ proceeds similarly. The analogous quadratic equation is:

$$y^2 + 0.152y - 0.00122 = 0$$

(where y is the concentration of free Na^+ ion) based on a K of 1/6.6 instead of 1/111.6. Solving the new equation gives an Na^+ ion concentration of $\boxed{0.0076 \text{ M}}$.

11-65 The problem really asks for a calculation of the solubility of AgCl in a solution that is 1.00 M in Cl^- ion. Let S equal this solubility. For every mole per liter of AgCl that dissolves, either an Ag^+ ion or an AgCl_2^- ion[17] forms. In mathematical form:

$$S = [\text{Ag}^+] + [\text{AgCl}_2^-]$$

---

[17] Dichloroargenate(I) ion.

Obviously, this assumes that all of the dissolved silver is present as one of two ions. The solubility equilibrium for AgCl assures that:

$$K_{sp} = 1.6 \times 10^{-10} = [Ag^+][Cl^-]$$

as long as solid silver chloride is present. The formation equilibrium for the $AgCl_2^-$ complex ion means:

$$\frac{[AgCl_2^-]}{[Ag^+][Cl^-]^2} = K_f = 1.8 \times 10^5$$

Solve these equations for $[Ag^+]$ and $[AgCl_2^-]$ and substitute the results into the expression for S:

$$S = \frac{1.6 \times 10^{-10}}{[Cl^-]} + (1.8 \times 10^5)(1.6 \times 10^{-10})[Cl^-]$$

Assume that the concentration of Cl^- ion is so large at 1.00 M that it is not substantially reduced by reaction with Ag^+ ion. If $[Cl^-] = 1.00$ M, then:

$$S = \frac{1.6 \times 10^{-10}}{1.00} + (1.8 \times 10^5)(1.6 \times 10^{-10})[1.00] = 2.8 \times 10^{-5} \text{ M}$$

The assumption that $[Cl^-]$ is only negligibly reduced from its original value is vindicated by this low solubility—only about 6×10^{-5} M of Cl^- ion is tied up in the complex. The solubility of AgCl in this solution is more than double the solubility of AgCl in pure water, which equals 1.3×10^{-5} M (by a computation like the one in **11-25**). Hence, AgCl(s) is $\boxed{\text{more soluble}}$ in 1.00 M NaCl than in pure water.

In 0.100 M NaCl, in which $[Cl^-]$ is 0.100 M, the equation for S becomes:

$$S = \frac{1.6 \times 10^{-10}}{0.100} + (1.8 \times 10^5)(1.6 \times 10^{-10})[0.100] = 2.9 \times 10^{-6} \text{ M}$$

AgCl is $\boxed{\text{less soluble}}$ in dilute NaCl than in pure water.

Tip. The reversal is remarkable. Complexation plays a potent role in determining solubilities.

11-67 $CuSO_4$ dissolves to give $Cu^{2+}(aq)$ and $SO_4^{2-}(aq)$. The aquated Cu^{2+} ion exists as the $Cu(H_2O)_4^{2+}$ complex ion. This ion acts as a Brønsted-Lowry acid according to the equation:

$$\boxed{Cu(H_2O)_4^{2+}(aq) + H_2O(l) \rightleftharpoons H_3O^+(aq) + Cu(H_2O)_3OH^+(aq)}$$

The solution is $\boxed{\text{acidic}}$ because the K_a for the complex exceeds K_b for the SO_4^{2-} ion. An equivalent answer is the equation:

$$Cu^{2+}(aq) + 2\,H_2O(l) \rightleftharpoons H_3O^+(aq) + (CuOH)^+(aq)$$

11-69 The computation is like other computations of the pH of solutions of weak acids.[18] The coordinated cobalt(II) ion is acidic:

	$Co(H_2O)_6^{2+}$	$+H_2O \rightleftharpoons$	$H_3O^+(aq) +$	$Co(H_2O)_5OH^+$
Init. Conc. (mol L^{-1})	0.10		small	0
Change in Conc. (mol L^{-1})	$-x$		$+x$	$+x$
Equil. Conc. (mol L^{-1})	$0.10 - x$		x	x

The equilibrium expression is:

$$\frac{[H_3O^+][Co(H_2O)_5OH^+]}{[Co(H_2O)_6^{2+}]} = K_a = 3 \times 10^{-10} = \frac{x^2}{0.10 - x}$$

Solve the equation to obtain $x = 5.5 \times 10^{-6}$. The concentration of H_3O^+ is 5.5×10^{-6} M, and the pH is therefore $\boxed{5.3}$.

11-71 The reaction: $Pt(NH_3)_4^{2+}(aq) + H_2O(l) \rightleftharpoons H_3O^+(aq) + Pt(NH_3)_3NH_2^+(aq)$ causes the acidity of the solution. The mass-action expression for this reaction is:

$$\frac{[Pt(NH_3)_3NH_2^+][H_3O^+]}{[Pt(NH_3)_4^{2+}]} = K_a$$

The concentration of the $Pt(NH_3)_3NH_2^+$ ion equals the concentration of the H_3O^+ ion in solution as long as this reaction is the only source of either ion. The concentration of the $Pt(NH_3)_4^{2+}$ ion equals 0.15 M minus the concentration of H_3O^+ ion. The H_3O^+ concentration can be calculated from the pH given in the problem:

$$[H_3O^+] = 10^{-4.92} = 1.20 \times 10^{-5} \text{ M}$$

Substitute this and the other concentrations in the K_a-expression:

$$K_a = \frac{(1.20 \times 10^{-5})(1.20 \times 10^{-5})}{0.15 - 1.20 \times 10^{-5}} = \boxed{9.6 \times 10^{-10}}$$

---

[18]Such as problem 10-29a.

11-73 The problem concerns the fate of $Pb^{2+}(aq)$ in a solution adjusted to pH 13.0 by the addition of NaOH. The precipitation reaction:

$$Pb^{2+}(aq) + 2\,OH^-(aq) \rightleftharpoons Pb(OH)_2(s)$$

tends to reduce the concentration of $Pb^{2+}(aq)$. The K for this reaction is the reciprocal of the K_{sp} of $Pb(OH)_2(s)$. It is quite large (K_{sp} is 4.2×10^{-15} so $1/K_{sp}$ is 2.38×10^{14}). This apparently means that 1.00 M Pb^{2+} ion gives a precipitate of $Pb(OH)_2(s)$ at pH 13.0. But there is a complication. The equilibrium:

$$Pb^{2+}(aq) + 3\,OH^-(aq) \rightleftharpoons Pb(OH)_3^-(aq)$$

ties up Pb^{2+} ion in a *soluble* form, thereby opposing precipitation of $Pb(OH)_2(s)$. The K for this reaction is K_f for $Pb(OH)_3^-(aq)$ and is large (4×10^{14}). Whether $Pb(OH)_2$ precipitates depends how the competition between these reactions plays out.

Suppose $Pb(OH)_2(s)$ *does* precipitate. Then, at pH 13.0, where $[OH^-] = 0.10$ M, the concentration of Pb^{2+} must fulfill the equation:

$$K_{sp} = 4.2 \times 10^{-15} = [Pb^{2+}][OH^-]^2 = [Pb^{2+}](0.10)^2$$

This means $[Pb^{2+}]$ is locked at 4.2×10^{-13} M if solid $Pb(OH)_2$ is present. The mass-action expression for the complexation equilibrium is:

$$K_f = 4 \times 10^{14} = \frac{[Pb(OH)_3^-]}{[Pb^{2+}][OH^-]^3}$$

Substitute $[OH^-] = 0.10$ and $[Pb^{2+}] = 4.2 \times 10^{-13}$ M, and solve for $[Pb(OH)_3^-]$. The answer is 0.17 M. Any concentration of Pb^{2+} ion that exceeds this threshold value causes precipitation. Because 1.00 M exceeds 0.17 M, $Pb(OH)_2(s)$ precipitates in the case defined in this problem. At equilibrium, $[Pb^{2+}]$ equals $\boxed{4.2 \times 10^{-13} \text{ M}}$, and the concentration of $Pb(OH)_3^-$ equals $\boxed{0.17 \text{ M}}$.

An initial concentration of 0.050 M Pb^{2+}, is less than the precipitation threshold of 0.17 M. There is no precipitate of $Pb(OH)_2(s)$. The K_{sp} equilibrium is *not* in effect. Essentially all of the Pb^{2+} ion is tied up in the complex, making the concentration of $Pb(OH)_3^-$ equal $\boxed{0.050 \text{ M}}$. Put this value into the K_f mass-action expression:

$$K_f = 4 \times 10^{14} = \frac{[Pb(OH)_3^-]}{[Pb^{2+}][OH^-]^3} = \frac{0.050}{[Pb^{2+}](0.10)^3}$$

Solving gives the concentration of free $Pb^{2+}(aq)$ as $\boxed{1 \times 10^{-13} \text{ M}}$.

11-75 The first reaction and its mass-action expression are:

$$Cl_2(g) \rightleftharpoons Cl_2(aq) \qquad \frac{[Cl_2]}{P_{Cl_2}} = K$$

Note that the solvent does not appear in the mass-action expression. Both the equilibrium concentration of the dissolved chlorine and the partial pressure of the gaseous chlorine are given in the problem. Hence:

$$K_1 = \frac{[Cl_2]}{P_{Cl_2}} = \frac{0.061}{1.00} = 0.061$$

The second reaction and its mass-action expression are:

$$Cl_2(aq) + H_2O(l) \rightleftharpoons H^+(aq) + Cl^-(aq) + HOCl(aq) \qquad \frac{[HOCl][H^+][Cl^-]}{[Cl_2]} = K_2$$

K_2 is readily calculated because $[H^+]$ must be equal to $[Cl^-]$ and to $[HOCl]$, based on the 1-to-1-to-1 stoichiometry of the reaction:

$$K_2 = \frac{[HOCl][H^+][Cl^-]}{[Cl_2]} = \frac{(0.030)(0.030)(0.030)}{(0.061)} = \boxed{4.4 \times 10^{-4}}$$

11-77 a) The decomposition reaction is $NH_4HS(s) \rightleftharpoons NH_3(g) + H_2S(g)$. One mole of gaseous ammonia forms per mole of gaseous hydrogen sulfide. Assuming that the product mixture follows Dalton's law, then the equilibrium partial pressure of each gas equals half the total pressure at equilibrium:

$$P_{NH_3} = P_{H_2S} = \frac{0.659 \text{ atm}}{2}$$

Insert these values into the mass-action expression and compute K:

$$K = P_{NH_3} P_{H_2S} = \left(\frac{0.659}{2}\right)^2 = \boxed{0.109}$$

b) Bringing the equilibrium partial pressure of $NH_3(g)$ to 0.750 atm will, by LeChatelier's principle, reduce the partial pressure of $H_2S(g)$. It will not change the equilibrium constant. Hence:

$$P_{H_2S} = \frac{K}{P_{NH_3}} = \frac{0.109}{0.750} = \boxed{0.145 \text{ atm}}$$

Tip. In the first part, the 2 is used twice: once as a divisor and once as an exponent. This is all right.

11-79 The K for this reduction of nickel oxide to nickel relates the partial pressures of $CO(g)$ and $CO_2(g)$. The nickel and nickel(II) oxide are pure solids and do not enter the mass-action expression:

$$\frac{P_{CO_2}}{P_{CO}} = K = 255.4$$

The total pressure equals the sum of the two partial pressures:

$$P_{CO_2} + P_{CO} = 2.50 \text{ atm}$$

Once the system is at equilibrium, the two equations relating the two partial pressures must be satisfied simultaneously. Therefore:

$$\frac{P_{CO_2}}{(2.50 - P_{CO_2})} = 255.4$$

Solving gives P_{CO_2} equal to $\boxed{2.49 \text{ atm}}$. Substitution back into the mass-action equation gives P_{CO} equal to $\boxed{9.75 \times 10^{-3} \text{ atm}}$.

11-81 Equal chemical amounts of PCB-2 and PCB-11 exist in solution in some volume V of water. Then, the same volume V of octanol is added. The treatment with octanol does nothing to alter the total chemical amount of either PCB but redistributes both. Once the redistribution comes to equilibrium:

$$n_{\text{PCB-2}} = [2]_{(aq)}\, V + [2]_{(oct)}\, V \quad \text{and} \quad n_{\text{PCB-11}} = [11]_{(aq)}\, V + [11]_{(oct)}\, V$$

where the bracketed numbers stand for the concentrations of the PCB-2 and PCB-11 and the "(aq)" and "(oct)" refer to water and octanol solutions. Setting the two chemical amounts equal to each other and dividing through by V gives:

$$[2]_{(aq)} + [2]_{(oct)} = [11]_{(aq)} + [11]_{(oct)}$$

The mass-action expressions for the partition of the PCB's between the solvents are:

$$K_2 = \frac{[2]_{(oct)}}{[2]_{(aq)}} \quad \text{and} \quad K_{11} = \frac{[11]_{(oct)}}{[11]_{(aq)}}$$

from which it follows that:

$$[2]_{(oct)} = K_2[2]_{(aq)} \quad \text{and} \quad [11]_{(oct)} = K_{11}[11]_{(aq)}$$

Substitution of these expressions into the first equation yields:

$$[2]_{(aq)} + K_2[2]_{(aq)} = [11]_{(aq)} + K_{11}[11]_{(aq)}$$

Solve this equation for the ratio of the concentration of the two PCB's in the water phase:

$$\frac{[2]_{(aq)}}{[11]_{(aq)}} = \frac{1 + K_{11}}{1 + K_2} = \frac{1 + 1.26 \times 10^5}{1 + 3.98 \times 10^4} = \boxed{3.17}$$

Tip. Try finding the ratio of the amounts of the two PCB's in the octanol. To start, divide the equation for $[2]_{(oct)}$ in terms of K_2 by the similar expression for $[11]_{(oct)}$ in terms of K_{11}:

$$\frac{[2]_{(oct)}}{[11]_{(oct)}} = \frac{K_2 [2]_{(aq)}}{K_{11} [11]_{(aq)}} = \frac{K_2}{K_{11}} \left(\frac{[2]_{(aq)}}{[11]_{(aq)}} \right)$$

The quantity in parentheses was just determined. Substitute for it and insert the K's:

$$\frac{[2]_{(oct)}}{[11]_{(oct)}} = \frac{K_2}{K_{11}} \left(\frac{1 + K_{11}}{1 + K_2} \right) = \frac{3.98 \times 10^4}{1.26 \times 10^5} \left(\frac{1 + 1.26 \times 10^5}{1 + 3.98 \times 10^4} \right) = 1.00$$

Although the amount of PCB-2 entering the water is triple the amount of PCB-11, so little of either leaves the octanol that the relative amounts of the two in the octanol remain essentially unchanged.

11-83 a) The chemical amount of I_2 initially in the 0.100 L of aqueous solution is:

$$0.100 \text{ L} \times \left(\frac{2 \times 10^{-3} \text{ mol } I_2}{1 \text{ L}} \right) = 2 \times 10^{-4} \text{ mol } I_2$$

Shaking the solution with 0.025 L of CCl_4 allows the I_2 to distribute itself between the two phases. Assume that this distribution comes to equilibrium. If y is the amount of I_2 that dissolves into the CCl_4, then the amount of I_2 left in the aqueous phase is $(2 \times 10^{-4} - y)$ mol. At equilibrium:

$$[I_2]_{(aq)} = \left(\frac{2 \times 10^{-4} - y}{0.100} \right) \text{ mol L}^{-1} \quad \text{and} \quad [I_2]_{(CCl_4)} = \left(\frac{y}{0.025} \right) \text{ mol L}^{-1}$$

The mass-action expression for this system is:

$$\frac{[I_2]_{(CCl_4)}}{[I_2]_{(aq)}} = K = 85 \qquad \text{from which:} \qquad \frac{(2 \times 10^{-4} - y)/0.100}{y/0.025} = 85$$

The last equation is easily solved for y, which equals 1.91×10^{-4} mol. Remember that y is the amount of I_2 that transfers to the CCl_4. and not a concentration. By subtraction, the amount remaining in the water is 0.09×10^{-4} mol. The fraction remaining equals the amount remaining divided by the original amount:

$$f = \frac{0.09 \times 10^{-4}}{2 \times 10^{-4}} = 0.045 = \boxed{0.04}$$

b) The first extraction with 0.025 L of CCl_4 leaves only 0.045 (4.5%) of the I_2 in the water. Another extraction with a fresh 0.025 L of CCl_4 will leave only 0.045 of that 0.045. The fraction remaining after these successive treatments is:

$$f = 0.045 \times 0.045 = \boxed{0.002}$$

c) From text Example 11-3,[19] the fraction of I_2 remaining in the water after a single 0.050 L extraction is 0.023, which is substantially larger than 0.0020.

Tip. It is about 11 times more efficient to extract the iodine with two half-sized portions of CCl_4 rather than one large portion. In general, it is more efficient to use several small portions of solvent, rather than one or two big ones in performing separations by extraction.

11-85 a) The numbers to plot are the second-to-last and last rows of the following table. Put $\ln K$ on the vertical axis and $1/T$ on the horizontal axis.

Temp. (K)	276.8	288.3	298	308	323.2
Equil. Constant K	1160	841	689	533	409
$1/T$ (K^{-1})	0.00361	0.00347	0.00336	0.00325	0.00309
$\ln K$	7.06	6.73	6.54	6.28	6.01

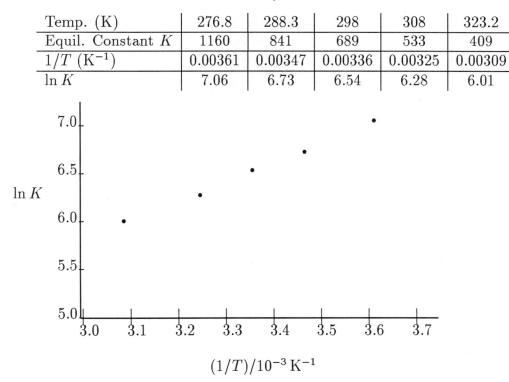

$(1/T)/10^{-3}\,\mathrm{K}^{-1}$

b) A form of the van't Hoff equation is:

$$\ln K = \frac{-\Delta H^\circ}{R}\left(\frac{1}{T}\right) + \frac{\Delta S^\circ}{R}$$

---

[19]Text page 365.

Comparing this equation to the equation of a straight line ($y = mx + b$) reveals that the slope m of the straight line that results from plotting $\ln K$ versus $1/T$ equals $-\Delta H°/R$. The slope m of the best straight line in the preceding graph comes out to 2020.3 K. Therefore:

$$\Delta H° = -mR = -(2020.3 \text{ K})(8.315 \text{ J K}^{-1}\text{mol}^{-1}) = \boxed{-16.8 \times 10^3 \text{ J mol}^{-1}}$$

11-87 The mercury(I) ion exists as Hg_2^{2+} in aqueous solution, according to the Table 11-2 and **11-29**. The dissolution equilibrium is:

$$Hg_2Cl_2(s) \rightleftharpoons Hg_2^{2+}(aq) + 2\,Cl^-(aq) \qquad K_{sp} = [Hg_2^{2+}][Cl^-]^2$$

11-89 The dissolution of barium sulfate is represented:

$$BaSO_4(s) \rightleftharpoons Ba^{2+}(aq) + SO_4^{2-}(aq) \qquad [Ba^{2+}][SO_4^{2-}] = K_{sp} = 1.1 \times 10^{-10}$$

If the equilibrium concentration of Ba^{2+} equals x, then the concentration of SO_4^{2-} is also x, as long as there are no additional sources (or sinks) of either ion. Then $x^2 = 1.1 \times 10^{-10}$ so that $x = [Ba^{2+}] = \boxed{1.1 \times 10^{-5} \text{ M}}$. This concentration is too low for any bad effects on patients drinking the suspension of $BaSO_4(s)$.

11-91 It takes 860 mL of 0.0050 M NaF to make 1.00 L of solution when mixed with 140 mL of 0.0010 M $Sr(NO_3)_2$. Just after mixing, but before any precipitation reactions occur, the concentrations of Sr^{2+} and F^- ions are:

$$[Sr^{2+}] = \left(\frac{140}{1000}\right)0.0010 = 1.4 \times 10^{-4} \text{ M} \qquad [F^-] = \left(\frac{860}{1000}\right)0.0050 = 4.3 \times 10^{-4} \text{ M}$$

The mixture will tend to precipitate SrF_2 only if the initial reaction quotient Q_0 exceeds K_{sp} for the reaction:

$$SrF_2(s) \rightleftharpoons Sr^{2+}(aq) + 2\,F^-(aq)$$

The value of Q_0 is computed by substitution of the initial concentrations into the mass-action expression:

$$Q_0 = [Sr^{2+}]_0[F^-]_0^2 = (1.4 \times 10^{-4})(4.3 \times 10^{-4})^2 = 2.6 \times 10^{-9}$$

Because Q_0 is less than K_{sp} (which equals 2.8×10^{-9}), there is $\boxed{\text{no precipitate of } SrF_2}$ at equilibrium.

Tip. The volumes of differing aqueous solutions are not in general additive (that is, 160 mL of solution A mixed with 840 mL of solution B does *not* always give 1000 mL of solution). However, these solutions are so dilute that additivity can be assumed.

11-93 Two solubility equilibria are going on simultaneously:

$$AgBr(s) \rightleftharpoons Ag^+(aq) + Br^-(aq) \qquad K_{sp} = [Ag^+][Br^-] = 7.7 \times 10^{-13}$$
$$CuBr(s) \rightleftharpoons Cu^+(aq) + Br^-(aq) \qquad K_{sp} = [Cu^+][Br^-] = 4.2 \times 10^{-8}$$

Divide the K_{sp} expression for the second by the K_{sp} for the first:

$$\frac{[Cu^+][Br^-]}{[Ag^+][Br^-]} = \frac{4.2 \times 10^{-8}}{7.7 \times 10^{-13}} = 5.45 \times 10^4$$

A single solution at equilibrium can have only one concentration of Br^- ion (or any other ion). Hence the $[Br^-]$ in the numerator equals the $[Br^-]$ in the denominator. Cancellation then gives:

$$\frac{[Cu^+]}{[Ag^+]} = \boxed{5.5 \times 10^4}$$

11-95 Only when Q exceeds K_{sp} can precipitation begin. Calculate the concentration of Ag^+ that will just suffice to precipitate each salt by setting $Q = K_{sp}$. The equilibria and K_{sp}-expressions are:

$$AgCl(s) \rightleftharpoons Ag^+(aq) + Cl^-(aq) \qquad\qquad [Ag^+][Cl^-] = 1.6 \times 10^{-10}$$
$$Ag_2CrO_4(s) \rightleftharpoons 2\,Ag^+(aq) + CrO_4^{2-}(aq) \qquad [Ag^+]^2[CrO_4^{2-}] = 1.9 \times 10^{-12}$$

Inserting the given concentrations of Cl^- and CrO_4^{2-} ions and solving the concentration of Ag^+ ion reveals that precipitation of $AgCl(s)$ requires 1.6×10^{-9} M Ag^+ and that precipitation of $AgCrO_4(s)$ requires is 2.8×10^{-5} M Ag^+. The second is much larger. Thus, $\boxed{AgCl(s)}$ will precipitate first.

Solid Ag_2CrO_4 just starts to come down when $[Ag^+] = 2.8 \times 10^{-5}$ M. Use the K_{sp} expression to calculate $[Cl^-]$ when $[Ag^+] = 2.8 \times 10^{-5}$ M:

$$[Cl^-] = \frac{K_{sp}}{[Ag^+]} = \frac{1.6 \times 10^{-10}}{2.8 \times 10^{-5}} = 5.7 \times 10^{-6} \text{ M}$$

The fraction of Cl^- remaining is:

$$f_{Cl^-} = \frac{[Ag^+]}{[Ag^+]_0} = \frac{5.7 \times 10^{-6} \text{ M}}{0.100 \text{ M}} = \boxed{5.7 \times 10^{-5}}$$

11-97 Magnesia dissolves in water and raises the pH by generating $OH^-(aq)$ ion:

$$MgO(s) + H_2O(l) \rightleftharpoons Mg^{2+}(aq) + 2\,OH^-(aq) \quad K = [Mg^{2+}][OH^-]^2$$

The pH of the solution at 25°C is 10.16. At this temperature, the sum of the pH and pOH of an aqueous solution equals 14.00. This means that the pOH of the saturated solution of magnesia equals 3.84. By the definition of pOH: $[OH^-] = 10^{-3.84} = 1.44 \times 10^{-4}$ M. Assume that dissolution of magnesia is the predominant source of OH^- in the solution, far outpacing the autoionization of water. Then:

$$2\,[Mg^{2+}] = [OH^-]$$

Substitute to obtain the value of K:

$$K = [Mg^{2+}][OH^-]^2 = \frac{[OH^-]}{2}[OH^-]^2 = \frac{(1.44 \times 10^{-4})^3}{2} = 1.5 \times 10^{-12}$$

The solubility of the MgO equals the final concentration of $Mg^{2+}(aq)$:

$$[Mg^{2+}] = \frac{[OH^-]}{2} = \boxed{7.2 \times 10^{-5} \text{ mol L}^{-1}}$$

11-99 Silver ion is complexed strongly by ammonia:

$$Ag^+(aq) + 2\,NH_3(aq) \rightleftharpoons Ag(NH_3)_2^{2+}(aq) \qquad K_f = 1.7 \times 10^7$$

When $AgBr(s)$ is placed in aqueous ammonia, a new dissolution reaction:

$$AgBr(s) + 2\,NH_3(aq) \rightleftharpoons Ag(NH_3)_2^{2+}(aq) + Br^-(aq) \qquad K = K_f K_{sp}$$

replaces the "regular" dissolution reaction. The new dissolution reaction equals the sum of the regular K_{sp} reaction and the complexation reaction. Its K is 1.7×10^7 times larger than the K_{sp}, which explains the increase in solubility of AgBr when ammonia is present.

11-101 a) The aqueous HCl reacts with $CdS(s)$ to give both $CdCl_4^{2-}(aq)$ and $H_2S(aq)$:

$$\boxed{CdS(s) + 2\,H_3O^+(aq) + 4\,Cl^-(aq) \rightleftharpoons CdCl_4^{2-}(aq) + H_2S(aq) + 2\,H_2O(l)}$$

The $HCl(aq)$ acts on the $CdS(s)$ by removing both sulfide ion (as H_2S) and cadmium ion (as the tetrachloro complex). The problem states that some $CdS(s)$ remains, so the above equation is an accurate description of the final equilibrium.

b) The equilibrium in the preceding part can be constructed as the sum of four reactions. The first represents the dissolution of $CdS(s)$ in pure water; the second and third represent the acid-base reactions ions HS^- and OH^- (both are bases) with water; the fourth represents the complexation of Cd^{2+} by Cl^- ions:

$$CdS(s) + H_2O(l) \rightleftharpoons Cd^{2+}(aq) + OH^-(aq) + HS^-(aq) \qquad K_1$$
$$HS^-(aq) + H_2O(l) \rightleftharpoons H_2S(aq) + OH^-(aq) \qquad K_2 = K_w/K_{a1}$$
$$2\,OH^-(aq) + 2\,H_3O^+(aq) \rightleftharpoons 4\,H_2O(l) \qquad K_3 = (1/K_w)^2$$
$$Cd^{2+}(aq) + 4\,Cl^-(aq) \rightleftharpoons CdCl_4^{2-}(aq) \qquad K_4 = K_f$$

where K_{a1} is the K_a for the first stage of the acid ionization of $H_2S(aq)$. The equilibrium constants of the four reactions are numbered for identification. The desired constant is the product of the four constants because the equation of interest is the sum of the four equations:

$$K = K_1 K_2 K_3 K_4 = K_1 \left(\frac{K_w}{K_{a1}}\right)\left(\frac{1}{K_w}\right)^2 (K_f)$$

The text[20] lists K_1 as 7×10^{-28} and K_{a1} as 9.1×10^{-8}. The problem gives K_f as 800. Insert these numbers and the well-known value of K_w into the preceding:

$$K = (7 \times 10^{-28})\left(\frac{1.0 \times 10^{-14}}{9.1 \times 10^{-8}}\right)\left(\frac{1}{1.0 \times 10^{-14}}\right)^2 (8 \times 10^2) = 6.2 \times 10^{-4}$$

Thus, at equilibrium:

$$\frac{[CdCl_4^{2-}][H_2S]}{[H_3O^+]^2[Cl^-]^4} = K = \boxed{6 \times 10^{-4}}$$

c) Let S equal the molar solubility of the $CdS(s)$ in 6 M HCl. Assume that at equilibrium the cadmium in solution is all in the form of $CdCl_4^{2-}(aq)$, and that the sulfur in solution is all in the form of $H_2S(aq)$. Then:

$$S = [CdCl_4^{2-}] = [H_2S]$$

Now, every Cd^{2+} ion that goes into solution consumes 4 Cl^- ions, and every S^{2-} ion that goes into solution consumes 2 H_3O^+ ions. This means:

$$[H_3O^+] = 6 - 2S \qquad \text{and} \qquad [Cl^-] = 6 - 4S$$

The 6 comes from the original concentration of HCl, which was 6 M. Substitution in the mass-action expression derived in the preceding parts gives:

$$\frac{[CdCl_4^{2-}][H_2S]}{[H_3O^+]^2[Cl^-]^4} = \frac{S^2}{(6 - 2S)^2(6 - 4S)^4} = 6.2 \times 10^{-4}$$

---

[20]In Table 11-4, text page 383, and in Example 11-10 on the same page.

Assuming that S is negligible compared to 6 M in this equation does not lead to an acceptable solution, as it is easy to confirm. However, S must lie between 0 and 6 mol L^{-1}. Guess that S equals 1.0 mol L^{-1}. Then the left side equals 3.9×10^{-3}, which exceeds 6.2×10^{-4}. Guess a smaller S, such as 0.6 mol L^{-1}. The left side then equals 9.3×10^{-5}, which is less than 6.2×10^{-4}. Further adjustments in the range between 0.6 and 1.0 give improved fits. The value $S = 0.8$ fits the equation fairly well. The solubility of CdS(s) in 6 M HCl is thus apparently 0.8 mol L^{-1}.

Now check assumptions. It was assumed that all of the sulfur in solution was in the form of H$_2$S(aq). Therefore, the concentration of H$_2$S(aq) should equal 0.8 M. But this exceeds the solubility of H$_2$S at room temperature, which is only 0.1 M.[21] Adding 6 M HCl to CdS(s) therefore forces H$_2$S out of solution. Gaseous H$_2$S bubbles out until the concentration of H$_2$S(aq) falls to 0.1 M. The previous equation is replaced by:

$$\frac{S\,(0.1)}{(6 - 2S)^2(6 - 4S)^4} = 6.2 \times 10^{-4}$$

Solving this new equation (by approximation), gives S, the molar solubility of CdS(s), equal to $\boxed{1 \text{ mol L}^{-1}}$.

11-103 The solution is prepared by mixing 0.020 mol of CuCl$_2$ and 0.100 mol of NaCN in 1.0 L of water. The original concentration of Cu^{2+} ion is therefore 0.020 M, and the original concentration of CN$^-$ is 0.100 M. These concentrations do not last long. The Cu^{2+}(aq) ion and CN$^-$(aq) soon combine to form a complex ion:

$$\text{Cu}^{2+}(aq) + 4\,\text{CN}^-(aq) \rightleftharpoons \text{Cu(CN)}_4^{2-}(aq) \qquad \frac{[\text{Cu(CN)}_4^{2-}]}{[\text{Cu}^{2+}][\text{CN}^-]^4} = K_f = 2 \times 10^{30}$$

The very large K_f means that nearly all of the Cu^{2+} ion is tied up as complex. At equilibrium then, $[\text{Cu(CN)}_4^{2-}] = 0.020$ M. Complexation reduces the concentration of CN$^-$ ion, which is in excess, from its original 0.100 M to $0.100 - 4(0.020) = 0.020$ M because 1 mol of Cu^{2+} accounts for 4 mol of CN$^-$(aq). The values:

$$[\text{CN}^-] = 0.020 \text{ M} \quad \text{and} \quad [\text{Cu(CN)}_4^{2-}] = 0.020 \text{ M}$$

might now be substituted into the K_f mass-action expression and used to compute an equilibrium concentration of Cu^{2+} ion. However, CN$^-$(aq) also hydrolyzes (reacts with H$_2$O) to give HCN(aq):

$$\text{CN}^-(aq) + \text{H}_2\text{O}(l) \rightleftharpoons \text{HCN}(aq) + \text{OH}^-(aq) \qquad \frac{[\text{HCN}][\text{OH}^-]}{[\text{CN}^-]} = \frac{K_w}{K_a} = 2.03 \times 10^{-5}$$

---

[21]Aqueous H$_2$S is saturated at a concentration of 0.1 M. See text Example 11-10, text page 383.

This reaction lowers the concentration of $CN^-(aq)$ from 0.020 M. Let x equal the concentration of CN^- that reacts in this way. Then:

$$\frac{[HCN][OH^-]}{[CN^-]} = 2.03 \times 10^{-5} = \frac{x^2}{0.020 - x}$$

Solving gives x equal to 6.27×10^{-4} M. The correct equilibrium concentration of CN^- is therefore $0.020 - 6.27 \times 10^{-4} = 0.0194$ M. Put this value into the K_f expression and solve for $[Cu^{2+}]$:

$$\frac{[Cu(CN)_4^{2-}]}{[Cu^{2+}][CN^-]^4} = 2 \times 10^{30} = \frac{(0.020)}{[Cu^{2+}](0.0194)^4} \qquad [Cu^{2+}] = \boxed{7 \times 10^{-26} \text{ M}}$$

Tip. If the hydrolysis of CN^- ion is ignored, the answer is 6×10^{-26} M—barely a significant difference.

11-105 a) Text Example 11-13[22] gives K_{a1} for $Fe(H_2O)_6^{3+}$ as 7.7×10^{-3}. The problem gives K_{a2} as 2.0×10^{-5}. Because K_{a2} is about 400 times smaller than K_{a1}, the second stage of the acid ionization has essentially $\boxed{\text{no effect}}$ on the pH. The pH of the 0.100 M solution of $Fe(NO_3)_3$ is 1.62, based solely on the first H^+-donation.[23] This corresponds to $[H_3O^+] = 2.4 \times 10^{-2}$ M. Thus, considering only the first stage:

$$[H_3O^+] = [Fe(H_2O)_5(OH)^{2+}] = 2.4 \times 10^{-2} \text{ M}$$

The K_a- expression for the *second* stage is:

$$K_{a2} = 2.0 \times 10^{-5} = \frac{[H_3O^+][Fe(H_2O)_4(OH)_2^+]}{[Fe(H_2O)_5OH^{2+}]}$$

Substitute the concentrations obtained from the first-stage-only calculation into this expression. Solving the resulting equation shows that the $[Fe(H_2O)_4(OH)_2^+]$ ion has a concentration of only $\boxed{2.0 \times 10^{-5} \text{ M}}$. This concentration is negligible compared to 0.024 M, the concentration of $Fe(H_2O)_5OH^{2+}(aq)$. The $[H_3O^+]$ that arises in the second stage is likewise negligible compared to 0.024 M.

b) The question refers to the dissociation of the $Fe(OH)_2^+$ complex ion. Dissociation proceeds through steps that are the reverse of the steps for formation:

$$Fe(OH)_2^+(aq) \rightleftharpoons Fe(OH)^{2+}(aq) + OH^-(aq) \qquad \frac{[Fe(OH)^{2+}][OH^-]}{[Fe(OH)_2^+]} = K_{d1}$$

$$Fe(OH)^{2+}(aq) \rightleftharpoons Fe^{3+}(aq) + OH^-(aq) \qquad \frac{[Fe^{3+}][OH^-]}{[Fe(OH)^{2+}]} = K_{d2}$$

---

[22]Text page 388.
[23]Computed in text Example 11-13.

These equilibria have constants labeled with d's for dissociation. Divide the first of the two mass-action equations into the equation $K_w = [H_3O^+][OH^-]$. The result is:

$$\frac{K_w}{K_{d1}} = \frac{[H_3O^+][Fe(OH)_2^+]}{[Fe(OH)^{2+}]}$$

The right-hand side of this equation is identical to the mass-action expression for K_{a2} in the previous part except that the chemical formulas in the K_{a2} expression show associated H_2O molecules explicitly. That is, $Fe(OH)_2^+$ is the same as $Fe(H_2O)_4(OH)_2^+$, and $Fe(OH)^{2+}$ is the same as $Fe(H_2O)_5OH^{2+}$. Hence, $K_{a2} = K_w/K_{d1}$. Similarly, $K_{a1} = K_w/K_{d2}$. Numerical values for K_{a1} and K_{a2} are given in the problem and in text Example 11-13. Substitution gives:

$$K_{d1} = \frac{K_w}{K_{a2}} = \frac{1.0 \times 10^{-14}}{2.0 \times 10^{-5}} = 5.0 \times 10^{-10} \qquad K_{d2} = \frac{K_w}{K_{a1}} = \frac{1.0 \times 10^{-14}}{7.7 \times 10^{-3}} = 1.3 \times 10^{-12}$$

Because the two-step dissociation of the complex exactly reverses the two-step formation of the complex, K_f of the complex equals the reciprocal of the product of K_{d1} and K_{d2}:

$$K_f = \frac{1}{K_{d1}K_{d2}} = \frac{1}{(5.0 \times 10^{-10})(1.3 \times 10^{-12})} = \boxed{1.5 \times 10^{21}}$$

11-107 Water is a better solvent for more covalent and ionic substances, and carbon tetrachloride (CCl_4) is a better solvent for covalent compounds. On this basis:
a) The polar covalent compound methanol (CH_3OH) has a larger concentration in $\boxed{H_2O}$.
b) The covalent compound hexachloroethane (C_2Cl_6) has a larger concentration in $\boxed{CCl_4}$.
c) Bromine (Br_2), which has a covalent bond, has a larger concentration in $\boxed{CCl_4}$.
d) The ionic compound sodium chloride ($NaCl$) has a larger concentration in $\boxed{H_2O}$.

11-109 Use ΔG_f°'s from text Appendix D to calculate ΔG° for the reaction, and from ΔG° obtain the equilibrium constant for the reaction:

$$\Delta G^\circ_{298.15} = 1 \underbrace{(0.00)}_{Ni(s)} + 1/2 \underbrace{(0.00)}_{O_2(g)} - 1 \underbrace{(-211.7)}_{NiO(s)} = 211.7 \text{ kJ}$$

$$\ln K_{298.15} = \frac{-\Delta G^\circ_{298.15}}{RT} = \frac{-(211.7 \times 10^3 \text{ J mol}^{-1})}{(8.315 \text{ J K}^{-1}\text{mol}^{-1})(298.15 \text{ K})} = -85.44$$

$$K = 7.8 \times 10^{-38}$$

The partial pressure of oxygen is related to K through the mass-action expression for this reaction. Because two of the three substances involved in the reaction are pure solids, this expression is very simple: $\sqrt{P_{O_2}} = K$. It follows that the equilibrium pressure of oxygen is the square of the equilibrium constant. It equals $\boxed{6.1 \times 10^{-75} \text{ atm}}$. The decomposition of $NiO(s)$ to its elements is *very* slight at room temperature!

11-111 The reaction is the sublimation of water, $H_2O(s) \rightleftharpoons H_2O(g)$. The K_{273} for this reaction equals the vapor pressure of ice at 273.15 K, which is 0.0060 atm. Ice at its freezing point sublimes spontaneously whenever the pressure of $H_2O(g)$ is less than 0.0060 atm. Cold ice sublimes less easily. Use the van't Hoff equation to calculate K for sublimation at $-15°C$ (258.15 K)

$$\ln\left(\frac{K_{258}}{K_{273}}\right) = \frac{-\Delta H_{sub}}{R}\left(\frac{1}{258.15 \text{ K}} - \frac{1}{273.15 \text{ K}}\right)$$

$$\ln\left(\frac{K_2}{0.0060}\right) = \frac{-50.0 \times 10^3 \text{ J mol}^{-1}}{8.315 \text{ J K}^{-1}\text{mol}^{-1}}\left(\frac{1}{258.15 \text{ K}} - \frac{1}{273.15 \text{ K}}\right)$$

$$K_{258} = 0.0017$$

Therefore ice at 258.15 sublimes spontaneously whenever the partial pressure of $H_2O(g)$ is $\boxed{\text{less than 0.0017 atm}}$.

11-113 The molar mass of codeine ($C_{18}H_{21}NO_3$) equals 299.370 g mol^{-1}. At room temperature, the molal solubility of this substances is:

$$\frac{1.00 \text{ codeine}}{120 \text{ mL water}} \times \left(\frac{1 \text{ mL water}}{1 \text{ cm}^3 \text{ water}}\right) \times \left(\frac{1 \text{ cm}^3 \text{ water}}{1.00 \text{ g water}}\right) \times \left(\frac{1000 \text{ g water}}{1 \text{ kg water}}\right)$$

$$\times \left(\frac{1 \text{ codeine}}{299.37 \text{ g codeine}}\right) = \boxed{0.0278 \text{ mol kg}^{-1}}$$

At 80°, the same mass of codeine dissolves in half the amount of water, so the molal solubility is double what it is at room temperature: $\boxed{0.0557 \text{ mol kg}^{-1}}$. The dissolution reaction is driven to the right (favoring the dissolved codeine) by the increase in temperature. The reaction is therefore $\boxed{\text{endothermic}}$; the "heat term" for the dissolution reaction is on the left.

11-115 The problem combines a calculation on a formic acid/formate buffer (Chapter 10) with a solubility calculation for the slightly soluble salt CaF_2 (Chapter 11). Start by treating the two calculations separately: compute the pH of the buffer and assume that the pH remains unchanged as the CaF_2 dissolves. This method neglects the effect of the F^- ion, a weak base, on the pH. It also assumes that the autoionization of water is negligible. These points will have to be checked.

The addition of 50.0 mL of 0.15 M HNO_3 to 100.0 mL of 0.12 M NaHCOO creates the buffer. The nitric acid converts 0.0075 mol of $HCOO^-$ ion to $HCOOH$ and leaves 0.0045 mol of $HCOO^-$ ion unreacted. The concentrations of these two species immediately after the conversion, but before either interacts further are:

$$[HCOOH] = \frac{0.0075 \text{ mol}}{0.150 \text{ L}} = 0.050 \text{ M} \qquad [HCOO^-] = \frac{0.0045 \text{ mol}}{0.150 \text{ L}} = 0.030 \text{ M}$$

Both concentrations change slightly as the formic acid/formate equilibrium takes effect. This equilibrium generates x mol L^{-1} of H_3O^+ as shown in the following:

	$HCOOH(aq)$	$+H_2O(aq) \rightleftharpoons$	$HCOO^-(aq) +$	$H_3O^+(aq)$
Init. Conc. (M)	0.50	—	0.030	small
Change in Conc. (M)	$-x$	—	$+x$	$+x$
Equil. Conc. (M)	$0.50 - x$	—	$0.030 + x$	x

Taking K_a from text Table 10-2[24] and substituting in the mass-action expression gives:

$$K_a = 1.77 \times 10^{-4} = \frac{[HCOO^-][H_3O^+]}{[HCOOH]} = \frac{(0.030 + x)x}{(0.050 - x)}$$

Solving for x is routine. The $[H_3O^+]$ equals 2.95×10^{-4} M.

Dissolution of $CaF_2(s)$ generates $F^-(aq)$ ion. Some of this ion reacts with water to give $HF(aq)$:

$$F^-(aq) + H_2O(l) \rightleftharpoons HF(aq) + OH^-(aq) \qquad \frac{[HF][OH^-]}{[F^-]} = K_b$$

Once equilibrium is established, the K_b equation holds and therefore the following, which is obtained by divided the K_b equation into the usual K_w equation also holds:

$$K_a = \frac{[H_3O^+][F^-]}{[HF]} = 6.6 \times 10^{-4}$$

By previous assumption, the concentration of hydronium ion is fixed at 2.95×10^{-4} M by the formic acid/formate buffer. Therefore:

$$\frac{2.95 \times 10^{-4}[F^-]}{[HF]} = 6.6 \times 10^{-4} \qquad \text{which means:} \qquad \frac{[F^-]}{[HF]} = 2.237$$

Now, CaF_2 dissolves:

$$CaF_2(s) \rightleftharpoons Ca^{2+}(aq) + 2F^-(aq) \qquad K_{sp} = [Ca^{2+}][F^-]^2 = 3.9 \times 10^{-11}$$

---

[24]Text page 323.

Obviously, the dissolution reaction generates F^- ion and Ca^{2+} ion in a 2 : 1 ratio. But it is *not* true at equilibrium that the concentration of F^- ion equals twice that of Ca^{2+} ion. Some F^- ion reacts with water to give HF. What *is* true is:

$$2[Ca^{2+}] = [F^-] + [HF]$$

Eliminating [HF] between this and the equation for the ratio [F^-]/[HF] gives:

$$[F^-] = 2[Ca^{2+}] - \frac{[F^-]}{2.237} \quad \text{which rearranges to:} \quad [F^-] = 1.382[Ca^{2+}]$$

Insertion of the equation for $[F^-]$ in terms of $[Ca^{2+}]$ into the K_{sp} expression gives:

$$[Ca^{2+}](1.382)^2[Ca^{2+}]^2 = 3.9 \times 10^{-11}$$
$$[Ca^{2+}] = 2.73 \times 10^{-4} \text{ M}$$

One mole of Ca^{2+} is present in solution for every one mole of CaF_2 that dissolved. Therefore, the molar solubility of CaF_2 in the buffer equals $\boxed{2.7 \times 10^{-4} \text{ mol L}^{-1}}$.

Now, check on the assumptions. A side calculation gives $[F^-]$ as 3.8×10^{-4} M, and [HF] as 1.7×10^{-4} M. These concentrations are both small compared to the concentrations of $HCOO^-$ and HCOOH. Therefore the HF/F^- equilibrium can affect the pH only negligibly. The water autoionization is similarly drowned out as a source of H_3O^+ by the large concentration of HCOOH.

Tip. It is instructive to set up the equations for an exact treatment[25] to solve this problem. First, list all the species present in the solution. In this case, nine different species exist at equilibrium (in addition to H_2O itself):

$$Na^+, H_3O^+, Ca^{2+}, NO_3^-, OH^-, F^-, HF, HCOO^-, \text{and HCOOH}$$

Next, establish all possible relationships among the concentrations of the species. The electrical neutrality of the solution requires that:

$$[Na^+] + [H_3O^+] + 2[Ca^{2+}] = [NO_3^-] + [OH^-] + [F^-] + [HCOO^-]$$

The Na^+ and NO_3^- ions do not react with other species in the solution, and it is easy to obtain their final concentrations:

$$[Na^+] = 0.080 \text{ M} \quad [NO_3^-] = 0.050 \text{ M}$$

Four mass-action relationships are satisfied at equilibrium:

$$1.77 \times 10^{-4} = \frac{[HCOO^-][H_3O^+]}{[HCOOH]} \quad 6.6 \times 10^{-4} = \frac{[F^-][H_3O^+]}{[HF]}$$
$$3.9 \times 10^{-11} = [Ca^{2+}][F^-]^2 \quad 1.0 \times 10^{-14} = [H_3O^+][OH^-]$$

---

[25]Text Section 10-7.

Finally, two material-balance relationships exist:

$$[HF] + [F^-] = 2[Ca^{2+}] \qquad [HCOOH] + [HCOO^-] = 0.080$$

There are now seven equations in seven unknowns, and the problem becomes "merely" to solve the simultaneous equations. Computer programs can be used for this, but mechanization ignores the chemical insights provided by the non-exact method. Resort to the exact method to check assumptions or when confusion threatens to sink all efforts. In this case it is worthwhile to list all equilibrium concentrations and confirm that they satisfy the preceding set of simultaneous equations.

Chapter 12

Electrochemistry

Balancing Oxidation-Reduction Equations

In oxidation-reduction (redox) reactions, the oxidation number of at least one element changes. The criteria for balance in redox equations are the same as in other chemical equations:

- **The number and kind of atoms shown on the left must equal the number and kind shown on the right;**

- **The net charge shown on the left must equal the net electrical charge shown on the right.**

Two situations are common:

1. The formulas of all of the reactants and products are known. The whole task is to find a set of coefficients. This is really a mathematical exercise. The only chemical knowledge required is familiarity with chemical symbolism.

2. Some of the reactants and products are not specified. Balancing the equation then requires **completion** (the insertion of additional reactants or products). In aqueous solution, the inserted species are most often H_2O, $H_3O^+(aq)$, and $OH^-(aq)$ on one side or the other of the equation.

When completing equations, bear these facts in mind:

1. The formula $H^+(aq)$ is equivalent to $H_3O^+(aq)$. The two are both valid representations of the hydrogen ion in aqueous solution. The second is the first with an H_2O added in.

2. In acid solution the concentration of $OH^-(aq)$ is low. It is therefore not realistic to use it as a reactant in a balanced equation. Use H_2O instead. If $OH^-(aq)$

339

is formed in an oxidation-reduction reaction in acid solution, then it will react with the plentiful $H_3O^+(aq)$ ion to give water. Recognize this fact in balanced equations.

3. In basic solution, $H_3O^+(aq)$ is scarce and should not appear as a reactant. Instead use H_2O. If $H_3O^+(aq)$ forms as a product, then it will react with the plentiful $OH^-(aq)$ to give water. Recognize this in balanced equations.

The text gives a six-step method to balance the oxidation-reduction equations in either case. These steps are built around the concept of **half-reactions**. A half-reaction is represented by a half-equation. In a half-equation, appear explicitly as reactants (reduction half-equations) or as products (oxidation half-equations). To arrive at an overall equation, write a balanced oxidation half-equation that releases the same number of electrons that is absorbed by a balanced reduction half-equation. Then add the two. The electrons cancel out. See **12-1** and **12-3**. Certain errors are common in balancing redox equations:

- Inserting H_2, O_2, H_2O_2 and other species in which H and O have oxidation numbers other than $+1$ and -2 respectively in order to complete half-equations. This always causes failure.

- Failing to recognize that two or three or even more elements in a compound can change oxidation number in the same reaction.

- Not understanding that a single substance can *disproportionate*, that is, be both oxidized and reduced in the same reaction.

Disproportionation

Some species can reduce and oxidize themselves. This means that some of the molecules (or atoms or ions) of the species lose electrons to give one set of products and other identical molecules (or atoms or ions) gain electrons to give different products. See **12-5c**. The phenomenon is called **disproportionation**. For example, the $Cu^+(aq)$ ion disproportionates to give $Cu(s)$ and $Cu^{2+}(aq)$ ion. Once it is understood that the same species can go both ways, balancing the equations presents no special problems. The oxidation half-equation and the reduction half-equation show the same species on the left. The coefficients of these species are added together when the half-equations are combined to make a full equation.

Electrochemical Cells

Oxidation-reduction reactions can occur without direct contact between the reactants. In an **electrochemical cell**, oxidation and reduction go on at **electrodes** that are well separated in space. One electrode collects the electrons from the species being oxidized. The electrons pass through a wire and are delivered to the species being reduced at a second electrode. The oxidation and reduction half-reactions are said to occur in two compartments of the cell. The reduction half-reaction occurs at the **cathode** and the oxidation half-reaction occurs at the **anode**. The electrodes are in contact with an **electrolyte**, often an aqueous solution.

The transfer of electrons in an electrochemical cell quickly stops unless some means of maintaining electrical neutrality within the electrolyte is provided. The design of electrochemical cells therefore includes a **salt bridge** to allow the transfer of counterions from the vicinity of one electrode to the vicinity of the other. The diagram shows the essential parts of an electrochemical cell:

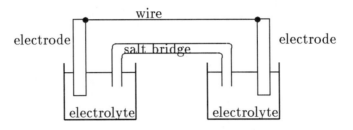

Suppose the electrode on the left is a piece of metallic magnesium dipping into a solution of $Mg(NO_3)_2$ and that the electrode on the right is a piece of metallic silver dipping into a solution of $AgNO_3$. At the moment a circuit is completed by attaching the wire, magnesium begins to be oxidized. Atoms of magnesium go into solution in the anode compartment (on the left) as Mg^{2+} ions. Simultaneously, NO_3^- ions flow from right to left in the salt bridge to maintain electrical neutrality in the anode compartment (see **12-9**). Silver ions are reduced in the cathode compartment; metallic silver plates out on the cathode.

Suppose that the two solutions start with concentrations of 1.00 M. This changes as electrons flow. The concentration of Mg^{2+} ion in the left compartment increases; the concentration of Ag^+ ion in the right compartment decreases. Electrons flow from left to right through the wire If an **ammeter** is inserted into the line connecting the two electrodes, the direction of this flow can be confirmed and its magnitude measured.

Galvanic and Electrolytic Cells

If a **voltmeter** is spliced into the wire connecting the electrodes in the above diagram, it is possible to measure the **difference in electrical potential $\Delta\mathcal{E}$** between the electrodes. Differences in electrical potential are measured in **volts** and are often called **voltages**. They are easily measured and are related to the tendency of the reaction spontaneously to proceed. This is quite different from the size of the electrical current through the wire, which tells the rate of the reaction. A positive difference in potential means there is an intrinsic "push" for the transfer of electrons through the wire from one side to the other: oxidation in one compartment and reduction in the other compartment tend spontaneously to occur.

An electrochemical cell operating spontaneously is a **galvanic cell**. Applying a sufficiently large outside voltage will force a spontaneous reaction to run in reverse. An electrochemical cell in which a non-spontaneous reaction is being forced to occur by the application of a potential difference from outside the cell is an **electrolytic cell**. Such a cell uses electrical energy to carry out chemical reactions that would not otherwise occur.

Both galvanic and electrolytic cells employ oxidation-reduction reactions. For example, the overall reaction:

$$Cu^{2+}(aq) + Zn(s) \rightarrow Cu(s) + Zn^{2+}(aq)$$

is the sum of the half-reactions:

$$Cu^{2+}(aq) + 2e^- \rightarrow Cu(s) \qquad \text{reduction}$$
$$Zn(s) \rightarrow Zn^{2+}(aq) + 2e^- \qquad \text{oxidation}$$

Breaking a balanced redox equation down into balanced half-equations is an important skill:

1. Determine the oxidation numbers of all elements in all compounds to learn which elements are oxidized and which reduced.

2. Write down an oxidation half-equation and a reduction half-equation separately.

3. Insert electrons so that the two sides of the resultant half-equations have the same net charge. See **12-5**.

The following are useful generalizations about half-equations:

• The oxidation half-reaction always involves the *loss* of electrons. It always takes place at the *anode* of an electrochemical cell.
• The reduction half-reaction always involves the *gain* of electrons. It always takes place at the *cathode* of an electrochemical cell.
• Electrons liberated at the anode pass through the outside circuit and are taken up by the reduction half-reaction at the cathode. The number of electrons liberated by the oxidation must exactly equal the number absorbed by the reduction.

In solving problems, it is vital to know the definitions of cathode and anode. Thus, in **12-1** or **12-3**, the only way to know which half-reaction occurs at which electrode is to recognize which is an oxidation and to know that oxidation always occurs at the anode:

Oxidation $\Longleftrightarrow$ Anode   Reduction $\Longleftrightarrow$ Cathode

In these pairings, the terms beginning with consonants (R and C) go together, and the terms beginning with vowels (O and A) go together. Do not bother to learn the plus/minus polarity of cathode and anode in electrochemical cells. These signs depend on whether the cell is a galvanic or an electrolytic cell.

Faraday's Laws

When channeled through an outside circuit, electrons are easily regulated. Their rate of flow (the **electric current**) is readily measured with an ammeter and can be varied at will by changing the electrical resistance of the outside circuit. Electric current is measured in amperes (A). A current of one ampere equals one coulomb flowing through an electrical circuit every second:

$$I = \frac{Q}{t}$$

See **12-17b** and **12-89** for uses of this definition.

In many problems, the electron can be treated as just another chemical reactant. See **12-11** and **12-13**. **Faraday's laws** extend the rules of stoichiometry to the electron as a reactant or product. The laws are:

1. In any cell the mass of a given substance produced or consumed at an electrode is proportional to the quantity of electrical charge Q passed through the cell.

2. Equivalent masses of different substances are produced or consumed at an electrode by the passage of a given quantity of electrical charge through the cell.

Although electrons cannot be weighed out, they are easily tracked because they are charged:

- **A mole of electrons equals 96485.31 coulomb.**

The **Faraday constant** $\mathcal{F}$ is defined as the charge per mole of electrons:

$$\mathcal{F} = 96485.31 \text{ C mol}^{-1}$$

Numerous problems use Faraday's laws. The following points arise in their use:

- The quantity of charge passing through a cell is the average current times the time it flows. If t is in seconds (s) and I in amperes (A), then their product is the quantity of electricity (total electrical charge) passing through the cell in coulombs (C). See **12-13b**.

- To convert a quantity of electricity from electrical units (coulombs) to chemical units (moles) use the Faraday constant. Memorize that 1 mole equals 96485 C.

- Balanced half-equations show the chemical changes at the electrodes of a cell. Treat the electron in a half-equation just like any other reactant or product. In solving stoichiometry problems, use half-equations on the same basis as full equations; treat electrons just like any other chemical substance. See **12-11**, **12-13c**, **12-19**, and **12-53**.

- An **equivalent mass** of a substance is the amount produced or consumed in an electrochemical cell by the passage of 1 mol of electrons (96485 C). The following table shows the relationship of molar mass and equivalent mass:

Reaction	Molar Mass	Equivalent Mass
$Ag^+(aq) + e^- \rightarrow Ag(s)$	107.9 g mol^{-1}	107.9 g per equiv
$Cu^{2+}(aq) + 2e^- \rightarrow Cu(s)$	63.54	31.8
$Al^{3+}(aq) + 3e^- \rightarrow Al(s)$	26.98	9.0
$2\,Cl^-(aq) \rightarrow Cl_2(g) + 2e^-$	70.91	35.5

Problem **12-15** involves equivalent masses.

For examples of the use of Faraday's laws see **12-17**, **12-69**, and **12-71**.

The Gibbs Free Energy and Cell Voltage

Usually, the only kind of work that chemical systems can do on their surroundings is pressure-volume work.[1] Electrochemical cells are the major exception. They are able to produce or consume **electrical work**, in addition to pressure-volume work. A galvanic cell is not an equilibrium system. The difference in electrical potential, or voltage, between its two electrodes ($\Delta\mathcal{E}$) represents the intrinsic thermodynamic push in the cell to come to equilibrium. Completing an electrical circuit between the electrodes allows the spontaneous flow of electrons from anode to cathode. If suitably harnessed, this flow can perform work. The amount of electrical work possible when an amount of charge $\mathcal{Q}$ moves through a potential difference $\Delta\mathcal{E}$ equals

$$w_{\text{elec}} = -\mathcal{Q}\Delta\mathcal{E}$$

See **12-73**. This equation defines the joule in terms of electrical quantities. The work w is in joules when $\mathcal{Q}$ is in coulombs (C) and $\Delta\mathcal{E}$ is in volts (V):

- **A joule (J) equals the energy required to push one coulomb of charge through a potential difference of one volt, 1 joule = 1 volt·coulomb**

---

[1]Text page 285.

Combining the equation for w_{elec} with the definition of electrical current I gives:

$$w_{elec} = -It\Delta\mathcal{E}$$

This equation shows that a joule equals an ampere·volt·second.

A spontaneous redox reaction taking place in a galvanic cell generates a measurable voltage between the electrodes of an electrochemical cell. This $\Delta\mathcal{E}$ is related to the free-energy change of the reaction in the cell:

$$\boldsymbol{\Delta G = -n\mathcal{F}\Delta\mathcal{E}}$$

Again, $\mathcal{F}$ is the Faraday constant, and n is the chemical amount of electrons transferred. The ΔG comes out in joules if $\Delta\mathcal{E}$ is in volts, $\mathcal{F}$ is in C mol^{-1}, and n is in moles. However, ΔG comes out in J mol^{-1} if n is regarded as the number of moles of electrons transferred per mole of chemical reaction. Either interpretation may be used. The difference amounts to the difference between an extensive and an intensive property.[2] Running a redox reaction in an industrial-scale cell transfers kilomoles of electrons. The same reaction in a single laboratory-scale cell transfers only millimoles of electrons. The *extensive* property ΔG (measured in J) is large in the first case but small in the second. The molar ΔG, an *intensive* property measured in J mol^{-1}, is the same in the two cases.

In an electrolytic cell, the chemical reaction is non-spontaneous. This corresponds to a positive ΔG and therefore to a negative $\Delta\mathcal{E}$. A negative $\Delta\mathcal{E}$ means a voltage must be impressed from outside to force the reaction to run. The reverse reaction has a positive $\Delta\mathcal{E}$ and runs spontaneously.

Current does not need to flow when the voltage of a galvanic cell is measured. If no current flows, then no chemical reaction occurs. The potential difference between the electrodes tells the *tendency* or *drive* for the chemical reaction to occur. It is truly well-named.

If current is allowed to flow, chemical change take place. Only then does a galvanic cell have even a chance of doing any electrical work on its surroundings. If the current flows infinitely slowly, the chemical change goes reversibly and can produce the maximum electrical work. Recall that the work produced by a system is $-w$ because $+w$ is defined as the work absorbed by a system. Then:

$$-w_{elec,\,max} = -\Delta G = n\mathcal{F}\Delta\mathcal{E}$$

This equation is used to solve **12-19**. The operation of a practical cell is always irreversible and produces *less* than $-w_{max}$:

$$-w_{elec,\,irrev} < -\Delta G$$

---

[2]The distinction is discussed on text page 204.

If some galvanic cell has a ΔG of -25 kJ, then, by the above equations, it can produce a maximum of $+25$ kJ of electrical work. In actual operation, *less* than 25 kJ of electrical work appears in the surroundings. Depending on how the output of the cell was used, as little as 0 J of electrical work might appear. All of free energy change might be diverted to simple resistive heating, for example.

Standard States and Cell Voltages

The above equation applies to cell reactions going on at any combination of T and P as long as T and P are constant. It certainly also applies when the products and reactants in the equation are in standard states at a temperature of 298.15 K and a pressure of 1 atm. Under these restricted circumstances, the superscript zero (naught) is added to the symbols:

$$\Delta G^\circ = -n\mathcal{F}\Delta\mathcal{E}^\circ$$

Solutes are in standard states when their concentration in an ideal solution is 1 M at 1 atm. Gases are in standard states when their partial pressures as ideal gases are 1 atm.

The above equation means that measurement of $\Delta\mathcal{E}^\circ$ values gives equilibrium constants:

$$\boldsymbol{\Delta G^\circ = -RT \ln K = -n\mathcal{F}\Delta\mathcal{E}^\circ} \quad \text{from which} \quad \ln K = \frac{n\mathcal{F}\Delta\mathcal{E}^\circ}{RT}$$

This relationship is used in **12-43**. In practice, $\Delta\mathcal{E}^\circ$ is measured in volts. Multiplication by $\mathcal{F}$ C mol^{-1}) and n, the number of moles of electrons transferred per mole of reaction as written, then gives $-\Delta G^\circ$ in J mol^{-1}. Division of this value by $-R$ (in J K^{-1}mol^{-1}) and T in kelvins gives $\ln K$ for the reaction as written.

Standard Reduction Potentials

Text Appendix E tabulates **standard reduction potentials** $\mathcal{E}^\circ$. Each standard potential is associated with a half-equation. The potentials are thus for **half-cells.** A half-cell can be visualized as one electrode compartment and its contents. Often it is a piece of metal dipping into a solution containing positive ions of that metal. **Example:** $Cu(s)$ dipping in a solution containing $Cu^{2+}(aq)$.

The structure of the table in text Appendix E must be understood:

- The larger the reduction potential of a species, the greater is its relative tendency to be reduced.

- As reduction potentials, $\mathcal{E}^\circ$'s refer only to half-reactions that gain electrons. Half-reactions that lose electrons have "oxidation potentials."

- A species on the *left* in a given half-reaction will spontaneously oxidize any species on the *right* of a half-reaction located *below* it in the table.[3] See **12-29a** and **12-79**.

- The standard reduction potentials are *intensive* properties. They are not affected when the coefficients in the half-reaction are changed. **Example:** The half-reaction $Cu^{2+}(aq) + 2e^- \rightarrow Cu(s)$ has the same $\mathcal{E}^\circ$ as $2\,Cu^{2+}(aq) + 4e^- \rightarrow 2\,Cu(s)$

- A standard half-cell reduction potential $\mathcal{E}^\circ$ cannot be directly measured. Listings of standard potentials are derived by measuring $\Delta\mathcal{E}^\circ$ values for different pairings of half-reactions and arbitrarily assigning a $\mathcal{E}^\circ$ of exactly 0 V to the *reference* half-reaction: $H_3O^+(aq) + e^- \rightarrow 1/2\,H_2(g) + H_2O(l)$.

The standard potential difference developed by a galvanic cell equals the *difference* between the standard reduction potential of the half-reaction taking place at the cathode and that of the half-reaction taking place at the anode:

$$\Delta\mathcal{E}^\circ = \mathcal{E}^\circ(\text{cathode}) - \mathcal{E}^\circ(\text{anode})$$
$$\Delta\mathcal{E}^\circ = \mathcal{E}^\circ(\text{reduction}) - \mathcal{E}^\circ(\text{oxidation})$$

Obtain $\Delta\mathcal{E}^\circ$'s as follows:

1. Balance the overall oxidation-reduction equation.

2. Break down the equation into a balanced reduction half-equation and a balanced oxidation half-equation.

3. Find these half-equations in Appendix E and note their standard reduction potentials. The oxidation half-reaction will be the *reverse* of one of the tabulated half-reactions. **Caution:** Find the correct half-equations, including the physical state of all reactants and products. Some half-equations resemble each other fairly closely, but have quite different standard reduction potentials.

4. Subtract the tabulated $\mathcal{E}^\circ$ for the half-reaction taking place at the anode (the oxidation) from the tabulated $\mathcal{E}^\circ$ for the half-reaction taking place at the cathode (the reduction).

This procedure is followed in **12-21**, **12-23b**, **12-35**, **12-45**, and **12-49**.

---

[3]This assumes that both species are present in standard states at the same temperature.

Addition and Subtraction of Half-Cell Reactions

As just pointed out, when two half-equations are combined to give a whole equation, in which no electrons appear explicitly, then the $\mathcal{E}°$ values are combined by simple subtraction:

$$\Delta\mathcal{E}° = \mathcal{E}°(\text{cathode}) - \mathcal{E}°(\text{anode})$$

If $\Delta\mathcal{E}°$ comes out negative, then the reaction is **non-spontaneous** as written and the reverse reaction is spontaneous.

When two half-equations are combined to make a new *half-equation* the calculation is not so simple. Then:

$$\mathcal{E}_3° = \frac{n_1\mathcal{E}_1° - n_2\mathcal{E}_2°}{n_3}$$

where the subscripts 1 and 2 refer to the reduction potentials and quantities of electrons in the two half-equations being combined, and where subscript 3 refers to the resultant half-equation. See **12-31a**. If this equation is used when half-reactions combine to make a whole reaction, it turns out that $n_1 = n_2 = n_3$, and the simple subtractive relationship already cited is the result. See **12-31**.

Oxidizing and Reducing Agents

Reduction potential diagrams help in comparing the reactivities of different chemical species. These diagrams consist of a list of species containing a particular element, either by itself or in various combination with hydrogen, oxygen or both. The species are written in a row in order of decreasing oxidation state of the subject element. Lines connecting pairs of species stand for a properly balanced half-equation representing the reduction of the species on the left to the one on the right. Standard reduction potentials appear above the lines.

The reduction potentials are for half-reactions taking place in water. The pH of the solution affects many reduction potentials sharply.[4] Reduction potential diagrams therefore generally indicate whether the solution is standard acidic solution (pH 0) or standard basic solution (pH 14).

Be able to determine the oxidation numbers of all the elements appearing in a reduction potential diagram, write out balanced half-equations for all possible reductions involving the species in the diagram, and derive reduction potentials for all possible reductions. Good balanced half-equations take into account the relative availability of $H_3O^+(aq)$ or $OH^-(aq)$. For example, the reduction potential diagram for oxygen in acid (at pH 0) appears on text page 418. Not shown there is the

---

[4]Text Figure 23-8 (text page 817) plots the dependence of some reduction potentials on pH.

reduction potential diagram for oxygen in base (at pH 14):

$$O_2 \xrightarrow{-0.076 \text{ V}} HO_2^- \xrightarrow{0.87 \text{ V}} OH^- \qquad \text{in base (pH 14)}$$

Three different species appear in this diagram. Therefore, three different reductions are possible. Balance the two half-reactions for which the reduction potential appears in the diagram this way:

$$O_2(g) + H_2O(l) + 2\,e^- \rightarrow HO_2^-(aq) + OH^-(aq) \qquad \mathcal{E}° = -0.076 \text{ V}$$
$$HO_2^-(aq) + H_2O(l) + 2\,e^- \rightarrow 3\,OH^-(aq) \qquad \mathcal{E}° = 0.87 \text{ V}$$

Another half-equation represents the direct reduction of O_2 to OH^-. Balance it:

$$O_2(g) + 2\,H_2O(l) + 4\,e^- \rightarrow 4\,OH^-(aq) \qquad \mathcal{E}° = \frac{2(-0.076) + 2(0.87)}{4} = 0.40 \text{ V}$$

Note that in base $H_3O^+(aq)$ is scarce and therefore does not appear in the balanced half-equations. The direct reduction of O_2 to OH^- could have been represented explicitly in the reduction potential diagram by a lines connecting the two and having "0.40 V" written above it.

The half-reactions represented in a reduction potential diagram can also be combined to give *whole* reactions. Recall that the $\Delta\mathcal{E}°$ for a whole reaction is:

$$\Delta\mathcal{E}° = \mathcal{E}°(\text{cathode}) - \mathcal{E}°(\text{anode})$$

Thus, in base, $OH_2^-(aq)$ oxidizes itself with a $\Delta\mathcal{E}°$ of $0.87 - (-0.076) = 0.97$ V. The products are $OH^-(aq)$ and $O_2(g)$. This is a disproportionation.

- **A species in a reduction potential diagram disproportionates spontaneously whenever a reduction potential on a line leading from it to the right exceeds a reduction potential on a line leading from it to the left.**

The reduction potential diagram for oxygen in base lists different voltages than the diagram for oxygen in acid. For example, the half-equations and reduction potentials for the reduction of O_2 to H_2O_2 at pH 14 and pH 0 are:

$$O_2(g) + H_2O(l) + 2\,e^- \rightarrow HO_2^-(aq) + OH^-(aq) \qquad \mathcal{E}° = -0.076 \text{ V} \quad \text{pH 14}$$
$$O_2(g) + 2\,H_3O^+(aq) + 2\,e^- \rightarrow H_2O_2(aq) + 2\,H_2O(l) \qquad \mathcal{E}° = 0.69 \text{ V} \quad \text{pH 0}$$

Why is it easier to reduce $O_2(g)$ in acid than in base? In base, the half-reaction generates OH^- against an outside OH^- concentration of 1 M. Making the solution acidic reduces the OH^- concentration in the surroundings and so favors the products

(LeChatelier's principle). This is reflected in the more positive reduction potential. As a basic solution is acidified, OH^- eventually disappears from among the products, and the best representation becomes the acid half-reaction, which has H_3O^+ on the left-hand side.

The voltage for the $O_2(g)$ to $HO_2^-(aq)$ reduction at pH 14 is computed by putting $[H_3O^+] = 10^{-14}$ M in the Nernst equation (see next section) for the half-reaction in acid. The voltage at pH 0 is computed by putting $[OH^-] = 10^{-14}$ M in the Nernst equation for the half-reaction in base. In reduction potential diagrams, take care to represent the correct predominant species according to the pH.

Reduction potential diagrams are a compact means of communicating quantities of chemical information. Often, diagrams for elements in a single group in the periodic table are displayed one after another to highlight trends.[5]

Concentration Effects and the Nernst Equation

Differences in potential calculated from the data in text Appendix E are correct only for cells in which all of the reactants and products are present in standard states (activity = 1). This is because potential difference depend on the activity of the reactants and products in the cell reaction as well as their identity. To compute $\Delta\mathcal{E}$ (without a superscript) requires knowledge of the concentrations (in mol L^{-1}) of all the solutes taking part in the chemical reaction in the cell and the partial pressures (in atm) of all the gases. Accept these numbers as equal to the activities of the solutes and gases.[6] Assume also that the activities of solids, liquids, and solvents equal 1. Then apply the **Nernst equation:**

$$\Delta\mathcal{E} = \Delta\mathcal{E}° - \frac{RT}{n\mathcal{F}}\ln Q$$

where Q is the reaction quotient of the reaction. . The subtracted term on the right side of the equation corrects for the non-1 activities of the reactants and products when they are in non-standard states. Suppose a galvanic cell employs the reaction:

$$Mg(s) + 2\,Ag^+(aq) \rightarrow Mg^{2+}(aq) + 2\,Ag(s)$$

The standard potential difference $\Delta\mathcal{E}°$ is $\mathcal{E}°$(cathode) minus $\mathcal{E}°$(anode). It equals 3.175 V. The form of the reaction quotient is:

$$Q = \frac{[Mg^{2+}]}{[Ag^+]^2}$$

---

[5]See page 574 of this Guide.

[6]Exact calculations would require activity coefficients of these substances as well. See text page 362–4.

The Nernst equation for this cell therefore is:

$$\Delta \mathcal{E} = 3.175 \text{ V} - \left(\frac{RT}{2\mathcal{F}}\right) \ln \frac{[\text{Mg}^{2+}]}{[\text{Ag}^+]^2}$$

Note that n equals 2 because 2 mol of electrons is transferred in the balanced equation. If $[\text{Mg}^{2+}] = [\text{Ag}^+] = 1$ M, Q is equal to 1 and $\ln Q$ equals zero. Under these conditions, the experimental $\Delta \mathcal{E}$ (as read from a voltmeter) equals $\Delta \mathcal{E}^\circ$, which is 3.175 V. If the anode and cathode are wired together, electrons spontaneously flows through the wire as the reaction proceeds. The concentration of Mg^{2+} increases, and the concentration of Ag^+ decreases, causing Q to increase and $\Delta \mathcal{E}$ to dwindle from 3.175 V. As electrons continues to flow, $\Delta \mathcal{E}$ ultimately trickles all the way down to 0.00 V as an ever-larger correction factor is subtracted from 3.175 V.

At $\Delta \mathcal{E} = 0$ the cell is *dead* as far as further release of chemical free energy by this reaction is concerned. The ΔG of the system is zero because $\Delta \mathcal{E}$ is zero. The free energy of the cell is at a minimum. A dead galvanic cell is at equilibrium. Meanwhile, neither $\Delta \mathcal{E}^\circ$ nor ΔG° has changed.

At equilibrium the reaction quotient Q equals K, the equilibrium constant. Measurement of $\Delta \mathcal{E}^\circ$'s of cells gives K's because:

$$0.00 = \Delta \mathcal{E}^\circ - \frac{RT}{n\mathcal{F}} \ln K \quad \text{from which} \quad \ln K = \frac{n\mathcal{F}}{RT} \Delta \mathcal{E}^\circ$$

The calculation of the K of a redox reaction is illustrated in **12-43** and **12-45**.

In problem solving, a convenient form of the Nernst equation is:

$$\Delta \mathcal{E} = \Delta \mathcal{E}^\circ - \left(\frac{0.0592 \text{ V}}{n}\right) \log_{10} Q \qquad \text{at 298.15 K}$$

This equation combines numerical values of R, T and $\mathcal{F}$ and converts from natural logarithms to base-10 logarithms. The solution to **12-35** shows how "0.0592 V" arises. This specialized form should *not* be used unless the temperature is 298.15 K. Like ΔG, $\Delta \mathcal{E}$ is strongly dependent on temperature. See **12-93**.

The Nernst equation applies to half-reactions and half-cell potentials just as it does to whole reactions. The half-reaction: $\text{H}_3\text{O}^+(aq) + e^- \rightarrow 1/2\,\text{H}_2(g) + \text{H}_2\text{O}(l)$ has the Nernst equation:

$$\mathcal{E} = \mathcal{E}^\circ - \frac{RT}{\mathcal{F}} \ln \left(\frac{P_{\text{H}_2}^{1/2}}{[\text{H}_3\text{O}^+]}\right)$$

Note the use of $\mathcal{E}^\circ$ instead of $\Delta \mathcal{E}^\circ$. This exact expression is used in **12-63**.

The operation of the **pH meter** illustrates the dependence of cell voltage on concentration. A reference electrode and an $\text{H}_3\text{O}^+(aq)$-sensitive electrode are placed in the aqueous solution under investigation. The observed cell potential depends linearly on $[H_3O^+]$ in the solution, according to the Nernst equation. At 25°C:

$$\Delta\mathcal{E} = \Delta\mathcal{E}° \, (\text{ref}) - 0.0592 \log[H_3O^+]$$

where $\Delta\mathcal{E}° \, (\text{ref})$ is a constant that depends on which half-reaction is selected for use as a reference. See **12-47**.

Batteries and Fuel Cells

Galvanic cells (**batteries**) have many important applications. **Primary cells** generate electrical work but cannot be recharged. **Secondary cells**, or accumulators, are cells that generate electrical work and can also be recharged by using electrical energy from an external source to drive the oxidation-reduction reaction by which they were discharged in reverse.

"Re-charging" does not involve putting electrons back into a cell. A cell does not in fact lose any electrons even if completely discharged. Instead it means forcing the electrons in the cell back to their former positions of higher chemical potential energy.

Important non-rechargeable galvanic cells are the **Leclanché cell** (see **12-85**), the **alkaline dry cell**, and the **zinc-mercuric oxide cell**.

Rechargeable Batteries

The **lead storage battery** is a rechargeable battery. As it is discharged the following takes place:

$$Pb(s) + SO_4^{2-}(aq) \rightarrow PbSO_4(s) + 2\,e^- \qquad \text{anode}$$
$$PbO_2(s) + SO_4^{2-}(aq) + 4\,H_3O^+(aq) + 2\,e^- \rightarrow PbSO_4(s) + 6\,H_2O \qquad \text{cathode}$$
$$Pb(s) + PbO_2(s) + 2\,SO_4^{2-}(aq) + 4\,H_3O^+(aq) \rightarrow 2\,PbSO_4(s) + 6\,H_2O(l) \quad \text{total}$$

Other rechargeable batteries are the **nickel-cadmium cell**, which uses a basic electrolyte, and the **sodium-sulfur cell**.

Fuel Cells

Unlike a battery, which is a closed system to which additional reactants cannot be added, a **fuel cell** continuously "burns" fuel electrochemically. Fuel cells offer a theoretical advantage over traditional means of using fuels. When a fuel is burned in the usual way, the conversion of the heat released, $-\Delta H$, to work encounters thermodynamic limitations on its efficiency. In a fuel cell, the free energy of the chemical reaction is converted directly to electrical energy, and this limitation is evaded.

Corrosion and Its Prevention

Corrosion in metals can very often be understood as the operation of short-circuited electrochemical cells, in which stress or exposure to water and air creates cathodic and anodic regions in a single piece of metal and permits the oxidation and dissolution of the metal. **Passivation** protects metals against corrosion. Often, a thin coating of a metal oxide forms on the surface of the metal and slows further electrochemical reactions. Another way of preventing corrosion is to use a **sacrificial anode** made out of a metal that is oxidized preferentially to the metal being protected. See **12-61**.

Electrolysis of Water and Aqueous Solutions

In pure water, H_3O^+ is not in a standard state (1 M) with activity 1. It has a much lower activity because its concentration is only 10^{-7} M. This lowers $\mathcal{E}$ for the reduction of H_3O^+ from 0.00 V, the standard potential, to -0.414 V. That is:

$$2\,H_3O^+(aq)(10^{-7}\ \text{M}) + 2\,e^- \rightarrow H_2(g)\,(1\ \text{atm}) + 2\,H_2O \qquad \mathcal{E} = -0.414\ \text{V}$$

It lowers $\mathcal{E}$ for the reduction of oxygen to water by the same amount: from a standard 1.229 V to 0.815 V:

$$1/2\,O_2(g)\,(1\ \text{atm}) + 2\,H_3O^+(aq)(10^{-7}\text{M}) + 2\,e^- \rightarrow 3\,H_2O(l) \qquad \mathcal{E} = 0.815\ \text{V}$$

The two lowered reduction potentials were calculated using the Nernst equation. Both results make sense with respect to LeChatelier's principle. If reactant $H_3O^+(aq)$ is present in less than 1 M (standard) concentration, then "drive to the right" is decreased.

Subtract the second of these half-equations from the first:

$$H_2O(l) \rightarrow H_2(g)\,(1\ \text{atm}) + 1/2\,O_2(g)\,(1\ \text{atm}) \quad \Delta\mathcal{E}^\circ = -0.414 - (0.815) = -1.229\ \text{V}$$

The negative $\Delta\mathcal{E}^\circ$ confirms that water does not spontaneously decompose to $H_2(g)$ and $O_2(g)$ at ordinary conditions.[7] The decomposition of water requires an outside voltage of 1.229 V (the **decomposition potential**) in an electrolytic cell to force it to occur. When such a cell is constructed and operated, $H_2(g)$ forms at the cathode and $O_2(g)$ at the anode.

These facts have important applications in the electrolysis of aqueous solutions:

- A species in neutral aqueous solution can be reduced only if its reduction potential exceeds -0.414 V.[8] If this condition is not met, then $H_3O^+(aq)$ is reduced instead of the species in question.

---

[7]The superscript reappears because all the reactants and products are in standard states.

[8]The number -0.50 for example is *less* than -0.414 because of the minus sign.

- A species in neutral aqueous solution can be oxidized only if its reduction potential is *less* than 0.815 V. Otherwise $H_2O(l)$ is oxidized to $O_2(g)$ instead.

These criteria are applied in **12-63**.

Detailed Solutions to Odd-Numbered Problems

12-1 The problem requires *completion* as well as balancing. This means the insertion of H_2O, OH^- ion and H_3O^+ ion on one or both sides of the equation. Follow the six-step procedure outlined in text Section 12-1.

a) The following gives the results of each step. Indications of state such as (aq) or (s) are omitted until the end.

1. $VO_2^+ \rightarrow VO^{2+}$ $\qquad\qquad\qquad\qquad\qquad\qquad\qquad$ $SO_2 \rightarrow SO_4^{2-}$
2. $VO_2^+ \rightarrow VO^{2+}$ $\qquad\qquad\qquad\qquad\qquad\qquad\qquad$ $SO_2 \rightarrow SO_4^{2-}$
3. $VO_2^+ \rightarrow VO^{2+} + H_2O$ $\qquad\qquad\qquad\qquad\quad$ $2\,H_2O + SO_2 \rightarrow SO_4^{2-}$
4. $2\,H_3O^+ + VO_2^+ \rightarrow VO^{2+} + 3\,H_2O$ $\qquad$ $6\,H_2O + SO_2 \rightarrow SO_4^{2-} + 4\,H_3O^+$
5. $e^- + 2\,H_3O^+ + VO_2^+ \rightarrow VO^{2+} + 3\,H_2O$ $\quad$ $6\,H_2O + SO_2 \rightarrow SO_4^{2-} + 4\,H_3O^+ + 2\,e^-$
6. $2\,e^- + 4\,H_3O^+ + 2\,VO_2^+ \rightarrow 2\,VO^{2+} + 6\,H_2O$ $\quad$ $6\,H_2O + SO_2 \rightarrow SO_4^{2-} + 4\,H_3O^+ + 2\,e^-$

$$\boxed{2\,VO_2^+(aq) + SO_2(g) \rightarrow 2\,VO^{2+}(aq) + SO_4^{2-}(aq)}$$

Step 5 shows that VO_2^+ is reduced (electrons appear on the left) and that SO_2 is oxidized (electrons appear on the right). In the final line, addition of the oxidation half-equation to the reduction half-equation has led, as planned, to the removal of e^-'s. The algebraic combining of terms removes six H_2O's and four H_3O^+'s.

Use the same method in the other parts of the problem.

b) $Br_2(l) + SO_2(g) + 6\,H_2O(l) \rightarrow 2\,Br^-(aq) + SO_4^{2-}(aq) + 4\,H_3O^+(aq)$

c) $Cr_2O_7^{2-}(aq) + 3\,Np^{4+}(aq) + 2\,H_3O^+ \rightarrow 2\,Cr^{3+}(aq) + 3\,NpO_2^{2+}(aq) + 3\,H_2O(aq)$

d) $5\,HCOOH(aq) + 2\,MnO_4^-(aq) + 6\,H_3O^+(aq) \rightarrow 5\,CO_2(g) + 2\,Mn^{2+}(aq) + 14\,H_2O(l)$

e) $3\,Hg_2HPO_4(s) + 2\,Au(s) + 8\,Cl^-(aq) + 3\,H_3O^+(aq) \rightarrow 6\,Hg(l) + 3\,H_2PO_4^-(aq) + 2\,AuCl_4^-(aq) + 3\,H_2O(l)$

Tip. The determination of oxidation numbers[9] is *not* necessary in balancing redox equations. However, oxidation numbers can come in handy in checking results. For example, in part a), vanadium is reduced from the +5 to the +4 state and sulfur is oxidized from the +4 to the +6 state. This requires a gain of one electron per vanadium atom and a loss of two electrons per sulfur atom, confirming the results of step 5.

---

[9]Oxidation numbers are discussed text page 85-86.

12-3 Follow the six-step method given in the text. The steps are the same as those used in **12-1** except that now OH^- and H_2O are used to balance hydrogen in step 4.
a) The following lists the results of each step. Indications of state such as (aq) or (s) are omitted until the end.

1. $Cr(OH)_3 \rightarrow CrO_4^{2-}$ $\qquad\qquad\qquad\qquad\qquad\qquad\qquad\qquad$ $Br_2 \rightarrow Br^-$
2. $Cr(OH)_3 \rightarrow CrO_4^{2-}$ $\qquad\qquad\qquad\qquad\qquad\qquad\qquad\qquad$ $Br_2 \rightarrow 2\,Br^-$
3. $H_2O + Cr(OH)_3 \rightarrow CrO_4^{2-}$ $\qquad\qquad\qquad\qquad\qquad\qquad$ $Br_2 \rightarrow 2\,Br^-$
4. $5\,OH^- + H_2O + Cr(OH)_3 \rightarrow CrO_4^{2-} + 5\,H_2O$ $\qquad\qquad$ $Br_2 \rightarrow 2\,Br^-$
5. $5\,OH^- + H_2O + Cr(OH)_3 \rightarrow CrO_4^{2-} + 5\,H_2O + 3\,e^-$ $\quad$ $2\,e^- + Br_2 \rightarrow 2\,Br^-$
6. $10\,OH^- + 2\,H_2O + 2\,Cr(OH)_3 \rightarrow 2\,CrO_4^{2-} + 10\,H_2O + 6\,e^-$ $\quad$ $6\,e^- + 3\,Br_2 \rightarrow 6\,Br^-$

$$\boxed{10\,OH^-(aq) + 2\,Cr(OH)_3(s) + 3\,Br_2(aq) \rightarrow 2\,CrO_4^{2-}(aq) + 6\,Br^-(aq) + 8\,H_2O(l)}$$

In the final line, addition of the oxidation half-equation to the reduction half-equation has led, as planned, to the removal of e^-'s. Combining like terms then removes two H_2O's.

Use the same method in the other parts of the problem.
b) $ZrO(OH)_2(s) + 2\,SO_3^{2-}(aq) \rightarrow Zr(s) + 2\,SO_4^{2-}(aq) + H_2O(l)$
c) $7\,HPbO_2^-(aq) + 2\,Re(s) \rightarrow 7\,Pb(s) + 2\,ReO_4^-(aq) + H_2O(l) + 5\,OH^-(aq)$
d) $4\,HXeO_4^-(aq) + 8\,OH^-(aq) \rightarrow 3\,XeO_6^{4-}(aq) + Xe(g) + 6\,H_2O(l)$
e) $N_2H_4(aq) + 2\,CO_3^{2-}(aq) \rightarrow N_2(g) + 2\,CO(g) + 4\,OH^-(aq)$

12-5 The problem requires a reveral of the steps given for the completion and balancing of redox equations.

a) $\qquad\qquad\qquad\qquad\qquad Fe^{2+}(aq) \rightarrow Fe^{3+}(aq) + e^-$ $\qquad\qquad$ (oxidation)
$\qquad H_2O_2(aq) + 2\,H_3O^+(aq) + 2e^- \rightarrow 4\,H_2O(l)$ $\qquad\qquad$ (reduction)
b) $\qquad\qquad 5\,H_2O(l) + SO_2(aq) \rightarrow HSO_4^-(aq) + 3\,H_3O^+(aq) + 2\,e^-$ (oxidation)
$\qquad MnO_4^-(aq) + 8\,H_3O^+(aq) + 5\,e^- \rightarrow Mn^{2+}(aq) + 4\,H_2O(l)$ $\qquad$ (reduction)
c) $\qquad\qquad\qquad\qquad\qquad ClO_2^-(aq) \rightarrow ClO_2(g) + e^-$ $\qquad\qquad$ (oxidation)
$\qquad ClO_2^-(aq) + 4\,H_3O^+(aq) + 4\,e^- \rightarrow Cl^-(aq) + 6\,H_2O(l)$ $\qquad$ (reduction)

Tip. The last reaction is a disproportionation.

12-7 The oxidation state of nitrogen both increases and decreases in this reaction, which is a disproportionation. The two half-reactions are:
$HNO_2(aq) + 4\,H_2O(l) \rightarrow NO_3^-(aq) + 3\,H_3O^+(aq) + 3\,e^-$ $\quad$ (oxidation)
$HNO_2(aq) + H_3O^+(aq) \rightarrow NO(g) + 2\,H_2O(l)$ $\qquad\qquad$ (reduction)

Combining the two half-equations gives the overall equation:[10]

$$3\,HNO_2(aq) \rightarrow NO_3^-(aq) + 2\,NO(g) + H_3O^+(aq).$$

12-9 Electrons flow from the left electrode to the right as Cr(II) is oxidized to Cr(III). In the salt bridge, negative ions flow from right to left and positive ions from left to right.

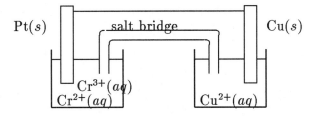

12-11 Formation of 1 mol of Sn(s) from $Sn^{4+}(aq)$ ions requires 4 mol of electrons: $Sn^{4+}(aq) + 4\,e^- \rightarrow Sn(s)$. Hence:

$$N_{Sn} = 6.95 \times 10^4\ C \times \left(\frac{1\ mol\ e^-}{96\,485.3\ C}\right) \times \left(\frac{1\ mol\ Sn}{4\ mol\ e^-}\right) = \boxed{0.180\ mol\ Sn}$$

Tip. Note the careful use of the word "maximum" in the statement of the problem. No Sn(s) is necessarily formed by the passage of the current because other electrolytic processes could take place to carry the current.

12-13 At the anode, Zn(s) is being oxidized to $Zn^{2+}(aq)$; at the cathode, $Cl_2(g)$ is being reduced to $Cl^-(aq)$.

a) $\boxed{Zn(s) + Cl_2(g) \rightarrow Zn^{2+}(aq) + 2\,Cl^-(aq)}$.

b) An ampere is a coulomb per second. Hence:

$$Q = \left(\frac{0.800\ C}{1\ s}\right) \times 25.0\ min \times \left(\frac{60\ s}{1\ min}\right) = \boxed{1.20 \times 10^3\ C}$$

In moles:

$$n_{e^-} = \left(\frac{0.800\ C}{1\ s}\right) \times 25.0\ min \times \left(\frac{60\ s}{1\ min}\right) \times \left(\frac{1\ mol\ e^-}{96485\ C}\right) = \boxed{0.0124\ mol\ e^-}$$

c) Two moles of electrons are produced by oxidation of one mole of Zn(s). Passage of 0.0124 mol of electrons therefore means 0.00622 mol of Zn(s) is oxidized. This is a loss of $\boxed{0.407\ g}$ (using a molar mass of 65.38 g mol^{-1} for zinc) from the zinc anode.

---

[10]This reaction has a $\Delta\mathcal{E}^\circ$ equal to 0.05 V, from the data in text Figure 21-17 (text page 774).

d) Two moles of electrons is equivalent in this reaction to one mole of $Cl_2(g)$. Passage of 0.0124 mol of electrons accordingly requires reduction of 0.00622 mol of $Cl_2(g)$. Use the ideal-gas law to calculate the volume of this chemical amount of chlorine at 25°C (298 K) and a pressure of 1 atm:

$$V_{Cl_2} = \frac{n_{Cl_2} RT}{P} = \frac{(0.00622 \text{ mol})(0.08206 \text{ L atm mol}^{-1}\text{K}^{-1})(298 \text{ K})}{1 \text{ atm}} = \boxed{0.152 \text{ L}}$$

Tip. In fact, it would be very difficult to pass the relatively large current cited in this problem (0.800 A) through any actual salt bridge of a convenient size. The point of the problem however is the computation of the quantities, not the engineering of the cell.

12-15 Calculate the ratio of the chemical amounts of oxygen and copper generated by the operation of the cell. The molar mass of oxygen is 32.0 g mol^{-1}, and the molar mass of copper is 63.54 g mol^{-1}. The cell therefore forms 0.500 mol of O_2 as it forms 1.00 mol of Cu. A balanced half-equation for the oxidation of water to gaseous oxygen is $3 H_2O(l) \rightarrow 1/2 O_2(g) + 2 H_3O^+(aq) + 2 e^-$. This equation states that the production of 1/2 mol of O_2 releases 2 mol of electrons. Hence:

$$\frac{0.500 \text{ mol } O_2}{1.00 \text{ mol Cu}} \times \left(\frac{2 \text{ mol } e^-}{1/2 \text{ mol } O_2} \right) = \frac{2.00 \text{ mol } e^-}{1 \text{ mol Cu}}$$

The copper starts off in the $\boxed{+2 \text{ oxidation state}}$ and is reduced as follows: $Cu^{2+}(aq) + 2e^- \rightarrow Cu(s)$.

12-17 a) In the electrolysis of molten KCl, the half-reactions are:

$$K^+(l) + e^- \rightarrow K(l) \qquad \qquad \text{reduction} \quad \text{cathode}$$

$$Cl^-(l) \rightarrow 1/2\frac{1}{2} Cl_2(g) + e^- \quad \text{oxidation} \quad \text{anode}$$

The sum of these two half-equations represents the overall cell reaction:

$$K^+ + Cl^- \rightarrow K(l) + 1/2 Cl_2(g)$$

b) A current of 2.00 A is a current of 2.00 C s^{-1}; 500 hours is 1.80×10^4 s. The amount of electricity passing through the cell equals the average current multiplied by the duration of the flow or 3.60×10^4 C. Since 9.6485×10^4 C is a mole of electrons, 0.373 mol of electrons passes through the cell. As one mole of electrons passes through the cell, one mole of K forms, and one half mole of Cl_2 forms. Therefore, 0.373 mol of K and 0.0187 mol of Cl_2 are generated during the 5.00 hr electrolysis run. These amounts are $\boxed{13.2 \text{ g}}$ of Cl_2 ($\mathcal{M} = 70.91$ g mol^{-1}) and $\boxed{14.6 \text{ g}}$ of K ($\mathcal{M} = 39.102$ g mol^{-1}).

Tip. A tacit assumption is that the cell has at least 27.8 g of KCl in it.

12-19 Compute the chemical amount of electrons transferred by use of the ratio furnished by the balanced half-equation $Ag^+ + e^- \rightarrow Ag(s)$:

$$n_{e^-} = 1.00 \text{ g Ag} \times \left(\frac{1 \text{ mol Ag}}{107.9 \text{ g Ag}} \right) \times \left(\frac{1 \text{ mol } e^-}{1 \text{ mol Ag}} \right) = 0.00927 \text{ mol } e^-$$

The maximum electrical work *produced* by the cell equals $-w_{elec,max}$ because positive work is by convention work that is absorbed. This maximum is attained if the cell is operated reversibly.[11] Thus:

$$-w_{elec,max} = -w_{elec,rev} = -\Delta G = n\mathcal{F}\Delta\mathcal{E}$$

In this case, all concentrations remain at standard-state values so the equation can be modified:

$$-w_{elec,max} = -w_{elec,rev} = -\Delta G^\circ = n\mathcal{F}\Delta\mathcal{E}^\circ$$

Substitution gives:

$$-w_{elec,max} = (0.00927 \text{ mol})(96485 \text{ C mol}^{-1})(1.03 \text{ V}) = \boxed{921 \text{ J}}$$

12-21 a) A brief answer to the question is the representation $Co|Co^{2+}\|Br_2|Br_2$. The full balanced half-equations and the equation for the overall reaction are:

anode: $\qquad\qquad\qquad\qquad Co(s) \rightarrow Co^{2+}(aq) + 2\,e^-$

cathode: $\qquad\qquad\qquad Br_2(l) + 2\,e^- \rightarrow 2\,Br^-(aq)$

overall reaction: $\qquad Co(s) + Br_2(l) \rightarrow Co^{2+}(aq) + 2\,Br^-(aq)$

b) The $\Delta\mathcal{E}^\circ$ of the cell equals the standard reduction potential at the cathode minus the standard reduction potential at the anode:

$$\Delta\mathcal{E}^\circ = \mathcal{E}^\circ(\text{cathode}) - \mathcal{E}^\circ(\text{anode}) = 1.065 - (-0.28) = \boxed{1.34 \text{ V}}$$

12-23 a) The $In^{3+}|In$ half-reaction must be the reduction because metallic indium plates out when the cell runs. The $Zn^{2+}|Zn$ half-reaction accordingly proceeds as an oxidation:

anode: $\quad Zn(s) \rightarrow Zn^{2+}(aq) + 2\,e^- \qquad\qquad$ cathode: $\quad In^{3+}(aq) + 3\,e^- \rightarrow In(s)$

b) The standard potential difference of the cell is 0.425 V. According to text Appendix E, the standard reduction potential at the zinc anode is -0.763 V. Hence:

$$0.425 \text{ V} = \mathcal{E}^\circ(\text{cathode}) - \mathcal{E}^\circ(\text{anode}) = \mathcal{E}^\circ(\text{cathode}) - (-0.763 \text{ V})$$

Solving gives $\mathcal{E}^\circ(\text{cathode})$ equal to $\boxed{-0.338 \text{ V}}$.

---

[11]Text page 412.

12-25 Powdered metallic aluminum should act as a reducing agent . A reducing agent is itself oxidized as it acts. The +3 oxidation state of Al is well-known. It is the obvious product when aluminum gives up electrons. There are no common negative oxidation states of Al; such states would have to result if Al served as an oxidizing agent. Finally, according to text Appendix E the reduction of Al^{3+} to $Al(s)$ has a large negative $\mathcal{E}^\circ$. If Al^{3+} is hard to reduce, then $Al(s)$ is easy to oxidize. Not only is powered aluminum a reducing agent, it is a powerful one.

12-27 The stronger an oxidizing agent is, the easier it is to reduce (see **12-25**). The more powerful oxidizing agent will have the algebraically larger reduction potential. For these two elements, the standard reduction potentials are:

$$Cl_2(g) + 2\,e^- \rightarrow 2\,Cl^-(aq) \qquad \mathcal{E}^\circ = +1.3583 \text{ V}$$
$$Br_2(l) + 2\,e^- \rightarrow 2\,Br^-(aq) \qquad \mathcal{E}^\circ = +1.065 \text{ V}$$

Since the Cl_2—Cl^- couple has a larger $\mathcal{E}^\circ$ than the Br_2—Br^- couple, $Cl_2(g)$ is the stronger oxidizing agent and (probably) the better disinfectant.

12-29 a) The problem is to find the strongest oxidizing agent under standard acidic conditions in this group of six species: $Co(s)$, $Ag^+(aq)$, $Cl^-(aq)$, $Cr(s)$, $BrO_3(aq)$ and $I_2(s)$. Compare the ease of reduction of the six by examining their standard reduction potentials. The metals $Co(s)$ and $Cr(s)$ and the $Cl^-(aq)$ ion are *hard* to reduce: negative oxidation states of the metals are rare, and the $Cl^{2-}(aq)$ ion is unknown. These three are immediately eliminated. Text Appendix E gives these reduction half-equations and standard potentials for the other three:

$$Ag^+(aq) + e^- \rightarrow Ag(s) \qquad\qquad \mathcal{E}^\circ = 0.7996 \text{ V}$$
$$BrO_3^-(aq) + 6\,H^+(aq) + 5\,e^- \rightarrow 1/2\,Br_2(l) + 3\,H_2O(l) \qquad \mathcal{E}^\circ = 1.52 \text{ V}$$
$$I_2(s) + 2\,e^- \rightarrow I^-(aq) \qquad\qquad \mathcal{E}^\circ = 0.535 \text{ V}$$

The strongest oxidizing agent is evidently BrO_3^- ion .

b) To find the strongest reducing agent, compare the ease of oxidation of the species. These half-equations and standard reduction potentials come from text Appendix E:

$$Co^{2+}(aq) + 2\,e^- \rightarrow Co(s) \qquad \mathcal{E}^\circ = -0.28 \text{ V}$$
$$Cr^{2+}(aq) + 2\,e^- \rightarrow Cr(s) \qquad \mathcal{E}^\circ = -0.557 \text{ V}$$
$$Cr^{3+}(aq) + 3\,e^- \rightarrow Cr(s) \qquad \mathcal{E}^\circ = -0.74 \text{ V}$$
$$Cl_2(g) + 2\,e^- \rightarrow 2\,Cl^-(aq) \qquad \mathcal{E}^\circ = 1.358 \text{ V}$$
$$I_2(s) + 2\,e^- \rightarrow I^-(aq) \qquad \mathcal{E}^\circ = 0.535 \text{ V}$$

The species under comparison are on the *right* sides of these half-equations. $\boxed{\text{Cr}(s)}$ is the best reducing agent since it is more easily oxidized than $\text{Co}(s)$, $\text{Cl}_2(g)$, or $\text{I}_2(s)$. Oxidation of the very poor reducing agent $\text{Ag}^+(aq)$ would give Ag^{2+}; oxidation of the very poor reducing agent $\text{BrO}_3^-(aq)$ would give $\text{BrO}_4^-(aq)$. No half-equations involving these species are given in Appendix E.

c) The standard reduction potential of $\text{Co}^{2+}(aq)$ ion (-0.28 V) is less than that of $\text{Pb}^{2+}(aq)$ ion (-0.1263 V), but higher than that of $\text{Cd}^{2+}(aq)$ ion (-0.4026 V). Therefore $\boxed{\text{Co}(s)}$ will reduce $\text{Pb}^{2+}(aq)$ ion, but not reduce $\text{Cd}^{2+}(aq)$ under standard acidic conditions.

12-31 a) The standard potential for the half-reaction: $\text{Mn}^{3+}(aq) + 3\,e^- \to \text{Mn}(s)$ is *not* equal to the simple sum of the half-cell potentials of the half-reactions:

$$\text{Mn}^{2+}(aq) + 2e^- \to \text{Mn}(s) \qquad \mathcal{E}° = -1.029 \text{ V}$$
$$\text{Mn}^{3+}(aq) + e^- \to \text{Mn}^{2+}(aq) \qquad \mathcal{E}° = 1.51 \text{ V}$$

even though the target half-reaction *is* the sum of these two half-reactions. Instead the potential is a *weighted average:*

$$\mathcal{E}_3° = \frac{n_1 \mathcal{E}_1° + n_2 \mathcal{E}_2°}{n_3}$$

where the three subscripted n's are the numbers of electrons transferred in the two half-reactions being combined and in the half-reaction that results. Substitution gives:

$$\mathcal{E}° = \frac{2(-1.029 \text{ V}) + 1(1.51 \text{ V})}{3} = \boxed{-0.183 \text{ V}}$$

b) The disproportionation: $3\,\text{Mn}^{2+}(aq) \to \text{Mn}(s) + 2\,\text{Mn}^{3+}(aq)$ combines the reduction of $\text{Mn}^{2+}(aq)$ to $\text{Mn}(s)$ and the oxidation of $\text{Mn}^{2+}(aq)$ to $\text{Mn}^{3+}(aq)$. It is represented by the second of the following half-equations subtracted from the first:

$$2\,e^- + \text{Mn}^{2+}(aq) \to \text{Mn}(s) \qquad \mathcal{E}° = -1.029 \text{ V}$$
$$2\,\text{Mn}^{3+}(aq) + 2\,e^- \to 2\,\text{Mn}^{2+}(aq) \qquad \mathcal{E}° = 1.51 \text{ V}$$

The coefficients in the second half-equation are all twice the coefficients appearing in text Appendix E. Because the final disproportionation reaction is a whole reaction, not a half-reaction, this doubling can be ignored in the computation of $\Delta\mathcal{E}°$:

$$\Delta\mathcal{E}° = \mathcal{E}°(\text{reduction}) - \mathcal{E}°(\text{oxidation}) = -1.029 - 1.51 = -2.539 \text{ V}$$

Solutions of $\text{Mn}^{2+}(aq)$ $\boxed{\text{do not disproportionate}}$ to $\text{Mn}(s)$ and $\text{Mn}^{3+}(aq)$; the standard potential difference for the process is negative.

Tip. To confirm why this calculation succeeds, write a weighted-average formula like the one in the preceding part, and use it to compute $\Delta\mathcal{E}^\circ$. Watch what happens when the values of n_1, n_2, and n_3 are substituted:

$$\Delta\mathcal{E}^\circ = \frac{n_1\mathcal{E}_1^\circ - n_2\mathcal{E}_2^\circ}{n_3} = \frac{2(-1.029) - 2(1.51)}{2} = -1.029 - 1.51 = -2.539 \text{ V}$$

In combining half-reactions to give a whole reaction n_1, n_2, and n_3 are *always* equal to each other. This is *never* the case when combining half-reactions to give another half-reaction.

12-33 a) The disproportionation of $Br_2(l)$ in acid is represented:

$$6\,Br_2(l) + 18\,H_2O(l) \rightarrow 2\,BrO_3^-(aq) + 10\,Br^-(aq) + 12\,H_3O^+(aq)$$

for which the standard potential difference is:

$$\begin{aligned}
\Delta\mathcal{E}^\circ &= \mathcal{E}^\circ(\text{reduction}) - \mathcal{E}^\circ(\text{oxidation}) \\
&= \mathcal{E}^\circ(Br_2|Br^-) - \mathcal{E}^\circ(BrO_3^-|Br_2) \\
&= 1.065 - 1.52 = -0.46 \text{ V}
\end{aligned}$$

The negative $\Delta\mathcal{E}^\circ$ means Br_2 will $\boxed{\text{not disproportionate}}$ to Br^- and BrO_3^- under standard acidic conditions.

b) The reduction giving Br_2 has a larger reduction potential than the one giving Br^- ion. Hence $\boxed{Br^-}$ is more easily oxidized than Br_2 and must be a better reducing agent under standard acidic conditions

---

12-35 The overall reaction in this cell is:

$$2\,Cr^{2+}(aq) + Pb^{2+}(aq) \rightarrow Pb(s) + 2\,Cr^{3+}(aq)$$

and the standard potential difference is:

$$\begin{aligned}
\Delta\mathcal{E}^\circ &= \mathcal{E}^\circ(\text{reduction}) - \mathcal{E}^\circ(\text{oxidation}) \\
&= \mathcal{E}^\circ(Pb^{2+}|Pb) - \mathcal{E}^\circ(Cr^{3+}|Cr^{2+}) \\
&= -0.1263 - (-0.41) = 0.28 \text{ V}
\end{aligned}$$

The standard reduction potentials are correct only if all reactants and products are in standard states. This is rarely the case. In this cell, none of the solute concentrations equals 1 M. The Nernst equation is the means of correcting standard potential differences to account for non-standard concentrations. It turns $\Delta\mathcal{E}^\circ$'s into $\Delta\mathcal{E}$'s:

$$\Delta\mathcal{E} = \Delta\mathcal{E}^\circ - \frac{RT}{n\mathcal{F}}\ln Q \qquad \text{from which} \qquad \Delta\mathcal{E} = \Delta\mathcal{E}^\circ - \frac{0.0592 \text{ V}}{n}\log Q \quad (\text{at } 25^\circ\text{C})$$

where Q is the reaction quotient. For this particular reaction and initial set of conditions:

$$Q = \frac{[Cr^{3+}]^2}{[Cr^{2+}]^2[Pb^{2+}]} = \frac{(0.0030)^2}{(0.15)(0.20)^2}$$

and n equals 2. Substitution of these values gives:

$$\Delta\mathcal{E} = 0.2837\ V - \frac{0.0592\ V}{2}\log\left(\frac{(0.0030)^2}{(0.15)(0.20)^2}\right) = \boxed{0.37\ V}$$

Tip. Confirm that "0.0592" is correct:

$$\frac{RT}{n\mathcal{F}}\ln Q = \frac{RT}{n\mathcal{F}}2.303\log Q = \frac{(8.315\ J\ K^{-1}mol^{-1})(298.15\ K)}{n(96485\ C\ mol^{-1})}2.303\log Q$$

$$= \frac{(8.315\ V\ C\ K^{-1}mol^{-1})(298.15\ K)}{n(96485\ C\ mol^{-1})}2.303\log Q = \frac{0.0592\ V}{n}\log Q$$

12-37 The Nernst equation works for half-reactions as well as for whole reactions. The half-reaction indicated by the cell notation is:

$$Cr^{3+}(aq) + e^- \rightarrow Cr^{2+}(aq) \qquad \mathcal{E}° = -0.41\ V$$

for which:

$$\mathcal{E} = \mathcal{E}° - \frac{RT}{n\mathcal{F}}\ln Q = -0.424\ V - \left(\frac{0.0592\ V}{1}\right)\log\left(\frac{0.0019}{0.15}\right) = \boxed{-0.31\ V}$$

Tip. By using "0.0592 V", one automatically assumes that the temperature is 25°C. Platinum is included in the notation for the half-cell because Pt is an electrode. It conducts electrons in or out but is not changed chemically.

12-39 The I_2/I^- half-reaction is at the cathode, the site of reduction. Hence the H_3O^+—H_2 half-reaction is an oxidation (at the anode). The overall reaction is:

$$2\ H_2O(l) + I_2(s) + H_2(g) \rightarrow 2\ H_3O^+(aq) + 2\ I^-(aq)$$

for which $\Delta\mathcal{E}°$ is 0.535 V. This is calculated by combining the standard reduction potentials of the half-reactions:

$$\Delta\mathcal{E}° = \mathcal{E}°(\text{cathode}) - \mathcal{E}°(\text{anode}) = 0.535(I_2|I^-) - (0.000)(H_3O^+|H_2) = 0.535\ V$$

The measured cell voltage depends on the concentrations and partial pressures of reactants and products according to the Nernst equation:

$$0.841\ V = 0.535\ V - \left(\frac{0.0592\ V}{2}\right)\log\left(\frac{[H_3O^+]^2[I^-]^2}{P_{H_2}}\right)$$

The $[I^-]$ equals 1.00 M; the P_{H_2} equals 1 atm. Substitution gives:

$$0.841 - 0.535 = -\frac{0.0592 \text{ V}}{2} \log \left(\frac{[H_3O^+]^2(1.00)^2}{1.00} \right)$$

Solving for $\log H_3O^+$ gives -5.17. The pH equals $- \log[H_3O^+]$, so it is $\boxed{5.17}$.

12-41 a) In the equation:

$$3 \, HClO_2(aq) + 2 \, Cr^{3+}(aq) + 12 \, H_2O(l) \rightarrow 3 \, HClO(aq) + Cr_2O_7^{2-}(aq) + 8 \, H_3O^+(aq)$$

Cr^{3+} is oxidized to $Cr_2O_7^{2-}$, and $HClO_2$ is reduced to $HClO$ under standard acidic conditions. Text Appendix E gives these half-equations and reduction potentials:

$$Cr_2O_7^{2-}(aq) + 14 \, H_3O^+(aq) + 6 \, e^- \rightarrow 2 \, Cr^{3+}(aq) + 21 \, H_2O(l) \qquad \mathcal{E}^\circ = 1.33 \text{ V}$$
$$3 \, HClO_2(aq) + 6 \, H_3O^+(aq) + 6 \, e^- \rightarrow 3 \, HClO(aq) + 9 \, H_2O(l) \qquad \mathcal{E}^\circ = 1.64 \text{ V}$$

If the first half-reaction occurs at the anode and the second at the cathode, then the difference between the standard potentials is positive:

$$\Delta \mathcal{E}^\circ = \mathcal{E}^\circ(\text{cathode}) - \mathcal{E}^\circ(\text{anode}) = 1.64 - 1.33 = \boxed{0.31 \text{ V}}$$

b) The concentration of Cr^{3+} ion is related to the measured cell potential by the Nernst equation. For the above overall reaction, the Nernst equation at 25°C becomes:

$$0.15 \text{ V} = 0.31 \text{ V} - \frac{0.0592 \text{ V}}{6} \log \left(\frac{[HClO]^3[Cr_2O_7^{2-}][H_3O^+]^8}{[HClO_2]^3[Cr^{3+}]^2} \right)$$

Insert the various concentrations, which are given in the problem (recall a pH of 0 means a $[H_3O^+]$ of 1.00 M):

$$0.15 \text{ V} = 0.31 \text{ V} - \frac{0.0592 \text{ V}}{6} \log \left(\frac{(0.20)^3(0.80)(1.00)^8}{(0.15)^3[Cr^{3+}]^2} \right)$$

Solving for the concentration of Cr^{3+} is straightforward:

$$\frac{6(0.15 - 0.31)}{0.0592} = - \log \left(\frac{1.896}{[Cr^{3+}]^2} \right) \qquad \text{hence} \qquad [Cr^{3+}] = \boxed{1 \times 10^{-8} \text{ M}}$$

12-43 The object is to compute the K of this reaction:

$$3 \, HClO_2(aq) + 2 \, Cr^{3+}(aq) + 12 \, H_2O(l) \rightarrow 3 \, HClO(aq) + Cr_2O_7^{2-}(aq) + 8 \, H_3O^+(aq)$$

It is known that $\Delta\mathcal{E}°$ equals 0.31 V. The K and $\Delta\mathcal{E}°$ are related as follows:

$$\ln K = \frac{n\mathcal{F}\Delta\mathcal{E}°}{RT} = \frac{6(9.6485 \times 10^4 \text{ C mol}^{-1})(0.31 \text{ V})}{(8.315 \text{ J K}^{-1}\text{mol}^{-1})(298.15 \text{ K})} = 72.4 \qquad K = \boxed{3 \times 10^{31}}$$

The equilibrium constant is so very large that the reaction goes to completion for all practical purposes: $HClO_2$ and Cr^{3+} react until one of the other is all used up. The two chromium-containing ions are the only colored species present. To determine the color of the solution, assume complete reaction, and calculate whether Cr^{3+} ion is all consumed or is in excess. There are 2.00 mol of $HClO_2$ and 1.0 mol of Cr^{3+} ion, and the two react in a 3-to-2 ratio. The green Cr^{3+} ion is the limiting reactant. Once all of the green ion is used up, the solution will be $\boxed{\text{orange}}$.

12-45 In a disproportionation reaction, a single species is simultaneously oxidized and reduced. In this instance, In^+ ion is oxidized to In^{3+} ion and also reduced to elemental In. Calculate the $\Delta\mathcal{E}°$ that would be developed if the overall reaction took place in an electrochemical cell. Then get K from $\Delta\mathcal{E}°$:

cathode:	$2\,In^+(aq) + 2\,e^- \rightarrow 2\,In(s)$	$\mathcal{E}° = -0.21$ V
anode:	$In^{3+}(aq) + 2\,e^- \rightarrow In^+(aq)$	$\mathcal{E}° = -0.40$ V
cell reaction:	$3\,In^+(aq) \rightarrow In^{3+}(aq) + 2\,In(s)$	$\Delta\mathcal{E}° = 0.19$ V

$$\ln K = \frac{n\mathcal{F}\Delta\mathcal{E}°}{RT} = \frac{2(9.6485 \times 10^4 \text{ C mol}^{-1})(0.19 \text{ V})}{(8.315 \text{ J K}^{-1}\text{mol}^{-1})(298.15 \text{ K})} = 15 \qquad K = e^{15} = \boxed{3 \times 10^6}$$

12-47 At the anode, H_2 is oxidized to H_3O^+; at the cathode, H_3O^+ to reduced to H_2. The sum of the oxidation and reduction is:

$$H_2(g, \text{anode}) + 2\,H_3O^+(aq, \text{cathode}) \rightarrow H_2(g, \text{cathode}) + 2\,H_3O^+(aq, \text{anode})$$

The two sides of this equation are identical, so $\Delta\mathcal{E}°$, the *standard* potential difference, equals zero. The observed non-zero Δ is caused by the unequal concentrations of H_3O^+ on the two sides. Write the Nernst equation for this reaction, which transfers 2 mol of electrons for every 1 mol of H_2:

$$0.150 \text{ V} = 0.00 \text{ V} - \frac{0.0592 \text{ V}}{2} \log \left(\frac{P_{H_2, \text{cathode}}[H_3O^+]^2_{\text{anode}}}{P_{H_2, \text{anode}}[H_3O^+]^2_{\text{cathode}}} \right)$$

$$0.150 \text{ V} = -\frac{0.0592 \text{ V}}{2} \log \left(\frac{(1.00)[H_3O^+]^2_{\text{anode}}}{(1.00)(1.00)^2_{\text{cathode}}} \right)$$

Solve for the H_3O^+ concentration at the anode. The answer is 0.00293 M. The pH at the anode is therefore $\boxed{2.53}$.

Buffer action in the solution surrounding the anode depends on the reaction:

$$HA(aq) + H_2O(l) \rightleftharpoons H_3O^+(aq) + A^-(aq) \qquad K_a = \frac{[H_3O^+][A^-]}{[HA]}$$

According to the problem, the concentrations of A^- and HA at equilibrium in the buffer solution are both equal t0 0.10 M at the same time the pH is 2.53. Substitution in the K_a expression gives:

$$K_a = \frac{[H_3O^+][A^-]}{[HA]} = \frac{(0.00293)(0.10)}{(0.10)} = \boxed{0.0029}$$

12-49 a) The cell reaction can be broken down into half-reactions:

cathode:	$Br_2(l) + 2\,e^- \rightarrow 2\,Br^-(aq)$	$\mathcal{E}° = 1.065$ V
anode:	$2\,H_3O^+(aq) + 2\,e^- \rightarrow H_2(g) + 2\,H_2O(l)$	$\mathcal{E}° = 0.000$ V

The standard potential difference is then:

$$\Delta\mathcal{E}° = \mathcal{E}°(\text{cathode}) - \mathcal{E}°(\text{anode}) = 1.065 - 0.000 = \boxed{1.065 \text{ V}}$$

b) The concentration of Br^- in the cell is related to the measured cell potential and other concentrations (and partial pressures) by the Nernst equation. For this cell (at 25°C) the Nernst equation becomes:

$$1.710 \text{ V} = 1.065 \text{ V} - \left(\frac{0.0592}{2}\right) \log\left(\frac{[Br^{2-}]^2[H_3O^+]^2}{P_{H_2}}\right)$$

At pH 0 the concentration of H_3O^+ is 1.00 M. The partial pressure of H_2 is 1.0 atm. Substitution of these values gives:

$$1.710 \text{ V} = 1.065 \text{ V} - \left(\frac{0.0592 \text{ V}}{2}\right) \log\left(\frac{[Br^-]^2(1.00)^2}{(1.0)}\right)$$

Solve for the concentration of bromide ion. The result, to two significant figures, is $\boxed{1.3 \times 10^{-11} \text{ M}}$.

c) The dissolution of $AgBr(s)$ is governed by a K_{sp} expression:

$$AgBr(s) \rightleftharpoons Ag^+(aq) + Br^-(aq) \qquad K_{sp} = [Ag^+][Br^-]$$

The $Br^-(aq)$ ion in the cell is at equilibrium with $AgBr(s)$ and $Ag^+(aq)$ ion. The concentrations of both ions are known: $[Br^-]$ was computed in the previous part, and $[Ag^+]$ is 0.060 M. Hence:

$$K_{sp} = (0.060)(1.27 \times 10^{-11}) = \boxed{7.6 \times 10^{-13}}$$

12-51 The half-reactions in a lead-acid cell are:

cathode: $PbO_2(s) + SO_4^{2-}(aq) + 4\,H_3O^+(aq) + 2\,e^- \rightarrow PbSO_4(s) + 6\,H_2O(l)$

anode: $Pb(s) + SO_4^{2-}(aq) \rightarrow PbSO_4(s) + 2\,e^-$

The standard reduction potentials are 1.685 V (cathode) and −0.356 V (anode). The potential difference is $1.685 - (-0.356) = \boxed{2.041\ \text{V}}$. When electrochemical cells are connected in series, their voltages add. The voltage generated by six lead-acid cells connected in series equals $6(2.041\ \text{V}) = \boxed{12.25\ \text{V}}$.

Tip. The standard 12-volt battery used in cars consists of six 2-volt lead-acid cells connected in series.

12-53 The "quantity of charge furnished" means the quantity of charge the battery can force through an outside resistance before its voltage falls to zero.

a) Oxidation of 1 mol of spongy lead requires that 2 mol of electrons pass through the circuit: $Pb(s) + SO_4^{2-}(aq) \rightarrow PbSO_4(s) + 2\,e^-$. Hence:

$$10 \times 10^3\ \text{g Pb} \times \left(\frac{1\ \text{mol Pb}}{207\ \text{g Pb}}\right) \times \left(\frac{2\ \text{mol } e^-}{1\ \text{mol Pb}}\right) \times \left(\frac{9.65 \times 10^4\ \text{C}}{1\ \text{mol } e^-}\right) = \boxed{9.3 \times 10^6\ \text{C}}$$

b) The maximum amount of electrical work that a battery can perform depends on the maximum amount of charge that it pushes through the outside circuit and the voltage difference between its two electrodes. The maximum work by the battery on the surroundings is:

$$-w_{\text{elec,max}} = \mathcal{Q}\Delta\mathcal{E}$$

Again, the work out of a system has a negative sign. This maximum work is obtained only if the discharge of the battery proceeds reversibly. Assume that the voltage of this lead-acid battery does not change as it is discharged. Writing the Nernst equation for the reaction (which is the sum of the half-reaction in the preceding problem) shows that this is tantamount to assuming that the concentration of sulfuric acid in the battery stays constant. Then:

$$-w_{\text{elec,max}} = (9.3 \times 10^6\ \text{C})(12\ \text{V}) = 1.1 \times 10^8\ \text{V C} = \boxed{1.1 \times 10^8\ \text{J}}$$

12-55 The amounts of Pb and PbO_2 in the battery diminish during discharge. Without reactants, the battery cannot function. Therefore, simply replacing the dilute H_2SO_4 with concentrated H_2SO_4 is not enough to recharge the battery. The $PbSO_4$ that accumulates must be removed and fresh PbO and PbO_2 added.

12-57 The maximum amount of electrical work done in an electrochemical reaction at constant T and P equals the change in free energy for that reaction. This quantity is in turn related to the potential difference and the amount of electricity passing the cell:

$$w_{elec,max} = \Delta G = -Q\Delta\mathcal{E}$$

If all of the gases in this fuel cell are kept at a pressure of 1 atm and the temperature is 298.15 K, then all reactants and products of the cell process are in standard states. The preceding equation becomes:

$$w_{elec,max} = \Delta G^\circ = -Q\Delta\mathcal{E}^\circ$$

Break the equation for the reaction in the fuel cell down into half-equations and look up their standard reduction potentials in text Appendix E:

anode: $\qquad H_2(g) + 2\,H_2O(l) \to 2\,H_3O^+(aq) + 2e^- \qquad \Delta\mathcal{E}^\circ = 0.00$ V

cathode: $\qquad 1/2\,O_2(g) + 2\,H_3O^+(aq) + 2\,e^- \to 3\,H_2O(l) \qquad \Delta\mathcal{E}^\circ = 0.229$ V

The standard potential difference of the cell is clearly 1.229 V. Now for Q. It equals the amount of electricity transferred as the fuel cell generates one gram of water:

$$Q = 1\text{ g } H_2O \times \left(\frac{1\text{ mol } H_2O}{18.015\text{ g } H_2O}\right) \times \left(\frac{2\text{ mol } e^-}{1\text{ mol } H_2O}\right) \times \left(\frac{9.6485 \times 10^4\text{ C}}{1\text{ mol } e^-}\right)$$

$$= 1.0711 \times 10^4 \text{ C}$$

The maximum electrical work produced in the surroundings per gram of water generated by this reaction equals:

$$-w_{elec,max} = Q\mathcal{E}^\circ = (1.0711 \times 10^4\text{ C})(1.229\text{ V}) = 1.316 \times 10^4 \text{ J}$$

At 60% efficiency only six-tenths of this maximum work flows. This is $\boxed{+7900\text{ J g}^{-1}}$.

12-59 Iron is oxidized, and water is reduced:

$$Fe(s) + 2\,H_2O(l) \to Fe^{2+}(aq) + 2\,OH^-(aq) + H_2(g)$$

The reaction has the standard potential difference:

$$\Delta\mathcal{E}^\circ = \mathcal{E}^\circ(\text{reduction}) - \mathcal{E}^\circ(\text{oxidation}) = -0.8277 - (-0.409) = \boxed{-0.419\text{ V}}$$

where the $\mathcal{E}^\circ$'s come from text Appendix E. The negative $\Delta\mathcal{E}^\circ$ means the reaction is not spontaneous with the products and reactants in standard states. But recall LeChatelier's principle. If the pH lowered from 14 to about 7 (lowering the concentration of OH^- by 7 orders of magnitude), the reaction will become spontaneous. Also, a low concentration of Fe^{2+} ion (relative to 1.00 M) and a low partial pressure (relative to 1.00 atm) of gaseous H_2 favor spontaneity. Since all three products would have far lower than standard concentrations (or pressures) under practical circumstances, corrosion of iron might well tend to occur by this reaction.

Tip. The question of the rate of corrosion is not considered.

12-61 In a remote theoretical sense, metallic sodium could be used as a sacrificial anode—it is far more easily oxidized than iron according to the standard reduction potentials (see text Appendix E). In practice, however, the ocean water would very rapidly oxidize it. Metallic sodium in fact reacts violently with H_2O to form $H_2(g)$. Sodium would be a ⟨bad sacrificial anode⟩ to protect the hull of a ship.

12-63 a) The product at the cathode could be either gaseous hydrogen from the reduction of 1.0×10^{-5} M $H_3O^+(aq)$ or metallic nickel from the reduction of 1.00 M $Ni^{2+}(aq)$. Direct observation of an operating cell would of course settle the issue at a glance.[12] The reduction potential for 1 M $Ni^{2+}(aq)$ to $Ni(s)$ is -0.23 V, as tabulated in text Appendix E. The reduction potential for H_3O^+/H_2O must be adjusted from the tabulated value of 0.00 V because the $[H_3O^+]$ is not 1 M (standard), but equals 1.0×10^{-5} M. Apply the Nernst equation (at 25°C) to do this:

$$\mathcal{E} = \mathcal{E}° - \frac{0.0592 \text{ V}}{1} \log \left(\frac{P_{H_2}^{1/2}}{[H_3O^+]} \right) = 0.0 - (0.0592 \text{ V}) \log \left(\frac{1}{10^{-5}} \right) = -0.296 \text{ V}$$

This reduction potential is algebraically less than the -0.23 V for the reduction for $Ni^{2+}(aq)/Ni(s)$. Therefore ⟨nickel⟩ forms first.

b) A current of 2.00 amperes for 10 hours is a current of 2.00 C s^{-1} for 36000 s. Therefore 7.20×10^4 C passes through the cell. The mass of Ni deposited is:

$$7.20 \times 10^4 \text{ C} \times \left(\frac{1 \text{ mol } e^-}{96\,485 \text{ C}} \right) \times \left(\frac{1 \text{ mol } Ni(s)}{2 \text{ mol } e^-} \right) \times \left(\frac{58.71 \text{ g Ni}}{1 \text{ mol Ni}} \right) = \boxed{21.9 \text{ g Ni}}$$

The volume of the electrolyte has to be so large that removal of 21.9 g of nickel does not lower the concentration of $Ni^{2+}(aq)$ to the point that H_3O^+ starts to be reduced.

c) If the pH is 1.0, then $[H_3O^+]$ is 0.10 M. The Nernst equation for the reduction of H_3O^+ to $H_2(g)$ at 1 atm and 25°C becomes:

$$\mathcal{E} = 0.00 - 0.0592 \text{ V} \log \left(\frac{1.00}{1 \times 10^{-1}} \right) = -0.0592 \text{ V}$$

---

[12]The candidates for reduction at the *cathode* are both *cations*. In electrolytic cells, positively charged cations migrate toward the negatively charged cathode.

At this pH, $\boxed{H_2(g)}$ rather than $Ni(s)$ tends to form at the cathode.

Tip. The results make sense with respect to LeChatelier's principle. In the first part, the highly dilute $H_3O^+(aq)$ is harder to reduce than the $Ni^{2+}(aq)$. In the second part, the higher concentration of $H_3O^+(aq)$ raises its reduction potential enough that it is easier to reduce than the $Ni^{2+}(aq)$, the concentration of which is not changed.

12-65 When Drano acts, aluminum is oxidized and water is reduced:

$$\boxed{2\,Al(s) + 6\,H_2O(l) + 2\,OH^-(aq) \rightarrow 2\,Al(OH)_4^-(aq) + 3\,H_2(g)}$$

12-67 The desired reaction equals the reverse of the non-spontaneous reaction that is forced to occur by the application of the outside voltage. It is:

$$\boxed{NaClO_3(aq) + 3\,H_2(g) \rightarrow NaCl(aq) + 3\,H_2O(l)}$$

12-69 a) Compute the chemical amounts of zinc and nickel(II) available for the reaction:

$$n_{Zn} = 32.68 \text{ g Zn} \times \left(\frac{1 \text{ mol Zn}}{65.39 \text{ g Zn}}\right) = 0.4998 \text{ mol Zn}$$

$$n_{Ni^{2+}} = 0.575 \text{ L Ni}^{2+} \text{ solution} \times \left(\frac{1.00 \text{ mol Ni}^{2+}}{1 \text{ L solution}}\right) = 0.575 \text{ mol Ni}^{2+}$$

The balanced equation $Zn(s) + Ni^{2+}(aq) \rightarrow Zn^{2+}(aq) + Ni(s)$ has Zn and Ni^{2+} reacting in a 1-to-1 ratio, so $\boxed{\text{zinc}}$ is the limiting reactant.

b) The cell is discharged (stops generating a potential difference) when the zinc is all gone:

$$0.4998 \text{ mol Zn} \times \left(\frac{2 \text{ mol } e^-}{1 \text{ mol Zn}}\right) \times \left(\frac{9.6485 \times 10^4 \text{ C}}{1 \text{ mol } e^-}\right) \times \left(\frac{1 \text{ s}}{0.0715 \text{ C}}\right) = \boxed{1.35 \times 10^6 \text{ s}}$$

c) The cell produces as many moles of new $Ni(s)$ as it consumes moles of $Zn(s)$:

$$0.4998 \text{ mol Ni} \times \left(\frac{58.70 \text{ g Ni}}{1 \text{ mol Ni}}\right) = \boxed{29.34 \text{ g Ni}}$$

d) The reaction has reduced 0.4998 mol of Ni^{2+} ion when it comes to a stop for want of zinc. There remains $0.575 - 0.4998 = 0.075$ mol of $Ni^{2+}(aq)$. This remaining nickel ion is still dissolved in the original 575 mL of solution. Its concentration is $0.075 \text{ mol}/0.575 \text{ L} = \boxed{0.13 \text{ mol L}^{-1}}$.

12-71 Set up a string of unit factors:

$$Q = 1.83 \text{ g Zn} \times \left(\frac{1 \text{ mol Zn}}{65.38 \text{ g Zn}} \right) \times \left(\frac{2 \text{ mol } e^-}{1 \text{ mol Zn}} \right) \times \left(\frac{9.6485 \times 10^4 \text{ C}}{1 \text{ mol } e^-} \right)$$

$$\times \left(\frac{100 \text{ C total}}{0.25 \text{ C in meter}} \right) = \boxed{2.16 \times 10^6 \text{ C total}}$$

Tip. The unit-factors from the molar mass of Zn and the stoichiometry of the half-reaction are routine. The factor derived from the Faraday constant is new in Chapter 12, but of the same type as the others. The fourth unit factor is slightly creative in defining total coulombs as different from coulombs passing through the electric meter.

12-73 In the conversion $Al_2O_3 \rightarrow Al$, aluminum passes from the $+3$ oxidation state to the 0 oxidation state; 3 mol of electrons is transferred for every 1 mol of aluminum formed. The total charge Q transferred for the year's supply of Al is:

$$1.5 \times 10^{10} \text{ kg} \times \left(\frac{1 \text{ mol Al}}{0.02697 \text{ kg}} \right) \times \left(\frac{3 \text{ mol } e^-}{1 \text{ mol Al}} \right) \times \left(\frac{9.65 \times 10^4 \text{ C}}{1 \text{ mol } e^-} \right) = 1.61 \times 10^{17} \text{ C}$$

The work required to transfer charge through this electrolysis cell depends on the amount of charge and its difference in potential:

$$w_{\text{elec}} = -Q \, \Delta \mathcal{E} = (1.58 \times 10^{17} \text{ C})(-5.0 \text{ V}) = 8.05 \times 10^{17} \text{ J}$$

where the negative potential difference reflects the fact that this is an electrolytic cell. Then:

$$\text{cost} = 8.05 \times 10^{17} \text{ J} \times \left(\frac{1 \text{ kWh}}{3.6 \times 10^6 \text{ J}} \right) \times \left(\frac{\$0.10}{1 \text{ kWh}} \right) = \$1.5 \times 10^{10} = \boxed{\$22 \text{ billion}}$$

12-75 a) The problem describes in detail the identity and state of the reacting species in two half-cells connected by a salt bridge and a wire to make a galvanic cell. Translate the descriptions into balanced reduction half-equations:

$$O_2(g) + 2 \, H_3O^+(aq) + 2e^- \rightarrow H_2O_2(aq) + 2 \, H_2O(l) \qquad \mathcal{E}° = 0.682 \text{ V}$$
$$MnO_2(s) + 4 \, H_3O^+(aq) + 2e^- \rightarrow Mn^{2+}(aq) + 6 \, H_2O(l) \qquad \mathcal{E}° = 1.208 \text{ V}$$

All reactants and products in the cell are in standard states. The quoted standard reduction potentials apply without correction to this cell. In a galvanic cell the two half-cells must combine to give a potential difference greater than 0.00 V. This means that reduction occurs in the half-cell with the algebraically larger $\mathcal{E}°$. Hence, the first half-reaction is the oxidation, and the second is the reduction. The overall reaction is the second half-reaction minus the first:

$$H_2O_2(aq) + MnO_2(s) + 2\,H_3O^+(aq) \rightarrow O_2(g) + Mn^{2+}(aq) + 4\,H_2O(l)$$

b) Since all participating species are in standard states:

$$\text{cell voltage} = \Delta\mathcal{E}° = \mathcal{E}°(\text{cathode}) - \mathcal{E}°(\text{anode}) = 1.208 - 0.682 = \boxed{+0.526 \text{ V}}$$

Tip. The schematic representation of this galvanic cell is:

$$C(graphite)|MnO_2(s)|Mn^{2+}(aq)\|H_2O_2(aq), H_3O^+(aq)|O_2(g)|Pt(s)$$

The gaseous oxygen is bubbled over the platinum electrode in such a way that it keeps in electrical contact with both the electrode and the solution. The solid MnO_2 must have electrical contact with the graphite electrode for the cell to work. Imposing an outside DC voltage exceeding 0.526 V (and having the negative wire attached to the Pt electrode) would in theory cause the cell to operate in reverse, as an electrolytic cell. The schematic representation of the cell would then be:

$$Pt(s)|O_2(g)|H_2O_2(aq), H_3O^+(aq)\|Mn^{2+}(aq)|MnO_2(s)|C(graphite)$$

12-77 a) A piece of metallic iron at the bottom of a solution of $Fe^{2+}(aq)$, removes unwanted $Fe^{3+}(aq)$ by reacting with it:

$$\boxed{Fe(s) + 2\,Fe^{3+}(aq) \rightarrow 3\,Fe^{2+}(aq)}$$

This redox reaction is a "re-proportionation", that is, the reverse of a disproportionation. It combines the oxidation of $Fe(s)$ to $Fe^{2+}(aq)$ with the reduction of $Fe^{3+}(aq)$ to $Fe^{2+}(aq)$. The reaction is represented by the first of the following half-equations subtracted from the second:

$$2\,e^- + Fe^{2+}(aq) \rightarrow Fe(s) \qquad \mathcal{E}° = -0.409 \text{ V}$$
$$2\,Fe^{3+}(aq) + 2\,e^- \rightarrow 2\,Fe^{2+}(aq) \qquad \mathcal{E}° = 0.770 \text{ V}$$

The standard potential difference is:

$$\Delta\mathcal{E}° = \mathcal{E}°(\text{reduction}) - \mathcal{E}°(\text{oxidation}) = .770 - (-.409) = \boxed{1.179 \text{ V}}$$

b) The $\Delta\mathcal{E}°$ for the reaction of $Mn(s)$ with $Mn^{3+}(aq)$ to give $Mn^{2+}(aq)$ is positive. This was established in **12-31** when a negative $\Delta\mathcal{E}°$ for the *reverse* reaction was obtained. Therefore, $\boxed{\text{put } Mn(s) \text{ in}}$ a solution of $Mn^{2+}(aq)$ to minimize the concentration of $Mn^{3+}(aq)$.

12-79 The $Fe^{3+}|Fe^{2+}$ half-reaction has a standard reduction potential of 0.770 V. If we model the insoluble Mn(III) and Mn(IV) compounds by MnO_2, we are dealing with a reduction potential of 1.208 V at pH 0. A reducing agent such as $\boxed{Br^-(aq)}$ would reduce MnO_2 at pH 0 without reducing the Fe^{3+} because the 1.065 V reduction potential for Br_2/Br^- is less than 1.208 V but more than 0.770 V. Deviations from standard conditions of course affect potentials. This fact and the different rates at which reactions might proceed in a restoration operation make the suggestion of Br^- very tentative.

12-81 The cell reaction is $2\,Ag^+(aq) + Zn(s) \rightarrow 2\,Ag(s) + Zn^{2+}(aq)$. The standard potential difference for the reaction is:

$$\Delta\mathcal{E}^\circ = \underbrace{(0.7996)}_{Ag^+|Ag} - \underbrace{(-0.7628)}_{Zn^{2+}|Zn} = 1.5624 \text{ V}$$

since Ag^+ is obviously reduced (cathode) and Zn oxidized (anode). The required voltage of 1.50 V is somewhat less than 1.5624 V. Even if 1 M solutions of $Ag^+(aq)$ and $Zn^{2+}(aq)$ were available, using them would not give the required voltage. The concentrations of one or both of the available solutions must be adjusted (by dilution) so that the cell generates 1.50 V. The Nernst equation for this cell relates its actual to its standard potential difference:

$$1.50 = 1.5624 - \frac{0.0592}{2}\log\left(\frac{[Zn^{2+}]}{[Ag^+]^2}\right) \qquad \text{which gives} \qquad \frac{[Zn^{2+}]}{[Ag^+]^2} = 10^{2.108} = 128$$

This equation must be satisfied to generate the required voltage. If $[Ag^+]$ equals 0.010 M, then $[Zn^{2+}]$ must equal 0.0128 M. There are many other combinations of concentrations that give the required potential difference, but this one is probably the simplest to prepare.

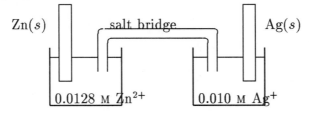

12-83 A gas confined at high pressure expands spontaneously to a lower pressure if a path is available. In a pressure cell, the free-energy decrease accompanying such a spontaneous expansion can appear as electrical work. In this case, the gas is Cl_2. At the cathode, $Cl_2(g)$ is reduced to 1 M $Cl^-(aq)$. At the anode, 1 M $Cl^-(aq)$ is oxidized to $Cl_2(g)$:

$$1/2\,Cl_2(g) + e^- \rightarrow Cl^-(aq) \qquad \text{cathode}$$
$$Cl^-(aq) \rightarrow 1/2\,Cl_2(g) + e^- \qquad \text{anode}$$

The cathode reaction *removes* $Cl_2(g)$ so the reduction must occur in the half-cell held at the *higher* pressure of $Cl_2(g)$, the 0.50 atm half-cell. The overall reaction is:

$$Cl_2\ (0.50\ \text{atm}) \rightarrow Cl_2\ (0.01\ \text{atm})$$

The $\Delta\mathcal{E}°$, the standard potential difference for this reaction, equals 0.000 V. The Nernst equation at 25°C gives the actual potential difference:

$$\Delta\mathcal{E} = \Delta\mathcal{E}° - \frac{0.0592\ \text{V}}{2} \log\left(\frac{0.010}{0.50}\right) = 0.000 - (-0.050) = \boxed{0.050\ \text{V}}$$

12-85 The reactions in the cell are:

anode: $\quad Zn(s) \rightarrow Zn^{2+}(aq) + 2\,e^-$

cathode: $\quad 2\,MnO_2(s) + 2\,NH_4^+(aq) + 2\,e^- \rightarrow Mn_2O_3(s) + 2\,NH_3(aq) + H_2O(l)$

Electrons are released at the anode and taken up at the cathode. Thus, electrons flow from the Zn electrode to the graphite electrode, at which they reduce MnO_2. The stud on the top of a common flashlight battery is positive, and the bottom is negative.

12-87 a) Oxygen is reduced to water at the cathode:

$$O_2(g) + 4\,H_3O^+ + 4\,e^- \rightarrow 6\,H_2O(l)$$

The nickel does not react. This is implied by the H_3O^+ and O_2 in the shorthand representation of the cell.

b) The overall reaction in the cell consumes ethanol and oxygen to give CO_2 and water:

$$C_2H_5OH(l) + 3\,O_2(g) \rightarrow 2\,CO_2(g) + 3\,H_2O(l)$$

This reaction transfers 12 mol of e^- when run as written. Compute its $\Delta G°$ and use it to get $\Delta\mathcal{E}°$. Obtain the necessary $\Delta G_f°$'s from text Appendix D:

$$\Delta G° = 2\ \underbrace{(-394.36)}_{CO_2(g)} + 3\ \underbrace{(-237.18)}_{H_2O(l)} - 1\ \underbrace{(-174.89)}_{C_2H_5OH(l)} = -1325.37\ \text{kJ}$$

$$\Delta\mathcal{E}^{\circ} = \frac{-\Delta G^{\circ}}{n\mathcal{F}} = \frac{-(-1325.37 \times 10^3 \text{ J})}{(12 \text{ mol})(96\,485 \text{ C mol}^{-1})} = 1.1447 \text{ J C}^{-1} = \boxed{1.1447 \text{ V}}$$

c) The standard potential difference equals the standard reduction potential at the cathode minus the standard reduction potential at the anode. The $\mathcal{E}^{\circ}$ at the cathode is for the reduction of gaseous oxygen to water at pH 0. Thus:

$$\Delta\mathcal{E}^{\circ} = \mathcal{E}^{\circ}(H_3O^+, O_2|H_2O) - \mathcal{E}^{\circ}(CO_2, H_3O^+|C_2H_5OH)$$
$$1.1447 \text{ V} = 1.229 \text{ V} - \mathcal{E}^{\circ}(CO_2, H_3O^+|C_2H_5OH)$$
$$\mathcal{E}^{\circ}(CO_2, H_3O^+|C_2H_5OH) = \boxed{+0.084 \text{ V}}$$

12-89 a) At the anode of the first cell, water is oxidized: $6\,H_2O(l) \rightarrow O_2(g) + 4\,H_3O^+ + 4\,e^-$. Water is far more easily oxidized than $Pt(s)$, the other possible candidate. In the second cell, metallic nickel is oxidized: $Ni(s) \rightarrow Ni^{2+}(aq) + 2\,e^-$. Nickel is more easily oxidized than water (by comparison of their standard reduction potentials).

b) The charge $\mathcal{Q}$ passing through the cells equals the current times the elapsed time. Compute this in coulombs and use the Faraday constant to convert to moles:

$$n_{e^-} = 0.10 \text{ C s}^{-1} \times 10 \text{ hr} \times \left(\frac{3600 \text{ s}}{1 \text{ hr}}\right) \times \left(\frac{1 \text{ mol } e^-}{9.6485 \times 10^4 \text{ C}}\right) = 0.0373 \text{ mol } e^-$$

Next, set up unit-factors from the half-equations. Treat the electron like any other product or reactant. At the anode in the first cell:

$$m_{O_2} = 0.0373 \text{ mol } e^- \times \left(\frac{1 \text{ mol } O_2}{4 \text{ mol } e^-}\right) \times \left(\frac{32.00 \text{ g } O_2}{1 \text{ mol } O_2}\right) = \boxed{0.30 \text{ g } O_2}$$

In the second cell:

$$m_{Ni^{2+}} = 0.0373 \text{ mol } e^- \times \left(\frac{1 \text{ mol } Ni}{2 \text{ mol } e^-}\right) \times \left(\frac{58.69 \text{ g } Ni}{1 \text{ mol } Ni}\right) = \boxed{1.1 \text{ g } Ni}$$

12-91 a) Text Appendix E gives these standard reduction potentials:

$$Co^{2+}(aq) \rightarrow Co(s) \qquad \mathcal{E}^{\circ} = -0.28 \text{ V}$$
$$Sn^{2+}(aq) \rightarrow Sn(s) \qquad \mathcal{E}^{\circ} = -0.1364 \text{ V}$$

$Sn^{2+}(aq)$ is clearly easier to reduce under standard conditions. Both ions have non-standard concentrations, but because their concentrations are equal and both are $+2$ ions, the effect on the reduction potential is the same for both. Therefore $\boxed{\text{Sn}(s) \text{ appears first}}$ if a mixture of 0.10 M CoCl_2 and 0.10 M SnCl_2 is electrolyzed.

b) The decomposition potential is the minimum $\Delta\mathcal{E}$ that drives a non-spontaneous electrochemical reaction. Start by establishing the overall reaction that takes place when an outside potential is applied. At the cathode, $\text{Sn}^{2+}(aq)$ ion is reduced. At the anode, water is oxidized to $\text{O}_2(g)$.[13] Hence the overall reaction:

$$\text{Sn}^{2+}(aq) + 3\,\text{H}_2\text{O}(l) \rightarrow \text{Sn}(s) + \frac{1}{2}\,\text{O}_2(g) + 2\,\text{H}_3\text{O}^+(aq)$$

The standard potential difference of this reaction is:

$$\Delta\mathcal{E}° = \mathcal{E}°(\text{Sn}^{2+}|\text{Sn}) - \mathcal{E}°(\text{O}_2|\text{H}_2\text{O}) = -0.1364 - 1.229 = -1.3654 \text{ V}$$

None of the reactants or products (except water) is in a standard state in this cell. Use the Nernst equation to obtain $\Delta\mathcal{E}$ for the concentrations that prevail:

$$\Delta\mathcal{E} = \Delta\mathcal{E}° - \frac{RT}{n\mathcal{F}}\ln Q = -1.3654 \text{ V} - \frac{0.0592 \text{ V}}{2}\log\left(\frac{P_{\text{O}_2}^{1/2}[\text{H}_3\text{O}^+]^2}{[\text{Sn}^{2+}]}\right) \quad (\text{at } 25° \text{ C})$$

Assume that the $\text{O}_2(g)$ is generated at a partial pressure of 1.00 atm and that the concentration of H_3O^+ is 1.0×10^{-7} (pH 7.00). Then:

$$\Delta\mathcal{E} = -1.3654 \text{ V} - \frac{0.0592 \text{ V}}{2}\log\left(\frac{(1.00)^{1/2}[1.0 \times 10^{-7}]^2}{[0.10]}\right) = -0.9806 \text{ V}$$

The decomposition potential is therefore $\boxed{0.981 \text{ V}}$.

c) The balanced equation for the electrolysis that involves Co^{2+} ion is written by simply replacing Sn with Co throughout the previous equation:

$$\text{Co}^{2+} + 3\,\text{H}_2\text{O}(l) \rightarrow \text{Co}(s) + 1/2\,\text{O}_2(g) + 2\,\text{H}_3\text{O}^+(aq)$$

The $\Delta\mathcal{E}°$ for this reaction is:

$$\Delta\mathcal{E}° = \mathcal{E}°(\text{Co}^{2+}|\text{Co}) - \mathcal{E}°(\text{O}_2|\text{H}_2\text{O}) = -0.28 - 1.229 = -1.509 \text{ V}$$

The reduction of Co^{2+} cannot start until the potential difference for reduction of Sn^{2+} has risen high enough to make the $\Delta\mathcal{E}$'s of the two reactions equal. Write the Nernst equation for the two reactions and set the $\Delta\mathcal{E}$'s equal:

$$-1.509 \text{ V} - \frac{0.0592 \text{ V}}{2}\log\left(\frac{P_{\text{O}_2}^{1/2}[\text{H}_3\text{O}^+]^2}{[\text{Co}^{2+}]}\right) = -1.3654 \text{ V} - \frac{0.0592 \text{ V}}{2}\log\left(\frac{P_{\text{O}_2}^{1/2}[\text{H}_3\text{O}^+]^2}{[\text{Sn}^{2+}]}\right)$$

---

[13] As shown in text Section 12-7, $\text{Cl}^-(aq)$ has a lesser tendency to be oxidized than $\text{H}_2\text{O}(l)$.

Simplify as follows:[14]

$$-0.1436 \text{ V} = \frac{0.0592 \text{ V}}{2} \log \left(\frac{P_{O_2}^{1/2}[\text{H}_3\text{O}^+]^2}{[\text{Co}^{2+}]} \right) - \frac{0.0592 \text{ V}}{2} \log \left(\frac{P_{O_2}^{1/2}[\text{H}_3\text{O}^+]^2}{[\text{Sn}^{2+}]} \right)$$

$$-4.851 = \log \left(\frac{P_{O_2}^{1/2}[\text{H}_3\text{O}^+]^2}{[\text{Co}^{2+}]} \right) - \log \left(\frac{P_{O_2}^{1/2}[\text{H}_3\text{O}^+]^2}{[\text{Sn}^{2+}]} \right)$$

$$-4.851 = \log \left(\frac{[\text{Sn}^{2+}]}{[\text{Co}^{2+}]} \right) \quad \text{which gives} \quad 1.408 \times 10^{-5} = \frac{[\text{Sn}^{2+}]}{[\text{Co}^{2+}]}$$

The concentrations of the two metal ions start equal, so only $\boxed{0.000014}$ of the Sn^{2+} remains when the Co^{2+} finally can plate out.

12-93 Calculate $\Delta H°$ and $\Delta S°$ for the reaction:

$$\text{Cu}(s) + 2\,\text{Ag}^+(aq) \rightarrow \text{Cu}^{2+}(aq) + 2\,\text{Ag}(s)$$

taking the needed $\Delta H_f°$ and $S°$ data from text Appendix D:

$$\Delta H° = \underbrace{(64.77)}_{\text{Cu}^{2+}(aq)} + 2\underbrace{(0.00)}_{\text{Ag}(s)} - \underbrace{(0.00)}_{\text{Cu}(s)} - 2\underbrace{(105.58)}_{\text{Ag}^+(aq)} = -146.39 \text{ kJ}$$

$$\Delta S° = \underbrace{(-99.6)}_{\text{Cu}^{2+}(aq)} + 2\underbrace{(42.55)}_{\text{Ag}(s)} - \underbrace{(33.15)}_{\text{Cu}(s)} - 2\underbrace{(72.68)}_{\text{Ag}^+(aq)} = -193.01 \text{ J K}^{-1}$$

Now, eliminate $\Delta G°$ between the two equations:

$$\Delta G° = \Delta H° - T\Delta S° \quad \text{and} \quad -n\mathcal{F}\Delta\mathcal{E}° = \Delta G°$$

Solving the resulting equation for $\Delta\mathcal{E}°$ gives:

$$\Delta\mathcal{E}° = \frac{-\Delta H°}{n\mathcal{F}} + T\frac{\Delta S°}{n\mathcal{F}}$$

This equation shows the dependence of the standard potential difference of *any* cell on temperature. If $\Delta S°$ of the cell reaction is positive, then raising the constant temperature T of the system raises the cell voltage; if $\Delta S°$ is negative, then raising T lowers the cell voltage. The $\Delta S°$ of the cell in this problem is negative, so the cell voltage $\boxed{\text{decreases with increasing } T}$.

Tip. It was not necessary to compute $\Delta H°$ to get the answer in this problem. Also, the way the voltage of the cell changes with temperature is of interest. Take the

---

[14]Recall that $\log a - \log b = \log(a/b)$.

derivative of $\Delta\mathcal{E}^\circ$ with respect to T in the preceding equation, and substitute the various values:

$$\frac{d(\Delta\mathcal{E}^\circ)}{dT} = \frac{\Delta S^\circ}{n\mathcal{F}} = \frac{-193.01 \text{ J K}^{-1}}{2(9.6485 \times 10^4 \text{ C})} = -1.00 \times 10^{-3} \text{ V K}^{-1}$$

The voltage drops by a millivolt for every kelvin that the temperature increases.

12-95 The problem is to estimate the solubility of AgBr in 0.10 M aqueous NaBr. It is possible to look up the K_{sp} of AgBr in text Table 11-3[15] and use the method of **11-25**, but it is more sporting to use the data given in the problem, which include:

$$\text{Ag}^+(aq) + e^- \rightarrow \text{Ag}(s) \qquad\qquad \mathcal{E}^\circ = 0.7996 \text{ V}$$
$$\text{AgBr}(s) + e^- \rightarrow \text{Ag}(s) + \text{Br}^-(aq) \qquad \mathcal{E}^\circ = 0.0713 \text{ V}$$

Combine these two half-equations to give the dissolution equation:

$$\text{AgBr}(s) \rightarrow \text{Ag}^+(aq) + \text{Br}^-(aq)$$

Since AgBr must end up on the left side of the dissolution equation, the first half-equation was subtracted from the second. The standard potential difference of this reaction is:

$$\Delta\mathcal{E}^\circ = \mathcal{E}^\circ(\text{cathode}) - \mathcal{E}^\circ(\text{anode}) = 0.0713 - 0.7996 = -0.7283 \text{ V}$$

Write the Nernst equation at 25°C for the dissolution:

$$\Delta\mathcal{E} = -0.7283 \text{ V} - \frac{0.0592 \text{ V}}{1} \log\left([\text{Ag}^+][\text{Br}^-]\right)$$

The solubility is measured at equilibrium, so $\Delta\mathcal{E}$ equals zero. Insert 0.10 M, which is a given, as the concentration of Br$^-$ ion. Then:

$$0 = -0.7283 \text{ V} - (0.0592 \text{ V}) \log\left([\text{Ag}^+][0.10]\right) \quad \text{from which:} \quad [\text{Ag}^+] = 5 \times 10^{-12} \text{ M}$$

The solubility of the AgBr equals the equilibrium concentration of Ag$^+$ ion because AgBr is the only source of Ag$^+$ ion and all dissolved silver is in the form of Ag$^+$ ion. The estimated solubility is therefore $\boxed{5 \times 10^{-12} \text{ mol L}^{-1}}$.

---

[15]Text page 374.

Chapter 13

Chemical Kinetics

Chemical kinetics is the study of the **rates** and **mechanisms** of chemical reactions. Most chemical reactions do not occur as represented in chemical equations but instead by a series of simpler steps that add together to give the net equation. A **mechanism** is a sequence of such **elementary steps**.

Rates of Chemical Reactions

As a chemical reaction progresses, the concentrations or partial pressures of the reactants fall, and the concentrations or partial pressures of the products increase. The **average reaction rate** equals the change in the concentration of a reactant or product divided by the time interval Δt over which it occurs. The **instantaneous rate** is the rate obtained in the limit as smaller and smaller values of Δt are considered.

The instantaneous rate of the general chemical reaction:

$$a\text{A} + b\text{B} \rightarrow c\text{C} + d\text{D}$$

is determined by monitoring the disappearance of a reactant or the appearance of a product. It is defined as:

$$\text{rate} = -\frac{1}{a}\frac{d[\text{A}]}{dt} = -\frac{1}{b}\frac{d[\text{B}]}{dt} = +\frac{1}{c}\frac{d[\text{C}]}{dt} = +\frac{1}{d}\frac{d[\text{D}]}{dt}$$

In this equation, the first d in the last term stands for the coefficient of the product D in the chemical equation, and the others are part of the derivative. Because the concentration of A falls as the reaction proceeds, the time rate of change of this concentration, $d[\text{A}]/dt$, is negative. The same holds for all reactants. The minus signs are put in to keep the rate positive.

- **The instantaneous rate of a reaction changes from moment to moment.**

The **initial rate** of a reaction is the rate at the moment that the reactants are mixed. It is the rate at $t = 0$.

The experimental determination of reaction rates requires a thermostat, because rates are quite sensitive to temperature, a clock to measure the time, and a means to monitor the concentration of at least one product or reactant. In some cases it is possible to *quench* (stop) a reaction by quickly cooling it and then to analyze for the concentration of a reactant or product. An alternative is to monitor some physical property as a continuous function of time. The total pressure is often used. See **13-50**. The color of the reaction mixture is another property frequently monitored.

Rate Laws

Rates of reactions almost always change with time. Most reactions slow down, but reactions that start slow and pick up speed are also known. One reason for a reaction to slow down is that any observed rate of a reaction is really a **net rate**, the difference between the rate of the forward reaction and of the reverse reaction. If pure reactants are mixed, reverse reaction cannot occur at first because no products are available to react. Back-reaction takes hold only later.

Reaction Order

The forward rate of a reaction may change with time even if back-reaction is prevented by quick removal of the products. In general the forward rate depends on the concentrations of the reactants. The dependence is expressed in the **rate law** of the chemical reaction. For the general chemical equation written above, the common form of the rate law is:

$$\text{rate} = -\frac{1}{a}\frac{d[\text{A}]}{dt} = k[\text{A}]^m[\text{B}]^n$$

The exponents m and n may equal positive or negative whole numbers or fractions or zero. A species whose concentration has an exponent of zero has no effect on the rate of the reaction (although some of it does have to be present if it is a reactant). A species with a negative exponent in a rate law makes the reaction go more slowly as its concentration is raised. See **13-6** for an illustration of both of these points.

The constant of proportionality k is the **rate constant**. Rate constants depend strongly on the temperature. See text Section 13-5.[1] The exponents m and n determine the **order** of the reaction: m is the order with respect to reactant A, and n is the order with respect to reactant B. The **overall order** of the reaction is the sum of all the exponents in the rate law.

---

[1] Text page 468.

The Units of k

The overall order of a reaction determines the units of its rate constant. A rate (not rate constant) is measured in units of moles per liter per second (mol $L^{-1}s^{-1}$). In the rate law for a first-order reaction, k is multiplied by a concentration (mol L^{-1}). The units of k must be s^{-1} to make the rate come out as required. In a second-order reaction k is multiplied by *two* concentrations (with overall units $mol^2 L^{-2}$). The units of k now must be L $mol^{-1}s^{-1}$ to make the units of the rate come out right. See **13-7b**.

Determining a Rate Law

Rate laws are experimental results. In one kind of experiment aimed at establishing rate laws, the concentrations of all but one reactant are held constant in a series of runs while the initial rate is measured with different concentrations of the one reactant under study. The initial rate is the rate observed before back-reaction that might occur among the products has a chance to confuse the issue. If doubling the concentration of the reactant under test doubles the initial rate, then the reaction is first-order in the test reactant. If the same doubling quadruples the rate, then the reaction is second order in that one reactant. See **13-5** and **13-7c**.

Integrated Rate Laws

The instantaneous rate of consumption (or production) of a reactant (or product) changes as a chemical reaction proceeds. An **integrated rate law** takes account of this and still expresses the concentration of the product or reactant directly as a function of time. Although there are mathematical methods by which to integrate nearly all of the many experimental rate laws that occur, the necessary calculus can be quite complex. Fortunately, first-order and second-order reactions are the most common and have simple integrated forms. It is best simply to learn them.

   • **First-order Integrated Rate Law.** Consider the reaction: A $\rightarrow$ products. Assume that it is first order. Let c equal the concentration of A at any time t after the reaction starts. Let c_0 equal the original concentration of A (the concentration at time zero). Then:

$$c = c_0 e^{-kt} \qquad \ln c - \ln c_0 = -kt \qquad \ln\left(\frac{c}{c_0}\right) = -kt$$

These all equations convey the same relationship. Because the change in $\ln c$ over equal intervals of time is constant in a first-order reaction, a plot of $\ln c$ against time will be a straight line. The rate constant k is then the negative slope of this line. See **13-13**.

The **half-life $t_{1/2}$** of a first-order reaction is the time required for the concentration of a reactant to decrease to half of its initial value. Big rate constants mean fast reactions. Fast reactions require little time to consume half the initial concentration of a reactant. Thus the half-life is inversely proportional to the rate constant:

$$\text{half life} = t_{1/2} = \frac{\ln 2}{k}$$

See **13-13** and **13-15**.

• **Second-order Integrated Rate Law.** For the second-order reaction:

$$2\,A \rightarrow \text{products} \qquad \text{the rate law is} \qquad \text{rate} = -\frac{1}{2}\frac{d[A]}{dt} = k[A]^2$$

and the integrated rate law is:

$$\frac{1}{c} = \frac{1}{c_0} + 2kt$$

where c is the concentration of A at any time t and c_0 is its original concentration. This integrated rate equation predicts that a plot of $1/c$ versus time will be a straight line. It is possible to predict the concentrations of reactants and products at any time given the rate constant and integrated rate law for a reaction (see **13-9** and **13-15**). It is also possible to figure out the time required for some specified change in concentration (**13-17** and **13-62**).

Integrated rate laws for reactions of the same order may differ somewhat if the stoichiometries of the reactions differ. The integrated rate law for the second order reaction $A + B \rightarrow$ products is:

$$\frac{1}{c} = \frac{1}{c_0} + kt$$

where c and c_0 refer to the concentration of A. The factor of 2 is missing here because the coefficient of 2 is missing in the balanced equation. This point is important in **13-17** and **13-69**.

Reaction Mechanisms

A reaction mechanism consists of a sequence of **elementary steps** that tell exactly what molecules must collide and in what sequence to convert the reactants into products. A mechanism also includes an indication of the rates of all the steps. This indication consists of a numerical rate constant for each step or a word or two (fast, slow, very fast, and so forth) giving a rough idea of the relative rates of the steps. Only the actual rate constants themselves are a complete statement of the rates and relative rates of the steps.

A mechanism may include **intermediates**. Intermediates are species produced in a early step and later consumed. Mechanisms may also include **catalysts**, which are species that come from outside, join in an early step in the mechanism and are regenerated in their original form by a later step. The sum of all of the steps of a reaction mechanism equals the balanced chemical reaction. See **13-21**, **13-57**, and **13-59**.

Elementary Reactions

An elementary step is also called an elementary reaction. Elementary reactions consist of the collision *in a single event* of the reactant particles to give the product particles. Elementary reactions have **molecularity**. The collision of two molecules to give products is a **bimolecular** elementary reaction. The simultaneous collision of three particles is a **termolecular** elementary reaction. An **unimolecular** reaction involves only a single reactant molecule. Only elementary reactions have molecularity. The order of elementary reactions equals their molecularity. The vast majority of chemical reactions proceed by way of mechanisms that are the sums of several elementary steps. The concept of molecularity does not apply to such non-elementary reactions.

Rate Constants and the Equilibrium Constant

In any mechanism, the product of the rate constants for the forward steps divided by the product of the rate constants for the reverse steps is equal to the equilibrium constant of the overall reaction. See **13-51**.

Rate Determining Steps

If one of the elementary steps in a mechanism is much slower than the rest, it acts as a bottle-neck and determines the overall rate of reaction. With this in mind, the procedure for writing the rate law predicted by a mechanism is:

1. Write the rate law for the slow step. This requires nothing more than taking the coefficients of the species in the slow step as the exponents in the rate law.

2. Use algebra to eliminate the concentrations of any intermediates that may occur in the slow-step rate law. Intermediates often appear in mechanisms but may never appear in experimental rate laws. Deducing a rate law from a mechanism is a waste of effort unless the result is in a form directly comparable with experiment.

Many mechanisms have one or more fast equilibria preceding the slow step. Problems **13-25** and **13-57** show how to get rid of concentrations of intermediates in writing rate laws from mechanisms.

The Steady-State Approximation

If all the steps in a mechanism have about the same rate, then no single one of them is rate-determining. The **steady-state approximation** is useful in such cases. In this approximation, it is assumed that the concentrations of intermediates remain constant over most of the course of the reaction. Suppose an intermediate is produced by the first step of a mechanism and consumed by the second. The net rate of change of its concentration equals its rate of production minus its rate of consumption. In the steady-state approximation this net rate of change is set equal to zero. See **13-33**. The net rate of change may involve several different terms, as in **13-57**. In every case, the steady-state approximation furnishes an equation giving the concentrations of intermediates in terms of the concentrations of non-intermediates and rate constants. In this way it allows the elimination of the concentrations of intermediates as a rate law is deduced from a mechanism. See **13-67**.

Chain Reactions

A chain reaction proceeds through a series of elementary steps, some of which are repeated many times. The repetition arises when a step in a mechanism consumes a reactive intermediate but simultaneously generates another intermediate. Reaction of this intermediate then generates some more of the first intermediate. Chain reactions have three stages: **initiation**, the production of the first intermediate; **propagation**, the use-one, make-one weaving of the chain, and **termination**, in which two reactive intermediates collide and thereby take each other out of the chain-generating scheme. A reaction in which exactly one new intermediate is generated for every one used proceeds at a steady rate (see **13-59**). A **branching chain reaction** features an increase in the number of intermediates and accelerates as it proceeds. In the branching chain reaction diagrammed in Figure 14-10,[2] which is a nuclear reaction, the intermediate is the neutron.

Effect of Temperature on Reaction Rates

Molecules must collide in order to react, but not all collisions lead to reactions. Most molecular collisions are not energetic enough. Only collisions for which the collision energy exceeds some minimum can result in reaction. The minimum is the **activation energy E_a** of the reaction.

The **Arrhenius equation** gives the temperature dependence of rate constants (and thus of reaction rates themselves):

$$k = Ae^{-E_a/RT}$$

---

[2]Text page 509.

The activation energy has the same units as RT; kJ mol^{-1} are common. The pre-exponential factor A is a constant having the same units as k. See **13-39a**. If E_a were zero, then the exponent in the Arrhenius equation would equal zero and, because any number raised to the zero power is 1, k would equal A.

The following form of the Arrhenius equation is useful for comparing rate constants at different temperatures:

$$\ln \left(\frac{k_1}{k_2} \right) = + \frac{E_a}{R} \left(\frac{1}{T_2} - \frac{1}{T_1} \right) = - \frac{E_a}{R} \left(\frac{1}{T_1} - \frac{1}{T_2} \right)$$

where k_1 is the rate constant at T_1 and k_2 is the rate constant at T_2. Note that if k_1 and k_2 are exchanged on the left in the above equation, then the signs in front of (E_a/R) reverse.

Clearly, measuring the rate constant at just two temperatures could give the activation energy. In practice, rate constants are measured at several different temperatures and the activation energy is determined from the slope of the line in a plot of $\ln k$ versus $1/T$ (see **13-35**).

Reactions have activation energies because formation of a product requires more than a mere encounter between the reactant molecules. As reactant molecules approach each other along a **reaction path**, they must possess sufficient energy to scale an energy barrier and become an unstable **activated complex**. Products form from this **transition state**. The activation energy for a forward reaction is in general different from the activation energy for the reverse reaction. The difference between $E_{a,f}$ and $E_{a,r}$ equals the thermodynamic ΔE of the reaction.

The activation energy of an elementary step is always positive. For overall reactions consisting of two or more elementary steps, E_a can be negative as well as positive. A negative E_a means that the observed rate constant diminishes with increasing temperature. This occurs when a fast exothermic equilibrium (with equilibrium constant K) precedes the rate-determining step (with rate constant k) in the reaction's mechanism. The observed rate constant is such a case equals Kk. See **13-25**. Although k rises with temperature, K for an exothermic reaction gets smaller and can outweigh the increase in k in determining the behavior of Kk.

Reaction Dynamics

Results from kinetic theory[3] on the rate of collisions among like molecules in the gaseous phase provide insight into the rates of gas-phase reactions. Rate constants can be predicted knowing the diameter d and molar mass $\mathcal{M}$ of the molecules that

---

[3]See text Section 4-6, text page 119.

are colliding and the activation energy of a collision:

$$k = 2Pd^2 N_0 \sqrt{\frac{\pi RT}{\mathcal{M}}} e^{-E_a/RT}$$

The factor P is a steric factor to account for the fact that some of the collisions that are energetic enough to cause reaction are ineffective because the molecules collide in the wrong orientation. The units of this expression require some care (see **10-43**).

Kinetics of Catalysis

A catalyst is a substance that enters into a chemical reaction and speeds it up, but is not itself consumed in the reaction. Catalysts do not appear in the balanced overall equation representing a reaction, but do appear in the rate law. Catalysts accelerate chemical reactions by providing new reaction pathways that have lower activation energies.

An **inhibitor** is the opposite of a catalyst. It *slows* the rates of chemical reactions. See **10-67**. In a sense a catalyst and intermediate are opposites, too. An intermediate is generated within a reaction's mechanism and consumed there; it may *not* appear in the rate law for the reaction. A catalyst joins a mechanism from without and is regenerated before the mechanism is over; it *may* appear in the rate law for the reaction. See **13-21** and **13-65**.

In **homogeneous catalysis** the catalyst is present in the same phase as the reactants. In **heterogeneous catalysis** the reactants are in one phase and the catalyst is in another phase.

Enzyme Catalysis

Enzymes catalyze many chemical reactions that take place in living organisms. They are high-molecular-mass proteins that bind at their **active sites** to reactant molecules (called **substrates** of the enzyme) and promote specific transformations of the reactant molecules. The kinetics of enzyme-catalyzed reactions involve no kinetic concepts that do not also appear in non-enzyme catalysis. Problem **13-67** shows how the rate of an enzyme-catalyzed reaction is lowered by the presence of an inhibitor.

Detailed Solutions to Odd-Numbered Problems

13-1 Text Figure 13-3[4] shows the concentration of NO in a system as a function of time. The instantaneous rate of production of NO equals the slope of the tangent to

---

[4]Text page 449.

this line. The slope of the tangent changes constantly. Sketch in the tangent line at t equal to 200 s. Reading from the graph (as shown in the Figure for $t = 150$ s) gives $\boxed{5.3 \times 10^{-5} \text{ mol L}^{-1}\text{s}^{-1}}$ as the instantaneous rate.

13-3 The rate of a reaction is expressed in terms of the rate of disappearance of a reactant, or formation of a product:

$$\text{rate} = \boxed{-\frac{1}{1}\frac{d[N_2]}{dt} = -\frac{1}{3}\frac{d[H_2]}{dt} = +\frac{1}{2}\frac{d[NH_3]}{dt}}$$

13-5 a) The way in which the rate of the reaction is affected by changes in concentration of the two reactants gives the order of the reaction. This reaction is first order in H_2 because the rate varies as to the first power of the concentration of H_2 and second order in NO because the rate varies as to the second power of the concentration of NO. The rate expression is

$$\boxed{\text{rate} = k[H_2][NO]^2}$$

The units of the rate constant k are $\boxed{\text{L}^2\,\text{mol}^{-2}\,\text{s}^{-1}}$ This can also be written as $\text{M}^{-2}\text{s}^{-1}$. The reaction has an overall order of 3.

b) Multiplying $[H_2]$ by 2 would double the rate, and multiplying $[NO]$ by 3 would increase the rate by a factor of 9. The combined effect would be to increase the rate by a $\boxed{\text{factor of 18}}$.

13-7 a) Compare the second and third data lines in the table, and assume that the difference that is shown is the *only* difference between the two runs. When $[C_5H_5N]$ is held constant and $[CH_3I]$ is doubled, the rate doubles. The reaction is first order in CH_3I. Next, compare the first and second lines in the table. When both concentrations are doubled, the rate is increased by a factor of 4. Half of this is due to the change in the concentration of CH_3I, so the other half is due to the change in the concentration of C_5H_5N. Hence:

$$\boxed{\text{rate} = k[C_5H_5N][CH_3I]}$$

b) Calculate k by substituting the data from any one of the three data points into the rate equation. Using the first line of the table:

$$7.5 \times 10^{-7} \text{ mol L}^{-1}\text{s}^{-1} = k(1.00 \times 10^{-4} \text{ mol L}^{-1})(1.00 \times 10^{-4} \text{ mol L}^{-1})$$

Hence k is $\boxed{75 \text{ L mol}^{-1}\text{s}^{-1}}$.

c) Substitute the known k and the two given concentrations into the rate equation. Remember that "M" stands for mol L^{-1}.

$$\text{rate} = (75 \text{ mol}^{-1}\text{L s}^{-1)})(5.0 \times 10^{-5} \text{ mol L}^{-1})(2.0 \times 10^{-5} \text{ mol L}^{-1})$$

$$= \boxed{7.5 \times 10^{-8} \text{ mol L}^{-1}\text{s}^{-1}}$$

13-9 The problem asks for the time it will take for the pressure to fall to half of its initial value. This is exactly the half-life. In a first-order process the half-life depends solely upon the rate constant. In this case:

$$t_{1/2} = \frac{\ln 2}{k} = \frac{0.69315}{2.2 \times 10^{-5} \text{ s}^{-1}} = \boxed{3.2 \times 10^4 \text{ s}}$$

Tip. Recall that partial pressures of gases are directly proportional to their concentrations (assuming ideality).

13-11 The decomposition of benzenediazonium chloride is first-order, following the integrated rate law:

$$P_{C_6H_5N_2Cl} = (P_{C_6H_5N_2Cl})_0 \, e^{-kt}$$

Express the elapsed time in seconds and insert the time, together with the initial partial pressure and the rate constant, into the preceding:

$$P_{C_6H_5N_2Cl} = (0.0088 \text{ atm})e^{(-4.3\times10^{-5} \text{ s}^{-1})(3.60\times10^4 \text{ s})} = \boxed{0.0019 \text{ atm}}$$

13-13 This first-order reaction follows the integrated rate law:

$$[C_2H_5Cl] = [C_2H_5Cl]_0 e^{-kt}$$

Divide both sides of this equation by the original concentration of the C_2H_5Cl, and take the natural logarithm of both sides:

$$\ln \frac{[C_2H_5Cl]}{[C_2H_5Cl]_0} = -kt$$

Substitution of two concentrations and the time gives:

$$\ln \frac{0.0016 \text{ M}}{0.0098 \text{ M}} = -k(340 \text{ s})$$

which is easily solved to give k equal to $\boxed{5.3 \times 10^{-3} \text{ s}^{-1}}$.

13-15 The integrated rate law for this very fast second-order association of iodine atoms is:

$$\frac{1}{[\text{I}]} = \frac{1}{[\text{I}]_0} + 2kt$$

Substitute the given concentration, rate constant and time:

$$\frac{1}{[\text{I}]} = \frac{1}{1.00 \times 10^{-4}\ \text{M}} + 2(8.2 \times 10^9\ \text{M}^{-1}\text{s}^{-1})(2.0 \times 10^{-6}\ \text{s})$$

Solving gives $[\text{I}] = \boxed{2.34 \times 10^{-5}\ \text{M}}$.

13-17 The reaction is the neutralization of $\text{OH}^-(aq)$ with $\text{NH}_4^+(aq)$. Aqueous acid-base reactions are generally fast. This reaction is no exception, as shown by the huge room-temperature rate constant of $k = 3.4 \times 10^{10}\ \text{M}^{-1}\ \text{s}^{-1}$. The answer will be a very short time. If 1.00 L of 0.0010 M NaOH and 1.00 L of 0.0010 M NH_4Cl are mixed, then *after* the mixing, but *before* the reaction can start each reactant has a concentration of 5.0×10^{-4} M. The kinetics are second-order overall:

$$\text{rate} = \frac{-d[\text{OH}^-]}{dt} = k[\text{OH}^-][\text{NH}_4^+]$$

Throughout the reaction $[\text{OH}^-] = [\text{NH}_4^+]$. Let this concentration be represented by c. Then:

$$\frac{-dc}{dt} = kc^2$$

Integrating this equation and inserting the initial condition gives:

$$\frac{1}{c} = \frac{1}{c_0} + kt$$

This equation does not include the factor of 2 that appears in text equation 13-4 because the stoichiometry of the reaction lacks that factor. For c_0 equal to 1.0×10^{-5} M and k equal to $3.4 \times 10^{10}\ \text{M}^{-1}\text{s}^{-1}$, the equation becomes:

$$\frac{1}{1.0 \times 10^{-5}\ \text{M}} = \frac{1}{5.0 \times 10^{-4}\ \text{M}} + (3.4 \times 10^{10}\ \text{M}^{-1}\text{s}^{-1})t$$
$$9.8 \times 10^4\ \text{M}^{-1} = (3.4 \times 10^{10}\ \text{M}^{-1}\text{s}^{-1})t$$
$$t = \boxed{2.9 \times 10^{-6}\ \text{s}}$$

13-19 a) Two particles collide in an elementary reaction. The reaction is therefore bimolecular. Its rate law is: $\boxed{\text{rate} = k[\text{HCO}][\text{O}_2]}$.

b) Three particles collide in an elementary reaction. The reaction is therefore termolecular: $\boxed{\text{rate} = k[CH_3][O_2][N_2]}$.

c) A single particle decomposes spontaneously in an elementary reaction. The reaction is unimolecular: $\boxed{\text{rate} = k[HO_2NO_2]}$.

13-21 a) The first step is unimolecular; the three subsequent steps are bimolecular. This is determined simply by counting the number of interacting particles on the left sides of the four equations. Molecularity has meaning only in reference to elementary reactions.

b) The overall reaction is the sum of the steps: $H_2O_2 + O_3 \rightarrow H_2O + 2\,O_2$.

c) The intermediates are O, ClO, CF_2Cl, and Cl. These species are produced in the course of the reaction and later consumed.

Tip. The CF_2Cl_2 is a catalyst. It reacts in an early stage in the mechanism but is later regenerated. The overall reaction could not proceed according to this mechanism without the presence and participation of CF_2Cl_2, but the compound is neither consumed nor produced.

13-23 The equilibrium constant of this elementary reaction equals the ratio of the rate constant of the forward reaction to the rate constant of the reverse reaction:

$$K = \frac{k_f}{k_r} = 5.0 \times 10^{10} = \frac{1.3 \times 10^{10}\ \text{L mol}^{-1}\text{s}^{-1}}{k_r}$$

Solving for k_r gives $0.26\ \text{L mol}^{-1}\text{s}^{-1}$. The reaction in question is in fact just the reverse of the original elementary reaction, so the answer is $\boxed{0.26\ \text{L mol}^{-1}\text{s}^{-1}}$.

13-25 a) The rate-limiting elementary step in a mechanism determines the overall reaction rate. In this case, the slow step is: $C + E \rightarrow F$. A preliminary version of the rate law is:

$$\text{rate} = k_2[C][E]$$

Unfortunately, the expression involves the concentration of C, an intermediate. This is unacceptable. To eliminate [C] in the rate law, consider how C is formed. It arises in the first step of the mechanism, a fast equilibrium. For that first step:

$$k_1[A][B] = k_{-1}[C][D]$$

Solve this equation for the concentration of C and substitute into the preliminary rate law:

$$\boxed{\text{rate} = \frac{k_1 k_2}{k_{-1}}\frac{[A][B][E]}{[D]}}$$

The overall reaction is the sum of the two steps: $\boxed{A + B + E \rightarrow D + F}$.

Tip. In this reaction system, the accumulation of one product (D) slows down the reaction, but the accumulation of the other product (F) does not.

b) The overall reaction in this case is: $\boxed{A + D \rightarrow B + F}$. For the two fast equilibria:

$$k_1[A] = k_{-1}[B][C] \quad \text{and} \quad k_2[C][D] = k_{-2}[E]$$

The last, slow step is the rate-determining step:

$$\text{rate} = k_3[E]$$

But E is an intermediate and its concentration may not appear in the final rate expression. To eliminate [E], solve the second of the preceding pair of equations of [E] and substitute:

$$\text{rate} = \frac{k_2 k_3}{k_{-2}}[C][D]$$

This expression *still* contains the concentration of an intermediate (C now). Eliminate [C] by solving the first of the pair for [C] and substituting:

$$\boxed{\text{rate} = \frac{k_1 k_2 k_3}{k_{-1} k_{-2}} \frac{[A][D]}{[B]}} = k_{\text{expt}} \frac{[A][D]}{[B]}$$

where the experimental k is the algebraic composite of the several step-wise rate constants. The reaction is first order in both A and D, is -1 order in B and first order overall.

13-27 The reaction of HCl with propane is first order in propane and third order in HCl. This is an experimental fact. Mechanism (a) proposes the rapid formation of the intermediate H from 2 HCl's followed by slow combination of the H with CH_3CHCH_2. This mechanism thus predicts second-order kinetics in HCl and must be wrong.

Mechanism (b) proposes *two* fast equilibria. The first involves HCl only and the second involves HCl and propane. The slow step is the chemical combination of the two intermediates that these equilibria produce. The rate is accordingly proportional to the concentrations of all the reactants in the two fast equilibria. HCl occurs three times among these reactants and propane occurs once so $\boxed{\text{mechanism (b) fits}}$ with the observed rate law.

Mechanism (c) involves HCl in the fast production of two different intermediates which then slowly combine to give the product (and to regenerate some HCl). It predicts second-order kinetics in HCl (2 HCl's consumed to furnish reactants for the slow step). It is therefore not consistent with observation.

Tip. A correct mechanism for a reaction must predict the observed rate law; a mechanism that correctly predicts the observed rate law may *still* be wrong. In other words, predicting the rate law is necessary but not sufficient for correctness in a mechanism.

13-29 The three mechanisms predict the three rate laws:

$$\text{rate} = k_1[NO_2Cl] \qquad \text{rate} = \frac{k_1 k_2}{k_{-1}} \frac{[NO_2Cl]^2}{[Cl_2]} \qquad \text{rate} = \frac{k_1 k_2 k_3}{k_{-1}k_{-2}} \frac{[NO_2Cl]^2}{[NO_2]}$$

The cluster of constants in the second and third rate laws can be regarded as a new single rate constant "k_{expt}." Only $\boxed{\text{mechanism (a)}}$ is consistent with experiment.

13-31 The overall reaction is $A + B + E \rightarrow D + F$. It proceeds first by an equilibrium between A and B giving D and the intermediate C, and then by the consumption of C in reaction with E to give F. The rate of appearance of C equals $k_1[A][B]$, and the rate of *disappearance* of C equals $k_{-1}[C][D] + k_2[C][E]$. This latter is a sum because C disappears both by back reaction (with D to give A plus B) and by *further* reaction (with E to give F). If the concentration of intermediate C is constant (the steady state approximation), the rate of disappearance of C must equal its rate of appearance:

$$k_1[A][B] = k_{-1}[C][D] + k_2[C][E]$$

The rate of the reaction can be expressed in terms of the rate of appearance of a product, for example, F:

$$\text{rate} = \frac{d[F]}{dt} = k_2[E][C]$$

Solve the steady-state equation for [C] and substitute the answer in the preceding:

$$\text{rate} = \boxed{\frac{k_1 k_2[A][B][E]}{k_2[E] + k_{-1}[D]}}$$

If $\boxed{k_2[E] \text{ is much smaller than } k_{-1}[D]}$ then this expression becomes:

$$\text{rate} \approx \left(\frac{k_1 k_2}{k_{-1}}\right) \frac{[A][B][E]}{[D]}$$

The second-step rate constant k_2 is much smaller than k_{-1} when the first step of the reaction is a fast equilibrium.

13-33 The reaction is the decomposition of nitryl chloride $2\,NO_2Cl \rightarrow 2\,NO_2 + Cl_2$. The mechanism involves a equilibrium breakdown of the reactant to NO_2 plus Cl followed by reaction of the atomic chlorine with a second molecule of NO_2Cl to generate the products. The change in the concentration of Cl with time is:

$$\frac{d[\text{Cl}]}{dt} = k_1[\text{NO}_2\text{Cl}] - k_{-1}[\text{Cl}][\text{NO}_2] - k_2[\text{Cl}][\text{NO}_2\text{Cl}]$$

because the rate of change of the concentration of Cl equals the rate of its production minus the rate of its consumption. The steady-state approximation is that $d[\text{Cl}]/dt = 0$. If so, then:

$$k_1[\text{NO}_2\text{Cl}] - k_{-1}[\text{Cl}][\text{NO}_2] - k_2[\text{Cl}][\text{NO}_2\text{Cl}] = 0$$

and:

$$[\text{Cl}] = \frac{k_1[\text{NO}_2\text{Cl}]}{k_{-1}[\text{NO}_2] + k_2[\text{NO}_2\text{Cl}]}$$

The rate of the overall reaction equals the rate of the final elementary step, which generates the two products:

$$\text{rate} = \frac{d[\text{Cl}_2]}{dt} k_2[\text{NO}_2\text{Cl}][\text{Cl}]$$

Substitute the expression for the concentration of the Cl into this equation:

$$\boxed{\text{rate} = \frac{d[\text{Cl}_2]}{dt} \frac{k_1 k_2[\text{NO}_2\text{Cl}]^2}{k_{-1}[\text{NO}_2] + k_2[\text{NO}_2\text{Cl}]}}$$

Tip. The answer differs from the rate expression given for the same reaction in **13-29**. It becomes equal to that expression as the second term in the denominator becomes large compared to the first. This happens if k_2 is large compared to k_{-1} or if $[\text{NO}_2]$ is small, as it would be in the initial stages of the reaction.

Tip. Lewis structures for NO_2Cl (nitryl chloride) are given in **3-63**, and the second-order formation of nitryl fluoride is mentioned on text page 450 and used as an example on text page 462.

13-35 a) The rate constant of an elementary reaction depends on the absolute temperature and activation energy E_a according to the Arrhenius equation:

$$k = Ae^{-E_a/RT}$$

where A is a constant. Take the natural logarithm of both sides:

$$\ln k = \ln A - \frac{E_a}{RT}$$

This means that a plot of $\ln k$ versus the reciprocal of T should be a straight line with a slope of $-E_a/R$ and an intercept (when $1/T = 0$) of $\ln A$. Two points determine a line. A quick way to estimate E_a is to select any two of the four data points in the problem (such as the first two), insert the values in the above equation:

$$\ln(5.49 \times 10^6) = \ln A - \frac{E_a}{(5000 \ K)R} \quad \text{and} \quad \ln(9.86 \times 10^6) = \ln A - \frac{E_a}{(1000 \ K)R}$$

and then solve for E_a (by eliminating $\ln A$ between the equations). This procedure gives $E_a = 432$ kJ mol^{-1}. Note that E_a has the same units as RT. Selecting *another* pair of points (for example the second two) and doing the same thing gives a somewhat different answer: $E_a = 392$ kJ mol^{-1}. The discrepancy means that the experimental data do not fall exactly on a straight line. The best way to use all data is to perform a *least-squares fit,* mathematically determining the slope of the straight line that comes closest to all four data points. Many electronic calculators are equipped to complete the necessary calculations almost without effort. Based on the minimization of the sum of the squares of the deviations, E_a is $\boxed{425 \text{ kJ mol}^{-1}}$.

b) As $1/T$ goes to zero, $\ln k$ approaches $\ln A$. Recall that using just the first two data points gave $E_a = 432$ kJ mol^{-1}. Substituting this value and the k and T values of the first data point into the Arrhenius equation gives $\ln A$ equal to 25.9. Using an E_a of 392 kJ mol^{-1} and the third or fourth (k, T) pair makes $\ln A$ equal 25.5. The least-squares fitting gives $\ln A = 25.76$ so that $A = \boxed{1.54 \times 10^{11} \text{ L mol}^{-1}\text{s}^{-1}}$. This is the best answer. The units of A are always the same as the units of k.

13-37 (a) Calculating $\ln A$ for this reaction using the Arrhenius equation and the values of k and E_a given in the problem. The T is equal to 303.2 K (30.0°C). The (rearranged) Arrhenius equation for this case is:

$$\ln A = \ln k + \frac{E_a}{RT} = \ln(1.94 \times 10^{-4}) + \left(\frac{1.61 \times 10^5 \text{ J mol}^{-1}}{(8.315 \text{ J K}^{-1}\text{mol}^{-1})(303.2 \text{ K})}\right) = 55.31$$

The value of A itself might be calculated at this point, but it is not needed. Simply put $\ln A$ back into the Arrhenius equation with T equal to 313.2 (40.0°C):

$$\ln k = \ln A - \frac{E_a}{RT} = 55.31 - \left(\frac{1.61 \times 10^5 \text{ J mol}^{-1}}{(8.315 \text{ J K}^{-1}\text{mol}^{-1})(313.2 \text{ K})}\right) = -6.51$$

Taking the antilogarithm of -6.51 gives k equal to $\boxed{1.49 \times 10^{-3} \text{ L mol}^{-1}\text{s}^{-1}}$. A 10 K increase in temperature (fairly small) increases the rate constant of the reaction nearly 8-fold.

b) This reaction is second order, but the following reasoning applies to reactions of any order. The larger the rate constant of a reaction, the more rapidly it goes. Faster reactions require less time to reach any designated point in their progress. Increasing the temperature of this reaction from 30.0 to 40.0°C increases k from 1.94×10^{-4} to 14.9×10^{-4} L mol^{-1}s^{-1}, which is a factor of 7.68. The time to reach the half-way mark in the reaction is therefore reduced by a factor of 7.68. The 50 percent conversion requires only $\boxed{1.30 \times 10^3 \text{ s}}$ at 40.0°C instead of the 10,000 s it requires at 30.0°C.

13-39 a) Solve the Arrhenius equation for A and insert the quantities given in the problem:

$$A = \frac{k}{e^{-E_a/RT}} = \frac{0.41 \text{ s}^{-1}}{\exp\left(-1.61 \times 10^5 \text{ J mol}^{-1}/(8.315 \text{ J K}^{-1}\text{mol}^{-1})(600 \text{ K})\right)}$$

$$= \boxed{4.3 \times 10^{13} \text{ s}^{-1}}$$

b) Assume that neither the activation energy E_a nor the Arrhenius A changes with temperature. Solve the Arrhenius equation for k, set T equal to 1000 K, and insert these values:

$$k = 4.25 \times 10^{13} \text{ s}^{-1} \exp\left(\frac{-1.61 \times 10^5 \text{ J mol}^{-1}}{(8.315 \text{ J K}^{-1}\text{mol}^{-1})(1000 \text{ K})}\right) = \boxed{1.7 \times 10^5 \text{ s}^{-1}}$$

13-41 The activation energy is the difference in energy between the initial state and the activated complex. The activated complex is 3.5 kJ mol^{-1} higher in energy than the reactants, which are HO(g) plus HCl(g). The products, H$_2$O(g) plus Cl(g), are themselves 66.8 kJ mol^{-1} lower in energy than the reactants. It follows that the activated complex is 70.3 kJ mol^{-1} higher in energy than the products. To pass from the products to the activated complex requires $\boxed{70.3 \text{ kJ mol}^{-1}}$.

13-43 Equation 13-7,[5] relates the rate constant of a second-order reaction between like molecules to the molecular diameter and molar mass. Rewrite this equation inserting the steric factor P, which accounts for the fact that some collisions with sufficient energy may have the wrong orientation for successful reaction:

$$k = P2d^2 N_0 \sqrt{\frac{\pi RT}{M}} e^{-E_a/RT}$$

The resulting equation has the form of the Arrhenius equation. Term-by-term comparison with the Arrhenius equation identifies the Arrhenius A as follows:

$$A = P2d^2 N_0 \sqrt{\frac{\pi RT}{M}}$$

---

[5]Text page 472.

All of the quantities on the right side of this equation are available. Express them in SI base units, if they are not already in such units. This includes in particular the molar mass of NOCl, which is 0.06546 kg mol^{-1}. Substitute them into the equation and compute A:

$$A = (0.16)(2)(3.0 \times 10^{-10})^2(6.022 \times 10^{23})\sqrt{\frac{(3.1416)(8.315)(298)}{0.06546}} = 5.98 \times 10^6$$

The units in this computation were omitted. It is wise to repeat the algebra on the units alone:

$$\text{units of } A = m^2(mol^{-1})\sqrt{\frac{(\text{ J K}^{-1}mol^{-1})(K)}{kg\ mol^{-1}}} = (m)^2(mol^{-1})\sqrt{\frac{(kg\ m^2\ s^{-2})}{kg}}$$

$$= m^2\ mol^{-1}\ (m\ s^{-1}) = m^3\ mol^{-1}s^{-1}$$

The answer is therefore 6.0×10^6 m^3 mol^{-1}s^{-1}. The cubic meter is too large for use as a unit of volume in laboratory-scale chemistry. It equals 1000 L. Converting from cubic meters to liters gives a final answer of $\boxed{6.0 \times 10^9 \text{ L mol}^{-1}\text{s}^{-1}}$.

13-45 a) In this reaction, penicillin is S, the substrate, and penicillinase[6] is E, the enzyme. Write the Michaelis-Menten equation and insert the constants K_m and k_2 that are given in the problem:

$$\text{rate} = \frac{k_2[E]_0[S]}{[S] + K_m} = \frac{(2 \times 10^3 \text{ s}^{-1})(6 \times 10^{-7} \text{ mol L}^{-1})[S]}{[S] + 5 \times 10^{-5} \text{ mol L}^{-1}}$$

The rate of the reaction increases as [S], the concentration of penicillin, increases. The maximum rate will be reached when [S] is large compared to 5×10^{-5} mol L^{-1}. At this point, the denominator of the fraction essentially equals [S] and cancels out the [S] in the numerator:

$$(\text{rate})_{\text{max}} = \frac{(2 \times 10^3 \text{ s}^{-1})(6 \times 10^{-7} \text{ mol L}^{-1})}{1} = \boxed{1.2 \times 10^{-3} \text{ mol L}^{-1}\text{s}^{-1}}$$

b) Rewrite the Michaelis-Menten equation inserting the desired rate (which equals half the maximum rate), the given concentration of enzyme, and the two constants:

$$\text{rate} = \frac{(\text{rate})_{\text{max}}}{2} = \frac{1.2 \times 10^{-3} \text{ mol L}^{-1}\text{s}^{-1}}{2} = \frac{(2 \times 10^3 \text{ s}^{-1})(6 \times 10^{-7} \text{ mol L}^{-1})[S]}{[S] + 5 \times 10^{-5} \text{ mol L}^{-1}}$$

Solving for [S] gives the concentration of substrate (penicillin) required. The answer is $\boxed{5.0 \times 10^{-5} \text{ mol L}^{-1}}$.

---

[6]Names ending in "-ase" signify enzymes.

13-47 The rate expression for the process is: rate $= $ [hemoglobin][O_2]. Straightforward substitution of the given data yields the answer:

$$\text{rate} = (4 \times 10^7 \text{ L mol}^{-1}\text{s}^{-1})(2 \times 10^{-9} \text{ mol L}^{-1})(5 \times 10^{-5} \text{ mol L}^{-1})$$

$$= \boxed{4 \times 10^{-6} \text{ mol L}^{-1}\text{s}^{-1}}$$

13-49 a) For this first-order process, $\ln[\text{In}^+] = -kt + \ln[\text{In}^+]_0$. Prepare a table of the natural logarithm of the concentration of In^+ versus time:

t (s)	0	240	480	720	1000	1200	10000
$\ln[\text{In}^+]$	-4.80	-5.05	-5.30	-5.55	-5.80	-5.80	-5.80

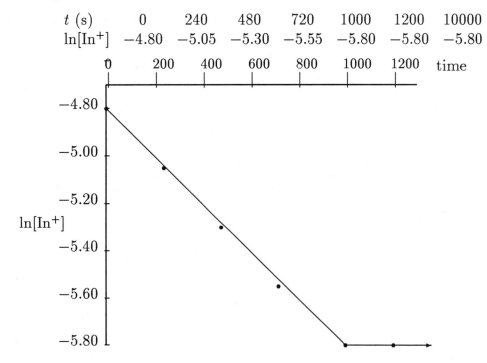

A common error is to make the values on the y-axis go up the axis as they become more negative. They should go down the axis as they become more negative. Then the plot consists of a straight line that slopes from northwest to southeast and levels off beyond 1000 s. The initial slope of the line is $-k$ and the intercept is $\ln[\text{In}^+]_0$. The slope can be read off the graph, but a least-squares analysis gives a somewhat better value. From such an analysis of the data, the slope is -1.01×10^{-3} s^{-1}. The rate constant is therefore $\boxed{1.01 \times 10^{-3} \text{ s}^{-1}}$.

b) The half-life can be determined from the rate constant,

$$t_{1/2} = \frac{\ln 2}{k} = \frac{0.6931}{1.01 \times 10^{-3} \text{ s}^{-1}} = \boxed{686 \text{ s}}$$

c) At 10000 seconds, the concentration of In^+ has persisted unchanged for 9000 seconds. The reaction has clearly reached equilibrium. The concentration of In^+ is 3.03×10^{-3} M, and the amount of In^+ in the 1.00 L solution is accordingly 3.03×10^{-3} mol. The equilibrium concentration of In^{3+} is calculated by subtracting this amount from the original amount of In^+, and using the stoichiometric relation between the In^+ and In^{3+} as follows:

$$n_{In^{3+}} = \frac{8.23 \times 10^{-3} - 3.03 \times 10^{-3}}{3} = 1.73 \times 10^{-3} \text{ mol}$$

The equilibrium concentration of In^{3+} therefore equals 1.73×10^{-3} M. Substitute in the usual mass-action expression:

$$K = \frac{[In^{3+}]}{[In^+]^3} = \frac{1.73 \times 10^{-3}}{(3.03 \times 10^{-3})^3} = \boxed{6.22 \times 10^4}$$

13-51 The reaction is: $OH^-(aq) + HCN(aq) \rightarrow H_2O(l) + CN^-(aq)$. This reaction is first-order in both OH^- and HCN, second-order overall:

$$\text{rate (forward)} = k_f[OH^-][HCN]$$

At equilibrium the forward rate is exactly equaled by the reverse rate, which, because the concentration of water is constant, depends solely on the concentration of CN^-:

$$\text{rate (reverse)} = k_r[CN^-]$$

It follows that:

$$k_f[OH^-][HCN] = k_r[CN^-] \quad \text{from which:} \quad \frac{k_r}{k_f} = \frac{[OH^-][HCN]}{[CN^-]}$$

Note that the units of the quantity on the right-hand side of the second equation are mol L^{-1}. Numerically, the right-hand side of the preceding equation equals the equilibrium constant K_b of the reaction:

$$CN^-(aq) + H_2O(l) \rightleftharpoons HCN(aq) + OH^-(aq)$$

The K_b of CN^- ion is related to K_a of its conjugate acid HCN[7]

$$K_a = \frac{K_w}{K_a} = \frac{1.0 \times 10^{-14}}{4.93 \times 10^{-10}} = 2.03 \times 10^{-5}$$

Therefore:

$$\frac{k_r}{k_f} = 2.03 \times 10^{-5} \text{ mol L}^{-1}$$

Substitution of 3.7×10^{-9} L $mol^{-1}s^{-1}$ for k_f gives $k_r = \boxed{7.5 \times 10^4 \text{ s}^{-1}}$.

---

[7]Taken from text Table 6-2.

13-53 Both reactions are third-order, but [M] can be treated as a constant since it is much larger than the original concentrations of the I and Br and is not changed by the progress of the reaction. The integrated rate laws for the two reactions are then:

$$\frac{1}{[\text{I}]} - \frac{1}{[\text{I}]_0} = 2k_\text{I}[\text{M}]t \quad \text{and} \quad \frac{1}{[\text{Br}]} - \frac{1}{[\text{Br}]_0} = 2k_\text{Br}[\text{M}]t$$

After one half-life, $[\text{I}] = 1/2\,[\text{I}]_0$ and $[\text{Br}] = 1/2\,[\text{Br}]_0$ so:

$$\frac{1}{[\text{I}]_0} = 2k_\text{I}[\text{M}]t_{1/2,\,\text{I}} \quad \text{and} \quad \frac{1}{[\text{Br}]_0} = 2k_\text{Br}[\text{M}]t_{1/2,\,\text{Br}}$$

Dividing the first equation by the second gives:

$$\frac{[\text{Br}]_0}{[\text{I}]_0} = \left(\frac{k_\text{I}}{k_\text{Br}}\right)\left(\frac{t_{1/2,\,\text{I}}}{t_{1/2,\,\text{Br}}}\right)$$

Substitute $[\text{I}]_0 = 2[\text{Br}]_0$ and $k_\text{I} = 3.0k_\text{Br}$:

$$\frac{1}{2} = 3.0\left(\frac{t_{1/2,\,\text{I}}}{t_{1/2,\,\text{Br}}}\right) \quad \text{from which} \quad \left(\frac{t_{1/2,\,\text{I}}}{t_{1/2,\,\text{Br}}}\right) = \frac{1}{6.0} = \boxed{0.17}$$

13-55 The two reactions are both reactions of hydrogen with a halogen to form a hydrohalic acid. However, the reactions must proceed by different mechanisms since the experimental rate laws are different:

$$\text{rate (bromine)} = k[\text{H}_2][\text{Br}_2]^{1/2} \quad \text{rate (iodine)} = k[\text{H}_2][\text{I}_2]$$

The currently accepted mechanisms for both reactions involve a fast equilibrium to split the halogen molecule into its two atoms followed by reaction of one atom with H_2. In the case of Br, this elementary process is rate-limiting, but in the case of I, it is fast. The third steps in the mechanisms differ.

$$\text{H} + \text{Br}_2 \rightarrow \text{HBr} + \text{Br} \quad \text{fast} \qquad \text{H}_2\text{I} + \text{I} \rightarrow 2\,\text{HI} \quad \text{rate-limiting}$$
for bromine reaction   for iodine reaction

Also, in the iodine reaction, the intermediate H_2I is thought to exist, whereas the analogous H_2Br does not appear in the proposed mechanism for the bromine reaction.

13-57 The reaction is the decomposition of ozone by light: $2\,\text{O}_3 + \text{light} \rightarrow 3\,\text{O}_2$. The mechanism involves the production of the intermediate O atom from O_3 (in the first step), and its consumption either to regenerate O_3 (the second step) or to make $2\,\text{O}_2$

in an encounter with an O_3 molecule (the third step). The change in the concentration of O with time is:

$$\frac{d[O]}{dt} = k_1[O_3] - k_2[O][O_2][M] - k_3[O][O_3]$$

This equation states that the rate of change of the concentration of O equals its rate of production minus its rate of consumption. The steady-state approximation is that [O] comes to a *steady* value. If [O] is steady it is unchanging, and $d[O]/dt = 0$. Then:

$$k_1[O_3] - k_2[O][O_2][M] - k_3[O][O_3] = 0$$

so that

$$[O] = \frac{k_1[O_3]}{k_2[O_2][M] + k_3[O_3]}$$

All of this concerns the intermediate. The rate of the overall reaction is:

$$\text{rate} = \frac{1}{3}\frac{d[O_2]}{dt} = k_3[O][O_3]$$

Insert the expression for the concentration of intermediate O into this rate law:

$$\boxed{\text{rate} = \frac{k_3 k_1[O_3]^2}{k_2[O_2][M] + k_3[O_3]}}$$

Divide the top and bottom of the fraction by k_3:

$$\text{rate} = \frac{k_1[O_3]^2}{(k_2/k_3)[O_2][M] + [O_3]}$$

Clearly, only the ratio k_2/k_3 affects the rate, not the individual values. This ratio tells how much of the intermediate O cycles back to O_3 relative to how much goes on to give the product.

13-59 The initiation step is $CH_3CHO \rightarrow CH_3 + CHO$. The propagation steps involve the intermediates CH_3 (the methyl radical) and CH_2CHO (the acetyl radical). The propagation step is the combination of the second and third elementary reactions in the problem. The CH_3 is consumed in the second reaction and regenerated in the third, which forms the product CO. This chain continues indefinitely to consume all available CH_3CHO except when cut by the termination reaction, which forms the by-product CH_3CH_3 when two methyl radicals encounter each other and react.

13-61 a) Assume that the rate constants of the reactions that occur in cooking vary with the temperature according to the Arrhenius equation. Since the food cooks two times faster at 112°C (385 K) than at 100°C (373 K), the rate constant must be twice as large at 385 K: $k_{385} = 2\,k_{373}$. An Arrhenius equation can be written for each of these k's. Hence:

$$Ae^{-E_a/385R} = 2\,Ae^{-E_a/373R}$$

Cancel out the A's, take the natural logarithm of both sides, and rearrange:

$$\frac{-E_a}{385\ \text{K}} - \frac{-E_a}{373\ \text{K}} = R\ln 2 = (8.315\ \text{J K}^{-1}\text{mol}^{-1})(0.6931)$$

Solving for E_a gives an activation energy equal to $\boxed{69.0\ \text{kJ mol}^{-1}}$.

b) The rates of the cooking reactions at 94.4°C (367.6 K) depend on the rate constant at that temperature. The following equation involving the rate constant for cooking at 367.6 K comes from dividing the Arrhenius equation at 367.6 K by the Arrhenius equation at 373 K:

$$\frac{k_{367.6}}{k_{373}} = \frac{e^{-E_a/367.6R}}{e^{-E_a/373R}}$$

Taking the natural logarithm of both sides gives

$$\ln\left(\frac{k_{367.6}}{k_{373}}\right) = \frac{-E_a}{367.6R} - \frac{-E_a}{373R} = \frac{-69.0 \times 10^3\ \text{J mol}^{-1}}{8.315\ \text{J K}^{-1}\text{mol}^{-1}}\left(\frac{1}{367.6} - \frac{1}{373}\right) = -0.339$$

where the value of E_a has been taken from part a). Take the antilogarithm of both sides:

$$\frac{k_{367.6}}{k_{373}} = e^{-0.339} = 0.712$$

Since the rate constant at 94.4°C is smaller by the factor 0.712 times than the rate constant at 100°C, the food takes $1/0.712 = 1.404$ times longer to cook. Instead of 10 minutes, the cooking requires $\boxed{14\ \text{minutes}}$.

13-63 The rate constants of the forward and reverse reactions in the equilibrium between gaseous hydrogen and iodine on the one side and gaseous hydrogen iodide on the other are related to the equilibrium constant by the equation:

$$K = \frac{k_f}{k_r}$$

Write Arrhenius equations for both the forward reaction and the reverse reaction:

$$k_f = Ae^{-E_{a,f}/RT} \qquad \text{and} \qquad k_r = Ae^{-E_{a,r}/RT}$$

Divide the first equation by the second and take the logarithm of both sides:

$$\ln\left(\frac{k_f}{k_r}\right) = \frac{-E_{a,f}}{RT} - \frac{-E_{a,r}}{RT} = \frac{1}{RT}(E_{a,r} - E_{a,f})$$

The left side of this equation equals $\ln K$, the logarithm of the equilibrium constant. But $\ln K$ of the reaction at the temperature T is related to the standard free-energy change at that temperature by the thermodynamic equation $\Delta G_T^\circ = -RT \ln K$. Substitution gives:

$$-\frac{\Delta G_T^\circ}{RT} = \frac{1}{RT}(E_{a,r} - E_{a,f})$$

Multiply both sides by $-RT$ (at the constant temperature T):

$$\Delta G_T^\circ = (E_{a,f} - E_{a,r})$$

Assuming that ΔH° and ΔS° do not change going from 298 to 1000 K, the standard free-energy change of this reaction at 1000 K is:

$$\Delta G_{1000}^\circ = \Delta H^\circ - 1000\,\Delta S^\circ$$

Use the standard enthalpies of formation of the reactants and products to compute ΔH°:

$$\Delta H^\circ = 2\underbrace{(26.48)}_{\text{HI}(g)} - 1\underbrace{(62.44)}_{\text{I}_2(g)} - 1\underbrace{(0.00)}_{\text{H}_2(g)} = -9.48 \text{ kJ}$$

Do the same with absolute entropies to get ΔS°:

$$\Delta S^\circ = 2\underbrace{(206.48)}_{\text{HI}(g)} - 1\underbrace{(260.58)}_{\text{I}_2(g)} - 1\underbrace{(130.57)}_{\text{H}_2(g)} = 21.81 \text{ J K}^{-1}$$

Combining these values gives ΔG° at 1000 K:

$$\Delta G_{1000}^\circ = -9.48 \times 10^3 \text{ J} - (1000 \text{ K})(21.81 \text{ J K}^{-1}) = -31.29 \times 10^3 \text{ J}$$

Now, two of the three quantities in the relationship $\Delta G_{1000}^\circ = (E_{a,f} - E_{a,r})$ are known. Substitute them:

$$-31.29 \text{ kJ mol}^{-1} = (165 \text{ kJ mol}^{-1} - E_{a,r})$$

Solving gives $\boxed{196 \text{ kJ mol}^{-1}}$ as the activation energy of the reverse reaction.

An easy way to obtain k_r is to calculate the equilibrium constant K_{1000} and use the known k_f in the equation: $K = k_f/k_r$ Thus:

$$\ln K = -\frac{\Delta G^\circ}{RT} = -\frac{-31.29 \times 10^3 \text{ J mol}^{-1}}{8.315 \text{ J K}^{-1}\text{mol}^{-1}(1000 \text{ K})} = 3.763 \quad \text{and} \quad K = 43.08$$

$$k_r = \frac{k_f}{K} = \frac{240 \text{ L mol}^{-1}\text{s}^{-1}}{43.08} = \boxed{5.6 \text{ L mol}^{-1}\text{s}^{-1}}$$

13-65 The CF_2Cl_2 enters into the chemical reaction and presumably speeds it up, but is not itself consumed in the reaction. It does not appear in the balanced overall equation representing the reaction because it is consumed in one step of the mechanism but regenerated later. It is a $\boxed{\text{catalyst}}$.

13-67 The mechanism of enzyme catalysis, as modified to allow for the action of an inhibitor, is:

$$
\begin{array}{lll}
E + S \rightleftharpoons ES & \text{fast equilibrium} & k_1 \text{ and } k_{-1} \\
ES \rightarrow E + P & \text{slow} & k_2 \\
E + I \rightleftharpoons EI & \text{fast equilibrium} & k_3 \text{ and } k_{-3}
\end{array}
$$

This is a case of competitive inhibition. The inhibitor (symbolized I) competes with the substrate S in binding to the enzyme E. The generation of product P is thereby slowed because P forms only through the complex ES. Follow the general pattern of the derivation in text Section 13-7, but allow for the complication of the inhibitor. Label the total concentration of enzyme $[E]_0$. The enzyme is present in one of three states: free, bound to the inhibitor, or bound to the substrate:

$$[E]_0 = [E] + [EI] + [ES]$$

Write the equilibrium expression for the third step of the mechanism, letting k_3/k_{-3} equal K_3:

$$K_3 = \frac{[EI]}{[E][I]}$$

Solve this expression for [EI], substitute into the first equation, and solve for [E] as follows:

$$[E]_0 = [E] + K_3[E][I] + [ES]$$

$$[E]_0 = [E]\Big(1 + K_3[I]\Big) + [ES]$$

$$[E] = \frac{[E]_0 - [ES]}{1 + K_3[I]}$$

Now, make the steady-state approximation for ES, the intermediate. This approximation is that the ES is generated as fast as it is consumed—its concentration does not change with time:

$$0 = \frac{d[ES]}{dt} = k_1[E][S] - k_{-1}[ES] - k_2[ES]$$

Solve for [ES] and substitute in with the previous expression for [E]. The result is:

$$[ES] = \frac{k_1[E]_0[S]}{k_1[S] + \Big(1 + K_3[I]\Big)(k_{-1} + k_2)}$$

The rate of the reaction equals the rate of the slow step, which equals $k_2[ES]$. Therefore:

$$\text{rate} = k_2[ES] = \frac{k_2 k_1 [E]_0 [S]}{k_1 [S] + \left(1 + K_3[I]\right)(k_{-1} + k_2)} = \frac{k_2 [E]_0 [S]}{[S] + K_m(1 + K_3[I])}$$

where K_m is defined as $(k_{-1} + k_2)/k_1$. This K_m is called the Michaelis-Menten constant.[8] Any concentration of the inhibitor increases the denominator of this expression and lowers the initial rate of the reaction.

13-69 The reaction is the gas-phase combination of H_2 and I_2 to give HI. The $\Delta H°$ of this change equals -9.48 kJ when two moles of HI are produced. See **13-63**. Obtain the initial rate of the reaction by substituting the rate constant and the initial concentrations of H_2 and I_2 into the rate equation:

$$\text{rate} = k[H_2][I_2] = 0.0242 \text{ M}^{-1}\text{s}^{-1}[0.081 \text{ M}][0.036 \text{ M}] = 7.06 \times 10^{-5} \text{ M s}^{-1}$$

The rate of formation of HI is twice this rate (because HI has a coefficient of 2 in the balanced reaction). The rate of absorption of heat equals the rate at which HI forms times the heat absorbed per mole of HI:

$$\frac{2(7.06 \times 10^{-5} \text{ mol HI})}{\text{L s}} \times \left(\frac{-9.48 \text{ kJ}}{2 \text{ mol HI}}\right) = \boxed{-6.7 \times 10^{-4} \text{ kJ L}^{-1}\text{s}^{-1}}$$

13-71 Substitution of the initial concentrations of NO and O_3 into the second-order rate law gives the desired initial rate of the reaction. Use the ideal-gas law to obtain these concentrations from the mole fractions of the two gases:

$$\left(\frac{n}{V}\right)_{NO} = \frac{P_{NO}}{RT} = \frac{0.00057(3.26 \text{ atm})}{(0.08206 \text{ L atm mol}^{-1}\text{K}^{-1})(500 \text{ K})} = 4.53 \times 10^{-5} \text{ mol L}^{-1}$$

$$\left(\frac{n}{V}\right)_{O_3} = \frac{P_{O_3}}{RT} = \frac{0.00026(3.26 \text{ atm})}{(0.08206 \text{ L atm mol}^{-1}\text{K}^{-1})(500 \text{ K})} = 2.07 \times 10^{-5} \text{ mol L}^{-1}$$

Then proceed with the substitution:

$$\text{rate} = k[NO][O_3]$$
$$= (7.6 \times 10^7 \text{ L mol}^{-1}\text{s}^{-1})(4.53 \times 10^{-5} \text{ mol L}^{-1})(2.07 \times 10^{-5} \text{ mol L}^{-1})$$
$$= \boxed{0.071 \text{ mol L}^{-1}\text{s}^{-1}}$$

---

[8]Introduced (although not named) on text page 478.

Chapter 14

Nuclear Chemistry

Atoms are composed of protons and neutrons (in the nucleus) surrounded by extra-nuclear electrons. Ordinary chemistry derives from the transfer and sharing of electrons among atoms. Nuclear chemistry is concerned with changes in the nuclei of atoms.

Mass-Energy Relationships in Nuclei

Many kinds of atoms (**radioactive** atoms) spontaneously change by emitting sub-atomic particles from their nuclei, by absorbing subatomic particles into their nuclei, or by having their nuclei split into two smaller nuclei. The masses of subatomic particles are quite small (at most about 10^{-27} kg). Convenient study of nuclear changes requires the definition of a new small unit of mass:

- The **atomic mass unit** is defined as 1/12 of the mass of a single neutral ^{12}C atom.

One mole of ^{12}C contains Avogadro's number of ^{12}C atoms and weighs exactly 12 g. Therefore:

$$1 \text{ u} = \left(\frac{1}{12}\right)\frac{12 \text{ g}}{6.022137 \times 10^{23}} = 1.660540 \times 10^{-24} \text{ g}$$

Most subatomic particles have masses of at least 1 u. The electron and positron are exceptions. The following summarizes the names, symbols, and exact masses of the most important subatomic particles:[1]

---

[1]See Text Table 14-1, text page 491.

Particle	Symbols	Mass (u)
Proton	$_1^1p$, $_1^1H^+$	1.00727647
Neutron	$_0^1n$	1.00866490
Electron	$_{-1}^0e^-$	0.00054857990
Positron	$_1^0e^+$	0.00054857990

The Einstein Mass-Energy Relationship

The mass of an atom is *not* equal to the sum of the masses of the protons, neutrons and electrons that make it up. Instead, for all elements, the mass of the atom is *less* than the mass of the component electrons, protons and neutrons. A change in mass is related to a change in energy by the Einstein equation:

$$\Delta E = c^2 \Delta m$$

where c equals 2.99792458×10^8 m s^{-1}, the speed of light in a vacuum. This equation applies to all nuclear reactions and to ordinary chemical reactions as well. See **14-63**. The Einstein equation is used less often for conversions of between Δm and ΔE than the following equivalencies, which are derived from it (as in **14-3**):

$$1 \text{ u} = 1.492419 \times 10^{-10} \text{ J}$$

$$1 \text{ u} = 931.494 \times 10^6 \text{ eV} = 931.494 \text{ MeV}$$

The **electron-volt, (eV)**, a non-SI unit of energy, equals the energy required to accelerate an electron through a potential difference of one volt. Energy changes in nuclear reactions are traditionally expressed in eV's or in MeV's. The **MeV** (mega-electron-volt) equals a million electron-volts. See **14-7** and **14-49**.

Since the formation of an atom from its component parts has a negative Δm, a nuclear reaction such as:

$$20\,_1^1H + 20\,_0^1n \rightarrow \,_{20}^{40}Ca$$

releases energy. See **14-3**. The quantity of energy released tells how tightly the **nucleons** (protons and neutrons) in the product nucleus are bound.

The **binding energy E_B** of a nucleus equals the negative of the energy change of its formation from its component nucleons. The **binding energy per nucleon** is the binding energy of a nucleus divided by the sum of the number of neutrons and protons in the nucleus. See **14-3**. The binding energy per nucleon is greatest for ^{56}Fe.

The preceding equation is a **nuclear equation**. Nuclear equations resemble ordinary chemical equations in requiring balancing. The criteria for balance however differ:

- In a balanced nuclear equation the sum of the left subscripts (the atomic numbers) must be same on the two sides of the equation.

- The sum of the left superscripts (mass numbers) must be the same on the two sides of the equation. See **14-3** and **14-57**.

The electron in nuclear equations is assigned an atomic number (left subscript) of -1 because the loss of an electron by a nucleus increases the atomic number by 1. The positron is assigned an atomic number of $+1$ because the loss of a positron by a nucleus decreases the atomic number by 1.

Nuclear Decay Processes

Spontaneous nuclear decay occurs only if the energy of the products is less than the energy of the reactants. Given the Einstein equation this implies that the total mass of the products must be less than the mass of the reactants:

$$\Delta E < 0 \quad \text{and} \quad \Delta m < 0$$

These criteria for spontaneity are equivalent to the thermodynamic criterion (at constant temperature and pressure) $\Delta G < 0$. In nuclear decay, a parent nucleus gives rise to a daughter nucleus, and energy leaves the system as kinetic energy of emitted particles and as electromagnetic radiation.

Although radioactive decay involves the conversion of one kind of nucleus into another, it is more convenient to use the masses of whole atoms (nucleus plus electrons) in calculations. Whole-atom masses are given in tables.[2] The following are important processes of nuclear decay:

- **Beta Decay.** This is the decay mode of **proton-deficient** nuclides. Such nuclides have fewer protons that stable nuclides of the same mass number. In beta decay, a neutron converts into a proton, emitting an electron (beta particle) and a massless **antineutrino** ($\bar{\nu}$). This process increases the number of protons in the nucleus: the atomic number Z of the nucleus goes up by one, but its mass number A stays the same. For example Si-31 turns into P-31 by beta decay:

$$^{31}_{14}\text{Si} \rightarrow {}^{31}_{15}\text{P} + {}^{0}_{-1}e^- + \bar{\nu}$$

The criterion for beta decay is:

$$m\,(\text{daughter atom}) - m\,(\text{parent atom}) = \Delta m < 0$$

The masses in this equation are whole-atom masses, not the masses of bare nuclei. The mass of the emitted electron is *not* explicitly added in with the mass of the products because it is already counted in the mass of the neutral daughter atom. (see **14-37** and **14-65**). If Δm is negative, then ΔE of the nuclear reaction is also

---

[2]Text Table 14-1, text page 491, for example.

negative. The antineutrino and the $_{-1}^0 e^-$ share in carrying off a total of $-\Delta E$ in energy. If the antineutrino should get zero kinetic energy, then the $_{-1}^0 e^-$ will have its maximum kinetic energy, $-\Delta E$.

• **Positron Emission.** The is the decay mode of **neutron-deficient** nuclides. Such nuclides have fewer neutrons that stable nuclides of the same A. In positron emission, the nucleus emits a positron ($_1^0 e^+$) and a neutrino and thereby converts a proton into a neutron. The mass number A of the nuclide remains unchanged, but the atomic number Z decreases by one. For example:

$$_{21}^{40}\text{Sc} \rightarrow {}_{20}^{40}\text{Ca} + {}_1^0 e^+ + \nu$$

The criterion for positron decay is:

$$m \, (\text{daughter atom}) - m \, (\text{parent atom}) + 2 \, m_e = \Delta m < 0$$

where m_e equals 0.00054858 u, the mass of the electron (and positron). The $2 \, m_e$ term appears because the use of whole-atom masses leaves one electron and one positron unaccounted for. See **14-7**, **14-45**, and **14-49**. The energy equivalent of two electron masses (2×0.00054858 u) equals 1.0220 MeV.[3] This is a lot of energy and neglecting it gives seriously wrong answers. See **14-49b**.

• **Electron Capture.** A neutron-deficient nucleus may capture one of its extra-nuclear electrons, converting a proton into a neutron. For example:

$$_4^7\text{Be} + {}_{-1}^0 e^- \rightarrow {}_3^7\text{Li} + \nu$$

This process achieves the same change as positron emission, that is, Z decreases by one while A stays the same. The energy (mass) criterion for electron capture is less stringent than for positron emission:

$$m \, (\text{daughter atom}) - m \, (\text{parent atom}) = \Delta m < 0$$

The additional $2 \, m_e$ term that appears in the case of positron emission is now absent. The mass of the electron on the left-hand side of the nuclear equation does not appear explicitly because it is included in the mass of the neutral parent atom. Nothing but a neutrino is emitted.

• **Alpha Decay.** An alpha particle is the nucleus of a helium atom, the $_2^4\text{He}^{2+}$ ion. A nucleus may emit an alpha particle, $_2^4\text{He}$, decreasing Z by 2 and A by 4. Thus:

$$_{88}^{226}\text{Ra} \rightarrow {}_{86}^{222}\text{Rn} + {}_2^4\text{He}$$

The mass (energy) criterion for α decay is:

$$m \, (\text{daughter atom}) + m({}_2^4\text{He atom}) - m \, (\text{parent atom}) = \Delta m < 0$$

---

[3]Using the unit-factor 931.494 MeV u^{-1}.

Alpha decay occurs mostly in the heaviest nuclides (Z greater than 83). See **14-65a**.

Many problems use the above relationships. Problem **14-45** varies the usual pattern by giving ΔE for two processes and the mass of the parent atom and asking for the mass of the daughter atom. Problem **14-13** uses none of the formal criteria but requires the ability to apply the fundamental criterion for spontaneous decay, a decrease in mass.

In nuclear decay problems, use the conversion factor 931.494 MeV u^{-1} to switch between mass units and energy units at will. As a practical matter, it may be easier to do arithmetic on the masses in u, just because the numbers are smaller, and convert to MeV as a last step. In computing energy and mass differences, do not round off prematurely. Carry extra digits through the intermediate stages, and round off at the last step.

Kinetics of Radioactive Decay

Nuclear decay processes occur according to first-order kinetics. Hence for all types of decay:

$$N = N_0 e^{-kt}$$

where N_0 is the number of nuclei originally present, N is the number present at time t and k is the rate constant. The half-life of a reaction is:

$$t_{1/2} = \frac{\ln 2}{k}$$

Radioactivity is measured by detecting the high-energy particles that the decay process produces. The decay rate at any moment is the **activity A** of the sample, in terms of the number of disintegrations per second.[4] The activity depends both on the number of atoms of the radioactive nucleus and the rate constant of the decay:

$$A = kN$$

The units on both sides of this definition are reciprocal time (typically s^{-1}). These definitions are used in **14-19, 14-21**, and **14-35**. **Specific activity** is the activity per given mass of sample. See **14-25** and **14-53**. The unit of activity is the **becquerel (Bq)**, defined as one radioactive disintegration per second (1 Bq = 1 s^{-1}). One **curie (Ci)** equals 3.7×10^{10} Bq. The activity and specific activity themselves decay with time:

$$A = A_0 e^{-kt}$$

---

[4]This activity is a "radioactivity." It has nothing to do with the thermodynamic activity a of a substance.

Once A and k are known the number of radioactive nuclei at any time can be calculated. See **14-21**.

In solving problems:

- Knowing the half-life is equivalent to knowing the rate constant for the decay because the product of the two is $\ln 2$ (0.69315). The units of k and $t_{1/2}$ are each other's reciprocals.

- Avoid confusing the units of time. Convert all half-lives (rate constants) in the same problem to the same unit of time (reciprocal time), preferably seconds (reciprocal seconds). See **14-35**.

- All of the concepts of chemical kinetics (Chapter 11) apply. In particular, the steady-state approximation[5] is prominent when a nuclide decays in a series of steps. See **14-65b**.

- Many problems involve radioactive dating. Dating using ^{14}C is a major example. These problems offer only one additional facet that first-order chemical kinetics do not: *activities* (the actual current rate of decay) rather than concentrations of reacting species are given.

Radiation in Biology and Medicine

The becquerel and curie measure the radioactivity of a species in terms of the number of nuclear decay events per second. Different kinds of events emit different kinds of particles. These particles differ in energy and penetrating power. In biology and medicine, the important factor is not the number of disintegrations per second in a radioactive source but the quantity of energy deposited in living tissue. See **14-35**. The **rad** is therefore defined as the amount of radiation that deposits 10^{-2} J in a 1 kg mass.

The **rem** is the unit of radioactive **dosage.** It is the number of rads absorbed by tissue multiplied by a "fix-up factor," the **relative biological effectiveness**, that accounts for variables such as the type of tissue and type of radiation and the dose rate.

Nuclear Fission

Nuclear fission is the splitting apart of a nucleus into less massive nuclei. Fission is exothermic—the binding energy per nucleon in the products exceeds binding energy per nucleon in the reactants. The fission of several elements, in particular uranium,

---

[5]See text page 465.

is induced by the absorption of neutrons. See **14-41**. When fission generates two or more new neutrons, the nuclear reaction is a branching chain reaction. It can lead to an explosion. In nuclear power reactors, neutron generation is moderated by the presence (in control rods) of ^{112}Cd or ^{10}B, which capture neutrons well. When the rate of neutron production equals the rate of consumption in a fission reactor, energy is generated at a controlled rate. The products of fission of ^{235}U include 34 different elements. Some are in radioactive isotopes that present disposal problems.

Nuclear Fusion and Nucleosynthesis

In nuclear fusion, light nuclides react to form heavier ones. The fusion of light atoms is thermodynamically spontaneous ($\Delta m < 0$), but has a very large activation energy. Fusion reactions are called thermonuclear because colliding atoms must have high kinetic energy (high T). Fusion reactions in stars lead to nucleosynthesis of elements. Hydrogen burning in stars proceeds by the reaction series:

$$2\,{}^{1}_{1}\text{H} \rightarrow {}^{2}_{1}\text{H} + {}^{0}_{1}e^{+} + \nu$$
$$\ {}^{2}_{1}\text{H} + {}^{1}_{1}\text{H} \rightarrow {}^{3}_{2}\text{He} + \gamma$$
$$2\,{}^{3}_{2}\text{He} \rightarrow {}^{4}_{2}\text{He} + 2\,{}^{1}_{1}\text{H}$$

where γ stands for a gamma ray. See **14-43**.

14-1 Use these tests for balance in nuclear equations: the sums of the left superscripts on the two sides of the equation must be equal; the sums of the left subscripts on the two sides of the equation must also be equal.

    a) $2\,{}^{12}_{6}\text{C} \rightarrow {}^{23}_{12}\text{Mg} + {}^{1}_{0}n$ **b)** ${}^{15}_{7}\text{N} + {}^{1}_{1}\text{H} \rightarrow {}^{12}_{6}\text{C} + {}^{4}_{2}\text{He}$ **c)** $2\,{}^{3}_{2}\text{He} \rightarrow {}^{4}_{2}\text{He} + 2\,{}^{1}_{1}\text{H}$

14-3 a) The total binding energy of the calcium-40 nuclide equals the negative of the energy equivalent of the change in mass that occurs with formation of the nuclide from its component particles:

$$20\,{}^{1}_{1}\text{H} + 20\,{}^{1}_{0}n \rightarrow {}^{40}_{20}\text{Ca}$$

$$\Delta m = m[{}^{40}_{20}\text{Ca}] - 20\,m[{}^{1}_{1}\text{H}] - 20\,m[{}^{1}_{0}n]$$
$$\Delta m = 39.962589 - 20(1.00782504) - 20(1.00866497) = -0.367209 \text{ u}$$

The energy equivalent of 1 u is 931.494 MeV. Therefore:

$$\Delta E = -0.367209 \text{ u} \times \left(\frac{931.494 \text{ MeV}}{1 \text{ u}}\right) = -342.053 \text{ MeV per atom}$$

The binding energy E_B is the negative of this, $\boxed{342.053 \text{ MeV}}$. To compute ΔE in joules, convert Δm to kilograms and substitute it into the Einstein equation:

$$\Delta E = c^2 \Delta m$$

$$= (2.9979246 \times 10^8 \text{ m s}^{-1})^2(-0.367209 \text{ u}) \times \left(\frac{1.660540 \times 10^{-27} \text{ kg}}{1 \text{ u}}\right)$$

$$= -5.48030 \times 10^{-11} \text{ J}$$

The binding energy of a mole of atoms equals the negative of this answer, which is for a single atom, multiplied by Avogadro's number:

$$(+5.48030 \times 10^{-11} \text{ J}) \times (6.022137 \times 10^{23} \text{ mol}^{-1}) = \boxed{3.30031 \times 10^{10} \text{ kJ mol}^{-1}}$$

Since $^{40}_{20}\text{Ca}$ has 40 nucleons, the E_B per nucleon is $342.053/40 = \boxed{8.55132 \text{ MeV}}$.

Tip. This is the value graphed in text Figure 14-2.

b) The Δm between the product $^{87}_{37}\text{Rb}$ and the reactants $50\,^1_0n$ and $37\,^1_1\text{H}$ equals -0.8135878 u. By computations like those in part a), the E_B per atom of $^{87}_{37}\text{Rb}$ is $\boxed{757.852 \text{ MeV}}$. The binding energy per mole is $\boxed{7.3122 \times 10^{10} \text{ kJ mol}^{-1}}$, and the binding energy per nucleon is E_B divided by 87 or $\boxed{8.7109 \text{ MeV}}$.

c) Uranium-238 has 92 protons and 146 neutrons. Compute Δm for the making of the atom by adding up the mass of 92 hydrogen atoms (not protons) and 146 neutrons and subtracting the result from the mass of the U-238 atom. The answer equals -1.934194 u. Convert this to a ΔE to obtain E_B per atom, which is $\boxed{1801.69 \text{ MeV}}$ or $\boxed{17.384 \times 10^{10} \text{ kJ mol}^{-1}}$. The binding energy per nucleon is $\boxed{7.57013 \text{ MeV}}$.

Tip. To save time in calculations observe that, according to the Einstein equation:

$$\Delta m = 1 \text{ u per atom} \quad \text{implies} \quad \Delta E = 8.9875519 \times 10^{10} \text{ kJ mol}^{-1}$$

14-5 The mass of a single ^8Be atom equals 8.0053052 u whereas the mass of two ^4He atoms equals twice 4.0026033 u, which is 8.0052066 u. Because the particles have a larger mass when organized as a ^8Be atom, the ^8Be atom is less stable than the pair of ^4He atoms. In other terms, the Δm for the nuclear reaction: $^8_4\text{Be} \rightarrow 2\,^4_2\text{He}$ is negative. The Δm is $\boxed{-9.86 \times 10^{-5} \text{ u}}$.

14-7 Represent the decay as:

$$^8_5\text{B} \rightarrow {}^8_4\text{Be} + {}^0_1e^+ + \nu$$

The difference in mass between the two sides of the reaction appears from this equation to be:

$$\Delta m = \underbrace{8.0053052}_{^8\text{Be}} + \underbrace{0.00054858}_{^0_1e^+} - \underbrace{8.024612}_{^8\text{B}} = -0.0187582 \text{ u ?}$$

using the numbers found in text Table 14-1[6] for the boron atom, beryllium atom, and positron. At this point a trap looms. The tabulated masses include the electrons that surround the nuclei of the atoms. Beryllium has only four electrons, but boron has five. The computation at this stage has wrongly allowed one electron to vanish. This is put right by including another electron mass with the mass of the products. The true change in mass is:

$$\Delta m = 8.0053052 + 2(0.00054858) - 8.024612 = -0.01820964 \text{ u}$$

This mass change has an energy equivalent of -16.962 MeV. Therefore $\boxed{16.962 \text{ MeV}}$ is released by the reaction of 1 atom of ^{8}B.

Tip. The trap is avoided by balancing the original equation as to charge. The positive charge on the positron cannot appear without compensation. Therefore Be must be a negative ion:

$$^8_5\text{B} \rightarrow {}^8_4\text{Be}^- + {}^0_1e^+ + \nu$$

14-9 Positron emission (loss of ${}^0_1e^+$) and electron capture by a nucleus always lower the atomic number by one; electron emission (loss of ${}_{-1}^0e^-$) raises the atomic number by one.

a) ${}^{39}_{17}\text{Cl} \rightarrow {}^{39}_{18}\text{Ar} + {}_{-1}^0e^- + \bar{\nu}$ **b)** ${}^{22}_{11}\text{Na} \rightarrow {}^{22}_{10}\text{Ne} + {}^0_1e^+ + \nu$

c) ${}^{224}_{88}\text{Ra} \rightarrow {}^{220}_{86}\text{Rn} + {}^4_2\text{He}$ **d)** ${}^{82}_{38}\text{Sr} + {}_{-1}^0e \rightarrow {}^{82}_{37}\text{Rb} + \nu$

14-11 The ^{19}Ne nucleus is neutron deficient and decays by converting a proton to a neutron. The ^{23}Ne nucleus is too rich in neutrons and decays by converting a neutron to a proton. Thus, the decay of ^{19}Ne proceeds by $\boxed{\text{positron emission}}$ to give ^{19}F, and the decay of ^{23}Ne proceeds by $\boxed{\text{electron emission (beta decay)}}$ to give ^{23}Na.

14-13 The electrically neutral neutron decays to a positive proton, a negative electron (${}_{-1}^0e^-$), and an antineutrino. Determine the change in mass and find the equivalent change in energy:

$$\Delta E = (1.007276 + 0.00054858 - 1.008665) \text{ u} \times \left(\frac{931.494 \text{ MeV}}{1 \text{ u}}\right) = -0.7824 \text{ MeV}$$

---

[6]Text page 491.

The electron has a maximum kinetic energy of $\boxed{0.7824 \text{ MeV}}$.

Tip. This is a maximum because the other products may also carry off some kinetic energy.

14-15 The answer requires the translation of the written description of the process into nuclear equations: $\boxed{^{30}_{14}\text{Si} + ^{1}_{0}n \rightarrow ^{31}_{14}\text{Si} \rightarrow ^{31}_{15}\text{P} + ^{0}_{-1}e^{-} + \bar{\nu}}$.

14-17 $^{210}_{84}\text{Po} \rightarrow ^{206}_{82}\text{C} + ^{4}_{2}\text{He}$ $^{4}_{2}\text{He} + ^{9}_{4}\text{Be} \rightarrow ^{12}_{6}\text{C} + ^{1}_{0}n$

14-19 Compute how many atoms are present in 0.0010 g of ^{209}Po:

$$0.0010 \text{ g} \times \left(\frac{1 \text{ mol Po}}{209 \text{ g Po}}\right) \times \left(\frac{6.02214 \times 10^{23} \text{ atom Po}}{1 \text{ mol Po}}\right) = 2.88 \times 10^{18} \text{ atom } ^{209}\text{Po}$$

The activity A, which equals the instantaneous rate of disintegration in the polonium, depends on the number of atoms present:

$$A = -\frac{dN}{dt} = kN$$

The k in this equation is a first-order rate constant. It equals $\ln 2/t_{1/2}$ where $t_{1/2}$ is the half-life of the decay. Substituting:

$$A = \left(\frac{\ln 2}{t_{1/2}}\right) N = \left(\frac{0.6931}{103 \text{ yr}}\right) 2.88 \times 10^{18} = 1.94 \times 10^{16} \text{ yr}^{-1}$$

Convert to a per minute basis as follows:

$$A = 1.94 \times 10^{16} \text{ yr}^{-1} \times \left(\frac{1 \text{ yr}}{365.2 \text{ day}}\right) \times \left(\frac{1 \text{ day}}{1440 \text{ min}}\right) = 3.7 \times 10^{10} \text{ min}^{-1}$$

That is, $\boxed{3.7 \times 10^{10} \text{ atoms}}$ of polonium decay per minute.

14-21 a) At the moment of preparation, the sample has initial activity (rate of decay) A_0. The activity is directly proportional to N_0, the number of atoms of O-19 that are initially present:

$$A_0 = kN_0$$

The rate constant k is $(\ln 2/29 \text{ s})$, which equals 0.0239 s^{-1}. Then:

$$N_{\text{O-19}} = \frac{A}{k} = \frac{2.5 \times 10^4 \text{ s}^{-1}}{0.0239 \text{ s}^{-1}} = 1.046 \times 10^6 \text{ atom} = \boxed{1.0 \times 10^6 \text{ atom } ^{19}\text{O}}$$

b) The number of O-19 atoms that remains at any time t is:

$$N = N_0 e^{-kt}$$

In this case t is 2.00 min or 120 s. Substitution of k and N_0 gives:

$$N = (1.046 \times 10^6) \exp\left(-(0.0239 \text{ s}^{-1})(120 \text{ s})\right) = \boxed{5.9 \times 10^4 \text{ atom}}$$

14-23 First calculate N, the number of atoms in 44 mg of ^{219}At:

$$N = 44 \times 10^{-3} \text{ g} \times \left(\frac{1 \text{ mol}}{219 \text{ g}}\right) \times \left(\frac{6.022 \times 10^{23} \text{ atoms}}{1 \text{ mol}}\right) = 1.21 \times 10^{20} \text{ atoms}$$

Then compute the activity of this amount of astatine using the half-life as given:

$$A = kN = \left(\frac{\ln 2}{t_{1/2}}\right) N = \left(\frac{0.6931}{54 \text{ s}}\right)(1.21 \times 10^{20}) = \boxed{1.6 \times 10^{18} \text{ s}^{-1}}$$

14-25 The activity of ^{14}C decays according to the same law as the number of atoms of ^{14}C:

$$A = A_0 e^{-kt} \quad \text{which gives:} \quad -kt = \ln\left(\frac{A}{A_0}\right)$$

For the papyrus, A is 9.2 disintegrations $\text{min}^{-1}\text{g}^{-1}$ and A_0 is 15.3 disintegrations $\text{min}^{-1}\text{g}^{-1}$. Also, k is $\ln 2/t_{1/2}$ or $1.21 \times 10^{-4} \text{ yr}^{-1}$. Substituting gives:

$$-(1.21 \times 10^{-4} \text{ yr}^{-1})\, t = \ln\left(\frac{9.2 \text{ min}^{-1}\text{g}^{-1}}{15.3 \text{ min}^{-1}\text{g}^{-1}}\right) \quad \text{so that:} \quad t = \boxed{4.2 \times 10^3 \text{ yr}}$$

Tip. It is assumed that the activity of ^{14}C in the biosphere has not changed since Egyptian times.

14-27 For every atom of U-238 that decays, eight atoms of He are produced. Compute the number of moles of He (using the ideal-gas law) per gram of rock:

$$n_{\text{He}} = \frac{PV}{RT} = \frac{(1 \text{ atm})(9.0 \times 10^{-8} \text{ L})}{(0.08206 \text{ L atm mol}^{-1}\text{K}^{-1})(273.15 \text{ K})} = 4.015 \times 10^{-9} \text{ mol}$$

The number of atoms of U-238 that has decayed is therefore:

$$4.015 \times 10^{-9} \text{ mol He} \times \left(\frac{6.022 \times 10^{23} \text{ atom He}}{1 \text{ mol He}}\right) \times \left(\frac{1 \text{ atom } ^{238}\text{U}}{8 \text{ atom He}}\right) = 3.02 \times 10^{14} \text{ atom}$$

The number of atoms of ^{238}U still present per gram of rock is also readily computed:

$$2.0 \times 10^{-7} \text{ g } ^{238}\text{U} \times \left(\frac{1 \text{ mol}}{238.0 \text{ g}}\right) \times \left(\frac{6.022 \times 10^{23} \text{ atom}}{1 \text{ mol}}\right) = 5.06 \times 10^{14} \text{ atom}$$

The initial number of atoms of U-238 equals the number of those decayed plus those remaining. It is 8.08×10^{14} atoms per gram of rock. The decay process is described by the equation:

$$\ln \left(\frac{N}{N_0} \right) = -kt = - \left(\frac{\ln 2}{t_{1/2}} \right) t$$

Substitution gives:

$$\ln \left(\frac{5.06 \times 10^{14}}{8.08 \times 10^{14}} \right) = - \left(\frac{\ln 2}{4.47 \times 10^9 \text{ yr}} \right) t$$

Solving for t gives the approximate age of the rock: $\boxed{3.0 \text{ billion years}}$.

14-29 First-order kinetics govern the radioactive decay of both ^{235}U and ^{238}U:

$$N(^{235}\text{U}) = N_0(^{235}\text{U})e^{-kt} \quad \text{and} \quad N(^{238}\text{U}) = N_0(^{238}\text{U})e^{-kt}$$

In the beginning, the two isotopes were equally abundant. Assume that the current difference in abundance is caused entirely by the faster decay of U-235. (The half-life of ^{235}U is briefer by a factor of about 6.) The current ratio of abundances is 137.7 to 1. In equation form:

$$N_0 \left(^{238}\text{U} \right) = N_0 \left(^{235}\text{U} \right) \quad \text{and} \quad N \left(^{238}\text{U} \right) = 137.7 N \left(^{235}\text{U} \right)$$

This means that:

$$\frac{137.7}{1} = \frac{N_0 \left(^{238}\text{U} \right) e^{-k_{238}t}}{N_0 \left(^{235}\text{U} \right) e^{-k_{235}t}} = \frac{e^{-k_{238}t}}{e^{-k_{235}t}}$$

where k_{238} is the rate constant for the decay of ^{238}U and k_{235} is the rate constant for the decay of ^{235}U. Take the natural logarithm of both sides of the equation:

$$\ln 137.7 = (-k_{238}t) - (-k_{235}t) = t \left(k_{235} - k_{238} \right)$$

For each isotope $k = \ln 2 / t_{1/2}$ so:

$$\ln 137.7 = t \left(\frac{\ln 2}{t_{1/2,235}} - \frac{\ln 2}{t_{1/2,238}} \right)$$

The half-lives of the isotopes are 7.13×10^8 years and 4.51×10^9 years respectively. Substitution in the equation gives t equal to $\boxed{6.0 \times 10^9 \text{ years}}$. The supposed supernova occurred about 1.5 billion years before the estimated time of the formation of the solar system.

14-31 Positron emission is accompanied by emission of a neutrino:
$$^{11}_{6}C \rightarrow ^{11}_{5}B + ^{0}_{1}e^{+} + \nu \qquad ^{15}_{8}O \rightarrow ^{15}_{7}N + ^{0}_{1}e^{+} + \nu$$

14-33 Assume that all of the ^{11}C and ^{15}O atoms decay before any are excreted or else that equal chemical amounts of the two radioactive nuclides are excreted. O-15 deposits 1.74 times more energy per kilogram of body mass than the C-11 because its positrons, which are emitted in equal number, are on the average more energetic by the factor $1.72/0.99 = \boxed{1.74}$.

14-35 a) Determine the number of atoms of I-131 ingested:

$$5.0 \times 10^{-6} \text{ g } ^{131}I \left(\frac{1 \text{ mol } ^{131}I}{131 \text{ g } ^{131}I}\right) \left(\frac{6.022 \times 10^{23} \text{ atoms } ^{131}I}{1 \text{ mol } ^{131}I}\right) = 2.3 \times 10^{16} \text{ atoms } ^{131}I$$

Express the half-life of I-131 in seconds: $8.041 \text{ days} \times 86400 \text{ s day}^{-1} = 6.947 \times 10^5$ s. The ingested activity is then:

$$A = kN = \left(\frac{\ln 2}{t_{1/2}}\right) N$$
$$= \left(\frac{0.6931}{6.947 \times 10^5 \text{ s}}\right) (2.3 \times 10^{16}) = 2.3 \times 10^{10} \text{ s}^{-1} = \boxed{2.3 \times 10^{10} \text{ Bq}}$$

b) Compute the energy emitted in the first second by the decay of I-131. Use the fact that a becquerel (Bq) is a decay event per second:

$$\left(\frac{2.3 \times 10^{10} \text{ events}}{1 \text{ s}}\right) \times \left(\frac{0.40 \text{ MeV}}{1 \text{ event}}\right) \times \left(\frac{1.602 \times 10^{-13} \text{ J}}{1 \text{ MeV}}\right) = 1.474 \times 10^{-3} \text{ J s}^{-1}$$

This equals 0.147 cJ per second. Since the victim weighs 60 kg and all of the energy is deposited within the victim's body, he or she absorbs 0.00245 cJ per kg of body mass in the first second. A millirad equals 0.0010 cJ per kg of body mass, so the radiation absorbed dose in the first second is $\boxed{2.4 \text{ millirad}}$.

c) The rate of dosage diminishes as the I-131 decays. Assume a constant activity equal to merely half the initial activity (the activity falls to this value after 8.041 days). Even at the artificially low rate of 1.25 millirad per second, the LD_{50} dose of 500 rad is delivered in a frighteningly brief time:

$$500 \text{ rad} \times \left(\frac{1 \text{ s}}{1.25 \times 10^{-3} \text{ rad}}\right) \times \left(\frac{1 \text{ day}}{86400 \text{ s}}\right) = 4.6 \text{ day}$$

The true dose in the first 4.6 days exceeds 500 rad, and dosage continues at a significant rate after the first 4.6 days. The actual dose will therefore $\boxed{\text{surely be lethal}}$.

Tip. Compare closely to **14-36**. The inhaled plutonium in that problem is equal in mass to the radioactive iodine ingested here, but is probably not a lethal dose. Be sure to see why.

14-37 a) The balanced nuclear reaction is $^{90}_{38}\text{Sr} \rightarrow \ ^{90}_{40}\text{Zr} + 2\ _{-1}^{0}e^- + 2\ \bar{\nu}$. The version $^{90}_{38}\text{Sr} \rightarrow \ ^{90}_{40}\text{Zr}^{2+} + 2\ _{-1}^{0}e^- + 2\ \bar{\nu}$ shows an exact charge balance between the two sides (in the right superscripts), in addition to balance as to Z (left subscripts) and A (left superscripts).

b) The overall nuclear reaction is two consecutive beta decays. The change in mass in beta decay is:

$$\Delta m = m\,(\text{daughter atom}) - m\,(\text{parent atom})$$

To get the Δm of the overall reaction, simply subtract the mass of an atom of ^{90}Sr from the mass of an atom of ^{90}Zr. The masses of the beta particles are automatically accounted for when this is done. The required isotopic masses are listed in the problem. The result is a Δm of -0.0030 u. The corresponding energy (taking 1 u as equivalent to 931.494 MeV) is $\boxed{2.8 \text{ MeV}}$.

c) This part is just like **14-19**. Compute the number of atoms in 1.00 g of ^{90}Sr:

$$N_{\text{Sr-90}} = 1.00 \text{ g} \times \left(\frac{1 \text{ mol}}{89.9073 \text{ g}}\right) \times \left(\frac{6.02214 \times 10^{23} \text{ atom}}{1 \text{ mol}}\right) = 6.698 \times 10^{21} \text{ atom } ^{90}\text{Sr}$$

The activity A, which is the instantaneous rate of disintegration of the Sr-90 depends on the number of Sr-90 atoms present:

$$A = -\frac{dN}{dt} = kN$$

The k in this equation equals $\ln 2/t_{1/2}$ where $t_{1/2}$ is the half-life. Substituting:

$$A = \left(\frac{\ln 2}{t_{1/2}}\right) N = \left(\frac{0.6931}{28.1 \text{ yr}}\right) (6.698 \times 10^{21}) = 1.65 \times 10^{20} \text{ yr}^{-1}$$

This is the number of disintegrations per year at the moment that the ^{90}Sr is released. The problem asks for the activity on a per second basis:

$$A = 1.65 \times 10^{20} \text{ yr}^{-1} \times \left(\frac{1 \text{ yr}}{365.2 \text{ day}}\right) \times \left(\frac{1 \text{ day}}{86400 \text{ s}}\right) = \boxed{5.23 \times 10^{12} \text{ s}^{-1}}$$

d) The activity of the Sr-90 falls off with time as the number of Sr-90 atoms persisting in the 1.00 g sample diminishes. The relationship is:

$$A = A_0 e^{-kt} = A_0 \exp\left(-\frac{\ln 2\, t}{t_{1/2}}\right)$$

Substituting the initial activity from part c), the specified time of 100 yr and the half-life of 28.1 yr gives

$$A = (5.23 \times 10^{12} \text{ s}^{-1}) \exp\left(\frac{-0.6931(100 \text{ yr})}{28.1 \text{ yr}}\right) = \boxed{4.44 \times 10^{11} \text{ s}^{-1}}$$

Tip. The activity of the isotope falls off to about 8.5% of its original value in 100 years. The problem could have asked for activities in becquerels (Bq) rather than disintegrations per second (s^{-1}). The conversion from s^{-1} to Bq is particularly simple: the answers are 5.23×10^{12} Bq and 4.44×10^{11} Bq.

14-39 The lighter isotopes of uranium happen to decay faster than the heavier isotopes. The quicker breakdown of the light isotopes leaves heavy isotopes behind, causing the average atomic mass of the uranium to $\boxed{\text{increase}}$ with time. This assumes that none of the lighter isotopes are the products of decay of the heavier isotopes.

14-41 The change in mass when one atom of U-235 gains a neutron and then undergoes fission as specified in the problem is the mass of the products minus the mass of the reactants:

$$\Delta m = 1\underbrace{(93.919)}_{^{94}\text{Kr}} + 1\underbrace{(138.909)}_{^{139}\text{Ba}} + 3\underbrace{(1.0086649)}_{^{1}_{0}n} - 1\underbrace{(235.043925)}_{^{235}\text{U}} - 1\underbrace{(1.0086649)}_{^{1}_{0}n}$$

$$= -0.1986 \text{ u}$$

Convert to kJ mol^{-1} using the mass-energy equivalence established in **14-3a**. The result is -1.785×10^{10} kJ mol^{-1}. The problem asks for the energy change per gram of ^{235}U:

$$-1.785 \times 10^{10} \text{ kJ mol}^{-1} \times \left(\frac{1 \text{ mol }^{235}\text{U}}{235.04 \text{ g }^{235}\text{U}}\right) = -7.59 \times 10^{7} \text{ kJ g}^{-1}$$

The energy released in the surroundings is the negative of the energy change of the system. It is $\boxed{+7.59 \times 10^{7} \text{ kJ g}^{-1}}$.

14-43 When a positron and electron meet, they annihilate each other to generate gamma radiation:

$$^{0}_{1}e^{+} + ^{0}_{-1}e^{-} \rightarrow 2\gamma$$

The mass of the electron and positron both equal 0.00054858 u. Therefore, Δm for the annihilation reaction is -0.00109716 u and:

$$\Delta E = -0.00109716 \text{ u} \times \left(\frac{931.494 \text{ MeV}}{1 \text{ u}}\right) = -1.02200 \text{ MeV}$$

Because neither particle had an kinetic energy, only this amount of energy appears in the surroundings /14-9.02200 MeV borne by two gamma rays of equal energy. The energy of each gamma ray is therefore $\boxed{0.51100 \text{ MeV}}$.

14-45 a) The beta decay of ^{64}Cu is:

$$^{64}_{29}\text{Cu} \rightarrow {}^{64}_{30}\text{Zn} + {}^{0}_{-1}e^- + \bar{\nu}$$

For this process:

$$\Delta m = m({}^{64}_{30}\text{Zn}) - m({}^{64}_{29}\text{Cu})$$

The ΔE for the process is given as -0.58 MeV. Convert this to atomic mass units and use it in the previous equation:

$$-0.58 \text{ Mev} \times \left(\frac{1 \text{ u}}{931.494 \text{ MeV}}\right) = m({}^{64}_{30}\text{Zn}) - m({}^{64}_{29}\text{Cu})$$

$$-0.000623 \text{ u} = m({}^{64}_{30}\text{Zn}) - 63.92976 \text{ u}$$

Hence the daughter ^{64}Zn weighs $\boxed{63.92914 \text{ u}}$.

b) This nuclear reaction produces ^{64}Ni from ^{64}Cu:

$$^{64}_{29}\text{Cu} \rightarrow {}^{64}_{28}\text{Ni} + {}^{0}_{1}e^+ + \nu$$

Its Δm is:

$$\Delta m = \left(m({}^{64}_{28}\text{Ni}) - m({}^{64}_{29}\text{Cu})\right) + 2\left(m({}^{0}_{1}e^+)\right)$$

The last term must be included because a positron is lost from the daughter-parent atom pair and the neutral daughter atom has one fewer electrons than the parent atom. The masses of a positron and an electron are equal. The value of ΔE is given as -0.65 MeV. Convert this energy to atomic mass units and set it equal to Δm in the previous equation:

$$-0.65 \text{ Mev} \times \left(\frac{1 \text{ u}}{931.494 \text{ MeV}}\right) = \left(m({}^{64}_{28}\text{Ni}) - m({}^{64}_{29}\text{Cu})\right) + 2\,(0.00054858 \text{ u})$$

Solving for the difference in mass between daughter and parent gives -0.00179 u. The parent ^{64}Cu weighs 63.92976 u, so the daughter ^{64}Ni weighs $\boxed{63.92797 \text{ u}}$

14-47 Only an element having Z larger by 2 can decay directly to Ac by alpha emission. Since Ac is element 89, this would be element 91, $\boxed{\text{protactinium}}$. Element 91 is in fact named as the parent of actinium ("proto-actinium"). Only an element with Z less by 1 can decay directly to Ac by beta emission. This would be element 88, which is $\boxed{\text{radium}}$. The fact that compounds of radium contain no actinium tends to rule out beta emission by radium as a significant source of actinium.

14-49 a) The formation of a ^{30}P atom is represented $15\,^1_1\text{H} + 15\,^1_0n \to\,^{30}_{15}\text{P}$. Note that the mass of the electrons is included in the mass of the ^1_1H atoms. The mass of the product is 29.97832 u, and the mass of the reactants is 15(1.00782504 + 1.00866497) or 30.24735 u. The difference between these masses is -0.26903 u, which is a difference in energy of -250.60 MeV. The binding energy equals the negative of this figure; the binding energy per nucleon is then $+250.60/30$ or $\boxed{8.353 \text{ MeV per nucleon}}$.

b) The equation for position emission by ^{30}P is: $^{30}_{15}\text{P} \to\,^{30}_{14}\text{Si} +\,^0_1e^+ + \nu$. The change in mass in this process is:

$$\Delta m = [m(^{30}_{14}\text{Si}) - m(^{30}_{15}\text{P})] + 2\,(0.00054858 \text{ u})$$

$$\Delta m = 29.97376 - 29.97832 + 0.0010972 = -0.003463 \text{ u}$$

The negative sign means that the process is spontaneous. The change in energy of the system is also negative:

$$\Delta E = (-0.003463 \text{ u}) \times \left(\frac{931.494 \text{ MeV}}{1 \text{ u}}\right) = -3.226 \text{ MeV}$$

The kinetic energy of the products equals $-\Delta E$ and is distributed among them. The positron has its maximum kinetic energy when the other decay products get none. This maximum is $\boxed{3.226 \text{ MeV}}$.

c) The quick way to get the fraction of ^{30}P atoms left after 450 s is to recognize that 450 s equals three 150 s half-lives. The fraction is then obviously $(1/2)^3$ or $\boxed{1/8}$. The rate constant is $\ln 2/t_{1/2}$; it is $\boxed{4.62 \times 10^{-3} \text{ s}^{-1}.}$

Tip. The energy equivalent of the mass of the positron and the extra electron in the mass-change equation is 1.178 MeV. Omitting this term introduces an unacceptable error (about 32%).

14-51 The activity of the gallium-67 has decayed to 5.0% of the initial activity when the number of atoms of Ga-67 reaches 5.0% of the original number of atoms: $N = 0.050N_0$. Insert this relationship into the first-order decay law:

$$\ln \frac{N}{N_0} = -kt \qquad \text{to obtain} \qquad \ln 0.05 = -kt$$

The rate constant k equals $\ln 2$ divided by the half-life, which is given as 77.9 hours. Therefore:

$$\ln 0.05 = -\left(\frac{\ln 2}{77.9 \text{ hr}}\right) \qquad \text{from which } t = \boxed{337 \text{ hr}}$$

14-53 The activity of ^{14}C equals the rate constant for its decay multiplied by the number of ^{14}C atoms present. Consider 1.00 g of carbon from the biosphere. The total number of carbon atoms equals:

$$N_{\text{tot}} = 1.00 \text{ g} \times \left(\frac{1 \text{ mol}}{12.01115 \text{ g}}\right) \times \left(\frac{6.022 \times 10^{23} \text{ atom}}{1 \text{ mol}}\right) = 5.014 \times 10^{22} \text{ atom}$$

The number of atoms of ^{14}C per gram of carbon is given by:

$$N_{\text{C-14}} = \frac{A}{k} = A\left(\frac{t_{1/2}}{\ln 2}\right)$$

where $t_{1/2}$ is the half-life of the ^{14}C and A is its specific activity. Converting $t_{1/2}$ from years to minutes so that the units of A cancel gives:

$$N_{\text{C-14}} = \left(\frac{15.3 \text{ min}^{-1}}{1 \text{ g}}\right) \times \left(\frac{3.014 \times 10^9 \text{ min}}{\ln 2}\right) = \boxed{6.65 \times 10^{10} \text{ atom file1}}$$

b) The preceding result means that 1.00 g of ordinary carbon contains 6.65×10^{10} atoms of ^{14}C. The number of atoms of C of all isotopes in 1.00 gram of C is 5.014×10^{22} atoms, which is vastly larger. The required fraction is the first number divided by the second. The answer is $\boxed{1.32 \times 10^{-12}}$.

14-55 Compute the mass of K-40 that was present in the rock at formation. In the following, the first unit-factor is valid because atoms of K-40 and Ar-40 have essentially the same mass. The second unit-factor deals with the fact that only 10.7%[7] of the original K-40 gives Ar-40:

$$0.42 \text{ mg } ^{40}\text{Ar} \times \left(\frac{1 \text{ mg } ^{40}\text{K by EC}}{1 \text{ mg } ^{40}\text{Ar}}\right) \times \left(\frac{100 \text{ mg } ^{40}\text{K orig}}{10.7 \text{ mg } ^{40}\text{K by EC}}\right) = 3.925 \text{ mg } ^{40}\text{K orig}$$

Since K-40 decays by first-order kinetics we can write:

$$m_{\text{K-40}} = (m_{\text{K-40}})_{\text{orig}} e^{-kt}$$

where the use of masses instead of number of atoms is valid because the number of atoms of K-40 is directly proportional to its mass. Take the logarithm of both sides and substitute the two masses and the rate constant. The rate constant k equals $\ln 2$ divided by the half-life of ^{40}K, which is available in text Table 14-2:

$$\ln\left(\frac{m_{\text{K-40}}}{(m_{\text{K-40}})_{\text{orig}}}\right) = -kt = \ln\left(\frac{1.00 \text{ mg}}{3.925 \text{ mg}}\right) = \left(\frac{-\ln 2}{1.28 \times 10^9 \text{ yr}}\right)t \qquad t = \boxed{2.5 \times 10^9 \text{ yr}}$$

---

[7]See text Table 14-2. This proportion of K-40 decays by electron capture (EC).

14-57 The incomplete nuclear equation: $^{10}_{5}\text{B} + ^{1}_{0}n \rightarrow ? + ^{4}_{2}\text{He}$ must be balanced by the insertion of a single symbol for the question mark. The mass number on this symbol must be 7 and the atomic number 3. The element of atomic number 3 is lithium. Hence the other atom that is formed is a $\boxed{^{7}_{3}\text{Li}}$ atom.

14-59 Use unit factors to obtain the energy released by the small A-bomb:

$$1.2 \text{ kg U} \times \left(\frac{1 \text{ mol U}}{0.238 \text{ kg U}}\right) \times \left(\frac{2 \times 10^{13} \text{ J}}{1 \text{ mol U}}\right) \times \left(\frac{1 \text{ ton TNT}}{4 \times 10^{9} \text{ J}}\right) = \boxed{2.5 \times 10^{4} \text{ ton TNT}}$$

14-61 The earth orbits the sun at a radius R of 1.50×10^{8} km. Imagine a sphere of this radius surrounding the sun. The surface area of this immense sphere is $4\pi R^2$. Radiation from the sun streams out in all directions, cutting through this sphere. Call the 6371 km radius of the earth r. From the sun, the earth appears as a tiny disk of area πr^2. This disk, minuscule in comparison to the area of the big sphere, intercepts a fraction f of the total radiation in proportion to the area of the big sphere that it covers:

$$f = \frac{\pi r^2}{4\pi R^2} = \frac{1}{4}\left(\frac{r}{R}\right)^2 = \frac{1}{4}\left(\frac{6371 \text{ km}}{1.50 \times 10^{8} \text{ km}}\right)^2 = 4.5 \times 10^{-10}$$

The surface area of the hemisphere of the earth that is exposed to the sun's rays at any time equals $2\pi r^2$. Assume that the radiant flux of 0.135 J s^{-1}cm^{-2} is the *average* value over the earth's exposed hemisphere. Then the earth receives the radiant power:

$$P(\text{earth}) = (0.135 \text{ J s}^{-1}\text{cm}^{-2}) \times 2\pi(6.371 \times 10^{8} \text{ cm})^2 = 3.44 \times 10^{17} \text{ J s}^{-1}$$

The total power output of the sun is $P(\text{earth})$ divided by the fraction of the sun's radiant power that hits the earth:

$$P(\text{sun}) = \frac{P(\text{earth})}{f} = \frac{3.44 \times 10^{17} \text{ J s}^{-1}}{4.5 \times 10^{-10}} = 7.6 \times 10^{26} \text{ J s}^{-1}$$

The mass equivalent of energy is given by the equation $\Delta E = c^2 \Delta m$. The sun emits 7.6×10^{26} J each second, so its ΔE is negative. The equivalent change in mass per second is:

$$\Delta m = \frac{\Delta E}{c^2} = \frac{-7.6 \times 10^{26} \text{ J s}^{-1}}{(3.0 \times 10^{8} \text{ m s}^{-1})^2} = \boxed{-8.5 \times 10^{9} \text{ kg s}^{-1}}$$

Tip. The sun burns (in the thermonuclear sense) 8.5 million metric tons of matter per second. How long can it keep burning? Its mass equals 1.99×10^{30} kg, so the sun is good for 2.3×10^{20} s (7.4 trillion years) at the current rate.

14-63 a) The $\Delta H°$ of the reaction $N_2H_4(l) + O_2(g) \rightarrow N_2(g) + 2\,H_2O(g)$ is the sum of the $\Delta H_f°$'s of the products less the sum of the $\Delta H_f°$'s of the reactants, according to Hess's law:

$$\Delta H° = 2\underbrace{(-241.82)}_{H_2O(g)} + 1\underbrace{(0.00)}_{N_2(g)} - 1\underbrace{(0.00)}_{O_2(g)} - 1\underbrace{(-50.63)}_{N_2H_4(l)} = \boxed{-534.27 \text{ kJ}}$$

where the numbers come from Appendix D (recall that elements in standard states have $\Delta H_f°$'s of zero).

b) The change in the standard internal energy during the constant-pressure reaction is:

$$\Delta E° = \Delta H° - P\Delta V$$

The $P\Delta V$ term equals $\Delta n_g RT$, if it is assumed that the gases in the reaction are ideal and that the volume of the liquid hydrazine is negligible.[8] As a mole of $N_2H_4(l)$ burns in $O_2(g)$ at 298.15 K, Δn_g of the system equals $+2$ mol, making $\Delta n_g RT$ equal to 4.96 kJ. Hence, for the combustion of one mole of $N_2H_4(l)$:

$$\Delta E° = \Delta H° - P\Delta V = -534.27 - 4.96 = \boxed{-539.23 \text{ kJ}}$$

c) Use $\Delta E°$ of the reaction as ΔE in the Einstein equation:

$$\Delta m = \frac{\Delta E}{c^2} = \frac{-539.23 \times 10^3 \text{ J}}{(2.9979 \times 10^8 \text{ m s}^{-1})^2} = \boxed{-5.9998 \times 10^{-12} \text{ kg}}$$

Tip. The chemical reaction of 32 g of hydrazine and 32 g of oxygen occasions a mass loss by the system of about 6 nanograms. This is too small to detect.

14-65 a) To produce ^{228}Ra, the ^{232}Th must emit an alpha particle in the first step of its decay:

$$^{232}_{90}\text{Th} \rightarrow \, ^{228}_{88}\text{Ra} + \, ^{4}_{2}\text{He}$$

In the next step, the ^{228}Ra emits a beta particle (an electron) to give ^{228}Ac:

$$^{228}_{88}\text{Ra} \rightarrow \, ^{228}_{89}\text{Ac} + \, ^{0}_{-1}e^- + \bar{\nu}$$

The decay products over the two steps are an alpha particle, an ^{228}Ac atom, an electron and an anti-neutrino. These products have the same total mass as an ^{228}Ac (228.03117 u) plus a $^{4}_{2}$He atom (4.0026033 u). The sum is 232.03377 u. The mass of the beta particle (electron) is included in the mass tabulated for a neutral actinium atom, and the anti-neutrino is massless. The mass of the reactant is 232.038054 u

[8]This is the same kind of assumption made in **7-43** and **8-65**.

so Δm for the process is -0.00428 u. This mass has an energy equivalent of -3.98 MeV. Hence, $\boxed{3.98 \text{ MeV}}$ is the energy lost by the system consisting of one Th atom and carried away as kinetic energy by the decay particles.

b) The thorium decays to radium in a first-order process, and the radium goes on to decay to actinium in another first-order process. Let k_1 be the rate constant for the first step and k_2 the rate constant for the second. When the radium is present in a steady-state amount then:

$$\frac{dN_{\text{Ra}}}{dt} = 0 = k_1 N_{\text{Th}} - k_2 N_{\text{Ra}} \quad \text{from which} \quad N_{\text{Ra}} = \frac{k_1}{k_2} N_{\text{Th}}$$

For each first-order process, the rate constant is $\ln 2$ divided by the half-file $t_{1/2}$. Applying this fact to the previous equation:

$$N_{\text{Ra}} = \left(\frac{t_{1/2,2}}{t_{1/2,1}}\right) N_{\text{Th}} = \left(\frac{6.7 \text{ yr}}{1.39 \times 10^{10} \text{ yr}}\right) N_{\text{Th}} = (4.8 \times 10^{-10}) N_{\text{Th}}$$

The number of ^{228}Ra nuclei equals $\boxed{4.8 \times 10^{-10}}$ of the number of ^{232}Th nuclei.

Chapter 15

Quantum Mechanics and Atomic Structure

Wave Motion and Light

A beam of light is electromagnetic radiation. Light consists of electric and magnetic fields (symbolized by E and H respectively) oscillating perpendicular to the direction in which the beam propagates and perpendicular to each other. Several parameters characterize electromagnetic radiation:

• **Wavelength λ.** The distance between successive peaks in the intensity of the oscillating electric field (or magnetic field) at any instant equals λ, the wavelength of the radiation.

• **Frequency ν.** Electromagnetic radiation is a periodic (cyclic) disturbance. As radiation travels past a given point, a maximum in the electric field is followed by a minimum, then another maximum, and so forth. The number of complete cycles occurring each second is the frequency of the radiation, represented by the symbol ν, the Greek nu. The units of ν are reciprocal seconds, s^{-1}. A reciprocal second is also called a Hertz, Hz. Sometimes the unit "cps" for "cycles per second" is encountered. One cps equals one reciprocal second.

The *reciprocal* of the frequency of a wave is the **period** of the wave. The period is the time required for the wave to complete one cycle. See **15-1**.

• **Amplitude** and **Intensity.** The amplitude of any wave disturbance is the maximum size of the excursion the wave makes during its oscillation. In classical wave theory, the **intensity I** of a wave is proportional to the square of its amplitude. For electromagnetic waves:

$$I \propto (E_{max}^2 + H_{max}^2)$$

• **Speed of propagation.** A traveling wave completes ν oscillations per second and with each oscillation advances by a distance λ, the distance between its successive crests. It follows that:

$$\text{speed of propagation} = \lambda \nu$$

This equation applies to all kinds of traveling waves. It is used in **15-1, 15-7, 15-65**, and numerous other problems. The speed of propagation of electromagnetic waves (more simply, **the speed of light**) in a vacuum is a constant c:

$$c = 2.99792458 \times 10^8 \text{ m s}^{-1} \text{ exactly} \quad \text{and} \quad c = \lambda \nu$$

 The second equation is the basis for **15-3** and **15-3**. It is also used in **15-15** and other problems. It state a simple inverse: the higher the frequency of light, the shorter the wavelength. It should be memorized.

 • **Color.** Visible light has wavelengths ranging from approximately 4×10^{-7} to 7×10^{-7} m (equivalent to 400 to 700 nm)[1]. The frequency of visible light is on the order of 10^{14} s^{-1}. Different wavelengths of light are perceived in different colors. Learn the order of the colors in the visible spectrum from long wavelength to short wavelength (that is, from low frequency to high frequency): red, orange, yellow, green, blue, indigo, violet. The initial letters spell out the name ROY G. BIV.

The Experimental Basis of Energy Quantization

Classical physics (19th century physics) treated electricity, magnetism, and electromagnetic radiation separately from **mechanics**, which dealt with the motions of particles and their interactions. Certain concepts of classical mechanics are required to understand quantum mechanics.

 • The **kinetic energy E_K** is the energy associated with the motion of a particle. For a single particle:

$$KE = E_K = \frac{p^2}{2m} = \tfrac{1}{2}mv^2$$

  where p, the product of m and v, is the **momentum** of the particle (m is its mass, and v is its velocity).

 • The **potential energy $V(x)$**, is the energy associated with the position of a particle. The "(x)" attached to the symbol V emphasizes the dependence of V on a coordinate that tells location. In three dimensions, the potential energy is represented $V(x, y, z)$.

---

[1]See text Figure 15-3, text page 526.

- The **total energy** of a particle is the sum of its potential and kinetic energies:

$$\text{total energy} = KE + PE$$
$$E = E_\text{K} + V(x)$$

Classical physics is highly successful in describing many phenomena. However, it makes wrong theoretical predictions in the study of radiation from a blackbody and the photoelectron effect.

- **Blackbody radiation.** A heated object such as an iron bar or tungsten light-bulb filament emits radiation with a broad range of wavelengths. A **blackbody** is an idealized version of such an emitter. Classical physics predicts that the intensity of the light having wavelength λ emitted by a blackbody depends directly on the absolute temperature and inversely on the fourth power of λ:

$$I = \left(\frac{8\pi R}{N_0}\right)\left(\frac{T}{\lambda^4}\right)$$

At long wavelengths, this formula agrees with observation. At short wavelengths, it gives values that are far too large. As λ gets smaller and smaller, the predicted intensity goes to infinity. This impossible theoretical result was called the **ultraviolet catastrophe.**

- **The photoelectric effect.** When a beam of light falls onto a metal or other material, "photoelectrons" are often ejected from the surface. Classical physics predicts that the maximum kinetic energy of photoelectrons should depend on the intensity of the beam of light. Instead, the maximum kinetic energy depends on the frequency of the light:

$$E_\text{K (max)} = \tfrac{1}{2}m_e v^2 = h(\nu - \nu_0)$$

where h is a constant and ν_0 is a threshold frequency that is characteristic of the metal being illuminated. Light of frequency less than ν_0 does not cause the emission of photoelectrons, no matter how great its intensity.

Quantum Explanations Succeed Where Classical Theory Fails

The theoretical difficulties presented by blackbody radiation and the photoelectric effect are resolved by the introduction of a central idea of quantum mechanics:

- **Energy is not continuous, but instead is quantized in discrete packets.**

Radiation transmits energy. The quantum idea applied to radiation (light) means that the energy of light is absorbed, emitted, or converted in individual packets, or quanta. The quantum of radiation is the **photon**. It is the particle of light. Its energy is directly related to its frequency:

$$E = h\nu = \frac{hc}{\lambda} \qquad \text{where} \qquad h = 6.626076 \times 10^{-34} \text{ J s}$$

The h is called **Planck's constant**. The equation is the Planck equation. When a body emits or absorbs a photon, the observed frequency of the photon gives on the change in the energy of the body. The Planck equation then takes the form:

$$\Delta E = h\nu = \frac{hc}{\lambda}$$

These equations occur repeatedly in problems. See **15-17, 15-19, 15-21**, and **15-63**. Problem **15-65** features both emission and absorption of a photon.

Planck's constant is extremely small. Consequently, the "grain-size" of particles of light is small. For many purposes, light is as good as continuous.

Photons, the quanta of light, have zero rest mass but *do* have momentum p (see **15-89**). Their total energy is:

$$E = pc$$

The preceding equations give the energy of light first in terms of frequency (a wave property) then in terms of momentum (a particle property):

- **In quantum theory, light has both wave and particle character.**

Planck "averted" the ultraviolet catastrophe by postulating that the oscillators that absorb and emit light at the surface of a blackbody have discrete energy states (discrete energy levels) and gain or lose energy only in amounts that correspond exactly to differences between pairs of levels. Putting the notion of the ultimate graininess of energy into the equations describing blackbody radiation gives predictions in accord with experiment.

The quantum explanation of the photoelectric effect treats light as a stream of photons (or particles or packets), each carrying a quantity of energy proportional to the frequency (not the intensity) of the light and given by the Planck equation $E = h\nu$. Liberation of a photoelectron requires a minimum energy to overcome the attractions holding it to its atom. This accounts for the existence of a minimum or threshold frequency for the liberation of photoelectrons. The **work function Φ** is the name for the minimum energy. Any extra energy from the photon can manifest itself as kinetic energy of the photoelectron:

$$E_{\text{K (max)}} = h(\nu - \nu_0) = h\nu - \Phi$$

See **15-11, 15-13**, and **15-69**.

Experiments Showing the Quantization of Energy in Atoms

Atomic Spectra

Both atoms (and molecules[2]) can emit and absorb light. Gaseous atoms of a particular element emit or absorb only at specific wavelengths. In other words: gaseous atoms emit or absorb only certain discrete, sharply defined frequencies of light. These emissions and absorptions are studied with a **spectrograph**, which is a device that uses a prism or other analyzer to spread out a beam of light in space according to the frequencies it contains. The spectrum recorded for atoms of a particular element consists of many discrete lines. Atoms of different elements atoms give different spectra. Atomic spectra characterize the elements as completely as fingerprints identify people.

The observation of lines in the spectra from single atoms strongly implies quantization of the energy of the atoms.

Spectroscopic lines follow patterns. Early spectroscopists discovered empirical formulas to predict the frequencies of related sets (series) of emission lines. For example, the formula:

$$\nu = \left(\frac{1}{4} - \frac{1}{n^2}\right) 3.29 \times 10^{15} \text{ s}^{-1} \qquad n = 3, 4, 5, \ldots$$

correctly predicts the frequencies of lines in the Balmer series, a group of lines in the visible portion of the emission spectrum of atomic hydrogen.

The Franck-Hertz Experiment

In this experiment, electrons move through a tube filled with gaseous atoms. No energy is transferred to the atoms until the speed (and kinetic energy) of the electrons reaches a certain minimum. Then the electrons transfer nearly all of their energy to the atoms. Additional distinct thresholds for the transfer of energy are observed when the kinetic energy of the incoming electrons is raised further. Clearly, the atoms accept energy only in "chunks" of certain sizes. Collisions that successfully transfer energy raise the energy of the atom from its **ground state**, which is its state of minimum energy, to an **excited state** . The relaxation of atoms from excited states back to the ground state emits light of a frequency that correspond very closely to the (known) amount of energy transferred in the collisions that created them. See **15-21**.

---

[2]See text Chapter 16.

Photoelectron Spectroscopy and Binding Energies

In photoelectron spectroscopy, high-frequency radiation ejects electrons from atoms, and the kinetic energy of the ejected electrons is measured. The difference between the energy supplied by the radiation and the observed kinetic energy of the ejected electrons equals the binding energy of the ejected electrons:

$$\text{Binding Energy} = h\nu - E_{\text{K}}(\text{ejected electron})$$

Highly energetic radiation eject electrons from even very tightly bound states in an atom. The experiment shows that:

- Electrons in atoms have definite, discrete energies. These are their binding energies.

- More than one electron may have the same energy in an atom. See **15-23**.

- Electrons in atoms are organized in shells or groups according to their binding energy.

Text Figure 15-37 shows the binding energy of the higher-energy electrons of the first 97 elements in the periodic table. Study its organization. For example, element 30 (zinc) has the binding energies of its 30 electrons organized into seven distinct groups.

Photoelectron spectroscopy differs fundamentally from the Franck-Hertz experiment, in which the collisions of relatively low-energy electrons excite electrons on atoms, but do *not* remove them.

The Bohr Model of the Atom

An important step in understanding quantization in atoms was Niels Bohr's model of the hydrogen atom. This model also applies to ions of other elements that, like hydrogen, have only one electron. The postulates of the Bohr model are:

1. The single electron revolves around the nucleus in a circular orbit.

2. Coulomb (electrostatic) attraction between the negatively-charged electron and the positively-charged nucleus provides the force needed to sustain the circular orbital motion of the electron.

3. The only orbits allowed are those for which the angular momentum of the electron is an integer n times the constant $h/2\pi$: That is, **angular momentum is quantized in units of $h/2\pi$:**

$$\text{angular momentum} = m_e v r = n\left(\frac{h}{2\pi}\right) \quad n = 1, 2, 3, \dots$$

The integer n is called a **quantum number**. The units of angular momentum are J s, the units of h. Since h is so small, the quantum of angular momentum is small. See **15-73**.

4. Electrons make transitions from one orbit to another only by absorbing or releasing a quantity of energy equal to the energy difference between the two orbits. When this energy is absorbed or emitted in the form of light:

$$\Delta E = h\nu$$

These postulates combine with the laws of classical mechanics and electrostatics to give two major results:

- **The radius of the hydrogen atom is quantized:**

$$r_n = \frac{\epsilon_0 n^2 h^2}{\pi Z e^2 m_e} \frac{n^2}{Z} = \left(\frac{\epsilon_0 h^2}{\pi e^2 m_e}\right) \frac{n^2}{Z} = \frac{n^2}{Z}(0.529 \times 10^{-10} \text{ m}) = \frac{n^2}{Z} a_0$$

where m_e and e are the mass and charge of the electron, ϵ_0 is a constant called the permittivity of free space, n is the quantum number, a whole number greater than or equal to one, and Z is the atomic number of the nucleus. The grouping of constants in the large parentheses is called a_0, the **Bohr radius**. It is a distance that occurs repeatedly in discussions of phenomena on the atomic level. The Bohr radius and other distances on the atomic scale are often expressed in a non-SI unit much favored by chemists, the Ångstrom (Å).

1 Ångstrom = 1.00 × 10⁻¹⁰ meter

- **The energy of the hydrogen atom is quantized:**

$$E_n = -\frac{Z^2 e^4 m_e}{8\epsilon_0^2 n^2 h^2} = -\left(\frac{e^4 m_e}{8\epsilon_0^2 h^2}\right)\frac{Z^2}{n^2} = -\frac{Z^2}{n^2}(2.18 \times 10^{-18} \text{ J})$$

The grouping of constants in the large parentheses has fundamental importance on the atomic scale. It therefore is used to define the **rydberg (Ry)**, a non-SI unit of energy:

1 rydberg = 2.18 × 10⁻¹⁸ joule

Start a study of these two equations by confirming the numerical values of a_0 and the rydberg. Insert the various constants in the preceding equations and check the cancellation of their units. Note that ϵ_0 equals 8.854×10^{-12} C² J⁻¹m⁻¹. Next, set Z equal to 1. Then r has a minimum value of 0.529×10^{-10} m (a_0, the Bohr radius) when $n = 1$. The radius of the atom gets bigger without bound as n steps up to

$2, 3, \ldots$. The energy E likewise has its minimum when $n = 1$, but this value is *less than zero*. The $n = 1$ state is the ground state of the atom. It has the lowest **allowed energy.** The energy rises toward a maximum of zero as n goes to infinity. Each different whole-number value of n greater than 1 corresponds to a different excited state of the atom.

Changes in the energy of a Bohr atom or ion are computed by subtracting the initial energy from the final energy:

$$\Delta E = E_{\text{f}} - E_{\text{i}} = \left[-\left(\frac{e^4 m_e}{8 \epsilon_0^2 h^2} \right) \frac{Z^2}{n_{\text{f}}^2} \right] - \left[-\left(\frac{e^4 m_e}{8 \epsilon_0^2 h^2} \right) \frac{Z^2}{n_{\text{i}}^2} \right]$$

$$= -\left(\frac{e^4 m_e}{8 \epsilon_0^2 h^2} \right) \left[\frac{Z^2}{n_{\text{f}}^2} - \frac{Z^2}{n_{\text{i}}^2} \right]$$

A negative change corresponds to a *loss* of energy. In spectroscopy, atoms change their energy by emitting or absorbing photons.

- In emission, atoms give off photons to carry away energy. The energy of an emitted photon equals $-\Delta E$ of the atom. See **15-19**, **15-21**, and **15-25**.

- In absorption, atoms take up energy supplied by photons from outside. The energy of the absorbed photon equals $+\Delta E$ of the atom.

- In both emission and absorption the frequency of the photon equals

$$\nu = \frac{|\Delta E|}{h}$$

Infinite n in a Bohr atom or ion corresponds to complete removal of the electron. The ionization energy of an atom is the minimum energy needed fully to remove an electron.[3] Removing an electron from an atom of hydrogen in its ground-state means raising it from $n = 1$ to $n = \infty$. This requires 2.18×10^{-18} J of energy.

Note that $n = 0$ corresponds to having never put an electron into the atom or ion. Only positive, integral values of n are allowed in the preceding equations.

The atomic number Z equals 1 for hydrogen, 2 for the one-electron He^+ ion, 3 for the Li^{2+} ion, 8 for O^{7+} (see **15-79**) and so forth. See **15-25** and **15-27** for a comparison of the results of the Bohr model for hydrogen and an one-electron ion. See **15-87** for an approximate treatment of alkali-metal atoms as one-electron atoms.

---

[3]See text page 57.

Waves, Particles and the Schrödinger Equation

Louis de Broglie suggested in 1924 that if light has both wave-like and particle-like properties, then perhaps matter (such as electrons, protons or even baseballs) has a wave aspect. The DeBroglie equation relates the wavelength of any moving particle to its momentum:

$$\lambda = \frac{h}{p} = \frac{h}{mv}$$

Because Planck's constant h is very small, the wavelength of most moving objects is too short to measure. The wavelength only becomes important for subatomic particles which have small masses. See **15-31**.

The Heisenberg Uncertainty Principle

The presence of a wave aspect in matter means that the positions and momenta of particles cannot be determined precisely. Instead there is a built-in uncertainty. In the measurement of positions, the uncertainty is on the order of the wavelength λ of the particle. For momenta it is on the order of h/λ . More accurately:

$$(\Delta p)(\Delta x) \geq \frac{h}{4\pi}$$

where Δp and Δx are the uncertainties in momentum and position respectively. Problem **15-33** requires nothing more than substitution in this inequality and some attention to the units. Problem **15-75** is a more complex calculation applying the uncertainty principle to an important case.

The Schrödinger Equation

Erwin Schrödinger generalized the DeBroglie wave relation to apply to bound particles such as electrons held in atoms. His contribution, a differential equation, is the fundamental equation of **quantum mechanics**. Study these points concerning this equation:

- Particles bound in different systems have different forms of the Schrödinger equation. The text considers only two bound systems: the particle in a box and the hydrogen atom. Many others exist.

- Because the Schrödinger equation is a differential equation, its solutions are mathematical functions, not numbers.

- Solutions to the Schrödinger equation are called **wave functions**, and are symbolized by the Greek letter ψ.

- An accurate wave function tells all that can be known about the behavior of the particle or particles it describes.

- For any bound system, a family of wave functions, not just one, satisfies the Schrödinger equation. Each describe a different allowed state of the system. For example, in **15-77**, substituting the positive integers for n in the wave function:

$$\psi(x) = \sqrt{\frac{2}{L}} \sin\left(\frac{n\pi x}{L}\right) \quad n = 1, 2, \ldots$$

gives the (infinite) family of different solutions.

- For every different ψ that satisfies the Schrödinger equation for a given system, there is a corresponding energy E. These are the **allowed energies** of the system. Most energies are *not* allowed. For the wave function just quoted, the allowed energies are:

$$E_n = n^2 \frac{h^2}{8mL^2}$$

- The wave function that corresponds to the lowest allowed energy is the ground state of the system. All others are excited states.

- The value of a wave function ψ changes from point to point in space. Regions of space in which ψ is positive have **positive phase**. Regions in which ψ is negative have **negative phase**. Regions where ψ passes through zero and changes sign are **nodes**. Finding a wave function's nodes helps in visualizing its shape. See **15-43**.

- The square of a wave function of a bound particle, ψ^2, evaluated at some point, equals the *probability* of finding the particle within a small volume about that point. See **15-43**. Thus, in **15-80**, which gives a ψ and asks for probabilities at several points, the proper procedure is to square the ψ, evaluate ψ^2 at the several points, and then multiply each of the results, which have units of m^{-3}, by the volume of the small sphere that is specified.

- A plot of ψ^2 against the space coordinates gives a picture of the **probability density distribution** of a bound particle that is following ψ. Plots like this appear in text Figure 15-24.

The Particle in a Box

A simple case in which the Schrödinger equation can be solved is that of a particle confined in a one-dimensional box. The wave functions that describe this system are all sine functions. They are:

$$\psi(x) = \sqrt{\frac{2}{L}} \sin\left(\frac{n\pi x}{L}\right) \qquad E_n = \frac{n^2 h^2}{8mL^2} \qquad n = 1, 2, \ldots$$

where n is an integer, m is the mass of the particle and L is the length of the box.

The particle-in-a-box problem has slight practical application (however, see **15-35**). Its main value is to illustrate how the wave property of a confined particle is accounted for. Key points are:

• In a one-dimensional box, the wave functions that satisfy the Schrödinger equation are characterized by *one* quantum number. In two-dimensional and three dimensional boxes the solutions are characterized by *two* and *three* quantum numbers, respectively.

• If the box is big (L large) or if the particle is massive (m large), the allowed energies of the particle are small. More importantly, the *differences* between neighboring values of allowed energy are small. When the allowed energies are closely spaced, the quantization of the energy is not apparent. This is the case with objects bigger than atoms.

• The quantum number alone is often used as a short-hand designation. Thus, the $n = 2$ wave function for a particle in a box is referred to as ψ_2 or simply "the $n = 2$ function".

The Hydrogen Atom

When the Schrödinger equation is set up and solved for the case of a single electron (charge $-e$ and mass m_e) bound in the vicinity of a central nucleus of charge $+Ze$, a set of wave functions characterized by *three* quantum numbers (n, ℓ, m) arises. There are three quantum numbers because the hydrogen atom is three-dimensional. These wave functions are **orbitals.** They are important in the theory of chemical bonding.

The solution of the hydrogen atom is in terms of **spherical coordinates r, θ,** and ϕ, with the origin at the nucleus, instead of x, y, and z, the familiar Cartesian coordinates. Spherical coordinates simplify the mathematics of the solution. They are related to Cartesian coordinates by the equations:

$$x = r \sin\theta \cos\phi \qquad y = r \sin\theta \sin\phi \qquad z = r \cos\theta$$

See text Figure 15-21.[4] The resultant wave functions can all be broken down into parts:

• A **radial** part, $R_{n\ell}(r)$, which depends only the distance from the nucleus, as given by the coordinate r. The form of $R(r)$ is determined by the first two quantum

---

[4]Text page 550.

numbers, n and ℓ, which is why they appear as subscripts on the symbol for the function.

• An **angular** part $\chi(\theta, \phi)$, which depends on the two angles θ and ϕ and is specified by the second and third quantum numbers, ℓ and m. See **15-43** and text Table 15-2.

The three quantum numbers, which arise naturally in the mathematics of the solution of the Schrödinger equation, occur in *sets* such as $\left(n = 3, \ell = 2, m = 0\right)$. See **15-37**. Each set signifies a different wave function. Table 15-2[5] gives *some* of the wave functions that satisfy the Schrödinger equation for the hydrogen atom. As the table shows, solutions to the Schrödinger equation are often complex and tiresome to write out in full. They are frequently designated by the quantum numbers alone. The set of quantum numbers $(3, 2, 0)$ means $\psi(3, 2, 0)$ and:

$$\psi(3, 2, 0) = R(3d)\chi(d_{z^2}) = \frac{4}{81\sqrt{30}}\left(\frac{Z}{a_0}\right)^{\frac{3}{2}} \sigma^2 \exp(-\sigma/3)\left(\frac{5}{16\pi}\right)^{\frac{1}{2}}(3\cos^2\theta - 1)$$

where the radial and angular functions on the right are copied from text Table 15-2.[6] More importantly, each set of three quantum numbers encodes the characteristic *energy* and characteristic *shape* of its wave function. Sets are *not* constructed by taking just any values for n, ℓ and m. Instead, possible values of the three are interlocked in a strict pattern:

• The principal quantum number n may have any integral value from $+1$ to infinity. The quantum number n gives the energy for a one-electron atom:

$$E_n = -\frac{1}{n^2}\left(\frac{Z^2 e^4 m_e}{8\epsilon_0 h^2}\right) \qquad \text{for one-electron atoms}$$

The principal quantum number also tells the number of **nodes** the wave function has:

$$\text{Number of nodes} = (n - 1)$$

The nodes of orbitals are a valuable physical reference for study. In the hydrogen atom wave functions, the nodes are always either **radial** or **angular**. Radial nodes are spheres of different radius with the nucleus at their center. They are called radial because their locations are specified by values of the coordinate r. Angular nodes are planes or curved surfaces. They are specified by values of the angular coordinates θ and ϕ. See **15-43**. Angular nodes always contain the nucleus. Radial nodes encircle

---

[5]Text page 554.
[6]The σ in the functions is discussed below.

the nucleus but never contain it. Keeping track of the nodes helps to solve problems. See **15-37**.

The more nodes in an orbital, the higher its energy.

• The **angular momentum quantum number** ℓ has allowed values ranging from 0 to $(n-1)$. The quantum number n imposes a ceiling on ℓ, which may *not* take on just any value. For example, when n is 1, then ℓ may equal 0 only.

The value for ℓ always equals the number of angular nodes in the orbital. The presence of angular nodes lends a distinct shape to orbitals, so ℓ really tells the shape of the orbital. When $\ell = 0$, the orbital has no angular nodes. All the nodes are automatically radial, and all orbitals with $\ell = 0$ are therefore **spherically symmetrical**. When $\ell = 1$, the orbital has 1 angular node. When $\ell = 2$, it has 2 angular nodes, and so forth.

In referring to orbitals, the numerical value of ℓ is replaced by a letter:

value of ℓ	0	1	2	3	4	...
code letter	s	p	d	f	g	rest of alphabet

An (n, ℓ) combination is referred to by affixing the letter for the ℓ value to the integer n. **Examples:** $2s$, $3d$. See **15-39**.

All s orbitals are spherically symmetrical. All p orbitals have 1 angular node, a plane. All d orbitals have 2 angular nodes. The shapes of s, p and d orbitals are important in chemistry. Study text Figures 15-23, 15-24, and 15-25.[7] Focus on the symmetry of the shapes. The three $2p$ orbitals are identical except for orientation. They differ by having their "lobes" directed along the three Cartesian coordinates.

• The **magnetic quantum number** m governs the behavior of the atom in external magnetic fields. It completes the description of an orbital. For a given value of ℓ, m ranges from $-\ell$ through 0 and up to $+\ell$. The number of possible values for m equals $(2\ell + 1)$. For example, an f orbital ($\ell = 3$) has 7 different possible values of m: $-3, -2, -1, 0, 1, 2, 3$. There is only one s orbital (because $\ell = 0$), but there are 3 p orbitals: p_x, p_y, p_z, and 5 d orbitals: d_{xy}, d_{xz}, d_{yz}, $d_{x^2-y^2}$, d_{z^2}. In these designations, the subscripts are Cartesian coordinates that tell the orientation of the single angular node (in the p case) and the two angular nodes (in the d case). They should be learned. To help, notice that the first three labels comprise every possible combination of two of the three Cartesian coordinates. In the last two labels, all of the coordinates are squared. In the hydrogen atom, orbitals having the same n and ℓ are equivalent except in orientation. Thus, m, which distinguishes among the members of the set, can be interpreted as telling the orientation in space of the orbital's angular nodes. It is indeed sometimes called the space quantum number.

---

[7]Text page 555–7.

• The rules setting the possible values of the three quantum numbers generate n^2 wave functions for a given n:

n	ℓ	label	Number of Orbitals	Total No. of Orbitals
1	0	$1s$	1	1
2	0	$2s$	1	
2	1	$2p$	3	4
3	0	$3s$	1	
3	1	$3p$	3	
3	2	$3d$	5	9
4	0	$4s$	1	
4	1	$4p$	3	
4	2	$4d$	5	
4	3	$4f$	7	16
n				n^2

Studying the One-Electron Wave Functions

To understand the wave functions in text Table 15-2, take them apart and study them piece by piece. Bear in mind that the wave functions *already* are in two parts. A complete wave function is the product of an angular part and a radial part.

Select a radial function (for example, $R(3s)$) for study. Simplify matters by setting $Z = 1$, which means dealing with the hydrogen atom and not one of the one-electron ions. Copy the function from text Table 15-2 and insert the numerical constants:

$$R(3s) = \frac{2}{81\sqrt{3}} \left(\frac{1}{a_0}\right)^{3/2} (27 - 18\sigma + 2\sigma^2)e^{-\sigma/3}$$

$$= (3.7051 \times 10^{13} \text{ m}^{-3/2})(27 - 18\sigma + 2\sigma^2)e^{-\sigma/3}$$

Prepare a table showing the values at different r's of the several terms in the function:

r (Å)	σ	$(27 - 18\sigma + 2\sigma^2)$	$e^{-\sigma/3}$	$R(r)$ (m$^{-3/2}$)
0.0000	0.0000	27.00	1.0000	10.0×10^{14}
0.2645	0.5000	18.50	0.8465	5.80×10^{14}
0.5290	1.0000	11.00	0.7165	2.92×10^{14}
0.7935	1.5000	4.50	0.6065	1.01×10^{14}
1.0060	1.9019	0.00	0.5305	0.0
2.0000	3.7807	-12.46	0.2836	-1.31×10^{14}
3.0000	5.6711	-10.76	0.1510	-0.602×10^{14}
3.7548	7.0980	0.00	0.0938	0.0
10.000	18.904	401.4	1.834×10^{-3}	0.272×10^{14}
100.00	189.04	6.810×10^4	4.302×10^{-28}	1.09×10^{-9}

Note that the variable σ has no units. It is the distance r divided by the Bohr radius a_0, which is 0.529 Å.

The term in the third column causes the overall function (last column) to become negative between $r = 1.006$ and 3.755 Å. The distances at which $R(3s)$ changes sign (goes from positive to negative phase and back again) are the locations of the two radial nodes.

Despite the fact that the term in the third column gets large with increasing r, the exponential term in the fourth column gets so small that it forces the overall value of $R(r)$ toward zero at large r.[8] Study the above numbers in combination with text Figure 15-23. Note that $1.9a_0$ and $7.1a_0$ equal 1.00 Å and 3.75 Å respectively.

Finally, the unit of $R(r)$ is meters to the minus three-halves. The physical meaning of this unit is hardly obvious. However, when squared, it becomes m^{-3}, and if m^{-3} is multiplied by a volume in m^3, the result is a pure number that is a probability (see the following). It is worthwhile to graph $R(3s)$ and other radial functions.

Sizes and Shapes of Orbitals

The wave nature of the electron makes it impossible to know exactly where an electron inside an atom is located. It even makes it impossible to know where "inside" leaves off and "outside" begins. This is the meaning of the Heisenberg uncertainty principle. Instead it is necessary to speak of probabilities.

The probability p of finding an electron at a point (r, θ, ϕ) in a hydrogen atom is:

$$p(r, \theta, \phi) = \left(\psi(r, \theta, \phi)\right)^2 = \left(R(r)\right)^2 \left(\chi(\theta, \phi)\right)^2$$

Different wave-functions give different probability distributions. Graphs of $\left(R(r)\right)^2$ and $\left(\chi(\theta, \phi)\right)^2$ convey the shape and size of orbitals. In studying these graphs, focus first on the number and orientation of the nodes of each orbital. The following generalizations hold:

- An orbital with quantum numbers n and ℓ has ℓ angular nodes and $n - \ell - 1$ radial nodes.

- At large r, all orbitals go to zero—the probability of finding the electron becomes very small far away from the nucleus.

- Close to the nucleus (as r approaches zero), all orbitals go to zero except s orbitals. An electron in an s orbital has a finite probability of being found right at the nucleus.

---

[8]This behavior resembles the behavior of the Maxwell-Boltzmann distribution function. See page 79 of this Guide.

- The size of orbitals of the same ℓ (for example all s orbitals) increases with increasing n. Thus, the $3p$ orbitals are bigger than the $2p$.

- The size of orbitals decreases with increasing Z. See **15-15**.

- For a given value of n the orbital size *decreases* with increasing ℓ. The $3d$ orbital is smaller than the $3p$, which is smaller than the $3s$.

A good quantitative measure of the size of an orbital is the average distance of an electron from the nucleus when it is in that orbital. This quantity is labeled $\bar{r}_{n,\ell}$:

$$\bar{r}_{n,\ell} = \frac{n^2 a_0}{Z}\left[1 + \frac{1}{2}\left(1 - \frac{\ell(\ell+1)}{n^2}\right)\right]$$

Study of this formula confirms the last three generalizations about size in the preceding list. See **15-79** for use of the formula.

Electron Spin

The electron behaves like a tiny magnet. In classical theory, this magnetism is explained by imagining the electron to be a ball of charge spinning about its own axis like a top. Two directions of spin are possible. They are identified with two possible values of the **spin quantum number**:

$$m_s = +\tfrac{1}{2} \quad \text{or} \quad -\tfrac{1}{2}$$

The electron spin quantum number adds a *fourth* quantum number to n, ℓ, and m, the three quantum numbers that characterize the wave function of a bound electron.

Many-Electron Atoms and the Periodic Table

When an atom contains more than one electron, the Schrödinger equation becomes too complicated for easy use. The usual description of the **electronic structure** of many-electron atoms avoids the complexity of an exact solution. and relies on an approximate solution.

In the **self-consistent field (SCF) orbital approximation**, each electron in the many-electron atom is described by a set of four quantum numbers: n, ℓ, m, m_s that specify an approximate one-electron orbital. These approximate orbitals are called **Hartree orbitals** after the developer of the SCF method. They resemble the hydrogen-atom orbitals. Consequently, hydrogen-atom orbitals offer a guide to the **electron configurations** of all other atoms. Certain rules govern how Hartree orbitals accommodate electrons:

- The **Pauli Exclusion principle** states that no two electrons in an atom may have the same set of four quantum numbers. An orbital, which is fully specified by a set of three quantum numbers, therefore holds at most two electrons, one with $m_s = +1/2$ and the other with $m_s = -1/2$.

- **Hund's rule** states that when electrons are added to a group of orbitals of equal energy they half-fill every orbital in the group before starting to pair up.

The ground-state electron configurations of all the elements in the periodic table are obtained by filling the Hartree orbitals in order of increasing energy while heeding the Pauli principle and Hund's rule. This embodies the **aufbau principle** (building-up principle). The following points are important in practicing aufbau:

- The order of filling the orbitals is:

$$1s \quad 2s \quad 2p \quad 3s \quad 3p \quad 4s \quad 3d \quad 4p \quad 5s \quad 4d \quad 5p \quad 6s \quad 4f \quad 5d \quad 6p \quad 7s \dots$$

In these designations, the numbers are n (the principal quantum number) and the letters tell the value of ℓ (the angular momentum quantum number) as explained on page 437 of this Guide. The s designations refer to just one orbital, but the p's refer to sets of 3 orbitals, and the d's and f's to sets of 5 and 7 orbitals because m is deliberately left unstated. For Hartree orbitals in general:

$$E_{ns} < E_{np} < E_{nd} \text{ for a given } n$$

- Elements are classified as s-block, p-block, d-block, or f-block depending on whether the latest electrons added in aufbau went into s, p, d, or f orbitals.
- A **shell** of orbitals in a many-electron atom is a set of orbitals with similar energies. A **subshell** is a group of orbitals with the same n and ℓ quantum numbers. Consider for example the $1s$, the $2s$, and $2p$ subshells in an atom of carbon. The $1s$ subshell, which contains a single orbital, is lower in energy than all other orbitals and is a distinct shell by itself. The $2s$ subshell (which contains one orbital) and the $2p$ subshell (which contains three orbitals) are relatively close together in energy and comprise a second shell.
- Orbitals of similar energy often are just orbitals with the same n. Sometimes however orbitals of different n are in the same shell. In cobalt, the $4s$, $3d$ and $4p$ subshells are in the same shell. See text Figure 15-37,[9] which displays a plot of the orbital energies of the elements.
- Electrons in the **outermost shell** of an atom determine the chemical behavior of that atom. These are the **valence electrons** of the atom. The valence electrons occupy the **valence orbitals**.

---

[9]Text page 571.

• When two electrons in an atom have the same spin quantum number their spins are **parallel** and the two electrons are **unpaired** (with each other). Unpaired electrons must be in different orbitals, according to the Pauli principle. Two electrons in the same orbital must have spin quantum numbers $+\frac{1}{2}$ and $-\frac{1}{2}$. The two are then **paired**.

• If all the electrons in an atom are paired, then the atom is **diamagnetic**. If one or more electrons are not paired, then their spins add together instead of cancelling each other out. An atom with one or more unpaired electrons is **paramagnetic**. Electron spin is related to the magnetism of the electron; paired electrons compensate for each other's magnetism. Diamagnetism and paramagnetism are physically observable properties.

Learn to write a correct representation of the ground-state electron configuration of any atom or ion. See **15-45** and **15-47**. In writing electron configurations:

- The number of electrons in each subshell is shown as a right superscript (and is often mistaken for an exponent). The maximum superscript is 2 for an *s* subshell (1 *s* orbital), 6 for a *p* subshell (3 *p*-orbitals), 10 for a *d* subshell (5 *d*-orbitals, and 14 for an *f* subshell (7 *f*-orbitals).

- Noble-gas electron configurations are abbreviated by writing the symbol of the noble gas in brackets as shown in **15-45** and **15-51**.

- The order of filling of the subshells in the neutral atoms need not be memorized. It can be determined from the pattern of the different blocks of elements in the periodic table.

- Nineteen *d*-block and *f*-block elements present exceptions to the usual pattern. These anomalous ground-state electron configurations are given in text Figure 15-33.[10] They are best memorized.

Another important exercise is writing the valence electron configuration of an element given its position in the periodic table. The number of the period that the element occupies gives the maximum value of n in such a configuration. For non-transition elements, the number of valence electrons equals the group number (number at head of the column in the periodic table).

Many problems are based on proposed violations of the rules that relate the quantum numbers of the electrons in many-electron atoms. See **15-53**.

---

[10]Text page 568.

Shielding Effects

Sizes of Atoms and Ions

Atomic size (the radii of ions and of metallic atoms) *decreases* with increasing atomic number across a period and *increases* going down a group in the periodic table. The *rate* of increase in size going down a group diminishes sharply once the $3d$ orbitals start to be filled. As Z increases across the transition metals, any one $3d$ electron is only imperfectly shielded from the attraction of the larger nuclear charge by the other electrons in the same subshell. Atomic size therefore contracts. The same effect occurs with p subshells and is responsible for the general contraction in size going across the table. The intervention of the first block of 10 elements in which d-orbitals are filled makes immediately subsequent elements smaller than they otherwise would have been.

    Isoelectronic species have the same electron configuration but different nuclear charge. S^{2-}, Cl^-, Ar, and K^+ are isoelectronic. The smallest species among an isoelectronic group is the one with the highest Z. See **15-85**.

Experimental Measures of Orbital Energies

The binding energies of the electrons in every orbital in every subshell and shell of the atoms of every element has been measured by photoelectron spectroscopy. The measurements confirm the order of energy of the Hartree orbitals.

Periodic Trends in Ionization Energies and Electron Affinities

Ionization energies are also measured by photoelectron spectroscopy. The positively charged ion produced by ionization can itself lose an electron. Thus an atom with n electrons has a series of n ionization energies, each one larger than the previous: $IE_1, IE_2, IE_3 \ldots$.

    In general the first ionization energy *increases* from left to right across a period (row) in the periodic table and *decreases* from top to bottom within a group (column). See **15-59**. Consider the IE_1 of lithium. It is much less than the IE_1 of He because the single $2s$ electron in Li "feels" a low effective nuclear charge. (see **15-87**). It is strongly shielded by the $1s$ electrons from the attraction of the nucleus. The IE_1 of beryllium is greater than that of lithium, which is consistent with the "increase-across" trend. This general trend is explained by increasing nuclear charge of the elements moving across a period holding the electrons more tightly. However, the IE_1 of boron is *less* than that of beryllium. This departure is explained by noting that the $2p$ subshell is farther out from the nucleus and is more effectively shielded from the nucleus by the $2s$ electrons and so more readily lost. Other variations disturb the general trends in IE_1. For example, nitrogen has a higher IE_1 than its neighbor to the right, oxygen. In every case, an explanation can be advanced in terms of the shell structure of the atoms.

The periodic trends in *EA* parallel the changes in IE_1. Both atomic properties increase from left to right in a period and from bottom to top in a group.

If a second electron joins an originally neutral atom, there is a second electron affinity: EA_2 is *always* negative—the -1 ion formed by the atom's gain of the first electron repels additional negative charge.

Detailed Solutions to Odd-Numbered Problems

15-1 The speed of propagation of a wave equals the product of its frequency and wavelength. A wave-crest hits the beach once every 3.2 s, which means that slightly less than one-third of a wave reaches the beach per second. The frequency of the waves equals the reciprocal of 3.2 s, or 0.3125 s^{-1}. Thus:

$$\text{speed} = \nu\lambda = \frac{1}{3.2 \text{ s}}(2.1 \text{ m}) = \boxed{0.66 \text{ m s}^{-1}}$$

Tip. The interval between recurring identical portions of a wave is called the **period** of the wave. The period equals the reciprocal of the frequency.

15-3 The speed of propagation of electromagnetic radiation through a vacuum is c, a quantity that is very well known. As with all traveling waves, the speed of propagation of the FM radio signal equals the product of its wavelength and frequency: $c = \lambda\nu$. Hence:

$$\lambda = \frac{c}{\nu} = \frac{2.998 \times 10^8 \text{ m s}^{-1}}{9.86 \times 10^7 \text{ s}^{-1}} = \boxed{3.04 \text{ m}}$$

15-5 a) Start with the same relationship as in the preceding, but solve for frequency:

$$\nu = \frac{c}{\lambda} = \frac{2.998 \times 10^8 \text{ m s}^{-1}}{6.00 \times 10^2 \text{ m}} = \boxed{5.00 \times 10^5 \text{ s}^{-1}}$$

b) The time for a wave to travel some distance d equals the distance divided by the speed of the wave. These electromagnetic waves advance at the known speed c. Hence:

$$t = \frac{d}{c} = \frac{8.0 \times 10^{10} \text{ m}}{3.00 \times 10^8 \text{ m s}^{-1}} \times \left(\frac{1 \text{ min}}{60 \text{ s}}\right) = \boxed{4.4 \text{ min}}$$

15-7 The wavelength of the sound waves can be determined from the frequency and speed of propagation of the waves:

$$\lambda = \frac{\text{speed}}{\nu} = \frac{343.5 \text{ m s}^{-1}}{261.6 \text{ s}^{-1}} = \boxed{1.313 \text{ m}}$$

The time to travel 30.0 m is:

$$t = \frac{d}{\text{speed}} = \frac{30.0 \text{ m}}{343.5 \text{ m s}^{-1}} = \boxed{0.0873 \text{ s}}$$

15-9 Blue light has a higher frequency than green light (see text Figure 15-3[11]). The photons of blue light are therefore more energetic than the photons of green light. Since the work function of the surface of the potassium is the same for both colors of light, the $\boxed{\text{electrons ejected by blue light}}$ have higher average kinetic energy.

15-11 Combine the relationships $c = \lambda\nu$ and $E = h\nu$ to obtain an equation for the wavelength of light of specified energy. Then insert the energy needed to eject electrons from a surface of cesium metal:

$$\lambda = \frac{c}{\nu} = \frac{hc}{E} = \frac{(6.626 \times 10^{-34} \text{ J s})(3.00 \times 10^8 \text{ m s}^{-1})}{3.43 \times 10^{-19} \text{ J}} = 5.80 \times 10^{-7} \text{ m}$$

This light is yellow (see text Figure 15-3). $\boxed{\text{Yellow light}}$ and light of shorter wavelength (green, blue, violet, ultraviolet, etc.) can eject electrons from cesium in the photoelectric experiment.

For selenium, the work function is bigger (more energy is required to eject electrons):

$$\lambda = \frac{hc}{E} = \frac{(6.626 \times 10^{-34} \text{ J s})(3.00 \times 10^8 \text{ m s}^{-1})}{9.5 \times 10^{-19} \text{ J}} = 2.1 \times 10^{-7} \text{ m}$$

This wavelength is well into the $\boxed{\text{ultraviolet}}$.

15-13 a) The maximum kinetic energy of an electron ejected from a chromium surface equals the difference between the energy of the incident photons and the work function of chromium:

$$E_{\text{Kmax}} = h\nu - \Phi = \frac{hc}{\lambda} - \Phi$$

$$E_{\text{Kmax}} = \frac{(6.626 \times 10^{-34} \text{ J s})(3.00 \times 10^8 \text{ m s}^{-1})}{2.50 \times 10^{-7} \text{ m}} - 7.21 \times 10^{-19} \text{ J} = \boxed{0.74 \times 10^{-19} \text{ J}}$$

---

[11]Text page 526.

b) The kinetic energy of a particle depends on its speed and mass: $E_K = \frac{1}{2}mv^2$. Solve for v, and insert the kinetic energy computed in the previous part and the known mass of the electron:

$$v = \sqrt{\frac{2E_K}{m}} = \sqrt{\frac{2(0.74 \times 10^{-19} \text{ J})}{9.109 \times 10^{-31} \text{ kg}}} = \boxed{4.0 \times 10^5 \text{ m s}^{-1}}$$

Tip. This speed is less than 0.2% of the speed of light, so relativistic effects (not mentioned in the text) are safely ignored.

15-15 The wavelength 671 nm is 6.71×10^{-7} m. Text Figure 15-3 shows that light of this wavelength is $\boxed{\text{red}}$.

15-17 The frequency corresponding to the transition energy is computed using a re-arranged version of the Planck equation $\Delta E = h\nu$:

$$\nu = \frac{\Delta E}{h} = \frac{3.6 \times 10^{-19} \text{ J}}{6.626 \times 10^{-34} \text{ J s}} = 5.43 \times 10^{14} \text{ s}^{-1}$$

The wavelength equals the speed of propagation divided by the frequency:

$$\lambda = \frac{c}{\nu} = \frac{2.9979 \times 10^8 \text{ m s}^{-1}}{5.43 \times 10^{14} \text{ s}^{-1}} = \boxed{5.5 \times 10^{-7} \text{ m}}$$

According to text Figure 15-3, this light is $\boxed{\text{green}}$.

15-19 a) The energy change of an atom and the wavelength λ of the radiation it emits are inversely related:

$$-\Delta E = \frac{hc}{\lambda}$$

The λ is 589.3 nm (or 5.893×10^{-7} m). Substitution for λ, h, and c gives:

$$\Delta E = \frac{(6.626 \times 10^{-34} \text{ J s})(2.998 \times 10^8 \text{ m s}^{-1})}{5.893 \times 10^{-7} \text{ m}} = \boxed{-3.371 \times 10^{-19} \text{ J}}$$

The negative sign means that the final energy of the sodium atom is less than its original energy.

b) A mole of sodium atoms consists of Avogadro's number of sodium atoms. The energy change per mole is:

$$\Delta E = \left(-3.371 \times 10^{-19} \frac{\text{J}}{\text{atom}}\right) \times \left(6.022 \times 10^{23} \frac{\text{atom}}{\text{mol}}\right) = \boxed{-2.030 \times 10^5 \text{ J mol}^{-1}}$$

c) The sodium arc light emits 1000 W (watt) of radiant energy; 1000 W equals 1000 $J \, s^{-1}$. Assume that all of this energy is carried by photons at the D-line. Then:

$$\frac{1.000 \times 10^3 \text{ J emitted}}{1 \text{ s}} \times \left(\frac{1 \text{ mol Na atoms}}{+2.030 \times 10^5 \text{ J emitted}} \right) = \boxed{4.926 \times 10^{-3} \text{ mol s}^{-1}}$$

15-21 a) The observed voltage of the first excitation threshold in the Franck-Hertz experiment on Na atoms equals 2.103 V. This means that an electron accelerated across a potential difference of 2.103 V is just energetic enough to transfer a quantum of energy to an Na atom, which does not accept smaller quanta of energy. The change in the energy of the Na that accepts the quantum of energy equals the product of the charge on the electron and the accelerating voltage:

$$\Delta E = (2.103 \text{ V})(1.602177 \times 10^{-19} \text{ C}) = 3.3694 \times 10^{-19} \text{ V C} = 3.3694 \times 10^{-19} \text{ J}$$

Later, the excited Na atom emits a quantum of light to relax to its original state. Its ΔE during the relaxation equals the negative of its ΔE during the excitation. The wavelength of the light that it emits is:

$$\lambda = \frac{hc}{-\Delta E} = \frac{(6.626 \times 10^{-34} \text{ J s})(2.9979 \times 10^8 \text{ m s}^{-1})}{-(-3.3694 \times 10^{-19} \text{ J})} = \boxed{5.895 \times 10^{-7} \text{ m}}$$

Tip. The calculated wavelength corresponds to 5895 Å, very close to the two closely spaced wavelengths (called a doublet) that are quoted for Na atoms in the problem.

15-23 a) The photoelectron spectroscopy experiment measures the kinetic energies of electrons that are ejected from atoms by the absorption of high-frequency photons. The difference between the energy of an incoming photon and the kinetic energy of an outgoing electron equals the binding energy of the electron. Multielectron atoms have many different binding energies.[12]. Some electrons are closer to the nucleus (more tightly bound) and others are farther away. Compute the binding energy of this electron by subtraction, as just discussed:

$$BE = h\nu - E_{\text{K electron}}$$

$$= \frac{hc}{\lambda} - E_{\text{K electron}}$$

$$= \frac{(6.626 \times 10^{-34} \text{ J s})(2.9979 \times 10^8 \text{ m s}^{-1})}{584.4 \times 10^{-10} \text{ m}} - (11.7 \text{ eV}) \left(\frac{1.602 \times 10^{-19} \text{ J}}{1 \text{ eV}} \right)$$

$$= \boxed{1.52 \times 10^{-18} \text{ J}}$$

The answer is equivalent to 9.52 eV.

---

[12]See text Figure 15-37.

15-25 The B^{4+} ion is a hydrogen-like ion. Like H, it has only one electron, but its atomic number Z equals 5. Answer the questions about it by substitution into these equations:

$$r_n = \frac{n^2}{Z}(5.29 \times 10^{-11} \text{ m}) \quad \text{and} \quad E_n = -(2.18 \times 10^{-18} \text{ J})\frac{Z^2}{n^2}$$

The radius for the state $n = 3$ in B^{4+} equals:

$$r_3 = \frac{3^2}{5}(5.29 \times 10^{-11} \text{ m}) = \boxed{9.52 \times 10^{-11} \text{ m}}$$

The energy for the same case ($n = 3$ and $Z = 5$) equals:

$$E_3 = -\frac{5^2}{3^2}(2.18 \times 10^{-18} \text{ J}) = -6.06 \times 10^{-18} \text{ J}$$

The negative of this answer is the energy needed to strip the electron away from a single B^{4+} ion in the $n = 3$ state. For a mole of B^{4+} ions, this energy must be multiplied by Avogadro's number:

$$E = \frac{+6.06 \times 10^{-18} \text{ J}}{\text{atom}} \times \left(\frac{6.022 \times 10^{23} \text{ atom}}{\text{mol}}\right) = \boxed{3.65 \times 10^6 \text{ J mol}^{-1}}$$

The energy change in a B^{4+} ion undergoing a $3 \rightarrow 2$ transition is the difference between the energies of the two states. The two energies are:

$$E_3 = -\frac{5^2}{3^2}(-2.18 \times 10^{-18} \text{ J}) = -\frac{25}{9}(-2.18 \times 10^{-18} \text{ J})$$

$$E_2 = -\frac{5^2}{2^2}(-2.18 \times 10^{-18} \text{ J}) = -\frac{25}{4}(-2.18 \times 10^{-18} \text{ J})$$

The difference equals the final energy minus the initial:

$$\Delta E = E_2 - E_3 = \left(\frac{-25}{4} - \frac{-25}{9}\right)(2.18 \times 10^{-18} \text{ J}) = -7.57 \times 10^{-18} \text{ J}$$

This is the energy change of the ion. The energy gained by the surroundings in the form of a photon equals the negative of this. Dividing this energy by h gives the frequency of the photon:

$$\nu = \frac{-\Delta E}{h} = \frac{-(-7.57 \times 10^{-18} \text{ J})}{6.626 \times 10^{-34} \text{ J s}} = \boxed{1.14 \times 10^{16} \text{ s}^{-1}}$$

The wavelength of the emitted photon is computed as follows:

$$\lambda = \frac{hc}{-\Delta E} = \frac{(6.626 \times 10^{-34} \text{ J s})(2.9979 \times 10^8 \text{ m s}^{-1})}{7.57 \times 10^{-18} \text{ J}} = \boxed{2.63 \times 10^{-8} \text{ m}}$$

15-27 The problem could be solved by substitution into the equations of the Bohr model. The fact that the $3 \to 2$ emission in hydrogen occurs at 656.1 nm is then not needed. A quicker solution uses this fact.

The allowed energy levels in hydrogen-like atoms are characterized by the factor $-Z^2/n^2$ multiplied by a constant. The $3 \to 2$ transition for the neutral H atom $(Z = 1)$ corresponds to an energy jump proportional to $(1^2/2^2 - 1^2/3^2)$, which equals 0.13889. For the Li^{2+} ion, $Z = 3$, and the $3 \to 2$ energy jump is proportional to $3^2/2^2 - 3^2/3^2$, or 1.2500. The energy jump is bigger in the Li^{2+} transition by the factor $(1.2500/0.13889) = 9.000$. Therefore the wavelength of the emitted light in the Li^{2+} transition is 9.000 times *shorter* than 656.1 nm. This is $656.1/9.000 = \boxed{72.90 \text{ nm}}$. Light of this wavelength is in the $\boxed{\text{ultraviolet}}$.

15-29 The wave in a guitar string is a standing wave. Its allowed wavelength satisfies the equation:

$$\frac{n\lambda}{2} = L$$

where L is the length of the string and n is an integer.

a) The first harmonic has $n = 1$: Solving the above equation gives:

$$\lambda_1 = \frac{2L}{1} = \frac{2(50 \text{ cm})}{1} = \boxed{100 \text{ cm}}$$

By substitution of $n = 3$, the wavelength λ_3 of the third harmonic is $\boxed{33 \text{ cm}}$.

b) The number of nodes in a standing wave is always one less than the number of the harmonic. Thus the 1st harmonic has 0 nodes, the 2nd harmonic has 1 node, and the 3rd harmonic has $\boxed{2 \text{ nodes}}$.

15-31 The deBroglie wavelength λ of an object is given by $\lambda = h/p$ where p is the momentum of the object. The momentum of an object equals its mass multiplied by its velocity.

a) For an electron moving at 1.00×10^3 m s^{-1}:

$$\lambda_e = \frac{h}{p} = \frac{h}{m_e v} = \frac{6.626 \times 10^{-34} \text{ J s}}{(9.11 \times 10^{-31} \text{ kg})(1.00 \times 10^3 \text{ m s}^{-1})} = \boxed{7.27 \times 10^{-7} \text{ m}}$$

b) For a proton moving at the same speed:

$$\lambda_p = \frac{h}{m_p v} = \frac{6.626 \times 10^{-34} \text{ J s}}{(1.673 \times 10^{-27} \text{ kg})(1.00 \times 10^3 \text{ m s}^{-1})} = \boxed{3.96 \times 10^{-10} \text{ m}}$$

c) A speed of 75 km hr^{-1} is equivalent to 20.8 m s^{-1} (multiply by 1000 m km^{-1} and then divide by 3600 s hr^{-1}). A 145-g baseball has a mass of 0.145 kg. Then:

$$\lambda_{\text{ball}} = \frac{h}{m_{\text{ball}}v} = \frac{6.626 \times 10^{-34} \text{ J s}}{(0.145 \text{ kg})(20.8 \text{ m s}^{-1})} = \boxed{2.2 \times 10^{-34} \text{ m}}$$

15-33 a) According to the Heisenberg uncertainty principle:

$$(\Delta x)(\Delta p) \geq \frac{h}{4\pi} \quad \text{hence} \quad \Delta p_{\text{min}} = \frac{h}{4\pi}\left(\frac{1}{\Delta x}\right)$$

The minimum uncertainty in the momentum of the electron is then:

$$\Delta p_{\text{min}} = \frac{6.626 \times 10^{-34} \text{ J s}}{4\pi(1.0 \times 10^{-9} \text{ m})} = \frac{6.626 \times 10^{-34} \text{ kg m}^2 \text{s}^{-2} \text{ s}}{4\pi(1.0 \times 10^{-9} \text{ m})} = 5.27 \times 10^{-26} \text{ kg m s}^{-1}$$

The minimum uncertainty in the velocity can be computed using the fact that the momentum is the product of the velocity and the mass of the electron:

$$\Delta v_{\text{min}} = \frac{\Delta p_{\text{min}}}{m} = \frac{(5.27 \times 10^{-26} \text{ kg m s}^{-1})}{9.11 \times 10^{-31} \text{ kg}} = \boxed{5.8 \times 10^4 \text{ m s}^{-1}}$$

b) The minimum uncertainty in the momentum is the same as in part a. The computation of the minimum uncertainty of the velocity must use the mass of a helium atom.[13] Conversion from atomic mass units to kilograms gives the m_{He} as 6.647×10^{-27} kg. Substitution then gives:

$$\Delta v_{\text{min}} = \frac{\Delta p_{\text{min}}}{m_{\text{He}}} = \frac{5.27 \times 10^{-26} \text{ kg m s}^{-1}}{6.647 \times 10^{-27} \text{ kg}} = \boxed{7.9 \text{ m s}^{-1}}$$

15-35 The allowed energies of a particle in a one-dimensional box are:

$$E_n = n^2\left(\frac{h^2}{8mL^2}\right)$$

This is text equation 15-18.[14] The particle is a particle of known mass (the electron), and the length of the box is 1.34 Å. If all quantities are expressed in SI units:

$$E_n = n^2\left(\frac{(6.626 \times 10^{-34} \text{ J s})^2}{8(9.109 \times 10^{-31} \text{ kg})(1.34 \times 10^{-10} \text{ m})^2}\right) = n^2(3.36 \times 10^{-18} \text{ J})$$

---

[13]Text Table 14-1, text page 491.
[14]See text page 549.

Substitution of $n = 1, 2, 3$ gives:

$$E_1 = \boxed{3.36 \times 10^{-18} \text{ J}} \qquad E_2 = \boxed{13.4 \times 10^{-18} \text{ J}} \qquad E_3 = \boxed{30.2 \times 10^{-18} \text{ J}}$$

To excite an electron from $n = 1$ (the ground state) to $n = 2$ (the first excited state) requires energy equal to the difference between E_1 and E_2. If this energy is supplied by one photon, then the wavelength of the photon must be:

$$\lambda = \frac{hc}{E_2 - E_1} = \frac{(6.626 \times 10^{-34} \text{ J s})(2.9979 \times 10^8 \text{ m s}^{-1})}{(13.4 \times 10^{-18} - 3.36 \times 10^{-18}) \text{ J}} = \boxed{1.97 \times 10^{-8} \text{ m}}$$

This wavelength, 197 Å, occurs in the ultraviolet region of the spectrum.

Tip. In the formula for the allowed energies, the quantum number n appears in combination with a quantity characterizing the universe (Planck's constant h), a quantity characterizing the particle (m), and a quantity characterizing the box (L).

15-37 Combination (a) is not allowed because ℓ must be less than n; it may never equal n. Combination (c) has $m > \ell$, which is not allowed. Combination (d) has $\ell < 0$, which is not allowed. $\boxed{\text{Only combination (b) is allowed}}$.

Tip. The rules are easier to apply when physical significance is attached. $(n - 1)$ equals the total number of all nodes of the wave function. The quantum number ℓ tells the number of angular nodes. Since the number of angular nodes cannot exceed the total number of nodes, ℓ may equal $(n - 1)$ at most. Its *minimum* is zero because "−1 angular nodes" has no physical meaning. The third quantum number m is related to the orientation of the angular nodes and may be negative.

15-39 a) $4p$ **b)** $2s$ **c)** $6f$

15-41 The total number of nodes equals one less than the quantum number n. The number of angular nodes equals the quantum number ℓ, which is obtained by decoding the s, p, d, f notation, as explained on page 437: **a)** $4p$: 2 radial nodes, 1 angular node. **b)** $2s$: 1 radial node, 0 angular nodes. **c)** $6f$: 2 radial nodes, 3 angular nodes.

15-43 The wave function $\psi(2p_z)$ is the product of a *radial* part $R(2p)$ and an angular part $\chi(p_z)$. The two parts of the functions are given in Table 15-2.[15] The total wave function equals:

$$\psi(2p_z) = \left(\frac{1}{32\pi}\right)^{1/2} \left(\frac{Z}{a_0}\right)^{3/2} \cos\theta \left(\frac{Zr}{a_0}\right) \exp\left(-Zr/2a_0\right)$$

---

[15]Text page 554.

The radial part of the function contributes the exponential dependence on r, and the angular part the $\cos\theta$ dependence. This particular orbital is not a function of the third coordinate ϕ. The probability of finding an electron at a point defined by a set of coordinates (r, θ, ϕ) depends on the square of this function:

$$\text{probability}(r, \theta, \phi) \propto \left(\psi(2p_z)\right)^2 \frac{1}{32\pi}\left(\frac{Z}{a_0}\right)^3 \cos^2\theta\left(\frac{Z^2 r^2}{a_0^2}\right)\exp\left(-Zr/a_0\right)$$

The question now becomes: what values of r and θ make the quantity on the right equal zero? Clearly, this happens when $r = 0$ (at the nucleus). It also happens when $\theta = \pi/2 = 90°$ (because $\cos^2\pi/2 = 0$), and when $\theta = 3\pi/2 = 270°$ (because $\cos^2 3\pi/2 = 0$). Consult text Figure 15-21[16] to confirm that the coordinate θ starts at zero along the $+z$ axis and equals either $\pi/2$ or $3\pi/2$ at right angles to the z axis, in the xy plane. Therefore, the probability equals zero at all points in the xy plane. This plane is a nodal plane.

Writing out and squaring the full wave function is not really necessary since the angular part alone controls the angular nodes of all the functions in Table 15-2. Thus, the square of the d_{xz} orbital has a $\sin^2\theta\cos^2\theta\cos^2\phi$ angular dependence. This function goes to zero whenever $\theta = \pi/2$ or $\theta = 3\pi/2$ (in the xy plane) and whenever $\phi = \pi/2$ or $\phi = 3\pi/2$ (in the yz plane). The $\boxed{xy\ \text{plane}}$ and $\boxed{yz\ \text{plane}}$ are the two angular nodes of the d_{xz} orbital.

The square of the $d_{x^2-y^2}$ orbital has a $\sin^4\theta\cos^2 2\phi$ angular dependence. This trigonometric function goes to zero at these values of ϕ:

$$\phi = \pi/4 \ (45°) \qquad \phi = 3\pi/4 \ (135°) \qquad \phi = 5\pi/4 \ (225°) \qquad \phi = 7\pi/4 \ (315°)$$

These first and third values of ϕ define a plane containing the z-axis and half-way between the x and y axes. The second and fourth values of ϕ define a plane containing the z-axis and at right angles to the first plane. These are the nodal planes of the $d_{x^2-y^2}$ orbital. The function also goes to zero at $\theta = 0$. This happens only along the z axis, the line at the intersection of the two nodal planes just identified.

15-45 The ground-state electron configurations are:
   a) C $1s^2 2s^2 2p^2$ **b)** Se $1s^2 2s^2 2p^6 3s^2 3p^6 3d^{10} 4s^2 4p^4$ **c)** Fe $1s^2 2s^2 2p^6 3s^2 3p^6 3d^6 4s^2$
The use of the bracketed symbol of a noble gas to represent the electron configuration of that element shortens the notation for most configurations:
   a) C [He]$2s^2 2p^2$ **b)** Se [Ar]$3d^{10} 4s^2 4p^4$ **c)** Fe [Ar]$3d^6 4s^2$

---

[16]Text page 550.

15-47 The ground-state configuration of an ion derives from the ground-state configuration of the atom. In the case of a negative ion, add electrons to available orbitals in order of energy. In the case of positive ions, remove electrons starting with the highest-energy occupied orbitals:

Be^+	$1s^2 2s^1$	C^-	$1s^2 2s^2 2p^3$	Ne^{2+}	$1s^2 2s^2 2p^4$	Mg^+	$[Ne]3s^1$
P^{2+}	$[Ne]3s^2 3p^1$	Cl^-	$[Ne]3s^2 3p^6$	As^+	$[Ar]3d^{10}4s^2 4p^2$		
I^-	$[Kr]4d^{10}5s^2 5p^6$						

All of these electron configurations are ground-state (lowest energy) configurations. Be^+, C^-, Ne^{2+}, Mg^+, P^{2+} and As^+ all have at least one unpaired electron (they have incomplete subshells) and should be paramagnetic. The Cl^- and I^- ions are diamagnetic.

15-49 a) The atom has 49 electrons (36 represented by [Kr] and 13 represented by superscripts). It is an $\boxed{\text{indium}}$ atom.

b) The ion has 18 electrons, and a charge of -2. The atomic number of its nucleus must be 16. It is therefore $\boxed{S^{2-}}$.

c) The ion has 21 electrons, and a charge of $+4$. The atomic number of its nucleus must be 25. It is therefore $\boxed{Mn^{4+}}$ ion.

15-51 As a halogen this element has a ground-state electron configuration of the form $...ns^2 np^5$. The next p-subshell after the $6p$ (used in the sixth row of the periodic table) is the $7p$. Accordingly, the electron configuration of the element would be $[Rn]5f^{14}6d^{10}7s^2 7p^5$ where [Rn] stands for the configuration of the first 86 electrons. Since the configuration represents 117 electrons, Z equals $\boxed{117}$.

15-53 If only one electron could occupy each orbital in many-electron atoms, then the configurations $1s^1$ and $1s^1 2s^1 2p^3$ and $1s^1 2s^1 2p^3 3s^1 3p^3$ would be closed-shell electron configurations. Atoms with $Z = \boxed{1,5,9}$ respectively would have these ground-state electron configurations.

15-55 a) A ground-state $\boxed{\text{K atom}}$ should have a larger radius than a ground-state Na atom. In K atoms the outermost electron occupies a $4s$ orbital, but in Na atoms the outermost electron occupies a closer $3s$ orbital.

b) The $\boxed{\text{Cs atom}}$ is larger than the Cs^+ ion. As a Cs^+ ion gains an electron to produce a Cs atom, the electron is accommodated in the more distant $n = 6$ shell.

c) The Rb^+ ion and the Kr atom are isoelectronic. The larger species is the one with smaller nuclear charge: $\boxed{Kr}$.

d) A Ca atom has two $4s$ electrons and a K atom has one $4s$ electron. The outermost electrons are in the same shell but Ca has a larger nuclear charge, contracting the electron cloud. Hence $\boxed{\text{potassium}}$ is larger.

e) The Cl^- ion and the Ar atom are isoelectronic. The larger species is the one with smaller nuclear charge: $\boxed{Cl^-}$.

15-57 a) The $\boxed{S^{2-}}$ ion should be larger than the O^- ion. Its outermost electrons occupy the $n = 3$ level whereas in O^- ion the outermost electrons are in the closer $n = 2$ level.

b) The $\boxed{Ti^{2+}}$ ion is larger than the Co^{2+} ion because the two have their outermost electrons in the same level, and Ti^{2+} has a smaller nuclear charge.

c) The $\boxed{Mn^{2+}}$ ion is larger than the Mn^{4+} ion because outermost electrons (those farthest away) are lost in going from the $+2$ to $+4$ ion.

d) The $\boxed{Sr^{2+}}$ ion is larger than the Ca^{2+} ion according to the trend to larger size going down the periodic table.

15-59 Use the periodic trends in ionization energy discussed on text pages 571–72.
a) $\boxed{Sr}$ has a higher first ionization energy than Rb.
b) $\boxed{Rn}$ has a higher IE_1 than Po.
c) $\boxed{Xe}$ has a higher IE_1 than Cs.
d) $\boxed{Sr}$ has a higher IE_1 than Ba.

15-61 Use the periodic trends in electron affinity discussed on text page 573 and shown in Figure 15-39.
a) $\boxed{Cs}$ has a larger *EA* than Xe.
b) $\boxed{F}$ has a larger *EA* than Pm.
c) $\boxed{K}$ has a larger *EA* than Ca.
d) $\boxed{At}$ has a larger *EA* than Po.

15-63 Convert the ionization energy of cesium from kilojoules per mole to joules per atom. This is done by multiplying it by 1000 J kJ^{-1} (to get to joules per mole) and then dividing by 6.022×10^{23} mol^{-1} (Avogadro's number). The answer is 6.239×10^{-19} J. Next, use the relationship $\Delta E = hc/\lambda$ to compute the wavelength that corresponding to this energy:

$$\lambda = \frac{hc}{\Delta E} = \frac{(6.626 \times 10^{-34} \text{ J s})(2.9979 \times 10^8 \text{ m s}^{-1})}{6.239 \times 10^{-19} \text{ J}} = 3.184 \times 10^{-7} \text{ m} = \boxed{318.4 \text{ nm}}$$

This wavelength is in the $\boxed{\text{near ultraviolet}}$ region of the electromagnetic spectrum.

15-65 The wavelength is the speed of the wave divided by its frequency:

$$\lambda = \frac{\text{speed}}{\nu} = \frac{343 \text{ m s}^{-1}}{440 \text{ s}^{-1}} = 0.780 \text{ m}$$

Dividing the distance by the speed gives the time. It takes the sound wave $\boxed{0.0292 \text{ s}}$ to travel 10.0 m.

15-67 As a blackbody is heated, the wavelength at which the maximum intensity is emitted becomes shorter, which explains the change from red to orange in the perceived color. Also, the intensity of the emitted radiation become larger at all wavelengths. A white-hot object emits strongly across a wide band of visible wavelengths (see the shape of the curve plotted in Figure 15-5.[17] As rising T shifts the wavelength of maximum intensity into the yellow and green, which are near the center of the visible range, the intensities at all other wavelengths are simultaneously great enough that the object appears white.

15-69 The energy of the photon is sufficient to overcome the work function of the nickel surface (pry an electron loose) and to impart a kinetic energy as large as 7.04×10^{-19} J to the ejected electron. It is known that:

$$\frac{hc}{\lambda} = \Phi + \tfrac{1}{2}mv^2$$

hence:

$$\frac{(6.626 \times 10^{-34} \text{J s})(3.00 \times 10^8 \text{ m s}^{-1})}{131 \times 10^{-9} \text{ m}} = \Phi + 7.04 \times 10^{-19} \text{ J}$$

$$\Phi = \boxed{8.1 \times 10^{-19} \text{ J}}$$

15-71 The Lyman series is emitted as hydrogen atoms undergo transitions from various excited states to the ground state. The energies of the emitted photons are:

$$E_n(\text{from H}) = Z^2 \left(\frac{1}{n_f^2} - \frac{1}{n_i^2} \right) \text{Ry} = (1^2) \left(\frac{1}{1} - \frac{1}{n_i^2} \right) \text{Ry}$$

where 1 Ry[18] equals 2.18×10^{-18} J, and n_i is the quantum number of the excited state. To be absorbed by a ground-state He$^+$ ion, the energy of the incoming photon must exactly equal the energy it takes to raise the electron from the $n = 1$ state to the $n = 2, 3, 4 \ldots$ state. These energies are:

$$E_n(\text{from He}^+) = Z^2 \left(\frac{1}{n_i^2} - \frac{1}{n_f^2} \right) \text{Ry} = (2^2) \left(\frac{1}{1} - \frac{1}{n_f^2} \right) \text{Ry}$$

---

[17]Text page 528.
[18]This unit of energy is introduced on text page 539.

Is there any combination of positive whole numbers for n_i and n_f that makes these two energies equal? To find out, subtract the first from the second and set the difference equal to zero. That is,

$$4\left(1 - \frac{1}{n_f^2}\right) - \left(1 - \frac{1}{n_i^2}\right) = 0 \quad \text{from which} \quad 3 + \frac{1}{n_i^2} = \frac{4}{n_f^2}$$

The left side of the second expression is always between 3.25 and 3 as n_i takes on its possible values; the right side is always equal to or less than 1 as n_f takes on *its* possible values. Since no combination of allowed n's makes the equation valid, the answer to the question is $\boxed{\text{no}}$.

15-73 The m of the earth and its v and r in its orbit around the sun are all given in SI units. The angular momentum of the earth is the product of the three values. The angular momentum of the earth in its orbit around the sun is quantized in units of $h/2\pi$: $mvr = nh/2\pi$. Thus:

$$mvr = 2.7 \times 10^{40} \text{ kg m}^2 \text{ s}^{-1} = n \left(\frac{6.626 \times 10^{-34} \text{ J s}}{2\pi}\right)$$

Solving for n gives the desired answer: $\boxed{2.6 \times 10^{74}}$. Note that one J s is the same as one kg m^2 s^{-1}, so n is unitless. Since n is truly huge, $+1$ in n has $\boxed{\text{no effect}}$ on the angular momentum.

15-75 a) The uncertainty in the kinetic energy of the electron (call it ΔE_K) is 0.02×10^{-19} J, the range of the values given in the problem. The uncertainty in the momentum of the electron is related to that of the kinetic energy by:

$$\Delta p = \Delta(mv) = m\Delta v = m\sqrt{\frac{2\,\Delta E_K}{m}}$$

The derivation of this equation employs $E_K = \frac{1}{2}mv^2$, the definition of kinetic energy. Compute Δp:

$$\Delta p = (9.11 \times 10^{-31} \text{ kg})\sqrt{\frac{2(0.02 \times 10^{-19} \text{ J})}{9.11 \times 10^{-31} \text{ kg}}} = 6 \times 10^{-26} \text{ kg m s}^{-1}$$

From the Heisenberg uncertainty principle:

$$\Delta x \geq \frac{h/4\pi}{\Delta p}$$

it follows that the *minimum* uncertainty in the position of the electron is:

$$\Delta x_{\min} = \frac{h/4\pi}{\Delta p} = \frac{(6.626 \times 10^{-34} \text{ J s})/4\pi}{6 \times 10^{-26} \text{ kg m s}^{-1}} = \boxed{9 \times 10^{-10} \text{ m}}$$

b) The mass of a helium atom is 6.647×10^{-27} kg.[19] The uncertainty in the momentum of a helium atom with the same ΔE_K as the electron in the previous part is

$$\Delta p = (6.647 \times 10^{-27} \text{ kg}) \sqrt{\frac{2(0.02 \times 10^{-19} \text{ J})}{6.647 \times 10^{-27} \text{ kg}}} = 5 \times 10^{-24} \text{ kg m s}^{-1}$$

Note that the uncertainty is larger for a helium atom than for an electron because of the larger mass of the helium atom. The minimum uncertainty in the position of the helium atom will be proportionately smaller:

$$\Delta x_{min} = \frac{h/4\pi}{\Delta p} = \frac{6.626 \times 10^{-34} \text{ J s}/4\pi}{5 \times 10^{-24} \text{ kg m s}^{-1}} = \boxed{1 \times 10^{-11} \text{ m}}$$

15-77 The wave functions for the particle in a box are sketched in text Figure 15-20b.[20] The $n = 2$ wave function has exactly one node, half-way between $x = 0$ and $x = L$. This node occurs because $\sin(2\pi x/L)$ equals zero when x equals $L/2$. The function has maxima at $x = L/4$ and $3L/4$. The figure clearly reveals the symmetry of the function. The probability of finding the particle at any point x between 0 and L is equal to ψ^2 evaluated at x. A plot of ψ^2 versus x has a zero at $x = L/2$ and symmetrical maxima at $x = L/4$ and $3L/4$ (see Figure 15-20c). Passing from $x = 0$ to $x = L/4$ covers one-fourth of the area under the curve defined by ψ^2, the probability distribution function. Since the particle must be in the box somewhere, the answer is $\boxed{1/4}$.

Tip. The same answer comes from writing out the particle-in-a-box function for $n = 2$, squaring it, and integrating from $x = 0$ to $x = L/4$:

$$\text{probability} = \int_0^{L/4} \left(\sqrt{\frac{2}{L}} \sin \frac{2\pi x}{L} \right)^2 dx$$

In doing this it is helpful to know that:

$$\int \sin^2 y \, dy = 1/2y - 1/4 \sin 2y$$

15-79 The $3d_{xy}$ orbital in O^{7+} ion has the $\boxed{\text{same shape}}$ as the $3d_{xy}$ orbital in an H atom. Both have two nodal planes at right angles. The $3d_{xy}$ orbital is much smaller in O^{7+} than in H because of the larger nuclear charge in O^{7+}.

---

[19]See text Table 14-1, text page 491.
[20]Text page 548.

Tip. It is instructive to compute the average distance of the electron in these two $3d$ orbitals:

$$\bar{r}_{n,\ell} = \frac{n^2 a_0}{Z}\left[1 + \frac{1}{2}\left(1 - \frac{\ell(\ell+1)}{n^2}\right)\right]$$

$$\bar{r}_{3,2} = \frac{3^2 a_0}{Z}\left[1 + \frac{1}{2}\left(1 - \frac{2(2+1)}{3^2}\right)\right] = \frac{9 a_0}{Z}\left[\frac{21}{18}\right] = \frac{10.5 a_0}{Z}$$

For H, the average distance is 5.55 Å; for O^{7+}, the average distance is 0.694 Å.

15-81 This atom of sodium is in an $\boxed{\text{excited state}}$. It can lose energy in a variety of ways to end up ultimately in its ground state, which is represented $[Ne]3s^1$.

15-83 In chromium(IV) oxide, the Cr^{4+} ion has the ground-state electron configuration: $\boxed{[Ar]3d^2}$. The neutral Cr atom has lost its $4s$ electron and three of its five $3d$ electrons. The two remaining $3d$ electrons are unpaired (in the ground state) so CrO_2 has $\boxed{\text{two}}$ unpaired spins per Cr atom.

15-85 The smallest by far is the hydrogen-like Co^{25+} ion. The rest of the order follows from periodic trends.[21]

$$\boxed{Co^{25+} < F^+ < F < Br < K < Rb < Rb^-}$$

15-87 The first ionization energy of Li(g) is 520×10^3 J mol^{-1}. Divide this by N_0 to put it on a per atom basis. The result is 8.635×10^{-19} J per atom. It requires this much energy to extract the $2s$ electron from lithium. The energy of an electron in an atom in a Hartree orbital is given by the formula:

$$E_n = -\frac{Z_{\text{eff}}}{n^2}\text{Ry} = \left(-2.18 \times 10^{-18}\ \text{J}\right)\frac{Z_{\text{eff}}^2}{n^2}$$

where Z_{eff} is the effective nuclear charge experienced by the electron and n is its quantum number. For the lithium $2s$ electron:

$$-(-2.18 \times 10^{-18}\ \text{J})\frac{Z_{\text{eff}}^2}{2^2} = 8.635 \times 10^{-19}\ \text{J}\quad\text{from which}\quad Z_{\text{eff}} = \boxed{1.26}$$

The true Z of lithium is 3. The inner two electrons, which are located mainly between the $2s$ electron and the nucleus, shield the influence of the nucleus considerably, but the screening is imperfect because Z_{eff} exceeds 1.

---

[21]Or text Figure 15-34, text page 569.

For Na ($Z = 11$), the $3s$ electron is lost, and the equation becomes:

$$-(-2.18 \times 10^{-18} \text{ J})\frac{Z_{\text{eff}}^2}{3^2} = \frac{496 \times 10^3 \text{ J mol}^{-1}}{6.022 \times 10^{23} \text{ mol}^{-1}} \quad \text{from which} \quad Z_{\text{eff}} = \boxed{1.84}$$

For K ($Z = 19$), the $4s$ electron is lost:

$$-(-2.18 \times 10^{-18} \text{ J})\frac{Z_{\text{eff}}^2}{4^2} = \frac{419 \times 10^3 \text{ J mol}^{-1}}{6.022 \times 10^{23} \text{ mol}^{-1}} \quad \text{from which} \quad Z_{\text{eff}} = \boxed{2.26}$$

15-89 The fall of photons on a surface exerts a pressure. Use the deBroglie relation, which gives p as a function of λ, to compute the momentum of the photons as they approach the surface:

$$p = \frac{h}{\lambda} = \frac{6.626 \times 10^{-34} \text{ J s}}{550 \times 10^{-9} \text{ m}} = 1.205 \times 10^{-27} \text{ kg m s}^{-1}$$

Each photon bounces away from the wall with momentum of the same magnitude but opposite sign. The change in momentum per collision is:

$$\Delta p = p_2 - p_1 = (1.205 \times 10^{-27}) - (-1.205 \times 10^{-27}) = 2.410 \times 10^{-27} \text{ kg m s}^{-1}$$

The number of photons colliding per second equals the given power of the laser (1.0 watt $= 1.0$ J s^{-1}) divided by the energy delivered per photon. The energy transported by one 550 nm photon is:

$$E = \frac{hc}{\lambda} = \frac{(6.626 \times 10^{-34} \text{ J s})(3.00 \times 10^8 \text{ m s}^{-1})}{550 \times 10^{-9} \text{ m}} = 3.614 \times 10^{-19} \text{ J}$$

Clearly, it requires lots of such photons to transport 1.0 J s^{-1}:

$$\left(\frac{1.0 \text{ J}}{1 \text{ s}}\right) \times \left(\frac{1 \text{ photon}}{3.614 \times 10^{-19} \text{ J}}\right) = \frac{2.77 \times 10^{18} \text{ photons}}{\text{second}}$$

As developed in Section 4-5,[22] the *total* force equals the change in momentum per collision multiplied by the number of collisions per second:

$$F = (2.410 \times 10^{-27} \text{ kg m s}^{-1}) \times (2.77 \times 10^{18} \text{ s}^{-1}) = 6.67 \times 10^{-9} \text{ kg m s}^{-2}$$

Pressure is defined as force divided by area. The area of the circular wall is πr^2 where r is its radius. Hence:

$$P = \frac{F}{A} = \frac{F}{\pi r^2} = \frac{6.67 \times 10^{-9} \text{ kg m s}^{-2}}{\pi (0.10 \times 10^{-3} \text{ m})^2} = \boxed{0.21 \text{ Pa}}$$

---

[22]Text page 115.

15-91 The problem asks for the lattice energy of KCl(s). This is the ΔE of the reaction:

$$KCl(s) \rightarrow K^+(g) + Cl^-(g) \qquad \Delta E = \text{lattice energy}$$

The best strategy is to discover a series of steps having known ΔE's that add up to the desired overall change. The sum of the ΔE's of the steps then equals the desired ΔE. Obvious steps are the vaporization of the solid KCl to give gaseous atoms of K and Cl:

$$KCl(s) \rightarrow K(g) + Cl(g) \qquad \Delta E = +653 \text{ kJ}$$

and then the transfer of electrons to create ions:

$$K(g) \rightarrow K^+(g) + e^- \qquad \Delta E = IE_1 = 418.8 \text{ kJ}$$
$$Cl(g) + e^- \rightarrow Cl^-(g) \qquad \Delta E = -EA = -349 \text{ kJ}$$

Since these equations add up to the specified process, simply add their ΔE's:

$$\text{lattice energy} = \Delta E = 418.8 + (-349) + 653 = \boxed{722 \text{ kJ}}$$

Tip. The literature value for the lattice energy of KCl(s) equals 715 kJ mol^{-1}.

Chapter 16

Quantum Mechanics and Molecular Structure

Molecular Orbitals in Diatomic Molecules

An atomic orbital (abbreviated AO) *localizes* the probability of finding its occupying electrons in regions close to the atom's nucleus. For example, the average distance of the $1s$ electron from the nucleus in hydrogen is only 0.8 Å.[1] In contrast, **molecular orbitals (MO's)** in general *delocalize* the probability of finding the electrons that occupy them over all the atoms of a molecule. Like atomic orbitals, molecular orbitals have a variety of shapes, sizes, and energies. An MO is constructed mathematically by mixing together the atomic orbitals of the atoms that make up the molecule. The process is **linear combination** of atomic orbitals, or **overlap** of atomic orbitals.

The simplest molecules contain just two atoms. The process of linear combination of atomic orbitals is fairly simple for diatomic molecules. A linear combination of two $1s$ atomic orbitals on a pair of neighboring atoms A and B is:

$$\psi(\text{bonding}) = C_A \psi_{1s}^A + C_B \psi_{1s}^B$$
$$\psi(\text{antibonding}) = C_A \psi_{1s}^A - C_B \psi_{1s}^B$$

where C_A and C_B are constants telling the degree of mixing, that is, the proportion of each parent's character that the daughter molecular orbital possesses. If the two parent atoms are identical, then C_A equals C_B on the basis of symmetry. If not, the constants differ. See **16-55**. As the preceding equations indicate, two orbitals on two neighboring atoms can combine in at least two ways:

• **Bonding.** The first combination places much electron probability density between the nuclei of the atoms. This tends to counteract internuclear repulsions and

---

[1] Verify this statement by using the formula on page 440 of this Guide.

461

leads to bonding between the atoms. The energy of a daughter molecular orbital that is a bonding MO is *less* than the energy of either of its AO parents.

• **Antibonding.** The second combination of the atomic orbitals (the one with the minus sign) places a node (a region of zero electron probability density) between the atoms. Its energy is higher than the energy of the parent atomic orbitals. Electrons in antibonding molecular orbitals are distributed in space in such a way that they *actively oppose* the continued existence of the molecule.

Study these additional points concerning molecular orbitals:

• Molecular orbitals are designated by Greek letters. The letters σ and π are analogous to the designations s and p used for atomic orbitals. A σ MO has zero angular nodes (planes) containing the internuclear axis. A π MO has one angular node containing this axis. See text Figures 16-4 and 16-5.[2]

• The parentage of the MO (for example, a $1s$ or $2p$ atomic orbital) is indicated by a subscript to the right of the Greek letter.

• MO's may be non-bonding as well as bonding and antibonding. Antibonding orbitals are indicated with a superscript * to the right of the Greek letter. Nonbonding orbitals are indicated with a superscript nb. Bonding molecular orbitals have no special designation.

• The relative occupancy by electrons of MO's of the different types determines bond order:

$$\text{Bond Order} = \tfrac{1}{2}(\text{No. } e^-\text{'s in bonding MO's}) - \tfrac{1}{2}(\text{No. } e^-\text{'s in antibonding MO's})$$

See **16-1**, **16-3**, and **16-57**. Nonbonding electrons do not affect the bond order. Determining a bond order requires assigning all electrons to a bonding, an antibonding or a nonbonding orbital. Sharing electrons between atoms does not by itself guarantee bonding since the electrons might be shared in antibonding molecular orbitals.

• Orbitals are conserved. The number of MO's formed in linear combination of AO's equals the number of AO's combined. If 10 atomic orbitals are mixed together, then 10 MO's result.

• *Correlation diagrams* show relative energies. Different MO's have different energies. A correlation diagram (such as text Figures 16-3 and 16-6) tells the order of energy and the parentage of a molecule's molecular orbitals.[3] The text does not cover the derivation of these diagrams, which should be taken as givens. However, the correlation diagrams for heteronuclear diatomic molecules show the energy levels of the more electronegative atom displaced downward because that atom attracts valence electrons more strongly.

---

[2] On text pages 586–7.
[3] It "co-relates" the MO's to their AO parents.

• Electron configurations are possible for MO's. A molecular electron configuration is like an atomic electron configuration but uses molecular orbitals instead of atomic orbitals.

To write a ground-state molecular electron configuration:

1. Consult the appropriate correlation diagram. The text gives eight such diagrams in Chapter 16.[4]

2. Put the available electrons into the MO's starting at lowest energy. Feed in electrons in accord with the Pauli principle and Hund's rules. That is, use only two electrons per orbital and put single electrons into orbitals of equal energy as unpaired electrons rather than pairing them in the same orbital.[5]

3. Write down each MO's designation and show the number of electrons occupying it with a right superscript. See **16-1** and **16-11**.

4. Treat only the *valence* electrons. Non-valence electrons are always equally distributed between bonding and antibonding orbitals.

Like atomic orbitals, MO's hold 0, 1 or 2 electrons. As with atomic orbitals, a single MO designation (example: π_{2p}) can refer to a *set* of two or more orbitals which have the same energy. This explains notation like $(\pi_{2p})^4$ in **16-9**. This particular symbolism means that four electrons occupy a set of two π MO's of equal energy deriving from the overlap of four $2p$ atomic orbitals. Similar notation appears in **16-23** and text Table 16-2.[6]

Molecular Orbitals in Polyatomic Molecules

The text describes bonding in polyatomic molecules by employing localized orbitals between particular pairs of atoms for σ bonds and for some π bonds. It uses delocalized molecular orbitals as necessary in the description of other π bonds.

In the localized orbital approach, the valence atomic orbitals of atoms with two or more bonds are **hybridized** to form new atomic orbitals. Hybridization involves the same technique of linear combination that was used to construct any other molecular orbital except:

In hybridization, all of the starting orbitals belong to the same atom; all of the resultant orbitals belong to that same atom.

---

[4]Text Figures 16-3, 16-6a, 16-6b, 16-8, 16-10, 16-16, 16-17, and 16-18.
[5]See text page 564.
[6]Text page 589.

Hybridization is an attempt to explain the geometrical shapes that are matters of experimental fact in molecules. Different hybrid combinations give different geometrical shapes. For example, suppose that one $2s$ and three $2p$ orbitals ($2p_x$, $2p_y$, $2p_z$) on the same atom are hybridized. The linear combinations are:

$$\psi_1 = 1/2(s + p_x + p_y + p_z)$$
$$\psi_2 = 1/2(s + p_x - p_y - p_z)$$
$$\psi_3 = 1/2(s - p_x + p_y - p_z)$$
$$\psi_4 = 1/2(s - p_x - p_y + p_z)$$

Taken together, the set of four hybrid daughter wave-functions ($\psi_1 \ldots \psi_4$) describes the same electron probability density as the four parents. This can be proved by squaring ψ_1, ψ_2, ψ_3, and ψ_4 and adding them together:

$$\psi_1^2 + \psi_2^2 + \psi_3^2 + \psi_4^2 = s^2 + p_x^2 + p_y^2 + p_z^2$$

However, unlike the four parents, the daughters in the new hybrid set of orbitals all have the same shape. They are exactly equivalent, except in their orientations. Each has one prominent lobe that points toward one of the four corners of a tetrahedron. They are called sp^3 orbitals because they arise from the mixing of one s and 3 p orbitals. **Caution:** the superscript in the symbol "sp^3" does not refer to the number of electrons occupying the orbital. Instead it refers to the number of parents of p character. An sp^3 orbital containing two electrons would be denoted by $(sp^3)^2$. This notation is confusing but is deeply entrenched and must be learned.

An atom has sp^3 hybridization when it has a steric number (SN) of 4.

Other important linear combinations are:

$1\ s + 1\ p \rightarrow 2\ sp$ orbitals	Linear	(used when $SN = 2$)
$1\ s + 2\ p \rightarrow 3\ sp^2$ orbitals	Trigonal	(used when $SN = 3$)

A σ bond results from the end-to-end overlap of hybrid orbitals (or atomic orbitals) along a line connecting two atoms. A 109.5° angle (the tetrahedral angle) is predicted between atoms bonded to a central atom that is using sp^3 hybrids. The predicted angles for sp^2 and sp hybrids are 120° and 180° respectively.

Once the framework of a molecule is set up using the suitable hybrid orbitals for σ bonds, the remaining orbitals may, within limits imposed by their symmetry, mix together to form π bonds, The correlation diagrams (such as text Figure 16-16 and 16-17) that result show only π orbitals. See **16-23**. The main problem in using these diagrams is making sure that the right number of electrons is used.

● All valence electrons do not necessarily take part in π bonding.

The number of electrons used in a π-system equals the number of valence electrons minus the number used in σ bonding minus the number in lone pairs. See **16-23** and **16-57**.

Quantum Calculations for Molecules

Use of the **variational principle** allows the calculation of the properties of molecules from the Schrödinger equation. The principle states

1. Only the exactly correct wave function for a molecule gives the true minimum energy of the molecule when inserted into the Schrödinger equation

2. Approximate wave functions all give higher energies.

3. The lower the energy, then the better the approximation.

This allows the step-by-step improvement of approximate wave functions by systematically altering them and re-computing the energy. The best wave function for the molecule gives the lowest energy. This computation-intensive process has been greatly facilitated in recent years by the availability of high-speed computers.

General Aspects of Molecular Spectroscopy

The energies of molecules, like the energies of atoms, are quantized. Molecules possess modes for the storage of energy, based on the relationship among atoms, that atoms do not have. This makes the pattern of the allowed energy states for molecules much more complex than for atoms. Charting the intensity and wavelength of the photons emitted or absorbed by molecules provides important information about the arrangement of the atoms (the bonding and shape) in the molecules.

The absorption of light by a sample causes the amount of light that it transmits to decrease. In a spectrophotometer, an absorbing sample is placed in a cell and subjected to a beam of light of specific wavelength λ. The **transmittance** of a sample equals the intensity of the departing beam divided by the intensity of the incoming beam of light. It is hard to measure these intensities directly. In a practical instrument the incoming beam is split. Half is passed through a cell containing the sample, and the other half through a reference cell. Then:

$$T = \frac{I_S}{I_R}$$

where the subscript S refers to light departing from the sample cell and R refers to light departing from the reference cell. The reference cell is the same as the sample cell but contains no sample.

According to the **Beer-Lambert law**, when light of a given wavelength passes through an absorber, the intensity of the light decreases logarithmically as the path length increases arithmetically:

$$-\log T = -\log \left(\frac{I_S}{I_R} \right) = a\ell$$

The proportionality constant in this equation is called the **absorption coefficient** or the **extinction coefficient**. The negative logarithm of the transmittance of a sample is called its **absorbance**. The Beer-Lambert law in terms of the absorbance is:

$$A = a\ell$$

If the absorber is in solution, then the absorption coefficient a changes if concentration c is changed. The a in the Beer-Lambert law is then replaced by the product of c and the **molar absorption coefficient** ϵ:

$$A = c\epsilon\ell$$

The absorbance A has no units. The units of a are cm^{-1} if ℓ is measured in centimeters. The units of ϵ are then $L\ mol^{-1}cm^{-1}$. The text does not include any problems using these relationships.

Molecular Spectroscopy

Molecular spectroscopy is the study of electromagnetic radiation (light) as it is absorbed or emitted by molecules passing from one allowed energy state to a second. The frequency of the light involved in such transitions depends on the difference in energy between the energy levels (states):

$$|\Delta E| = h\nu$$

The absolute value is taken because ΔE is negative in emission and positive in absorption.

Different types of spectroscopy are distinguished based on the magnitude of the ΔE and the physical basis of the interaction between photon and molecule.

• **Nuclear magnetic resonance (NMR) spectroscopy.** Many nuclei have a property called nuclear spin. When such nuclei are placed in a magnetic field, two or more quantized nuclear spin states of different energies appear. Nuclei change spin state by absorbing or emitting photons having energies between 2×10^{-5} and $20 \times 10^{-5}\ kJ\ mol^{-1}$. The corresponding frequencies and wavelengths are on the order of 50×10^6 and $500 \times 10^6\ s^{-1}$ and 6 to 0.6 m. These are radio waves. The exact energy of an absorption or emission (called a resonance) depends on the chemical environment of the nucleus, that is, on the distribution of electrons (including valence electrons) surrounding it and on the identity and location of neighboring nuclei. Chemically equivalent nuclei have the same **chemical shift**, because their surroundings are equivalent, and show NMR peaks at the same frequency. Chemical shifts are measured relative to easily observed resonance frequencies in the NMR spectra of agreed-upon reference compounds.

The NMR experiment allows identification of bonded groups in a molecule and measurement of distance between nuclei.

• **Rotational spectroscopy.** Molecules can tumble and turn in space, and the energies of these motions are quantized. Transitions between rotational quantum states involve energy differences ranging from 0.001 to 0.1 kJ mol^{-1}. The corresponding frequencies and wavelengths are 2.5×10^9 and 2.5×10^{11} s^{-1} and 120 mm to 1.2 mm. These occur in the microwave and short-wave-radio region of the electromagnetic spectrum.

Rotational energy depends on the molecular **moments of inertia**. A moment of inertia is measured relative to a specified axis. It gives the tendency of an object to persist in rotating about that axis. In general, molecules have three moments of inertia. Because they have a simple linear structure, diatomic molecules are a special case. A diatomic molecule containing atoms of mass m_1 and m_2 separated by distance R_e has only one non-zero moment of inertia I:

$$I = \left(\frac{m_1 m_2}{m_1 + m_2} \right) R_e^2 = \mu R_e^2 \qquad \text{diatomic molecule}$$

The quantity involving the masses is the **reduced mass μ** unboldmath of the molecule. See **16-27**, **16-59**, and **16-61** for computations of reduced mass. A reduced mass mingles the masses of two objects. It is their product over their sum. Note that the units of this combination are mass units (and not reciprocal mass or mass squared). Checking the units prevents errors in using the reduced mass. Typical values for molecular moments of inertia are on the order of 10^{-45} kg m^2. See **16-27**. Note that the kg m^2 is the SI unit for moments of inertia.

The quantization of the rotational motion of a heteronuclear diatomic molecule is in terms of the **rotational quantum number J**:

$$E(\text{rotational}) = \frac{h^2}{8\pi^2 I} J(J+1) \qquad J = 0, 1, 2, \ldots$$

According to this formula, the rotational energies of such a molecule grow farther and farther apart as J increases.[7] However, the rotational spectrum of such a molecule consists of lines separated by equal intervals of frequency. Each line arises as the molecule changes from an initial state to a final state with a J that is exactly 1 larger or 1 smaller: $\Delta J = \pm 1$. Larger changes in J are forbidden. Measurements of rotational spectra give bond lengths in molecules (see **16-29**).

• **Vibrational Spectroscopy.** Chemical bonds can be compared to springs connecting atoms. The atoms in molecules vibrate ceaselessly about some equilibrium or home position. The energies of these molecular vibrations are quantized. Transitions between allowed vibrational states range in energy from about 2 to 40 kJ

---

[7]See text Figure 16-27, text page 610.

mol^{-1}. These energies correspond to frequencies and wavelengths between 5×10^{12} and 100×10^{12} s^{-1} and 60000 and 3000 nm respectively. This is in the infra-red region of the spectrum.

A restoring force opposes the stretching or compression of a chemical bond from its equilibrium length. The **force constant k** tells the strength of the restoring force. If a bond is a spring, then the force constant is its stiffness. A typical molecular force constant is 500 newton per meter. If a bond with a k of 500 N m^{-1} is stretched 0.1 Å(1×10^{-11} m), the restoring force is 5×10^{-9} N.

The force constants of the bonds of a molecule and the masses they connect determine the allowed vibrational frequencies of that molecule. In the case of a diatomic molecule, the relationship has a simple form: The frequency of the vibrational stretch of the bond

$$\nu = \frac{1}{2\pi}\sqrt{\frac{k}{\mu}} \qquad \text{in a diatomic molecule}$$

Recall that μ is the reduced mass of the diatomic molecule. The allowed vibrational energies of a heteronuclear[8] diatomic molecule are then:

$$E(\text{vibrational}) = h\nu(v + \tfrac{1}{2}) \qquad\qquad v = 0, 1, 2, \ldots$$

where v (the letter vee) is the **vibrational quantum number**. This formula means that vibrational energy states are uniformly spaced and, remarkably, implies that when $v = 0$ the molecule still has some vibrational energy, the **zero-point energy**. Transitions between vibrational states are allowed only if $\Delta v = \pm 1$. Observations of such transitions allow determination of ν and computation of the force constants of bonds (see **15-37**). **Caution:** Sloppiness in writing the symbols ν (nu, a frequency) and v (vee, a quantum number) causes confusion in using the preceding formulas.

• **Electronic Spectroscopy.** The total quantized energy of a molecule is approximately equal to the sum of contributions from three kinds of motion:

$$E(\text{total}) = E(\text{electronic}) + E(\text{vibrational}) + E(\text{rotational})$$

The electronic contribution is the largest. Changes in the electron configuration of molecules generally require (or release) from 50 to 500 kJ mol^{-1}. These energies are on the order of the strengths of chemical bonds. Frequencies of light that supply (or take away) this amount of energy per photon range from 10^{14} to 10^{15} s^{-1}. The corresponding wavelength range is 2400 nm to 240 nm, which covers the *visible* region of the spectrum and goes well into the ultra-violet. The text considers electronic excited states in a separate section.

---

[8]Diatomic molecules having two identical atoms vibrate, but do not absorb or emit radiation.

Molecular Energy Levels and Their Thermal Occupation

Molecules can get into higher-energy rotational, vibrational, and electronic state (excited states) by taking up energy transferred to them by the random chaotic jostling of other molecules. This is thermal excitation. The probability P_i that a molecule is in an excited state of energy E_i on this basis depends on the temperature:

$$P_i \propto \exp(-E_i/RT) \qquad \text{or:} \qquad P_i \propto \exp(-E_i/k_B T)$$

where the first equation uses a molar energy basis and the second a molecular energy.[9] Among many molecules, P_i equals the fraction occupying energy state i. The ratio of the occupation fraction for two states i and j is:

$$\frac{P_i}{P_j} = \exp\left(\frac{-(E_i - E_j)}{RT}\right) \qquad \text{or} \qquad \frac{P_i}{P_j} = \exp\left(\frac{-(E_i - E_j)}{k_B T}\right)$$

At room temperature RT equals about 2.5 kJ mol^{-1}. At room temperature, many excited rotational states, a few excited vibrational states (see **16-35**), and essentially no excited electronic states are occupied.

Solving problems in molecular spectroscopy

- Pay heed to the units. Work exclusively in SI units and convert at the end if necessary. The units of moments of inertia (kg m^2) and reduced mass (kg) sometimes cause trouble—moments in u Å^2 or g Å^2 and masses in atomic mass units or grams are used by mistake. Follow the pattern in **16-59**.

- Distinguish between patterns in the *energy levels* of molecules and patterns in the *frequencies* of the transitions connecting those energy levels. In **16-29**, the observed spectroscopic lines in the pure rotational spectrum of a diatomic molecule are uniformly spaced by frequency. The rotational energy levels of this molecule are *not* uniformly spaced.

Excited Electronic States

Visible and ultraviolet light can excite molecules into higher electronic quantum states. These states have distributions of electron density that differ significantly from those of the ground state. This causes molecules in excited states to have sharply different properties. The new distributions are described by electron configurations in which molecular orbitals (MO's) of higher energy are occupied by electrons transferred from lower energy MO's. For example, a π electron in a C=C double bond can

---

[9]Compare to the different versions of the formulas for the average speeds of molecules in a gas on page 80 of this Guide.

be excited from a bonding MO to an antibonding MO (a $\pi \rightarrow \pi^*$ transition). Transitions of this type occur at wavelengths well into the ultraviolet in isolated double bonds.

If a compound absorbs light of a visible wavelength (such as yellow), then the light that goes on through has the color that is the **complement** of the color absorbed (the complement of yellow is violet). Complementary colors lie opposite each other in the **color wheel** (Figure 16-35, text page 616). The colors on the color wheel are given by the mnemonic Roy G. Biv[10] written around the circumference of a six-segmented circle (indigo and violet share a segment).

The Fate of Excited Electronic States

Molecules in excited electronic states experience a variety of fates:

- Some molecules quickly reemit a photon, and return to their ground state. This is **fluorescence**, and occurs within about 10^{-9} s.

- Some molecules manage to cross from a **singlet** excited state to a long-lived **triplet** excited state. A singlet state has all of the electron spins paired; a triplet state has two unpaired electron spins. Transition from a triplet excited state to a singlet ground state is slow and leads to **phosphorescence** in the sample, which glows for seconds or even minutes.

- Some molecules relax from the excited state to the ground state in a cascade of rotational and vibrational transitions that release its extra energy to the surroundings as heat.

- Some excited molecules suffer the rupture of chemical bonds, and never return to their ground states. They react chemically.

Photochemistry is the study of reactions that follow the electronic excitation of molecules by light. For example, **photodissociation**, the breaking of bonds under the influence of light, may follow exposure of a compound to light of suitable frequency.

New Developments in Spectroscopy

Three new spectroscopic methods are discussed:

- **Time-resolved spectroscopy.** In this method, molecules are excited by pulses of light and the emitted photons subsequently identified both by wavelength and time delay. The input pulses have durations of 10^{-12} to 10^{-15} seconds. Very rapid changes in the nature of the molecular excited state can be followed.

---

[10]See page 426 in this Guide.

- **Cooling in supersonic jets.** Cold, isolated molecules are procured for study by mixing the sample with a carrier gas and forcing the mixture through a nozzle at supersonic speed. Adiabatic cooling leads to low effective temperatures downstream in the jet. This puts nearly all molecules in the ground state and reduces thermal blurring. More details appear in the spectrum.

- **Matrix isolation.** The sample is prepared at low concentration in a cold solid (the matrix), and excited with tightly focused laser light. Very sensitive detectors enable measurement of signals from one molecule at a time.

Illustration: Conjugated Systems

Conjugated molecules contain sets of molecular orbitals that span two or more double or triple bonds that occur close to each other in the molecule. Examples are 1,3-butadiene and benzene. Simple theory views 1,3-butadiene, which contains four C atoms in a row, as having two double bonds separated by one single bond: C=C—C=C. In actuality, the double bonds interact. MO theory views the multiple bonds as a set in which four electrons occupy two of the π MO's that arise from mixing the four $2p_z$ orbitals on the carbon atoms. The resulting system is a **conjugated π electron system.**

Benzene C_6H_6 is another example of conjugation. Simple theory suggests alternating single and double bonds around the six-membered ring in benzene. The actual bonding, in which all six C—C bonds are equivalent, is better understood to arise from the occupancy of three of the six MO's derived from mixing the six $2p_z$ orbitals on the six C atoms. Study text Figure 17-6.[11] See also **17-25**.

Conjugation reduces the energy difference between the electronic ground state and the lowest electronic excited state of molecules. It lowers the energy of excited-state π orbitals more than it lowers the energy of ground-state π orbitals. As a result, a sufficiently large conjugated system experiences a shift of its first electronic absorption from the ultra-violet into the visible region of the spectrum. The substance takes on color. The observed color is the complement of the color that the molecules absorb. Use the color wheel[12] to predict complementary colors.

The **fullerenes** present a third example of conjugation. Molecules of buckminsterfullerene (C_{60}) have a cage structure maintained by 90 σ bonds. Added to these bonds is a conjugated π system derived from 60 p orbitals contributed by the 60 carbon atoms in the cage. See **16-43**. Other fullerenes (such as C_{70}) also possess conjugated π systems.

---

[11]Text page 637.
[12]See Text page 616.

Illustration: Global Warming and the Greenhouse Effect

The atmosphere has a layered structure. The **troposphere** extends from the ground to about 12 km, the **stratosphere** from 12 to 50 km, the **mesosphere** from 50 to 85 km, and the **thermosphere** from 85 km out. Temperature falls with altitude in the troposphere and mesosphere, but rises with altitude in the stratosphere and thermosphere.

In the thermosphere, photodissociation of oxygen by the reaction:

$$O_2 \rightarrow 2\,O \qquad \Delta E_d = 496 \text{ kJ mol}^{-1}$$

absorbs much high-energy ultraviolet radiation (UV light with wavelengths less than 200 nm) from incoming sunlight. Some high-energy radiation penetrates to the stratosphere to cause the photodissociation of O_2 to O at that altitude. In the stratosphere, atomic O reacts further to form ozone:

$$O_2 + O \rightarrow O_3^* \qquad \text{then:} \qquad O_3^* + M \rightarrow O_3 + M$$

where the second equation shows the transfer of energy from the excited-state ozone (O_3^*) by collision with a non-reacting molecule. Without this step, O_3^* immediately dissociates. Thus sunlight converts O_2 to O_3 in the stratosphere. However, other sunlight breaks down O_3 in the stratosphere. Light of wavelength 200 to 350 nm is sufficiently energetic to photodissociate ozone:

$$O_3 + h\nu \rightarrow O_2 + O \qquad \Delta E_d = 106 \text{ kJ mol}^{-1}$$

Ultraviolet light of these somewhat longer wavelengths is not energetic enough to dissociate O_2, but is still harmful on the ground. Absorption by ozone in the stratosphere filters most of it from sunlight.

The stratospheric steady-state of ozone production and consumption is interfered with by chemicals released from the surface. Chlorofluorocarbons used as refrigerants are unreactive enough to mix into the stratosphere. Once there, they are photodissociated (see **16-41** and **16-45**) to release Cl, a **radical** (a species with an odd number of electrons). The Cl soon reacts with O or O_2 to give the radical ClO, which catalyzes the destruction of stratosphere ozone. The proposed mechanism is:

$$2\,ClO + M \rightarrow ClOOCl + M$$
$$ClOOCl + h\nu \rightarrow ClOO + Cl$$
$$ClOO + M \rightarrow Cl + O_2 + M$$
$$2 \times \left(Cl + O_3 \rightarrow ClO + O_2 \right)$$

This catalytic cycle is broken when Cl reacts with other species to form **reservoir molecules** such as HCl and $ClONO_2$. Circumstances that release Cl from these reservoirs cause intense episodes of ozone depletion. Such circumstances arise during the Antarctic winter and result in annual "ozone holes" in that region.

Photochemical reactions in the troposphere create photochemical smog. "NO_x" stands for interconverting mixtures of the nitrogen oxides NO, NO_2, and N_2O_4; these oxides are formed when air (N_2 plus O_2) is heated. Pollutant ozone is then produced photochemically: $NO_2 + h\nu \rightarrow NO + O$, followed by $O + O_2 + M \rightarrow O_3 + M$. Ozone in the troposphere is harmful in itself and also oxidizes organic compounds to produce irritants.

Oxides of sulfur are a more wide-spread atmospheric pollutant. They arise from combustion of sulfur-containing impurities in fuels. The SO_2 that is the immediate product is eventually oxidized to SO_3; SO_3 reacts with water to form H_2SO_4.

Acid rain is precipitation (it includes fog and snow) made unduly acidic by nitric acid (from NO_2 pollution) and sulfuric acid (from SO_2 pollution). Acid rain damages forests and lakes. The removal of SO_2 emissions from smokestacks is essential control measure—straightforward acid-base chemistry suggests the use of $CaO(s)$ to absorb SO_2.

Certain gases in the troposphere absorb infrared radiation emitted by warm surface of the earth. Since the IR radiation otherwise would go off into space, the effect is like than of a greenhouse. Water, carbon dioxide and methane are all effective absorbers; man-made increases in carbon dioxide and methane concentrations may lead to undesirable global warming through the **greenhouse effect.**

Detailed Solutions to Odd-Numbered Problems

16-1 **a)** Fluorine (F_2) is a homonuclear diatomic molecule with $Z = 9$. It has 14 valence electrons. The F_2^+ ion has only 13 valence electrons. Use the correlation diagram in text Figure 16-6b[13] to get the energetic order of valence molecular orbitals in both species. The MO's are then filled in order of increasing energy with the available valence electrons:

$$F_2 \quad (\sigma_{2s})^2(\sigma_{2s}^*)^2(\sigma_{2p})^2(\pi_{2p})^4(\pi_{2p}^*)^4$$
$$F_2^+ \quad (\sigma_{2s})^2(\sigma_{2s}^*)^2(\sigma_{2p})^2(\pi_{2p})^4(\pi_{2p}^*)^3$$

b) The F_2 molecule has two more bonding than antibonding electrons. Its bond order equals $\boxed{1}$; F_2^+ ion has three more bonding than antibonding electrons. Its bond order is $\boxed{3/2}$.

c) The molecule of F_2 has no unpaired electrons; accordingly, F_2 is $\boxed{\text{diamagnetic}}$. The F_2^+ ion has an odd number of electrons. Because a π_{2p}^* electron is unpaired, F_2^+ is $\boxed{\text{paramagnetic}}$.

---

[13]Text page 588.

d) The $\boxed{F_2^+}$ ion has a larger bond order and therefore requires $\boxed{\text{more energy}}$ to dissociate than the F_2 molecule.

16-3 The valence-electron configuration of S_2 should be like that of O_2 except in using $n = 3$ orbitals: $(\sigma_{3s})^2(\sigma_{3s}^*)^2(\sigma_{3p})^2(\pi_{3p})^4(\pi_{3p}^*)^2$. The bond order is $\boxed{2}$, and the molecule should be $\boxed{\text{paramagnetic}}$ (two unpaired electrons).

16-5 In each case, count the valence electrons. This result and the charge on the species identify the column of the periodic table in which the element is located. All the configurations involve MO's from the $n = 2$ shell and therefore involve elements in the second row of the periodic table. The bond order is half the number of bonding electrons minus half the number of antibonding electrons:

a) F_2, bond order 1.   **b)** N_2^+, bond order 5/2.   **c)** O_2^-, bond order 3/2.

16-7 Check unpaired valence electrons:
a) diamagnetic.   **b)** paramagnetic.   **c)** paramagnetic.

16-9 Nitrogen is more electronegative than carbon. The energies of its atomic orbitals are *lowered* in the correlation diagram in text Figure 16-18[14] relative to the energies of the corresponding orbitals of the carbon atom. A CN molecule has 9 valence electrons. The ground-state valence configuration is: $(\sigma_{2s})^2(\sigma_{2s}^*)^2(\pi_{2p})^4(\sigma_{2p})^1$. The bond order of the molecule is $\boxed{5/2}$, Its unpaired electron causes $\boxed{\text{paramagnetism}}$.

16-11 The valence-electron configurations are:

CF $(\sigma_{2s})^2(\sigma_{2s}^*)^2(\sigma_{2p})^2(\pi_{2p})^4(\pi_{2p}^*)^1$ and CF^+ $(\sigma_{2s})^2(\sigma_{2s}^*)^2(\sigma_{2p})^2(\pi_{2p})^4(\pi_{2p}^*)^0$

Removing an electron from the π_{2p}^* orbital in the CF molecule gives the CF^+ ion. The loss of an antibonding electron increases the bond order from 5/2 to 3. The bond $\boxed{\text{strengthens}}$.

16-13 The electron configuration for HeH^- would be $(\sigma_{1s})^2(\sigma_{2s}^*)^2$. The ion has a bond order of $\boxed{\text{zero}}$ and should be $\boxed{\text{unstable}}$.

16-15 The central N atom in NH_2^- is surrounded by eight valence electrons (5 from the N, 1 each from the H's and 1 for the overall negative charge). The valence orbitals of the N atom are $\boxed{sp^3}$ hybridized. Two of the hybrids overlap in σ bonds with $1s$ orbitals on the two H atoms. The other two contain lone pairs. The molecular ion is $\boxed{\text{bent}}$ with an H—N—H angle far less than 180°.

Tip. Experimentally, the angle equals 106.7° (even less than 109.5°).

---

[14]Text page 554.

16-17 In all of these species, the hybridization on the central atom follows from the number of lone pairs plus bonded pairs surrounding the central atom (the steric number SN). The molecular geometry depends on the hybridization, but the shapes of molecules are named only with reference to actual atoms.

a) The central C atom in CH_4 has SN 4. This atom is $\boxed{sp^3}$ hybridized, and the molecule is $\boxed{\text{tetrahedral}}$.

b) The central C atom has SN 2 in CO_2 and is $\boxed{sp}$ hybridized. The molecule is $\boxed{\text{linear}}$.

c) The central O atom has SN 3 in OF_2 and is $\boxed{sp^3}$ hybridized. Two of the hybrid orbitals on the O atom accommodate lone pairs of electrons, and two overlap with orbitals on the fluorine atoms. The molecule is $\boxed{\text{bent}}$.

d) The central C atom in CH_3^- has SN 3 and is $\boxed{sp^3}$ hybridized. One of the four hybrid orbitals contains a lone pair of electrons. The other three overlap with $1s$ orbitals of the three hydrogen atoms. The molecular ion is $\boxed{\text{pyramidal}}$.

e) The central Be atom in BeH_2 has SN 2 and is $\boxed{sp}$ hybridized. The molecule is $\boxed{\text{linear}}$.

16-19 The ClO_3^+ and ClO_2^+ ions have 24 and 18 valence electrons respectively. The central Cl atom in ClO_3^+ has SN 3 and therefore three $\boxed{sp^2}$ hybrid orbitals overlapping with orbitals from the oxygen atoms. It has a $\boxed{\text{trigonal planar}}$ geometry. The central Cl atom in ClO_2^+ likewise has a set of three $\boxed{sp^2}$ hybrid orbitals, but only two overlap with orbitals on oxygen atoms. The third sp^2 orbital contains a lone pair. The ClO_2^+ molecular ion is $\boxed{\text{bent}}$. The central chlorine atoms in the following Lewis structures are shown with expanded octets. Other resonance structures can be drawn; these particular structures minimize formal charges. Compare to **3-61**.[15]

---

[15]Page 67 of this Guide.

16-21 The central nitrogen atom in the orthonitrate ion can attain an octet by forming four single bonds, one to each of the four oxygen atoms. We expect $\boxed{sp^3}$ hybridization on the N atom and a $\boxed{\text{tetrahedral}}$ geometry.

Tip. The Lewis structure of this ion puts a +3 formal charge on the central nitrogen atom. No wonder the ion is unfamiliar.

16-23 Like the CO_2 molecule, the azide ion N_3^- is linear and has 16 valence electrons. The correlation diagram of text Figure 16-16[16] applies. Two N—N σ bonds result from overlap of sp hybrid orbitals on the central N atom with $2p_z$ orbitals on the two outer N atoms. These bonds use 4 electrons. Lone pairs in each of the $2s$ orbitals of the outer N atoms use another 4 electrons. The π orbitals pictured in text Figure 16-15 accommodate the remaining eight valence electrons. Four of these are in two bonding π orbitals that lie perpendicular to each other. The other four are in the two nonbonding π orbitals. Thus: $(\pi)^4(\pi^{nb})^4$, which means a total of two π bonds. The overall bond order of the molecule is 4: (2 σ bonds plus 2 π bonds). The two N-to-N linkages are identical; each has bond order 2. All of the electrons are paired so the compound is $\boxed{\text{diamagnetic}}$.

The N_3 molecule has 15 valence electrons. It derives from N_3^- by the loss of an electron. The loss comes from the highest energy molecular orbital which is (by consulting text Figure 16-16) a nonbonding MO. N_3 is $\boxed{\text{bound}}$ with an overall bond order of 4, just like N_3^-. Unlike N_3^-, N_3 has an unpaired electron and is $\boxed{\text{paramagnetic}}$.

The N_3^+ ion has 14 valence-electrons. It derives from N_3^- by the loss of two nonbonding π electrons. The N_3^+ molecular ion is therefore $\boxed{\text{bound}}$ with bond order 4. There are two unpaired electrons in the set of π^{nb} orbitals so N_3^+ is $\boxed{\text{paramagnetic}}$, too.

Tip. The odd electron in N_3 is accommodated in one of the nonbonding orbitals pictured in text Figure 16-15c. Compare the "location" of the odd electron with the location of the odd electron in the Lewis structures of the related compound ONO, which are discussed in **3-65**. Also, compare to these structures while solving **16-24**.

16-25 The Lewis structure of acetaldehyde is:

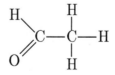

The hybridization on the —CH_3 (methyl) carbon atom is $\boxed{sp^3}$; the hybridization

---

[16]Text page 597.

on the carbonyl C is $\boxed{sp^2}$. In addition to the σ bonds from the overlap of these hybrid orbitals, two electrons in a π orbital bond the carbonyl C to the O atom. A π antibonding orbital spanning the same two atoms remains empty. The three groups bonded to the carbonyl C lie in a plane with bond angles near $\boxed{120°}$. The geometry at the methyl C atom is approximately tetrahedral, with all six H—C—H and H—C—C angles near $\boxed{109.5°}$.

16-27 The rotational energy of diatomic molecules such as NO is given by:

$$E_{rot} = \left(\frac{h^2}{8\pi^2 I}\right) J(J+1)$$

where I is the moment of inertia of the molecule and J is the rotational quantum number $(J = 0, 1, 2, \dots)$. The rotational energies in the ground state and first excited states of such molecules equal:

$$E_{rot,\,J=0} = \left(\frac{h^2}{8\pi^2 I}\right)(0) = 0 \qquad \text{and} \qquad E_{rot,\,J=1} = \left(\frac{h^2}{8\pi^2 I}\right)(2)$$

To use these formulas, obtain data on the $^{14}N^{16}O$ molecule[17] and compute I:

$$
\begin{aligned}
I = \mu R_e^2 &= \frac{m_N m_O}{m_N + m_O} R_e^2 \\
&= \left(\frac{(14.00307400 \text{ u})(15.9949146 \text{ u})}{(14.00307400 + 15.9949146) \text{ u}}\right)\left(1.154 \times 10^{-10} \text{ m}\right)^2 \times \left(\frac{1 \text{ kg}}{6.022137 \times 10^{26} \text{ u}}\right) \\
&= 1.651103 \times 10^{-46} \text{ kg m}^2
\end{aligned}
$$

Then the desired ΔE is:

$$
\begin{aligned}
E_{J=1} - E_{J=0} &= \frac{h^2}{8\pi^2 I}(2) - 0 \\
&= \frac{(6.626076 \times 10^{-34} \text{ J s})^2}{4\pi^2(1.651103 \times 10^{-46} \text{ kg m}^2)} = \boxed{6.736 \times 10^{-23} \text{ J}}
\end{aligned}
$$

This equals 40.56 J mol^{-1}.

16-29 a) The spacing of the spectroscopic lines in the pure rotational spectrum of a diatomic species is *uniform* with a frequency separation equal to $h/4\pi^2 I$ where I is the moment of inertia of the species. Average the two differences in frequency to

---

[17] The mass data come from text Table 14-1 (text page 491); the bond length comes from text Table 3-6 (text page 79).

estimate this separation for ^{12}C—^{16}O. It equals 1.155×10^{11} s^{-1}. Then compute the moment of inertia by solving for I as follows:

$$1.155 \times 10^{11} \text{ s}^{-1} = \left(\frac{h}{4\pi^2 I}\right) \qquad I = \left(\frac{6.626 \times 10^{-34} \text{ J s}}{4\pi^2 (1.155 \times 10^{11} \text{ s}^{-1})}\right) = \boxed{1.45 \times 10^{-46} \text{ kg m}^2}$$

b) The energy of a rotational state is given by:

$$E_{\text{rot}} = \left(\frac{h^2}{8\pi^2 I}\right) J(J+1) \quad \text{which is the same as} \quad E_{\text{rot}} = \frac{h}{2}\left(\frac{h}{4\pi^2 I}\right) J(J+1)$$

For ^{12}C—^{16}O, the factor $h/4\pi^2 I$ equals 1.155×10^{11} s^{-1}. Insert it, $h/2$, and the different J's in the preceding:

$$E_{\text{rot}} = \frac{6.626 \times 10^{-34} \text{ J s}}{2}(1.155 \times 10^{11} \text{ s}^{-1})J(J+1)$$

$$E_{J=1} = \boxed{7.65 \times 10^{-23} \text{ J}} \quad E_{J=2} = \boxed{23.0 \times 10^{-23} \text{ J}} \quad E_{J=3} = \boxed{45.9 \times 10^{-23} \text{ J}}$$

c) The mass of a ^{12}C atom equals 12.0000 u, and the mass of an ^{16}O atom equals 15.994915 u. Convert these masses to kilograms by dividing by 6.022137×10^{26} u kg^{-1}. The moment of inertia depends on the reduced mass and bond distance:

$$I = \mu R_e^2 = \frac{m_O m_C}{m_O + m_C} R_e^2$$

Solve for the bond distance R_e and substitute the masses and moment of inertia:

$$R_e = \sqrt{(1.45 \times 10^{-46} \text{ kg m}^2)\left(\frac{(2.656 \times 10^{-26}) + (1.993 \times 10^{-26}) \text{ kg}}{(2.656 \times 10^{-26})(1.993 \times 10^{-26}) \text{ kg}^2}\right)}$$

$$= 1.13 \times 10^{-10} \text{ m} = \boxed{1.13 \text{ Å}}$$

16-31 In vibrational spectra, only transitions between adjacent vibrational states are allowed. Therefore the change in the vibrational quantum number $(v_2 - v_1)$ equals $+1$ in this weak absorption by Li$_2$. The change in vibrational energy equals the energy of the final state minus the energy of the initial state:

$$\Delta E_{\text{vib}} = h\left(\frac{1}{2\pi}\right)\sqrt{\frac{k}{\mu}}$$

where k is the desired force constant and μ is the reduced mass of Li$_2$. The change in vibrational energy is related to the wavelength of the absorbed light λ by $\Delta E_{\text{vib}} = hc/\lambda$. Substitute this relation into the preceding:

$$\frac{hc}{\lambda} = h\left(\frac{1}{2\pi}\right)\sqrt{\frac{k}{\mu}}$$

The wavelength of the absorption line is quoted in the problem as 2.85×10^{-5} m. From Table 14-1, the mass of ^{7}Li is 7.016005 u; the reduced mass of Li$_2$ is half of this or 3.508003 u. Converting the reduced mass to kilograms gives 5.8253×10^{-27} kg. Insert this and the other quantities in the preceding equation, solve for k, and complete the arithmetic:

$$k = \left(\frac{2\pi c}{\lambda}\right)^2 \mu = \left(\frac{2\pi(2.9979 \times 10^{10} \text{ m s}^{-1})}{2.85 \times 10^{-5} \text{ m}}\right)^2 (5.8253 \times 10^{-27} \text{ kg}) = \boxed{25.4 \text{ kg s}^{-2}}$$

Note that Planck's constant h cancels out of the calculation.

Tip. Confirm that a kg s^{-2} (kilogram per square second) equals a N m^{-1} (newton per meter).

16-33 The frequency of infrared light absorbed by the "signature" C—H stretch is given by $\nu = (1/2\pi)\sqrt{k/\mu}$. Also, $\nu = c/\lambda$. Eliminate ν between these equations and solve for the force constant k:

$$k = \left(\frac{2\pi c}{\lambda}\right)^2 \mu$$

Next, insert the given values of λ and μ. The reduced mass μ is taken as the mass of the H atom, which equals the molar mass of H divided by Avogadro's number:

$$k = \left(\frac{2\pi(2.9979 \times 10^8 \text{ m s}^{-1})}{3.4 \times 10^{-6} \text{ m}}\right)^2 \frac{0.001008 \text{ kg mol}^{-1}}{6.022 \times 10^{23} \text{ mol}^{-1}} = \boxed{510 \text{ kg s}^{-2}}$$

16-35 The ratio of the fraction of molecules occupying two quantum states of different energies is given by:

$$\frac{P_i}{P_j} = \exp\left(-(E_i - E_j)/k_B T\right)$$

The problem asks for this ratio for the vibrational ground state and first excited state in N$_2$ at 450 K. Let the j-th state be $v = 0$, the ground state, and let the i-th state be $v = 1$, the first excited vibrational state. The difference between the energies of these states is just h times the natural oscillation frequency (vibrational frequency)[18] of the system:

$$E_1 - E_0 = h\nu = (6.626 \times 10^{-34} \text{ J s})(7.07 \times 10^{13} \text{ s}^{-1}) = 4.685 \times 10^{-20} \text{ J}$$

Obtain the desired ratio by substitution in the equation for the distribution:

$$\frac{P_i}{P_j} = \exp\left(\frac{-(4.685 \times 10^{-20} \text{ J})}{(1.3808 \times 10^{-23} \text{ J K}^{-1})(450 \text{ K})}\right) = \boxed{0.00053}$$

---

[18]See text page 580.

16-37 The lowest unoccupied molecular orbital of ethylene is a π^* antibonding orbital. The electron that is gained when the $C_2H_4^-$ ion is formed from C_2H_4 goes into this orbital. An additional antibonding electron means that the bond order $\boxed{\text{decreases}}$ in $C_2H_4^-$ relative to C_2H_4.

16-39 The color of a substance is the complement of the color of the light that the substance absorbs. The complementary color of orange is blue. We would therefore expect absorption $\boxed{\text{around 450 nm}}$.

16-41 The bond dissociation energy of Cl—F equals 252 kJ mol^{-1}, according to text Table 3-2.[19] Express this energy change in joules per molecule by dividing it by Avogadro's number and calculate the corresponding wavelength:

$$\lambda = \frac{hc}{\Delta E} = \frac{(6.626 \times 10^{-34} \text{ J s})(2.9979 \times 10^8 \text{ m s}^{-1})}{(2.52 \times 10^5 \text{ J mol}^{-1})/(6.0221 \times 10^{23} \text{ mol}^{-1})} = \boxed{4.75 \times 10^{-7} \text{ m}}$$

Wavelengths equal to or shorter than 475 nm can dissociate ClF at room temperature.

16-43 The 60 carbon atoms in fullerene must form 120 bonds in order to satisfy the octet rule for every atom. The reasoning is that each C must have four bonds, but each bond is shared by two C's: $4 \times 60/2 = 120$. The structure of fullerene has 60 vertices (the 60 C atoms), 32 faces (20 hexagonal and 12 pentagonal), and 90 edges. These facts can be obtained by inspection.[20] The 90 edges must consist of 60 single bonds and $\boxed{30}$ double bonds. There is no other way to account for 120 shared pairs of electrons. One simple rule for placing the 30 double bonds is to put them at all $\boxed{\text{edges that join two hexagonal faces}}$.

16-45 Divide 440 kJ, the energy of a mole of bonds, by Avogadro's number to obtain 7.31×10^{-19} J, the energy change in the dissociation of one bond. Calculate the wavelength corresponding to this energy change as follows:

$$\lambda = \frac{hc}{\Delta E} = \frac{(6.626 \times 10^{-34} \text{ J s})(2.9979 \times 10^8 \text{ m s}^{-1})}{7.31 \times 10^{-19} \text{ J}} = \boxed{2.72 \times 10^{-7} \text{ m}}$$

Light of wavelengths shorter than this supplies more than enough energy to break C—F bonds; light of longer wavelengths cannot dissociate these bonds.

Tip. The 272 nm light also suffices to dissociate the C—C bonds and C—Cl bonds in the chlorofluorocarbon because the energies of these bonds are substantially less than the energy of the C—F bond. See text Table 7-3.[21] Also, the problem is nearly identical to **16-41**.

---

[19]See text page 67.

[20]Inspect a soccer ball. Also, the following formula relates the number of vertices, edges and faces of any polyhedron: $V + F = E + 2$.

[21]Text page 228.

16-47 The best Lewis structures are a resonance pair:

$$\left[\quad :\ddot{O} - \ddot{O} = \ddot{O}: \quad \longleftrightarrow \quad :\ddot{O} = \ddot{O} - \ddot{O}: \quad \right]$$

The central O atom has a lone pair in both structures. The VSEPR model assigns the central O atom $\boxed{SN\,3}$ and thereby predicts $\boxed{sp^2}$ hybridization and an angle of (approximately) 120° at the central O atom: the ozone molecule is $\boxed{\text{bent}}$. There are two electrons in a bonding π orbital formed from the three $2p_z$ orbitals perpendicular to the molecular plane. The non-bonding and antibonding orbitals in this π system[22] are unoccupied. The total bond order is 3, which is a bond order of $\boxed{3/2}$ for each O-to-O linkage.

16-49 a) Refer to text Figure 16-6a to identify the five occupied valence molecular orbitals in the N_2 molecule:

$$N_2 \quad (\sigma_{2s})^2 (\sigma_{2s}^*)^2 (\pi_{2p_x})^2 (\pi_{2p_y})^2 (\sigma_{2p_z})^2$$

The highest-energy occupied orbital is a σ_{p_z} MO, derived from $2p_z$-$2p_z$ overlap. Its shape is shown in Figure 16-4a.[23] The next two highest occupied MO's are a pair of π bonding MO's of equal energy and identical shape that differ only in orientation. Their shape is shown in Figure 16-5a.[24] One is from $2p_x$-$2p_x$ overlap, and the other is from $2p_y$-$2p_y$ overlap. Occupation of these two π orbitals by four electrons furnishes a cylindrical muff of electron density surrounding the σ bond between the two N atoms. At lower energy come a σ^* orbital derived from antibonding overlap of the 2s orbitals (text Figure 16-4b) and a σ orbital from the bonding overlap of the same orbitals (text Figure 16-4a).

b) Since the highest occupied molecular orbital of N_2 is a bonding orbital, the removal of one electron from N_2 will decrease the bond order, and $\boxed{\text{lengthen}}$ the N-to-N bond.

16-51 The correlation diagram in Figure 16-6a[25] gives the ground-state configuration $(\sigma_{2s})^2 (\sigma_{2s}^*)^2 (\pi_{2p_x})^1 (\pi_{2p_y})^1$ when used for B_2 with its six valence electrons. There are two unpaired electrons in this configuration. The diagram in Figure 16-6a on the other hand implies the configuration $(\sigma_{2s})^2 (\sigma_{2s}^*)^2 (\sigma_{2p_z})^2$. There are no unpaired electrons in this configuration. Diagram 16-6b is not consistent with the fact that B_2 is paramagnetic.

---

[22]The system is diagrammed in text Figure 16-17, text page 598.
[23]Text page 586.
[24]Text page 587.
[25]Text page 588.

16-53 a) Consider the molecular orbital energy-level diagrams for H_2 and O_2 (text Figures 16-3 and 16-6b respectively). The ionization energy equals the energy required to remove the highest energy electron from a gaseous molecule or atom. In the case of H compared to H_2, the $1s$ electron of the atom is higher in energy than a σ_{1s} electron of the molecule. Therefore, it requires less energy to remove the atomic $1s$ electron than a molecular σ_{1s} electron. In the case of O versus O_2, the atomic $2p$ electron lies lower in energy than the molecular π_{2p}^* electron. Consequently, it requires more energy to ionize O than O_2.

b) The highest occupied molecular orbital of F_2 is the π_{2p}^* orbital (see **16-1** and text Figure 16-6b). It is higher in energy than the atomic $2p$ orbital. Therefore, the F_2 molecule should have a lower ionization energy than the F atom.

16-55 The molecular orbital and the square of the molecular orbital for the ground state of the heteronuclear molecule are:

$$\psi = C_A\psi_A + C_B\psi_B \quad \text{and} \quad \psi^2 = C_A^2\psi_A^2 + 2C_AC_B\psi_A\psi_B + C_B^2\psi_B^2$$

where the C's are constants. The *square* of the wave-function is given because it is the quantity that is related to the probability of finding the electron. Neglecting the overlap of the two orbitals means neglecting the cross-term in the squared wave-function:

$$\psi^2 \approx C_A^2\psi_A^2 + C_B^2\psi_B^2$$

If the electron spends 90 percent of its time in orbital ψ_A then $C_A^2 = 9C_B^2$. Also, the electron must be either on atom A or atom B so $C_A^2 + C_B^2 = 1$. Solution of the two simultaneous equations gives $C_A = \boxed{0.949}$ and $C_B = \boxed{0.316}$.

16-57 a) Nitramide has 24 valence electrons. It must have one double bond somewhere if the octet rule is obeyed.[26] If the structure is non-planar, this double bond is strongly localized to the —NO_2 portion of the molecule. The two electrons occupy a π orbital derived from $2p_z$ orbitals on the N atom and the two O atoms bonded to it. Inclusion in the π system of orbitals and electrons from the other N atom would require coplanarity of the H_2N— and —NO_2 portions of the molecule. If the two portions are not coplanar, then overlap and effective mixing of p-orbitals is not possible and the N—N bond order is $\boxed{1}$.

b) If the nitramide molecule were planar, the four $2p_z$ orbitals present on the two nitrogen atoms and two oxygen atoms after completion of the σ bonding could overlap to form one π, two π^{nb}, and one π^* MO's. Four electrons would occupy this π system. Two of the electrons would be in the bonding orbital, and the other two electrons

---

[26]Confirm this using the procedure on text page 59–60.

would be in the non-bonding orbitals. The resulting π system would possess a total a net of two bonding electrons across the four atoms involved. The bond order of the N—N bond would be 1 (from the σ interaction) plus 1/3 (from the π system) or $\boxed{4/3}$.

16-59 The moment of inertia of a diatomic molecule is $I = \mu R_e^2$ where R_e is the equilibrium bond distance and μ is the reduced mass. Substitution of the masses[27] of ^{1}H, ^{19}F and ^{81}Br into the formula for the reduced mass gives:

$$\mu_{HF} = \frac{(1.00782505)(18.9984033)}{1.00782505 + 18.9984033} = 0.957055 \text{ u}$$

$$\mu_{HBr} = \frac{(1.00782505)(80.91629)}{1.00782505 + 80.91629} = 0.99543 \text{ u}$$

In kilograms, the two reduced masses are:

$$\mu_{HF} = \boxed{1.5893 \times 10^{-27} \text{ kg}} \quad \text{and} \quad \mu_{HBr} = \boxed{1.6529 \times 10^{-27} \text{ kg}}$$

The equilibrium bond distances equal 0.926×10^{-10} m for HF and 1.424×10^{-10} m for HBr.[28] Using $I = \mu R_e^2$ gives these moments of inertia:

$$I_{HF} = 1.363 \times 10^{-47} \text{ kg m}^2 \quad \text{and} \quad I_{HBr} = 3.352 \times 10^{-47} \text{ kg m}^2$$

The rotational spectra of diatomic molecules consist of lines equally spaced in frequency with the separation between adjacent lines equal to $h/4\pi^2 I$. Therefore:

$$\left(\frac{h}{4\pi^2 I}\right)_{HF} = \frac{6.626 \times 10^{-34} \text{ J s}}{4\pi^2(1.363 \times 10^{-47} \text{ kg m}^2)} = \boxed{12.3 \times 10^{11} \text{ s}^{-1}}$$

$$\left(\frac{h}{4\pi^2 I}\right)_{HBr} = \frac{6.626 \times 10^{-34} \text{ J s}}{4\pi^2(3.352 \times 10^{-47} \text{ kg m}^2)} = \boxed{5.01 \times 10^{11} \text{ s}^{-1}}$$

The large gain in molecular mass from HF and HBr causes only a rather small gain in the reduced mass of the diatomic molecule. To understand why, picture the rotation of the diatomic molecules. In HF, the center of rotation is already very close to the F atom because F is 19 times heavier than H; the H does most of the moving about the center of rotation. Even a big increase in the mass of the heavy atom (replacement of F by Br) moves the center of mass only fractionally closer to the heavy atom.

---

[27]From Text Table 14-1.
[28]Text Table 3-6.

16-61 The reduced masses of the three diatomic molecules are:

$$\mu_{\text{NaH}} = \frac{(22.989770)(1.0078205)}{22.989770 + 1.00782505} = \boxed{0.9654994 \text{ u}}$$

$$\mu_{\text{NaCl}} = \frac{(22.989770)(34.9688527)}{22.989770 + 34.9688527} = \boxed{13.870686 \text{ u}}$$

$$\mu_{\text{NaI}} = \frac{(22.989770)(126.904477)}{22.989770 + 126.904477} = \boxed{19.463754 \text{ u}}$$

The reduced mass of NaD will also be needed:

$$\mu_{\text{NaD}} = \frac{(22.989770)(2.0141079)}{22.989770 + 2.0141079} = 1.8518678 \text{ u}$$

The force constant of the bond in a diatomic molecule is $k = \mu(2\pi\nu)^2$ where μ is the reduced mass and ν is the vibrational frequency. Therefore:

$$k_{\text{NaH}} = \frac{0.9654994 \text{ u}}{6.022137 \times 10^{26} \text{ u kg}^{-1}}(4\pi^2)(3.51 \times 10^{13} \text{ s})^2 = \boxed{78.0 \text{ kg s}^{-2}}$$

$$k_{\text{NaCl}} = \frac{13.870686 \text{ u}}{6.022137 \times 10^{26} \text{ u kg}^{-1}}(4\pi^2)(1.10 \times 10^{13} \text{ s})^2 = \boxed{110 \text{ kg s}^{-2}}$$

$$k_{\text{NaI}} = \frac{19.463754 \text{ u}}{6.022137 \times 10^{26} \text{ u kg}^{-1}}(4\pi^2)(0.773 \times 10^{13} \text{ s})^2 = \boxed{68.6 \text{ kg s}^{-2}}$$

Insert the reduced mass of NaD and the force constant of NaH into the formula for the vibrational frequency of a diatomic molecule. The reduced mass of NaD is calculated in atomic mass units in the preceding. Converting to kilograms gives $3.0751007 \times 10^{-27}$ kg. Then:

$$\nu = \frac{1}{2\pi}\sqrt{\frac{k}{\mu}} = \frac{1}{2\pi}\sqrt{\frac{78.0 \text{ kg s}^{-2}}{3.0751007 \times 10^{-27} \text{ kg}}} = \boxed{2.53 \times 10^{13} \text{ s}^{-1}}$$

Tip. The three bond distances were not needed.

16-63 The difference in energy ΔE between the $v = 1$ and $v = 0$ vibrational states of HgBr(g) equals $h\nu$ where ν is the given vibrational frequency of the molecule. The ratio of the occupation of the two states depends on $e^{-\Delta E/k_\text{B}T}$ and equals 0.127 at some temperature T. Compute T as follows:

$$\ln\left(\frac{P_1}{P_0}\right) = \frac{-\Delta E}{k_\text{B}T} \qquad \text{or, after substitution:} \qquad \ln(0.127) = \frac{-h\nu}{k_\text{B}T}$$

Solve for the temperature and evaluate:

$$T = \frac{-h\nu}{k_\text{B}\ln(0.127)} = \frac{-(6.626 \times 10^{-34} \text{ J s})(5.58 \times 10^{12} \text{ s}^{-1})}{(1.3807 \times 10^{-23} \text{ J K}^{-1})(-2.0636)} = \boxed{130 \text{ K}}$$

16-65 a) There are five C=C double bonds. The isomer to the left of the arrow has four *trans* C=C double bonds in the chain extending to the right from the six-membered ring. The double bond in the six-membered ring is also *trans* when the relative positions of the two largest groups, one of which is the long side-chain, are considered. Hence, this isomer has $\boxed{\text{five}}$ *trans* double bonds. The isomer to the right of the arrow is the same except that the second C=C double bond from the right end of the side chain is *cis*. This isomer has $\boxed{\text{four}}$ *trans* double bonds.

b) The absorption maximum would $\boxed{\text{shift to shorter wavelength}}$. Loss of the ring and the —CHO group would reduce the range of delocalization of electrons in a system of alternating single and double bonds because the ring contains a C=C double bond, and the —CHO group contains a C=O double bond.

16-67 a) The carbon atom in formaldehyde is sp^2 hybridized.

b) Formaldehyde has ten valence orbitals: three σ-orbitals formed by overlap of sp^2 orbitals on the C atom with $1s$ orbitals on the two H atoms and the $2p$-orbital on the O atom that points toward the carbon; three empty σ^* orbitals with the same parents; two lone-pair $2s$ and $2p$ orbitals on the O atom; one occupied π (bonding) orbital derived from the two remaining $2p$ orbitals, which are directed perpendicular to the plane of the molecule; one empty π^* (antibonding) orbital derived from the same parents.

c) The weaker transition at lower frequency is probably due to excitation of an electron from a lone-pair $2p$ orbital on the oxygen atom to the π^*-orbital.

16-69 Nitrogen dioxide is a radical. In the stratosphere, NO_2 would photodissociate to give NO which could catalyze the destruction of O_3. Plausible mechanisms are easy to write:

$$NO + O_3 \rightarrow NO_2 + O_2 \qquad NO_2 + O \rightarrow NO + O_2$$

In the troposphere, where the concentration of O_3 is small and the concentration of NO_2 is higher, NO_2 participates in the formation of O_3:

$$NO_2 + h\nu \rightarrow NO + O \qquad O + O_2 + M \rightarrow O_3 + M$$

Ozone is bad in the troposphere because of its high toxicity. Unfortunately, O_3 created in the troposphere is too reactive to diffuse up into the stratosphere, where it might do some good.

16-71 At thermal equilibrium, the rate of excitation from $v = 0 \rightarrow v = 1$ $\boxed{\text{equals}}$ the rate of the reverse process $v = 1 \rightarrow v = 0$. If this were not so, then the relative populations of the states would change, and the system would not be at equilibrium.

Assume that the rates of excitation and deexcitation depend solely on P_0 and P_1, the populations of the two states. Then:

$$\text{rate}_{0\rightarrow 1} = k_{0\rightarrow 1}P_0 = k_{0\rightarrow 1}\exp(-E_0/k_BT)$$
$$\text{rate}_{1\rightarrow 0} = k_{1\rightarrow 0}P_1 = k_{1\rightarrow 0}\exp(-E_1/k_BT)$$

where the k's are first-order rate constants. The two rates are equal. Set them equal to each other in the two equations and solve for the ratio of the k's:

$$\frac{k_{0\rightarrow 1}}{k_{1\rightarrow 0}} = \boxed{\exp\big((E_0 - E_1)/k_BT\big)}$$

16-73 The problem concerns the hydroxyl radical. This is *not* the species OH^-, which has 8 valence electrons, but is the neutral species OH, which has 7 valence electrons.
a) Convert the concentration of OH radicals to mol L^{-1}:

$$\frac{1 \times 10^7 \text{ molec.}}{\text{cm}^3} \times \left(\frac{1000 \text{ cm}^3}{L}\right) \times \left(\frac{1 \text{ mol OH}}{6.022 \times 10^{23} \text{ molec.}}\right) = 1.66 \times 10^{-14} \text{ mol L}^{-1}$$

Now use the ideal-gas equation and the definition of mole fraction:

$$P_{OH} = \left(\frac{n}{V}\right)RT = \left(\frac{1.66 \times 10^{-14} \text{ mol}}{L}\right)(0.082 \text{ L atm mol}^{-1}\text{K}^{-1})(298 \text{ K})$$
$$= \boxed{4 \times 10^{-12} \text{ atm}}$$

$$X_{OH} = \frac{P_{OH}}{P_{tot}} = \frac{4 \times 10^{-12} \text{ atm}}{1 \text{ atm}} = \boxed{4 \times 10^{-12}}$$

b) The OH radical reacts with NO_2 to give nitric acid:

$$OH + NO_2 \rightarrow HNO_3$$

The oxidation state of N $\boxed{\text{increases}}$ in this reaction. The HNO_3 interacts with atmospheric water and eventually comes to the surface in the form of acid rain.

Chapter 17

Bonding, Structure, and Reactions of Organic Molecules

Organic chemistry is the study of the reactions and properties of compounds containing carbon. Carbon has an intermediate electronegativity. Its atoms form four bonds. These may be to hydrogen, nitrogen, oxygen, or any of numerous other elements. Most importantly, carbon atoms bond well to other carbon atoms, a fact that allows them to string together in chains of essentially unlimited length.

Carbon forms triple bonds and double bonds as well as single bonds. The most common triple bonds are to other carbon atoms or to nitrogen. The most common double bonds are to other carbon atoms and to oxygen or sulfur. The most common single bonds are to other carbon atoms, to the halogens, and to hydrogen.

Bonding and Structure in Organic Molecules

The text describes the bonding in organic molecules in terms of localized molecular orbitals for σ bonds and either localized or delocalized molecular orbitals for π bonds. The approach is the one established in text Section 16-2. It is applied here in the following steps:

1. Write the Lewis electron-dot structure for the molecule. Take particular care to get the number of valence electrons right.

2. Determine the hybridization of every atom. To do this, use VSEPR theory[1] to obtain the steric number of each atom. Then read the hybridization of the

---

[1]Text pages 80–84.

atoms from the following table (see **16-17**):

Steric Number	Hybridization
2	sp
3	sp^2
4	sp^3

3. Place electron pairs in each localized (σ) molecular orbital. This constructs the single-bond framework of the molecule. The single-bond framework uses two valence electrons between each pair of bonded atoms. Total up the number of valence electrons used in this way.

4. Identify the p orbitals that were not used in hybridization. Combine them to form π molecular orbitals.

5. Place the valence electrons not used in step 3 into the π molecular orbitals. Problems **17-1** and **17-5** show the process and its results.

Isomerism

Isomers are substances having the same number and kinds of atoms but arranged differently. Isomers usually do not interconvert rapidly because interconversion requires the breaking and re-making of chemical bonds, a process with a high activation energy. Several types of isomerism are distinguished:

- In **constitutional isomerism,** the order of connections of the atoms differs between two molecules having the same formula, Different constitutional isomers have different atomic *skeletons* (review what is meant by a skeleton, on page 49 of this Guide). Problems **3-17** and **3-49** treat constitutional isomerism. The text discusses this kind of isomerism in Section 17-2[2] in connection with branched-chain alkanes. Constitutional isomerism is also called structural isomerism.

- In *cis-trans* **isomerism** (also called **geometrical isomerism**), two molecules have the same molecular formula and the same overall skeleton but differ in the spatial arrangement of their atoms. This kind of isomerism appears, in organic compounds, in association with double bonds. *Cis* means "on the near side," and *trans* means "on the far side."[3] In a *cis*-isomer, two identical groups are on the same side of a double bond. In a *trans*-isomer. two identical groups are on opposite sides of a double bond. The two sides of a double bond differ because:

---

[2]On text page 640.

[3]Remember: you need *trans*portation to get to the far side of town.

- **There is no free rotation about a double bond.**

Rotation around single bonds is unimpeded because σ overlap is cylindrically symmetrical. With double bonds, rotation requires breaking a π bond, a process that requires excitation of an electron to a higher-energy π^* orbital. Conversion of *cis* to *trans* isomers (or vice versa) is therefore slow. *Cis-trans* isomerism is common in coordination complexes (see text Section 18-3[4]).

- In **optical isomerism**, two compounds have the same molecular formula, the same overall skeleton, and the same spatial arrangement of their atoms, but differ because the molecules of one are the non-superimposable mirror images of the molecules of the other. A left hand and a right hand are non-superimposable mirror images of each other. Structures that are not superimposable on their mirror images are said to possess **chirality**. In organic chemistry the most common source of chirality is a carbon atom that is bonded to four different atoms or groups of atoms. Chirality is not unique to organic compounds. It is discussed further in connection with coordination complexes. See text Section 18-3.[5]

Conjugated Molecules

Conjugated molecules contain sets of molecular orbitals that span two or more double or triple bonds that occur close to each other in the molecule. Examples are 1,3-butadiene and benzene. See page 471 in this Guide. On text page 636, The text deliberately re-introduces material from text page 614-15 concerning conjugated molecules to put it into the context of organic chemistry. It then expands the discussion to cover the bonding in the benzene molecule.

Petroleum Refining and the Hydrocarbons

Hydrocarbons are compounds that contain only carbon and hydrogen. The primary source of hydrocarbons at this time is petroleum. The processing of petroleum yields fuels such as gasoline and starting materials for the petrochemical industry, which creates a large variety of useful products.

Hydrocarbons fall into four categories: **alkanes**, **alkenes**, **alkynes**, and the **aromatic** hydrocarbons. Carbon's unique propensity for forming stable chains allows the existence of straight-chain, branched-chain, and **cyclic**, or ring structures in all four categories of hydrocarbon compound. Hydrocarbons lacking any rings are **acyclic**. All aromatic hydrocarbons contain at least one ring.

---

[4]Text page 672.

[5]Text page 673. Chirality is also discussed on page 507 in this Guide.

Alkanes

Alkanes are **saturated hydrocarbons** because every carbon atoms in the molecule is linked at enough hydrogen atoms to give it four bonds to four different atoms—all C-to-C (and C-to-H) bonds in alkanes are single bonds. This gives alkanes the general formula C_nH_{2n+2}. In straight-chain alkanes (sometimes called normal alkanes), all the carbons except those on the ends of the chain are linked to two other C's and to two H's. The straight-chain alkanes are the major constituents of petroleum.

Branched-chain alkanes have the same general formulas as straight-chain alkanes, but now the chain of carbon atoms has branches so that at least one carbon atom is bound to three other carbon atoms. The names of the alkanes having from 1 to 15 C atoms are listed in text Table 17-1.[6]

- **Memorize the names of at least the first ten alkanes.**

Note that after $n = 4$ the names employ familiar stems followed by the suffix "-ane."

The free rotation about the C—C single bond allows many **conformations** for alkanes as various segments of the chain rotate into proximity with each other. Different conformations of a given carbon chain look different but are the same compound as long as they have the same sequence of atoms.

No amount of free rotation can convert one constitutional (structural) isomer into another. Instead, bonds must be broken and re-formed. Straight-chain and branched-chain alkanes with the same formula are structural isomers.

The standard (IUPAC) system of naming casts light on the nature of isomers by it way it deals with the problem of naming them. Follow these steps:

1. Identify the longest continuous chain of C atoms. Chain length can be concealed by writing structures in zig-zags on the page. Do not be deceived. What counts is *sequence*. Follow the chain around corners.

2. Number the atoms in the chain, starting from the end nearer any branches. The aim is to give the lowest possible number to the positions of side-groups. **Substituent groups** (or side-groups) are branches from the main chain. Alkyl side-groups are named by removing the suffix "-ane" from the name of the alkane and adding "-yl."

3. Write the name of the compound using a number for each side-group to tell where it is attached, followed by the name of each side-group and finally by the name for the main chain.

---

[6]Text page 638.

4. Side-groups enter a name in alphabetic order. The presence of identical side-groups at different locations on the chain is indicated by the appropriate multiplying prefix. See **17-9**.

The above procedure does not deal with every possible naming situation in the alkanes, but it does work with most common compounds. **Example:** Name the alkane:

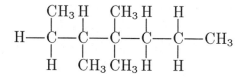

Solution: The name is 2,3,3-trimethylheptane. One common wrong answer is "1,2,3,3-tetramethylhexane." The methyl group that is apparently a side-group on the left-most carbon in the main line of the structure is really part of the chain. Once the longest chain is identified, the question turns to which direction to use for numbering. Numbering this chain from the left to the right, instead of the reverse, gives *smaller* prefix numerals in the name. That is, "2,3,3-" is preferable to "4,4,5-".

An increased fraction of branched-chain alkanes in a gasoline improves the smoothness of its combustion, a property that is measured by **octane number**.

Alkenes and Alkynes

Alkenes are hydrocarbons that contain at least one carbon-carbon double bond. Some of the C atoms in alkenes are sp^2 hybridized. See **17-15**. Alkenes are said to be **unsaturated** because they have room for more hydrogen; they react readily with H_2 to form alkanes. Alkenes have the generic formula C_nH_{2n}. If there are two double bonds in an alkene, it is a **diene** and has the generic formula C_nH_{2n-2}. Trienes have the formula C_nH_{2n-4}, and so forth. Hydrocarbons with multiple double bonds are **polyenes**. There is no free rotation about the C=C double bond. This allows the existence of *cis-trans* isomers.[7] *Cis* and *trans* isomers have their atoms connected in the same sequence but lying in different geometrical relationships. See **17-11**.

Alkenes are named similarly to alkanes. Always pick the longest chain *that includes the double bond*. The position of the double bond is signified in the name by the number of the carbon atom that is followed by the double bond in the structural formula. Number the atoms of the chain and designate side-groups so that this number is as small as possible. Then comes the root name of the parent hydrocarbon followed by the ending "-ene". See **17-13**. The trivial names "ethylene" and propylene" for ethene and propene respectively must be memorized. Alkenes do not occur in petroleum, but they can be produced from it by **cracking** reactions.

---

[7]See text Figures 17-3 (text page 634) and 18-12 (page 673).

These include **catalytic cracking** and **thermal cracking.** The two methods produce shorter-chain alkenes from the alkanes in the starting petroleum distillate (see **17-7**).

Alkynes contain C-to-C triple bonds. The simplest alkyne is ethyne (acetylene). Again, the chain selected for naming must include the triple bond. In naming these compounds, the position of the triple bond is indicated by a number, and the ending "-yne" is attached. When there is no ambiguity the number is omitted. For example, "1-propyne" and "2-propyne" are the same compound, propyne. The same policy of simplification applies to alkenes (above). There is only one possible propene, so numbers are not needed in its name.

Cyclic alkanes, alkenes, polyenes, alkynes, and polyalkynes are all possible. These hydrocarbons all contain at least one ring. They are named by using the prefix "cyclo" in conjunction with the appropriate root name for the number of carbons found in the ring. Thus, propane (C_3H_8) is an alkane and cyclopropane C_3H_6 is a cycloalkane It has the same formula as the alkene propene. Formation of a ring from an alkane reduces the number of H atoms by two. This makes cycloalkanes the isomers of alkenes. Several cyclic hydrocarbons are pictured in text Figure 17-14.[8] A cycloalkane is the subject of **17-5**, and naming cyclic alkanes and alkenes is part of **17-9**, **17-10**, and **25-21**. Also see **16-65**.

Do not forget to include cyclic structures when listing the possible isomers of a hydrocarbon which is, by its formula, unsaturated.

Aromatic Hydrocarbons

The **aromatic hydrocarbons** contain rings of carbon atoms in which delocalized π-electrons significantly increase the molecular stability. These compounds are also called **arenes.** Benzene C_6H_6 is the archetypal aromatic hydrocarbon. It has an unsaturated six-carbon ring with a system of six π-electrons.[9] Each C-to-C bond is effectively a 3/2 bond. The increase in molecular stability caused by this aromatic system in benzene is computed in **17-35**. It is a **resonance stabilization energy.** Petroleum contains useful quantities of benzene.

Substituent groups may be attached to the benzene ring. It is customary to omit the hydrogen atoms when drawing the structure of the benzene ring. Avoid the dual errors of not counting these H atoms in molecular formulas or forgetting to subtract them from molecular formulas when they are replaced by substituent groups. Thus, methylbenzene is C_6H_5—CH_3: one of the six hydrogen atoms of benzene is replaced by a methyl group. This compound is more commonly called **toluene.**

If two side-groups are attached to the benzene ring, *three* different isomers are

---

[8]Text page 646.
[9]See text page 637.

possible. Relative positions are distinguished by numbers. The 1,2-, 1,3-, and 1,4- di-substitution patterns are named *ortho, meta and para,* respectively. This means that there are three dimethylbenzenes. These isomeric compounds are called **xylenes**.

The **BTX aromatics** are a mixture of benzene, toluene, and *ortho-*, *meta-*, and *para*-xylene. These compounds are very important in gasoline because they significantly increase the octane number of the gasoline. They are made by **reforming reactions** from straight-chain alkanes separated from petroleum by distillation. Toluene and the xylenes can be converted to benzene by **hydrodealkylation** reactions. In such reactions an alkyl side-group is removed (hence "-dealkyl") and replaced by hydrogen.

Functional Groups and Organic Synthesis

The insertion of non-carbon atoms in a hydrocarbon chain or their attachment to such a chain creates sites of greater reactivity that are called **functional groups**. The structures of the common functional groups are given in text Table 17-3.[10]

Many functional groups can be regarded as deriving from a simple inorganic molecule by insertion of an alkyl group such as $-CH_3$ or an aromatic group such as $-C_6H_5$. The general symbol for an alkyl group is "R." See **17-21**. The general symbol for an aromatic group (an **aryl** group) is "Ar."

• **Alkyl halides** have the generic formula R—X. Formally, they are derivatives of the hydrohalic acids H—Cl, H—Br, and H—I.

• **Alcohols** R—OH and **phenols**, Ar—OH can be regarded as derived from water (H—OH) by replacing one H with an alkyl or aryl group. In alcohols, the —OH group must be attached to a carbon atom that is saturated, that is, a carbon atom with four single bonds. Alcohols are called **primary**, **secondary**, and **tertiary** according to the number of other carbon atoms bonded to the carbon atom that bears the —OH group. In phenols, the —OH group is attached to an aromatic group; the compound C_6H_5—OH is named phenol.

• **Ethers** all contain the R—O—R′ functional group where R and R′ are the same or different alkyl or aryl groups. They are like water (H—O—H) with both H atoms replaced. Neither of the carbon atoms attached to the ether O may itself be double-bonded to an O atom, because then the molecule would be an ester (see below).

• **Aldehydes** and **ketones** contain a carbonyl group   C=O. The aldehydes have at least one H bonded to the carbonyl carbon. The other group bonded to the carbonyl carbon is —R (an alkyl group) or An aromatic group. The simplest aldehyde is $H_2C=O$ (formaldehyde). The ketones have alkyl or aromatic groups replacing both H atoms in $H_2C=O$.

---

[10]Text page 647.

• **Carboxylic acids** have the formula R—COOH. They can be viewed as deriving from HO—COOH (carbonic acid) by the replacement of an —OH group with an —R group. They act as weak acids, forming salts upon neutralization with bases. Typical carboxylic acids include formic acid, in which R equals H, and acetic acid, in which R equals CH_3.[11]

• **Esters** derive from carboxylic acids when the acid reacts with an alcohol:

R—COOH (acid) + HO—R′ (alcohol) → R—COOR′ (ester) + H_2O

• **Amines** can be regarded as derivatives of ammonia NH_3 in which one, two or three H atoms are replaced by alkyl groups. They are sub-classified: RNH_2 is a primary amine, R_2NH is a secondary amine, and R_3N is a tertiary amine.

• **Amides** can also be regarded as derivatives of ammonia NH_3. In amides, either ammonia or a primary or secondary amine has condensed with a carboxylic acid:

R—COOH (acid) + NH_2—R′ (amine) → R—CONHR′ (amide) + H_2O

Detailed Solutions to Odd-Numbered Problems

17-1 The Lewis structure of acetaldehyde is:

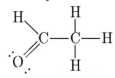

This structure has a total of 18 valence electrons. The *SN*'s (steric numbers) of the carbon atoms are 4 (for the methyl carbon) and 3 (for the carbonyl carbon). The *SN*'s of the other atoms are immaterial because each of them forms only one bond. The *SN* 4 on the methyl C means it has sp^3 hybridization; the *SN* 3 on the carbonyl C means it has sp^2 hybridization. Constructing the single bond framework of the molecule uses 12 valence electrons. At this point a *p* orbital on the carbonyl C contains a single electron and the three *p* orbitals on the oxygen contain 5 electrons. The *p* orbital on the C and a *p* orbitals on the O overlap to form a π bonding orbital. Two electrons occupy this orbital. The remaining 4 electrons remain as lone pairs on the O. A π^* antibonding orbital is created simultaneously with the π orbital spanning C and O, but it remains empty. The three groups bonded to the carbonyl C lie in a plane with bond angles near 120°. The geometry at the methyl C atom is approximately tetrahedral, with all six H—C—H and H—C—C angles near 109.5°.

[11]See **10-45** and **10-49**.

17-3 The Lewis structures for HCOOH and HCOO⁻ are:

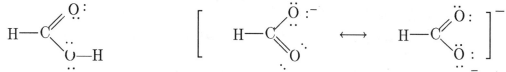

One resonance form is given for HCOOH, but two are given for the formate anion HCOO⁻. In formic acid, one oxygen atom is doubly bonded to the carbon atom, and the other is singly bonded. In the anion, there is some double-bond character in both C—O bonds. The carbon atom in HCOOH is sp^2 hybridized (*SN* 3), and the OH oxygen atom is sp^3 hybridized (*SN* 4). The immediate surroundings of the carbon atom have trigonal planar geometry, and the C—O—H group is bent. In the HCOO⁻ ion, the carbon atom and both oxygen atoms are sp^2 hybridized (*SN* 3), possessing a three-center four-electron π system. In HCOOH, π overlap occurs between orbitals on the carbon atom and only one oxygen atom. Both C-to-O bond lengths in the formate ion should lie somewhere between the value for the single bond (1.36 Å) and the value for the double bond (1.23 Å).

17-5 a) The ΔH_f° of gaseous cyclopropane equals the enthalpy change in forming $C_3H_6(g)$ in its standard state from $C(s)$ and $H_2(g)$ in their standard states: $3\,C(s) + 3\,H_2(g) \rightarrow C_3H_6(g)$. Imagine this reaction to proceed by the atomization of 3 mol of $C(s)$ and 3 mol of $H_2(g)$ to $C(g)$ and $H(g)$ followed by the formation of 3 mol of C—C bonds and 6 mol of C—H bonds. The sum of the ΔH°'s of these steps equals the desired ΔH_f°, by Hess's law. Use the bond enthalpies and enthalpies of atomization from Table 7-3.[12]

$$\Delta H_f^\circ = 3(716.682) + 6(217.96) + 3(-348) + 6(-413) = \boxed{-64 \text{ kJ}}$$

b) By Hess's law, the standard enthalpy change for the combustion of cyclopropane equals the enthalpy of formation of the products minus the enthalpy of formation of the reactants in the reaction:

$$C_3H_6(g) + \tfrac{9}{2}O_2(g) \rightarrow 3\,CO_2(g) + 3\,H_2O(g)$$

Thus:

$$\Delta H^\circ = -1959 \text{ kJ} = 3\underbrace{(-393.51)}_{CO_2(g)} + 3\underbrace{(-241.82)}_{H_2O(g)} - \tfrac{9}{2}\underbrace{(0)}_{O_2(g)} - 1\underbrace{(\Delta H_f^\circ)}_{C_3H_6(g)}$$

$$\Delta H_f^\circ = \boxed{53 \text{ kJ mol}^{-1}}$$

c) The calorimetric ΔH_f° of $C_3H_6(g)$ is considerably higher that the ΔH_f° based on bond enthalpies. It requires more energy to make $C_3H_6(g)$ than would be expected

---

[12]Text page 228.

on the basis of the formation of normal single bonds. Formation of the triangular ring of C atoms forces C—C—C bond angles of 60°, much smaller than the usual tetrahedral angle, 109.5°. The extra energy is the "strain enthalpy" of cyclopropane. It is about $\boxed{117 \text{ kJ mol}^{-1}}$.

17-7 a) The catalytic cracking reaction is:

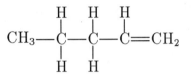

b) Another isomer of the alkene is 2-pentene, in which the double bond lies between the second and third carbon atoms in the chain.

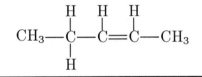

17-9 The structural formulas are:

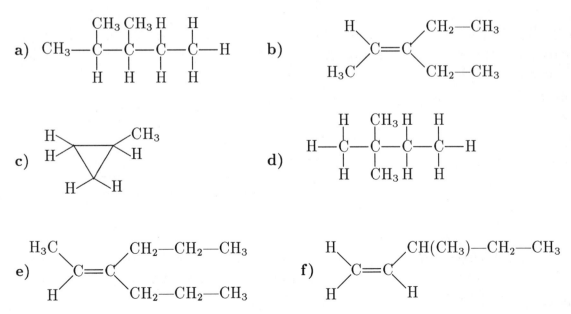

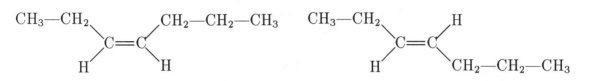

g) CH_3—$\overset{\overset{\displaystyle H}{|}}{\underset{\underset{\displaystyle CH_3}{|}}{C}}$—$\overset{\overset{\displaystyle H}{|}}{\underset{\underset{\displaystyle H}{|}}{C}}$—$\overset{\overset{\displaystyle C_2H_5}{|}}{\underset{\underset{\displaystyle H}{|}}{C}}$—$\overset{\overset{\displaystyle H}{|}}{\underset{\underset{\displaystyle H}{|}}{C}}$—$\overset{\overset{\displaystyle H}{|}}{\underset{\underset{\displaystyle H}{|}}{C}}$—$CH_3$ h) CH_3—$C{\equiv}C$—$\overset{\overset{\displaystyle CH_2CH_3}{|}}{\underset{\underset{\displaystyle H}{|}}{C}}$—$CH_2CH_2CH_3$

17-11 The *cis* isomers of 3-heptene is at the left and the *trans* is at the right:

$$CH_3{-}CH_2 \diagdown \qquad \diagup CH_2{-}CH_2{-}CH_3 \qquad\qquad CH_3{-}CH_2 \diagdown \qquad \diagup H$$
$$\qquad\qquad C{=}C \qquad\qquad\qquad\qquad\qquad\qquad C{=}C$$
$$H \diagup \qquad \diagdown H \qquad\qquad\qquad\qquad\qquad H \diagup \qquad \diagdown CH_2{-}CH_2{-}CH_3$$

17-13 a) 1,2-hexadiene   **b)** 1,3,5-hexatriene (or *trans*-1,3,5-hexatriene)
c) 2-methyl-1-hexene   **d)** 3-hexyne

17-15 Count the number of atoms linked to the C atom in question, and add the number of lone pairs.[13] The results equals the steric number *SN* of the C in question. If *SN* equals 2, the C is *sp* hybridized; if *SN* equals 3, the C is *sp²* hybridized; if *SN* equals 4, the C is *sp³* hybridized.

a) From left to right in the structure, the hybridization of the C atoms is: *sp²*, *sp*, *sp²*, *sp³*, and *sp³*.

b) All C atoms are *sp²* hybridized.

c) The two C atoms involved in the double bond are *sp²* hybridized. The rest are *sp³* hybridized.

d) From the left, the hybridization is *sp³*, *sp³*, *sp*, *sp*, *sp³*, and *sp³*.

17-17 a) This is an esterification. It formally resembles an acid-base neutralization with the alcohol playing the part of the base:

$$CH_3CH_2CH_2CH_2OH + CH_3COOH \rightarrow CH_3COOCH_2CH_2CH_2CH_3 + H_2O$$

The organic product is the ester $\boxed{\text{butyl acetate}}$.

b) The reaction is a dehydration: $H_3C{-}COO^- NH_4^+ \rightarrow H_3C{-}CO{-}NH_2 + H_2O$.

c) The H atom attached to the O atom and one of the H atoms on the neighboring C atom are removed: $H_3CCH_2CH_2OH \rightarrow H_3CCH_2CHO + H_2$. The organic product

---

[13]Lone pairs are quite uncommon on C atoms in organic compounds.

is propanal (also called propionaldehyde).

d) $H_3CCH_2CH_2CH_2CH_2CH_2CH_3 + 11\,O_2 \rightarrow 7\,CO_2 + 8\,H_2O.$

17-19 a) Brominate ethylene to give 1,2-dibromoethane. Then dehydrobrominate the 1,2-dibromoethane. This means: treat ethylene with bromine so that the bromine molecule adds across the double bond. Then treat the product in such a way that hydrogen bromide is abstracted:

$$CH_2{=}CH_2 + Br_2 \rightarrow CH_2BrCH_2Br \quad \text{then} \quad CH_2BrCH_2Br \rightarrow CH_2{=}CHBr + HBr$$

b) Treat 1-butene with water (in the presence of H_2SO_4):
$CH_3CH_2CH{=}CH_2 + H_2O \rightarrow CH_3CH_2CH(OH)CH_3.$
c) Treat propene with water to give 2-propanol, and then dehydrogenate over a catalyst (such as metallic copper):

$$CH_3CH{=}CH_2 + H_2O \rightarrow CH_3CH(OH)CH_3 \text{ and } CH_3CH(OH)CH_3 \rightarrow CH_3COCH_3 + H_2$$

17-21
$$\underset{\underset{R}{|}}{\overset{\overset{R}{|}}{R{-}C}}{-}OH + HO{-}\overset{\overset{O}{\|}}{\underset{\underset{H}{|}}{C}}{-}R' \rightarrow HOH + \underset{\underset{R}{|}}{\overset{\overset{R}{|}}{R{-}C}}{-}O{-}\overset{\overset{O}{\|}}{C}{-}R'$$

17-23 It is clear that the molar ratio of the ethylene to the ethylene dichloride is 1-to-1, based on the formulas of the two compounds. Use this ratio to compute the amount of ethylene required to make the 6.26×10^9 kg of ethylene dichloride. Then:

$$6.26 \times 10^9 \text{ kg } C_2H_4Cl_2 \times \left(\frac{1 \text{ mol } C_2H_4Cl_2}{0.09896 \text{ kg } C_2H_4Cl_2}\right) \times \left(\frac{1 \text{ mol } C_2H_4}{1 \text{ mol } C_2H_4Cl_2}\right)$$
$$\times \left(\frac{0.02805 \text{ kg } C_2H_4}{1 \text{ mol } C_2H_4}\right) = 1.77 \times 10^9 \text{ kg } C_2H_4$$

This is 11.2% of the 15.87×10^9 kg total annual production of ethylene. The mass of the chlorine is 4.49×10^9 kg , by a similar computation.

17-25 The six π molecular orbitals of pyridine arise as a combination of the six p_z orbitals of the nitrogen and five carbons atoms in the ring:

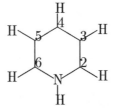

According to the problem, MO's that put electron density onto the N atom will be lower in energy in pyridine than comparable orbitals in benzene. Thus, molecular orbitals that have the N-atom p_z orbital among their parents will be lower in energy. Let the N atom occupy position 1 in a numbering scheme that goes counter-clockwise around the ring (see above). Also, refer to Figure 17-6,[14] in which the six π MO's of benzene are sketched. The strongly bonding and strongly antibonding (the highest and lowest) MO's in Figure 17-6 both have parentage that includes the p_z orbital on atom 1. These two molecular orbitals are therefore *lowered* in energy in pyridine relative to benzene. One of the two weakly bonding molecular orbitals in benzene has $p_z(N)$ parentage, but the other does not. Instead, its parentage includes p_z-orbitals from carbon atoms 2 through 5. The first of the two weakly bonding MO's (located on the left in Figure 17-6) is therefore lowered in energy in pyridine relative to benzene, but the energy of the second is not affected. Similarly, the two weakly antibonding MO's in benzene are split in energy. The one that has some $p_z(N)$ parentage (to the right in Figure 17-6) is lowered, but the other is unchanged. The final result is an energy-level diagram for pyridine with six different π-orbital energies, four lower than the corresponding benzene π-orbitals, and two unchanged in energy.

17-27 Cyclodecene has a 10-membered ring and one C-to-C double bond. The end of every line in the following is understood to be occupied by an H. Every intersection is understood to be occupied by a C:

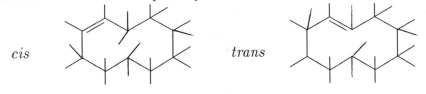

cis   *trans*

17-29 Dehydrogenation is the removal of hydrogen H_2 from a compound; dehydration is the removal of water H_2O. Ethane CH_3CH_3 is dehydrogenated to ethylene CH_2CH_2, and ethanol CH_3CH_2OH is dehydrated to ethylene.

17-31 The unsaturated hydrocarbon must have a straight-chain skeleton of six carbon atoms because it gives straight-chain hexane when reduced with hydrogen gas. Oxidation at the double bond splits it to a four-carbon acid (butanoic acid) and a two-carbon acid (acetic acid). The double bond is therefore at the 2 position: CH_3—CH_2—CH_2—CH=CH—CH_3. The compound is 2-hexene. This compound has *cis* and *trans* isomers, but the available data do not allow a decision about which isomer is present. The balanced equations are:

$$C_6H_{12}(l) + H_2(g) \rightarrow C_6H_{14}(l)$$

---

[14]Text page 637.

$$27\,H_3O^+(aq) + 9\,MnO_4^-(aq) + 5\,C_6H_{12}(l)$$
$$\rightarrow 9\,Mn^{2+}(aq) + 5\,C_4H_8O_2(aq) + 5\,C_2H_3O_2(aq) + 43\,H_2O(l)$$

17-33 a) The combustion reactions are:

$$C_7H_{16}(l) + 11\,O_2(g) \rightarrow 7\,CO_2(g) + 8\,H_2O(g)$$
$$C_8H_{18}(l) + 25/2\,O_2(g) \rightarrow 8\,CO_2(g) + 9\,H_2O(g)$$

The formulas of heptane and isooctane, which are liquids at room conditions, come from the text.[15] Take ΔH_f° data from text Appendix D as needed for the reactants and products other than the two fuels. Combine the ΔH_f°'s to get the ΔH°'s of the two combustions:

$$\Delta H_1^\circ = 8\underbrace{(-241.82)}_{H_2O(g)} + 7\underbrace{(-393.51)}_{CO_2(g)} - 11\underbrace{(0)}_{O_2(g)} - 1\underbrace{(-187.82)}_{heptane(l)} = \boxed{-4501.31\ \text{kJ}}$$

$$\Delta H_2^\circ = 9\underbrace{(-241.82)}_{H_2O(g)} + 8\underbrace{(-393.51)}_{CO_2(g)} - \frac{25}{2}\underbrace{(0)}_{O_2(g)} - 1\underbrace{(-224.13)}_{isooctane(l)} = \boxed{-5100.33\ \text{kJ}}$$

b) The combustion of 1 mol of isooctane clearly produces much more heat (ΔH_2°) than the combustion of 1 mol of heptane (ΔH_1°). The question however concerns the comparison of the ΔH°'s per gallon of the two fuels:

$$\frac{-4501.31\ \text{kJ}}{1\ \text{mol heptane}} \times \left(\frac{1\ \text{mol heptane}}{100.21\ \text{g heptane}}\right) \times \left(\frac{453.59\ \text{g}}{1\ \text{lb}}\right) \times \left(\frac{5.71\ \text{lb}}{1\ \text{gal}}\right) = \frac{-1.16 \times 10^5\ \text{kJ}}{1\ \text{gal}}$$

$$\frac{-5100.31\ \text{kJ}}{1\ \text{mol octane}} \times \frac{1\ \text{mol octane}}{114.23\ \text{g octane}} \times \frac{453.59\ \text{g}}{1\ \text{lb}} \times \frac{5.77\ \text{lb}}{1\ \text{gal}} = \frac{-1.17 \times 10^5\ \text{kJ}}{1\ \text{gal}}$$

The $\boxed{\text{isooctane}}$ produces slightly more heat per gallon.

Tip. Compare the ΔH° for the combustion of isooctane computed here to the ΔH° computed in **8-69**. The reason for the difference between the two is the difference in the physical states of the products.

17-35 a) The ΔH° of the reaction $6\,C(g) + 6\,H(g) \rightarrow C_6H_6(g)$ equals the ΔH_f°'s of the products minus the ΔH_f°'s of the reactants:

$$\Delta H^\circ = 1\underbrace{(82.93)}_{C_6H_6(g)} - 6\underbrace{(716.682)}_{C(g)} - 6\underbrace{(217.96)}_{H(g)} = \boxed{-5524.92\ \text{kJ}}$$

---

[15]Text Table 17-1, text page 638.

Neither of the ΔH_f°'s of the reactants equals zero because gaseous monatomic H and gaseous C are *not* the standard states of these two elements. The answer is an experimental result; it is based on experimental ΔH_f°'s and Hess's law.

b) The formation of 1 mol of $C_6H_6(g)$ with alternating double and single bonds requires the formation of 6 mol of C—H bonds, 3 mol of C=C bonds, and 3 mol of C—C bonds. The negative of the sum of the bond dissociation enthalpies of these bonds is $\boxed{-5367 \text{ kJ}}$.

c) The resonance stabilization in 1 mol of benzene equals the difference between the ΔH° expected and the ΔH° observed: $(-5367) - (-5524.92) = \boxed{158 \text{ kJ}}$.

17-37 Conjugation of multiple bonds (π-delocalization) tends to increase the wavelength of the absorbed light in electronic transitions. Hence, we expect cyclohexene to absorb at $\boxed{\text{shorter}}$ wavelengths than benzene because the π bonding is localized in cyclohexene and delocalized in benzene.

Chapter 18

Bonding in Transition Metals and Coordination Complexes

The Chemistry of the Transition Metals

The main transition group includes the fourth-period elements Sc through Zn, the fifth-period elements Y through Cd, and the sixth-period elements Lu through Hg. In these d-block elements, the $3d$, $4d$, or $5d$ subshell is either partly or just filled. The text presents a section on the physical properties and another on the chemical properties of the d-block metals. Study this information with a copy of the periodic table in hand. The properties are:

- **Ionization Energies.** The first and second ionization energies generally increase going across the periods. As an exercise, plot the first IE's as a function of Z, noting the various irregularities and the broad trend.[1] The IE's in the sixth period start lower (at Lu) but rise higher (at Au) than in the fourth and fifth periods.

- **Atomic Radius.** The line of the trend across all three periods resembles a droopy clothesline: high at the left, sagging in the middle, getting high again at the right. See text Figure 18-1. Elements in the fourth transition period are smaller that the elements below them in the table, but the atomic sizes of transition metals in the fifth are similar to those of their congeners in the sixth period. The lanthanide contraction[2] causes this similarity. The increasing nuclear charge during the filling of the $4f$ subshell makes the atomic radii of the sixth-period transition metals smaller than they would otherwise be.

---

[1]Data are available in text Appendix F and text Table 18-1.
[2]See text Section 15-8, text page 570.

- **Strength of Metallic Bonding.** Bonding in the elemental transition metals reaches maximum strength in the middle of each transition series as shown by their melting points, boiling points, and $\Delta H°$'s of atomization. Bonding strength rises at first as electrons added during aufbau go into d-derived molecular orbitals of mainly bonding character. After the mid-point of the d-block, the added electrons go into MO's of mainly anti-bonding character, and the strength of the bonding decreases.[3] See **18-45**.

- **Hydration Enthalpy of $M^{2+}(g)$ Ions.** The $\Delta H°$'s for the reaction $M^{2+}(g) \rightarrow M^{2+}(aq)$, where M^{2+} is a transition-metal ion, become generally less going across the table. This property is introduced because exceptions to the trend can be explained elegantly by crystal-field theory, the theory of the bonding between molecules like water and transition metal ions (see text Section 18-4).

- **Variability of Oxidation States.** Nearly all the transition elements form compounds in two or more different oxidation states. Among the **early** transition metals (the first five in each period), the maximum oxidation number equals the number of valence electrons. The maximum oxidation numbers of the **late** transition elements (those past the middle of the period) are less. The elements with the most variability in oxidation number are toward the center of each period.

- **Importance of Higher Oxidation States for Heavier Transition Metals.** Compounds having a transition element in a high oxidation state grow more prevalent moving down the periodic table. This trend is the exact reverse of the trend among the main-group elements.

- **Covalent Versus Ionic Bonding.** Compounds having transition metals in high oxidation states tend to be covalent; compounds having a transition metal in a low oxidation state tend to be ionic. See **18-1**.

- **Acid/Base Behavior of Oxides.** Oxides in which transition metals are in high oxidation states are acid anhydrides; oxides in which transition metals are in low oxidation states are base anhydrides. See **18-73**.

- **Ease of Oxidation.** The transition metals of the fourth period, with the exception of Cu, have negative reduction potentials under standard acidic conditions—they tend to be oxidized easily. The oxidation of the bulk metals is often greatly slowed by **passivation**, the formation of a protective coating of the initial products of oxidation. The early transition elements in the fifth and sixth periods

---

[3]Metallic bonding is discussed in text Section 19-3, text pages 708–10.

also have negative standard reduction potentials, but the late elements, culminating in gold, have large positive reduction potentials and resist oxidation.

The Formation of Coordination Complexes

Complexes are compounds or ions built up from simpler species than can exist independently. Coordination complexes contain transition metal atoms (the **central metal**) bonded to two to eight surrounding ions or neutral molecules (the **ligands**). Think of the bonding in acid-base terms. The ligands have unshared electron pairs which they donate, acting as Lewis bases. The metal atoms have vacant d-orbitals (as well as vacant s- and p-orbitals) and accept electron pairs, acting as Lewis acids. The resulting coordinate covalent bonds[4] are intermediate in character between ionic and covalent. The bonded ligands are in the **coordination sphere** of the central metal. The number of ligand-to-metal bonds is the **coordination number** (CN) of the metal in the complex.

Ions of transition metals coordinate many ligands strongly, giving directional bonds. The resulting complexes have wide range of structures, colors, and magnetic properties.

Attachment of a ligand to a metal occurs at a ligating (donor) atom. Ligands having only one ligating atom (examples: Cl^-, NH_3) are **monodentate** ligands. Ligands having and using two or more ligating sites are bidentate, tridentate, and so forth.

Chelation

Chelates are complexes in which a single ligand is attached to the same central atom by two or more donors. In complexes containing exclusively monodentate ligands, the number of ligands equals the coordination number. Polydentate ligands (such as $H_2\ddot{N}CH_2CH_2\ddot{N}H_2$ where the dots represent electron pairs ready for donation) can span the distance between two points of coordination. See **18-23**. The resulting complexes are chelates. Chelates are more stable than non-chelate complexes with the same kind of donor atom. See **18-75**.

Coordination Alters Properties

Coordination complexes persist in aqueous solution. See **18-39**. The solutions have characteristic conductivities, colors, magnetism, and reactivities which are not the simple sums of the conductivities, colors, magnetism and reactivities of the parts. A coordinated ligand reacts differently from an uncoordinated ligand. For example, a

---

[4]See page 246 of this Guide.

basic ligand (like NH_3) can behave as an acid when coordinated (see **11-71**), and coordinated Cl^- ion is often *not* precipitated by $Ag^+(aq)$. Instead it is held tightly in the coordination sphere. See **18-53**.

Formulas and Names of Coordination Compounds

The overall charge on a coordination complex is the algebraic sum of the charges of its coordinated parts. Ligands are often neutral (e.g. H_2O, NH_3, CO) or negatively charged (Cl^-, CN^-, NCO^-, and so forth). Thus, the compound $K_2[PtCl_6]$ contains the $[PtCl_6]^{2-}$ complex ion, which consists of a Pt(IV) atom and four Cl^- ions. The formulas of coordination complexes are (usually) enclosed in brackets to recognize their independent existence. Do not confuse these brackets with brackets referring to the concentrations of dissolved species. Thus $[Ag(NH_3)_2]^+$ is a formula of a complex ion, and "$[[Ag(NH_3)_2]^+]$" stands for the concentration of that species in solution. Chemists avoid doubling brackets.[5] The meaning of single brackets is usually clear from the context.

The names of coordination complexes derive from the names of the ligands with prefixes to indicate how many ligands are present, and include the name of the central metal. A complex may be an ion or a neutral molecule; if it is ionic, its compound is named according to the pattern used for simple ionic compounds once the complex ion is named.

- Within the complex, the ligands are named first, followed by the metal ion. If more than one type of ligand is present, negatively charged ligands are listed first, followed by neutral ones. Within these categories ligands are listed in alphabetic order.

- The names of anionic ligands end with the letter o, whereas neutral ligands are usually called by the names of the molecules. Exceptions are H_2O (aqua), CO (carbonyl), and NH_3 (ammine).

- When several ligands of a particular kind are present, the Greek prefixes di-, tri-, tetra-, penta-, and hexa- are used to give the count. Thus the ligands in $[Co(NH_3)_4Cl_2]^+$ are designated "dichlorotetraammine." If the ligand itself contains a Greek prefix, use the prefixes *bis* (2), *tris* (3), *tetrakis* (4) to indicate the number of ligands present. The name "ethylenediamine" already contains the term "di" so "bis(ethylenediamine)" is used to indicate two ethylenediamine ligands.

- If the complex is an anion, attach the suffix "-ate" to the name of the metal. Unfamiliar terms like "zincate", "cobaltate", "rhodate" result.

---

[5]For example, in text Section 11-8, text page 384.

- The oxidation number of the metal is written in Roman numerals following the name of the metal. Thus $[Fe(CN)_6]^{3-}$ is named hexacyanoferrate(III) ion.

Example: A coordination complex has the formula $[Co(NH_3)_5Cl]Cl_2$. Which atom is the central atom, what is the charge on the complex ion, what is the oxidation number of the central atom, and what is the name of the compound? **Solution:** The transition metal cobalt is the central atom, and the ligands are ammonia and chloride ion. Since two chloride ions (charge of -1 each) are needed to balance its charge, the complex ion (in brackets) must have a charge of $+2$. Cobalt is in the $+3$ oxidation state because there are three -1 chlorides in the formula to be countered electrically, and the ammonia molecules are neutral. The name of the compound is chloropentaamminecobalt(III) chloride.

Familiarity with nomenclature is tested in two ways: by giving names and asking for formulas and by giving formulas and asking for names. See **18-11, 18-13,** and **18-49**.

Ligand Substitution Reactions

The concept of coordination organizes an immense number of different compounds as different ligands linked in varied combinations with different central metals. A major aspect of the chemical reactivity of complexes is ligand substitution, in which one ligand leaves the coordination sphere and another replaces it. **Labile** complexes undergo rapid substitution of one ligand for another. **Inert** complexes are those in which ligand substitution proceeds very slowly or not at all. See **18-15**. Kinetic stability (inertness) is *not* the same as thermodynamic stability, and lability is not the same as thermodynamic reactivity. The rate and fundamental tendency of ligand substitution reactions are often pH-dependent, as in **18-17**.

Structure of Coordination Complexes

The number of bonds formed by the central atom to its surrounding ligands is its coordination number (CN). The geometrical structure of coordination complexes can be predicted from their CN's using VSEPR theory with some modifications:

- Coordination numbers of 2, 4 and 6 are more common than any others. Other CN's (3, 5 and 7 or more) also occur.

- When CN is 2, the L—M—L group is usually linear.

- Most CN 6 complexes have octahedral structures.

- Complexes with CN 4 are most often **square-planar** (having the four ligands in a flat square around the central metal) or tetrahedral. Between these two, tetrahedral coordination predominates except that the square planar geometry is preferred in complexes of Pd(II) and Pt(II) and complexes of Au(III), Ir(I) and Rh(I).

In reading and writing structural formulas of coordination complexes, remember that coordination is to the central metal. Solid or dotted lines drawn from ligand to ligand (as in text Figure 18-12b[6] are *not* bonds but merely suggest the geometry of the coordination sphere. The compound pictured in the figure has exactly 4 coordinate bonds and no ligand-to-ligand bonds.

Isomerism

Isomers are substances having the same number and kinds of atoms but arranged differently. See **3-65**. In **geometrical isomerism** in coordination complexes, the ligands are arrayed in different relative positions about the central metal atom. Two ligands which are near to each other, on the same side of a metal atom, are *cis*. Two ligands which are far apart (on opposite sides of a metal atom) are *trans*. In square-planar complexes, geometrical isomerism is possible when the formula is Ma_2b_2 or Ma_2bc where a, b, and c are different ligands. See **18-21a**. In octahedral complexes, any given site ("o" in the following figure) has one site that is *trans* to it (labeled "t") and four equivalent sites that are *cis* to it (labeled "c"). The c sites are equivalent, although they do look different as usually sketched (compare to **18-21b**):

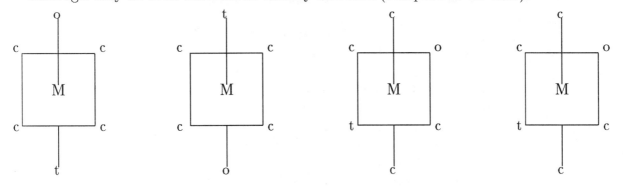

Chiral Structures

A more subtle type of isomerism occurs when two structures are each other's mirror images. See **18-21c** and **18-23**. Non-superimposable mirror-image structures are **enantiomers**. Enantiomers have the same relationship to each other as the left and right hand. If the source of mirror-image isomerism is the three-dimensional distribution of other, bonded atoms about a single atom, then that atom is a **chiral center**. Study the structure of the Co(III)-EDTA complex ion in text Figure 18-15.[7]

---

[6]Text page 673.
[7]Text page 674.

This structure has an enantiomer. The ligand EDTA has a backbone connecting the two N donors and four arms connecting the N donors to the four O donors. The ligand wraps around the central Co atom. The enantiomeric structure has one chelate arm leading from the front nitrogen atom to the *top* of the octahedron (instead of the bottom), and one chelate arm leading from the back nitrogen to the *bottom* of the octahedron (instead of the top). The wrapping of the chelate rings then has an opposite twist or chirality about the central cobalt atom, which is a chiral center.

In the tetrahedral geometry, optical isomerism with the metal as chiral center is possible only if there are four different ligands. See **18-55**. In the square-planar geometry, mirror-image isomers with the central metal as the chiral center is not possible. Both optical and geometrical isomerism occur in **18-23**.

A common task is to sketch the geometry of a complex, including all isomers, given its formula. Attack such problems as follows:

1. Determine the *CN* of the metal and the geometry of the complex.

2. Sketch a starting-point structure.

3. Test for geometrical isomerism by moving ligands around while systematically checking pair-wise relationships of all the ligands. Do the variant structures have the same *cis* and *trans* relationships as in the original sketch? If so, then the new structure is not a geometrical isomer of the original. If not, then it is.

4. Check for optical isomerism by drawing a vertical line next to the original sketch and using it as a mirror to create a mirror image of the original. Remember that atoms *near* the mirror line are still near it in the reflection on its other side. Atoms far from the mirror remain far from it in the reflection. An effective alternative tactic is to turn the original drawing over and trace it through the back of the paper. The result is the mirror image of the original.

5. Check whether the new structure is superimposable on the original. Rotate it so as to put as many ligands as possible in the same positions as in the original sketch. This requires some practice in three-dimensional visualization. The best way to practice is to perform the rotation stepwise, drawing helper sketches at each step. If the mirror image structure cannot be superimposed upon the original by any combination of rotations, then the two are genuine enantiomers.

Bonding in Coordination Complexes

A successful bonding theory must explain: a) the *geometry* of coordination complexes; b) the often striking *colors* of the complexes; c) the variation in the *magnetism* of the complexes (paramagnetism, caused by unpaired electrons, versus diamagnetism, when all electrons are paired).

The text presents two bonding theories. They are **crystal-field theory** and **ligand-field theory.** The "field" in these names refers to the electrostatic influence exerted by the ligands upon the central metal atom.

Crystal-Field Theory

In crystal-field theory, the bonding between metal and ligands is modeled as ionic. The ligands are approximated by point charges that can be maneuvered at various distances from and geometries around the central metal. The negative charges are brought up toward the central-metal atom or ion, and the effects on the valence orbitals of the metal are studied. The symmetry of the electrostatic field determines the form of these effects. If the symmetry of the ligands is **octahedral,** then the metal's five nd orbitals are split into two groups, the t_{2g} and e_g. The t_{2g} contains *three* of the d-orbitals (t for triple) and is lower in energy than the two orbitals of the e_g level. The degree of splitting is the **crystal-field splitting** and is symbolized Δ_o where the subscript refers to the octahedral field.

Splitting occurs because the 5 d-orbitals have different spatial distributions. Some orbital lobes point at the ligands, and some point between them. Problem **18-58** tests understanding of the reasons for d-orbital splitting by asking the energetic fate of the p orbitals in fields of various symmetries.

An octahedral field gives one kind of splitting pattern. If the symmetry of the crystal field is tetrahedral or square planar instead of octahedral, then *different* splitting patterns occur. See Figure 18-19.[8] Figures depicting these patterns are **crystal-field splitting diagrams.** The magnitudes of the crystal-field splittings are symbolized Δ_t (subscript t for tetrahedral) and Δ_1, Δ_2 and Δ_3 (for square planar). A square planar field creates three intervals of splitting because it is inherently less symmetrical than the tetrahedral or octahedral fields. A completely asymmetric arrangement of six ligands would split the five d-orbitals to five different energies.

The magnitude of the splitting depends on the strength of the perturbing crystal field, which in turn depends on the identity of the ligands. The magnitude of the splitting varies considerably from ligand to ligand, but is on the order a few hundred kilojoules per mole.

- **Crystal-field splittings are on the general order of the strength of chemical bonds.**

The **crystal-field stabilization energy (CFSE)** is the energy gained by a complex in which electrons occupy the lower-energy d orbitals relative to what the

---

[8]Text page 680.

energy would be if there were no crystal-field splitting. Text Table 18-4[9] gives the CFSE's for all possible high-field and low-field occupancies of the d-orbitals in an octahedral field. In this table, negative CFSE's correspond to greater stability.

Understand the method by which CFSE's are calculated:

In an octahedral field, every electron in a t_{2g} orbital stabilizes the complex by 2/5 Δ_o. Every electron in an e_g orbital destabilizes the complex by 3/5 Δ_o.

The crystal-field stabilization energy contributes to the stability of the complex. See **18-27**. As **18-35** and **18-63** establish, the CFSE is an important factor in the stability of complexes.

The many colors exhibited by coordination complexes are interpretable in terms of the crystal-field theory. **Strong field** ligands split the d-orbitals far apart. Their Δ is big. It requires *more* energy to excite a d-electron from the low-lying set of d-orbitals to an orbital of higher energy. Complexes with strong field ligands therefore absorb light more toward the blue end of the visible spectrum. Complexes with weak-field ligands absorb more toward the red (low frequency) end of the spectrum. The **spectrochemical series** is an empirical ordering of common ligands according to the strength of the field that they exert:

weak-field   strong-field

$$I^- < Br^- < Cl^- < F^- < OH^- < H_2O < NH_3 < en < CO \approx CN^-$$

To understand the visible spectra of coordination complexes, recall the colors of the spectrum. From low-frequency to high-frequency they are:

Red Orange Yellow Green Blue Indigo Violet

Figure 16-35[10] curves this line of colors into a circle, the color wheel. When a compound absorbs light of a given color, it removes those frequencies from the spectrum. The color perceived is the given color's **complement**, which is on the opposite side of the color wheel. This idea in important in **18-31**, **18-33**, **18-37**, and **18-39**.

The splitting of the metal d orbitals in the crystal-field gives rise to two possible types of d-electron configurations. **Low-spin** configurations occur when Δ is large and the d-electrons remain paired in the orbitals split to lower energy. High-spin configurations occur when Δ is small, and the d-electrons remain unpaired because the energy it would require to pair them exceeds Δ, the energy required to push them up to occupy the d-orbitals that are split to higher energy. See **18-25**. The correspondence in terminology is:

---

[9]Text page 677.

[10]Text page 616. See also ROY G. BIV on page 426 of this Guide.

strong-field $\Longleftrightarrow$ low-spin   **weak-field $\Longleftrightarrow$ high-spin**

The magnetic properties of complexes depend on the number of unpaired electrons and are successfully predicted by crystal-field theory. The success of the predictions extends not only to paramagnetism (all high-spin and many low-spin complexes) versus diamagnetism (some low-spin complexes) but also to the number of unpaired electrons.

Ligand-Field Theory

In this theory the ligands are no longer regarded as point negative charges exerting an electrostatic field as they surround the central metal atom. Instead, the ligands are allowed to have orbitals. Molecular orbitals are constructed from the metal valence orbitals and the ligand orbitals. Orbital correlation diagrams that are similar in principle to the correlation diagrams of Chapter 14 result. Compare text Figures 16-16 and 18-22.[11] The ligand-field theory is intrinsically more realistic than crystal-field theory. It adds to crystal-field theory an understanding of the *variations* in field strength of the ligands. Also, the correlation diagrams of ligand-field theory include as a component the crystal-field splitting diagrams previously discussed. Hence, they do not contradict but include and expand upon crystal-field theory.

Ligands may, if they have properly arrayed orbitals, receive electrons from as well as give electrons to the metals. This is **π-back-bonding** and is associated with an increase in the field strength exerted by the ligand. Common ligands that can back-bond are the CN^- ion and the CO molecule.

Organometallic Compounds and Catalysis

Organometallic compounds have bonds between metal and carbon atoms. The carbon monoxide molecule ($:C\equiv O:$) has a lone pair at both ends. It bonds through its carbon end to many metals. It donates π electron probability density to the metal and accepts electron probability density back into its empty π^* (antibonding) orbitals.

Special stability occurs among organometallics when the central metal is surrounded by 18 valence electrons. See **18-41** and **18-67**. Attaining 18 electrons means the metal has a closed-shell electron configuration. The rule of 18 is thus conceptually like the rule of 8 (the octet rule) and similarly subject to violation. The special stability is associated with 18 valence electrons is as important for organometallics as the octet rule is for compounds of main-group elements. The rule of 18 explains the dimeric structures of $Mn_2(CO)_{10}$ and $Co_2(CO)_8$ in contrast to monomeric $Ni(CO)_4$.

Sandwich compounds are organometallic compounds in which a metal atom or ion is situated between two planar aromatic rings, as in $Fe(C_5H_5)_2$. Ligands of the

---

[11]On text pages 597 and 683 respectively.

type found in this compound can donate more than one electron pair (see **18-67**). The electrons are typically π-electrons from a molecular orbital like the one illustrated for benzene in text Figure 17-6.[12] The rule of 18 still works in sandwich compounds.

Many transition metals catalyze processes giving valuable organic compounds. Catalysis proceeds with intermediate formation of metal-carbon bonds. An example is the **Monsanto process** for the production of acetic acid from methanol and carbon monoxide. The process involves an organometallic complex of rhodium (see text Figure 18-27).

Coordination Complexes in Biology

Many biologically important compounds contain some derivative of **porphine,** a cyclic organic compound in which four N-donors lie in a plane, surrounding and coordinating a metal ion. The heme portion of hemoglobin contains an iron ion coordinated by a porphine-derived tetradentate ligand. Chlorophyll contains a different porphine derivative coordinated to Mg^{2+} ion. Vitamin B_{12} contains a corrin ring (somewhat similar to porphine) coordinated to a central cobalt atom.

Detailed Solutions to Odd-Numbered Problems

18-1 a) The more water-soluble of the two compounds is $\boxed{PtF_4}$. It has Pt in the lower oxidation state and should be more ionic than PtF_6 and therefore more soluble.

b) The more volatile is $\boxed{PtF_6}$, which is the more covalent of the two.

18-3
$$V_{10}O_{28}^{6-}(aq) + 16\,H_3O^+(aq) \to 10\,VO_2^+(aq) + 24\,H_2O$$

The vanadium is in the $\boxed{+5}$ oxidation state both before and after this acid-base reaction. The oxide of vanadium in which V is in this oxidation state is $\boxed{V_2O_5}$.

18-5 The reduction of titanium(IV) oxide to titanium(III) oxide with hydrogen is represented:
$$2\,TiO_4 + 5\,H_2 \to Ti_2O_3 + 5\,H_2O$$

The $\boxed{product}$, an oxide in which Ti is in a lower oxidation state, should be more basic.

18-7 Methylamine is a $\boxed{monodentate}$ ligand that binds to a central metal ion by donating a lone pair of electrons from the $\boxed{N\ atom}$. This is the only lone pair in the molecule.

---

[12]Text page 637.

18-9 a) $[V(NH_3)_4Cl_2]$ The V atom is in the $\boxed{+2}$ oxidation state.

b) $[Mo_2Cl_8]^{4-}$ The Mo atom is in the $\boxed{+2}$ oxidation state.

c) $[Co(H_2O)_2(NH_3)Cl_3]^-$ The Co atom is in the $\boxed{+2}$ oxidation state.

d) $[Ni(CO)_4]$ The Ni atom is in the $\boxed{0}$ oxidation state.

18-11 **a)** $Na_2[Zn(OH)_4]$ **b)** $[Co(H_2NCH_2CH_2NH_2)_2Cl_2]NO_3$
c) $[PtBr(H_2O)_3]Cl$ **d)** $[Pt(NH_3)_4(NO_2)_2]Br_2$

18-13 **a)** Ammonium diamminetetraisothiocyanatochromate(III)
b) Pentacarbonyltechnetium(I) iodide
c) Potassium pentacyanomanganate(IV)
d) Tetraammineaquachlorocobalt(III) bromide

Tip. In part b, the "iso" in the name indicates that the donor atom in the thiocyanate ligand is the N, as in Cr—NCS. If the formula were $NH_4[Cr(NH_3)_2(SCN)_4]$, the linkage would be Cr—SCN and the name would be thiocyanato. Compare to **18-57**. Inorganic chemists are sometimes careless about the order of symbols in ligands.

18-15 If the ligands of a complex ion can be rapidly substituted for by other ligands, the complex ion is labile. A complex that is thermodynamically stable and labile will persist indefinitely as it rapidly exchanges ligands with the surroundings (usually the solvent). The occurrence of exchange can be verified by isotopic labeling experiments.

18-17 The dissociation of the tetraamminecopper(II) ion can be represented:

$$[Cu(NH_3)_4]^{2+}(aq) \rightarrow Cu^{2+}(aq) + 4\,NH_3(aq)$$

Taking data from Appendix D:

$$\Delta G^\circ = 1\underbrace{(65.49)}_{Cu^{2+}(aq)} + 4\underbrace{(-26.50)}_{NH_3(aq)} - 1\underbrace{(-111.07)}_{[Cu(NH_3)_4]^{2+}(aq)} = \boxed{+70.56 \text{ kJ}}$$

Because ΔG° exceeds zero, the reaction is not spontaneous as written. The tetraamminecopper(II) ion is thermodynamically $\boxed{\text{stable}}$ with respect to this reaction. Under acidic conditions, the dissociation reaction is:

$$[Cu(NH_3)_4]^{2+}(aq) + 4\,H_3O^+(aq) \rightarrow Cu^{2+}(aq) + 4\,NH_4^+(aq) + 4\,H_2O(l)$$

$$\Delta G^\circ = 1\underbrace{(65.49)}_{Cu^{2+}(aq)} + 4\underbrace{(-79.31)}_{NH_4^+(aq)} + 4\underbrace{(-237.18)}_{H_2O(l)} - 1\underbrace{(-111.07)}_{[Cu(NH_3)_4]^{2+}(aq)} - 4\underbrace{(-237.18)}_{H_3O^+(aq)}$$

$$= \boxed{-140.68 \text{ kJ}}$$

Hence $[Cu(NH_3)_4]^{2+}(aq)$ reacts $\boxed{\text{spontaneously}}$ with water at pH 0.

Tip. LeChatelier's principle applies. Strong acid (pH 0) favors substitution because the product NH_3 is removed by reaction with H_3O^+ ion.

18-19 The four substances all dissolve in water to make 0.010 M solutions. The more ions per mole of solute then the greater the conductivity of the solution at a given concentration of solute. Hence in order of increasing conductivity:

$$[Cu(NH_3)_2Cl_2] \quad < KNO_3 \quad < Na_2[PtCl_6] \quad < [Co(NH_3)_6]Cl_3$$

$$0 \text{ ions} \qquad\quad 2 \text{ ions} \qquad 3 \text{ ions} \qquad\quad 4 \text{ ions}$$

18-21 a) There are two isomers (*cis* and *trans*) of $[Pt(NH_3)_2BrCl]$. Neither is optically active:

b) The $[Co(CN)_3(H_2O)_2Cl]^{2-}$ ion has three possible isomers. None of the three isomers is optically active.

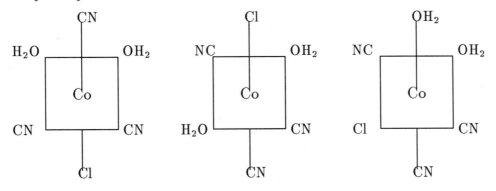

c) $[V(C_2O_4)_3]^{3-}$ ion is enantiomeric with two possible optical isomers. In the following the oxalato ligand (^-OOC—COO^-) is abbreviated as "O—·—O":

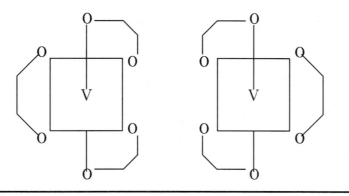

18-23 There are three isomeric $[Fe(en)_2Cl_2]^+$ complexes. They are the *trans*-dichloro-bis(ethylenediamine)iron(III) ion (at the left in the following) and the two mirror-image *cis*-dichlorobis(ethylenediamine)iron(III) ions. All involve octahedral coordination about the Fe atom. The NH_2—CH_2CH_2—NH_2 ligand is abbreviated en. This bidentate ligand ligates through its two —NH_2 groups. It can span an edge of the Fe octahedron but not opposite corners. In the following it is represented "N—·—·—N":

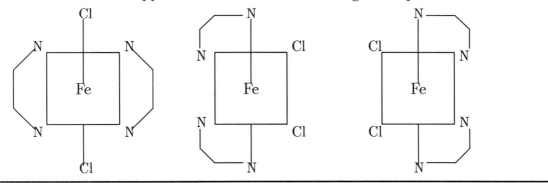

18-25 Strong-field octahedral complexes have a large splitting between the t_{2g} and e_g sets of orbitals; weak-field complexes have a small splitting between the t_{2g} and e_g orbitals. When the splitting energy Δ_o exceeds the pairing energy of the electrons, electrons pair up in the t_{2g} level and fill it completely before occupying the e_g level. Otherwise, electrons remain unpaired as long as possible.

a) The electron configuration of Mn^{2+} is $[Ar]3d^5$. It has $\boxed{5}$ unpaired electrons in a weak field and $\boxed{1}$ unpaired electron in a strong field:

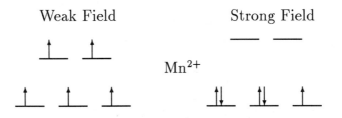

b) Zn^{2+} ion has the electron configuration $[Ar]3d^{10}$. It has $\boxed{zero}$ unpaired electrons in both weak and strong fields:

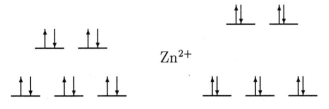

c) Cr^{3+} ion has configuration $[Ar]3d^3$. $\boxed{3}$ unpaired electrons in both weak and strong fields:

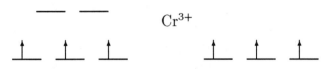

d) Mn^{3+} ion has configuration $[Ar]3d^4$. It has $\boxed{4}$ unpaired electrons in weak fields and $\boxed{2}$ in strong fields:

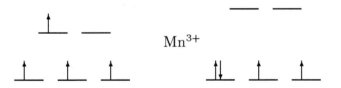

e) Fe^{2+} ion has configuration $[Ar]3d^6$. It has $\boxed{4}$ unpaired electrons in weak fields and $\boxed{zero}$ in strong fields:

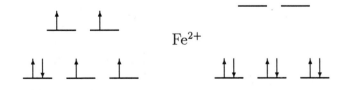

18-27 The ground-state Fe^{3+} ion has 5 d-electrons. In the strong octahedral field exerted by six CN^- ligands, the d-electron configuration is $(t_{2g})^5(e_g)^0$. All of the d-electrons are in the three t_{2g} orbitals; only $\boxed{one\ electron}$ can be unpaired. In the weak field exerted by six H_2O ligands, the d-electron configuration is $(t_{2g})^3(e_g)^2$. All $\boxed{five}$ d-electrons remain unpaired.

Each of the five d-electrons in t_{2g} levels stabilizes $[Fe(CN)_6]^{3-}$ by $2/5\Delta_o$. The total crystal-field stabilization energy equals $5 \times -2/5\Delta_o$ or $\boxed{-2\Delta_o}$. See text Figure 18-4.

The two e_g electrons in the weak-field $[Fe(H_2O)_6]^{3+}$ complex contribute $+3/5\Delta_o$ each to the CFSE, and the three t_{2g} electrons contribute $-2/5\Delta_0$ each for a total CFSE of $\boxed{zero}$.

18-29 In an octahedral field, d^3 systems are particularly stable because the t_{2g} set is half filled whether the ligands are strong-field or weak-field. In d^8 octahedral systems, the t_{2g} set is completely filled, and the e_g set is half filled whether the ligand is strong-field or weak-field. This also promotes stability. Octahedral d^5 systems would be expected to be particularly stable in complexes of weak-field ligands. The configuration of the d-electron system is then $(t_{2g})^3(e_g)^2$, in which the t_{2g} and e_g sets of orbitals are half filled. Octahedral d^6 systems would be expected to be particularly stable in complexes of strong-field ligands. The configuration of the d-electron system is then $(t_{2g})^6(e_g)^0$ in which the t_{2g} orbitals are a filled set.

18-31 The color perceived in a solution is the complement of the color of light absorbed. A colorless ion (such as $[Zn(H_2O)_6]^{2+}$) $\boxed{\text{does not absorb}}$ a significant amount of visible light.

18-33 A solution of $[Fe(CN)_6]^{3-}$ ion transmits red light. Assuming that it absorbs any visible light, the complex ion must absorb light in the green portion of the spectrum. See text Figure 16-35.[13] According to Figure 15-3,[14] green light has a wavelength of about $\boxed{5 \times 10^{-7} \text{ m}}$. Using the relationship $\Delta E = hc/\lambda$ with Planck's constant and the speed of light in the proper units gives an energy difference of 4×10^{-19} J. This is equivalent to 240 kJ mol^{-1}. Assume that absorption of the light excites a single electron from a low-lying t_{2g} orbital to an e_g level. Then Δ_o equals this energy difference, and is also about $\boxed{240 \text{ kJ mol}^{-1}}$.

18-35 The hexacyanoferrate(III) ion has a d^5 configuration on the central ion that is split by a strong octahedral field. As Table 18-4 shows, the crystal field stabilization energy for the resulting $(t_{2g})^5$ configuration is $-2\Delta_o$. The value of Δ_o is 240 kJ mol^{-1} (see preceding problem) so the CFSE equals $\boxed{480 \text{ kJ mol}^{-1}}$.

18-37 a) Use the color wheel in text Figure 16-35. The complement of blue-violet is $\boxed{\text{orange-yellow}}$.

---

[13] Text page 616.
[14] Text page 526.

b) The absorbed light is orange-yellow with a wavelength λ of maximum absorption near $\boxed{600 \text{ nm}}$. See text Figure 15-3. The experimentally observed transition turns out to be at 575 nm.

c) Cyanide ion is a strong-field ligand, and water is an weak-field ligand. Replacing coordinated water molecules with cyanide ions increases the crystal-field splitting. Increasing the splitting increases the frequency of the light that is absorbed, and causes a $\boxed{\text{decrease}}$ in the wavelength of maximum absorption.

18-39 a) In an aqueous solution of $Fe(NO_3)_3$, the Fe^{3+} ion is coordinated to six water molecules. The weak field of these ligands allows the high-spin electron configuration $(t_{2g})^3(e_g)^2$ on the central Fe^{3+} ion. In the case of $[Fe(CN)_6]^{3-}$, the strong field exerted by the CN^- ligands forces the electron configuration $(t_{2g})^5(e_g)^0$. Replacing the weak-field ligand water with the weak-field ligand fluoride should not change the $(t_{2g})^3(e_g)^2$ configuration. The absorption of light by the fluoride complex ion should therefore resemble that by the aqua complex. The solution of $K_3[FeF_6]$ should be $\boxed{\text{pale}}$.

b) The ground-state electron configuration of Hg^{2+} ion is $[Xe]4f^{14}5d^{10}$. A full subshell of 10 d-electrons means that electronic transitions in which electrons are redistributed among d-orbitals are not possible. Such transitions are mostly responsible for the colors of coordination complexes. A solution of $K_2[HgI_4]$ should therefore be $\boxed{\text{colorless}}$.

18-41 Common ligands donate a pair of electrons to the central metal atom of a complex except that hydrogen donates only a single electron. The number of valence electrons provided by a metal atom is given by its position in the periodic table.

The compound $[Cr(CO)_4]$ has a total of $\boxed{14}$ valence electrons at the metal, 6 from the Cr atom and 8 from the four CO ligands. The compound $[Os(CO)_5]$ has a total of $\boxed{18}$ valence electrons at the metal: 8 from the Os atom and 10 from the ligands. The compound $[H_2Fe(CO)_4]$ has a total of $\boxed{18}$ valence electrons at the metal: 2 from the H atoms, 8 from the Fe atom, and 8 from the four CO ligands. The compound $K_3[Fe(CN)_5CO]$ contains the $[Fe(CN)_5CO]^{3-}$ ion. The Fe(II) ion contributes 6 valence electrons, and the six ligands contribute 12 for a total of $\boxed{18}$ valence electrons about the Fe(II).

18-43 1.00 mol of vitamin B_{12s} contains 1.00 mol of Co. 1.00 mol of cobalt weighs 58.93 g, but this is only 4.43% of the mass of a mole of the vitamin. The mass of 1.00 mol of the vitamin is therefore larger than that of a mole of cobalt by the factor 100/4.43:

$$\left(\frac{100}{4.43}\right) \times 58.93 \text{ g mol}^{-1} = \boxed{1330 \text{ g mol}^{-1}}$$

18-45 Zinc has the lowest melting point and lowest boiling point of the fourth-period transition metals. This element has a d^{10} configuration. Bonding in metals, including elemental zinc, arises from the combination of valence atomic orbitals into delocalized bonding and antibonding molecular orbitals. A complete d subshell means that the antibonding orbitals deriving from d orbitals are completely occupied. This makes the bonding in the metal weaker, which implies a lower melting point and boiling point than other metals.

18-47 Electronegativity is a measure of the ability of an atom in a compound to draw electrons to itself. Mo(VI), the highest oxidation state of Mo has (formally at least) the most positive charge and should therefore have the highest electronegativity.

Tip. The electronegativity values referred to in the problem were computed by the method of Pauling and equal 2.16 for Mo(II), 2.19 for Mo(III), 2.24 for Mo(IV), 2.27 for MO(V), and 2.35 for Mo(VI).[15]

18-49 In $[Ru_2(NH_3)_6Br_3](ClO_4)_2$, all six ammonia ligands are neutral, each bromide ion has a -1 oxidation state and each perchlorate ion has a -1 charge. Because the sum of the oxidation states must equal zero (the overall charge of the complex), the oxidation state of the ruthenium is 2.5.

Tip. One of the ruthenium atoms may be in the $+2$ oxidation state, and the other in the $+3$ oxidation state.

18-51 Ligands are electron-pair donors. On the basis of simple electrostatics, it is much harder for a positively charged species to donate electron pairs than for a neutral or negatively charged species to do so.

18-53 Water coordinated to the central Cr is more tightly bonded and harder for a dehydrating agent to remove than water held in the solid as water of crystallization. Compound 1 loses two moles of H_2O per mole; it therefore has two waters of crystallization. The other water is in the coordination sphere: $[Cr(H_2O)_4Cl_2]Cl \cdot 2H_2O$. This ion has octahedral coordination about the central Cr atom. Compound 2 loses only one mole of H_2O per mole so it has only one water of hydration: $[Cr(H_2O)_5Cl]Cl_2 \cdot H_2O$. Compound 2 also has an octahedral structure. Compound 3 loses no water of hydration, so it must have all six water molecules coordinated: $[Cr(H_2O)_6]Cl_3$; compound 3 has an octahedral structure.

Solutions of silver nitrate precipitate AgCl only with chloride ions that are not in the coordination sphere. Therefore, compound 1 gives one mole of AgCl per mole

[15] A. L. Allred, *J. Inorg. Nucl. Chem.*, **1961**, *17*, 215.

of complex; compound 2 gives two moles of AgCl per mole of complex; compound 3 gives three moles of AgCl per mole of complex. For compound 1:

$$100.0 \text{ g} \times \left(\frac{1 \text{ mol}}{266.44 \text{ g}} \right) \times \left(\frac{1 \text{ mol AgCl}}{1 \text{ mol}} \right) \times \left(\frac{143.32 \text{ g AgCl}}{1 \text{ mol AgCl}} \right) = \boxed{53.79 \text{ g AgCl}}$$

Similar 100.0 g samples of compounds 2 and 3, which have the same molar mass, give respectively twice and three times as much AgCl: $\boxed{107.6 \text{ g}}$ and $\boxed{161.4 \text{ g}}$.

18-55 Tetrahedral structures do not display *cis-trans* isomerism because each corner of a tetrahedron is the same distance from the other three corners. In this case:

Tetrahedral structures exhibit mirror-image isomerism if they have four different atoms attached to the central atom. Because the complex [CoCl₂(en)] has two identical ligands, it does not exhibit mirror-image isomerism. The tetrahedral complex [CoClBr(en)] also lacks *cis-trans* isomers, as explained above. It has no optical isomers because the two ends of the en ligand are equivalent.

18-57 a) In the preferred formulation, the Pt in the [Pt(en)₂(SCN)₂]²⁺ cation is in the +4 oxidation state, while the Pt in the [PtBr₂(SCN)₂]²⁻ anion is in the +2 oxidation state. These complexes form a 1-to-1 ionic compound with the correct empirical formula. The Pt(IV) has 6 *d*-electrons, and is surrounded by 6 donors. If all 6 *d*-electrons are in the t_{2g} level (strong-field), the ion has no unpaired electrons and is diamagnetic. The Pt(II) has 8 *d*-electrons and is surrounded by 4 donors. The anion can therefore also have all of its electrons paired and be diamagnetic. By contrast, if the molecular formula were [PtBr(en)(SCN)₂], then all Pt atoms would be in the +3 oxidation state and have 7 *d*-electrons. This compound would be paramagnetic.

b) Bis(ethylenediamine)dithiocyanatoplatinum(IV) dibromodithiocyanatoplatinate(II).

18-59 The cyanide ligand has a −1 charge. This means that in [Mn(CN)₆]⁵⁻ the oxidation state of the Mn must be $\boxed{+1}$. A +1 added to 6(−1) gives the observed −5 charge on the complex ion. Similarly, the oxidation states of Mn in [Mn(CN)₆]⁴⁻ and [Mn(CN)₆]³⁻ are $\boxed{+2}$ and $\boxed{+3}$, respectively. The ground-state electron configurations of the manganese ions are:

$$\text{Mn}^{+1} \quad [\text{Ar}]3d^6 \qquad\qquad \text{Mn}^{+2} \quad [\text{Ar}]3d^5 \qquad\qquad \text{Mn}^{+3} \quad [\text{Ar}]3d^4$$

The low-spin (strong-field) d-orbital occupancy diagrams for each complex are:

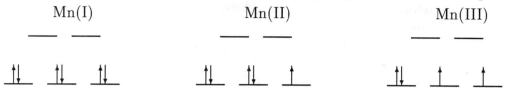

Mn(I)   Mn(II)   Mn(III)

18-61 In the list of the ionic radii, minima occur for the ions with 3 and 8 d-electrons. These are the V^{2+} and Ni^{2+} ions. Presumably the oxides are high-field, $\boxed{\text{low-spin}}$ complexes in which the 4th and 9th electrons are accommodated in a higher energy orbital, increasing the radius.

18-63 The central cobalt(II) in $[CoCl_4]^{2-}$ ion has a total of $\boxed{7}$ d-electrons. The Co atom (electron configuration $[Ar]3d^7 4s^2$) loses the two valence s-electrons in becoming the Co(II). The tetrahedral field of the 4 chloride ligands splits the d-orbitals of Co(II) into e_g orbitals (at *lower* energy) and t_{2g} orbitals (at *higher* energy). The ground-state electron configuration of the seven electrons is $\boxed{(e_g)^4 (t_{2g})^3}$. This configuration results regardless of the strength of the ligand field: only high-spin states are possible for a d^7 species in a tetrahedral field. The tetrahedral complex has crystal-field stabilization energy (CFSE) equal to $-6/5\Delta_t$; this stabilization favors the tetrahedral complex.

18-65 The cesium ions each have a $+1$ charge, and the fluoride ions each have a -1 charge. Thus, copper is in the $\boxed{+4}$ oxidation state. There are six monodentate ligands attached to the Cu(IV), so the most likely geometry about the central metal atom will be $\boxed{\text{octahedral}}$. The ground-state electron configuration of Cu(IV) is $[Ar]3d^7$. In a weak octahedral field, the d-elect configuration would become $\boxed{(t_{2g})^5 (e_g)^2}$. This high-spin configuration is far more likely than the low-spin $(t_{2g})^6 (e_g)^1$ configuration because the F^- ion is a weak-field ligand.

18-67 The cyclopentadienyl ion ($C_5H_5^-$) has a -1 charge and is a six-electron donor. The neutral C_6H_6 molecule is also a six-electron donor. The other ligands are all two-electron donors.

The Co atom in $[Co(C_5H_5)_2]^{2+}$ has 6 $3d$ electrons and shares 12 more from the two ligands. It sees a total of $\boxed{18}$ valence electrons.

The Fe atom in $[Fe(C_5H_5)(CO)_2Cl]$ also sees a total of $\boxed{18}$ valence electrons: the Fe(II) starts with 6, the $C_5H_5^-$ donates 6, the Cl^- and each CO donate 2.

The Mo in $[Mo(C_5H_5)_2Cl_2]$ sees a total of $\boxed{18}$ valence electrons: the Mo(IV) starts with 2 and each $C_5H_5^-$ contributes 6. The Cl^- ligands donate 2 electrons each.

The Mn in the complex $[Mn(C_5H_5)(C_6H_6)]$ sees a total of $\boxed{18}$ valence electrons: the Mn(I) ion has 6, the $C_5H_5^-$ contributes 6, and the C_6H_6 also contributes 6.

18-69 The rule of 18 holds that special stability is conferred on organometallic compounds having a central metal atom surrounded by 18 electrons. To relate this to ligand-field theory, study the MO diagram in Figure 18-22.[16] The nine molecular orbitals of lowest energy in the diagram are all either bonding or nonbonding. These MO's hold 18 electrons. Any complex that has electrons in just these MO's has zero antibonding electrons. In $Cr(CO)_6$, the ground-state valence electron configuration is:

$$\sigma_s^2 \sigma_p^6 \sigma_d^4 (d_{xy}^{nb})^2 (d_{xz}^{nb})^2 (d_{yz}^{nb})^2$$

with no electrons in antibonding orbitals. Additional electrons would go into antibonding orbitals and decrease the stability of the species.

18-71 Transition metal coordination compounds play a central role in biology. In hemoglobin, for example, iron is essential in binding O_2 for transport to cells and CO_2 for transport from cells. Cobalt is the central metal atom in vitamin B_{12}, a coordination compound that plays a vital role in metabolism.

Some of these biological complexes are colored because of the closely spaced d-levels that allow absorption of visible light. Ligand exchange lets these complexes act as catalysts for chemical reactions in living organisms, in analogy to the catalysis developed in synthetic chemistry. Finally, the existence of more than one oxidation state lets some complexes act as oxidizing or reducing agents in biochemical processes.

18-73 Manganese and chlorine appear in the same group ("Gruppe VII") in Mendeleev's early periodic table mainly because of the similarity in their compounds in the +7 oxidation state. Both elements form heptaoxides (Cl_2O_7 and Mn_2O_7) that are liquids are room conditions (indicating covalent bonding) and that react with water to give strong acids (perchloric acid $HClO_4$ and permanganic acid $HMnO_4$). Both of these acids are powerful oxidizing agents, and many perchlorate salts resemble permanganate salts closely.

Among the lower oxides of manganese, MnO is basic, Mn_2O_3 is weakly basic, MnO_2 is feebly acidic, and MnO_3 is more strongly acidic. A similar trend is found among the oxides of chlorine. Manganic acid (H_2MnO_4) resembles chloric acid $HClO_3$ in that it disproportionates to an acid in a higher oxidation state ($HMnO_4$) and an oxide in a lower oxidation state (MnO_2). However manganates (salts of manganic acid) resemble sulfates and chromates much more than chlorates.

Manganese and chlorine both have seven valence electrons, but these include d electrons in the case of Mn. The availability of additional low-lying d-orbitals in Mn leads

---

[16]Text page 683.

to metallic bonding in that element and enormous differences between it and Cl.

18-75 The problem requires the comparison of the two reactions:

$$Cd^{2+}(aq) + 4\,NH_2CH_3 \rightarrow [Cd(NH_2CH_3)_4]^{2+}(aq) \quad (1)$$
$$Cd^{2+}(aq) + 2\,en \rightarrow [Cd(en)_2]^{2+}(aq) \qquad (2)$$

where en stands for ethylenediamine $H_2NCH_2CH_2NH_2$. For the two:

$$-RT\ln K_1 = \Delta H_1^\circ - T\Delta S_1^\circ \qquad \text{and} \qquad -RT\ln K_2 = \Delta H_2^\circ - T\Delta S_2^\circ$$

Subtracting the second equation from the first gives:

$$RT\ln K_2 - RT\ln K_1 = \Delta H_1^\circ - \Delta H_2^\circ + T\Delta S_2^\circ - T\Delta S_1^\circ$$

According to the problem, the ΔH°'s of the two reactions are about equal. Therefore:

$$RT\ln K_2 - RT\ln K_1 \approx T\Delta S_2^\circ - T\Delta S_1^\circ$$

$$R\ln\left(\frac{K_2}{K_1}\right) \approx \Delta S_2^\circ - \Delta S_1^\circ$$

Judging from the number of particles present among the reactants, $\boxed{\Delta S_2^\circ}$ is less negative than ΔS_1°. This makes the right-hand side of the preceding equation positive. and means that K_2 exceeds K_1. The second reaction lies farther to the right at equilibrium because of its larger K: the chelate effect stabilizes the $\boxed{[Cd(en)_2]^{2+}}$ complex.

Chapter 19

Structure and Bonding in Solids

Solids are classified into two kinds, depending on the range of microscopic order within them: **crystalline solids** possess long-range order into their structures, whereas **amorphous solids** do not. This chapter considers both kinds of solids and also **liquid crystals**, states of matter intermediate between solid and liquid.

Crystal Symmetry and the Unit Cell

The most fundamental classification of crystals considers the number and kind of their **symmetry elements.** Elements of symmetry include:

- **Rotational axes.** If the rotation of a crystal about an axis "takes it into itself" (makes its new appearance indistinguishable from its previous appearance), then the axis is an axis of symmetry. A C_2 symmetry axis brings the crystal into itself at 180° intervals; a C_4 axis does so at 90° intervals; a C_n axis does so at $360°/n$ intervals.

- **Mirror planes.** Imagine a plane slicing through a crystal. If reflection across the plane, exchanging left for right and right for left, leaves the crystal unchanged in appearance, then the plane is a mirror plane.

- **Centers of inversion.** A crystal has a center of inversion if an imagined operation of "pulling itself through" its central point leaves it unchanged in appearance. If the central point is taken as the origin $(0, 0, 0)$, a crystal with a center of inversion has an inverted feature at $(-x, -y, -z)$ for every regular feature at (x, y, z).

Symmetry operations may exist in any object (see **19-1**). This includes molecules (see **19-3**).

The long-range order in a crystal means that sites with identical surroundings recur in a regular pattern. The array of all such points (called **lattice points**) in a given crystal is the **crystal lattice**. Like physical crystals themselves, crystal lattices, which are mathematical constructions, can be characterized by their symmetry. There are only seven different possible combinations of symmetry elements in crystals lattices, the seven **crystal systems**. Each crystal system has an essential or minimum symmetry and a descriptive name: **hexagonal, cubic, tetragonal, trigonal, orthorhombic, monoclinic,** and **triclinic**.

- Hexagonal, tetragonal, trigonal, and monoclinic crystals have the minimum and essential symmetries: 6-fold, 4-fold, 3-fold and 2-fold axis, respectively.

- Cubic, orthorhombic, and triclinic crystals have the respective minimum essential symmetries: a set of four independent 3-fold rotations, a set of three mutually perpendicular 2-fold rotations, none.

Crystal systems have more symmetry that just the essential minimum. Study text Table 19-1[1] closely and sketch the relationship of the symmetry elements in the different crystal systems. The text identifies *all* of the symmetry elements for the cubic system.[2] An obvious exercise is to make such a listing for the other six crystal systems.

Unit Cells in Crystals

A crystal can be thought of as built from blocks of a unit structure, its **unit cell**. Unit cells are imaginary "building bricks" stacked side by side in three-dimensional space. All unit cells have twelve edges, eight corners, and three pairs of parallel faces.[3] This shape allows them to stack together snugly top to bottom, side to side, and front to back. Of course, no little cells walls actually exist inside crystals: unit cells are constructions of the mind. As such, they are chosen as **the smallest units that retain all the symmetry of the crystal system**. A **primitive** unit cell contains one lattice point. Unit cells containing two or more lattice points are **non-primitive**.

Usually, the walls of a unit cell will slice through one or more atoms. When portions of atoms are excluded just outside one wall, the loss is compensated for inside the unit cell since a similar portion is automatically included just *inside* the opposite wall. The structure of an entire crystal is given by the shape and contents of its unit cell.

There are seven types of unit cell, one for each of the crystal systems. Each type has the same symmetry as one of the seven crystal systems. A set of six **cell constants**, three edge lengths (**a, b,** and **c**) and three angles (**α, β,** and **γ**) between

---

[1] Text page 696.
[2] In text Figure 19-3, text page 695.
[3] Unit cells are parallelepipeds.

the edges specify the size and shape of a unit cell. The angles are *not* necessarily 90°. Unit cells with higher symmetry have conditions relating the values of a, b and c and restricting the angles α, β, and γ to certain values, usually 90°. The symmetry-imposed conditions on the cell constants in the seven crystal systems should be memorized:

hexagonal	$a = b$	$\alpha = \beta = 90°$, $\gamma = 120°$
cubic	$a = b = c$	$\alpha = \beta = \gamma = 90°$
tetragonal	$a = b$	$\alpha = \beta = \gamma = 90°$
trigonal	$a = b = c$	$\alpha = \beta = \gamma \neq 90°$
orthorhombic	none	$\alpha = \beta = \gamma = 90°$
monoclinic	none	$\alpha = \gamma = 90°$
triclinic	none	none

The axial lengths of unit cells range upward from about 4 Å. Although cell axes may be long relative to the diameter of atoms (**19-11a**), they are always small compared to the physical crystal itself (**19-11b**). Most unit cells are buried deep in the crystalline interior. In the interior, every unit cell is just like the next only moved over ("translated") in one of three directions by the length of one of the three cell edges. Therefore, the surroundings of the eight corners of every unit cell are identical. These corners are the lattice points. They are at the intersections of a regularly-spaced three dimensional lattice, or grid, of lines running along the edges of the unit cells.

- **Every lattice point has identical surroundings.**

Often, it is important to know how many lattice points there are per unit cell. A unit cell has eight corners but shares each one with seven other cells. Its corners alone contribute $1/8 \times 8 = 1$ lattice point. See **19-19**. Some lattices have more than one lattice point per unit cell and still conform to the symmetry of one of the seven crystal systems. The additional lattice points appear:

- At the exact center of the unit cell. This is **body-centering.** A body-centered cell has two lattice points: one from its eight corners, and the other at the body center.

- At the centers of all six faces. This is **face-centering.** A face-centered unit cell has four lattice points. See **19-19**. Understanding the face-centered structure is essential in **19-59**.

- At the centers of just one of the three sets of parallel faces. This is **side-centering.** A side-centered unit cell has two lattice points.

Unit cell constants are sometimes called **lattice parameters** because they tell the spacing of the lattice points.

Scattering of X-rays by Crystals

X-rays have wavelengths λ on the order of the distances between layers of atoms in crystals. When a beam of x-rays strikes a layer of atoms in a crystal, it interacts with the electrons of the atoms and is **scattered** in all directions. The atoms lie in a regular array. Some of the scattered x-rays from one layer of atoms in a set of equally spaced layers may **constructively interfere** with scattered x-rays from the next layer, to be reinforced in phase by more scattered x-rays from the *next* layer, and so forth. The result is a scattered beam of x-rays striking out from the crystal at an angle to the incoming beam. Many scattered beam are possible because many different sets of layers of atoms exist in a crystal.[4]

Each scattered beam is called a **reflection.** The **Bragg law** is the criterion for scattering:

$$n\lambda = 2d \sin \theta$$

In this important equation, n is an integer, λ is the wavelength of the x-rays, d is the perpendicular distance between the layers of atoms, and θ is one-half of the angle between the incident beam of x-rays and the reflected beam. The angle θ is the **Bragg angle.**

In common applications, λ is known and 2θ (and thus θ) is measured. The order of the reflection is often given explicitly. See **19-5** and **19-7**.

Example: Suppose that x-rays of wavelength 1.54 Å are scattered from a set of evenly-spaced layers of atoms in a crystal with a spacing of 4.62 Å. Compute the Bragg angle of all the resultant reflections. **Solution:** From the Bragg equation with $\lambda = 1.54$ Å and $d = 4.62$ Å:

$$\theta = \sin^{-1} \left(\frac{n\lambda}{2d} \right) = \sin^{-1} (n/6.00)$$

Systematic substitution for the integer n gives:

$n = 1$	$\theta = 9.59°$	$n = 4$	$\theta = 41.81°$
$n = 2$	$\theta = 19.47$	$n = 5$	$\theta = 56.44$
$n = 3$	$\theta = 30.00$	$n = 6$	$\theta = 90.00$

The list ends at $n = 6$ because higher n's give $\sin \theta > 1.0$, but numbers exceeding 1.0 do not have an inverse sine. Every reflection in the list is found scattered at an angle 2θ relative to the incident beam. Compare to **19-9**. For $n = 6$ the angle 2θ is 180°. The $n = 6$ reflection scatters from this crystal right back into the incident beam.

---

[4]Just as a field of corn has more rows than the ones that the planting machine followed. Such "angle-rows" are often visible from a car passing on a road that borders at an angle to the field.

X-rays are not unique in being scattered by matter. Both neutrons and electrons have wave-like properties, as predicted by the DeBroglie equation (Chapter 12):

$$\lambda = \frac{h}{p} = \frac{h}{mv}$$

where p is the momentum of the particle. When beams of neutrons or electrons of the proper momentum impinge on crystals, they are scattered according to the Bragg equation. See **19-59**, which is like **19-5** except that the λ of the neutrons has to be calculated (using the DeBroglie equation) before applying the Bragg equation.

Crystal Structure

The volume V_c of a unit cell can be computed from its six cell constants:

$$\boldsymbol{V_c = abc\sqrt{1 - \cos^2 \alpha - \cos^2 \beta - \cos^2 \gamma + 2 \cos \alpha \, \cos \beta \, \cos \gamma}}$$

When one or more of the three angles equals 90°, the above formula is drastically simplified, since $\cos 90° = 0$. See **19-11a** and **19-47**.

The cubic crystal system is particularly important to understand because it underlies the solid-state structure of many simple substances. Three lattices have cubic symmetry: the **simple cubic (s.c.)**, the **face-centered cubic (f.c.c.)**, and the **body-centered cubic (b.c.c.)**. Some geometrical facts about the cube are useful:

- One parameter, the length of an edge a, fully characterizes a cube. This is the same as saying that one parameter, the length of a unit cell edge, fully characterizes a cubic lattice.

- The volume of a cube is equal to its edge cubed (a^3). Similarly the volume of a cubic unit cell is a^3.

- A line drawn diagonally across one face of a cube is the **face diagonal f.** The length of f is $\sqrt{2}a$.

- A line connecting the opposite corners of a cube is the **body diagonal b.** The length of b is $\sqrt{3}a$.

The density of a crystal is equal to its mass divided by its volume. In terms of the unit cell, the density of a crystal is the mass of the contents of the cell divided by the volume of the cell:

$$\text{density} = \rho = \frac{m \text{ (contents of cell)}}{V_c}$$

This leads to the useful expression:

$$\rho = \frac{n_c \mathcal{M}}{N_0 V_c}$$

in which n_c, which is always an integer, equals the number of formula units in the unit cell, $\mathcal{M}$ equals the molar mass of the formula units, N_0 is Avogadro's number, and V_c is the volume of the unit cell. If $\mathcal{M}$ is in g mol^{-1} and V_c is in cm^3, then the density comes out in g cm^{-3}. This equation is used in **19-5, 19-7, 19-9**, and **19-53**. One error in its use is failure to compute V_c properly. If cell constants are given in Å, then cell volume will come out in Å^3. Remember:

$$1 \text{ cm}^3 = 10^{24} \text{ Å}^3 \qquad \text{because} \qquad 1 \text{ cm} = 10^8 \text{ Å}$$

Another error in computing densities is to omit N_0. To understand the inclusion of N_0 in the formula, reason that $\mathcal{M}$ has units of g mol^{-1} and something has to be done to remove the mol^{-1} part because densities do not have moles in their units. Problem **19-15** gives additional perspective on why N_0 appears.

Even in simple substances, all of the atoms in a unit cell do *not* automatically sit at lattice points. See **19-17** and **19-19**. Atoms can be anywhere in the cell. Their locations are given by a set of three **fractional coordinates.** One of the corners of the unit cell serves as the origin and the three cell edges which intersect at the origin serve as axes. Any point inside the cell has coordinates (x, y, z), where x tells in units of a (that edge's length) how far away from the origin the point is along a line parallel to the first edge, and y and z do the same for the second and third edges. For example, an atom in a cubic cell might have the fractional coordinates (0.40, 0.15, 0.35). If the three cell edges were all equal to 4.0 Å, this atom would lie 1.60 Å out from the origin in a direction parallel to the a cell edge, 0.60 Å out parallel to b, and 1.40 Å out parallel to c.

The eight corners of the unit cell have the fractional coordinates:

$$(0,0,0) \quad (1,0,0) \quad (0,1,0) \quad (0,0,1) \quad (1,1,1) \quad (0,1,1) \quad (1,0,1) \quad (1,1,0)$$

These eight corners are exactly equivalent. In fractional coordinates, adding or subtracting 1 to any member of the triple does not create new or unique positions but instead refers to the *same* relative location in the *adjoining* unit cell.

Example: What are the fractional coordinates of atoms located at the face-centers of a cubic unit cell? **Solution:** The face centers all have one fractional coordinate equal to 0 and the other two equal to $\frac{1}{2}$. The correct answer:

$$(\tfrac{1}{2}, \tfrac{1}{2}, 0) \quad (\tfrac{1}{2}, 0, \tfrac{1}{2}) \quad (0, \tfrac{1}{2}, \tfrac{1}{2})$$

This answer, which is correct, often puzzles students who reason that a cube must have six face centers because it has six faces. The centers of the other three faces are:

$$(\tfrac{1}{2}, \tfrac{1}{2}, 1) \quad (\tfrac{1}{2}, 1, \tfrac{1}{2}) \quad (1, \tfrac{1}{2}, \tfrac{1}{2})$$

These coordinates do *not* specify unique, new points in the lattice. They instead specify points that are equivalent to the first three but translated by the length of a cell edge along z, y and x respectively. A cubic unit cell "owns" only three face centers because each of its six faces is shared 50:50 with a neighboring unit cell. See **19-25** for a related case. Fractional coordinates are used to advantage in **19-19** and **19-29**.

Atomic Packing in Crystals

In elemental metals there is only one kind of atom in the crystal. This makes for simplicity in the crystal structure. Many metals crystallize in cubic lattices in which the atoms are in contact with their neighbors and are located at each point of the lattice:

- In a simple cubic (s.c.) lattice, the distance between neighboring lattice points is a. The radii of identical metal atoms sitting at these points and in contact is $1/2a$.

- In a face-centered cubic (f.c.c.) lattice, nearest-neighbor identical atoms situated at the lattice points touch along a face diagonal. The distance between their centers is $\sqrt{2}a/2$, and the radius of the metal atoms is $\sqrt{2}a/4$. This statement applies only with all the atoms in the structure are identical. See **19-59** for a contrasting case.

- In a body-centered cubic (b.c.c.) lattice, nearest-neighbor identical atoms touch along a body diagonal. The distance between their centers is $\sqrt{3}a/2$; the radius of the metal atoms is accordingly $\sqrt{3}a/4$.

If atoms are hard spheres, then gaps must appear between them when they are packed together, like oranges in a crate, to form a crystal. A **close-packed** structure is a structure in which identical spheres (atoms) occupy the greatest possible fraction of the total space. The two most efficient methods for packing identical spheres are **cubic close-packing (c.c.p.)** and **hexagonal close-packing (h.c.p.)**. In both, the atoms occupy 74.0 percent of the volume and the other 26.0 percent is empty. In the close-packing of spherical, equal atoms, each has 12 nearest neighbors: six in the same plane, three in the plane below, three in the plane above.

- Cubic close-packing corresponds to setting down layers of identical spheres in the sequence *abcabcabc*..., where the letters refer to the offset in position of the higher layers relative to the first. The recurrence of *a* in the fourth position means that atoms in the fourth layer are positioned exactly above atoms in the first layer. In c.c.p., the lattice defined by points at centers of the atoms is cubic and face-centered.

- Hexagonal close-packing corresponds to the sequence of layers *ababab*.... The recurrence of *a* in the third position means that atoms in the third layer are positioned exactly above atoms in the first layer. In h.c.p., the lattice is hexagonal and primitive.

The gaps between close-packed atoms are **interstitial sites.** Octahedral interstitial sites are surrounded by six atoms. Tetrahedral interstitial sites are smaller and surrounded by only four atoms. See **19-49**. A face-centered cubic lattice has four octahedral sites and eight tetrahedral sites per unit cell. In other kinds of lattices there are other kinds of interstitial sites. An examination problem might refer to the geometry of a **cubic** site (surrounded by eight atoms) to test understanding.

Cohesion in Solids

The strengths of the forces that hold crystalline substances together vary enormously. Crystals are classified according to these forces. The categories are: **molecular, ionic, metallic,** and **covalent.**

Molecular crystals

Molecular crystals consist of molecules (two or more atoms held together by strong, directional covalent bonds) maintained in relative position in a lattice by weaker van der Waals (VdW) forces.

- VdW forces are attractions among complete molecules and neutral atoms and include dipole-dipole forces, dipole-induced dipole forces, and induced dipole-induced dipole forces. They are "non-bonded" interactions.

- VdW forces are not strong enough to bring two neighboring atoms closer together than the sum of their **van der Waals radii.**

- VdW radii are known for every element.

In molecular crystals, the molecules pack together in such a way as to minimize empty space but to avoid overlay of VdW radii of any atoms. The molecules are only slightly affected internally by interaction with their neighbors. Molecular crystals typically have low melting points, crystallize in lattices of lower symmetry, conduct electricity poorly, are easily deformed (are soft), and are diamagnetic.

Ionic crystals

Compounds among elements that differ considerably in electronegativity often form ionic crystals. They contain alternating cations and anions arranged on a lattice. Such crystals behave to a good approximation as if constructed of hard, charged spheres held in contact by electrostatic attractions. Such attractions are strong and nondirectional.

To maximize the attractions (between ions of unlike charge) and to minimize repulsions (between ions of like charge), the lattice of an ionic crystal has every positive ion surrounded by as many negative ions as possible and vice versa. Since fixed bond angles play no part in electrostatic attractions, ionic crystals favor densely packed structures. Ionic crystals are typically hard, high-melting, and brittle.

Each positive ion occupies an interstitial site in a lattice of negative ions and each negative ion occupies an interstitial site in a lattice of positive ions. The shape (octahedral, tetrahedral or other) and proportion of occupied sites depend on the relative size and charge of the ions.

In ionic crystals, the negative ions are nearly always larger than the positive ions. The cation/anion **radius ratio,** a number less than 1, determines the structure which charged hard spheres adopt. As the radius ratio gets larger, more anions can surround the cation and just touch it without bumping into each other. See **19-25** and **19-49**.

Learn these important structures for ions of equal charge:

- The **zinc blende** structure. When the radius ratio is less than 0.414, each ion is surround by four near neighbors of opposite charge.

- The **rock salt** structure. When the radius ratio is between 0.414 and 0.732, each ion is surrounded by six near neighbors of opposite charge.

- The **cesium chloride** structure. When the radius ratio is greater than 0.732. each ion surrounded by eight near neighbors of opposite charge. See **19-29**.

The structures are diagrammed in text Figures 19-19, 19-17, and 19-18.

Avoid the trap of thinking that every ion in an ionic crystal sits at a lattice point. In NaCl, the Na^+ ions sit at the lattice points at the corners and the face centers. The Cl^- ions sit at the body center and at the midpoints of the edges of a face-center cubic unit cell. These are not lattice points.

Metallic Crystals

In metallic crystals, valence electrons are delocalized in huge (crystal-sized) molecular orbitals. This results in cohesive forces called metallic bonds that are nondirectional. They involve the sharing of electrons among all the atoms in the crystal. The electrons occupy molecular orbitals arising from the overlap of atomic orbitals of similar energy on all of the atoms. For example, 1 mol of sodium metal has 6.02×10^{23} $3s$ orbitals. These orbitals overlap to make a **band** of molecular orbitals. The molecular orbitals comprise a band because they are so numerous and their energies are so nearly the same. The band has room for $2 \times 6.02 \times 10^{23}$ electrons and therefore is only half-full, under the Pauli principle. Metallic crystals can be regarded as a set of positive ions awash in an "electron sea" of electrons delocalized in this way.

The excellent electrical and thermal conductivities of metals are explained by the great mobility of electrons at the top of the sea of occupied levels. The uppermost occupied or half-occupied molecular orbital in the band is the **Fermi level.** Electrons at or near the Fermi level require only slight amounts of energy to excite them to occupy levels lying above.

The identical atoms in elemental metallic crystals often pack in the most efficient possible manner, which is cubic close packing or hexagonal close packing.

Covalent Crystals

In covalent (or **network**) crystals, the atoms are linked by strong, directional covalent bonds in a giant network. The crystals are hard and rigid, but conduct electricity poorly. Diamond, discussed in **19-45** and diagrammed in Figure 19-23[5] is a typical case. Its crystal structure consists of a face-centered cubic array of C atoms with additional C atoms occupying every other tetrahedral interstitial site. In this arrangement, each C atom is surrounded by 4 nearest neighbors situated perfectly for σ overlap between sp^3 hybrid orbitals.

Structures of the Elements

The elements provide examples of three of the four types of cohesion in solids:

- Atoms of nonmetallic elements at the top of the periodic table tend to use up all their bonding capacity in intramolecular bonds. They consequently form mainly molecular crystals. A trend away from the formation of multiple bonds and toward formation of chains and rings of atoms appears moving down the table.

---

[5]Text page 711.

- The structures of the metallic elements are those of metallic crystals, usually with one atom per lattice point.

- Group VII elements form covalently bound diatomic molecules that interact by VDW forces to form molecular crystals.

- In Group VI, oxygen forms diatomic molecules, and solid oxygen is a molecular crystal. Sulfur favors chains and rings so that its room-temperature solid is a crystal consisting of S_8 molecules. Selenium and tellurium form long spiral chains in the solid state; their crystals are covalent along the chains but molecular between them.

- In Group V, nitrogen solidifies as a molecular crystal (with N_2 units); phosphorus solidifies as white phosphorus (molecular crystal with P_4 units) or red or black phosphorus (network solids); solid As and Sb have network structures.

- Group IV elements are often on the borderline between metallic and covalent in the solid state.

Lattice Energies of Crystals

The **lattice energy** of a crystal is the energy decrease that results when the constituent atoms, molecules or ions condense from a gas to form the crystal at 0 K.

Lattice energies of molecular crystals can be estimated from the Lennard-Jones parameters of the molecules. They are much smaller than the lattice energies of other types of crystals. For example, compare the answer to **19-31a** or **19-33a** with the answer to **19-53a**.

In ionic crystals, Coulomb forces are the predominant type of interaction between the constituent particles, which are ions. Taking the Coulomb contribution to the lattice energy of as the whole lattice energy in an ionic crystal consisting of $+1$ and -1 ions gives:

$$\text{lattice energy} = \frac{e^2 N_0}{4\pi\epsilon_0 R_0} M$$

where e is 1.6021×10^{-19} C, where ϵ_0 is 8.854×10^{-12} C^2 J^{-1}m^{-1}, where R_0 is the *minimum* distance between the two ions, and where M is a special constant, the **Madelung constant**. The Madelung constant accounts for the fact that the electrostatic energy of a crystal is the sum of all possible pair-wise attractions and repulsions in the crystal. The value of the Madelung constant depends on the type of crystal lattice. Values have been computed, using the methods of infinite series, for all lattices. Table 19-3[6] gives some Madelung constants.

---

[6]Text page 714.

Calculations using the above formula give *theoretical* estimates of lattice energies. Experimental lattice energies are obtained using the **Born-Haber cycle**, an application of the first law of thermodynamics. It is impossible directly to decompose an ionic crystal into its constituent ions in their gaseous states. Fortunately, ΔE for this process, which defines the lattice energy, can be determined as the sum of several measurable ΔE's for steps that combine to give this reaction.

The solution to **19-33** shows the use of the Born-Haber cycle. Form an idea of the size of the different energies involved in the Born-Haber cycle. This will allow a common-sense approach to checking lattice energy calculations. Note that the lattice energy is an energy change, not an enthalpy change. The Born-Haber cycle must use ΔE values for all of its steps to compute a lattice energy properly. In practice, the difference between ΔH and ΔE is usually small. See **19-33**.

Many problems require some creative use of a Born-Haber cycle. **Example:** Estimate the enthalpy of formation of the hypothetical compound CaCl(s). Why does $CaCl_2(s)$ always form when Ca(s) and $Cl_2(g)$ react? **Solution:** The enthalpy of formation of CaCl(s) can only be estimated because the compound does not exist, and direct measurements are not possible. The key insight is to use the lattice energy of KCl to approximate the lattice energy of the hypothetical compound. First, either look up or calculate this lattice energy. Then apply the Born-Haber cycle to the formation of the hypothetical CaCl(s). The answer is a ΔH_f° of about -150 kJ mol^{-1}. The compound is favored thermodynamically relative to the elements! Why does it not form? The answer is in the second part of the question. The real compound, $CaCl_2(s)$, has a ΔH_f° of -795.8 kJ mol^{-1} (text Appendix D). It is even more favored.

Defects and Amorphous Solids

Real crystals are imperfect. They have finite extent; their atoms move around (vibrate), they include impurities; some atoms are always displaced from expected sites or missing entirely. Their impurities either take the places of "true" occupants at sites in the crystal or lodge in interstices between them.

- **Schottky defects** are atom or ion vacancies.

- **Frenkel defects** are displacements of atoms or ions from expected sites to neighboring interstitial sites.

- **Color centers** (F-centers) are the replacement of an anion in an ionic crystal by an electron. See text Figure 19-29. They lend striking colors to some specimens of ionic compounds.

Both Frenkel and Schottky defects in crystals are mobile. They jump from one lattice sit to another. Diffusion in crystalline solids is due mainly to the presence and mobility of defects.

Many solid compounds, called **nonstoichiometric** compounds, violate the law of definite proportions by having atoms in ratios that differ substantially from whole-number ratios. The existence of defects explains why. If some proportion of the positive ions in an ionic lattice is oxidized (meaning that some ions lose additional electrons), then there can be Schottky defects at some of the positive ion sites. The result is a ratio of positive ion to negative ion that is *less* than stoichiometric. The fraction of positive ions that is oxidized is related to the number of defects and the degree of nonstoichiometry in the compound. See **19-37** and **19-57**.

Amorphous Solids and Glasses

When a solid has so many defects that crystalline order is destroyed, then it is an **amorphous** solid or a **glass**. A strong tendency to form glasses is associated with the presence in a solid of long or irregularly shaped molecules than can easily become tangled. In principle however, any solid that can be liquefied can be prepared in an amorphous state, usually by cooling it very rapidly. Glasses soften and flow when heated instead of melting sharply at a defined temperature. This plasticity makes glasses very easy to fabricate into desired shapes.

Liquid Crystals

Liquid crystals are thermodynamically stable phases intermediate between liquid and solid. Heating an ordinary crystal of certain substances can disrupt the lattice well before orientational order among the molecules is broken down. The transition from solid to true liquid then can pass through several intermediate liquid-crystal states. Liquid crystals flow like liquids, but there is a degree of statistical order in the distribution of their molecules.

Liquid crystals usually form in substances having molecules with elongated rod-like shapes. They are solid-like in showing orientational ordering of these molecules, but liquid-like in the random distribution of the centers of the molecules. The exact nature of residual solid-state order differs among the types of liquid crystals. In **nematic phases**, molecules show a preferred orientation to a particular directions, but their centers are distributed at random. In **smectic phases**, orientational order is present and is joined by a tendency for molecules to lie in layers. In **cholesteric phases**, molecules show nematic ordering in layers, but the preferred direction of orientation changes from layer to layer.

Detailed Solutions to Odd-Numbered Problems

19-1 a) In an isosceles triangle one of the sides is non-equivalent to the other two, and no 3-fold rotation axis exists

b A 3-fold axis passes through the center of an equilateral triangle.

c) A 3-fold axis passes through the center of each of the four triangular faces in a tetrahedron and out through the opposite vertex.

d) A 3-fold axis passes along each of the four long diagonals of a cube, that is, from each vertex to the most distant opposite vertex.

19-3 The CCl_2F_2 molecule has two mirror planes. The first is defined by the two Cl atoms and the central C atom, and the second is defined by the two F atoms and the central C atom. The intersection of the two mirror planes coincides with a single 2-fold axis of rotation. This axis passes through the central C atom and bisects the angles defined Cl—C—Cl and F—C—F.

19-5 The Bragg law $n\lambda = 2d\sin\theta$ becomes in this case:

$$2(1.660\text{ Å}) = 2d\sin\left(\frac{54.70°}{2}\right) \quad \text{from which} \quad d = \frac{1.660\text{ Å}}{\sin 27.35°} = \boxed{3.613\text{ Å}}$$

19-7 The Bragg law $n\lambda = 2d\sin\theta$ becomes:

$$4(1.936\text{ Å}) = 2(4.950\text{ Å})\sin\theta$$

where $n = 4$ comes from the specification of fourth-order diffraction and 4.950 Å is the interplanar spacing. Solving gives θ equal to 51.46° and 2θ equal to $\boxed{102.9°}$.

Tip. The angle 128.54° $(180 - 51.46°)$ also fulfills the equation. This gives $2\theta = 257.1°$, which is equivalent to $-102.9°$. This corresponds to "reflection" from the other side of the layers of atoms.

19-9 Solve the Bragg law for 2θ and substitute the values given for this case of diffraction by crystalline LiCl:

$$2\theta = 2\sin^{-1}\left(\frac{n\lambda}{2d}\right) = 2\sin^{-1}\left(\frac{n\,2.167\text{ Å}}{2(2.570\text{ Å})}\right) = 2\sin^{-1}(0.4216n)$$

Inserting integers for n gives 2θ equal to $\boxed{49.87°}$ for $n = 1$ and 2θ equal to $\boxed{115.0°}$ for $n = 2$. Higher values of n lead to arguments of $\sin^{-1}$ that exceed 1.00. The inverse sine function is not defined in such cases. Consequently 2θ has only the four possible values.

19-11 a) The cell angles are all 90°, because the crystal is tetragonal. Then:

$$V_{cell} = abc = (223.5)^2(113.6) = \boxed{5.675 \times 10^6 \ \text{Å}^3}$$

b) The volume of the box-shaped crystal is likewise the product of the lengths of the three edges. It equals 3 mm³—small, but easily visible with the unaided eye. To compare the two volumes divide one by the other and use a suitable unit factor to make sure the units cancel out:

$$\frac{V_{crystal}}{V_{cell}} = \frac{3 \ \text{mm}^3}{5.675 \times 10^6 \ \text{Å}^3} \times \left(\frac{1 \ \text{Å}^3}{10^{-21} \ \text{mm}^3}\right) = \boxed{5 \times 10^{14}}$$

This ratio equals the number of units cells in the crystal.

19-13 Compute the mass of the contents of the unit cell and divide it by the volume of the cell to obtain the density of the cell. Since the substance consists of many copies of the unit cell side-by-side, this result is the density of the whole crystal. The mass of the contents of the unit cell equals twice the mass of a single formula unit of $Pb_4In_3B_{17}S_{18}$:

$$m_{contents} = 2 \times 1934.235 = 3868.47 \ \text{u}$$

The volume of the cell is:

$$V_c = abc\sqrt{1 - \cos^2\alpha - \cos^2\beta - \cos^2\gamma + 2\cos\alpha\cos\beta\cos\gamma}$$
$$= abc\sqrt{1\cos^2\beta} = abc\sin\beta$$
$$= (21.021 \ \text{Å})(4.014 \ \text{Å})(18.989 \ \text{Å})(0.9924) = 1582.46 \ \text{Å}^3$$

The density is then:

$$\rho = \frac{m}{V} = \frac{3868.47 \ \text{u}}{1582.46 \ \text{Å}^3} \times \left(\frac{10^{24} \ \text{Å}^3}{1 \ \text{cm}^3}\right) \times \left(\frac{1 \ \text{g}}{6.022 \times 10^{23} \ \text{u}}\right) = \boxed{4.059 \ \text{g cm}^{-3}}$$

The two unit-factors take the answer from an unfamiliar unit of density to a familiar one.

19-15 a) The volume of the cubical unit cell in elemental silicon is just the edge of the cell cubed. It equals $(5.431 \ \text{Å})^3$, which is 160.19 Å³. There are 10^8 Å in a centimeter and consequently 10^{24} Å³ in a cubic centimeter. Hence:

$$\frac{160.19 \ \text{Å}^3}{1 \ \text{unit cell}} \times \left(\frac{1 \ \text{cm}^3}{10^{24} \ \text{Å}^3}\right) = \boxed{1.602 \times 10^{-22} \frac{\text{cm}^3}{\text{unit cell}}}$$

b) The mass of the contents of the unit cell of crystalline silicon equals the volume of the unit cell multiplied by its density:

$$\frac{1.602 \times 10^{-22} \text{ cm}^3}{\text{unit cell}} \times \left(\frac{2.328 \text{ g Si}}{1 \text{ cm}^3}\right) = \boxed{\frac{3.729 \times 10^{-22} \text{ g Si}}{\text{unit cell}}}$$

c) The unit cell contains eight atoms of silicon for a total mass of 3.729×10^{-22} g. Consequently, a single atom has a mass of $\boxed{4.662 \times 10^{-23} \text{ g}}$.

d) One mole of silicon contains Avogadro's number of atoms of silicon. The molar mass of silicon is 28.0855 g mol^{-1}. Divide this molar mass by the single-atom mass of silicon to obtain Avogadro's number:

$$\frac{28.0855 \text{ g mol}^{-1}}{4.662 \times 10^{-23} \text{ g}} = \boxed{6.025 \times 10^{23} \text{ mol}^{-1}} = N_0$$

This is only about 0.05% larger than the accepted value.

19-17 The volume V_c of the unit cell in sodium sulfate equals the product of its three cell edges. a, b, and c. This follows because the term under the radical sign in text equation 19-2[7] equals 1. Hence V_c is 708.47 Å^3, which is 7.0847×10^{-22} cm^3. The volume of a mole of unit cells of sodium sulfate is Avogadro's number times the volume of one cell:

$$(7.0847 \times 10^{-22} \text{ cm}^3) \times (6.022 \times 10^{23} \text{ mol}^{-1}) = 426.6 \text{ cm}^3 \text{ mol}^{-1}$$

The density of a unit cell equals the density of the substance itself since a crystal consists of many unit cells stacked side by side. Multiplying the volume of a mole of unit cells by the density of the substance gives the mass of a mole of unit cells:

$$\left(\frac{426.6 \text{ cm}^3}{1 \text{ mol}}\right) \times \left(\frac{2.663 \text{ g}}{1 \text{ cm}^3}\right) = 1136.1 \text{ g mol}^{-1}$$

The molar mass corresponding to the formula Na_2SO_4 equals 142.04 g mol^{-1}. This is far less than 1136.1 g mol^{-1}. The unit cell must hold several formula units. Because 142.04 is almost exactly 1/8th of 1136.1, it follows that each unit cell contains $\boxed{8}$ Na_2SO_4 formula units.

Tip. The unit cell contains 56 atoms. Obviously, all of these atoms cannot reside at the corners of the unit cell. They are in fact distributed throughout the volume of the cell.

---

[7]Text page 701.

19-19 In this crystalline compound, rhenium atoms lie at the eight corners of the unit cell, and oxygen atoms lie at the 12 edges. Start by figuring out the number of atoms of Re and O per cell: each cell has 1 rhenium atom ($8 \times 1/8$) and 3 oxygen atoms ($12 \times 1/4$). The 1/8 and 1/4 appear because every corner of a unit cell is shared among a total of eight cells and every edge is shared among four cells. The ratio of these numbers gives the empirical formula because the compound is composed of many repeats of the unit cell. The chemical formula therefore is $\boxed{ReO_3}$.

Tip. Another way to explore the locations of the atoms is with fractional coordinates. The equivalent Re atoms have these coordinates:

$$(0,0,0) \quad (1,0,0) \quad (0,1,0) \quad (0,0,1) \quad (1,1,1) \quad (0,1,1) \quad (1,0,1) \quad (1,1,0)$$

These are the corners of the cube in text Figure 19-13.[8] The O atoms at the centers of the cell edges have these fractional coordinates:

$$\left(\tfrac{1}{2},0,0\right) \quad \left(0,\tfrac{1}{2},0\right) \quad \left(\tfrac{1}{2},1,0\right) \quad \left(1,\tfrac{1}{2},0\right)$$
$$\left(0,0,\tfrac{1}{2}\right) \quad \left(1,0,\tfrac{1}{2}\right) \quad \left(0,1,\tfrac{1}{2}\right) \quad \left(1,1,\tfrac{1}{2}\right)$$
$$\left(\tfrac{1}{2},0,1\right) \quad \left(0,\tfrac{1}{2},1\right) \quad \left(\tfrac{1}{2},1,1\right) \quad \left(1,\tfrac{1}{2},1\right)$$

Only three of these locations are distinct. The nine containing a 1 are translations ("one cell over") of these three.

19-21 a) A body-centered cubic structure means two Fe atoms per unit cell, one in the center of the cell and one at each of the eight corners of the cell (each corner atom is shared by seven neighboring cells). The two atoms have a total mass of 111.694 u and touch along the body diagonal of the cell, but not along the edges. Compute the volume of the unit cell by multiplying its mass by its density:

$$V_c = 111.694 \text{ u Fe} \times \left(\frac{1 \text{ g Fe}}{6.0221 \times 10^{23} \text{ u Fe}}\right) \times \left(\frac{1 \text{ cm}^3}{7.86 \text{ g Fe}}\right) = 2.36 \times 10^{-23} \text{ cm}^3$$

The edge a of the cubic unit cell is the cube root of the volume. It is 2.87×10^{-8} cm, which is 2.87 Å. The nearest-neighbor distance is one-half the body diagonal b of the unit cell. The body diagonal is related to the edge as follows:

$$b = \sqrt{3}e = \sqrt{3}(2.87 \text{ Å}) = 4.97 \text{ Å}$$

Hence nearest neighbors are $\boxed{2.48 \text{ Å}}$ apart.

b) The lattice parameter equals $\boxed{2.87 \text{ Å}}$, the cubic cell's edge. See above.

---

[8]Text page 702.

c) The atomic radius of Fe equals one quarter of the body diagonal of the unit cell because Fe atoms are "in contact" along this diagonal. It is therefore $\boxed{1.24 \text{ Å}}$.[9]

19-23 a) A body-centered cubic lattice has two lattice points per unit cell. In metallic sodium, one Na atom is associated with each lattice point to give $\boxed{\text{two}}$ Na atoms per cell.

b) Let r_{Na} equal the radius of the Na atom. In the crystal, Na atoms touch along the body diagonal b of the cubic cell, which has atoms at its corners and center. This means $4r_{Na} = b$. But b is $\sqrt{3}$ times the edge of the cell. Hence $4r_{Na} = \sqrt{3}e$. Cubing the equation gives:

$$64r_{Na}^3 = 3\sqrt{3}e^3$$

The volume of the cell V_c is e^3. The volume of a single Na atom is $4/3\pi(r_{Na})^3$, and, obviously, two Na atoms have twice this volume:

$$V_c = e^3 \quad \text{and} \quad V_{2Na} = 2 \times \left(\frac{4\pi(r_{Na})^3}{3}\right)$$

Solving these equations for e^3 and $(r_{Na})^3$ respectively and substituting into the equation that precedes them gives:

$$64\left(\frac{3V_{2Na}}{8\pi}\right) = 3\sqrt{3}V_c \quad \text{from which} \quad \left(\frac{V_{2Na}}{V_c}\right) = \frac{(3\sqrt{3})(8\pi)}{3(64)} = \boxed{0.680}$$

Tip. This is not the most efficient possible packing of spheres. If the spheres lie at the lattice points of a face-centered cubic lattice, then the most efficient packing (cubic close packing) is attained.

19-25 The atoms making up the simple cubic array are the host atoms. These atoms touch along the edges of the cubic unit cell. A guest interstitial atom sits hole at the center of the host unit cell. The body diagonal b of the host unit cell runs between two diagonally opposite host atoms and passes along the diameter of the guest atom. If the guest is as large as it can be without pushing the host atoms out of contact, then:

$$e = 2r_{host} \qquad b = 2r_{host} + 2r_{guest}$$

But the body diagonal of a cube is longer than the edge by a factor of $\sqrt{3}$: Hence:

$$2r_{host} + 2r_{guest} = \sqrt{3}\,(2r_{host}) \quad \text{so that:} \quad \frac{r_{guest}}{r_{host}} = \sqrt{3} - 1 = \boxed{0.732}$$

---

[9]This answer equals the value tabulated in Appendix F.

19-27 Use electronegativity differences and position in the periodic table.
a) $BaCl_2$–ionic   **b)** SiC–covalent   **c)** CO–molecular   **d)** Co–metallic.

19-29 In the simple cubic CsCl lattice, the positive ion has $\boxed{8}$ Cl^- ions as nearest neighbors. The second nearest neighbors are a set of $\boxed{6}$ Cs^+ ions, and the third nearest neighbors are a set of $\boxed{12}$ yet more distant Cs^+ ions.

To obtain this answer, imagine a Cs^+ ion at the center of a home cell that has Cl^- ions on its eight corners. These are the 8 nearest-neighbor Cl^- ions. The home cell has 6 faces and 12 edges. The 6 second-nearest neighbors are the Cs^+ ions at the centers of the 6 face-adjoining unit cells. The 12 third-nearest neighbors are the Cs^+ ions at the centers of the 12 edge-adjoining cells.

Tip. Avoid getting bogged down using messy sketches to count neighbors and decide which neighbors are nearer. A better way uses fractional coordinates. Define an origin $(0,0,0)$ at the Cs^+ ion at the center of the home cell. In fractional coordinates, the edges of the unit cell are used as units of length. This puts a Cl^- ion at coordinates $(\frac{1}{2}, \frac{1}{2}, \frac{1}{2})$. The cubic symmetry means that the x, y and z coordinates are equivalent and that the plus and minus directions on each coordinate are equivalent, too. Permuting equivalent fractional coordinates and changing the signs of the fractional coordinates therefore generate equivalent locations. For $(\frac{1}{2}, \frac{1}{2}, \frac{1}{2})$ the following eight sets of coordinates result from these operations:

$$\left(+\tfrac{1}{2},+\tfrac{1}{2},+\tfrac{1}{2}\right) \quad \left(-\tfrac{1}{2},+\tfrac{1}{2},+\tfrac{1}{2}\right) \quad \left(+\tfrac{1}{2},-\tfrac{1}{2},+\tfrac{1}{2}\right) \quad \left(+\tfrac{1}{2},+\tfrac{1}{2},-\tfrac{1}{2}\right)$$
$$\left(-\tfrac{1}{2},-\tfrac{1}{2},-\tfrac{1}{2}\right) \quad \left(+\tfrac{1}{2},-\tfrac{1}{2},-\tfrac{1}{2}\right) \quad \left(-\tfrac{1}{2},+\tfrac{1}{2},-\tfrac{1}{2}\right) \quad \left(-\tfrac{1}{2},-\tfrac{1}{2},+\tfrac{1}{2}\right)$$

These are the coordinates of the 8 nearest neighbors of the Cs^+ ion. The 6 second-nearest neighbors have these coordinates:

$$(0,0,1) \quad (0,1,0) \quad (1,0,0) \quad (0,0,-1) \quad (0,-1,0) \quad (-1,0,0)$$

And the 12 third-nearest neighbors have these coordinates:

$$(0,1,1) \quad (0,1,-1) \quad (0,-1,1) \quad (0,-1,-1) \quad (1,0,1) \quad (1,0,-1)$$
$$(-1,0,1) \quad (-1,0,-1) \quad (1,1,0) \quad (1,-1,0) \quad (-1,1,0) \quad (-1,-1,0)$$

In this description of CsCl, triples that contain any half-integers locate Cl^- ions, and triples that contain all whole numbers locate Cs^+ ions. Also, fractional coordinates make it easier to compute interatomic distances in a lattice. For a cubic crystal the distance d between two points (x_1, y_1, z_1) and (x_2, y_2, z_2) specified in fractional coordinates equals:

$$d = a\sqrt{(x_2 - x_1)^2 + (y_2 - y_1)^2 + (z_2 - z_1)^2}$$

where a equals the edge of the cubic cell.

19-31 The problem requests computation of the lattice energy of $RbCl(s)$. The electrostatic (Coulomb) lattice energy of a crystal is given by:

$$\text{Lattice energy} = \frac{N_0 e^2}{4\pi \epsilon_0 R_0} M$$

where M is the Madelung constant and R_0 is the distance between neighboring ions. Substitute the correct values for $RbCl(s)$. The Madelung constant quoted in the problem is for the rock-salt structure, which is the structure of $RbCl(s)$; R_0 equals the sum of the radii of the Rb^+ and Cl^- ions, which is 3.29 Å, or 3.29×10^{-10} m:

$$\text{Lattice energy} = \frac{(6.022 \times 10^{23} \text{ mol}^{-1})(1.602 \times 10^{-19} \text{ C})^2}{4\pi(8.854 \times 10^{-12} \text{ C}^2 \text{ J}^{-1}\text{m}^{-1})(3.29 \times 10^{-10} \text{ m})} 1.7476$$

$$= 738 \times 10^3 \text{ J mol}^{-1}$$

Reducing this energy by 10% to account for non-Coulomb effects gives $\boxed{664 \text{ kJ mol}^{-1}}$ for the dissociation energy of $RbCl(s)$. The experimental value is 680 kJ mol^{-1}.

19-33 a) The lattice energy of $LiF(s)$ equals the energy change of the reaction:

$$LiF(s) \rightarrow Li^+(g) + F^-(g) \qquad \text{lattice energy} = \Delta E$$

Direct experimental measurement of this ΔE is not possible. The Born-Haber cycle is a series of lesser steps taking place at 25°C that add up to the above change. The ΔE of each step *can* be measured. The steps are:

1. Decomposition of $LiF(s)$ to give $Li(s)$ and $F_2(g)$;

2. Vaporization of $Li(s)$ to $Li(g)$ and dissociation of $F_2(g)$ to $F(g)$;

3. Transfer of electrons from $Li(g)$ to $F(g)$ to give $F^-(g)$ ions and $Li^+(g)$ ions.

Apply the first law of thermodynamics to this cycle:

$$\Delta E_{\text{cycle}} = \Delta E_1 + \Delta E_2 + \Delta E_3$$

Evaluate ΔE_3 first. It equals the first ionization energy of $Li(g)$ minus the electron affinity of $F(g)$:

$$\Delta E_3 = 520 - 328 = 192 \text{ kJ mol}^{-1}$$

ΔE_2 is the energy change accompanying the vaporization of $Li(g)$ to $Li(s)$ plus the energy change accompanying the dissociation of $F_2(g)$ into atoms. The $\Delta H°$'s for

544

Chapter 19

these processes (text Appendix D) equal 159.37 kJ mol^{-1} and 78.99 kJ mol^{-1} respectively. The tabulated $\Delta H°$'s are not the same as the desired ΔE's, but are related as follows:

$$\Delta E = \Delta H° - RT\Delta n_{\mathrm{g}} = \Delta H° - (2.48 \text{ kJ mol}^{-1})\Delta n_{\mathrm{g}} \quad \text{(at 298 K)}$$

For $\mathrm{Li}(s) \to \mathrm{Li}(g)$, Δn_{g} is +1. It follows that:

$$\Delta E_{\mathrm{vap}} = 159.37 - (2.48)(1) = (159.37 - 2.48) \text{ kJ mol}^{-1}$$

For the dissociation $1/2\,\mathrm{F}_2(g) \to \mathrm{F}(g)$, Δn_{g} is +1/2. Therefore:

$$\Delta E_{\mathrm{dissoc}} = 78.99 - (2.48)(1/2) = (78.99 - 1.24) \text{ kJ mol}^{-1}$$

Adding the two parts of step 2:

$$\Delta E_2 = (159.37 - 2.48) + (78.99 - 1.24) = 234.64 \text{ kJ mol}^{-1}$$

The first step in the cycle was the decomposition of one mole of LiF(s) to form one mole of Li(s) and one-half mole of F$_2$(g):

$$\mathrm{LiF}(s) \to \mathrm{Li}(s) + \frac{1}{2}\mathrm{F}_2(g) \qquad \Delta H° = +615.97 \text{ kJ mol}^{-1}$$

This reaction is the exact "un-formation" of LiF. Its $\Delta H°$ is therefore the negative of the standard enthalpy of formation of LiF(s) from Appendix D. Again, compute the energy change from the energy change and the change in volume:

$$\Delta E_3 = \Delta H_3° - RT\Delta n_{\mathrm{g}} = +615.97 - (2.48)(1/2) = 614.73 \text{ kJ mol}^{-1}$$

Add up the ΔE's for the three steps in the cycle:

$$\Delta E_{\mathrm{cycle}} = 192 + 234.64 + 614.73 = \boxed{1041 \text{ kJ mol}^{-1}}$$

b) The computation of the Coulomb energy for LiF(s) follows the pattern of **19-31**:

$$\text{Lattice energy} = \frac{N_0 e^2}{4\pi\epsilon_0 R_0}M$$

where M is the Madelung constant and R_0 is the distance between neighboring positive and negative ions. For lithium fluoride R_0 is 2.014 Å, or 2.014×10^{-10} m.[10] Obtain the M for rock salt from Table 19-5. Then:

$$\text{Lattice energy} = \frac{(6.022 \times 10^{23} \text{ mol}^{-1})(1.602 \times 10^{-19} \text{ C})^2}{4\pi(8.854 \times 10^{-12} \text{ C}^2\text{ J}^{-1}\text{m}^{-1})(2.014 \times 10^{-10} \text{ m})}1.7476$$

$$= \boxed{1205 \times 10^3 \text{ J mol}^{-1}}$$

---

[10]The ionic radii of Li$^+$ and F$^-$ ions are 0.68 and 1.33 (text Appendix F). The ratio of these radii is 0.51, which confirms that LiF(s) adopts the rock-salt structure.

This (theoretical) Coulomb energy is about 15 percent larger than the (experimental) Born-Haber lattice energy. The discrepancy arises because the Coulomb calculation ignores non-Coulomb interactions.

Tip. In the first page, if the $\Delta n_g RT$ corrections in step 2 and 3 are omitted, the answer comes out 1046 kJ mol^{-1}. The difference is less than 1%. In view of the experimental uncertainty of many ΔE values, taking ΔE to equal ΔH is often defensible. Also, the cycle is performed at 298.15 K, but the lattice energy is defined at 0 K.

19-35 The presence of Frenkel defects will $\boxed{\text{not change}}$ the density of a crystal by a significant amount, because the vacancies at lattice sites are compensated for by interstitial atoms. Frenkel defects in large numbers might cause a small bulging of the crystal and a consequential decrease in its density.

19-37 a) A sample of 100 g of this iron(II) oxide contains 76.55 g of Fe and 23.45 g of O. This corresponds to 1.3707 mol of Fe and 1.4657 mol of O. Dividing one by the other gives the formulas $FeO_{1.0693}$ or $\boxed{Fe_{0.9352}O}$. It is improper to round off to the stoichiometric formula FeO. The experimental analysis is precise to four significant figures, and the chemical formula should have the same precision.

b) Let a equal the fraction of sites occupied by Fe^{3+} ions and b equal the fraction of sites occupied by Fe^{2+} ions. The Fe^{3+} ions that occur in Fe^{2+} sites compensate with their extra charge for missing Fe^{2+} ions elsewhere and make the compound as a whole electrically neutral. The average positive charge per site must be 2. Also the sum of a and b is 0.9352, as shown by the empirical formula. In equation form this means:

$$3a + 2b = 2 \quad \text{and} \quad a + b = 0.9352$$

Solution of these simultaneous equations gives $a = 0.1296$. This equals the fraction of sites occupied by Fe^{3+} ions. The fraction of the iron in the +3 state is the fraction of sites having +3 iron divided by the fraction having iron of either kind: $0.1296/0.9352 = \boxed{0.1386}$.

19-39 The entropy of the $\boxed{\text{isotropic phase}}$ (the liquid phase) exceeds the entropy in the smectic liquid crystal phase of a substance. Compare the degrees of order apparent in text Figure 19-32a and 19-32b. The enthalpy of the $\boxed{\text{isotropic phase}}$ exceeds that of the smectic phase because heating the smectic phase converts it to the isotropic phase.

19-41 The Bragg law $n\lambda = 2d \sin\theta$ becomes, in this case of first-order diffraction of water waves:

$$1(3.00 \text{ m}) = 2(5.00 \text{ m}) \sin\theta$$

Solving for θ gives 17.46° so 2θ is $\boxed{35°}$.

19-43 a) The unit-cell volume V_c of NaCl is the cell edge cubed or $\boxed{179.43 \text{ Å}^3}$.

b) The volume V_p of the primitive unit cell of NaCl equals one-fourth of the volume of the conventional unit cell or $\boxed{44.856 \text{ Å}^3}$.

c) For relatively long wavelength (2.2896 Å) x-rays:

$$N_{\text{beams}} = \frac{4}{3}\pi \left(\frac{2}{\lambda}\right)^3 V_p = \frac{4}{3}\pi \left(\frac{2}{2.2896 \text{ Å}}\right)^3 (44.856 \text{ Å}^3) = \boxed{125}$$

d) For short wavelength (0.7093 Å) x-rays, there are far more diffracted beams.

$$N_{\text{beams}} = \frac{4}{3}\pi \left(\frac{2}{\lambda}\right)^3 V_p = \frac{4}{3}\pi \left(\frac{2}{0.7093 \text{ Å}}\right)^3 (44.856 \text{ Å}^3) = \boxed{4212}$$

19-45 In diamond the C—C bond distance equals the distance between any two nearest-neighbor atoms. Reviewing the list of coordinates given in the problem shows one such pair of atoms is the C at $(0,0,0)$ and the C at $(\frac{1}{4},\frac{1}{4},\frac{1}{4})$. This is also clear in Figure 19-23 in the text. Other pairs of carbons are equally close but none is closer. These carbons are separated by one-fourth of the body diagonal of the unit cell. The body diagonal is $\sqrt{3}$ times the edge of the cell or $3.57\sqrt{3}$ Å. The bond distance is 1/4 of this or $\boxed{1.55 \text{ Å}}$.

19-47 a) The cell is monoclinic so two of the three cell angles automatically equal 90°.

b) The volume of the cell is:

$$V_c = abc\sqrt{1 - \cos^2\alpha - \cos^2\beta - \cos^2\gamma + 2\cos\alpha\cos\beta\cos\gamma}$$

Because both α and γ are 90°, this becomes (with the use of the trigonometric identity $\sin^2\beta + \cos^2\beta = 1$):

$$V_c = abc\sqrt{1 - \cos^2\beta} = abc\sin\beta = (11.04)(10.98)(10.92)\sin 96.73° = 1314.6 \text{ Å}^3$$

The volume equals 1.3146×10^{-21} cm³. The density is then computed as follows:

$$\rho = \frac{n_{\text{S atoms}}\mathcal{M}_S}{N_0 V_c} = \frac{48(32.066 \text{ g mol}^{-1})}{(6.022 \times 10^{23})(1.3146 \times 10^{-21} \text{ cm}^3)} = \boxed{1.944 \text{ g cm}^{-3}}$$

19-49 Any tetrahedral interstitial site can be viewed as occupying the center of a cube that has every other one of its eight corners occupied by spherical atoms of radius r_1. Let the edge of such a cube have length 1. Then the face diagonal f has length $\sqrt{2}$ and the body diagonal b has length $\sqrt{3}$. The four atoms at the alternate corners surround the center and touch each other along the face diagonals of the cube. Therefore:

$$2r_1 = \sqrt{2}$$

Let r_2 be the radius of a spherical atom placed at the interstitial site, the center of the cube. The largest such atom will just touch all four atoms at the corners. The body diagonal in that case equals the sum of the diameters of the two atoms:

$$2r_1 + 2r_2 = b = \sqrt{3}$$

Dividing the second equation by the first gives:

$$\frac{(r_1 + r_2)}{r_1} = \frac{\sqrt{3}}{\sqrt{2}} \quad \text{hence} \quad 1 + \frac{r_2}{r_1} = 1.225$$

Since r_2/r_1 is 0.225, the largest possible value for r_2 is $\boxed{0.225r_1}$.

19-51 Non-metals like the chalcogens and the halogens form molecular crystals in the their solid states; metals like the transition metals and the alkali metals form metallic crystals. Elements at the center of the periodic table (in Group IV) such as carbon and silicon form covalent crystals.

19-53 a) According to the equations developed in the text[11], the potential energy and intermolecular distance in a face-centered-cubic molecular crystal depend on the Lennard-Jones parameters for the molecules comprising the crystal as follows:

$$R_0 \approx 1.09\sigma \quad \text{and} \quad V_{\text{tot}} \approx -8.61\epsilon N_0$$

where σ is the first Lennard-Jones parameter, R_0 is the equilibrium spacing (at 0 K), V_{tot} is the total potential energy of the lattice, and ϵ is the other Lennard-Jones parameter. For N_2, σ is 3.70 Å.[12] Therefore, R_0 is about $\boxed{4.03 \text{ Å}}$.
For N_2, ϵN_0 is 0.790 kJ mol^{-1}. The potential energy of the lattice is accordingly -6.80 kJ mol^{-1}. This makes $\boxed{+6.80 \text{ kJ mol}^{-1}}$ a reasonable estimate of the lattice energy.

b) The density of a crystal is related to the volume of its unit cell by:

$$\rho = \frac{n_c \mathcal{M}}{N_0 V_c}$$

---

[11]Text page 713.
[12]See text Table 4-4, text page 127.

For $N_2(s)$, ρ is 1.026 g cm^{-3}. The crystal has four N_2 molecules per unit cell and each molecule has a mass of 28.014 g mol^{-1}. Solve the preceding for V_c and substitute:

$$V_c = \frac{n_c \mathcal{M}}{N_0 \rho} = \frac{4(28.014 \text{ g mol}^{-1})}{6.022 \times 10^{23} \text{ mol}^{-1}(1.026 \text{ g cm}^{-3})} = 181.36 \times 10^{-24} \text{ cm}^3$$

The edge of the cubic cell is the cube root of the volume of the cell. It equals $\boxed{5.660 \times 10^{-8} \text{ cm}}$ (5.660 Å). In a face-centered cubic cell, a nitrogen molecule lies at the center of every face of the cell and at every corner. The face diagonal is $5.660\sqrt{2}$ Å or 8.005 Å long. One-half of this is the distance from an N_2 at a face center to an N_2 at a face corner. This, the intermolecular distance, is $\boxed{4.002 \text{ Å}}$. This result is only about 0.7 percent less than the distance computed using the Lennard-Jones parameter. The good agreement tends to confirm the analysis in text section 19-4.

19-55 Sodium chloride is an ionic solid. If there are Schottky defects, a fraction of the Na^+ sites is vacant. To maintain electrical neutrality an equal fraction of the Cl^- sites must be vacant. The density of defect-free NaCl is 2.165 g cm^{-3}. Introducing 0.0015 mole fraction of Schottky defects reduces the chemical amount of NaCl per cm^3 to 0.9985 of what had been. Therefore, the mass of NaCl per cm^3 is 0.9985 of what it had been, or $\boxed{2.162 \text{ g cm}^{-3}}$.
Frenkel defects involve displacement from a regular lattice site to an interstitial site. No mass is removed from the crystal, so the density stays at $\boxed{2.165 \text{ g cm}^{-3}}$ as long as the volume of the crystal is not changed.

19-57 a) The binary compound is 28.31 percent O and 71.69 percent Ti by mass. 100 g of the compound contains 1.4973 mol of Ti and 1.7694 mol of O. The formula is $Ti_{0.8462}O$ where the $\boxed{0.8462}$ equals the ratio of 1.4973 to 1.7694.

b) Only 0.8462 of the stoichiometric quantity of Ti is present; 0.1538 of the total Ti^{2+} sites then must be vacant. Let a equal the fraction of Ti^{2+} sites with a Ti^{3+} occupying them, and b the fraction of sites with a Ti^{2+}. Clearly: $a+b = 0.8462$. The net positive charge per oxygen must be $+2$. Each Ti^{3+} contributes $+3$ and each Ti^{2+} contributes $+2$. Electrical neutrality requires $3a + 2b = 2$. Solution of the simultaneous equations gives b equal to $\boxed{0.5386}$ and a equal to $\boxed{0.3026}$. About 30% of the Ti^{2+} sites contain a Ti^{3+} ion, about 15% are vacant, and about 54% contain a Ti^{2+}.

19-59 a) Use the deBroglie relation to obtain the wavelength of the neutrons:

$$\lambda = \frac{h}{mv} = \frac{6.626 \times 10^{-34} \text{ J s}}{(1.6750 \times 10^{-27} \text{ kg})(2.639 \times 10^3 \text{ m s}^{-1})} = \boxed{1.499 \times 10^{-10} \text{ m}}$$

b) The edge length of the unit cell is the interplanar spacing of the planes doing the scattering. Compute it by solving the Bragg law for d and substituting:

$$a = d = \frac{n\lambda}{2\sin\theta} = \frac{2(1.499 \times 10^{-10}\text{ m})}{2\sin(36.26°/2)} = 4.817 \times 10^{-10}\text{ m} = \boxed{4.817\text{ Å}}$$

c) Sodium hydride adopts the rock-salt structure. Therefore, Na^+ and H^- ions touch along the edges of the unit cell. The Na^+ ions occupy the corners and center of each face of the cell, forming a pattern like the pattern of five dots on the face of die. Four H^- ions also lie in each face; they occupy the edges between Na^+ ions. The distance from the center of an Na^+ ion to the center of the adjoining H^- is therefore one-half of the edge of the unit cell. This equals $\boxed{2.409\text{ Å}}$.

d) As established in slightly different words in the preceding, the edge e of the unit cell is the sum of two Na^+ radii and two H^- radii:

$$2r_{H^-} + 2r_{Na^+} = 4.817\text{ Å}$$

Substitution of 0.98 Å for r_{Na^+} gives r_{H^-} equal to $\boxed{1.43\text{ Å}}$.[13]

19-61 Applying the rule of thumb assigns each water molecule a volume of 18 Å^3. The mass of a water molecule is 18.02 u so the density of water would be 18.02 u/18 Å^3. Convert this density to g cm^{-3}:

$$\frac{18.02\text{ u}}{18\text{ Å}^3} \times \left(\frac{1\text{ g}}{6.022 \times 10^{23}\text{ u}}\right) \times \left(\frac{10^{24}\text{ Å}^3}{1\text{ cm}^3}\right) = \boxed{1.7\text{ g cm}^{-3}}$$

The density based on the rule of thumb is much higher than the actual density of solid water (0.90 g cm^{-3}). The rule of thumb fails in this case because the hydrogen bonding in ice maintains an abnormally open structure.

---

[13]Text Appendix F tabulates a radius of 1.46 Å for H^- ion.

Chapter 20

Chemical Processes for the Recovery of Pure Substances

The Chemical Industry

In a **process-based** approach to chemistry, the interaction of the complete range of chemical principles is considered with respect to the operation of practical industrial processes that transform starting materials to desired products. Chemical processes required careful development from their original laboratory scale through a series of **pilot-plant** stages to a final production status. This also often involves a changeover from a **batch** to a **continuous** process, if the latter is feasible and more economical. See **18-1.** Starting materials should be easily accessible and transportable. Most starting materials have low free energies, a fact that gives particular important to high-free-energy materials like hydrocarbons from petroleum. See **18-15**. The best sources of the most important elements should be learned (see **18-3**).

Desired reactions for which ΔG is positive can be made to proceed by linking them to other reactions for which ΔG is negative.

An ideal chemical process gives the desired product in 100% yield, at a controllable, convenient rate, and in high purity. The proper disposal of waste is an integral part of the design of chemical processes. Chemical processes often involve many steps and intermediate chemical compounds. Most large-volume chemicals are little used by consumers, but they are essential in making the products that consumers do use.

The products of a practical process must be obtained at a sufficient yield, at a sufficient rate and either pure enough to use immediately or readily purified. Thermodynamics imposes fundamental limitations on the yield of chemical reactions. Pressure and temperature must be selected accordingly. But practical processes may not proceed arbitrarily slowly. Compromises sometimes must be made between the best conditions based on thermodynamics and conditions under which the process will go at an acceptable rate. Much effort goes into designing catalysts to speed up thermodynamically favored processes enough to make them practical. Finally, if side-reactions contaminate the desired product with by-products that are dangerous to handle or hard to remove or both, the conditions of the process may again have to be modified to eliminate the troublesome by-products.

Hydrogen

The production and uses of hydrogen exemplify process-based chemistry. The element is abundant in water, but the free energy of water is too low for economically direct conversion to gaseous hydrogen (with oxygen as by-product. Instead, the major source is the **reforming reaction** between water and methane (or another hydrocarbon) from natural gas (see **18-7**). This reaction produces **synthesis gas,** which is a mixture of hydrogen and carbon monoxide. Synthesis gas can be used as a mixture for a variety of syntheses. Further reaction of the carbon monoxide in synthesis gas with water in the exothermic **shift reaction:**

$$CO(g) + H_2O(g) \rightarrow CO_2(g) + H_2(g)$$

gives more hydrogen and some carbon dioxide, which finds uses of its own. See **18-8**.

Many of the uses of hydrogen are **captive**—that is, occurring directly after production. Hydrogen is used to make ammonia, methanol, or (in direct-reduction processes), to convert iron ore to iron. Another use is in the **hydrogenation** of **unsaturated** hydrocarbons, in which carbon-carbon double bonds are reduced to single bonds, and the compound is thereby saturated. Hydrogenation is applied to petroleum products as well as to edible oils (see **18-11**).

Extractive Metallurgy

Extractive metallurgy concerns the winning of useful metals from ores. Ores contain metals in positive oxidation states in compounds (sulfides, oxides) of low free energy. Reductions of ore to metal plus oxygen (or sulfur) are nonspontaneous (example: Fe_2O_3 to give Fe and O_2). Therefore, compensating spontaneous reactions (such as O_2 plus C to give CO_2) are run to furnish the needed free energy. **Pyrometallurgy** refers to recovery of metals at high temperature.

The extraction of copper from its sulfide ores requires ore benefaction and high-temperature reduction. The iron that is usually present in copper ores must also be removed. **Froth-flotation** of copper ores uses water, oil, and a detergent: oil-wetted ore particles rise in air-churned froth to separate from water-wetted rock. Roasting of enriched copper ores converts iron that is present to iron oxide, but leaves copper as copper(I) sulfide. Further reaction of Cu_2S with oxygen gives metallic Cu.

The **hydrometallurgical** method of separating copper from iron in ore uses aqueous redox chemistry and thereby avoids production of the pollutant $SO_2(g)$.

Iron and Steel

The winning of iron from ores is also a high-temperature reduction. Iron ores are reduced in a blast furnace by $CO(g)$ from coke ($C(s)$). Alumina (Al_2O_3) and silica (SiO_2) impurities are taken up to form slag upon addition of limestone ($CaCO_3$) as flux. A blast furnace has a charge of iron ore, coke and limestone. The charge meets a blast of pre-heated air from nozzles (**tuyeres**). Ensuing reactions produce impure iron (pig iron) and slag.

Steel (an Fe-C alloy with lower carbon content than pig iron) is made from pig iron by the **basic oxygen process**. In this process, blasts of pure oxygen are blown into molten pig iron through a **lance** and tuyere to remove excess C and impurities and make steel. Most carbon in the pig iron is removed as CO. Impurity silicon and phosphorus give SiO_2 and P_2O_5, which are acidic and react with the flux of CaO and MgO. The products of these acid-base reactions transfer to a **slag** phase and are easily removed. Dissolved gases (such as oxygen, nitrogen and hydrogen) are removed from the melt by **vacuum degassing** in which argon gas is blown through the molten steel as large pumps reduce the pressure above. The basic oxygen process is fast, flexible, and capable of close, computerized control. It is the dominant steel-making process today.

Electrometallurgy

The production of aluminum requires the reduction of aluminum ores that are of low free-energy. The aluminum oxide-containing ore bauxite is purified in the **Bayer process**, in which the Al_2O_3 dissolves in base, but impurities do not. The pure oxide is re-precipitated after filtration of the solution. Aluminum is then obtained in an electrolytic cell in the **Hall-Héroult process**. This involves dissolution of aluminum oxide in molten cryolite (Na_3AlF_6) at 950° and reduction at a steel cathode. See **20-43**.

Magnesium metal is produced by the electrochemical reduction of molten $MgCl_2$. Metals are refined as well as produced by electrometallurgical methods. Metals can be plated out electrochemically as coatings on other metals.

Detailed Solutions to Odd-Numbered Problems

20-1 A batch process is carried out by mixing reactants, letting them react, and then removing products in a series of steps. In a continuous process reactants are continuously added to the reaction vessel and the products are continuously removed. A continuous process requires more than one opening in the reaction vessel; a batch process can operate in a vessel with only one opening.

20-3 a) Sulfur occurs in useful amounts in $\boxed{\text{salt domes}}$ under the surface of the earth. The deposits formed when calcium sulfate reacted with natural gas and carbon dioxide. Interestingly, the reaction is catalyzed by enzymes that are secreted by bacteria:

$$CaSO_4(s) + CH_4(g) + CO_2(g) \rightarrow CaCO_3(s) + S(s) + 6\,H_2O(l)$$

b) A useful source of carbon is bituminous $\boxed{\text{coal}}$. Coal is formed from plant matter by bacterial reduction, heat, and pressure.

c) $\boxed{\text{Phosphate rock}}$ is the most useful source of phosphorus, It is $Ca_5(PO_4)_3F$ in a nearly pure condition.

d) A good source of calcium is $\boxed{\text{limestone}}$ or marble, both of which are principally composed of calcium carbonate.

20-5 A good source of small amounts of pure hydrogen is:

$$\boxed{Zn(s) + 2\,H_3O^+(aq) \rightarrow H_2(g) + Zn^{2+}(aq) + 2\,H_2O(l)}$$

Zinc is too expensive to use to make large amounts of hydrogen

---

20-7 The reaction in this problem is the endothermic reforming of methane:

$$CH_4(g) + H_2O(g) \rightarrow CO(g) + 3\,H_2(g)$$

This reaction becomes spontaneous at a high enough temperature. To compute that T, use the relationship $\Delta G = \Delta H° - T\Delta S°$. If $\Delta G°$ is equal to zero, then:

$$T = \frac{\Delta H°}{\Delta S°}$$

Text Appendix D supplies the standard entropies and standard enthalpies of formation of the reactants and products: Combine the data to get the $\Delta H°$ and $\Delta S°$ of the reaction:

$$\Delta H° = 3\underbrace{(0.00)}_{H_2(g)} + 1\underbrace{(-110.52)}_{CO(g)} - 1\underbrace{(-74.81)}_{CH_4(g)} - 1\underbrace{(-241.82)}_{H_2O(g)} = 206.11 \text{ kJ}$$

$$\Delta S° = 3\underbrace{(130.57)}_{H_2(g)} + 1\underbrace{(197.56)}_{CO(g)} - 1\underbrace{(186.15)}_{CH_4(g)} - 1\underbrace{(188.72)}_{H_2O(g)} = +214.40 \text{ J K}^{-1}$$

$$T = \frac{\Delta H^\circ}{\Delta S^\circ} = \frac{206.11 \times 10^3 \text{ J}}{214.40 \text{ J K}^{-1}} = \boxed{960 \text{ K}}$$

20-9 a) The equation is $3\,\text{Fe}(s) + 4\,\text{H}_2\text{O}(g) \rightarrow \text{Fe}_3\text{O}_4(s) + 4\,\text{H}_2(g)$.

b) Use the values of ΔH°_f in Appendix D to calculate ΔH° for the above reaction:

$$\Delta H^\circ = 1\,\underbrace{(-1118.4)}_{\text{Fe}_3\text{O}_4(s)} + 4\,\underbrace{(0)}_{\text{H}_2(g)} - 4\,\underbrace{(-241.82)}_{\text{H}_2\text{O}(g)} - 3\,\underbrace{(0)}_{\text{Fe}(s)} = \boxed{-151.1 \text{ kJ}}$$

The negative ΔH° means that the reaction is exothermic. Increasing the temperature lowers K and shifts the reaction from right to left. Thus, the yield of the reaction at equilibrium would $\boxed{\text{decrease}}$, although an elevated temperature is needed for the reaction to proceed rapidly.

Tip. This is a standard "catch" in designing industrial processes.

20-11 The balanced chemical equation for the conversion is:

$$\text{C}_{18}\text{H}_{32}\text{O}_2 + 2\,\text{H}_2(g) \rightarrow \text{C}_{18}\text{H}_{36}\text{O}_2(g)$$

Compute the chemical amount of H_2 needed:

$$500.0 \text{ g C}_{18}\text{H}_{32}\text{O}_2 \times \left(\frac{1 \text{ mol}}{280.45 \text{ g C}_{18}\text{H}_{32}\text{O}_2}\right) \times \left(\frac{2 \text{ mol H}_2}{1 \text{ mol C}_{18}\text{H}_{32}\text{O}_2}\right) = 3.5657 \text{ mol H}_2$$

Then use the ideal-gas equation:

$$V_{\text{H}_2} = \frac{n_{\text{H}_2} RT}{P} = \frac{(3.5657 \text{ mol})(0.082057 \text{ L atm mol}^{-1}\text{K}^{-1})(273 \text{ K})}{1 \text{ atm}} = \boxed{79.9 \text{ L}}$$

20-13 The balanced equation for the reduction of pyrolusite to manganese by aluminum is:

$$3\,\text{MnO}_2(s) + 4\,\text{Al}(s) \rightarrow 2\,\text{Al}_2\text{O}_3(s) + 3\,\text{Mn}(s)$$

Combine the standard enthalpies of formation and standard free energies of formation of the two reactants and two products (text Appendix D) to get the standard enthalpy change and standard free energy change of the reaction at 298.15 K:

$$\Delta H^\circ = 2(-1675.7) + 3(0) - 3(-520.03) - 4(0) = -1791.3 \text{ kJ}$$
$$\Delta G^\circ = 2(-1582.3) + 3(0) - 3(-465.17) - 4(0) = -1769.1 \text{ kJ}$$

These changes in enthalpy and free energy apply to the production of 3 mol of Mn. The values per mole of Mn are only one-third as large:

$$\Delta H^\circ = \boxed{-597.1 \text{ kJ mol}^{-1}} \qquad \text{and} \qquad \Delta G^\circ = \boxed{-589.7 \text{ kJ mol}^{-1}}$$

20-15 The balanced equation for the decomposition of mercury(II) oxide to its elements is:

$$2\,HgO(s) \rightarrow 2\,Hg(l) + O_2(g)$$

Combine the standard enthalpies of formation and absolute entropies of the reactants and products at 25° (text Appendix D) to get the standard enthalpy change and standard entropy change of the reaction:

$$\Delta H° = 2(0) + 1(0) - 2(-90.83) = +181.66\ kJ$$
$$\Delta S° = 1(205.03) + 2(76.02) - 2(70.29) = 216.49\ J\ K^{-1}$$

If $\Delta H°$ and $\Delta S°$ are independent of temperature, $\Delta G°$ for this reaction becomes equal to zero at the temperature fulfilling the equation:

$$T = \frac{\Delta H°}{\Delta S°} = \frac{181.66 \times 10^3\ J}{216.49\ J\ K^{-1}} = \boxed{839\ K}$$

At temperatures exceeding 839 K, $\Delta G°$ is negative.

Tip. Mercury(II) oxide, a red solid, indeed does decompose with gentle heating.[1]

20-17 The reduction of $WO_3(s)$ to $W(s)$ with gaseous hydrogen is:

$$WO_3(s) + 3\,H_2(g) \rightarrow W(s) + 3\,H_2O(g)$$

$$\Delta H° = 3(-241.82) + 1(0) - 1(-842.87) - 3(0) = 117.41\ kJ$$
$$\Delta S° = 3(188.72) + 1(32.64) - 1(75.90) - 3(130.57) = 131.19\ J\ K^{-1}$$

The reaction is endothermic. High temperatures favor the products in such a case. When K equals unity, $\Delta G°$ equals zero. The temperature that makes $\Delta G°$ equal zero (if $\Delta H°$ and $\Delta S°$ are temperature-independent) is:

$$T = \frac{\Delta H°}{\Delta S°} = \frac{117.41 \times 10^3\ J}{131.19\ J\ K^{-1}} = \boxed{895\ K}$$

20-19 According to its formula, chalcopyrite is 34.62% Cu. Similarly, covellite is 66.46% copper by mass, and bornite is 63.61% copper by mass. The mass of copper per 100 g of this ore is:

$$\left(\frac{34.62\ g\ Cu}{100\ g\ chalcopyrite} \times \frac{1.1\ g\ chalcopyrite}{100\ g\ ore}\right) + \left(\frac{66.46\ g\ Cu}{100\ g\ covellite} \times \frac{0.42\ g\ covellite}{100\ g\ ore}\right)$$
$$+ \left(\frac{63.61\ g\ Cu}{100\ g\ bornite} \times \frac{0.51\ g\ bornite}{100\ g\ ore}\right) = \frac{0.984\ g\ Cu}{100\ g\ ore}$$

---

[1]See text Figure 1-6, text page 9.

A metric ton of ore is 1000 kg of ore. This is 10^4 times more than 100 g of ore. Accordingly there is 0.98×10^4 g of copper per metric ton of ore, or $\boxed{9.8 \text{ kg}}$ of copper per metric ton of ore.

20-21 a) The balanced equations are:

$$(1) \quad Fe_2O_3(s) + 3\,CO(g) \rightarrow 2\,Fe(s) + 3\,CO_2(g)$$
$$(2) \quad Fe_3O_4(s) + 4\,CO(g) \rightarrow 3\,Fe(s) + 4\,CO_2(g)$$
$$(3) \quad FeCO_3(s) + CO(g) \rightarrow Fe(s) + 2\,CO_2(g)$$

b) The reactions each reduce 1 mol of an iron compound, but the yields are 2, 3, and 1 mol of Fe respectively. For each reaction, $\Delta G°$ is the $\Delta G_f°$ of the reactants minus the $\Delta G_f°$ of the products. Taking free energies of formation from Appendix D:

Equation No.	$\Delta G°$	$\Delta G°$ per mol Fe
(1)	-29.4 kJ	-14.7 kJ mol^{-1}
(2)	-13.3	-4.4
(3)	$+15.1$	$+15.1$

The first reaction has the most negative $\Delta G°$ per mol of Fe: $\boxed{Fe_2O_3}$ is the easiest compound to reduce with CO from the standpoint of thermodynamics. The easiest to reduce in practice depends on factors such as ease of handling and rate of reduction.

20-23 The half-equations for the Downs process for sodium are:

$$Cl^- \rightarrow \frac{1}{2}\,Cl_2(g) + e^- \quad \text{anode} \qquad\qquad Na^+ + e^- \rightarrow Na(l) \quad \text{cathode}$$

20-25 A steady current of 55,000 A for 24 h means that 4.75×10^9 C passes through a single cell. Dividing by the Faraday constant gives the chemical amount of electricity passing through the cell. It is 4.93×10^4 mol. It takes 3 mol of electrons to deposit 1 mol of Al. The theoretical yield of Al is therefore 1.64×10^4 mol, which is 4.43×10^5 g of Al per cell. There are 100 cells, so the total theoretical yield of Al is 100 times larger than for a single cell. It is $\boxed{4.4 \times 10^7 \text{ g}}$.

20-27 The Kroll process uses the reaction:

$$\boxed{TiCl_4(l) + 2\,Mg(s) \rightarrow Ti(s) + 2\,MgCl_2(s)}$$

The minimum mass of magnesium to produce 100 kg of titanium by this process is:

$$100 \text{ kg Ti} \times \left(\frac{1 \text{ kmol Ti}}{47.88 \text{ kg Ti}}\right) \times \left(\frac{2 \text{ kmol Mg}}{1 \text{ kmol Ti}}\right) \times \left(\frac{24.305 \text{ kg Mg}}{1 \text{ kmol Mg}}\right) = \boxed{102 \text{ kg Mg}}$$

20-29 7.32 g of zinc is to be coated onto the steel garbage can. This is 0.112 mol of Zn. Each mole of Zn requires 2 mol of electrons to plate it out, and a mole of electrons is 96,485 coulombs. The total charge passed through the cell is therefore 2.161×10^4 C. A current of 8.50 A means that 8.50 C passes through the cell every second. The time required to pass the required charge is:

$$t = \frac{Q}{I} = \frac{2.161 \times 10^4 \text{ C}}{8.50 \text{ C s}^{-1}} = 2.54 \times 10^3 \text{ s} = \boxed{42.4 \text{ min}}$$

20-31 Very reactive chemicals are chemicals of high free energy. A compound of high free energy will tend to react spontaneously with other substances. This is why, for example, there is very little elemental aluminum in the Earth's crust. Any elemental aluminum initially present has long since reacted to form compounds such as Al_2O_3 or $Al_2Si_2O_5 \cdot (OH)_4$ that have lower free energies.

a) Coal, hydrogen and air do not spontaneously react to form aspirin. This means that ΔG exceeds zero for this reaction. Aspirin has a higher free energy than its constituent carbon (from coal) hydrogen and oxygen (from air).

b) Pure water, sodium chloride and magnesium carbonate react spontaneously to form seawater. Thus, seawater is in a state of lower free-energy.

c) Carbon dioxide and water do not spontaneously react to form vitamin A and oxygen. This means that the vitamin A and oxygen lie in higher free-energy states than carbon dioxide and water.

20-33 The reaction of iron or zinc with sulfuric acid makes a poor large-scale preparation for hydrogen because zinc and iron are relatively expensive starting materials. They require a large input of energy to free them from their ores. A economical and useful large-scale preparation must use energetically cheaper starting materials.

20-35 Use all of the data in unit factors:

$$63 \times 10^9 \text{ cans} \times \left(\frac{0.355 \text{ L}}{1 \text{ can}}\right) \times \left(\frac{0.15 \text{ mol CO}_2}{1 \text{ L}}\right) \times \left(\frac{0.0440 \text{ kg CO}_2}{1 \text{ mol CO}_2}\right)$$
$$= \boxed{1.5 \times 10^8 \text{ kg CO}_2}$$

20-37 The problem requires classification of reactions into three categories: acid-base (see text page 172), redox (see text pages 174 and 402), and dissolution-precipitation (see text page 376).

a) This reaction is a $\boxed{\text{redox}}$ reaction. Sulfur is reduced from the +6 oxidation state (in $CaSO_4$) to the 0 oxidation state (in S). Methane is oxidized.

b) The hydrogenation of carbon monoxide is a $\boxed{\text{redox}}$ reaction. Hydrogen is oxidized from the 0 oxidation state (in H_2) to the +1 oxidation state (in CH_3OH). The reduction is the conversion of CO (C in the +2 oxidation state) to CH_3OH (C in the −2 oxidation state).

c) This hydrogenation of ethylene is a $\boxed{\text{redox}}$ reaction. Hydrogen is oxidized from the 0 oxidation state (in H_2) to the +1 oxidation state (in H_3CCH_3). The reduction is the conversion of $H_2C{=}CH_2$ (C in the −2 oxidation state) to $H_3C{-}CH_3$ (C in the −3 oxidation state).

d) This reaction is a $\boxed{\text{redox}}$ reaction. Iron is oxidized from the 0 oxidation state (in Fe) to the +2 oxidation state (in $FeSO_4$). The reduction which accompanies this oxidation involves converting H_2SO_4 (H in the +1 oxidation state) to H_2 (H in the 0 oxidation state).

20-39 The equilibrium $CO_2(g) + C(s) \rightarrow 2\,CO(g)$ is crucial in the reduction of iron ore in a blast furnace. At room temperature it lies far to the left, but at 1875° (2148 K) it is very much shifted to the right. To answer the question, compute $\Delta H°$ and $\Delta S°$ of this reaction from the data in text Appendix D and then use the relationship:

$$-RT \ln K_T = \Delta H° - T\Delta S° \qquad \text{which gives} \qquad \ln K_T = \frac{-\Delta H°}{RT} + \frac{\Delta S°}{R}$$

to solve for K at both of the temperatures. It is found that $\Delta H°$ is 172.47 kJ and $\Delta S°$ is 175.75 J K^{-1}. The answers are $K_{298} = \boxed{9.2 \times 10^{-22}}$ and $K_{2148} = \boxed{9.7 \times 10^4}$. The second K is a rough approximation because $\Delta H°$ and $\Delta S°$ do change between 298 and 2148 K.[2]

20-41 Iron ore is Fe_2O_3 admixed with silica, alumina and minor impurities; copper ore is usually a sulfide. Iron ore is reduced with $CO(g)$ in a blast furnace by reaction with coke (carbon) and a limited amount of air. Fluxes like CaO are added to make a fluid slag from the silica and alumina impurities. Blast furnace slag is used as road ballast and in making portland cement. The roasting of copper ores gives off sulfur dioxide, which must not be vented into the atmosphere and is often processed into sulfuric acid. Roasting reduces the copper only partially (to the copper(I) sulfide) and oxidizes impurities. Roasting of copper ores takes place at relatively low temperatures. Removal of impurities goes on at higher temperature (about 1100°C) but still below the 1535°C at which iron is reduced in a blast furnace. The flux in this step of copper recovery is $CaCO_3$, and the slag contains impurity iron. The copper is still combined chemically in Cu_2S. This compound is decomposed into impure metallic copper by heating it in air, which generates more $SO_2(g)$.

---

[2]The answer to this problem also appears on text page 742.

20-43 Both aluminum and magnesium are produced electrochemically. The starting material for the production of aluminum is a mixture of cryolite with bauxite. The mixture is melted and electrolyzed; aluminum is produced at the cathode. The commercial source of magnesium is seawater. Magnesium hydroxide is precipitated from seawater by treating it with a cheap base such as calcined (burnt) dolomite CaO·MgO. The $Mg(OH)_2$ is then converted to $MgCl_2$ by reaction with $HCl(aq)$. The $MgCl_2$ is melted and electrolyzed, producing chlorine gas at the anode and metallic magnesium at the cathode.

20-45 a) The balanced equation is:

$$4\,FeCr_2O_4(s) + 8\,Na_2CO_3(l) + 7\,O_2(g) \rightarrow 2\,Fe_2O_3(s) + 8\,Na_2CrO_4(l) + 8\,CO_2(g)$$

b) The standard enthalpies of formation (in $kJ\ mol^{-1}$) of these substances are:

$FeCr_2O_4(s)$	$Na_2CO_3(l)$	$O_2(g)$	$Fe_2O_3(s)$	$Na_2CrO_4(l)$	$CO_2(g)$
-1445	$(-1130.7 + \Delta H_{fus})$	0	-824.2	$(-1342 + \Delta H_{fus})$	-393.5

The entries for the two liquids require comment. The ΔH_f°'s of the two liquids exceed the ΔH_f°'s of their solids by amounts equal to the enthalpy of fusion of the solid. According to the problem, the two enthalpies of fusion are about the same. The two compensate for each other in the computation of ΔH° by Hess's law:

$$\Delta H^\circ = 2(-824.2) + 8(-1342 + \Delta H_{fus}) + 8(-393.5)$$
$$- 7(0) - 8(-1130.7 + \Delta H_{fus}) - 4(-1445) = \boxed{707\ kJ}$$

c) Leach the soluble sodium chromate away from the insoluble by-products by treating the product mixture with water. Evaporate the water. Heat the resulting solid sodium chromate with carbon (charcoal) to reduce it to chromium. Other products would include $CO_2(g)$ and $Na_2O(s)$, which could be taken up in a suitable acidic flux.

20-47 Compute the volume of zinc to be coated onto the steel, then the mass of the volume of zinc and the chemical amount in that mass. Take the density of zinc from text Appendix F:

$$V_{Zn} = \frac{(0.250 \times 10^{-3}\ m)(1.00\ m)(100\ m)}{1\ side} \times 2\ sides = 0.0500\ m^3$$

$$m_{Zn} = 0.0500\ m^3 \times \left(\frac{10^6\ cm^3}{m^3}\right) \times \left(\frac{7.133\ g\ Zn}{cm^3}\right) = 3.566 \times 10^5\ g$$

$$n_{Zn} = 3.566 \times 10^5\ g\ Zn \times \left(\frac{1\ mol\ Zn}{65.38\ g\ Zn}\right) = 5.454 \times 10^3\ mol\ Zn$$

Each mole of Zn requires 2 mol of electrons to plate it out, and a mole of electrons is 96485 C. Hence:

$$Q = 5.454 \times 10^3 \text{ mol Zn} \times \left(2 \text{ mol } e^- over 1 \text{ mol Zn}\right) \times \left(\frac{96485 \text{ C}}{\text{mol } e^-}\right) = 1.053 \times 10^9 \text{C}$$

The energy used in the plating operation is the amount of charge passed multiplied by the voltage that pushes it through the circuit. The voltage is 3.5 V in this case so the energy is 3.68×10^9 V C which is 3.68×10^9 J. Divide this amount of energy by 0.9 because the galvanizing is only 90% efficient. This raises the energy consumption to 4.09×10^9 J, equivalent to 1.14×10^3 kW hr, which costs $\boxed{\$114}$.

Tip. The problem draws on ideas covered in **20-29** and in **12-13**.

Chapter 21

Chemical Processes Based on Sulfur, Nitrogen, and Phosphorus

Sulfuric Acids and Its Uses

Sulfuric acid plays a central role in modern industrial society. Historically, it was produced in quantity in the **lead-chamber** process, in which oxides of nitrogen serve as oxygen carriers for the oxidation of SO_2 to SO_3. Currently, the **contact process**, which uses the direct oxidation of SO_2 by O_2 in the presence of a catalyst, followed by absorption of the SO_3 in previously produced H_2SO_4, is the main synthetic route to sulfuric acid.

Sulfur for sulfuric acid comes from metal sulfides (see **21-5**) or from deposits of elemental sulfur that are mined by the **Frasch process**. In this technique, super-heated water is pumped underground to melt the sulfur, which is then forced up from the sulfur-bearing formation with heated compressed air. Sulfur present as sulfides (mainly hydrogen sulfide) in natural gas and petroleum is another source of sulfur. These impurities must be removed in any case before combustion of the fuels to avoid objectionable smells and the release of SO_2 to the atmosphere. The sulfides are acidic and are extracted from the fuel by scrubbing it with a basic solution. Then, in the **Claus process,** the hydrogen sulfide is reacted with SO_2 to yield elemental sulfur.

Sulfuric acid is used in a large number of chemical processes. Most of its uses are indirect in that sulfur from the acid does not become a part of the product, but ends up as a sulfate waste or as **spent acid**. See **21-41**. The important chemical sodium sulfate, however, is synthesized from SO_2, NaCl, and O_2 in the **Hargreaves process**[1] and is a by-product of rayon manufacture.

An important application of sulfur chemistry is the processing of wood to wood pulp for paper and cardboard. In the **sulfate**, or **Kraft process**, an alkaline digestion

---

[1]See text Example 2-3, text page 43.

liquor containing $NaOH$ and Na_2S breaks down the lignin in the wood and allows separation of the cellulose. This pulp is further processed into paper and the digestion liquor is recycled. In the **sulfite process**, wood chips are digested in an acidic solution containing the hydrogen sulfite ion to free the cellulosic pulp.

Phosphorus Chemistry

Elemental phosphorus exists in several allotropic forms. White phosphorus consists of tetrahedral P_4 molecules. In red and black phosphorus, the atoms of the element are extensively bonded in network structures.[2]

The important oxides of phosphorus are P_4O_6 and P_4O_{10}. They are both acidic oxides and react with excess water to give H_3PO_3 (phosphorous acid) and H_3PO_4 (phosphoric acid) respectively.

The single most important use of sulfuric acid is in the production of phosphate fertilizers. Phosphate rock (principally $Ca_5(PO_4)_3F$) is treated with H_2SO_4 to give **superphosphate** fertilizer, which is a mixture of calcium dihydrogen phosphate $(Ca(H_2PO_4)_2 \cdot H_2O)$ and gypsum $(CaSO_4 \cdot 2H_2O)$. The same type of reaction using an excess of sulfuric acid produces **wet-process** phosphoric acid. The main use of wet-process phosphoric acid is in reaction with further phosphate rock to generate **triple superphosphate** fertilizer (see **21-43**). A higher grade of phosphoric acid than the wet-process product is afforded by the **furnace process**, in which phosphate rock is first reduced to elemental phosphorus. The P_4 is burned in air to P_4O_{10}, which is reacted with water to give H_3PO_4. The reaction of phosphoric acid with ammonia gives ammonium phosphate, the major phosphorus-containing fertilizer in current use. Sodium phosphates from the neutralization of phosphoric acid by sodium hydroxide are used in cleaning products and as builders in detergents.

Phosphate ions link together to form polyanions. Tetrahedral PO_4^{3-} groups join at their corners to give chains that are analogous to the polysilicates. See text Section 23-1.[3]

Example: Phosphoric acid (H_3PO_4) condenses to give the dimer diphosphoric acid ($H_4P_2O_7$) and the trimer tripolyphosphoric acid ($H_5P_3O_{10}$). Predict the formula of the polyphosphoric acid with four phosphates linked in a straight chain. Predict the formula of the polyphosphoric acid with four phosphates linked in a ring. **Solution:** Each upward step corresponds to adding H_3PO_4 and subtracting H_2O. This is equivalent to adding HPO_3. The next formula is $H_6P_4O_{13}$. Closing a tetraphosphate chain into a ring would involve one more condensation step and the concomitant loss of one more H_2O. The cyclic tetraphosphoric acid would have the formula $H_4P_4O_{12}$. This example is related to **23-1**.

---

[2]See text Figure 19-25, text page 712.

[3]Text page 810.

The existence of condensed phosphates is very important in biochemistry. The crucial biological molecules adenosine diphosphate and adenosine triphosphate, ADP and ATP, are derivatives of diphosphoric acid and triphosphoric acid respectively, just as adenosine monophosphate, AMP, is a derivative of phosphoric acid. See text Figure 25-17b.[4]

Nitrogen Fixation

Nitrogen is an essential nutrient for plants and animals. Although the element is abundant at or near the surface of the earth, most of the supply is chemically inaccessible because of the great stability of the triple bond in the N_2 molecule. The formation of compounds between molecular nitrogen and other elements is called **nitrogen fixation**. In nature, atmospheric nitrogen is fixed in the form of NO (which reacts immediately with O_2 to form NO_2) by the intense heat in lightning flashes. Nitrogen is also fixed by the action of certain bacteria that grow in association with the roots of some plants. Large supplies of nitrogen fertilizers, which are essential to high crop yields, required an economical means for the fixing of atmospheric nitrogen.

An early method of nitrogen fixation, the **electric-arc process**, fixed nitrogen by electrical discharge, but was expensive to operation and hard to regulate. In the **cyanamide**, or **Frank-Caro, process**, calcium carbide (CaC_2) was generated by heating the inexpensive materials lime (CaO) and coke (C) in an electric furnace. Calcium carbide was then reacted with nitrogen to give calcium cyanamide ($CaCN_2$) (see **21-15**), which was used directly as fertilizer or treated with water to give ammonia. The major source of industrial fixed nitrogen today is the **Haber-Bosch process**, the direct combination of N_2 and H_2 to yield NH_3 (see **9-29** and **9-61**). The practical process requires a compromise between low temperature, which favors high yield of ammonia at equilibrium, and high temperature, which increases the rate of attainment of equilibrium. The best operating temperature is between 700 and 900 K. A catalyst speeds up the reaction, and high pressure favors higher equilibrium yields of ammonia. The Haber-Bosch process is energy intensive and requires major capital investment to build large structures strong enough to withstand high pressure. See **21-17**.

Once nitrogen is fixed in the form of ammonia, a variety of products are easily formed by acid-base reactions for use as fertilizers. The oxidation of ammonia by the **Ostwald process** produces nitric acid (HNO_3). This conversion is performed in stepwise fashion, first to NO, and then to NO_2. The reaction of the NO_2 with water gives HNO_3 and NO, which is cycled back into the process. The major use of nitric acid is in reaction with ammonia to give ammonium nitrate (NH_3NO_3), which is used for fertilizer and as an industrial explosive.

---

[4]Text page 872.

Hydrazine (N_2H_4) is made from aqueous ammonia by oxidation with the aqueous hypochlorite ion ($ClO^-(aq)$) in the **Raschig synthesis.** Note that the formula of hydrazine equals twice the formula of ammonia with H_2 subtracted. Subtraction of H_2 corresponds to oxidation, and the oxidation number of nitrogen in hydrazine accordingly equals -2 (versus -3 in ammonia). Hydrazine is H_2NNH_2, and its conjugate acid is $H_2NNH_3^+$ ion, the hydrazinium ion.[5] Since both nitrogen atoms are basic, hydrazine can gain two hydrogen ions.

Oxoacids of Sulfur, Phosphorus, and Nitrogen

Recall that oxoacids are compounds having a nonmetal X bonded to one or more oxygen atoms which in turn may have hydrogen bonded to them. Their general formula is $XO_n(OH)_m$. The text compares and contrasts oxoacids of each of the three elements considered in this chapter. Six compounds are considered: nitric acid (HNO_3), nitrous acid (HNO_2), sulfuric acid (H_2SO_4), sulfurous acid (H_2SO_3), phosphoric acid (H_3PO_4, and phosphorous acid (H_3PO_3). Study text Figure 21-17,[6] and note the following points:

- The only strong acids in the group are sulfuric acid and nitric acid.

- Sulfurous acid at low pH is represented as $SO_2(aq)$ rather than as H_2SO_3.

- The phosphorus oxoacids are the poorest oxidizing agents among the six considered.

- Under acidic conditions, the best oxidizing agent in the group is nitrous acid. However, nitrous acid disproportionates spontaneously and rapidly (to nitric acid and nitrogen oxide). This makes it unsuitable for use in oxidations.

- The second-best theoretical (and best practical) oxidizing agent in the group is nitric acid.

- The oxidizing strengths of all the oxoacids are all lowered, but to different degrees, when the pH is raised to 14 and they are replaced by their basic forms.

- The lower oxidation states of oxo-compounds of sulfur and phosphorus (in species such as hypophosphite ion $H_2PO_2^-$ and sulfite ion (SO_3^{2-}) are good reducing agents at pH 14.

- Having an element in a high oxidation state does not automatically make an oxoacid or oxoanion a strong oxidizing agent. Thermodynamic stability changes with oxidation state in an irregular fashion.

---

[5]Just as ammonia is NH_3, and its conjugate acid is NH_4^+ ion, the ammonium ion.
[6]Text page 774.

Thermodynamic versus Kinetic Stability

Grasping the distinction between the thermodynamic stability of a compound and its kinetic stability is essential in understanding their actual, observed chemical behavior. This is their **descriptive chemistry.** Thus, NO_2^- is thermodynamically unstable in basic aqueous solution with respect to disproportionation but such solutions can be kept indefinitely because they are kinetically stable. Similarly, the compound hydrazine is thermodynamically unstable with respect to disproportionation (see **21-25**), but can be kept for long periods. Also, the oxidation of ammonia with oxygen is strongly favored thermodynamically at room temperature but does not proceed at any perceptible rate because no effective kinetic pathway is available. It is in fact difficult to burn ammonia in air even at higher temperatures although addition of a Pt catalyst speeds the reaction. See **22-9** for a case where thermodynamic feasibility does not guarantee economic feasibility.

Compare the distinction between kinetic and thermodynamic stability to the distinction between "lability" and "inertness" in coordination complexes in text Chapter 18.

How Oxidation State Influences the Acidity of Oxoacids

In general, higher oxidation states in the central atom X in oxoacids correspond to more covalent X—O bonds and stronger acids. This follows because a more positive X withdraws electron density from the oxygen atoms bonded to it causing the H atoms on the other side to be more easily released since there is less electron density to hold them. For a series of acids of the same general formula $XO_n(OH)_m$, the strongest acid (largest K_a) is the one with the largest n—the largest number of lone oxygen atoms attached to the central atom. Each increase in n increases the K_a by a factor of about 10^5. See **21-53**.

Detailed Solutions to Odd-Numbered Problems

21-1 The reaction is $SO_2(g) + 1/2\,O_2(g) \rightleftharpoons SO_3(g)$. Increasing the pressure would drive this equilibrium to the right. Decreasing the temperature would also favor the products (the reaction is exothermic). Finally, continuous removal of the SO_3 would induce the production of more product to replace it.

21-3 The problem proposes replacing H_2SO_5 in the oxidation of the pollutant SO_2 by H_2O_2:

$$SO_2(g) + H_2O_2(aq) \rightarrow HSO_4^-(aq) + H^+(aq)$$

Taking data from text Appendix D to compute $\Delta G°$ gives:

$$\Delta G° = 1 \underbrace{(0)}_{H^+(aq)} + 1 \underbrace{(-755.91)}_{HSO_4^-(aq)} - 1 \underbrace{(-134.03)}_{H_2O_2(aq)} - 1 \underbrace{(-300.19)}_{SO_2(g)} = -321.69 \text{ kJ}$$

The reaction is therefore spontaneous and would be thermodynamically $\boxed{\text{feasible}}$.

Tip. The powerful oxidizing agent H_2SO_5 mentioned in the problem can be regarded as containing S in the +8 oxidation state. A better view however is to see it as regular sulfuric acid with one of its O atoms replaced by a peroxide (O_2^{2-}) group. This gives it the formula $H_2SO_3(OO)$, keeps the sulfur in the +6 oxidation state, and explains the name: peroxymonosulfuric acid.

21-5 $ZnS(s) + 2 O_2(g) + H_2O(l) \rightarrow ZnO(s) + H_2SO_4(l)$. This reaction equals the reaction on text page 759[7] added to the reverse of the first reaction on text page 757.

21-7 Sodium sulfate is cheaply obtained by neutralizing sulfuric acid with sodium hydroxide. Sodium sulfide is then prepared by heating the sodium sulfate obtained in this way with carbon (in the form of coal or charcoal):

$$H_2SO_4(l) + 2 NaOH(s) \rightarrow Na_2SO_4(s) + 2 H_2O(l)$$
$$Na_2SO_4(s) + 2 C(s) \rightarrow Na_2S(s) + 2 CO_2(g)$$

21-9 This is the oxidation of phosphorus by sulfuric acid.
$5 H_2SO_4(aq) + 2 P(s) \rightarrow 5 SO_2(g) + 2 H_3PO_4(aq) + 2 H_2O(l)$.
Or: $10 H_2SO_4(aq) + P_4(s) \rightarrow 10 SO_2(g) + 4 H_3PO_4(aq) + 4 H_2O(l)$.

21-11 If *pairs* of phosphoric acid molecules react and split out water, the reaction must be $2 H_3PO_4(l) \rightarrow H_4P_2O_7(aq) + H_2O(l)$. The structure of $H_4P_2O_7$ (diphosphoric acid) is:

$$
\begin{array}{ccc}
\ddot{\text{O}}\text{--H} & & \ddot{\text{O}}\text{--H} \\
| & & | \\
\ddot{\text{O}}=\text{P}\text{--O--}\text{P}=\ddot{\text{O}} \\
| & & | \\
\text{:O--H} & & \text{:O--H}
\end{array}
$$

21-13 This is a straightforward conversion problem:

$$\frac{2 \times 10^8 \text{ kg N}}{1 \text{ year}} \times \left(\frac{1000 \text{ g N}}{1 \text{ kg N}}\right) \times \left(\frac{1 \text{ mol N}}{14.0 \text{ g N}}\right) \times \left(\frac{6.02 \times 10^{23} \text{ atom N}}{1 \text{ mol N}}\right)$$

$$\times \left(\frac{1 \text{ year}}{3.15 \times 10^7 \text{ s}}\right) = \boxed{\frac{3 \times 10^{26} \text{ atom N}}{\text{s}}}$$

---

[7]Under the heading "Sources for Sulfur and Sulfur Dioxide."

21-15 The first step is the thermal decomposition of calcium carbonate (limestone) to calcium oxide. The calcium oxide is then converted to calcium carbide by reaction with carbon (coke). Next the calcium carbide is reacted at high temperature with nitrogen to form calcium cyanamide. This is the actual nitrogen-fixing step. Finally the calcium cyanamide is treated with water to liberate ammonia. The four steps and their sum are:

$$CaCO_3(s) \rightarrow CaO(s) + CO_2(g)$$
$$CaO(s) + 3\,C(s) \rightarrow CaC_2(s) + CO(g)$$
$$CaC_2(s) + N_2(g) \rightarrow CaNCN(s) + C(s)$$
$$CaNCN(s) + 4\,H_2O(l) \rightarrow Ca(OH)_2(s) + CO_2(g) + 2\,NH_3(g)$$

$$CaCO_3(s) + 2\,C(s) + N_2(g) + 4\,H_2O(l) \rightarrow$$

$$\boxed{Ca(OH)_2(s) + 2\,CO_2(g) + CO(g) + 2\,NH_3(g)}$$

The standard enthalpy change of this reaction is obtained by combination of the proper ΔH_f°'s from text Appendix D. Assume the calcium carbonate to be in the form of calcite. The answer is $\boxed{374.4 \text{ kJ}}$. Therefore ΔH° equals 187.2 kJ mol per mol of $NH_3(g)$.

21-17 The advantage of working at higher pressures is the increase in yield: the equilibrium, $N_2(g) + 3\,H_2(g) \rightleftharpoons 2\,NH_3(g)$ shifts to the right with increasing pressure. The disadvantage of higher pressure is that stronger equipment must be constructed. This means more costly vessels, valves, and pipes.

The standard enthalpy change for the reaction is negative (-92.2 kJ) indicating that the reaction is exothermic. Therefore, equilibrium yield is also increased by lower temperature. The general rule is that reaction rates increase with increasing temperature. At fixed yield a higher temperature can be used if the pressure is higher, giving a faster reaction.

21-19 The three steps in the Ostwald process are:

$$4\,NH_3(g) + 5\,O_2(g) \rightarrow 4\,NO(g) + 6\,H_2O(g) \quad \Delta H^{\circ} = -905.48 \text{ kJ}$$
$$2\,NO(g) + O_2(g) \rightarrow 2\,NO_2(g) \qquad\qquad\qquad \Delta H^{\circ} = -114.14 \text{ kJ}$$
$$3\,NO_2(g) + H_2O(l) \rightarrow 2\,HNO_3(l) + NO(g) \quad \Delta H^{\circ} = -71.66 \text{ kJ}$$

All three steps are exothermic. Hence, for all three steps, a high equilibrium yield will be favored by $\boxed{\text{low temperature}}$.

21-21 Note the similarity of the two equations:

$$3\,HNO_3(aq) + Cr(s) + 3\,H_3O^+(aq) \rightarrow 3\,NO_2(g) + Cr^{3+}(aq) + 6\,H_2O(l)$$
$$3\,HNO_3(aq) + Fe(s) + 3\,H_3O^+(aq) \rightarrow 3\,NO_2(g) + Fe^{3+}(aq) + 6\,H_2O(l)$$

21-23 The compound has 24 valence electrons:

$$H_2\ddot{N} \diagdown \underset{\cdot\cdot}{\ddot{N}} {=} \underset{\cdot\cdot}{N} \diagup \ddot{N}H_2$$

21-25 One way to obtain the equation for the disproportionation reaction is to re-write one of the half-equations as an oxidation and combine it with the reduction:

reduction	$2\,N_2H_4 + 4\,H_2O + 4\,e^- \rightarrow 4\,NH_3 + 4\,OH^-$
oxidation	$N_2H_4 + 4\,OH^- \rightarrow N_2 + 4\,H_2O + 4\,e^-$
total	$3\,N_2H_4 \rightarrow N_2 + 4\,NH_3$

Note that the reduction half-equation has been multiplied by 2 to equalize the electron count. The disproportionation reaction involves neither OH^- nor H_3O^+ ion. Its potential is consequently independent of pH. Combine the two half-cell potentials listed in the problem to obtain the potential difference of the reaction. It is $+1.26$ V, indicating that 1 M aqueous hydrazine is thermodynamically $\boxed{\text{unstable}}$ under standard conditions with respect to disproportionation to nitrogen and ammonia. It does not follow that the reaction occurs at any appreciable rate.

21-27 "Completely innocuous in the atmosphere" makes it easy to determine the products. The equation is then balanced by the usual method (see text Section 12-1) for redox reactions. Each hydrazine gains 4 electrons to get to N(0) while each nitric acid loses 5 electrons to get to n(0). Note where these numbers appear in the answer:

$$5\,N_2H_4(l) + 4\,HNO_3(l) \rightarrow 7\,N_2(g) + 12\,H_2O(l)$$

Combining ΔH_f°'s according to Hess's law:

$$\Delta H^\circ = 12(-285.83) + 7(0) - 5(50.63) - 4(-174.10) = \boxed{-2986.71 \text{ kJ}}$$

21-29 The more electronegative the hydrogen-replacing group, the more the nitrogen lone pair is drawn toward it, and the poorer the electron-donating ability of the nitrogen atom. Hydroxylamine should be $\boxed{\text{less basic}}$ than ammonia because O is more electronegative than H. Experiment bears out this prediction.

21-31 A good tactic in working out a synthetic route to a target compound is to work backwards. N_2O_4 exists in equilibrium with NO_2. This equilibrium represents the last step in the synthesis:

$$2\,NO_2(g) \rightleftharpoons N_2O_4(g)$$

This exothermic reaction can be shifted to the right by cooling. Also, N_2O_4, a gas at room temperature, freezes out at $-11°C$. The gaseous NO_2 is prepared from ammonia as part of the Ostwald process for nitric acid:

$$4\,NH_3(g) + 5\,O_2(g) \rightarrow 4\,NO(g) + 6\,H_2O(g)$$
$$2\,NO(g) + O_2(g) \rightarrow 2\,NO_2(g)$$

Ammonia can be prepared directly from the elements: $N_2(g) + 3\,H_2(g) \rightarrow 2\,NH_3(g)$. Thus, the final four-step route to N_2O_4 from the elements:

$$N_2(g) + 3\,H_2(g) \rightarrow 2\,NH_3(g) \qquad 4\,NH_3(g) + 5\,O_2(g) \rightarrow 4\,NO(g) + 6\,H_2O(g)$$
$$2\,NO(g) + O_2(g) \rightarrow 2\,NO_2(g) \qquad 2\,NO_2(g) \rightarrow N_2O_4(s)$$

21-33 a) Tripling all coefficients in the second half-equation appearing in the problem and subtracting it from the first half-equation gives the following disproportionation:

$$4\,P_4(s) + 12\,OH^-(aq) + 12\,H_2O(l) \rightarrow 4\,PH_3(g) + 12H_2PO_2^-(aq)$$

The potential difference for this reaction is:

$$\Delta\mathcal{E}° = \mathcal{E}°(\text{reduction}) - \mathcal{E}°(\text{oxidation}) = -0.89 - (-2.05) = 1.16\ \text{V}$$

The positive standard potential difference means that P_4 $\boxed{\text{does}}$ disproportionate spontaneously at pH 14.

b) The stronger reducing agent is the more easily oxidized substance. P_4 is on the right side of the reduction half-equation with the more negative $\mathcal{E}°$. Hence $\boxed{P_4}$ is a stronger reducing agent than PH_3.

Tip. Insert the two half-reactions and their potentials into the right side of text Figure 21-17[8] to help visualize the relative power of P_4 and PH_3 as reducing agents.

21-35 Compute the standard free-energy changes of the two reactions using the ΔG_f° data in text Appendix D. The $\Delta G°$ of the oxidation of SO_2 by $O_2(g)$ according to the first equation in the problem is -70.89 kJ. The $\Delta G°$ for the oxidation of SO_2 by $NO_2(g)$ according to the second equation in the problem is -35.63 kJ. The

[8]Text page 774.

oxidation by O_2 has a larger thermodynamic driving force. Oxidation of SO_2 by NO_2 is however faster at low temperature, which explains the early success of the lead-chamber process for sulfuric acid.

Tip. The $\Delta H°$ and $\Delta S°$ data quoted in the text[9] can be very conveniently used in the equation $\Delta G° = \Delta H° - T\Delta S°$ to obtain $\Delta G°$ for the first reaction. The result equals the result using $\Delta G_f°$'s.

21-37 The reaction in the electrolysis unit combines the oxidation of sulfate ion and the reduction of hydronium ion:

$$2\,SO_4^{2-}(aq) \rightarrow S_2O_8^{2-}(aq) + 2\,e^- \qquad\qquad \text{oxidation}$$
$$2\,H_3O^+(aq) + 2\,e^- \rightarrow H_2(g) + 2\,H_2O(l) \qquad\qquad \text{reduction}$$
$$2\,H_3O^+ + 2\,SO_4^{2-}(aq) \rightarrow H_2(g) + S_2O_8^{2-}(aq) + 2\,H_2O(l) \qquad \text{overall}$$

The standard potential difference is:

$$\Delta\mathcal{E}° = \mathcal{E}°(\text{reduction}) - \mathcal{E}°(\text{oxidation}) = 0.0 - 2.0 = -2.0 \text{ V}$$

The Nernst equation for this reaction (at 25°C) is:

$$\Delta\mathcal{E} = -2.0 \text{ V} - \frac{0.0592 \text{ V}}{2} \log\left(\frac{P_{H_2}[S_2O_8^{2-}]}{[H_3O^+]^2[SO_4^{2-}]^2}\right)$$

Inserting the given pressure and concentrations gives:

$$\Delta\mathcal{E} = -2.0 - \frac{0.0592}{2} \log\left(\frac{(0.10)(0.50)}{(1.0)^2(1.0)^2}\right) = -1.96 \text{ V}$$

The electrolysis unit requires a minimum voltage of $+1.96$ V to force the reaction to run.

21-39 a) $Ba^{2+}(aq) + H_2SO_4(aq) + 2\,H_2O(l) \rightarrow BaSO_4(s) + 2\,H_3O^+(aq)$

b) $2\,NH_3(aq) + H_2SO_4(aq) \rightarrow 2\,NH_4^+(aq) + SO_4^{2-}(aq)$

c) $Cu(s) + 2\,H_2SO_4(aq) \rightarrow Cu^{2+}(aq) + SO_2(g) + SO_4^{2-}(aq) + 2\,H_2O(l)$

[9]Text page 758.

21-41 a) Compute the $\Delta H°$ and $\Delta S°$ of the reaction using data from text Appendix D:

$$\Delta H° = 2\underbrace{(-241.82)}_{H_2O(l)} +4\underbrace{(-296.83)}_{SO_2(g)} +1\underbrace{(-393.51)}_{CO_2(g)} -4\underbrace{(-395.72)}_{SO_3(g)} -1\underbrace{(-74.81)}_{CH_4(g)}$$

$$= -406.78 \text{ kJ}$$

$$\Delta S° = 2\underbrace{(188.72)}_{H_2O(l)} +4\underbrace{(248.11)}_{SO_2(g)} +1\underbrace{(213.63)}_{CO_2(g)} -4\underbrace{(256.65)}_{SO_3(g)} -1\underbrace{(186.15)}_{CH_4(g)}$$

$$= 370.76 \text{ J K}^{-1}$$

Use these numbers to obtain $\Delta G°_{1000}$ as follows:

$$\Delta G°_{1000} = \Delta H° - T\Delta S° = -406.78 \text{ kJ} - (1000 \text{ K})(0.37076 \text{ kJ K}^{-1}) = \boxed{-780 \text{ kJ}}$$

b) Compute the $\Delta H°$ and $\Delta S°$ for the critic's proposed reaction in the same way:

$$\Delta H° = 1\underbrace{(-241.82)}_{H_2O(l)} +1\underbrace{(-20.63)}_{H_2S(g)} +1\underbrace{(-393.51)}_{CO_2(g)} -1\underbrace{(-395.72)}_{SO_3(g)} -1\underbrace{(-74.81)}_{CH_4(g)}$$

$$= -185.43 \text{ kJ}$$

$$\Delta S° = 1\underbrace{(188.72)}_{H_2O(l)} 1\underbrace{(205.68)}_{H_2S(g)} +1\underbrace{(213.63)}_{CO_2(g)} -1\underbrace{(256.65)}_{SO_3(g)} -1\underbrace{(186.15)}_{CH_4(g)}$$

$$= 165.23 \text{ J K}^{-1}$$

From which:

$$\Delta G°_{1000} = -185.43 \text{ kJ} - (1000 \text{ K})(0.16523 \text{ kJ K}^{-1}) = \boxed{-351 \text{ kJ}}$$

c) Both reaction are spontaneous at 1000 K. Thermodynamics does not predict which, if either, will occur preferentially. Although the first is more favored than the second, the second could easily predominate if it were faster and if the reaction:

$$H_2S(g) + 3SO_3(g) \rightarrow 4SO_2(g) + H_2O(g)$$

were slow. Note that this reaction added to the second reaction gives the first. The best way to settle the issue is by experiment.

21-43 Superphosphate is $(Ca(H_2PO_4)_2 \cdot H_2O)_3 (CaSO_4 \cdot 2H_2O)_7$. This compound has a molar mass of 1961.4 g mol^{-1}. The mass percentage of P_2O_5 in superphosphate is:

$$\frac{3 \times 141.94 \text{ g mol}^{-1}}{1961.4 \text{ g mol}^{-1}} \times 100\% = \boxed{21.7\%}$$

where 141.94 g mol^{-1} is the molar mass of P_2O_5. This is about one-third the mass percentage of P_2O_5 in triple superphosphate.[10]

21-45 The fixing of nitrogen from available starting materials generally requires the input of energy. For example, the Haber-Bosch process requires high pressures of $N_2(g)$ and $H_2(g)$. It takes energy to produce these pressures. The first step in the Ostwald process for nitric acid requires heating to 800°C. In the 1970's, when the price of petroleum rose, so did the price of energy, and consequently, the cost of manufacturing ammonia, nitric acid, and fertilizer.

21-47 a) $6\,NO_2(g) + 4\,(NH_2)_2CO(aq) \rightarrow 7\,N_2(g) + 4\,CO_2(g) + 8\,H_2O(l)$

b) $6\,NO(g) + 2\,(NH_2)_2CO(aq) \rightarrow 5\,N_2(g) + 2\,CO_2(g) + 4\,H_2O(l)$

21-49 In the Raschig synthesis, hydrazine is produced from aqueous hypochlorite ion and ammonia:

$$2\,NH_3(aq) + OCl^-(aq) \rightarrow N_2H_4(aq) + H_2O(l) + Cl^-(aq)$$

A two-step mechanism that involves NH_2Cl as an intermediate is:

$$NH_3(g) + OCl^-(aq) \rightarrow NH_2Cl(aq) + OH^-(aq)$$
$$NH_2Cl(aq) + NH_3(aq) + OH^-(aq) \rightarrow N_2H_4(aq) + H_2O(l) + Cl^-(aq)$$

21-51 a) The molar free energies of formation (in kJ mol^{-1}) of the reactants and products are:

$$
\begin{array}{ccccccc}
 & 6\,Fe_2O_3(s) & + & N_2H_4(aq) & \rightarrow & 4\,Fe_3O_4(s) & + N_2(g) & +2\,H_2O(l) \\
\Delta G_f^\circ & -742.2 & & 128.1 & & -1015.5 & 0 & -237.18
\end{array}
$$

Combine these in the usual way to obtain ΔG° of the reaction:

$$\Delta G^\circ = 4(-1015.5) + 1(0) + 2(-237.18) - 6(-742.2) - 1(128.1) = \boxed{-211.3 \text{ kJ}}$$

b) The proposal is no good. When hydrazine reacts with rust it acts as a reducing agent. Under highly acidic conditions, hydrazine gains H^+ to form the hydrazinium ion $N_2H_5^+$ which is a good oxidizing agent. The suggested improvement would probably aggravate the rusting problem.

21-53 The structure is $\boxed{H_2PO(OH)}$, in which two of the hydrogen atoms are bound directly to the central P. The acid $HP(OH)_2$, which has two acidic hydrogen atoms, would be orders of magnitude weaker. See text Table 21-1. On this basis, $H_2PO(OH)$ should be $\boxed{\text{monoprotic}}$.

---

[10]See the example on page 26 in this Guide to review how this calculation works.

Chapter 22

Chemical Processes Based on Halogens and the Noble Gases

The halogens include chlorine, bromine, iodine and fluorine, and comprise Group VII of the periodic table. The noble gases are helium, neon, argon, krypton, and radon. They are Group VIII.

Chemicals from Salt

Glassmaking requires sodium carbonate (soda ash), and soapmaking requires a cheap base that is more soluble than calcium hydroxide (slaked lime). A third important process, the bleaching of cloth or fibers, requires a quick-acting bleach. Suitable chemicals for all these purposes come from sodium chloride, by a variety of production methods.

In the **Leblanc process** for sodium carbonate (**22-3**), sodium chloride was heated with sulfuric acid to form sodium sulfate. Heating sodium sulfate with a mixture of carbon (from coal) and limestone ($CaCO_3$) gives a mixture from which the water-soluble sodium carbonate could be extracted. The Leblanc process, which is no longer used, had two noxious by-products—CaS and HCl. Oxidation of the HCl from the Leblanc process to Cl_2 either by reaction with the mineral pyrolusite (MnO_2) or by the **Deacon process** changed the HCl from a nuisance to an asset because chlorine, in the form of **bleaching powder** CaCl(OCl) makes an excellent bleach for textiles.

The **Solvay process** for sodium carbonate supplanted the Leblanc process. In it, the net reaction is the conversion of NaCl and $CaCO_3$ to Na_2CO_3 and the by-product $CaCl_2$. See **2-53**. Ammonia and ammonium chloride are important intermediates that are recycled through this continuous process. Much Na_2CO_3 is now obtained by mining **trona**, and sodium hydroxide (formerly made by reacting Solvay Na_2CO_3 with $Ca(OH)_2$) is now largely produced electrolytically.

Sodium hydroxide and chlorine are produced from salt by the electrolysis of concentrated aqueous solutions. The by-product is hydrogen, which has many uses. In the **diaphragm cell** for the electrolysis of brine, the anode is made of titanium coated with a noble metal, and the cathode is steel. The two are separated by a diaphragm that prevents the mixing of the product gases Cl_2 (anode) and H_2 (cathode) but allows migration of ions to preserve electrical neutrality. Sodium hydroxide, a strong soluble base, has many uses in chemical and materials processing.

The Chemistry of Chlorine, Bromine and Iodine

This section treats the properties, reactions and compounds of the elements of Group VII, the halogens, with the exception of fluorine. One format for studying such material is to outline it under the following headings:

1. Occurrence and preparation of the element

2. Characteristic chemical and physical properties

3. Important compounds and types of compounds

4. Uses in society

The halogens exhibit many regular trends in their properties in accord with the predictions of the periodic law. All are reactive non-metals that serve as oxidizing agents. Their oxidizing strength decreases regularly going down the group. The solution chemistry of the halogens can in large part be summarized in a set of reduction potential diagrams. (see page 349 of this Guide). Many of the potentials come from text Appendix E.

Reduction Potential Diagrams of the Halogens
Acidic Solution (pH 0)

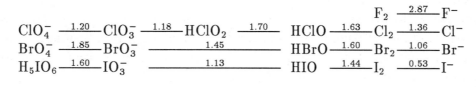

Basic Solution (pH 14)

Observe how much these diagrams can convey. They show, for example:

- The high reduction potential of iodine(VII) in acid solution. Iodine(VII) can oxidize $Mn^{2+}(aq)$ to $MnO_4^-(aq)$ because the reduction potential of the MnO_4^- to Mn^{2+} couple is only about 1.5 V.

- That F_2 oxidizes Cl^-. In addition the diagrams show that each elemental halogen will oxidize all halide ions (X^- ions) in compounds of the halogens beneath it in the periodic table. See **22-43**.

- That the chemistry of periodic acid involves complications not present in perbromic and perchloric acids. The formulas of the major compound of iodine(VII) both in acid and base show the expansion of the iodine coordination sphere to include additional water molecules.

Fluorine and Its Compounds

The chemistry of fluorine is discussed under a separate heading because of the extreme differences between it and the other halogens. These include:

- Fluorine exhibits only the 0 and -1 oxidation states unlike the other halogens, all of which can be oxidized to positive oxidation states ($+1$, $+3$, $+5$, and $+7$).

- Only fluorine of all the halogens has been shown to form compounds with any of the noble gases (the Group VIII elements).

- The dissociation energy of the F_2 molecule is abnormally low.

- The oxidizing ability of fluorine is extremely high. This is related to its high electronegativity, its low dissociation energy, and the strong bonds it forms with other elements.

In studying the descriptive material here, note how the theoretical concepts of previous chapters are put to work in systematizing great quantities of chemical information. For example, fluorine forms compounds with nearly every other element. It might be possible to memorize the ΔH_f° of all of these compounds. But linking the data to the periodic table allows useful *general* conclusions. Look for valid generalizations and avoid memorizing minutiae.

Fluorine atoms form weak bonds to other fluorine atoms but particularly strong bonds to most other kinds of atoms. These facts underlie the extraordinary chemical reactivity of F_2 and the difficulties encountered in the first attempts to prepare it in elemental form. Modern methods for preparing fluorine are based on the original 1886 electrolytic process, and employ KF dissolved in HF, both fluorides derived from natural sources in fluoroapatite or fluorite minerals. Fluorine displaces other halogens from their compounds and oxidizes elements in compounds exposed to it. Elemental fluorine is used in uranium production and isotope separation (see **4-51**), and in the manufacture of sulfur hexafluoride. Calcium fluoride is used in the steel industry, and hydrogen fluoride is used to make synthetic cryolite for aluminum production.

Most fluorine compounds are made directly or indirectly from hydrogen fluoride, a colorless hydrogen-bonded liquid. The compounds of fluorine with metals in their lower oxidation states are salts with strong ionic character. The higher oxidation states of many of the same elements yield volatile fluorides with significant covalent character (see **22-29**). Compounds of fluorine with the nonmetals possess a wide range of reactivities and molecular geometries. Some compounds, such as SF_6, are quite inert toward chemical reaction; others such as BF_3 are strong Lewis acids (see **3-27** and **22-27**) that readily accept a share in electron pairs from fluoride ions or other Lewis bases; still others, such as SbF_5, are strong fluorinating agents. One of the strongest of these agents, PtF_6, opened up modern research on the compounds of the noble gases by its reaction with xenon; xenon and krypton fluorides are made by direct reaction with fluorine or with other fluorinating agents. Xenon fluorides react with water to yield xenon trioxide.

Fluorinated hydrocarbons arise by the substitution of fluorine for hydrogen in hydrocarbons and often have greater thermal and chemical stability than the parent hydrocarbon. The release of chlorine upon eventual decomposition of **chlorofluo-rocarbons** in the outer atmosphere leads to depletion of the ozone in that region. The unique properties of fluorine shape the properties of the important fluorinated polymer Teflon (**22-55**).

Compounds of Fluorine and the Noble Gases

The noble gases are in Group 0 (VIII) of the periodic table. Their principal uses come from their physical properties. The unreactive character of these elements prevents their entry into chemical compounds with most other elements. Fluorine and oxygen form several compounds with Xe. Note that the ΔH_f°'s of all of the xenon fluorides are negative (see **22-35**). The compounds are thermodynamically stable with respect to their elements.

• **The problems in Chapter 22 all involve the application of previously studied principles to the particular cases of the halogens and noble gases.**

Detailed Solutions to Odd-numbered Problems

22-1 Use ΔG_f° data (text Appendix D) to compute $\Delta G_{298.15}^\circ$ for this reaction: The set-up is:

$$\Delta G_{298.15}^\circ = 2\underbrace{(-36.8)}_{\text{OCl}^-(aq)} + 1\underbrace{(-237.18)}_{\text{H}_2\text{O}(l)} - 2\underbrace{(-157.24)}_{\text{OH}^-(aq)} - 1\underbrace{(97.9)}_{\text{Cl}_2\text{O}(g)} = -94.2 \text{ kJ}$$

Use this value to compute K at this temperature:

$$\ln K_{298.15} = \frac{\Delta G_{298.15}^\circ}{-RT} = \frac{-94.2 \times 10^3 \text{ J}}{(8.315 \text{ J K}^{-1}\text{mol}^{-1})(298.15 \text{ K})} = 38.0 \quad K_{298.15} = \boxed{3.2 \times 10^{16}}$$

22-3 The overall reaction in the Leblanc process is:

$$2\,\text{NaCl}(s) + \text{H}_2\text{SO}_4(l) + 2\,\text{C}(s) + \text{CaCO}_3(s)$$
$$\rightarrow \text{Na}_2\text{CO}_3(s) + 2\,\text{CO}_2(g) + \text{CaS}(s) + 2\,\text{HCl}(g)$$

Combine the enthalpies of formation from text Appendix D in accord with Hess's law to compute the ΔH° of this process:

$$\Delta H^\circ = 1\underbrace{(-1130.68)}_{\text{Na}_2\text{CO}_3(s)} + 2\underbrace{(-393.51)}_{\text{CO}_2(g)} + 1\underbrace{(-482.41)}_{\text{CaS}(s)} + 2\underbrace{(-92.31)}_{\text{HCl}(g)}$$
$$- 2\underbrace{(-411.15)}_{\text{NaCl}(s)} - 1\underbrace{(-813.99)}_{\text{H}_2\text{SO}_4(l)} - 2\underbrace{(0.00)}_{\text{C}(s)} - 1\underbrace{(-1206.92)}_{\text{CaCO}_3(s)} = \boxed{+258.48 \text{ kJ}}$$

The LeBlanc process is $\boxed{\text{endothermic}}$.

22-5 The production of Cl_2 is tied to the production of NaOH by the stoichiometry of the reaction: $2\,\text{NaCl} + 2\,\text{H}_2\text{O} \rightarrow \text{Cl}_2 + 2\,\text{NaOH} + \text{H}_2$. According to this equation, 2 mol of NaOH is produced per 1 mol of Cl_2. The mass of 2 mol of NaOH is 80.0 g, and the mass of 1 mol of Cl_2 is 70.9 g. If there were no losses of either material during production, the ratio of their masses would always equal $\boxed{1.13 \text{ to } 1}$. Evidently, more NaOH is lost in practice than Cl_2, reducing this ratio to the 1.05 quoted in the problem.

22-7 The problem reverses the more common task of determining an equilibrium constant from ΔG° data. It gives an equilibrium constant (in slightly disguised form) and asks for an estimate of ΔG°. The reaction is: $\text{Br}_2(l) \rightleftharpoons \text{Br}_2(aq)$. At 25°C, 33.6 g of bromine dissolves in one liter of water. At saturation, the $l \rightleftharpoons aq$ equilibrium exists. The concentration of $\text{Br}_2(aq)$ at saturation is 0.210 M.[1] The mass-action expression

---

[1] Taking the molar mass of Br_2 as 159.82 g mol^{-1}.

for the reaction is: $K = [Br_2](aq)$, so K is 0.210. The equilibrium constant and $\Delta G°$ at this temperature are related:

$$\Delta G° = -RT \ln K = -(8.315 \text{ J K}^{-1}\text{mol}^{-1})(298.15 \text{ K}) \ln(0.210) = \boxed{3.87 \times 10^3 \text{ J}}$$

for the reaction as written, which is the reaction giving 1 mol of $Br_2(aq)$. The $\Delta G_f°$ for one mole of $Br_2(aq)$ is +3.93 kJ, according to text Appendix D.

Tip. *Some* Br_2 does dissolve despite the fact that $\Delta G°$ is positive.

22-9 For the reaction:

$$2\,Br^-(aq) + Cl_2(g) \rightleftharpoons Br_2(g) + 2\,Cl^-(aq)$$

$$\Delta G° = 2\underbrace{(-131.23)}_{Cl^-(aq)} + 1\underbrace{(\ 3.14)}_{Br_2(g)} - 2\underbrace{(-103.96)}_{Br^-(aq)} = -51.40 \text{ kJ}$$

where the numbers come from text Appendix D. Compute the equilibrium constant as follows:

$$\ln K_{298.15} = \frac{-\Delta G°_{298.15}}{RT} = \frac{-(-51.40 \times 10^3 \text{ J})}{(8.315 \text{ J K}^{-1}\text{mol}^{-1})(298.15 \text{ K})} = 20.7$$

from which K is $\boxed{1.0 \times 10^9}$. The $K_{298.15}$ is large enough to make the extraction of bromine from sea water by oxidation with $Cl_2(g)$ thermodynamically $\boxed{\text{feasible}}$.

Tip. K can also be obtained for the related reaction that gives liquid rather than gaseous bromine:

$$2\,Br^-(aq) + Cl_2(g) \rightleftharpoons Br_2(l) + 2\,Cl^-(aq)$$

Obtain $\Delta\mathcal{E}°$ for this reaction by combining the standard reduction potentials of its two half-reactions (Appendix E). The result is 0.293 V. Substitute this standard potential difference into text equation 12-8:[2]

$$\log K_{298.15} = \frac{n}{0.0592 \text{ V}}\Delta\mathcal{E}° = \frac{2}{0.0592 \text{ V}}(0.293 \text{ V}) = 9.90$$

It follows that $K_{298.15}$ equals 7.9×10^9.
Thermodynamic feasibility is not enough. If the chlorine is cheap and the reaction is fast, the process might also be economically feasible.

---

[2]Text page 420.

22-11 a) The balanced equation is: $Br^-(aq) + H_3PO_4(aq) \rightarrow H_2PO_4^-(aq) + HBr(g)$.
Other equations (ones that involve NaBr as a reactant and give PO_4^{3-} or Na_3PO_4 or
other forms of phosphate as product) are possible. This ionic equation is best.

b) The 100 mL of solution contains 0.050 mol $L^{-1} \times 0.100$ L $= 0.0050$ mol of $Br^-(aq)$.
According to the balanced equation, 0.0050 mol of $HBr(g)$ can form. If it does, and
if it behaves ideally, its volume is:

$$V_{HBr} = \frac{n_{HBr}RT}{P} = \frac{(0.0050 \text{ mol})(0.08206 \text{ L atm mol}^{-1}\text{K}^{-1})}{1 \text{ atm}} = \boxed{1.1 \text{ L}}$$

22-13 A reaction involving reactants and products in standard states is spontaneous
at constant T and P if $\Delta G° < 0$. Because $\Delta G° = -nF\Delta\mathcal{E}°$, a reaction is likewise
spontaneous if $\Delta\mathcal{E}° > 0$. If the first half-reaction in the problem runs as a reduction
(at the cathode) and the second as an oxidation (at the anode), then:

$$\Delta\mathcal{E}° = \mathcal{E}°(\text{cathode}) - \mathcal{E}°(\text{anode}) = 0.954 - (-0.25) = 1.20 \text{ V} > 0$$

The change is spontaneous as written. The balanced whole reaction is:

$$2\,ClO_2(g) + 2\,OH^-(aq) \rightarrow ClO_2^-(aq) + ClO_3^-(aq) + H_2O(l)$$

Thus, ClO_2 is $\boxed{\text{unstable}}$ in standard basic solution.

22-15 The standard reduction potential diagram for iodine at pH 14 shows that $I_2(s)$
disproportionates spontaneously to $IO^-(aq)$ and $I^-(aq)$:

$$I_2(s) + 2\,OH^-(aq) \rightarrow I^-(aq) + IO^-(aq) + H_2O(l)$$

This is because 0.535 V, the standard reduction potential for $I_2(s)$ to $I^-(aq)$, exceeds
0.45 V, the potential for $IO^-(aq)$ to $I_2(aq)$. But $IO^-(aq)$ itself disproportionates
spontaneously to $IO_3^-(aq)$ and $I^-(aq)$:

$$3\,IO^-(aq) \rightarrow IO_3^-(aq) + 2\,I^-(aq)$$

as shown by a comparison of the IO_3/IO^- and IO^-/I^- standard reduction potentials.
The overall balanced equation for the disproportionation is:

$$6\,OH^-(aq) + 3\,I_2(aq) \rightarrow IO_3^-(aq) + 5\,I^-(aq) + 3\,H_2O(l) \quad \Delta\mathcal{E}° = 0.49 - 0.14 = \boxed{0.35 \text{ V}}$$

b) The half-reaction: $IO_3^-(aq) + 3\,H_2O(l) + 6\,e^- \rightarrow I^-(aq) + 6\,OH^-(aq)$ is the sum of:

$$IO_3^- + 2\,H_2O + 4\,e^- \rightarrow IO^- + 4\,OH^- \quad \mathcal{E}° = 0.14 \text{ V}$$
$$IO^- + H_2O + 2\,e^- \rightarrow I^- + 2\,OH^- \quad \mathcal{E}° = 0.49 \text{ V}$$

The first half-reaction transfers 4 electrons, and the second transfers 2 electrons for an overall transfer of 6 e^-'s:

$$\mathcal{E}° = \frac{(4 \times 0.14) + (2 \times 0.49)}{6} = \boxed{0.26 \text{ V}}$$

22-17 a) $2\,F_2(g) + 2\,SrO(s) \rightarrow 2\,SrF_2(s) + O_2(g)$
b) $2\,F_2(g) + O_2(g) \rightarrow 2\,OF_2(g)$ **c)** $UF_4(s) + F_2(g) \rightarrow UF_6(g)$

22-19 Calculate the relative number of moles of Na and F. Their ratio is $1:3$. Enough hydrogen must be present to balance the -1 charges of the extra two fluoride ions. Therefore, the empirical formula is $\boxed{NaF\cdot 2HF}$.

22-21 The compound is 38.35% Cl and 61.65% F, which corresponds to the empirical formula ClF_3. The molecular formula is then Cl_nF_{3n} where n is an integer. To obtain the molecular formula, determine the molar mass. The molar mass of an ideal gas can be calculated from its measured vapor density ρ at any conditions of T and P. It can be shown from the ideal-gas equation and the definition of molar volume[3] that:

$$\mathcal{M}_{vap} = \frac{\rho_{vap} RT}{P}$$

Substitution of the P and T values in this case along with the vapor density give a molar mass of 92.5 g mol^{-1}. When $n = 1$, the above molecular formula gives exactly this molar mass. Hence the molecular formula is $\boxed{ClF_3}$.

22-23 The shapes of all of these compounds can be predicted using the valence shell electron pair repulsion (VSEPR) theory.[4]

a) OF_2 has a central O atom with a steric number of 3. This includes the two F atoms and one lone pair of electrons. The molecule is $\boxed{bent}$.

b) BF_3 has a central B atom with SN 3. The molecule is $\boxed{trigonal\ planar}$.

c) In BrF_3, the central Br has SN of 5. The molecule has a $\boxed{distorted\ T}$ shape.

d) In BrF_5, the central Br is surrounded by five F atoms and one lone pair for a SN of 6. The molecule has a $\boxed{square\ pyramidal}$ shape.

e) In IF_7, the central I is surrounded by seven F atoms. The SN of the central I is therefore 7. One arrangement that minimizes repulsions among seven electron

---

[3]See text page 29.
[4]See page 56 in this Guide.

pairs distributed about a center is a pentagonal bipyramid. See **3-73b**. This is the structure of IF_7.

f) In SeF_6, the central Se is surrounded by six F atoms and no lone pairs. The SN is 6, and the predicted geometry is octahedral.

22-25 a) Molecules of thionyl difluoride SOF_2 and selenyl difluoride $SeOF_2$ should have the same geometry because the central atoms of the two have the same number of valence electrons and the same number of bonded atoms. Use the VSEPR model to predict the molecular geometry. The central atoms of the molecules have SN 4 with one lone pair and three bonded atoms. The predicted molecular geometry is pyramidal.

b) Recall that Lewis bases are electron-pair donors and Lewis acids are electron-pair acceptors. In the reaction between SOF_2 and BF_3, the SOF_2 molecule donates a lone pair of electrons to the BF_3 molecule, and a coordinate covalent bond results:

$$F_3B + : SOF_2 \rightarrow F_3B—SOF_2$$

Thus, SOF_2 is acting as a Lewis base.

22-27 Fluorine chemists often think of fluoride-ion donors as bases and fluoride-ion acceptors as acids. The autoionization equilibrium in liquid BrF_3 is:

$$2\,BrF_3(l) \rightleftharpoons BrF_4^-(solv) + BrF_2^+(solv)$$

The reaction can be seen as a competition between two Lewis acids, BrF_3 and BrF_2^+, for a Lewis base, F^-. One side of the competition is the acceptance of an electron pair from the Lewis base F^- by the Lewis acid BrF_3 to create a bond:

$$F_3\ddot{Br} + (: \ddot{F} :)^- \rightarrow (F_3\ddot{Br}—\ddot{F} :)^-$$

Vying against this is the acceptance of an electron pair from the Lewis base F^- by the Lewis acid BrF_2^+:

$$(F_2\ddot{Br})^+ + (: \ddot{F} :)^- \rightarrow F_2\ddot{Br}—\ddot{F} :$$

Because BrF_2^+ is a much stronger Lewis acid than BrF_3, the first of these reactions lies far to the left at equilibrium and the second lies far to the right. Still, some autoionization occurs, just as in H_2O.

22-29 The molecule with the least ionic character, and weakest intermolecular interactions is expected to have the lowest boiling point. The ranking would therefore be $PtF_6 < PtF_4 < CaF_2$. In fact, PtF_4 decomposes upon melting, and so does not possess a normal boiling point.

22-31 The reaction of interest is: $C_2F_4(g) \rightarrow C(s) + CF_4(g)$. Combine the ΔH_f°'s to obtain ΔH°:

$$\Delta H^\circ = 1 \underbrace{(-925)}_{CF_4(g)} + 1 \underbrace{(0)}_{C(s)} - 1 \underbrace{(-651)}_{CF_4(g)} = -274 \text{ kJ}$$

Thus, if one mol of C_2F_2 were to explode, 274 kJ of heat would be released. The molar mass of C_2F_2 is 100 g mol^{-1}. The 1.00 kg of C_2F_4 therefore is 10.0 mol. Its explosion would release about $\boxed{2740 \text{ kJ}}$.

22-33 The redox reaction: $XeF_6(g) + 3 H_2(g) \rightarrow Xe(g) + 6 HF(g)$ has ΔH° equal to the sum of the ΔH_f°'s of the products minus the sum of the ΔH_f°'s of the reactants:

$$\Delta H^\circ = 1 \underbrace{\Delta H_f^\circ}_{Xe(g)} + 6 \underbrace{\Delta H_f^\circ}_{HF(g)} - 3 \underbrace{\Delta H_f^\circ}_{H_2(g)} - 1 \underbrace{\Delta H_f^\circ}_{XeF_6(g)}$$

Inserting data from the problem and text Appendix D gives:

$$-1282 \text{ kJ} = 1 \underbrace{(0)}_{Xe(g)} + 6 \underbrace{(-271.1)}_{HF(g)} - 3 \underbrace{(0)}_{H_2(g)} - 1 \underbrace{(\Delta H_f^\circ)}_{XeF_6(g)}$$

Solving gives the ΔH_f° for $XeF_6(g)$. It is $\boxed{-345 \text{ kJ mol}^{-1}}$.

b) From the preceding part and text Appendix D:

$$XeF_6(g) \rightarrow Xe(g) + 3 F_2(g) \qquad \Delta H^\circ = 345 \text{ kJ}$$

$$\frac{1}{2} F_2(g) \rightarrow F(g) \qquad\qquad\qquad \Delta H^\circ = 78.99 \text{ kJ}$$

Multiplying the second equation by 6 and adding it to the first gives:

$$XeF_6(g) \rightarrow Xe(g) + 6 F(g) \quad \Delta H^\circ = 818.94 \text{ kJ}$$

The ΔH° here is the enthalpy change to atomize 1 mol of $XeF_6(g)$. The average bond enthalpy of the Xe—F bond is 1/6 of this or $\boxed{136 \text{ kJ mol}^{-1}}$.

22-35 Calculate ΔH° for the hydrolysis:

$$XeF_6(g) + H_2O(l) \rightarrow XeOF_4(l) + 2 HF(g)$$

by combining ΔH_f° values according to Hess's law. The first and third in the following are given in the problem; the others are from text Appendix D:

$$\Delta H^\circ = 1 \underbrace{(148)}_{XeOF_4(l)} + 2 \underbrace{(-271)}_{HF(g)} - 1 \underbrace{(-298)}_{XeF_6(g)} - 1 \underbrace{(-285.83)}_{H_2O(l)} = 190 \text{ kJ}$$

The standard enthalpy change for the hydrolysis of $XeF_6(g)$ is $\boxed{190 \text{ kJ mol}^{-1}}$.

22-37 The balanced half-equations are:

$$Mn^{2+}(aq) + 12\,H_2O(l) \rightarrow MnO_4^-(aq) + 8\,H_3O^+(aq) + 5\,e^-$$
$$XeO_6^{4-}(aq) + 12\,H_3O^+(aq) + 8\,e^- \rightarrow Xe(g) + 18\,H_2O(l)$$

The oxidation of the Mn^{2+} releases $5\,e^-$ per ion, and the reduction of the XeO_6^{4-} consumes $8\,e^-$ per ion. The overall equation is accordingly:

$$8\,Mn^{2+}(aq) + 5\,XeO_6^{4-}(aq) + 6\,H_2O(l) \rightarrow 8\,MnO_4^-(aq) + 5\,Xe(g) + 4\,H_3O^+(aq)$$

22-39 Gaseous $\boxed{\text{chlorine}}$ is a stronger oxidizing agent than $Br_2(l)$ because the reduction potential for Cl_2/Cl^- is more positive than the reduction potential for Br_2/Br^-. The stronger oxidizing agent will be the stronger bleach.

22-41 (a) A current of $100\,000$ A corresponds to the passage of $100\,000$ C s^{-1}. There are $86\,400$ seconds in a day, so 8.64×10^9 C passes through each of the 250 cells daily. This equals 8.95×10^4 mol of electrons per cell (computed using the Faraday constant ($\mathcal{F} = 96485$ C mol^{-1}). As each mole of electrons that passes through a cell, $1/2$ mol of Cl_2 is produced, according to the half-reaction $Cl^- \rightarrow Cl_2 + 2\,e^-$. Therefore, 4.48×10^4 mol of Cl_2 is produced by one cell in a day. This is 3.17×10^6 g of Cl_2 per cell. The total production is 7.94×10^8 g of Cl_2 per day for the 250-cell plant. This is $\boxed{794 \text{ metric tons}}$ of Cl_2.

b) The power consumption in watts equals the operating potential in volts multiplied by the current. Each cell uses 3.5 V $\times 100\,000$ A $= 350\,000$ W. The set of 250 cells uses 250 times this: 8.75×10^7 W. Multiplying this power by the 24 hours in a day gives 2.10×10^6 kilowatt-hour as the daily energy consumption of the plant. Since a kWh is 3.600×10^6 J,[5] the daily energy consumption is $\boxed{7.56 \times 10^{12} \text{ J}}$.

c) The electric bill for one day at this plant equals $\boxed{\$105,000}$.

---

[5]See **12-73**.

22-43 a) The trend in acid strength among the hydrogen halides runs:

$$HI > HBr > HCl$$

Therefore HAt should be an even $\boxed{\text{stronger}}$ acid than HI.

b) The question concerns $\mathcal{E}^\circ$ for the half-reaction:

$$At_2(s) + 2\,e^- \rightarrow 2\,At^-(aq)$$

$At_2(s)$ oxidizes $Zn(s)$ to $Zn^{2+}(aq)$, and $\mathcal{E}^\circ$ for the $Zn^{2+}(aq)/Zn(s)$ half-reaction is -0.763 V. Therefore $\mathcal{E}^\circ$ for the $At_2(s)/At^-(aq)$ reduction exceeds -0.763 V. However, At_2 does not oxidize $Fe^{2+}(aq)$ to Fe^{3+}. Since the standard reduction potential for $Fe^{3+}(aq)$ to $Fe^{2+}(s)$ is 0.770 V, $\mathcal{E}^\circ$ for the reduction of $At_2(s)$ is less than 0.770 V. The range is therefore from $\boxed{-0.763 \text{ to } +0.770 \text{ V}}$. On the basis of the periodic trends among the halogens, we also expect $\mathcal{E}^\circ$ for At_2/At^- is be less than $\mathcal{E}^\circ$ for I_2/I^-. This would make the range -0.763 to 0.535 V.

c) $Cl_2(g)$ will easily oxidize $At^-(aq)$: $2\,At^-(aq) + Cl_2(g) \rightarrow 2\,Cl^-(g) + At_2(s)$.

d) Solid At_2 should have an even greater tendency to disproportionate in aqueous base than $I_2(s)$. Thus:

$$\boxed{6\,At_2(s) + 12\,OH^-(aq) \rightarrow 2\,AtO_3^-(aq) + 10\,At^-(aq) + 6\,H_2O(l)}$$

22-45 Fluorine chemists often think of fluoride-ion acceptors as Lewis acids and fluoride-ion donors as Lewis bases.[6] The reaction is:

$$K_2[NiF_6](s) + TiF_4(s) \rightarrow K_2[TiF_6](s) + NiF_2(s) + F_2(g)$$

TiF_4 accepts F^- ions. Thus, TiF_4 is acting as a $\boxed{\text{Lewis acid}}$.

22-47 The standard reduction potential $\mathcal{E}^\circ$ for the reduction of $F_2(g)$ to $F^-(aq)$ is 2.87 V. This is much more positive than the 0.815 V potential for the reduction of $O_2(g)$ to 10^{-7} M $OH^-(aq)$ ion. Because 1 M $F^-(aq)$ has a very much smaller tendency to be oxidized than 10^7 M $OH^-(aq)$, the product at the anode in the electrolysis of NaF is $O_2(g)$, not $F_2(g)$; $F_2(g)$ is $\boxed{\text{not obtained.}}$

22-49 The balanced equation: $4\,HF(aq) + SiO_2(s) \rightarrow SiF_4(g) + 2\,H_2O(l)$ shows that 4 mol of $HF(aq)$ is needed to dissolve 1 mol of quartz rock. The rock contains mere traces of gold (1 part in 10^5 by mass). Suppose enough quartz is dissolved to recover

---

[6]See problem **10-7** in this Guide.

1 troy ounce (31.3 g) of gold. This is 10^5 times 31.3 g of quartz. The cost of the required HF is:

$$31.3 \times 10^5 \text{ g quartz} \times \left(\frac{1 \text{ mol quartz}}{60.1 \text{ g}}\right) \times \left(\frac{4 \text{ mol HF}}{1 \text{ mol quartz}}\right) \times \left(\frac{20.0 \text{ g HF}}{1 \text{ mol HF}}\right)$$

$$\times \left(\frac{2 \text{ g HF sol'n}}{1 \text{ g HF}}\right) \times \left(\frac{1 \text{ L sol'n}}{1170 \text{ g sol'n}}\right) \times \left(\frac{\$0.25}{1 \text{ L sol'n}}\right) = \$1780$$

The solvent cost more that the \$350 the troy ounce of gold would bring. The method is a $\boxed{\text{losing proposition}}$. It breaks even on materials when the ore is 5.1 times richer in gold (5.1 is the ratio of \$1780 to \$350). For true feasibility the cost of labor and equipment would have to be covered, too.

22-51 Follow the rules in Chapter 2 for writing Lewis dot structures. The fluorine atoms in O_2F_2 are terminal:

$$:\ddot{F} - \ddot{O} - \ddot{O} - \ddot{F}:$$

The bond order of the O—F bond is 1, and the bond order of the O—O bond is 1. Each F—O—O bond angle is roughly 109° reflecting the steric number of four at each O atom. The analogous compound of oxygen and hydrogen is $\boxed{\text{hydrogen peroxide}}$.

22-53 Use the VSEPR model[7] to predict the molecular geometry. The Lewis structure of the SF_5^- ion is:

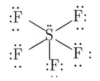

The central S atom in this ion has *SN* 6 with 1 lone pair and 5 bonded atoms; the molecular geometry is square pyramidal.

22-55 Teflon has the formula $(CF_2)_n$. It consists of many CF_2 monomer units linked together in a chain. The mass percentage of fluorine in the polymer is the same as the mass percentage of fluorine in the simplest repeating unit. The first value of mass percentage of fluorine (48.4%) was quite low compared to the actual mass percentage (76.0%) and gave the apparent formula $CF_{0.59}$. There should be no chlorine in Teflon, but under the circumstances it was wise to check.

---

[7]Text page 80.

22-57 The reaction of interest is: $Xe(g) + 2\,F_2(g) \rightarrow XeF_4(s)$. The given free energy of formation applies directly to this reaction. Use the relationship $\Delta G° = -RT \ln K$:

$$\ln K = \frac{\Delta G°}{-RT} = \frac{-134 \times 10^3 \text{ J}}{(8.315 \text{ J K}^{-1}\text{mol}^{-1})(298.15 \text{ K})} = 54.1 \quad \text{from which } K = \boxed{3 \times 10^{23}}$$

Chapter 23

Ceramic Materials

Naturally Occurring Minerals: the Lithosphere

The earth (approximate radius 6370 km) consists of the three distinct parts: **core** (thickness 3470 km), **mantle,** (2880 km) and **crust** (17 km). The thin crust is the **lithosphere.** The lithosphere is the region of contact between the cool atmosphere and the warm interior of the earth. It is in a state of continual chemical transformation.

Silicates

Compounds containing silicon and oxygen make up much of the earth's crust. The strong bond between Si and O is the basis for a large class of minerals called **silicates**.

The orthosilicate anion, SiO_4^{4-} has a central Si surrounded by four O's lying at the corners of a tetrahedron. It is the basic structural unit in silicates. The molar ratio of Si to O in this anion is $1:4$. Orthosilicate groups can link to each other by sharing O atoms. As several SiO_4^{4-} tetrahedra link together by sharing corners, the number of oxygen atoms per silicon diminishes.

Example: Determine the charge on the cyclosilicate ion in Figure 23-1c.[1] **Solution:** The ion contains three Si atoms. There are three links between SiO_4 tetrahedra in this cyclic structure. This reduces the number of oxygen atoms from 12, the number in $(SiO_4)_3$, to 9. In silicates the oxidation number of Si is $+4$ and of O -2. Hence, the ion is $Si_3O_9^{6-}$. Another approach notes that silicon atoms have a zero formal charge in silicates, that oxygen atoms with just one bond in a silicate always have a -1 formal charge, and that oxygen atoms that are shared between silicon atoms have two bonds and a formal charge of zero. Three oxygen atoms are shared between the SiO_4 tetrahedra in text Figure 23-1c, and six oxygen atoms on the perimeter of the

---

[1]Text page 810.

structure. The sum of the formal charges of all atoms is −6, hence the net charge on the ion is −6. Compare to **23-1**.

A wide variety of silicate structures occurs in nature, ranging from single tetrahedra to pairs, rings, chains, sheets, and three-dimensional networks of linked tetrahedra, with cations interspersed to balance the total charge in the crystal. The ratio of the number of silicon to oxygen atoms provides a means to classify silicate structures. The criteria for classification are summarized in Table 23-2.[2] The minimum Si : O ratio is 1 : 4; the maximum is 1 : 2. There are two complications:

- Not all of the oxygen atoms in a mineral are part of the silicate structural system. Some may be present in water of crystallization, for example. Such casual oxygen atoms are not counted in determining the Si : O ratio. See **23-3**.

- Aluminum can substitute for silicon in silicate structures. If it does, it must be counted as an Si atom in predicting silicate structures. See **23-5**. Note that aluminum in silicate minerals does not always replace silicon. Sometimes it is present only as a cation to maintain electrical neutrality.

Aluminosilicates

When an Al(III) replaces an Si(IV) in tetrahedral coordination in a silicate mineral, an **aluminosilicate** results. Because Al(III) contributes only three valence electrons instead of four, aluminosilicates must obtain one additional electron per Al(III) to achieve a bonding pattern equivalent to that in silicates. These electrons generally come from metal atoms. Accordingly, aluminosilicates generally contain cations interspersed in the vicinity of their aluminum atoms and sufficient in number to just balance the net negative charge on the aluminosilicate network or chain. **Feldspars** and **micas** are two important kinds of aluminosilicate minerals.

Clays and Zeolites

In **clay** minerals, water can be absorbed between sheets of bonded atoms and cause swelling. These materials are used to seal bore holes in the drilling of oil wells, in construction, and in pottery making. **Zeolites** are aluminosilicates with three-dimensional network structures that are open enough to allow the inclusion of small molecules and ions. As just explained, aluminosilicates always contain cations to maintain electrical neutrality. These cations often reside in the openings in the zeolite framework.

Zeolites are used for ion-exchange. An equilibrium is established as one type of positive ion becomes lodged in the pores and tunnels of the aluminosilicate structure

---

[2]Text page 810.

and another is released. In addition, small neutral molecules like water can be taken up and held in the internal cavities of zeolites. The pore size in zeolites can be adjusted so that small molecules pass through, but large molecules are stopped. Such zeolites serve as **molecular sieves.** Finally, zeolites can hold small molecules in favorable orientations for reaction, thus serving as catalysts.

Geochemistry

Geochemistry is the study of the chemistry of the earth. Understanding earth science requires use of nearly all the principles covered in the first half of the text. Thus, **23-9** could come from Chapter 7, **23-29** from Chapter 3, and **23-33** from Chapter 11.

The **geochemical cycle** starts with the crystallization of a **magma,** a molten silicate fluid, to form **igneous** rocks. **Sedimentary** rocks comes from the weathering, dissolution, and subsequent precipitation of minerals. Many of these minerals show strong **differentiation**, based on chemical factors, from the original material. Another type of change is the **metamorphosis** of sedimentary rocks under heat and pressure. High-grade metamorphism leads to remelting, and magma formation once again.

The text introduces a new device for displaying chemical data. It is the **potential versus pH diagram.** See text Figure 23-8.[3] The key points in understanding and using these diagrams are:

- Inconsistencies in plotting and scaling give confusing results. Always plot *reduction* potentials *(not* oxidation potentials) on the vertical axis. They appear as horizontal lines on the diagram. Then, the top of the diagram is favored in oxidizing environments, and the bottom in reducing environments. Remember that as pH increases (to the right on the horizontal scale) the concentration of $H_3O^+(aq)$ becomes less, (not more).

- The reduction potentials of half-reactions in which H_3O^+ or OH^- do not occur explicitly are pH independent.

- Breaks in potential versus pH lines occur when one half-reaction supersedes another. Thus, in **23-7**, the species $Co^{2+}(aq)$ is reduced in acid solution, but $Co(OH)_2(s)$ is reduced in basic solution. A bend in the potential versus pH curve marks the cross-over.

- The slope of the reduction potential versus pH line at 25°C equals -0.0592 V per pH unit. This derives from the logarithm term in the Nernst equation.

---

[3]Text page 817.

- The potential versus pH diagrams identify **stability regions** for the elements in different oxidation states. They help to decide which minerals will form under certain conditions (oxidizing versus reducing and acidic versus basic).

Clapeyron Equation

The Clapeyron equation states the effect of pressure on a chemical equilibrium:

$$\frac{dP}{dT} = \frac{\Delta H}{T \Delta V}$$

This equation gives the slopes of phase coexistence lines in the phase diagrams discussed in Chapter 5.

In using the Clapeyron equation, remember that it is valid *only* at combinations of temperature and pressure for which the system is at equilibrium. A common error is to substitute a temperature which does *not* fit the equation:

$$\Delta G = 0 = \Delta H - T \Delta S$$

and expect the Clapeyron equation to work. See **23-11**. As long as ΔG is 0 for a change taking place at constant temperature and pressure, the quotient ($\Delta H/T$) equals ΔS. An alternate version of the Clapeyron equation then is:

$$\frac{dP}{dT} = \frac{\Delta S}{\Delta V}$$

This form of the equation is used in **23-35**.

One hindrance to understanding the Clapeyron equation arises because it concerns equilibrium at constant temperature and pressure, yet the term dP/dT clearly involves changes in temperature and pressure. Imagine that a system is brought to equilibrium under constant temperature and pressure in separate experiments at many different (P, T) combinations. It would certainly be possible to plot these sets of values on a graph with P on the vertical axis and T on the horizontal axis. The term dP/dT equals the slope of the resulting line.

A second hindrance in applying the Clapeyron equation occurs because it has constantly been assumed in problems that the ΔS of a change does not depend strongly on the temperature. This is true, at least across limited ranges of temperature. Unfortunately, it encourages the incorrect conclusion that ΔS values are also only weakly dependent upon the *pressure*. If the change in the number of moles of gas Δn_g is non-zero in a process, then the ΔS of the process is *strongly* dependent upon pressure. See **23-9**.

The units of dP/dT are Pa K^{-1} (pascal per kelvin) in the SI system. These will be the units of the answer if ΔS (or $\Delta H/T$) is in J K^{-1} and ΔV is in m^3. Volume changes however nearly always derive from densities, which almost never are quoted with m^3 as part of their unit. Do not forget to convert if ΔV is in liters or cubic centimeters or if $(\Delta H/T)$ is in kJ K^{-1}. The unfamiliar Pa K^{-1} unit can readily be converted to atm K^{-1}, if desired. See **23-11**.

The Properties of Ceramics

Ceramics are synthetic materials that have as their essential components inorganic, nonmetallic materials. They are able to resist high temperatures and corrosion, and are stiff and hard, but also brittle, making them liable to failure when stressed. They are subject to **thermal shock**.

Composition and Structure of Ceramics

Ceramics employ a wide range of compounds. Useful ceramic bodies are nearly always a mixture of compounds. The classification of ceramics includes **silicate ceramics**, which all contain the SiO$_4$ group; **oxide ceramics,** in which some elements other than silicon combines with oxygen; and **nonoxide ceramics**, in which oxygen is replaced by another nonmetal as a principal component. Nonoxide ceramics are based on compounds like Si$_3$N$_4$ and SiC.

A **ceramic phase** is any portion of the whole ceramic body that is physically homogeneous and bounded by a surface that separates it from other parts. The **microstructure** of a ceramic piece can be quite complex, with grains of different phases, voids, and pores. Microstructure has a significant effect on the properties of the finished products. Ceramics are made by **firing** to a high temperature, which leads to partial melting, crystal phase changes, and to **sintering**, in which grains grow together and the ceramic body shrinks or **densifies.** The properties of a ceramic piece depend on its microstructure which in turn depends markedly on the exact conditions under which the piece was formed and fired. The biggest problem with ceramics as structural materials is inconsistent quality.

Silicate Ceramics

Many **silicate ceramics** use as their starting material natural clay minerals such as kaolin. Clays are aluminosilicate minerals that contain hydrated cations between the layers of an infinite sheet structure. Firing an object formed of clay drives away the water and, among other reactions and phase changes, leads to the formation of mullite (Al$_6$Si$_2$O$_{13}$), which lends strength to the ceramic. Compare the reaction:

$$3\,Al_2Si_2O_5(OH)_4(s) \rightarrow 3\,Al_6Si_2O_{13}(s) \text{ (mullite) } + 4\,SiO_2(s) + 6\,H_2O(g)$$

to the reaction in **23-13**. A **glaze** is a thin, glassy layer that coats the object, giving it strength and also often decorating it.

Glasses

Glasses are amorphous solids of variable composition. Different compositions of silicate glass are used for different applications. See **23-17**. Pure silica (SiO_2) forms a glass that has a high melting point. The melting point is lowered by the addition of sodium oxide and calcium oxide to give the **soda-lime** glass used in most applications. Incorporating boron oxide in silicate glass yields a glass with better resistance to thermal shock because it has a smaller coefficient of linear thermal expansion. Soda-lime glass has the approximate formula $Na_2O \cdot CaO \cdot 6SiO_2$. Its structure consists of SiO_4 tetrahedra linked in a random three dimensional network. The cations (Na^+ and Ca^{2+}) occupy voids in the network and maintain electrical neutrality. Glasses are **isotropic** in their physical properties (the properties are the same in all directions). They soften over a range of temperatures rather than melt sharply. This makes it possible to work glass and to **anneal** it. Annealing reduces internal stresses that freeze in when a glass body hardens rapidly.

Cements

Portland cement is another silicate ceramic. It is a complex mixture of compounds produced from oxides of calcium, aluminum, silicon, and iron, to which calcium sulfate is added. Portland cement is made from ground limestone mixed with aluminosilicates. Firing the mixture in a cement kiln causes many reactions from which the major product is tricalcium silicate (Ca_3SiO_5), and the minor product is tricalcium aluminate ($Ca_3Al_2O_6$). This mixed material, in the form of *clinkers,* is mixed with a small percentage of gypsum ($CaSO_4 \cdot 2H_2O$). Cement hardens shortly after being mixed with water because of the hydration of the tricalcium silicate. Successive reactions subsequently take place that give off heat and incorporate the water in microscopic needle-shaped hydrated crystals, which interlock, solidify the cement, and give it impressive strength. **Mortar** is sand mixed with portland cement. **Concrete** is a mixture of sand and an aggregate (crushed stone or pebbles).

Nonsilicate Ceramics

Oxide ceramics do not contain major amounts of silicon but instead have a metal in combination with oxygen. Their high melting points and resistance to corrosion make them good **refractories** for lining furnaces and chemical tank reactors. **High-density alumina** Al_2O_3, a material fabricated with minimal pore size, is useful in cutting tools and armor plating. **Magnesia** MgO combines very high thermal conductivity and heat resistance with very low electrical conductivity.

Some oxide ceramics have been found to be **superconducting,** losing all resistance to the flow of an electric current at or below a certain temperature. This temperature exceeds the boiling point of liquids nitrogen for certain formulations. The **1-2-3** compound $YBa_2Cu_3O_{(9-x)}$ is one such high-temperature superconductor. It is a mixed-valence compound that is a nonstoichiometric solid, with oxygen atoms missing from many sites in the ideal **perovskite** structure. See text Figure 23-18.[4]

Nonoxide ceramics have strong covalently bonded network structures. They include **silicon nitride** Si_3N_4, which can be made by reaction of silicon with nitrogen at high temperatures, or by reaction of $SiCl_4$ with NH_3. It is possible to alloy silicon nitride by replacing Si^{4+} with Al^{3+} while simultaneously compensating for the change in charge by inserting some O^{2-}'s in place of N^{3-}. This results in new ceramics called **sialons.** **Boron nitride** BN occurs in different structures that are related to the structure of graphite and diamond. It is used for high-temperature ceramic vessels. **Silicon carbide** SiC is an abrasive that is now used to reinforce other ceramics. Such **composite ceramics** are important new structural materials.

Detailed Solutions to Odd-Numbered Problems

23-1 A Lewis structure for the disilicate ion $Si_2O_7^{6-}$ is:

The six O atoms on the perimeter of the structure have three lone pairs and single bonds to an Si atom. All six have a formal charges of -1. The O atom linking the Si atoms has f.c. zero. The Si atoms also have f.c. zero. There are 56 valence electrons in the Lewis structure.

The $P_2O_7^{4-}$ and $S_2O_7^{2-}$ ions are isoelectronic (that is, they also have 56 valence electrons). The Lewis structure drawn above works for $P_2O_7^{4-}$ by simply replacing the Si's with P's. Both P's then have f.c. $+1$. Similarly, the Lewis structure appearing above works for $S_2O_7^{2-}$ ion by simply replacing the Si's with S's. Both S atoms then have f.c. $+2$. The analogous compound of chlorine is Cl_2O_7 (dichlorine heptaoxide). Again, the Si's in the above structure can be replaced, this time with Cl's. The two Cl atoms have f.c. $+3$.

---

[4]Text page 831.

Tip. Many additional resonances structures can be drawn if the octet rule is broken for the Si (or P or S) atoms in these structures. Such structures have one or more double bonds from O atoms to the central atoms. They increase the average bond order and lower the formal charge on the central atoms. See **3-61**.

23-3 In each example, determine the Si : O ratio for the network. Ignore oxygen atoms found in, for example (OH) groups. Then use text Table 23-2.[5]
a) Tetrahedra. Ca, +2; Fe, +3; Si, +4; O, −2.
b) Infinite sheets. Na, +1; Zr, +2; Si, +4; O, −2.
c) Pairs of tetrahedra. Ca, +2; Zn, +2; Si; +4; O, −2.
d) Infinite sheets. Mg, +2; Si, +4; O, −2; H, +1.

23-5 The problem is like the preceding except that Al atoms grouped in the formulas with the Si atoms are counted as Si atoms in determining the Si : O ratio.
a) Infinite network. Li, +1; Si, +4; Al, +3; O, −2.
b) Infinite sheets. K, +1; Al, +3; Si, +4; O, −2; H, +1.
c) Closed rings or infinite single chains. Al, +3; Mg, +2; Si, +4; O, −2.

23-7 The standard reduction potential for the half-reaction $Co^{2+}(aq) + 2\,e^- \rightarrow Co(s)$ equals −0.28 V at pH 0. As the pH rises the potential remains constant until, after a brief intermediate regime, cobalt(II) hydroxide precipitates, and the half-reaction $Co(OH)_2(s) + 2\,e^- \rightarrow Co(s) + 2\,OH^-(aq)$ more accurately describes matters. The standard potential for this half-reaction is −0.73 V. The Nernst equation for the new half-reaction is:

$$\mathcal{E} = \mathcal{E}^\circ - \frac{0.0592}{2}\log[OH^-]^2 = -0.73 - 0.0592\log[OH^-]$$

But

$$-0.0592\log[OH^-] = +0.0592\,pOH = 0.0592\,(14 - pH)$$

Therefore:

$$\mathcal{E} = -0.73\text{ V} + (0.0592\text{ V})(14 - pH) = (0.0988 - 0.0592\,pH)\text{ V}$$

This equation gives the half-cell potential as a function of pH. A plot of $\mathcal{E}$ versus pH is a straight line with a slope of −0.0592. If the line is extended back to pH 0, it intercepts the voltage axis at 0.0988 V. Of course, the actual behavior at low pH is governed by the −0.28 V half-reaction.

---

[5]Text page 810.

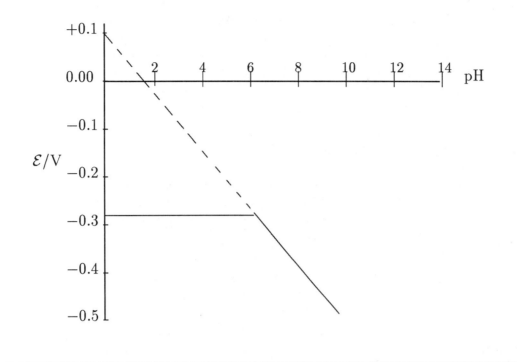

23-9 a) $CaCO_3(s) + SiO_2(s) \rightarrow CaSiO_3(s) + CO_2(g)$.

b)

$$\Delta H° = 1\underbrace{(-1634.94)}_{CaSiO_3(s)} + 1\underbrace{(-393.51)}_{CO_2(g)} - 1\underbrace{(-1206.92)}_{CaCO_3(s)} - 1\underbrace{(-910.94)}_{SiO_2(s)} = \boxed{+89.41 \text{ kJ}}$$

$$\Delta S° = 1\underbrace{(81.92)}_{CaSiO_3(s)} + 1\underbrace{(213.63)}_{CO_2(g)} - 1\underbrace{(92.9)}_{CaCO_3(s)} - 1\underbrace{(41.84)}_{SiO_2(s)} = \boxed{+160.8 \text{ J K}^{-1}}$$

c) Assume that $\Delta H°$ and $\Delta S°$ are independent of temperature. Then, when $\Delta G°$ is zero:

$$T = \frac{\Delta H°}{\Delta S°} = \frac{89.41 \times 10^3 \text{ J}}{160.8 \text{ J K}^{-1}} = \boxed{556 \text{ K}}$$

d) Although the $\Delta H°$ of this process is essentially independent of the pressure, the $\Delta S°$ *does* depend on pressure. Gaseous CO_2 has less entropy when released at a higher pressure. Hence ΔS for this reaction at 500 atm is less than $\Delta S°$, which is at 1 atm. Compressing 1 mol of $CO_2(g)$ from 1 atm to 500 atm changes the entropy by:

$$\Delta S(\text{pressure change}) = -R\ln\frac{P_2}{P_1} = -R\ln(500/1) = -51.67\text{J K}^{-1}$$

This formula is discussed in Chapter 8.[6] Therefore, replace 160.8 J K^{-1} with a ΔS of $160.8 + (-51.67)$ or 109.1 J K^{-1}. Write $\Delta G = \Delta H - T\Delta S$ at 500 atm, set ΔG equal to zero, and compute T. It is $\boxed{819 \text{ K}}$.

Tip. At high pressure, it takes a *higher* temperature to make the equilibrium constant of the reaction equal 1. This fits with the sense of LeChatelier's principle for this reaction: high pressure favors the reactants because their volume is less.

23-11 a) The conversion from calcite to aragonite is:

$$1 \text{ mol calcite} \rightarrow 1 \text{ mol aragonite}$$

The $\Delta H°$ for this change is the difference between the $\Delta H_f°$'s (text Appendix D) of calcite and aragonite. The $\Delta H°$ of the reaction is $\boxed{-0.21 \text{ kJ}}$. The $\Delta S°$ is the difference in the absolute entropies text (text Appendix D) of the two substances at 298.15 K. It equals $\boxed{-4.2 \text{ J K}^{-1}}$. The $\Delta G°_{298.15}$ of the reaction is $\boxed{+1.04 \text{ kJ}}$, by a similar calculation with $\Delta G_f°$'s, using the same source for data.

b) The $\Delta G°_{298.15}$ for the calcite$\rightarrow$aragonite conversion is positive. This means that $\boxed{\text{calcite}}$ is thermodynamically favored over aragonite at 25°C and 1 atm.

c) Assume that $\Delta H°$ and $\Delta S°$ depend only weakly on T. Then: $\Delta G_T° \approx \Delta H° - T\Delta S°$. When the two forms are equally favored, $\Delta G°$ equals zero. Setting $\Delta G°$ equal to zero gives $T = 50$ K. Below 50 K, $\boxed{\text{aragonite}}$ is favored. At temperatures above 50 K, calcite is favored (as long as the pressure is 1 atm).

d) Higher pressure favors the more dense form. The molar volume of aragonite is 34.16 cm^3 mol^{-1}. Aragonite is more dense than calcite, which occupies 36.94 cm^3 mol^{-1}. $\boxed{\text{Aragonite}}$ is favored at high pressure.

e) The Clapeyron equation gives the slope of the coexistence line for two phases in a P versus T phase diagram:

$$\frac{dP}{dT} = \frac{\Delta H}{T\Delta V}$$

For the transformation of 1 mol of calcite to 1 mol of aragonite, $\Delta H°$ equals -210 J. The ΔV for this transition is $(34.16 - 36.94)$ cm^3 or -2.78 cm^3 which is -2.78×10^{-6} m^3. The coexistence line for the two solids passes through the point defined by the (P, T) combination: $(P = 1 \text{ atm}, T = 50 \text{ K})$, according to the results of the previous part of this problem. Assume that the ΔH and ΔV values hold correct at

---

[6]See text page 259.

this low temperature. Then the slope of the coexistence line at this point is:

$$\frac{dP}{dT} = \frac{-210 \text{ J}}{(50 \text{ K})(-2.78 \times 10^{-6} \text{ m}^3)} = 1.5 \times 10^6 \text{ Pa K}^{-1}$$

Text Appendix B confirms that a J m^{-3} equals a pascal (Pa). One atmosphere is 101325 pascal, so an equivalent answer is $\boxed{15 \text{ atm K}^{-1}}$.

Tip. The answer is the rate of change of pressure with temperature at $P = 1$ atm and $T = 50$ K. As long as ΔH and ΔV are both independent of the constant T and the constant P as which the conversion takes place, the coexistence line is straight and has this slope.

23-13 Firing steatite (soapstone) drives out water:

$$\boxed{\text{Mg}_3\text{Si}_4\text{O}_{10}(\text{OH})_2(s) \rightarrow 3\,\text{MgSiO}_3(s) + \text{SiO}_2(s) + \text{H}_2\text{O}(g)}$$

23-15 The reaction for the preparation of the glass is:

$$\text{Na}_2\text{CO}_3(s) + \text{CaCO}_3(s) + 6\,\text{SiO}_2(s) \rightarrow \text{Na}_2\text{O·CaO·(SiO}_2)_6(s) + 2\,\text{CO}_2(g)$$

The problem is now a routine calculation in stoichiometry. The molar mass of the glass is 479 g mol^{-1}, which equals 0.479 kg mol^{-1}. The chemical amount of CO_2 produced is:

$$n_{\text{CO}_2} = 2.50 \text{ kg glass} \times \left(\frac{1 \text{ mol glass}}{0.479 \text{ kg glass}}\right) \times \left(\frac{2 \text{ mol CO}_2}{1 \text{ mol glass}}\right) = 10.44 \text{ mol}$$

Use the ideal-gas equation to compute the volume of the CO_2 at 0°C and 1 atm pressure:

$$V_{\text{CO}_2} = \frac{n_{\text{CO}_2} RT}{P} = \frac{(10.44 \text{ mol})(0.08206 \text{ L atm mol}^{-1}\text{K}^{-1})(273 \text{ K})}{1 \text{ atm}} = \boxed{234 \text{ L CO}_2}$$

23-17 Assume a sample of exactly 100 g of the soda-lime glass and calculate the chemical amount of each element which is present. This requires use of the molar masses of the several binary oxides. The following table summarizes the results:

Mass of Oxide	$\mathcal{M}$(g mol^{-1})	Amounts of Elements	
72.4 g SiO$_2$	60.08	1.205 mol Si	2.410 mol O
18.1 g Na$_2$O	61.98	0.5841 mol Na	0.2920 mol O
8.10 g CaO	56.07	0.1444 mol Ca	0.1444 mol O
1.00 g Al$_2$O$_3$	101.96	0.01962 mol Al	0.02942 mol O
0.20 g MgO	40.304	0.004962 mol Mg	0.004962 mol O
0.20 g BaO	153.33	0.001304 mol Ba	0.001304 mol O

The total chemical amount of oxygen in the sample equals the sum of all the listings for O in the right-most column. It is 2.882 mol. The chemical amounts of the various elements per mole of oxygen are their amounts in the above table divided by 2.882. After rounding to the correct number of significant digits, they are:

> 0.418 mol Si, 0.203 mol Na, 0.050 mol Ca, 0.0068 mol Al, 0.002 mol Mg, 0.0005 mol Ba .

Tip. Note how the proportions of the oxides in this glass differ from the nominal proportions in soda-lime glass.

23-19 The reaction for the production of tricalcium silicate is:

$$SiO_2(s, quartz) + 3\,CaO(s) \rightarrow (CaO)_3 \cdot SiO_2(s)$$

The enthalpy of the reaction is the sum of the enthalpies of formation of the products minus the sum of the enthalpies of formation of the reactants. Text Appendix D gives the ΔH_f°'s for $SiO_2(s, quartz)$ and $CaO(s)$. Then:

$$\Delta H^\circ = 1 \underbrace{(-2929.2)}_{(CaO)_3 SiO_2(s)} - 1 \underbrace{(-910.94)}_{SiO_2(s)} - 3 \underbrace{(-635.09)}_{CaO(s)} = \boxed{-113.0 \text{ kJ}}$$

23-21 The sum of the oxidations numbers of the atoms in the compound must equal zero. Assign the oxidation numbers -2 to oxygen, $+2$ to Ba, and $+3$ to Y. The copper must then have an oxidation number of $\boxed{7/3}$ to bring the sum to zero.

23-23 a) $\boxed{SiO_2(s, quartz) + 3\,C(s, graphite) \rightarrow SiC(s) + 2\,CO(g)}$.

b) Find necessary ΔH_f°'s in Appendix D. Then:

$$\Delta H^\circ = 1 \underbrace{(-65.3)}_{SiC(s)} + 2 \underbrace{(-110.52)}_{CO(g)} - 1 \underbrace{(-910.94)}_{SiO_2(s)} - 3 \underbrace{(0)}_{C(s)} = \boxed{624.6 \text{ kJ}}$$

c) Silicon carbide should, like diamond, be hard, high melting, and a poor conductor of electricity.

23-25 The reaction is: $SiC(s) + 2\,O_2(g) \rightarrow SiO_2(s, quartz) + CO_2(g)$. Refer to text Appendix D for the necessary values of ΔG_f°:

$$\Delta G^\circ = 1 \underbrace{(-394.36)}_{CO_2(g)} + 1 \underbrace{(-856.67)}_{SiO_2(s)} - 1 \underbrace{(-62.8)}_{SiC(s)} - 2 \underbrace{(0)}_{O_2(g)} = \boxed{-1188.2 \text{ kJ}}$$

The ΔG° is less than zero so the reaction is spontaneous. Thus, $SiC(s)$ is thermodynamically unstable in the presence of oxygen at standard conditions. The rate of this reaction is however vanishingly slow.

23-27 a) Apophyllite contains infinite sheets of silicate units. The Si and O are in the +4 and −2 oxidation states respectively. The oxidation states are F, −1; K, +1; Ca, +2. The water of hydration in the mineral has H in the +1 and O in the −2 oxidation states.

b) Rhodonite contains infinite single chains of SiO_4 units. The Ca and Mn both have +2 oxidation states.

c) Margarite is an aluminosilicate. It contains infinite sheets of aluminosilicate groups. The Ca and non-infinite-sheet Al are in the +2 and +3 oxidation states, respectively. The Si, Al, and O in the aluminosilicate framework are in the +4, +3 and −2 states; the hydroxide H and O are in the +1 and −2 states respectively.

23-29 Let x equal the oxidation number of the Fe, and write an equation to express the electrical neutrality of the compound:

$$\underbrace{2.36(2)}_{\text{Mg}} + \underbrace{0.48(x)}_{\text{Fe}} + \underbrace{0.16(3)}_{\text{Al}} + \underbrace{2.72(4)}_{\text{Si}} + \underbrace{1.28(3)}_{\text{Al}} + \underbrace{10(-2)}_{\text{O}} + \underbrace{2(-1)}_{\text{OH}} + \underbrace{0.32(2)}_{\text{Mg}} = 0$$

Solving gives $x = 3$; the oxidation state of the iron is $\boxed{3}$.

23-31 $CaAl_2Si_2O_8(s) + CO_2(aq) + 2\,H_2O(l) \rightarrow CaCO_3(s) + Al_2Si_2O_5(OH)_4(s)$.
Lowering the pH will solubilize the product $CaCO_3$, which dissolves readily in acid (see the following problem). Thus, we predict by LeChatelier's principle that lower pH $\boxed{\text{increases}}$ the extent of weathering.

23-33 a) The dissolution of $CaCO_3(s)$ in water involves the equilibria:

$$CaCO_3(s) \rightleftharpoons Ca^{2+}(aq) + CO_3^{2-}(aq) \qquad K_{sp} = 7.6 \times 10^{-9}$$
$$CO_3^{2-}(aq) + H_2O(l) \rightleftharpoons HCO_3^-(aq) + OH^-(aq) \qquad K_{b1} = 2.08 \times 10^{-4}$$
$$HCO_3^-(aq) + H_2O(l) \rightleftharpoons H_2CO_3(aq) + OH^-(aq) \qquad K_{b2} = 2.32 \times 10^{-8}$$

The law of mass action for the two acid-base equilibria gives:

$$K_{b1} = \frac{[OH^-][HCO_3^-]}{[CO_3^{2-}]} \quad \text{and} \quad K_{b2} = \frac{[OH^-][H_2CO_3]}{[HCO_3^-]}$$

Note that K_{b1} is K_w divided by the K_{a2} of H_2CO_3 and K_{b2} is K_w divided by K_{a1} of H_2CO_3.[7]

---

[7]These K's and the K_{sp} are from text Table 10-2 (text page 323) and 11-3 (text page 374).

Since the water is at pH 7, $[OH^-]$ is 1.0×10^{-7} M. Substitute this value into the two K_b expressions and rearrange:

$$[HCO_3^-] = (2080)[CO_3^{2-}] \quad \text{and} \quad [H_2CO_3] = (482)[CO_3^{2-}]$$

Let S equal the solubility of the $CaCO_3(s)$. Then S is equal to $[Ca^{2+}]$ and is also equal to the sum of the concentrations of the three carbon-containing species. This statement is a material-balance condition[8] for the carbonate. Expressed mathematically:

$$S = [Ca^{2+}] \quad \text{and} \quad S = [CO_3^{2-}] + [HCO_3^-] + [H_2CO_3]$$

Substitute the independent expressions for the bicarbonate and carbonic acid concentrations into the second equation to obtain:

$$S = [CO_3^{2-}](1 + 2080 + 482) = [CO_3^{2-}](2560)$$

Combine this equation with the K_{sp} expression for $CaCO_3(s)$ to calculate S:

$$8.7 \times 10^{-9} = [Ca^{2+}][CO_3^{2-}] = S\frac{S}{2560} \quad \text{which gives} \quad S = \boxed{0.0047 \text{ M}}$$

b) Decreasing the pH will $\boxed{\text{increase}}$ the solubility of $CaCO_3(s)$ in water. More carbonate ion is converted to bicarbonate ion or carbonic acid at lower pH.

c) The river's annual flow is 8.8×10^{12} L. This much water would, at equilibrium, dissolve 4.1×10^{10} mol of $CaCO_3$ ($\mathcal{M} = 100$ g mol^{-1}) which is $\boxed{4.1 \times 10^6 \text{ metric tons}}$.

Tip. Do not thoughtlessly assume that $HCO_3^-(aq)$ is the only carbon-containing species present in significant concentration. This assumption corresponds to leaving out the first and third terms in the parentheses in the equation for S and leads to an S of 0.0042 M, which is 10 percent less than the correct answer. A worse error is to ignore the acid-base interaction of the carbonate ion with the water entirely. This corresponds to omitting the second and third terms within the parentheses and gives an S of 9.3×10^{-5} M, about 50 times too low.

23-35 The balanced equation is:

$$\boxed{Mg_3Si_4O_{10}(OH)_2(s) + Mg_2SiO_4(s) \rightarrow 5\,MgSiO_3(s) + H_2O(g)}$$

Trial-and-error balancing gives this answer. Another approach is to note that Mg_2SiO_4 loses O^{2-} ions on a per silicon basis, and $Mg_3Si_4O_{10}(OH)_2$ gains O^{2-} ions on the same basis. This approach puts O^{2-} ion in this reaction in the role played by the electron

---

[8]Discussed on text page 349 and page 259 of this Guide.

in the standard method of balancing redox reactions. Arranging the gain of O^{2-} to equal the loss of O^{2-} rapidly gives a balanced equation:

O^{2-} loss $\qquad Mg_2SiO_4(s) \rightarrow MgSiO_3(s) + O^{2-} + Mg^{2+}$

O^{2-} gain $\qquad O^{2-} + Mg^{2+} + Mg_3Si_4O_{10}(OH)_2(s) \rightarrow 4\,MgSiO_3(s) + H_2O(g)$

b) Use LeChatelier's principle. Increasing the total pressure increases the activity of $H_2O(g)$ and shifts the reaction to the left. The products are $\boxed{\text{disfavored}}$.

c) The slope of the coexistence curve is given by the Clapeyron equation:

$$\frac{dP}{dT} = \frac{\Delta H}{T\Delta V} = \frac{\Delta S}{\Delta V}$$

For this reaction ΔV is clearly positive because the products include 1 mol of gas, which has a large volume, but the reactants include no gas. The ΔS of the reaction is positive, for the same reason. The slope of the curve is accordingly $\boxed{\text{positive}}$.

23-37 To impart a red color to the pot, the oxidation state of the iron must be high. The iron in iron oxides will be in a high oxidation state if bound to many oxide anions. Thus, an air-rich (oxygen-rich) atmosphere should be employed. To impart a black color to the pot, the oxidation state of the iron must be low. A smoky fuel-rich atmosphere has relatively little oxygen in it. In such an atmosphere the iron is not oxidized. To make a red pot use an air-rich atmosphere; to make a black pot use an fuel-rich atmosphere.

23-39 A 100.0 g sample of pure dolomite contains 45.7 g of $MgCO_3$ and 54.3 g of $CaCO_3$. This can be confirmed by dividing the masses of the two compounds by their molar masses (84.31 g mol^{-1} and 100.08 g mol^{-1}). The chemical amounts of the two equal to 0.542 mol. If the compounds are present in equal chemical amount the formula is "$(MgCO_3)_1 \cdot (CaCO_3)_1$" which is better written $MgCO_3 \cdot CaCO_3$ or $MgCaC_2O_6$, or $\boxed{MgCa(CO_3)_2}$.

23-41 The equation for the reaction of silicon nitride with hydrofluoric acid is

$$Si_3N_4(s) + 12\,HF(aq) \rightarrow 3\,SiF_4(g) + 4\,NH_3(aq)$$

In this case, no kinetic barrier exists to slow the reaction. Parts fabricated from silicon nitride are rapidly corroded by $HF(aq)$.

Chapter 24

Optical and Electronic Materials

Semiconductors

In the elemental state, silicon forms a covalently bonded network solid[1] in which all the electrons are localized between pairs of atoms. If a crystal of silicon contains N atoms, then it contains $4N$ valence orbitals. These valence orbitals mix to create two bands of very closely spaced orbitals that are separated by a gap. The $4N$ valence electrons possessed by the N Si atoms in a crystal occupy the lower-lying **valence band** of orbitals. This band is essentially completely filled. Consequently, although the electrons might trade places readily, a net transfer of electric charge cannot occur in the valence band. At somewhat higher energy than the valence band lies the conduction band. Electrons in the conduction band are highly mobile, but in pure silicon the conduction band is essentially empty. The result for pure silicon is a low electrical conductivity that can be increased by heating the solid or illuminating it (see **24-1**) thereby exciting electrons across the **band gap.** Other **intrinsic semiconductors** are made by combining elements to give crystals with the same number of valence electrons per atom as silicon (four). An important example is gallium arsenide (compare with **24-5**).

An **insulator** such as diamond has an even larger band gap than a semiconductor, so that its conduction band is completely unoccupied by electrons at ordinary temperatures. See **24-3**). Note the similarity of the formula given in **24-3** to the formula for the thermal occupation of molecular energy levels.[2]

Because the conduction band in a metal is only partly filled, electrons can change levels with a y minimal input of energy. The conductivities of silicon and other intrinsic semiconductors can be increased by **doping**, the deliberate addition of impurity atoms of selected types. If a Group V element is added to pure silicon, the excess

---

[1]The structure of silicon is the same as that of diamond, which is shown in text Figure 19-34 on text page 711.

[2]Text page 609.

electrons occupy levels just below the conduction band, and the conductivity is increased 1000-fold, giving a *n*-**type semiconductor.** If atoms of a Group III element substitute for silicon atoms, they make a *p*-**type semiconductor,** in which electrical conduction occurs by the hopping of positive holes in the valence band when a voltage is applied (see **24-15**).

Phosphors are wide band-gap materials with dopants selected to create new energy levels such that particular colors of light are emitted when elements previously excited to these levels fall back to the ground state (see **24-7**).

Photosynthesis

The text describes the process of photosynthesis in purple photosynthetic bacteria as a prelude to the study of photosynthesis in green plants. The steps in bacterial photosynthesis are given in the answer to **24-15**. The net effect of these steps is to harness the energy of the light to pump hydrogen ion from inside the cell to outside against the concentration gradient.

The process of photosynthesis in green plants is more complex than in the purple bacteria. Green plants employ two types of reaction center: photosystem I and photosystem II, which absorb light of different wavelengths. Unlike the purple bacteria, green plants employ water as a source of hydrogen and give off oxygen as a by-product of their photosynthesis.

Detailed Solutions to Odd-Numbered Problems

24-1 The band-gap energy equals the energy of light of wavelength 920 nm. Use Planck's relation to compute this energy:

$$E_g = h\nu = \frac{hc}{\lambda} = \frac{(6.626 \times 10^{-34} \text{ J s})(2.9979 \times 10^8 \text{ m s}^{-1})}{920 \times 10^{-9} \text{ m}} = \boxed{2.16 \times 10^{-19} \text{ J}}$$

24-3 The problem requires substitution in the formula for the number of electrons excited. The gap energy E_g is 8.7×10^{-19} J, which is equivalent to 5.24×10^5 J mol^{-1}. With T at 300 K and R equal to 8.315 J K^{-1}mol^{-1} the formula becomes:

$$n_e(4.8 \times 10^{15} \text{ cm}^{-3} \text{ K}^{-3/2})(300 \text{ K})^{3/2} \exp\left(\frac{-5.24 \times 10^5 \text{ J mol}^{-1}}{2(8.315 \text{ J K}^{-1}\text{mol}^{-1})(300 \text{ K})}\right)$$

$$= 6.1 \times 10^{-27} \text{ cm}^{-3}$$

In a 1.00-cm^3 diamond (pretty big for a diamond) at room temperature, only 6.1×10^{-27} electrons are excited to the conduction band; there are essentially $\boxed{\text{no electrons}}$ in the conduction band.

24-5 a) Phosphorus-doped silicon is an $\boxed{n\text{-type}}$ semiconductor because substitution of a P (five valence electrons) at an Si (four valence electrons) site populates the conduction band. The carriers of electric current are mobile electrons.
b) Zinc-doped indium antimonide is a $\boxed{p\text{-type}}$ semiconductor. Mobile holes are the charge carriers in this material.

24-7 Use the Planck equation to compute the wavelength that corresponds to a difference in energy of 2.9×10^{-19} J:

$$\lambda = \frac{hc}{\Delta E} = \frac{(6.626 \times 10^{-34} \text{ J s})(2.9979 \times 10^8 \text{ m s}^{-1})}{2.9 \times 10^{-19} \text{ J}} = \boxed{6.8 \times 10^{-7} \text{ m}}$$

Light of this wavelength is $\boxed{\text{red}}$.

24-9 At room temperature, zinc white does not absorb in the visible region, although it does absorb ultraviolet light. It appears white. When zinc white is heated, the absorption in the UV is shifted into the blue end of the visible region. Yellow is the color complement of blue, so the absorption of blue light makes the substance appear yellow. The shift into the blue from the UV is a shift to lower frequency and indicates a $\boxed{\text{decrease}}$ in band gap.

24-11 First, compute the energy of a single photon:

$$E = h\nu = \frac{hc}{\lambda} = \frac{(6.626 \times 10^{-34} \text{ J s})(2.9979 \times 10^8 \text{ m s}^{-1})}{4.30 \times 10^{-7} \text{ m}} = 4.62 \times 10^{-19} \text{ J}$$

The energy of 1.00 mol of photons equals this value multiplied by Avogadro's number. The answer is 278 kJ. The conversion ADP→ATP has $\Delta G°$ equal to $+34.5$ kJ. The 278 kJ from 1.00 mol of photons is 8.06 times larger than 34.5 kJ. Therefore 8.06 mol of ATP could be produced by 1.00 mol of 430 nm photons. This means that at most $\boxed{8 \text{ molecules}}$ of ATP are produced by a single 430 nm photon.
Tip. The value of $\Delta G°$ given in the problem is for "biochemical standard conditions." This kind of $\Delta G°$'s is often distinguished by adding a prime to the symbol: $\Delta G°'$. No $\Delta G°'$'s appear in text Appendix D. Under biochemical standard conditions the activity of H_3O^+ is defined as 1 at pH 7 (rather than at pH 0). Biochemical standard conditions provide a reference state for thermodynamic values that is closer to the conditions prevailing in living cells. The ΔG of the ADP $\rightarrow$ ATP reaction, like that of most biochemical reactions, is highly dependent on pH.

24-13 The hybridization of silicon atoms in Si(s) is sp^3, giving rise to a 3-dimensional network of tetrahedral silicon atoms that is just like the diamond structure of carbon. Graphite consists of parallel sheets of hexagonally arrayed σ bonded carbon atoms. Less directional bonds join the sheets. Graphite is an excellent electrical conductor in a direction parallel to the sheets as a result of extensive electron delocalization in the out-of-plane π-system . If silicon were to adopt the graphite structure, we would expect $\boxed{\text{high}}$ electrical conductivity.

24-15 The empty seat will appear to move to the right at a rate of one seat position per five minutes. The analogy with hole motion in p-type semiconductors is this: each seat is a lattice site, the empty seat is the hole, and the people are the electrons,

24-17 The steps in bacterial photosynthesis are:
1. The bacteria gather radiant energy using different "antenna molecules" that are excited to high-energy electronic states but quickly transfer the energy to neighboring molecules and on to the photosynthetic reaction center.
2. The reaction center contains four bacteriochlorophyll molecules: the "special pair" (symbolized $(BChl)_2$) and two others. It also contains two bacteriopheophytin molecules, two ubiquinone molecules (UQ), and an iron(II) ion. These species are arranged in a symmetrical fashion with the ubiquinone molecules, which include long conjugated chains of 50 carbon atoms, forming two branches (the A branch and B branch). The bacteriochlorophyll molecules in the special pair are pushed up into electronic excited states as they briefly trap the input energy. They then transfer an electron (and are themselves oxidized) to a bacteriopheophytin molecule.
3. The electron moves from the bacteriopheophytin to the ubiquinone molecule in the A branch.
4. The electron zips down the ubiquinone molecule and across to the ubiquinone molecule in the B branch. This UQ molecule is reduced to UQ^-.
5. The oxidized special pair is reduced to its original state by picking up an electron from a cytochrome protein (Cyt). The special pair is re-excited (with energy from another photon) and transfer a second electron to the same ubiquinone molecule in the B branch, forming UQ^{2-}, which, being a base, picks up hydrogen ions to form UQH_2.
6. The doubly reduced ubiquinone UQH_2 undocks from the reaction center as a fresh ubiquinone molecule comes in.
7. The UQH_2 reduces Cyt^+ protein back to Cyt. The location of the Cyt^- allows the by-product H^+ ion to be generated outside the cell wall.

The net effect is harness the energy of the light to pump hydrogen ion from inside the cell to outside against the concentration gradient.

Chapter 25

Polymeric Materials

Polymerization Reactions for Synthetic Polymers

Polymers are molecules built up by the linking together, in long chains, sheets, or three-dimensional networks, of many identical structural units called **monomer units.** Silicate minerals (Chapter 19) are polymers. Many man-made polymers are based on organic starting materials. The two major types of polymer growth are **addition polymerization** and **condensation polymerization.**

In addition polymerization, monomer units react to form a polymer chain without net loss of atoms. The polymerization of an alkenes such as ethylene is an addition polymerization. See **25-19**. Addition polymerization often proceeds by free-radical chain reaction[1] of molecules that have C=C double bonds. It is started by a suitable **initiator.** In other cases, addition polymerization proceeds by an ionic mechanism.

In condensation polymerization, a small molecule, often water, is split out. See **25-7**. In this type of polymerization, each molecule of monomer has (at least) two different functional groups that react to join the units together in a "head-to-tail" fashion. **Copolymers** form when chemically different monomers are mixed in the polymerization process. If the differing monomer units join the growing polymer chain at random, the result is a **random copolymer**. It is also possible to make **block copolymers**, in which long sequences of each type of monomer unit are chemically bonded to form the polymer chain, and **graft copolymers**, in which polymer chains of a second sort branch from a polymer chain of the first sort. The polymer chains growing from monomers with more than two reactive sites can be **cross-linked** into sheets and networks, which often have desirable physical properties.

Applications for Synthetic Polymers

Useful polymers occur in fibers, plastics, and elastomers (rubber).

---

[1]Chain reactions are introduced and discussed on text page 467 as part of Chapter 13.

- **Fibers.** The important natural fiber **cellulose** is a condensation polymer of glucose. See below. The —OH side-groups along the cellulose chain can be modified to give derivatives like guncotton, nitrocellulose (which is made into celluloid), and cellulose acetate. The first true synthetic polymeric fiber was nylon, a polyamide formed by the condensation of a dicarboxylic acid and a diamine.

- **Plastics.** Plastics are polymeric materials that can be molded or extruded into appropriate shapes and that harden upon cooling or solvent evaporation. Polyethylene, formed by the free-radical addition polymerization of ethylene, can be created as low-density polyethylene (LDPE), high-density polyethylene (HDPE), and linear low-density polyethylene (LLDPE). The difference concerns the number and length of side-chains projected from the polymer molecules. If a methyl group replaces one of the hydrogen atoms in every monomer unit of polyethylene, the result is polypropylene. The relative positions of the methyl groups attached to the carbon backbone may be **isotactic** (all on the same side), **syndiotactic** (alternating in a regular pattern), or **atactic** (distributed at random). Substitution of a chlorine atom for one of the hydrogen atoms in ethylene gives vinyl chloride, which polymerizes to polyvinyl chloride, another very useful plastic. **Plasticizers** are often added to polyvinyl chloride and other plastics to soften them and make them flexible.

- **Elastomers.** An elastomer is a plastic that can be deformed to a large extent and still recover its original form when the deforming stress is removed. Natural rubber is a polymer of **isoprene** (2-methylbutadiene), in which the geometry at the double bonds along the polymer chain is *cis*. If natural rubber is cross-linked by S—S bonds in a process called **vulcanization**, the product is harder, more resilient, and does not melt. It is this material that is used for automobile tires. Synthetic rubbers include neoprene, in which a chlorine atom replaces the methyl group in natural rubber, and copolymers of butadiene with styrene or butadiene with acrylonitrile.

Natural Polymers

The text identifies three important classes of natural polymers: polysaccharides, proteins, and nucleic acids.

Carbohydrates and Polysaccharides

Carbohydrates are organic compounds containing C, H, and O and no other elements. They have the general formula $C_n(H_2O)_m$. The complete removal of water from a carbohydrate (its dehydration) leaves pure carbon.

Simple sugars (**monosaccharides**) are a group of carbohydrates having the more restricted general formula $C_n(H_2O)_n$. The most important simple sugar is the 6-carbon sugar (or **hexose**) $C_6H_{12}O_6$ called glucose. Note that the ending "-ose" in a chemical name always signals a sugar. Glucose is an aldehyde at C-1 and has —OH groups on C-2 through C-6. The presence of four different side-groups at C-2 through C-5 causes chiral centers at these four atoms. The name D-glucose refers to specific chiralities at C-2 through C-5—isomers of glucose with different chiralities at one or more C atoms have different names. The aldehyde carbon (C-1) in glucose reacts readily with the —OH oxygen at C-5 to close a six-membered ring. Two cyclic forms of glucose result:

α-D-glucose (to the left) and β-D-glucose. The two differ in the chirality at C-1, which is non-chiral until the ring is closed. Compare to **25-11**.

A second important simple sugar is fructose, an isomer of glucose. Fructose is the sweetest of the simple sugars. The carbonyl group in fructose is at C-2 rather than C-1, making fructose a ketone. Fructose has chiral centers at C-3, C-4, and C-5; the name D-fructose refers to specific chiralities at all three centers. In solution, fructose forms both a five-membered ring (when C-2 links to the —OH at C-5) and a six-membered ring (when C-2 links to the —OH at C-6). In each case, two chiralities are possible at the C-2. This gives fructose a total of four possible cyclic forms.

Disaccharides and **polysaccharides** arise from the condensation polymerization of simple sugars. Each condensation step eliminates one molecule of water. Simple sugars always occur in some cyclic form in polysaccharides. Sucrose (cane sugar) is the most important disaccharide. It forms from the reaction of the —OH at C-1 of an α-D-glucose ring with the —OH on C-2 of a β-D-fructose five-membered ring. The result is elimination of H_2O and formation of a —O— link between the two cylic monomer unit. See text Figure 25-10b.[2] The edible polymer starch is poly-α-D-glucose where the condensation is between the C-1 hydroxide of one monomer unit and the C-4 hydroxide of the next. The inedible polymer cellulose is poly-β-D-glucose. with the same (C-1)—O—(C-4) linkage.

Proteins

In proteins, the polymer chain is formed of α-amino acid monomer units. Molecules of α-amino acids contain an amine (—NH_2) and carboxylic acid group (—COOH) attached to the same C atom. Amino acids are amphoteric. See **10-95**. All α-amino acids but glycine have a carbon atom that is a chiral center. This means they come in two forms mirror-image isomeric forms (D and L). Twenty different amino acids commonly occur in proteins. Natural proteins contain almost exclusively L-amino

---

[2]Text page 865.

acids. The properties of proteins are strongly affected by the nature of the side-groups (see **25-15**). In a protein, the amino group on one amino acid is linked to the acid group on the next in the **peptide linkage**, which is a amide group. Every time a link is forged, a molecule of water is split out. Creation of a peptide linkage is non-spontaneous at room conditions. but is driven in organisms by free energy from spontaneous reactions. Protein chains of up to about 50 amino-acid groups are often termed **polypeptides**. All polypeptide and protein chains have distinguishable ends. One end is the amino end, and one end is the acid end. There is essentially an infinite number of possible different protein molecules. Not only is the length of the polymer chain variable, but different amino acids may be linked in different orders.

Although the sequence of side-groups in proteins is exceedingly variable (**25-55**), this is only part of their structural variability. Protein structure also involves the way the chain is folded, coiled, looped back on itself by hydrogen bonds, or cross-linked to other protein chains. **Fibrous proteins** are structural materials that form sheets or fibers. They have regular three-dimensional structures. The polymeric chain may coil into a spiral from which the side-groups of the amino acids extend outward in a helical pattern. Such a right-handed coil is called the **α-helix**. It is maintained by hydrogen bonds. Alternatively, neighboring polypeptide chains may hydrogen-bond together in sheet-like structures. **Globular** proteins have chains that are folded irregularly into a more or less compact globular shape. The folding causes amino acids which are widely separated in the sequence along the chain to lie adjacent to each other in the globule. This kind of folding is vital in the functioning of hemoglobin, the oxygen-carrying molecule in the blood.

The same folding occurs in the structures of **enzymes**. Enzymes are globular proteins that catalyze particular reactions in the cell. The exact folding creates active sites at which steric and electronic factors combine to hold substrate molecules in such as way as to enhance the rates of specific reactions.

Nucleic Acids

Nucleic acids preserve and transmit information about the sequence of amino acids in proteins. They are nucleic because they occur in the nuclei of cells. They are acids because they are derivatives of the weak acid phosphoric acid. Nucleic acids are polymers. The backbone of the polymer chain consists of $—(HO)OPO_2—$ groups alternating with a cyclic sugar. In deoxyribonucleic acid (DNA), the sugar is deoxyribose. See text Figure 25-17.[3] Each sugar has as a side-group a nitrogenous base. In DNA, the bases are cytosine (C) guanine (G), adenine (A), and thymine (T). Expressed in somewhat different terms, a nucleic acid molecule is built by linking four different nucleotide units together in a long chain. A nucleotide consists of a base

---

[3]Text page 872.

plus a sugar plus a phosphate.

The order of the bases along the polymeric chain encodes the information that the nucleic acid maintains or transmits. A nucleic acid molecule that is only 10 nucleotide units long has $4^{10} = 1048580$ possible isomers based on the different sequences of the nucleotides. The concept is the same as the theme of **25-55**. The difference is that the subject now is nucleotides linking together, not amino acids.

In DNA, two polymeric strands intertwine in a double helix. The cystosine side-group on one strand links through hydrogen bonds to a guanine side-group on the other. Each adenine links through hydrogen bonds to a thymine. The base-pairings are quite specific: C...G A...T.

When the DNA double helix is unwound in the present of a supply of nucleotides and a suitable means for their delivery, each single strand serves as a template for the creation of a new complementary strand. The result is two new DNA molecules identical in their base sequences to each other and to their progenitor.

Detailed Solutions to Odd-Numbered Problems

25-1 The addition polymerization is $n\,Cl_2C{=}CH_2 \rightarrow \left(CCl_2{-}CH_2\right)_n$.

25-3 Addition polymerization does not split out any small molecules. From the formula of the polymer then, the starting monomer must be $\boxed{\text{formaldehyde}}$ ($H_2C{=}O$).

25-5 **a)** As glycine (NH_2CH_2COOH) polymerizes to the polypeptide, one molecule of $\boxed{\text{water}}$ is lost in the formation of each peptide bond.

b) The repeating structure in the polypeptide is:

$$-\overset{\displaystyle H}{\underset{}{N}}-\overset{\displaystyle H}{\underset{\displaystyle H}{C}}-\overset{\displaystyle O}{\underset{}{C}}-$$

25-7 The repeating unit in the polyamide has the formula $C_{12}H_{22}N_2O_2$. This formula is the sum of the molecular formulas of adipic acid and hexamethylenediamine minus twice the formula of water, a relationship that derives from the fact that the diacid and diamine polymerize by condensation with loss of one molecule of water for each unit added to the chain. The molar mass of the repeating unit is 226.32 g mol^{-1}. Then the chemical amount of the repeating unit that is needed is:

$$1.00 \times 10^6 \text{ g polymer} \times \left(\frac{1 \text{ mol of units}}{226.32 \text{ g polymer}}\right) = 4419 \text{ mol}$$

This means that the synthesis needs 4419 mol of adipic acid (which has a molar mass of 146.1 g mol^{-1}) and 4419 mol of hexamethylenediamine (molar mass 116.2 g mol^{-1}). These chemical amounts convert to $\boxed{646 \text{ kg}}$ of adipic acid and $\boxed{513 \text{ kg}}$ of hexamethylenediamine.

25-9 Polyethylene is formed by addition polymerization. This means that the mass of the monomer used to make the polymer equals the mass of the polymer that is formed. No mass is split out in the form of water or other by-product, as in condensation polymerization. Use this fact in a train of unit-conversions and then do an ideal-gas calculation:

$$4.37 \times 10^9 \text{ kg LDPE} \times \left(\frac{1 \text{ kg C}_2\text{H}_2}{1 \text{ kg LDPE}}\right) \times \left(\frac{1 \text{ mol C}_2\text{H}_2}{0.02805 \text{ kg C}_2\text{H}_2}\right) = 1.558 \times 10^{11} \text{ mol C}_2\text{H}_2$$

$$V_{\text{C}_2\text{H}_2} = \frac{n_{\text{C}_2\text{H}_2}RT}{P} = \frac{(1.558 \times 10^{11} \text{ mol C}_2\text{H}_2)(0.08206 \text{ L atm mol}^{-1}\text{K}^{-1})(273 \text{ K})}{1.00 \text{ atm}}$$

$$= \boxed{3.49 \times 10^{12} \text{ L C}_2\text{H}_2}$$

Tip. This equals 3.49 km³ (cubic kilometers!) of gaseous ethylene.

25-11 The compound β-D-galactose has $\boxed{\text{five}}$ chiral centers: It is an isomer of β-D-glucose. Note the different chirality at C-4.

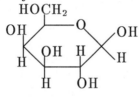

25-13 A tripeptide is a chain of three monomer units and has distinguishable ends. Any of the three kinds of building block can go in the first position, any of the three can go in the second position, and any of the three can go in the third position. There are accordingly $3^3 = \boxed{27}$ possible tripeptides.

25-15 The pentapeptide is:

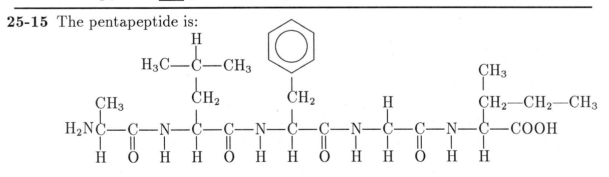

The pentapeptide has all non-polar side-groups. It should be more soluble in $\boxed{\text{octane}}$ than in water.

25-17 The empirical formula of phenylalanine is $C_6H_5CH_2CH(NH_2)COOH$,[4] which is equivalent to $C_9H_{11}NO_2$. The polypeptide forms with the removal of an H from the

[4]See text Table 25-3, text page 867.

amine end of the molecule and an OH from the carboxylic acid end. Except for the two monomer units at the two extreme ends, which are negligible, each phenylalanine loses one HOH as the polymer forms. The empirical formula of the polymer is therefore $\boxed{C_9H_9NO}$. The molar mass of a C_9H_9NO unit is 147.2 g mol^{-1}. If the molar mass of the polypeptide is 17500 g mol^{-1} it contains $17500/147.2 = \boxed{119}$ monomer units.

25-17 A catalyst is by definition not consumed in a reaction; it is taken up in one step of a mechanism but regenerated in a subsequent step. In the polymerization of acrylonitrile, the butyl lithium is consumed and the butyl anion is incorporated into the product. Doing this creates an new anion by which chain-building propagates. The butyl lithium is incorporated into the product. It is irrecoverable and hence not a catalyst.

25-21 Polyvinyl chloride is $-(CH_2-CHCl)_n$. The molar mass of its monomer unit is 65.50 g mol^{-1}. Polyvinyl chloride is formed by polymerization of ethylene dichloride CH_2ClCH_2Cl ($\mathcal{M} = 98.95$ g mol^{-1}) with the loss of one molecule of HCl per monomer unit added to the chain. The theoretical yield from 950 million pounds of monomer is therefore

$$950 \text{ million lb} \times \frac{62.50}{98.95} = 600 \text{ million lb}$$

If the actual yield of polymer is 500 million pounds, then the percent yield is $500/600 \times 100\% = \boxed{83\%}$. If the actual yield gets as high as 550 million pounds, the percent yield is $\boxed{92\%}$.

25-23 Hair consists of polymeric chains of amino acid cross-linked by —S—S— bridges. To curl hair, treat the hair with a reducing agent, which breaks some of the cross-links. Then arrange the strands of hair in the desired curls, and treat the hair with an oxidizing agent. The disulfide bridges then reform, but in different locations. The new cross-links maintain the curls.

25-25 Two kinds of amino acid can appear at each of the 22 positions in the polypeptide chain. The two ends of the chain are distinguishable. Therefore, there are $\boxed{2^{22}}$ or 4.194 million possible isomeric molecules.

25-27 The acid form of alanine can donate two hydrogen ions:

$$^+H_3NCH(CH_3)COOH(aq) + H_2O(aq) \rightleftharpoons {}^+H_3NCH(CH_3)COO^-(aq) + H_3O^+(aq)$$
$$^+H_3NCH(CH_3)COOH(aq) + H_2O(aq) \rightleftharpoons H_2NCH(CH_3)COOH(aq) + H_3O^+(aq)$$

The two K's are $K_1 = 10^{-2.3} = 5.0 \times 10^{-3}$ and $K_2 = 10^{-9.7} = 2.0 \times 10^{-10}$. Notice that the although both K's are K_a's, the reactions are competitive, not consecutive. The notation K_{a1} and K_{a2} is therefore avoided. Abbreviate the formulas of the three forms of alanine:

$$A \Rightarrow {}^{+}H_3NCH(CH_3)COOH \quad Z \Rightarrow {}^{+}H_3NCH(CH_3)COO^{-} \quad B \Rightarrow H_2NCH(CH_3)COOH$$

This makes it easier to write out the mass-action expressions for the above equilibria:

$$\frac{[H_3O^+][Z]}{[A]} = K_1 \quad \text{and} \quad \frac{[H_3O^+][B]}{[A]} = K_2$$

At pH 7 $[H_3O^+]$ equals 1.0×10^{-7} M, and the two expressions become:

$$\frac{(1.0 \times 10^{-7})[Z]}{[A]} = 5.0 \times 10^{-3} \quad \text{and} \quad \frac{(1.0 \times 10^{-7})[B]}{[A]} = 2.0 \times 10^{-10}$$

$$\frac{[Z]}{[A]} = 5.0 \times 10^4 \quad \text{and} \quad \frac{[B]}{[A]} = 2.0 \times 10^{-3}$$

a) The fraction[5] of the alanine in the Z-form equals the concentration of Z divided by the sum of the concentrations of all three forms:

$$f_Z = \frac{[Z]}{[A] + [Z] + [B]} = \frac{5.0 \times 10^4[A]}{[A] + (5.0 \times 10^4)[A] + (2.0 \times 10^{-3})[A]} = \boxed{0.99998}$$

Essentially $\boxed{\text{all}}$ of the molecules are in the zwitterion-form at pH 7.

b) The fraction of alanine in the B-form is the concentration of that form divided by the sum of the concentrations of all three forms:

$$f_B = \frac{[B]}{[A] + [Z] + [B]} = \frac{2.0 \times 10^{-3}[A]}{[A] + (5.0 \times 10^4)[A] + (2.0 \times 10^{-3})[A]} = \boxed{4.0 \times 10^{-8}}$$

---

[5]Compare to **10-91**.

Appendices

Answers to Odd-Numbered Problems, App. A

A-1 The trailing zeros in d) and e) must not be omitted when the number is put into scientific notation.
a) 5.82×10^{-5} **b)** 1.402×10^3
c) 7.93 **d)** -6.59300×10^3 **e)** 2.530×10^{-3} **f)** 1.47

A-3 a) 0.000537 **b)** $9,390,000$ **c)** -0.00247 **d)** 0.006020 **e)** $20,000.$

A-5 The number is 746 million kilograms or 746,000,000 kg.

A-7 a) Statistical methods for deciding when to omit a outlier are not developed in the Appendix. Instead the appeal is to use good judgment. The value 135.6 g is grossly out of line with the others.
b) The mean is 111.34 g **c)** The standard deviation is 0.22 g and the 98% confidence limit is

$$\text{confidence limit} = \pm\frac{t\sigma}{\sqrt{N}} = \pm\frac{2.57(0.22 \text{ g})}{\sqrt{6}} = \pm 0.23 \text{ g}$$

where t comes from Table A-2, text page A-4.

A-9 The measurement of mass in problem A-7 is more precise.

A-11 a) five **b)** three **c)** ambiguous (two or three significant figures) **d)** three **e)** four.

A-13 a) 14 L **b)** -0.0034°C **c)** 3.4×10^2 lb **d)** 3.4×10^2 miles
e) 6.2×10^{-27} J

A-15 Eight, $2\,997\,215.55$

A-17 a) -167.25 **b)** 76 **c)** 3.1693×10^{15} **d)** -7.59×10^{-25}

A-19 a) -8.40 **b)** 0.147 **c)** 3.24×10^{-12} **d)** 4.5×10^{13}

A-21 The area of the triangle is 337 cm². Three significant figures appear in the answer.

Answers to Odd-Numbered Problems, App. B

B-1 a) 6.52×10^{-11} kg **b)** 8.8×10^{-11} s **c)** 5.4×10^{12} kg m² s⁻³
d) 1.7×10^4 kg m² s⁻³ A⁻¹

B-3 a) 4983°C, but it is very hard to measure such a high temperature to ±1°C.
b) 37.0°C **c)** 111°C **d)** −40°C.

B-5 a) 5256 K **b)** 310.2 K **c)** 384 K **d)** 233 K.

B-7 a) 24.6 m s⁻¹ **b)** 1.15×10^3 kg m⁻³ **c)** 1.6×10^{-29} A s m
d) 1.5×10^2 mol m⁻³ **e)** 6.7 kg m² s⁻³ = 6.7 W.

B-9 One kW-hr is equal to 3.6×10^6 J. Hence, 15.3 kW-hr is 5.51×10^7 J.

B-11 The engine displacement is 6620 cm³ or 6.62 L.

Answers to Odd-Numbered Problems, App. C

C-1 The slope is 50 miles hr⁻¹.

C-3 a) The equation is in the required form with m (slope) equal to 4 and b (y-intercept) equal to −7.
    b) The equation is $y = 7/2x − 5/2$. The slope is 7/2, and the y-intercept is −5/2.
    c) The equation is $y = −2x + 4/3$. The slope is −2 and the y-intercept is 4/3.

C-5 The graph of y versus x for the equation $y = 2x^3 − 3x^2 + 6x − 5$ is not linear. The value of y rises from −45 at $x = −2$ and +11 at $x = +2$. The graph cuts the x-axis (has $y = 0$) at $x = 1$.

C-7 a) $x = −5/7$ **b)** $x = 3/4$ **c)** $x = 2/3$.

C-9 The answers are given to 4 significant figures. **a)** $x = 0.5447, −2.295$ **b)** $x = −0.6340, −2.366$ **c)** $x = +0.6340, +2.366$.

C-11 a) Assuming that x is small compared to 2.00 gives $x = 6.5 \times 10^{-7}$. There are also two complex roots.
    b) The best method of solution is graphical. There are three roots because this is a third-degree equation: $x = 4.07 \times 10^{-2}, 0.399, −1.011$.
    c) The only real root is $x = −1.3732$. It can be arrived at graphically. The other two roots are imaginary. They are of little interest in chemical applications.

C-13 a) 4.551 (the three significant figures appear in the mantissa) **b)** To help understand the significant figures, divide the exponent in the number by 2.302585093 to re-express it as a power of 10: $10^{-6.814}$. The "6" plays the role of the characteristic when the antilog is taken and the mantissa has three significant figures. Hence the answer has three significant figures: 1.53×10^{-7}.   **c)** 2.6×10^8 **d)** -48.7264

C-15 The answer is 3.015.

C-17 Few calculators accommodate a number with an exponent exceeding 99 in absolute value. To answer this problem, write

$$\log 3.00 \times 10^{121} = \log(3.00) + \log 10^{121} = 121 + 0.477 = 121.477$$

C-19 Simply change the characteristic from 0 to 7 or from 0 to -3 and add it to the same mantissa: $7 + 0.751 = 7.751$ and $-3 + 0.751 = -2.249$.

C-21 The problem is to find x in the equation $\log \ln x = -x$. One way to proceed is to guess an x, put it into the left side of the equation and see on a calculator if the indicated operations gives back the guess. Adjust the guess and repeat until satisfied. The answer is 1.086.

C-23 a) $8x$ **b)** $3\cos 3x - 8\sin 2x$ **c)** 3   **d)** $1/x$.

C-25 a) 20   **b)** $78125/7$ **c)** 0.0675.